auto-catalytic

endo-PG, PME

citrate release

Membrane damage → Ascorbate secretion.

O_2

Cu^{2+}

$^\circ OH$ (apo)

$O_2^{\circ -}$

pectic polysacch solubilisation

polysacch
scission

oligosacch
release

Ethylene formation

expansin

Softening

ANNUAL REVIEW OF
PLANT PHYSIOLOGY

ANNUAL REVIEW OF PLANT PHYSIOLOGY

VOLUME 38, 1987

WINSLOW R. BRIGGS, *Editor*

Carnegie Institution of Washington, Stanford, California

RUSSELL L. JONES, *Associate Editor*

University of California, Berkeley

VIRGINIA WALBOT, *Associate Editor*

Stanford University

ANNUAL REVIEWS INC.   4139 EL CAMINO WAY   PO BOX 10139   PALO ALTO, CALIFORNIA 94303-0897 USA

ANNUAL REVIEWS INC.
Palo Alto, California, USA

International Standard Serial Number: 0066–4294
International Standard Book Number: 0–8243–0638-4
Library of Congress Catalog Card Number: A-51-1660

Annual Review and publication titles are registered trademarks of Annual Reviews Inc.

Annual Reviews Inc. and the Editors of its publications assume no responsibility for the statements expressed by the contributors to this *Review*.

Typesetting by Kachina Typesetting Inc., Tempe, Arizona; John Olson, President Typesetting coordinator, Janis Hoffman

PRINTED AND BOUND IN THE UNITED STATES OF AMERICA

To Jean Heavener

One of the remarkable things about the production of the *Annual Review of Plant Physiology* is how smoothly the numerous chapters are drawn together to form each volume. The majority of the credit for this must go to the production editor. In our case we have been truly fortunate to have the services of Ms. Jean Heavener who served with the Annual Review staff since 1968, working on volumes 19–38 with us. During the preparation of this volume, Jean contracted an incurable cancer. It is with great sadness that we report her recent death. We have lost not only a wonderful editor but a cheerful person who improved each volume through gentle persuasion. We will miss her very much.

THE EDITORIAL COMMITTEE

For the convenience of readers, a detachable order form/envelope is bound into the back of this volume.

Annual Review of Plant Physiology
Volume 38, 1987

CONTENTS

PREFATORY

Living in the Golden Age of Biology, *Beatrice M. Sweeney*   1

MOLECULES AND METABOLISM

Photochemical Reaction Centers: Structure, Organization, and
    Function, *Alexander N. Glazer and Anastasios Melis*   11
Some Aspects of Calcium-Dependent Regulation in Plant
    Metabolism, *H. Kauss*   47
Some Molecular Aspects of Plant Peroxidase Biosynthetic
    Studies, *R. B. van Huystee*   205
Cellulose Biosynthesis, *Deborah P. Delmer*   259
Membrane-Proton Interactions in Chloroplast Bioenergetics:
    Localized Proton Domains, *Richard A. Dilley, Steven M.
    Theg, and William A. Beard*   347
Gibberellin Biosynthesis and Control, *Jan E. Graebe*   419

ORGANELLES AND CELLS

Phosphorylation of Proteins in Plants: Regulatory Effects and
    Potential Involvement in Stimulus/Response Coupling,
    R. Ranjeva and A. M. Boudet   73
Membrane Control in the Characeae, *Masashi Tazawa, Teruo
    Shimmen, and Tetsuro Mimura*   95
The Plant Cytoskeleton: The Impact of Fluorescence
    Microscopy, *Clive W. Lloyd*   119
Regulation of Gene Expression in Higher Plants, *Cris
    Kuhlemeier, Pamela J. Green, and Nam-Hai Chua*   221
Agrobacterium-Mediated Plant Transformation and Its Further
    Applications to Plant Biology, *Harry Klee, Robert Horsch,
    and Stephen Rogers*   467

TISSUES, ORGANS, AND WHOLE PLANTS

Fruit Ripening, *C. J. Brady*   155

Differentiation of Vascular Tissues, *Roni Aloni*   179

Plants in Space, *Thora W. Halstead and F. Ronald Dutcher*   317

POPULATION AND ENVIRONMENT

Genetics of Wheat Storage Proteins and the Effect of Allelic
Variation on Bread-Making Quality, *Peter I. Payne*   141

Plant Virus-Host Interactions, *Milton Zaitlin and Roger Hull*   291

Evolution of Higher-Plant Chloroplast DNA-Encoded Genes:
Implications for Structure-Function and Phylogenetic Studies,
Gerard Zurawski and Michael T. Clegg   391

INDEXES

Author Index   487

Subject Index   507

Cumulative Index of Contributing Authors, Volumes 30–38   516

Cumulative Index of Chapter Titles, Volumes 30–38   518

RELATED ARTICLES FROM OTHER *ANNUAL REVIEWS*

From the *Annual Review of Genetics,* Volume 20 (1986)

The Early Days of Yeast Genetics: A Personal Narrative, *Herschel Roman*
Molecular Genetics of Transposable Elements in Plants, *H.-P. Döring and P. Starlinger*
Pre-mRNA Splicing, *Michael R. Green*

From the *Annual Review of Microbiology,* Volume 40 (1986)

Recent Advances in Bacterial Ion Transport, *Barry P. Rosen*
Competition for Nodulation of Legumes, *D. N. Dowling and W. J. Broughton*
The Autotrophic Pathway of Acetate Synthesis In Acetogenic Bacteria, *Lars G. Ljungdahl*
Organization of the Genes for Nitrogen Fixation in Photosynthetic Bacteria and Cyanobacteria, *Robert Haselkorn*

From the *Annual Review of Biochemistry,* Volume 56 (1987)

Inositol Trisphosphate and Diacylglycerol: Two Interacting Second Messengers, *Michael J. Berridge*
Fractionation and Structural Assessment of Oligosaccharides and Glycopeptides by Use of Immobilized Lectins, *T. Osawa and T. Tsuji*
Inhibitors of the Biosynthesis and Processing of N-Linked Oligosaccharide Chains, *Alan D. Elbein*
Topography of Glycosylation in the Rough Endoplasmic Reticulum and Golgi Apparatus, *Carlos B. Hirschberg and Martin D. Snider*
Protein Glycosylation in Yeast, *M. A. Kukuruzinska, M. L. E. Bergh, and B. J. Jackson*
The Nucleus: Structure, Function, and Dynamics, *John W. Newport and Douglass J. Forbes*
Cyclic AMP and Other Signals Controlling Cell Development and Differentiation in *Dictyostelium, Günther Gerisch*
Enzymes of General Recombination, *Michael M. Cox and I. R. Lehman*
P450 Genes: Structure, Evolution, and Regulation, *Daniel W. Nebert and Frank J. Gonzalez*
Aminoacyl tRNA Synthetases: General Scheme of Structure-Function Relationships in the Polypeptides and Recognition of Transfer RNAs, *Paul Schimmel*
Intracellular Proteases, *Judith S. Bond and P. Elaine Butler*
Dynamics of Membrane Lipid Metabolism and Turnover, *E. A. Dawidowicz*
Biosynthetic Protein Transport and Sorting by the Endoplasmic Reticulum and Golgi, *Suzanne R. Pfeffer and James E. Rothman*
Ferritin: Structure, Gene Regulation, and Cellular Function in Animals, Plants, and Microorganisms, *Elizabeth C. Theil*

(continued) ix

Transfer RNA Modification, *Glenn R. Björk, Johanna U. Ericson, Claes E. D. Gustafsson, Tord G. Hagervall, Yvonne H. Jönsson, and P. Mikael Wikström*

From the *Annual Review of Phytopathology,* Volume 24 (1986)

Remote Sensing of Biotic and Abiotic Plant Stress, *Ray D. Jackson*
Biosynthesis and Functions of Fungal Melanins, *Alois A. Bell and Michael H. Wheeler*
Molecular Evolution of Plant RNA Viruses, *R. W. Goldbach*
Mechanisms of Resistance to Plant Viruses, *Fernando Ponz and George Bruening*
Phytoalexin Synthesis: Biochemical Analysis of the Induction Process, *Jürgen Ebel*
The Role of Pectic Enzymes in Plant Pathogenesis, *Alan Collmer and Noel T. Keen*
Tissue Culture and the Selection of Resistance to Pathogens, *Margaret E. Daub*

From the *Annual Review of Cell Biology,* Volume 2 (1986)

Core Particle, Fiber, and Transcriptionally Active Chromatin Structure, *D. S. Pederson, F. Thoma and R. T. Simpson*
Proton-Translocating ATPases, *Qais Al-Awqati*
Protein Import into the Cell Nucleus, *Colin Dingwall and Ronald A. Laskey*
Microtubule-Associated Proteins, *J. B. Olmsted*
Mechanism of Protein Translocation Across the Endoplasmic Reticulum Membrane, *Peter Walter and Vishwanath R. Lingappa*

From the *Annual Review of Biophysics and Biophysical Chemistry,* Volume 16 (1987)

Structure and Dynamics of Water Surrounding Biomolecules, *Wolfram Saenger*
The Thermodynamic Stability of Proteins, *John A. Schellman*
Biophysical Chemistry of Metabolic Reaction Sequences in Concentrated Enzyme Solution and in the Cell, *D. K. Srivastava and Sidney A. Bernhard*
An Introduction to Molecular Architecture and Permeability of Ion Channels, *George Eisenman and John A. Dani*
Permeation in Potassium Channels: Implications for Channel Structure, *Gary Yellen*
Calcium Channels: Mechanisms of Selectivity, Permeation, and Block, *Richard W. Tsien, Peter Hess, Edwin W. McCleskey, and Robert L. Rosenberg*
Absorption, Scattering, and Imaging of Biomolecular Structures with Polarized Light, *Ignacio Tinoco, Jr., William Mickols, Marcos F. Maestre, and Carlos Bustamente*
Thermodynamic Efficiency in Nonlinear Biochemical Reactions, *John Ross and Mark Schell*
Nuclear Magnetic Resonance and Distance Geometry Studies of DNA Structures in Solution, *Dinshaw J. Patel, Lawrence Shapiro, and Dennis Hare*
Peptides with Affinity for Membranes, *E. T. Kaiser and F. J. Kézdy*

Beatrice M. Sweeney

Ann. Rev. Plant Physiol. 1987. 38:1–9

LIVING IN THE GOLDEN AGE OF BIOLOGY

Beatrice M. Sweeney

Department of Biological Sciences, University of California, Santa Barbara, California 93106

What great good luck for a biologist to live between 1914 and 1986, to come of age at the same time as the science of biology! Our predecessors were the herbalists, the catalogers, a few physiologists working with the most Stone-Age tools. Now instruments can measure biological parameters in seconds, automatically programmed by computers. With the help of isotopes of phosphorus, carbon, and sulfur, we can find a single gene in the tangle of chromosomes or a single protein with the help of antibodies specific for that protein. We can see the structure of a cell at the nanometer level. With these tools and more, much about how plants and animals, bacteria and protists manage to live and reproduce has been discovered during my lifetime.

My own life has reflected the development of biology in the last 70 years. My passion for botany began before I can really remember, when I was so small that my eyes were level with the celandine's translucent leaves on the stone walls along the road when we walked to Cabot Woods. A half mile then seemed a very great distance, so far that my brother, a year younger, rode there in a wicker baby carriage. In the wood at that time, a stream ran at the bottom of a rounded ridge covered with oaks and underbrush. By the brook we gathered dog tooth violets in the spring, feeling their smooth, mottled leaves and nodding lily flowers, pulling them up to show their pale underground stems. We discovered hepaticas and bloodroot under the dry oak leaves on the ridge. In the hot days of summer, we found forget-me-nots blossoming by the brook. Later Cabot Woods was given to the city to be preserved, but the city crews came to clean it up, burning the hill and bulldozing the stream until there was nothing left.

By the time I was six, I must have had a good idea of the flowers of New England, because I found the plants in Pasadena excitingly different from those at home when we spent a winter there in 1920. In the garden were Cecil

1

Brunner doll roses and a thicket of pampas grass. My brother and I explored the Arroyo Seco, a wild tangle of rushes and chapparal with a small dirt path down the center. In the spring, we went for walks through the fields of lupines and poppies, fields long since solid with houses.

I lived my early scientific life entirely outside of school. Before I could read, I recorded the flowers I found in drawings, frustratingly primitive and inaccurate, later by dark and out-of-focus photographs made with my mother's ancient 4 × 5 view camera. Collecting and naming flowers was my greatest interest at that time. We had our first car then, driven by John, our Irish chauffeur. This dear man was very long-suffering with me and would stop the car as soon as I called out, so that I could look at a plant that I had spotted as I leaned out the car window. Cars then did not travel as fast as they do now. As I grew older, I learned the plants of the Adirondack forest where we spent the summers, the twin flowers, Indian pipe, and tawny hawkweed that I discovered while tramping in the mountains with my brother.

My mother took me to a biological supply house when I was probably no more than eight years old. This treasure house sold magnifying glasses with brass legs that came off, as well as "botany boxes," elongated lunch boxes with covers that folded back for collecting, and large sheets of grey blotting paper for drying plants. My taxonomic phase lasted until I was in my teens, but sooner than that I began to discover the microscopic world of cells and to wonder how cells worked.

My interest in physiology really began because of the rarity of yellow lady slippers and an advertisement in a plant catalog for roots of these plants. But we didn't have a bog in our garden. I described my rather unsuccessful efforts to create a bog in my back yard in an essay for school that was actually published, my first publication (7).

Perhaps today our children are so busy after school with music lessons, soccer, and gymnastics that there is no time to explore the world alone or to think. At school I had best friends and played on teams, but my time alone is what gave direction to my life.

Not until I was a freshman at Smith College did I take a formal course in biology: beginning botany. We looked in the microscope and I saw plants at another level, cells that did different things, cells with chloroplasts, xylem, epidermis with water-impermeable walls and stomata. At that time photosynthesis was taught as the transformation of CO_2 to carbohydrate by a single light step in which oxygen was evolved, perhaps with the formation of formaldehyde. It wasn't until I was in graduate school that I heard about Van Niel's deduction from observations of the sulfur bacteria that the oxygen evolved came from water!

Fortunately for me, I was admitted to the Special Honors Program as a junior at Smith. We were allowed to take fewer courses and to do small research projects in each, working independently with our professors. In botany, I did my first research under the direction of Miss Smith, a study of the effects of drugs on the rate of protoplasmic streaming in *Elodea* and *Nitella* (8). I worked in a laboratory in the greenhouse head-house, measuring rate of streaming with a stop watch. I had a key to the door, a microscope of my own, reagent bottles, cultures of *Nitella,* and an aquarium filled with *Elodea.* I was launched on my career.

My teachers in botany at college were fine women, but they were all spinsters almost without private lives. I resolved that I would be different, that I would marry and have children and be a scientist too. After all, men in science did not have to give up family life. After college I proceeded to marry and have children, and my life was pretty tempestuous for a number of years; but I didn't give up science. In this my graduate advisor, Kenneth Thimann, was a great help. He never in any way implied that I was inferior to his male graduate students. He quietly made it possible for me to continue my research during whatever time I could manage when I had my first and second children during graduate school. Once when he and I were working together on a paper and I hesitated to make suggestions, he told me "Two heads are better than one." Could he have considered our heads to be equal?

Folke Skoog was a postdoctoral fellow at Harvard at the time. He arranged double-blind tests for me for my research on the effects of auxin on pro-toplasmic streaming in *Avena* (18). We had many discussions about research while eating cherries and throwing the pits out the window at a chimney near by. At a AAAS meeting at Dartmouth, we listened together to the radio announcement of the beginning of the Second World War in Europe.

The state of the art in biology when I started graduate school in 1936 was as follows. Electron transport in photosynthesis and photophosphorylation had not yet been discovered. That chromosomes were equally divided among the progeny in cell division was understood; the genetic material was unknown, though Mendel's work had demonstrated its necessity. Microtubules were unknown—in fact, the electron microscope had not yet been invented. Mitochondria had been seen in the light microscope as barely visible elongated bodies. The citric acid cycle had just been discovered: I learned it for my orals. Plant hormones had been chemically characterized, and animal hormones were just being discovered. X-ray diffraction methods for determining molecular structure were not in general use, and there were of course no electronic calculators or computers to make analysis easy and quick. Cytoplasm was thought to be a gel-sol, essentially an aspic of enzymes and substrates.

Plant work was actually advanced compared to that in animals, plants being in some ways simpler to dissect at the physiological and biochemical level. For example photoperiodic control, discovered apropos flowering in plants in

1914 (19), was still very little known in animals. The circadian rhythms shown in the leaf sleep movements in leguminous plants were observed to continue in constant darkness in the 17th century (2), while animal circadian rhythms and the biological clock for which they are evidence began to be studied extensively only in the 1950s.

With a fresh PhD in my hand, I went with my physician husband to Rochester, Minnesota, so that he could be a Fellow there. We went in January. Snow was so deep in our backyard that the garbage pail was buried from sight; the temperature was −20°F, and the freshly washed diapers froze solid on the line. Our ears froze. There were no jobs available in botany at all. I worked half days at a farm that had been converted to a hormone-assay lab, learned to keep a rat colony and do animal surgery. I left by bus for the farm at 7:30 a.m., still night during winter, and returned at lunchtime to my two small kids. Finally I got a postdoctoral fellowship to study the effects of estrogens on respiration of rat uteri (10). The Warburg manometer was at first a mystery to me. Hardest was finding out the composition of Brodie's solution, but it finally turned up in one of the earliest of Warburg's papers.

Meanwhile the United States had entered the war. My husband left Rochester to be a flight surgeon in San Diego. We lived in La Jolla, near Scripps Institution of Oceanography, and I hurried there to find something to do in plant physiology. C. K. Tseng took me into his laboratory, where he was studying the agar-producing red alga, *Gelidium*. C. K. kept his plants in tanks of seawater bubbled with air, so that the lab resounded with this pleasant burbling. I measured photosynthetic gas exchange with an ancient Van Slyke apparatus (20) and for lunch ate Lipton's chicken soup with C. K., which he stirred with chopsticks.

During the 1950s, I shared a laboratory at Scripps Institution of Oceanography with Francis Haxo. He was interested in action spectra for photosynthesis in algae, including the red alga *Porphyridium*. The spectrum for this alga was peculiar in that those wavelengths absorbed by phycoerythrin were much more effective in photosynthesis than were the wavelengths absorbed by chlorophyll itself, yet it was known from fluorescence measurements that energy absorbed by the phycobilin pigments was transferred to chlorophyll. What was the matter with the light absorbed directly by chlorophyll *a*? Was it for some reason ineffective? Lawrence Blinks had the answer almost in his hand when he showed that when red wavelengths of light were exchanged for green light of the same effectiveness at steady state there was for a short time a peak of higher oxygen evolution. The explanation, however, did not come from a study of red algae but from Robert Emerson's careful measurements of photosynthesis of the green alga *Chlorella* at the red end of the spectrum. He noted that at wavelengths longer than 680 nm the efficiency of photosynthesis decreased a little faster than did the absorption of the chlorophyll. With brilliant intuition, Emerson irradiated *Chlorella* with two wavelengths at once. How he conceived this experiment is beyond understanding. The result, as you know, was his discovery that two photosystems with different pigment composition must be excited at the same time (3). Emerson immediately understood the explanation for the inefficiency of light absorbed by chlorophyll in red algae, where phycoerythrin is the light-harvesting pigment; this was in fact a much clearer case of the necessity for the "enhancement," as Emerson called it. How do I know he had understood? Because I went to see Emerson at Urbana just at this moment. He invited me to lunch with his family and after we had finished eating, he took me by the arm, led me into the living room, sat me down in a corner, drew up a chair, and started asking me questions about what we were doing at Scripps with the red algae—a very exciting experience for me, and a little scary.

It was by chance that I began to work with the dinoflagellates. It was difficult to get funds for equipment at Scripps in 1950. I wanted to work on photosynthesis in red and brown algae in different colors of light, for which I fancied I needed an integrating sphere, lights, and a monochromator. Marston Sargent overheard my complaints and said, "While you're waiting for money, Beazy, why don't you see if you can grow some of the dinoflagellates?" They had not been cultured at that time. It actually proved to be easy because fresh seawater and plankton samples were available at the end of the Scripps Pier. I could haul them up through a hatch with a bucket on a rope, as from a well. James Bonner wrote me a note praising the paper that described this research (11), such a splendid thing for a well-known scientist to do for a novice!

One of the dinoflagellates that I was able to culture was *Gonyaulax polyedra*. This dinoflagellate was bioluminescent in the laboratory! At that time the luminescence of *Renilla* had been shown to be inhibited by light, so I wanted to find out whether light might also inhibit the bioluminescence of *Gonyaulax*. A large number of tubes with samples from a culture were prepared and their light emission was measured at successive times with the photomultiplier photometer Jim Snodgrass had made for me. At first all went as expected: When samples from the constant light bank were darkened, the bioluminescence increased, reaching a maximum in 6 hr; but then it began to fall. Perhaps the cells were dying? Because I still had unused samples and I hate to waste anything, I continued to make measurements. I found that after about 24 hr in the dark, bioluminescence began again to increase, forming another peak and then decreasing once more. I presented this result at the conference on luminescence at Asilomar, California, that March, 1954 (6). There I met Woody Hastings, and we arranged to work together the following summer at my laboratory. He recognized right away that I had seen a circadian rhythm in the luminescence of *Gonyaulax*, because he was at the time an assistant professor at Northwestern where Frank Brown was experimenting with rhythms in crabs and potatoes. Well, I just never went back to the question of the spectral properties of algal photosynthesis.

Woody, Marlene Karakashian (an undergraduate at Northwestern), and I worked together at Scripps for three summers. During this time we characterized the circadian rhythm in stimulated bioluminescence in *Gonyaulax polyedra* (16) and the biochemistry of bioluminescence with respect to enzyme, substrate, oxygen requirement and lack of apparent cofactors (4). The structure of the substrate, a very unstable molecule, was only discovered recently.

This was a time of great excitement in our research. Data poured from our hands and interpretations were hammered out in late-night discussions while we waited for the time to make more readings of light emission. Perhaps the most exciting experiment was that determining the temperature effects on the period of the circadian rhythm in bioluminescence. Woody found out (I would never have dared to ask) that we could use all the small temperature-controlled rooms at the phytotron at California Institute of Technology in the late summer while the facility was being serviced. The three of us rented one bed at a graduate student house in Pasadena and occupied it in eight-hour shifts, each of us making measurements of the bioluminescence at the different temperatures for eight hours and then playing for eight hours. The first experiment failed because the lights in the temperature-controlled rooms, run at overvoltage, were bluer than our lights at Scripps. The extra blue light inhibited the mechanical stimulation of bioluminescence, as we learned later. The second experiment, using a lower irradiance, succeeded (5).

In the winters, beside doing experiments on bioluminescence and rhythms with *Gonyaulax,* I helped Francis Haxo with his measurements of the effect of different temperatures on photosynthesis of attached algae from arctic and temperate habitats. When Francis arranged to extend this study to the tropics with Pete Scholander's expedition in 1960, he asked me to come along. The Michael Pilson family, who were helping me look after my children, now four in number, agreed to take over for three months so that I could go. Twelve scientists traveled inside the Great Barrier Reef from Cairnes to Thursday Island in the 60-foot *Tropic Seas,* a crocodile-hunting boat skippered by Vince Vlasov. I occupied a bunk in the stern with the canned goods. When we reached Thursday Island, established ourselves in the CSIRO oyster laboratory, and waded out on the sand flat in front, we found *Acetabularia!*

For some time I had wanted to find out whether the nucleus was necessary for circadian timing. *Acetabularia* was the obvious organism to use because it can live and function normally for a long time after the nucleus is removed. However, I had no *Acetabularia* plants in La Jolla. Furthermore, no circadian rhythm was known in *Acetabularia* at that time. We had with us on Thursday Island only manometers to measure photosynthetic oxygen evolution, so I measured the oxygen evolution of *Acetabularia* at different times of day in natural light and in a makeshift artificial constant light. What luck! Rates of photosynthesis differed greatly between day and night, and this cycle continued in constant light. The nucleus can be removed by simply cutting off the basal part of an *Acetabularia* cell, so it was easy to test whether or not the rhythm continued in the absence of the nucleus. It did (17).

It's a wonderful thing to see something you've never seen before, the Great Barrier Reef or the ultrastructure of a cell you've been working with for a long time. In the light microscope, *Gonyaulax* is almost completely opaque. When I was at Yale I had the opportunity to see the details of the internal structure of *Gonyaulax* for the first time. Ben Bouck fixed and sectioned cells and we saw the chloroplasts, the mitochondria, and curious membrane-bound structures, square in cross sections, the trichocysts (1).

After this I learned to do electron microscopy myself, even freeze-fracture, a technique I found difficult. The limited number of cells one can examine in electron micrographs makes gathering quantitative data tiring and frustrating (13). However, some of the pictures, particularly of freeze-fractured cells, are truly beautiful.

It is wonderful, too, suddenly to understand something. The question of how a cell can keep time had long been in my mind. I was sitting at my typewriter composing a talk for an *Acetabularia* conference in Wilhelms-haven when all at once it came to me that membrane-bound organelles must be important in time-keeping, since not a single prokaryote is known to have a circadian rhythm. An exchange of ions or molecules must be taking place between cytoplasm and organelles in a controlled fashion, a feedback loop where the membrane must change the passage of ions and the passage of ions must change the membrane in a cycle (12). This might explain the peculiar observation in *Acetabularia* that, while the nucleus is clearly not necessary for timing, when it is present it can change timing. Of course this idea was not completely right, but it started a train of thought and a series of experiments both by myself and by others that confirmed the importance of membranes in generating time information in cells.

In 1967 I came to the University of California at Santa Barbara and the next year became an associate professor of biology, my first position as a real professor. I discovered the pleasures of working with students of my own— my graduate students and the undergraduates in the College of Creative Studies here. This institution is remarkable in being a small college embedded within a large university with all its resources. It was designed by Marvin Mudrick for the education of intellectually gifted students. In science, even the freshmen work in the faculty laboratories helping with research or even doing their own projects. I was very fortunate to have a number of these students in my laboratory, including several who are already well-known scientists. Their questions enlivened discussions and their ideas were an inspiration. To teach such students is pure fun.

During the days when the *R. V. Alpha Helix* was an active biological research ship, I was included in two expeditions to study bioluminescence, one to New Guinea and the other to South East Asia. These trips were productive for all of us. I studied the physiology of the *Noctiluca* with symbiotic green flagellates swimming in its vacuole (14) and brought back a culture of *Pyrocystis fusiformis,* a very large cell that proved useful for electrophysiology (21) and for a comparison of the cell cycle and the circadian cycle (15).

I have always enjoyed doing my own laboratory work, either alone or with one or two active collaborators—for example Francis Haxo and Woody Hastings, more recently Marie-Therese Nicolas and Goran Samuelsson (9). I find that I understand the advantages and flaws in an experimental technique when I do it myself, and most of my ideas have come when I am actually gathering data. Molecular biology, which has yielded so much information about the genes, has an inherent danger. It requires large research groups, as evidenced by the long list of authors on single papers, and thus requires large amounts of money for salaries and supplies. The senior scientist in such a group must thus act as administrator and fund raiser and will probably have difficulty finding time to work in the laboratory. We make a mistake when we convert our most able scientists into administrators and hence inactivate them.

People often ask me whether or not I have suffered from discrimination against women scientists. As a young girl, I really did not feel this. My family and my teachers were always encouraging, so I did not suffer any effects of discrimination until I had my PhD. The problems I met after that arose from attitudes that I had somehow acquired with respect to society's expectations of female behavior toward men. I instinctively felt that I should put men's interests ahead of my own. For example, I deferred to my husband in consenting to his taking a position at Mayo Clinic, in an isolated community where there was nothing for me to do scientifically. I deferred to my early collaborators in first authorship of the more important papers. I was satisfied with a research position when my children were small because I only wanted to be committed to work half-time. When they were grown and I could hold a full-time faculty position, it was difficult for me to get one. Universities considered that I had too many publications to be an assistant professor and too little teaching experience to be an associate professor. Finally I succeeded, but only barely—even in the 1960s when all universities were enlarging their faculties. The attitude of servility and guilt about pleasing ourselves that we learn as little girls is, in my experience, the greatest deterent to equal opportunity for women.

I advise young women in science to keep their maiden names, dare to be assertive, and refuse to get discouraged. They must keep learning and reading, especially if they have to take leave from science for family reasons. It may not be easy to combine a scientific career with a family but it can be done and it's worth all the effort. Research has sustained me through many otherwise unbearable times and given me much satisfaction, made my older years exciting. There is a temptation for women to settle for the family alternative as simpler. However, sooner or later the family will want a second income. A high percentage of all women do work at some point in their lives. When that point arrives it is crucial for a woman to be prepared to enter, or to resume, a challenging career.

As Goran Samuelsson said in introducing me recently at a seminar at the University of Umea, I have had a lifelong love affair with research.

Literature Cited

1. Bouck, G. B., Sweeney, B. M. 1966. The fine structure and ontogeny of trichocysts in marine dinoflagellates. *Protoplasma* 61:205–23
2. De Mairan, M. 1729. Observation botanique. *Hist. Acad. R. Sci., Paris,* p. 35
3. Emerson, R., Chalmers, R., Cederstrand, C. 1959. Some factors influencing the long-wave limit of photosynthesis. *Proc. Natl. Acad. Sci. USA* 43: 133–43
4. Hastings, J. W., Sweeney, B. M. 1957. The luminescent reaction in extracts of the marine dinoflagellate, *Gonyaulax polyedra. J. Cell Comp. Physiol.* 49: 209–26
5. Hastings, J. W., Sweeney, B. M. 1957. On the mechanism of temperature independence in a biological clock. *Proc. Natl. Acad. Sci. USA* 43:804–11
6. Haxo, F. T., Sweeney, B. M. 1955. Bioluminescence in *Gonyaulax polyedra.* In *The Luminescence of Biological Systems,* ed. F. H. Johnson, pp. 415-20. Washington, DC: AAAS
7. Marcy, B. 1931. In praise of bogs. In *Essays of Today,* ed. R. A. Witham, pp. 323–26. Cambridge: The Riverside Press
8. Marcy, B. 1937. Effect of ethylene chlorhydrin and thiourea on *Elodea* and *Nitella. Plant Physiol.* 21:207–12
9. Samuelsson, G., Sweeney, B. M., Matlick, H. A., Prezelin, B. B. 1983. Changes in photosystem II account for the circadian rhythm in photosynthesis in *Gonyaulax polyedra. Plant Physiol.* 73:329–31
10. Sweeney, B. M. 1944. The effect of estrone on anaerobic glycolysis of the uterus of the rat *in vitro. J. Lab. Clin. Med.* 29:957–62
11. Sweeney, B. M. 1954. *Gymnodinium splendens,* a marine dinoflagellate requiring vitamin B_{12}. *Am. J. Bot.* 41:821–24
12. Sweeney, B. M. 1974. A physiological model for circadian rhythms derived from the *Acetabularia* rhythm paradoxes. *Int. J. Chronobiol.* 2:95–110
13. Sweeney, B. M. 1976. Freeze-fracture studies of *Gonyaulax polyedra.* I. Membranes associated with the theca and circadian changes in the particles on one membrane face. *J. Cell Biol.* 68:451–61
14. Sweeney, B. M. 1976. *Pedinomonas noctilucae* (Prasinophyceae), the flagellate symbiotic in *Noctiluca* (Dinophyceae) in Southeast Asia. *J. Phycol.* 12:460–64
15. Sweeney, B. M. 1982. Interaction of the circadian cycle with the cell cycle in *Pyrocystis fusiformis. Plant Physiol.* 70:272–76
16. Sweeney, B. M., Hastings, J. W. 1957. Characteristics of the diurnal rhythm of luminescence in *Gonyaulax polyedra. J. Cell Comp. Physiol.* 49:115–28
17. Sweeney, B. M., Haxo, F. T. 1961. A persistence of a photosynthetic rhythm in enucleated *Acetabularia. Science* 134:1361–63
18. Sweeney, B. M., Thimann, K. V. 1938. The effect of auxin on protoplasmic streaming. II. *J. Gen. Physiol.* 21:439–161
19. Tournois, J. 1914. Études sur la sexualité du houblon. *Ann. Sci. Nat. Bot. Biol. Veg.* 19:49–191
20. Tseng, C. K., Sweeney, B. M. 1946. Physiological studies of *Gelidium cartilagineum.* I. Photosynthesis, with special reference to the carbon dioxide factor. *Am. J. Bot.* 33:706–15
21. Widder, E. A., Case, J. F. 1981. Bioluminescence excitation in a dinoflagellate. In *Bioluminescence, Current Perspectives,* ed. K. H. Nealson, pp. 125–32. Minneapolis: Burgess

Ann. Rev. Plant Physiol. 1987. 38:11–45
Copyright © 1987 by Annual Reviews Inc. All rights reserved

PHOTOCHEMICAL REACTION CENTERS: STRUCTURE, ORGANIZATION, AND FUNCTION

Alexander N. Glazer

Department of Microbiology and Immunology, University of California, Berkeley, California 94720

Anastasios Melis

Division of Molecular Plant Biology, 313 Hilgard Hall, University of California, Berkeley, California 94720

CONTENTS

INTRODUCTION 12
PHOTOSYSTEM II 13
   Photochemical Reaction Center Complex 13
   Chlorophyll a "Core" Antenna Complex 14
   Chlorophyll a/b Light-Harvesting Complex II (LHC II) 14
PHOTOSYSTEM I 16
   Photochemical Reaction Center Complex 16
   Chlorophyll a "Core" Antenna Complex 17
   Chlorophyll a/b Light-Harvesting Complex I (LHC I) 18
PHOTOSYSTEM STOICHIOMETRY 18
EXCITATION ENERGY DISTRIBUTION IN THE THYLAKOID
     MEMBRANE OF OXYGENIC PHOTOSYNTHESIS 19
   Cyanobacteria 19
   Wild-Type Higher Plant Chloroplasts 20
   Mutations Affecting the Development and/or Assembly of LHC II 21
REGULATION OF ELECTRON TRANSPORT 22
   The Effect of LHC II Phosphorylation on PSI 22
   The Physiological Significance of LHC II Phosphorylation 24
CYANOBACTERIA: OVERVIEW 24

11

0066-4294/87/0601-0011$02.00

PURPLE BACTERIA 25
   General Properties of Purple Bacterial Reaction Centers............................ 27
   Rhodopseudomonas viridis Reaction Center Complex................................... 28
   Rhodopseudomonas viridis B1015 Antenna Complex.................................... 29
   Organization of the Rhodopseudomonas viridis Photosynthetic Membrane 31
   Antenna Domain Sizes for Energy Transfer.. 32
GREEN BACTERIA 32
   Antenna Pigment Organization... 33
   Reaction Centers.. 34

INTRODUCTION

Solar energy, the ultimate source of all biological energy on this planet, is converted to chemical form by the process of photosynthesis. In higher plants, photosynthetic energy production for the reduction of energy-poor compounds like CO_2 and nitrate involves the elaborate processes of electron transport and coupled photophosphorylation that occur in chloroplasts. Membraneous organelles 5–10 μM in diameter, chloroplasts are enclosed within a double membrane, or envelope. The interior of the envelope space is divided between the amorphous stroma medium and the highly oriented membrane phase (39). The membrane phase shows a further differentiation into regions of stacked thylakoids (grana) and "stroma thylakoids" (interconnecting the grana) (39, 146).

Electron transport occurs in the photosynthetic membrane phase and requires coordinated interaction between a large number of electron carrier compounds and enzymatic proteins that facilitate the transfer of electrons (reducing power) from dissociated H_2O molecules to $NADP^+$. The electron transport components are highly organized in the thylakoid membrane and facilitate the transfer of electrons laterally in the plane of the membrane from the grana regions (appressed areas) to the stroma-exposed regions. Functionally, electron transport occurs from intermediate-to-intermediate in a sequential manner formulated as the Z-scheme of electron transport by Hill & Bendall over twenty-five years ago (81).

The overall process of electron transport from H_2O to $NADP^+$ is strongly endergonic and is realized through the consumption of light energy at two discrete electron transport steps occurring at Photosystem II and Photosystem I. The term photosystem (PS) defines a membrane-bound complex composed of specialized chlorophyll molecule(s) (which acts as the reaction center) and of specialized electron acceptor molecules. Tightly bound to the reaction center is a "core" chlorophyll a-containing complex that collects and transfers excitation energy to the photochemical reaction center. In addition, there is an accessory Chl a- and Chl b-containing light-harvesting complex (LHC). Each photosystem apparently constitutes a structurally independent thylakoid membrane protein complex.

Cyanobacteria carry out oxygen-evolving photosynthesis in much the same manner as do higher plant chloroplasts. The major difference is in the antenna pigment and thylakoid membrane organization. Phycobilisomes replace the chlorophyll a/b LHC antenna of PSII, and cyanobacterial PSI closely resembles the "core" PSI complex of higher plants. Appressed thylakoid regions (grana) do not form in cyanobacteria. The anoxygenic photosynthesis of the purple and green bacteria is performed by a single photosystem in each instance. The reaction center components in the purple bacteria show relationships to those of PSII, whereas in the green bacteria, depending on the particular organism, photosystems with similarities to either PSII or PSI have been found.

A treatise devoted entirely to the structure and function of photosynthetic membranes has recently appeared (171), as have more specialized reviews on photosynthetic electron transport (76), state transitions (61), and photoregulation (4).

PHOTOSYSTEM II

Photochemical Reaction Center Complex

The photochemical reaction center of PSII is a specialized chlorophyll species termed P680 owing to the absorbance change around 680 nm that occurs upon its photooxidation (49, 50). The primary electron acceptor of PSII is a pheophytin molecule (94, 95) that acts as a transitory electron transfer intermediate to the secondary electron acceptor. The latter is a specialized plastoquinone molecule termed Q_A (124, 176, 188), and it can be photoreduced to its plastosemiquinone anion form Q_A^-. When excitation energy from the light-harvesting pigment arrives at the reaction center, an endergonic charge separation occurs between P680 and pheophytin, followed by electron transfer to Q_A according to the reaction mechanism:

$$P680 \cdot Ph \cdot Q_A \xrightarrow[\text{energy}]{\text{light}} P680^* \cdot Ph \cdot Q_A \xrightarrow{<10 \text{ ps}} P680^+ \cdot Ph^- \cdot Q_A,$$

$$P680^+ \cdot Ph^- \cdot Q_A \xrightarrow{\sim 100} P680^+ \cdot Ph \cdot Q_A^- \xrightarrow{50} P680 \cdot Ph \cdot Q_A^- \xrightarrow{200} P680 \cdot Ph \cdot Q_A.$$

As a result, the light energy absorbed by PSII is stored as chemical energy in the form of an electron on Q_A. The midpoint oxidation-reduction potential ($E_{m,7}$) of the components involved is the following (P680/P680$^+$) = $\sim + 1.0$ V, (Ph$^-$/Ph) = -0.62 V, and (Q_A^-/Q_A) = -0.05 V. The energy contained in a red (680 nm) photon is equal to 1.82 eV, whereas the potential energy stored in the chemical pair P680$^+ \cdot$Ph$^-$ is equal to 1.62 eV. These figures demonstrate the highly efficient conversion of light energy to chemical energy by the photochemical reaction center.

Closely associated with the primary quinone acceptor Q_A of PSII is one atom of Fe(II), which does not participate directly as an intermediate in the electron-transport process. The Fe(II) is known to be magnetically coupled to the semiquinone anion form of $Q_A(Q_A^-)$ (58, 159) and is probably situated about 7 Å away from Q_A and from the secondary quinone Q_B (47). The oxidation of Fe(II) to Fe(III) by exogeneously added ferricyanide (150) coincides with the oxidation of a previously unidentified high-potential electron acceptor of PSII (Q_{400}; Ref. 87). The function of Fe(II) in the electron transfer process from Q_A^- to Q_B is unknown and most likely involves semiquinone anion stabilization in the thylakoid membrane (150, 205).

Polypeptides associated with the photochemical reaction center complex of PSII include the 32-kDa plastoquinone/herbicide-binding protein (136) (better known as Q_B protein) and possibly a 34-kDa protein (currently known as D_2) whose function remains unknown. By analogy with the structural organization of the L and M subunits in purple bacterial reaction centers, it has been postulated that the Q_B and D_2 proteins carry the photochemical reaction center components of PSII (8, 129, 183, 184). An interesting property of the 32-kDa protein of PSII is its relatively high turnover rate as evidenced in pulse-chase experiments (118, 119). Localization studies of the 32-kDa Q_B-binding protein revealed the presence of two distinct pools of this polypeptide. The major pool was localized in the grana partition region (about 70%) with the remainder in stroma-exposed lamellae (199). This distribution of an integral PSII polypeptide correlates with the relative amounts of $PSII_\alpha$ (in the membrane of the grana partition regions) and $PSII_\beta$ (in stroma-exposed lamellae) (122).

Chlorophyll a "Core" Antenna Complex

Closely associated with the photochemical reaction center of PSII are two polypeptides, with apparent molecular weights of 51 and 44 kDa, that collectively contain nearly 50 Chl a molecules (8, 31, 32, 65, 181, 206–208). Each polypeptide may be associated with up to 25 Chl a molecules, i.e. 1 Chl a molecule per about 2-kDa protein. Of the two polypeptides the 51 kDa forms the inner core antenna; it is directly associated with P680, so that excitation energy from the internal antenna Chl of the 44-kDa polypeptide must go through the Chl a molecules of the 51-kDa polypeptide in order to reach the photochemical reaction center (31, 207). Moreover, it was suggested that P680 and Ph are actually localized on the 51-kDa polypeptide (140), in effect making it the reaction center-binding protein. This view is in contrast with the concept mentioned earlier, and more work is needed to distinguish between the two alternatives.

Chlorophyll a/b Light-Harvesting Complex II (LHC II)

The accessory light-harvesting antenna of PSII contains Chl a and Chl b (Chl a/Chl b = 1.1) organized in two distinct Chl a/b LHC entities. The tightly bound LHC II complement contains about 80 ± 20 Chl (a + b) molecules (LHC II-intrinsic). The addition of LHC II-intrinsic to the Chl a core antenna of PSII results in a PSII functional antenna size, N_{PSII}, of about 120 Chl (a + b) molecules. This PSII configuration is evident in plants grown under high light intensity [N_{PSII} = 130 Chl, Chl a/Chl b = 6.7 (109)], in developmental mutants where only limited amounts of LHC II are assembled [N_{PSII} = 130 Chl, Chl a/Chl b = ~5 (1, 181)], and in PSII$_\beta$ in several plant species [N_β = 130 ± 30 (121, 123, 181)].

The peripheral LHC II [also referred to as mobile (170)] augments the light-harvesting antenna size of PSII by about 120 Chl (a + b) molecules (104, 123). Thus, the mature form of PSII residing in the membrane of the grana partition region contains, on the average, a total of 250 Chl (a + b) molecules in its light-harvesting antenna.

The organization of the LHC II-intrinsic and LHC II-peripheral in the thylakoid membrane is largely unknown. Based on work with membrane crystals of the LHC II, Kühlbrandt proposed that the basic *in situ* organization of the LHC II polypeptides is trimeric (98). Furthermore, assuming that each LHC II polypeptide contains about 13 Chl (a + b) molecules (29, 135),[1] i.e. 1 Chl per approximately 2 kDa of protein, it may be concluded that the formation and assembly of LHC II occurs in trimers, each containing about 39 Chl (a + b) molecules. Hence, the LHC II-intrinsic will contain 2 such trimers (about 80 Chl molecules) whereas, in fully developed higher plant chloroplasts, the LHC II-peripheral must contain a minimum of 3 trimers (about 120 Chl molecules). Moreover, evidence in the literature indicates the existence of as many as three distinct polypeptides (29, 27, and 25 kDa) associated with the light-harvesting antenna of PSII (29, 54, 170).

It is tempting to speculate that six 29-kDa polypeptides form the tightly bound LHC II-intrinsic, whereas both 27- and 25-kDa polypeptides are associated with the LHC II-peripheral. This hypothesis receives partial support from the kinetics of phosphorylation of the light-harvesting complex (101, 102, 107). It was demonstrated that the specific phosphorylation of the 25-kDa polypeptide was greater than that of the 27-kDa polypeptide (107). No phosphorylation of the 29-kDa polypeptide has been reported. It is of interest to observe that phosphorylation of the LHC II results in the functional dissociation of the LHC II-peripheral from the rest of the PSII complex, leaving the LHC II-intrinsic tightly bound on the PSII-core complex (48, 100, 107). Since phosphorylation of LHC II polypeptides is assumed to proceed from the outermost to the intermost subunits, it is implied that 27-kDa polypeptides may comprise the inner complement of the peripheral LHC II,

---

[1]Estimate of the Chl (a +b) content of LHC II from spinach (prepared by the method of Ref. 29, based on amino acid analysis) was 15.5 ± 1.0 per 28,500 daltons of protein (A. N. Glazer, unpublished data).

whereas 25-kDa polypeptides make up the outer complement of the peripheral LHC II. This hypothetical configuration of the light-harvesting accessory antenna of PSII bears some resemblance to the phycobilisome (PBS) of cyanobacteria. Structurally and functionally, the phycobilisome is divided into core-cylinder complexes composed of allophycocyanin (equivalent to the LHC II-intrinsic) and peripheral rods containing phycocyanin/phycoerythrin (equivalent to the LHC II peripheral). Unlike the PBS, which is a largely extrinsic complex bound to the protoplasmic surface of the thylakoid membrane of cyanobacteria, the LHC II polypeptides are transmembrane with a significant portion of the protein protruding from the lipid bilayer on both sides (7, 110).

It is known that polypeptides of the LHC II are coded for by nuclear genes. Work by Dunsmuir (55), among others, has uncovered the existence of five distinct Chl a/b LHC gene families. On the basis of nucleotide sequencing information, some heterogeneity in both length and polypeptide composition was predicted (55). However, the physiological significance of the multigene families is not known. Clearly, more research is required to reveal unequivocally the molecular and membrane level complexity of the accessory antenna of PSII. However, it is useful to bear in mind the structural-organizational and functional parallel between the cyanobacterial PBS and the higher plant LHC II.

PHOTOSYSTEM I

Photochemical Reaction Center Complex

The photochemical reaction center of PSI is a specialized chlorophyll dimer termed P700 because of the absorbance change at 700 nm associated with its photooxidation (82, 196). Recent evidence suggests that the structure of the primary acceptor of the photochemical reaction center I is a 13-hydroxy-20-chloro Chl a (51). Associated with the reaction center are a number of PSI electron acceptors (A_0, A_1, X_1, B, and A). The primary electron acceptor A_0 is believed to be a specialized reaction center-bound Chl a molecule (11, 80, 87a, 117, 164, 165). The secondary electron acceptor of PSI (A_1) could be a quinone molecule. Recent evidence in the literature suggests the presence of phylloquinone molecules that copurify with PSI preparations (162, 179). The subsequent electron acceptors X, B, and A are bound iron-sulfur centers (82, 113, 164); each center contains 4 Fe and 4 labile S atoms (112). When excitation energy from the light-harvesting pigment arrives at the reaction center, an endergonic charge separation reaction occurs between P700 and A_0, followed by further electron transfer steps, according to the reaction mechanism:

$$P700 \; A_0 \; A_1 \; XBA \xrightarrow[\text{energy}]{\text{light}} P700^* \; A_0 \; A_1 \; XBA \xrightarrow{\leq 10 \text{ ps}} P700^+ \; A_0^- \; A_1 \; XBA,$$

$$P700^+ A_0^- A_1 XBA \xrightarrow[\text{ps}]{\sim 100} P700^+ \; A_0 \; A_1^- \; XBA \longrightarrow P700^+ \; A_0 \; A_1 \; X^- BA.$$

Characteristic of the electron-acceptor molecules on the reducing side of PSI is the extremely negative oxidation-reduction potential ($E_{m,7}$) estimated as follows: $(P700/P700^+) = +0.48$ V, $(A_0^-/A_0) = \sim -1.2$ V, $(A_1^-/A_1) = \sim -1.0$ V, $(X^-/X) = \sim -0.73$ V, $(B^-/B) = -0.59$ V and $(A^-/A) = -0.55$ V (57).

Chlorophyll a "Core" Antenna Complex

The photochemical reaction center of PSI (P700) is associated with a 70-kDa tetramer (112, 191) which, in addition to P700, binds about 130 Chl a and 16 carotenoid molecules. Thus, each 70-kDa polypeptide may bind about 32 Chl a (approximately 1 Chl a per 2-kDa protein) and 4 carotenoid molecules. The 70-kDa tetramer along with the 130 Chl a molecules form the Chl a "core" antenna of PSI. The structural makeup of the photochemical reaction center of PSI within this tetramer is not totally clear. Evidence in the literature suggests that in intermittent light-developing chloroplasts and in Chl deficient mutants, i.e. under conditions of limited Chl biosynthesis, PSI is fully functional with only about 90 Chl a molecules in its light-harvesting antenna (121, 181). Similarly, treatments with detergents have successfully resulted in the isolation of a fully functional PSI complex containing only about 60 Chl a molecules (137, 138). It is tempting to speculate that the reaction center of PSI is stabilized by and/or bound to the inner 70-kDa dimer (13, 14), whereas the other 70-kDa dimer forms the "outer" core antenna complex of PSI.

The purified PSI complex contains a number of polypeptides, in addition to the 70-kDa protein. Nelson and co-workers identified subunits of 25, 20, 18, 16, and 8 kDa in native PSI preparations (13, 14). It is believed that subunits of 20 and 16 kDa are closely associated with the large 70-kDa complex (141). Lundell et al (112) defined the following subunit stoichiometry in native PSI preparations from *Synechococcus* 6301: [70 kDa]$_4$[18–20 kDa]$_1$[17.7 kDa]$_1$[16 kDa]$_1$[10 kDa]$_2$. Lagoutte et al (103) investigated the functional role of the minor 8-kDa band in PSI particles. They reported that both in vivo ^{35}S labeling and carboxymethylation with iodo[^{14}C]acetate showed that most cysteine residues are located in this 8-kDa polypeptide, which suggested that it might contain the iron-sulfur centers X, B, or A. The work of Bonnerjea et al (21) indicated the presence of the iron-sulfur centers B and A in a PSI complex containing only the 70-kDa and 19-KDa polypeptides. More conclusively, the work of Sakurai & San Pietro (160) suggested that Fe-S centers are covalently bound to the large 70-kDa PSI subunit(s) containing the reaction center P700.

Chlorophyll a/b Light-Harvesting Complex I (LHC I)

The accessory light-harvesting antenna of PSI (LHC I) contains 60–80 Chl a and Chl b molecules (Chl a/Chl b = 3). Thus, the total functional antenna Chl of PSI contains 200–210 Chl $(a+b)$ molecules, only slightly smaller than that of PSII (121, 123, 145). Polypeptides associated with LHC I migrate in the 20–25 kDa region (3, 5, 79) and are immunologically distinct from those of LHC II (106). The proteins associated with LHC I have been further fractionated by Triton into distinct complexes (105). The lighter sucrose gradient fraction contained two polypeptides with relative molecular masses of 20 and 23 kDa (LHC Ia) and an emission maximum at 730 nm. The heavier sucrose gradient fraction was enriched in a 20-kDa polypeptide (LHC Ib) with a fluorescence emission at 680 nm (105). By analogy with LHC II, it is of interest to ask whether LHC Ia and LHC Ib might respectively represent tightly bound and peripheral components of the accessory antenna of PSI.

PHOTOSYSTEM STOICHIOMETRY

A significant development in our understanding of the structural and functional organization of the chloroplast photosystems is the recognition that the stoichiometry of PSI and PSII complexes is markedly variable among different photosynthetic tissues (124). Cyanobacteria commonly display reaction center RCII/RCI ratios of 0.3–0.8 (114, 116, 126, 139) while higher plant chloroplasts from spinach and from other species grown under direct sunlight have a RCII/RCI reaction center ratio of 1.4–1.9 (121–127). Interestingly, higher plant chloroplast mutants deficient in the Chl a-b LHC display RCII/RCI > 2 (1, 65, 181) (Table 1).

The observation that RCII/RCI ratios are different from unity and show a large variation among different photosynthetic membranes is significant be-

Table 1 Photochemical apparatus organization in cyanobacteria, higher plant chloroplasts, and chlorophyll b mutants

	$\dfrac{\text{Chl } a}{\text{Chl } b}$	$\dfrac{\text{Chl}}{\text{PSI}}$	$\dfrac{\text{Chl}}{\text{PSII}}$	$\dfrac{\text{RCII}}{\text{RCI}}$
Synechococcus 6301	—	160	370	0.43
Higher plant chloroplasts, wild type	2.8	590	330	1.8
Tobacco Su/su	4.7	330	120	2.7
Barley chlorina f2	∞	300	100	3.0

cause it is contrary to the conventional prejudice (from an extremely literal interpretation of the Z-scheme) that the ratio of the two photosystems should always be equal to 1.0 (199a). This question should be viewed in terms of the optimal utilization of light in oxygenic photosynthesis, which requires approximately equal utilization of light by the two photoreactions, rather than equal photosystem stoichiometry. A summary of the structural and functional organization of the photochemical apparatus of oxygenic photosynthesis is given below.

EXCITATION ENERGY DISTRIBUTION IN THE THYLAKOID MEMBRANE OF OXYGENIC PHOTOSYNTHESIS

A meaningful evaluation of the balance between light absorption and electron transport in the two photosystems of cyanobacteria and higher plant chloroplasts would require a rigorous assessment of a number of parameters including the photosystem stoichiometry, size and composition of the light-harvesting pigments of each photosystem, light absorption in the grana versus that in the stroma-exposed thylakoids of chloroplasts, and light filtering through the leaf tissue. A limited assessment of the effect of these parameters is presented here.

Cyanobacteria

The photochemical apparatus organization and photosystem stoichiometry in the thylakoid membrane of *Synechoccus* 6301 is schematically shown in Figure 1 (114, 139). Each PSI complex contains a light-harvesting antenna of about 140 Chl a molecules, whereas PSII complexes contain only about 35 Chl a molecules. The lower stoichiometric quantities (RCII/RCI = 0.43, see Table 1) and the lower Chl a antenna size ($N_{II} = 35$, $N_I = 140$) of PSII in cyanobacteria are countered by the association of PSII with the phycobilisome (PBS). Each PBS contains approximately 400 bilins (67–69, 114) organized in the form of phycocyanin (PC) and allophycocyanin (AP) complexes in the peripheral rods and core cylinders, respectively (see Figure 1, also Ref. 67–69). Two PSII complexes, each functionally connected to one of the PBS core cylinders (115), compete for excitation energy from the same PBS. Figure 2 presents the absorbance spectrum of intact *Synechococcus* 6301 cells (solid line) and, upon deconvolution, the absorbance spectra of PBS-PSII (dashed line) and of PSI (dotted lines) in the cell. The integrated absorbance of light by the PBS-PSII complexes (area defined by the dashed trace) is approximately equal to the integrated absorbance of light by the PSI complexes in the cell (area defined by the dotted trace), supporting the notion that

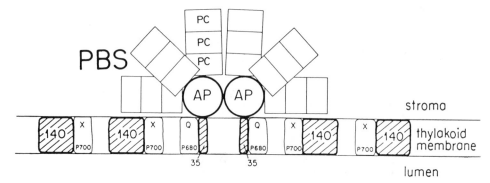

Figure 1 Photochemical apparatus organization and photosystem stoichiometry in *Synechococcus* 6301. The thylakoid membrane integral components include the RC complexes of PSI (P700 X) and PSII (P680 Q), each associated with a distinct Chl-protein light-harvesting complex. The light-harvesting antenna of PSI contains approximately 140 Chl molecules, that of PSII about 35. The PBS is loosely bound to the Chl-protein of two PSII complexes via the core cylinders. Each PBS contains about 400 bilin molecules organized in the form of phycocyanin (PC) and allophycocyanin (AP) complexes.

overall absorption of light by the PBS-PSII complexes in *Synechococcus* 6301 cells is equal to that of the PSI complexes.

The above analysis provides a rationale for understanding the low RCII/RCI < 1 stoichiometry in cyanobacteria and suggests that a photosystem stoichiometry need not be equal to 1.0 in order to ensure balanced absorption of light by PSII and PSI complexes in the thylakoid membrane.

Wild-Type Higher Plant Chloroplasts

A summary of the photochemical apparatus organization and photosystem stoichiometry in higher plant chloroplasts (spinach, pea, tobacco, barley) is presented in Table 2 (6, 123). About 65% of all Chl is associated with PSII and only about 35% is associated with PSI. However, PSII is handicapped in terms of light absorption at three levels (molecular, membrane, leaf). (*a*) Most of the Chl *b* in the chloroplast is associated with PSII; however, Chl *b* is not as efficient in light absorption as Chl *a*, which results in attenuation of light absorption by PSII (molecular level). (*b*) Further attenuation of light absorption by PSII occurs as a result of segregation of most of PSII in the grana partition regions where mutual pigment shading is greater than that in stroma-exposed lamellae (membrane level) (93a). (*c*) Attenuation of light absorption owing to light filtering through the leaf is more pronounced for PSII (leaf level). Hence, the average RCII/RCI = 1.7 in higher plant chloroplasts may reflect a response of the plant in countering the multilevel attenuation of light absorption by PSII and in establishing approximately equal absorption and utilization of light among the two photosystems (127).

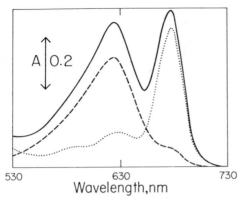

Figure 2   Absorbance spectra of *Synechococcus* 6301 cells (——) and the partial spectra of PBS-PSII (– –) and PSI (···) in the cell. The partial spectra were obtained upon deconvolution of the in vivo absorbance spectrum of *Synechococcus* 6301.

Mutations Affecting the Development and/or Assembly of LHC II

Some photosynthetic mutants are deficient in the peripheral Chl *a/b* LHC of PSII and PSI. Such mutants possess fully functional photochemical reaction centers and other electron transport intermediates. However, they show Chl *a*/Chl *b* ratios that are significantly higher than the corresponding wild type, and they often show altered thylakoid membrane ultrastructure. Since 80% or more of the Chl antenna of PSII is comprised of the Chl *a/b* LHC II, whereas only about 30% of the Chl antenna of PSI is Chl *a/b* LHC I, it follows that in LHC-deficient mutants the light-harvesting capacity of PSII is lowered substantially more than that of PSI (126).

  Since the mutation reduced primarily the light-harvesting capacity of PSII, elevated RCII/RCI ratios (Table 1) may be viewed as restoring the balance of light absorption between PSII and PSI. This is an example in which a specific mutation caused a disproportionate reduction in the number of the antenna Chl

Table 2   Photosystem stoichiometry and chlorophyll distribution in wild type higher plant chloroplasts

	$PSII_\alpha$	$PSII_\beta$	PSI
Stoichiometry	1.3	0.4	1.0
Antenna size Chl($a+b$)	230	100	200
Chl *a*	145	90	180
Chl *b*	85	10	20
Total Chl percent	56	7	37

molecules of PSII and in which the plasticity of the organelle (chloroplast) allowed an elevated RCII/RCI in the thylakoid membrane, in essence countering the effect of the mutation (126).

The guiding principle for understanding the wide variation in RCII/RCI ratios among different photosynthetic membranes is the following: Within the constraints dictated by the size and composition of the light-harvesting antenna of each photosystem, the thylakoid membrane of oxygenic photosynthesis will adjust and optimize the photosystem stoichiometry to ensure a balanced absorption of light by PSII and PSI complexes (116, 126).

REGULATION OF ELECTRON TRANSPORT

The regulation of the process of light absorption and the regulation of the rate of electron transport between the three main integral thylakoid membrane complexes has attracted considerable interest among investigators in photosynthesis. Since the beginning of the present decade, studies of two aspects of this regulation have progressed independently and almost simultaneously. The first involved the important discovery by Bennett (15) of a reversible phosphorylation of the LHC II of PSII. The physiological significance attributed to this phenomenon was the regulation of excitation energy distribution between PSII and PSI in higher plant chloroplasts (16). It was hypothesized that phospho-LHC II separates from the PSII complex in the grana partition region and upon migration into the non-appressed membrane region it couples energetically to PSI complexes where it serves as a PSI accessory antenna (12). The first part of this model has received considerable support from a variety of experimental approaches. The energetic coupling of phospho-LHC II with PSI, however, is currently a controversial issue; more research is required to provide unambiguous answers.

The second aspect of the regulation of chloroplast structure and function involved the discovery of changes in the PSII/PSI reaction center ratio in the thylakoid membrane of chloroplasts and cyanobacteria. These changes occurred as a response to variation in the quality of light during plant growth (71, 72, 116, 121, 125, 126, 139). Thus, it became evident that both higher plant chloroplasts and cyanobacteria will compensate for an imbalance in the rate of light absorption between the two photosystems by changing the relative concentration of the two photosystem complexes in the thylakoid membrane. The adjustment in the photosystem stoichiometry was necessary and sufficient to establish an overall balanced absorption of light by PSII and PSI (116, 126).

The Effect of LHC II Phosphorylation on PSI

Phosphorylation of the LHC II is believed to be the underlying mechanism of State transitions (77). Work by several investigators (15, 16, 85, 100, 101,

107, 175) has helped uncover the importance and consequences of LHC II phosphorylation with respect to the function of PSII. The physiological significance of the phenomenon of LHC II phosphorylation remains a matter of debate, however. Currently, this question impinges upon the fate of the phospho-LHC II in the stroma-exposed lamellae, where it presumably moves following the covalent binding of phosphate. There is controversy in the literature on whether phospho-LHC II regulates excitation distribution by reversibly binding to PSI. Results obtained from low temperature fluorescence emission spectra in control and phosphorylated membranes were presented in support of such a regulatory function (12, 61, 77). However, the interpretation of low temperature fluorescence emission spectra is ambiguous, because the technique is rather indirect. The few direct measurements assaying the absorption cross-section properties of PSI before and after phosphorylation are not all in agreement, either. The first such measurement by Haworth & Melis (78), assaying the rate of light absorption by PSI before and after phosphorylation, failed to show any differences. Telfer et al (180) used broad band 600-nm laser flashes and compared the flash saturation response of P700 (ΔA_{820}) before and after phosphorylation. Following phosphorylation of the LHC II (as well as in the absence of Mg^{2+} ions), they measured a significant enhancement in the amplitude of the ΔA_{820} signal at subsaturating flash intensities and a smaller relative increase of the ΔA_{820} signal at saturating flash intensities (180). The enhancement of the ΔA_{820} signal was interpreted as a PSI absorption cross-section increase following phosphorylation.

On the other hand, Kramer et al (97) presented detailed excitation spectra of PSI and PSII in green algae in State 1 (dark adapted) or State 2 (preilluminated). State 2 is widely assumed to correspond to the phosphorylation of the LHC II under in vivo conditions (77). These measurements were performed at liquid helium temperatures that allowed reasonable resolution of individual excitation bands at 650 nm (Chl *b*) and 670–690 nm (Chl *a*). Kramer et al (97) failed to detect any enhancement in excitation energy transfer from Chl *b* to PSI upon phosphorylation of the LHC II (State 2). More recently, Larsson et al (108) measured the light saturation curve of PSI electron transport before and after phosphorylation of the LHC II in isolated thylakoid membranes. They used 650-nm excitation arguing, correctly, that Chl *b* excitation ought to amplify the effect of a phospho-LHC II association with PSI. Based on the extent of LHC II phosphorylation in their samples, they expected a 30–40% change in the apparent absorption cross section of PSI. They reported an "upper limit" of an 8% change, i.e. well below what was expected on the basis of LHC II phosphorylation and probably within the standard error of their measurement (see also 85). Similarly, Deng & Melis (48) concluded that the significant loss of light-harvesting capacity by PSII, occurring upon phosphorylation of the LHC II (48, 100, 107,

175), is not accompanied by a gain in light-harvesting capacity by photosystem I.

The Physiological Significance of LHC II Phosphorylation

Recent work from several laboratories revealed the manner in which oxygen-evolving photosynthetic organisms, from cyanobacteria to higher plants, respond to environmental light-quality and light-intensity changes (4, 116, 126). The response involves changes in the stoichiometry of electron-transport complexes and changes in the Chl antenna of the two photosystems that are adjusted and optimized in order to ensure balanced electron flow under the particular plant growth conditions. Emphasis has been placed on light-quality changes during plant growth that simulate a State 1 $\longrightarrow$ State 2 transition, and vice versa (116, 126). It was suggested that state transitions, as measured by the change in the steady-state course of fluorescence yield and of the rate of oxygen evolution (20, 195), and the underlying protein phosphorylation of thylakoid membrane proteins in vivo (77) may constitute the initial events leading to change in the stoichiometry of electron-transport and Chl-protein complexes. The phosphorylation of the LHC II, in this case, could earmark subunits of this polypeptide for removal and degradation. It must be noted that environmental light conditions that favor phosphorylation of the LHC II also result in the removal of this complex from the thylakoid membrane. Is it then possible that phosphorylation of thylakoid membrane polypeptides serves as a signal in an elaborate regulatory mechanism designed to identify surplus polypeptides and/or complexes for further processing?

CYANOBACTERIA: OVERVIEW

The photochemical apparatus of cyanobacteria consists of three macro-molecular complexes: (a) the phycobilisome, which serves as the light-harvesting antenna for PSII. This antenna complex is unique to cyanobacteria and red algae, neither of which contains Chl a/b proteins. (b) PSI, resembling the "core" PSI complex of chloroplasts stripped of its LHC I antenna, and (c) PSII, which likewise resembles the PSII "core" complex of chloroplasts (28, 206–208). Recent comprehensive reviews have dealt with phycobilisome structure and function (67–70) and with the molecular genetics of phycobilisome components (28). A very detailed review of the current knowledge of the structure, function, and molecular genetics of the photosynthetic, electron, and proton transport components of the cyanobacterial thylakoid membrane has appeared (28). A complete analysis of the data on cyanobacteria would duplicate the latter review as well as some of the information already given on the photosystems of higher plants. Consequently, only the general conclusions are presented below.

A high degree of sequence homology is seen between the genes encoding polypeptide components of PSI, PSII, the ATPase complex, and cytochrome b_6-f in cyanobacteria and in higher plant chloroplasts (28). Moreover, to the degree that the spectroscopic signatures and prosthetic group compositions have been compared between cyanobacterial and chloroplast PSI and PSII reaction center complexes, no significant differences have emerged.

Spectroscopic (114–116) and fractionation (34, 35, 99) studies support the view that the phycobilisome is physically linked to PSII and transfers energy virtually exclusively to this photosystem. The physical link between the phycobilisome and the PSII complex is very readily disrupted. This is an interesting point because the link between LHC II and PSII is likewise weak. One can view the phycobilisome and LHC II as alternative light-absorbing modules grafted onto a common type of PSII core complex in different organisms. In each case the accessory antenna-PSII contact region might involve relatively limited protein-protein contacts. An inverse situation is discussed below for green bacteria, where the same type of light-harvesting module, the chlorosome, is grafted onto different types of reaction center complexes in different organisms.

PURPLE BACTERIA

The purple bacteria are divided into two families: the Chromatiaceae (purple sulfur bacteria) and Rhodospirillaceae (purple nonsulfur bacteria) (152). In both of these families, the entire photosynthetic apparatus is located in the cytoplasmic membrane. In most purple bacteria, growing photosynthetically, the cytoplasmic membrane is enlarged by infolding into the cytoplasm to form what is generally referred to as intracytoplasmic membranes (ICM). In thin section, these infoldings can appear as vesicles, tubes, or lamellar stacks. In a few cases [e.g. *Rhodopseudomonas (Rps.) gelatinosa]*, there are no infoldings and the photosynthetic apparatus is housed within the cytoplasmic membrane. When ICM are present, the photosynthetic apparatus is localized within these invaginations (53, 168).

Bacteriochlorophylls (BChl) *a* and *b* are the major light-harvesting pigments. Either one or the other BChl is present, but the nature of the BChl does not correlate with either family or genus (Table 3). Various carotenoids are also components of the photosynthetic apparatus and contribute both to light-harvesting (36) and photoprotection (74).

Purple bacteria can be divided into three categories on the basis of the types of their light-harvesting (LH) complexes. Group I organisms contain a single type of LH complex that is present at a constant ratio to the reaction center (RC) and is independent of growth conditions (166). For example, in *Rps. viridis* this is the complex designated B1015, and in *Rhodospirillum rubrum* it

Table 3 BChl a and b distribution among purple bacteria

Organism	BChl a^a	BChl b^a
Purple sulfur bacteria		
Chromatium D	+	
Ectothiorhodospira halochloris		+
Ectothiorhodospira abdelmalekii		+
Thiocapsa floridana	+	
Thiocapsa pfennigii		+
Purple nonsulfur bacteria		
Rhodobacter capsulatus[b]	+	
Rhodobacter sphaeroides[b]	+	
Rhodopseudomonas palustris	+	
Rhodopseudomonas viridis		+
Rhodopseudomonas sulfoviridis		+
Rhodospirillum rubrum	+	

[a]Characteristic absorption maxima of bacteriochlorophylls in living cells are 375, 590, 800–810, 830–890 nm for BChl a and 400, 605, 835–850, 1015–1035 nm for BChl b (151). The absorption maxima for BChl a and b in ether are at 775 nm and 790 nm, respectively.
[b]These species were formerly included in the genus *Rhodopseudomonas* (88).

is B880, where "B" refers to "bulk," or light-harvesting, BChl, (RC BChl is designated by the prefix "P") and the number refers to the absorption maximum in nanometers. The organisms mentioned above contain only this type of LH. Group II organisms contain complexes homologus to the ones found in group I bacteria and, in addition, complexes designated B800–850. Group II organisms include two extensively studied bacteria, *Rhodobacter (Rb.) capsulatus* and *Rb. sphaeroides*. Group III organisms (e.g. *Chromatium vinosum, Rps. acidophila* 7050) contain two complexes homologous to those seen in group II bacteria and in addition contain a third class of complexes designated B800–820. It is plausible to infer that group II- and group III-type LH systems evolved from group I-type LH systems by the consecutive addition to the periphery of the assembly of LH components absorbing at higher energies, leading ultimately in group III LH apparatus to the energy transfer sequence (1a):

$$B800–820 \longrightarrow B800–850 \longrightarrow B880 \longrightarrow RC$$

It is evident from the foregoing that group I organisms have the simplest photosynthetic apparatus, and it is a fortuitous circumstance that the first determination of the structure of a photosynthetic reaction center was performed on a complex isolated from *Rps. viridis,* a member of this group. The photosynthetic apparatus of this organism is discussed below in some detail.

General Properties of Purple Bacterial Reaction Centers

Most of the photosynthetic reaction center complexes from purple bacteria consist of one copy each of three membrane-spanning polypeptide subunits named H (heavy), M (medium), and L (light), on the basis of their apparent molecular weights as determined by SDS-polyacrylamide gel electrophoresis (44). This nomenclature is unfortunate, since the actual molecular weights of these polypeptides deviate significantly from the apparent values (Table 4). In fact, the H subunits are the shortest of the three polypeptides (130, 131, 201, 202, 209).

Detailed studies of the *Rps. sphaeroides* reaction center complex have revealed that it contains four molecules of BChl a, two molecules of bacteriophaeophytin (BPh a), one molecule of carotenoid (absent in reaction centers isolated from carotenoidless mutants), two molecules of ubiquinone-10 (Q_A and Q_B), and one atom of Fe^{2+} (59, 144). Reaction centers from other purple bacteria have a similar prosthetic group composition. The H subunit can be removed from the *Rb. sphaeroides* complex (44), or that of *Rs. rubrum* (200), and the residual LM complex retains the pigments and photochemical activity. The removal of H does perturb the electron transfer from Q_A to Q_B (44). Both Q_A and Q_B are magnetically coupled to the Fe^{2+}. However, the electron transfer characteristics of Fe-depleted reaction centers reconstituted with Mn^{2+}, Co^{2+}, Ni^{2+}, Cu^{2+}, and Zn^{2+} were the same as those of native complexes (43). Comparison of Fe-depleted and native reaction centers showed that the presence of a divalent metal ion in the Fe site is necessary to establish the native electron-transfer properties of Q_A. It is possible that the Fe^{2+} plays more than one role (see Ref. 44). The sequence of electron transfer events following excitation of the reaction center was delineated by a variety

Table 4 Apparent and actual molecular weights of the reaction center and light-harvesting complex polypeptides of *Rhodopseudomonas viridis*

	Apparent M_r^a	Number of residues	Actual M_r^b	Percent error in apparent M_r
Reaction center polypeptides				
M	28,000	323	35,902	−22.0
L	24,000	273	30,531	−21.4
H	35,000	258	28,345	+23.5
Polypeptides of the B1015 LH complex				
α	11,000	58	6,848	+60.6
β	8,000	55	6,138	+30.3
γ	4,000	36	4,200	− 4.8

[a]Determined from electrophoretic mobility of "denatured" polypeptides on polyacrylamide gels in the presence of sodium dodecylsulfate (182).
[b]Calculated from the amino acid sequence (24, 130, 131).

of spectroscopic studies (147) and is examined in conjunction with the consideration of the structure of the *Rps. viridis* reaction center.

Rhodopseudomonas viridis Reaction Center Complex

The landmark crystallographic studies of this reaction center (45, 46), coupled with the determination of the sequences of its constituent subunits (130, 131), have led to the description of this complex at nearly atomic resolution. In addition to the H, L, and M subunits, the complex contains a tightly bound four-heme cytochrome molecule. The L and M subunits each consist of five long transmembrane helices. The H subunit possesses one long transmembrane helix at its amino terminus. The cytochrome subunit appears to be attached to the complex at the periplasmic surface (46). The disposition of the prosthetic groups within the complex is shown in Figure 3. In contrast to *Rb. spheroides,* the *Rps. viridis* complex contains BChl *b* and BPh *b*; the tightly bound quinone (Q_A) is menaquinone-9 (163); the second quinone (Q_B) is more loosely bound and is missing from the crystals (45, 46). The location of the Q_B site was deduced from examination of the binding sites for three different competitive inhibitors of Q_B binding.

The "special pair" of BChls, postulated to function as the primary electron donor on the basis of spectroscopic results (142), is indeed present and is designated "P" (Figure 3). In P, the distance between the BChl *b*-Mg ions is

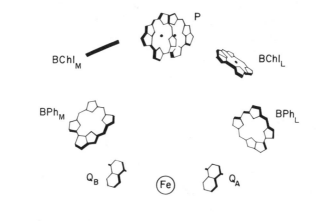

$$P^* BPh \xrightarrow{2-4\,ps} P^+ BPh^- \xrightarrow{200\,ps} P^+ Q_A^- Q_B \xrightarrow{10^{-4}\,s} P^+ Q_A Q_B^-$$

Figure 3   Approximate arrangement of the chromophores in the *Rps. viridis* reaction center (modified from Fig. 1 of Ref. 203). The abbreviations used are: BChl, bacteriochlorophyll *b*; P, "special pair" BChl dimer; BPh, bacteriophaeophytin *b*; Q_A, menaquinone; Q_B, ubiquinone. The subscripts L and M refer to the L and M subunits of the reaction center. The approximate position of the Q_B site was inferred from crystallographic studies of reaction center complexes with competitive inhibitors (45).

~7Å, and the angle between ring planes is ~15°. As shown in Figure 3, a BChl b is adjacent to each of the BChl b molecules of P at a Mg-Mg distance of ~13 Å and angles between ring planes of ~70°. Each of these monomeric BChl b is in close proximity to a BPh b with a distance between ring centers of ~11 Å and angles between ring planes of ~64°. The isoprenoid side-chain of the menaquinone-9 in the Q_A site approaches one of the BChl b molecules (BPh b_M) to within van der Waals distance. The Fe^{2+} approaches the menaquinone to within ~7 Å (45). Upon excitation of the reaction center, the electron moves consecutively from P* to BPh_L, to Q_A, and thence to Q_B. The kinetics of these steps are given in Figure 3. The photooxidized primary donor is reduced by the cytochrome c molecules in ~270 nanoseconds (84). Excitation of the reaction center by a second photon sends another electron by the same path to Q_B, generating the quinol Q_BH_2. The latter dissociates from the reaction center and carries electrons and protons to a cytochrome bc_1 complex (204).

It will be noted that the ring systems of the four BChl and two BPh pigments are related by a local twofold rotation axis that runs between the two BChls constituting P on the periplasmic side of the membrane and through the Fe on the cytoplasmic side of the membrane. At the level of pigment organization, therefore, there are two structurally equivalent branches that could be used for electron transfer across the membrane (45). The symmetry is broken by the distinctive protein environment of each of the pigments in the two branches. The absorption bands of the two BPh molecules are well resolved at low temperature (5 K) and are located at 530 and 545 nm, respectively. Only the 545-nm band is bleached at a short time after excitation. The bleaching of this band is presumed to reflect the reduction of BPh_L. The absorption band of the other BPh is only slightly affected (148). Analysis of the absorbance spectra of crystals of the *Rps. viridis* reaction center complex, taken with plane-polarized light, shows that the BPh absorbing at higher wavelength is the one closer to Q_A (211). These experiments establish that electron transfer proceeds only along the right-hand path (in Figure 3).

Rhodopseudomonas viridis B1015 Antenna Complex

Rps. viridis contains a single type of antenna complex. Separation of this complex in unperturbed form from the reaction center has yet to be achieved. The building block of the B1015 complex consists of polypeptides α, β, and γ in a 1:1:1: ratio (see Table 4 for M_r values). Per $\alpha\beta\gamma$, the complex contains 2 BChl b and 1 carotenoid (1,2-dihydrolycopene and 1, 2-dihydroneurosporene) (24, 37, 186). By comparison with the LH-type I complexes from other organisms, it is inferred that these pigments are carried by the α and β polypeptides. The sequences of the α, β, γ polypeptides are given in Figure 4. In common with the homologous polypeptides from other purple bacteria [e.g. the B870 α and β polypeptides of *Rs. rubrum* (25) and *Rp. gelatinosa* (213), the B800–850 polypeptides of *Rb. capsulatus* (177, 178, 209, 210)], the α and β polypeptides of *Rps. viridis* have a polar N-terminal domain, a hydrophobic central region, and a charged C-terminal domain (Figure 4). An $\alpha\beta$ pair is envisaged to serve as the basic structural unit of all the bacterial light-harvesting complexes (37, 213). The transmembrane orgnanization of such units has been examined by a wide range of techniques including antibody labelling (91, 93), surface-specific iodination (90), chemical cross-linking (111, 149), and susceptibility to degradation by

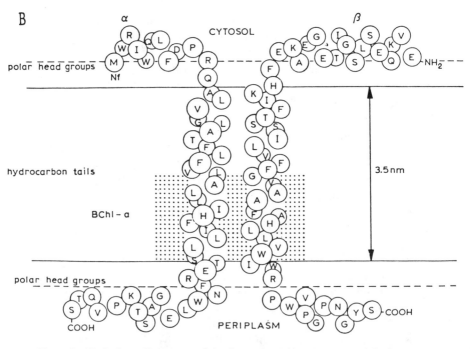

A
α ATEYRTASWKLWLILDPRR<u>VLTALFVYLTVIALLIHFGLL</u>STDRLNWWEFQRGLPKAA

β ADLKPSLTGLTEEEAKEFHG<u>IFVTSTVLYLATAVIVHYLVW</u>TAKPWIAPIPKGWV

γ YFAADGSVVPSISDWNLWVPLGILGIPTIWIAL$_V^T$YR

Figure 4 (A) Amino acid sequence of the three polypeptide components of the *Rps. viridis* light-harvesting complex: B 1015-α, B1015-β, and B1015-γ (24). The hydrophobic middle domain in the α and β subunits is underlined. (B) A proposed model for the disposition of the α and β polypeptides of the B890 light-harvesting complex of *Rs. rubrum* with respect to the photosynthetic membrane (from Ref. 27).

various proteases (26). The results have been summarized in several comprehensive recent reviews (9, 37, 52, 212, 213), which should be consulted for references to many additional studies. Models of the organization of these complexes, such as the one shown in Figure 4 for the B890 $\alpha\beta$ complex of *Rps. rubrum,* have been proposed on the basis of these studies.

Organization of the Rhodopseudomonas viridis Photosynthetic Membrane

Electron microscopy of thylakoid membranes from *Rps. viridis* at resolution of 16–20 Å reveals that this membrane contains extensive regular arrays of circular particles, 125–130 Å in diameter, that are arranged on a two-dimensional hexagonal lattice (66, 92, 133). The particles consist of a core 45 Å in diameter, surrounded by a ring ~20 Å wide in projection, and show an apparent sixfold rotational symmetry. This "crystalline" organization of the thylakoid membrane is shared by several other BChl *b*-containing purple bacteria (56) as well as the BChl *a*-containing *Rs. rubrum* (128, 143, 187), and it correlates with the presence of only one light-harvesting complex, LHI (56). The crystalline arrays are not seen in the thylakoids of those bacteria that possess both LHI and LHII. It is interesting to note that ATPase complexes are absent from the crystalline membrane regions (89).

Several lines of evidence indicate that in *Rps. viridis,* the circular particles, termed "photoreceptor units," consist of a reaction center surrounded by twelve LH $\alpha\beta\gamma$ complexes. Measurements on two-dimensional (132) and three-dimensional (45) crystals of isolated *Rps. viridis* reaction centers have shown that the cross section of the reaction center in the membrane plane is roughly elliptical, 4.5 × 6.0 nm on the average (45). The 45-Å diameter core of the photoreceptor units would therefore correspond to a single reaction center. Since at the molecular level, the reaction center cannot have sixfold symmetry, the Fourier maps of the photoreceptor units presumably represent averages over six very similar possible orientations of a structural unit without intrinsic rotational symmetry arranged into a hexagonal lattice (174). The *Rps. viridis* antenna consists of 24 BChl *a* per reaction center (22). On the assumption that each $\alpha\beta$ unit carries two BChl *b* molecules, the apparent sixfold symmetry suggests that the core is surrounded by a ring of six $\alpha_2\beta_2\gamma_2$ complexes (174). The inferred arrangement of LH polypeptides is supported by results of cross-linking experiments on thylakoid membranes. Tetramers ($\alpha_2\beta_2$) were found among the products of reaction with cross-linkers of 6-Å span width. This indicates that $\alpha\beta$ units must be closely associated with one another (111). Other cross-linking experiments have indicated a close approach of the reaction center polypeptide H and B1015-β (149). The approximate volume of the photoreceptor units is compatible with a molecular weight of ~360,000 calculated for a complex of $(\alpha_2\beta_2\gamma_2)_6$ with one reaction center.

Site-specific mutagenesis has thus far been applied only to the study of the assembly of the photoreceptor units of *Rb. capsulatus,* which contains both LHI and LHII complexes. Mutants in which light-harvesting complexes are missing still assemble fully functional reaction center complexes. The reverse is also true. Absence of reaction center polypeptides from the membrane is not necessary for the insertion of LHI and LHII into the membrane (30).

Antenna Domain Sizes for Energy Transfer

There is a general consensus that in the purple bacterial thylakoid membrane numerous photoreceptor units are interconnected and cooperate in the trapping of excitation quanta. The evidence for the "lake" model of energy transfer, in which a large number of antenna pigments serve many reaction centers, rests on measurements of the fluorescence yield as a function of the fraction of traps in the state P^+ (193), on the determination of the yield of BChl fluorescence as a function of the intensity of an exciting picosecond flash (10, 33, 189), and on the efficiency of singlet excitation by triplet states (134) or of singlet-singlet annihilation (192). Such measurements indicate in *Rs. rubrum,* for example, an antenna domain size of at least 1000 molecules of BChl (192). From an estimate of 35 BChl per P870 in *Rs. rubrum* (192), it is evident that at least 25 photoreceptor units must be connected. Recall that the thylakoid membrane of *Rs. rubrum* shows a hexagonal lattice of photoreceptor units similar to that described for *Rps. viridis.* The spectroscopic results indicate that the photosynthetic membrane in these bacteria should be viewed as containing clusters of photoreceptor units, varying in size and in dynamic equilibrium, and that the Fourier maps derived from electron microscopy represent mean positions of photoreceptor units averaged over many clusters.

In organisms such as *Rb. sphaeroides,* which contain both B875 and B800–850 complexes, the latter are believed to bind at the periphery of the B875-reaction center complexes (38, 134) and to bridge between the complexes (134). In *Rb. sphaeroides* the antenna domain size is even greater than that in *Rs. rubrum* (192).

GREEN BACTERIA

The green bacteria are subdivided into two families with overall physiological similarities to the purple sulfur and the purple nonsulfur bacteria, respectively. The Chlorobiaceae, of which the most extensively studied are the very similar organisms *Chlorobium limicola* and *Prosthecochloris aestuarii,* are unicellular strict anaerobes that utilize sulfur compounds as electron donors.

The Chloroflexaceae, exemplified by *Chloroflexus aurantiacus,* are faculta-
tively aerobic filamentous gliding bacteria that can utilize reduced carbon
compounds as electron donors. These two families are grouped under the
rubric of "green bacteria" because in both the major light-harvesting pigments, BChl *c, d,* or *e,* are assembled in chlorosomes. These large bag-like
structures are attached to regions of the cytoplasmic membrane that contain
reaction center complexes associated with additional antenna components.
From a variety of studies, it is evident that the chlorosome represents the
single significant common feature (albeit a morphologically striking one)
between the organisms belonging to the two families. In fact, 16S RNA
comparisons indicate that *Chloroflexus* is only very distantly related to any of
the other photosynthetic bacteria (169). Detailed examination of the
photosynthetic apparatus of the green bacteria (see Refs. 2, 18, 19 for recent
reviews) leads to the fascinating conclusion that the same type of light-
harvesting antenna, the chlorosome, is grafted onto membrane-bound reaction
centers of the purple bacterial type in the Chlorobiaceae and onto very
different reaction centers, in many ways resembling chloroplast PSI, in the
Chloroflexaceae.

Antenna Pigment Organization

Currently available data on the structure of the antenna and reaction center
complexes in the green bacteria are summarized in Table 5 and Figure 5. The
size of the chlorosomes and their content of BChl *c* is greater in *C. limicola*
than in *C. aurantiacus,* but the general features of these antenna assemblies
are the same in the two organisms. In the *Chlorobium* chlorosome, it is
estimated that there are 7 BChl *c* per 5.6-kDa polypeptide (198). This is a far
higher ratio of pigment to protein than is seen in any other Chl- or BChl-
containing antenna complex. Detailed models and supporting data have been
provided by Zuber and his colleagues (198) for the assembly of the BChl
c-polypeptide complexes forming the rod structures of the chlorosome core
(Figure 5). In these assemblies, it has been suggested that the polypeptide-
bound BChl *c* molecules interact with each other by intermolecular hydrogen
bonding between the -OH group of ring I and the C=O group of ring V (198).
An alternative proposed mode for BChl *a* oligomer formation involves hydrogen bonding between the -OH group of ring I and the magnesium atom of the
adjacent BChl *c* (167).

   As deduced from the structural data (Figure 5, Table 5), and from
spectroscopic studies, the energy transfer pathway in *Chlorobium* is:

BChl *c* complex → BChl *a* complex → BChl *a* complex → P840 complex.
λ_{max}749 nm λ_{max}794 nm λ_{max}804 nm
Chlorosome   Chlorosome   Baseplate   Membrane

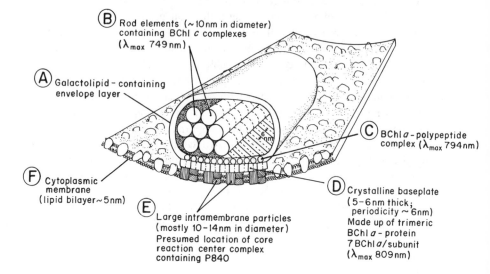

Figure 5   Diagrammatic representation of the organization of the photosynthetic apparatus in *Chlorobium limicola*, based on the data in References 63 and 173.

The corresponding pathway for *Chloroflexus* is (18, 19):

BChl *c* complex $\rightarrow$ BChl *a* complex $\rightarrow$ B806–865 $\rightarrow$ P870 complex.
λ_{max}740 nm λ_{max}790 nm λ_{max}865 nm
Chlorosome   Baseplate   Membrane   Membrane

The *Chloroflexus* B806–865 antenna complex contains a single polypeptide whose amino acid sequence is homologous to that of the α polypeptides of the antenna complexes of purple bacteria.

Reaction Centers

CHLOROBIUM   The prosthetic group composition of the *Chlorobium* reaction center complex is very similar to that of the purple bacteria except that one BChl *a* has been replaced by one BPh *a* and two polypeptides are present rather than three. ESR data suggest a dimer structure for P840$^+$ (reviewed in Ref. 18) and BPh *c* and an Fe-S protein ($E_m = -550$ mV) as primary and secondary electron acceptors, respectively. Three membrane-bound Fe-S centers, reducible by P840*, are present in *Chlorobium*. The properties of these centers are similar to those of Fe-S centers of PSI. The known steps in the electron transfer pathway in the reaction center complex of *Chlorobium* are:

Table 5   Some properties of the photosynthetic apparatus in *Chlorobium limicola* and *Chloroflexus aurantiacus*

Component	Chlorobium	Chloroflexus
Chlorosome	100×30 nm[a] (172)	100×260 nm[a] (173)
Envelope	Glycolipids; polypeptides (40, 83, 161)	Glycolipids; polypeptides of 11 and 18 kDa (60, 83, 161)
Rods	BChl *c*-polypeptide complex; BChl *a*-polypeptide complex (at base of chlorosome) BChl *c* : BChl *a* : RC[c] = 1000–1500 : 80 : 1 (153)	BChl *c*; β and γ-carotene, 7–8 BChl c/M_r 5,592 polypeptide[b], 16–300 BChl *c*/RC
Baseplate	BChl *a*-protein trimer[d] 7 BChl a/M_r 43,750	B790 BChl *a*-protein complex (17)
Cytoplasmic membrane		
Antenna complex	BChl *a*-polypeptide complex (63)	B806-865[e] containing 1 B808 BChl *a*, 2 B865 BChl *a*. BChl *a*: γ-carotene = 2 : 1 (190)
RC core complex	P840(E_m = +240 mV) tightly associated with 20 BChl *a*, BPh *c* and Fe-S center (g = 1.94) E_m = −550 mV. Polypeptides: 65 kDa; 32 kDa Rieske-type Fe-S protein; 24 kDa *c*-type cytochrome (86, 96)	P870 (E_m = +360 mV); 3 BChl *a*; 3 BPh *a*; 2 mena-quinone; polypeptides of 30 and 28 kDa (155, 156)

[a]Heterogeneous in size. Values given are at upper limit of size.
[b]The sequence of this polypeptide is (198) ATRGWFSESSAQVAQIGDIMFQGHWQWVSNAL-QATAAAVDNINRNATPGVR.
[c]RC = reaction center.
[d]The sequence of this protein and its three-dimensional structure have been determined (41, 120, 185).
[e]The sequence of the 4.9-kDa B806–B865 apopolypeptide (named B806–B865-α) is (197) MQPRSPVRTNIVIFTILGFVVALLIHFIVLSSPEYNWLSNAEGG.

$$P840{*}BPh\ c \rightarrow P840^+ \cdot BPh\ c^- \rightarrow P840^+ \cdot Fe - S^-.$$
$$-1200\ \text{mV} \qquad -800\ \text{mV} \qquad\qquad -550\ \text{mV}$$

CHLOROFLEXUS   The electron transfer pathway within *Chloroflexus* reaction centers is very similar to that defined for purple bacterial reaction centers (18, 19):

$$P870{*} \cdot BChl\ a/BPh\ a \rightarrow P870^+ \cdot BChl\ a/BPh\ c^- \rightarrow P870^+ \cdot MQ_A^- \rightarrow P870^+ \cdot MQ_B^-$$
$$-1000\ \text{mV} \qquad\qquad -800\ \text{mV} \qquad\qquad -50\ \text{mV} \qquad \sim 0\ \text{mV}$$

where M is menaquinone.

HELIOBACTERIUM CHLORUM   *H. chlorum* is a recently described strictly anaerobic photosynthetic bacterium that possesses the previously unknown BChl *g* as antenna and reaction center pigment (23, 64). The structure proposed for BChl *g* differs from that for BChl *b* in only one respect: A vinyl group in ring I of BChl *g* occupies the position of the acetyl group of BChl *b* (23). The absorption spectra of BChl *g* and *b* in dioxane do not differ significantly. However, reaction centers containing BChl *b* exhibit red absorption maxima at 960 nm (P960) (42), whereas the reaction center of *H. chlorum* is P798 (62). *H. chlorum* possesses a single undifferentiated plasma membrane containing the photosynthetic pigments, and no chlorosomes are present. Information on the photochemistry of *H. chlorum* reaction centers is still limited. The primary electron donor BChl has E_m = +225 mV; an acceptor Fe-S cluster (E_m = −510 mV), with properties similar to the Fe-S clusters of the acceptors of *Chlorobium,* has been identified (157).

It is interesting to note that phototrophic organisms, lacking in intracytoplasmic membranes but otherwise similar to each of the groups of photosynthetic bacteria, have now been described. As noted, *H. chlorum* has close functional similarities to green sulfur bacteria. A newly described BChl *a*-containing organism, *Heliothrix oregonensis,* which lacks chlorosomes and BChl *c,* has many characteristics in common with *Chloroflexus* (154). *Rhodospirillum tenue* is a purple nonsulfur bacterium lacking intracytoplasmic membranes (73, 194). *Gloeobacter violaceus,* a cyanobacterium lacking thylakoids (158), houses the photosynthetic apparatus in its cytoplasmic membrane to which its phycobilisomes are directly attached (75). Such organisms may be viewed either as ancestors of the more common organisms, with highly specialized intracytoplasmic membrane systems, or as derivatives that have lost their ability to elaborate such membranes or chlorosomes.

ACKNOWLEDGMENTS

We wish to thank Ms. Janice Quartieri for typing the manuscript. The authors' work has been supported by grants from the USDA-Competitive Research Grants Office and the NSF-Metabolic Biology Division (A.M.), and by N. S. F. DMB 85-18066 and N. I. H. GM 28994 grants (A. N. G.).

Literature Cited

1. Abadia, J., Glick, R. E., Taylor, S. E., Terry, N., Melis, A. 1985. Photochemical apparatus organization in the chloroplasts of two *Beta vulgaris* genotypes. *Plant Physiol.* 79:872–78

1a. Amesz, J. 1978. Fluorescence and energy transfer. In *The Photosynthetic Bacteria,* ed. R. K. Clayton, W. R. Sistrom, pp. 333–40. New York: Plenum. 946 pp.

2. Amesz, J. 1985. Photosynthesis: structure of the membrane and membrane proteins. *Prog. Bot.* 47:87–104

3. Anderson, J. M. 1984. A chlorophyll *a/b* protein complex of photosystem I. *Photobiochem. Photobiophys.* 8:221–28

4. Anderson, J. M. 1986. Photoregulation of the composition, function, and structure of thylakoid membranes. *Ann. Rev. Plant Physiol.* 37:93–136

5. Anderson, J. M., Brown, J. S., Lam, E., Malkin, R. 1983. Chlorophyll *b*: an

integral component of photosystem I of higher plant chloroplasts. *Photochem. Photobiol.* 38:205–10

6. Andersson, B., Anderson, J. M. 1980. Lateral heterogeneity in the distribution of chlorophyll-protein complexes of the thylakoid membranes of spinach chloroplasts. *Biochim. Biophys. Acta* 593:427–40

7. Andersson, B., Anderson, J. M., Ryrie, I. J. 1982. Transbilayer organization of the chlorophyll proteins of spinach thylakoids. *Eur. J. Biochem.* 123:465–72

8. Arntzen, C. J., Pakrasi, H. B. 1986. Photosystem II reaction center: Polypeptide subunits and functional cofactors. See Ref. 171, pp. 457–67

9. Bachofen, R., Viemken, V. 1986. Topology of the chromatophore membranes of purple bacteria. See Ref. 171, pp. 603–19

10. Bakker, J. G. C., van Grondelle, R., Den Hollander, W. T. F. 1983. Trapping, loss, and annihilation of excitations in a photosynthetic system. II. Experiments with the purple bacteria *Rhodospirillum rubrum* and *Rhodopseudomonas capsulata*. *Biochim. Biophys. Acta* 725:508–18

11. Baltimore, B., Malkin, R. 1980. On the nature of the intermediate electron acceptor (A_1) in the photosystem I reaction center. *Photochem. Photobiol.* 31:485–90

12. Barber, J. 1982. Influence of surface charges on thylakoid structure and function. *Ann. Rev. Plant Physiol.* 33:261–95

13. Bengis, C., Nelson, N. 1975. Purification and properties of the PSI reaction center from chloroplasts. *J. Biol. Chem.* 250:2783–88

14. Bengis, C., Nelson, N. 1977. Subunit structure of the chloroplast PSI reaction center. *J. Biol. Chem.* 252:4564–69

15. Bennett, J. 1977. Phosphorylation of chloroplast membrane polypeptides. *Nature* 269:344–46

16. Bennett, J., Steinback, K. E., Arntzen, C. J. 1980. Chloroplast phosphoproteins: Regulation of excitation energy transfer by phosphorylation of thylakoid membrane polypeptides. *Proc. Natl. Acad. Sci. USA* 77:5253–57

17. Betti, J. A., Blankenship, R. E., Natarajan, L. V., Dickinson, L. C., Fuller, R. C. 1982. Antenna organization and evidence for the function of a new antenna pigment species in the green photosynthetic bacterium *Chloroflexus aurantiacus*. *Biochim. Biophys. Acta* 680:194–201

18. Blankenship, R. E. 1985. Electron transport in green photosynthetic bacteria. *Photosynth. Res.* 6:317–33

19. Blankenship, R. E., Fuller, R. C. 1986. Membrane topology and photochemistry of the green photosynthetic bacterium *Chloroflexus aurantiacus*. See Ref. 171, pp. 390–99

20. Bonaventura, C., Myers, J. 1969. Fluorescence and oxygen evolution from *Chlorella pyrenoidosa*. *Biochim. Biophys. Acta* 189:366–83

21. Bonnerjea, J., Ortiz, W., Malkin, R. 1985. Identification of a 19 kDa-polypeptide as an Fe-S center apoprotein in the photosystem I primary electron acceptor complex. *Arch. Biochem. Biophys.* 240:15–20

22. Breton, J., Farkas, D. L., Parson, W. W. 1985. Organization of the antenna bacteriochlorophylls around the reaction center of *Rhodopseudomonas viridis* investigated by photoselection techniques. *Biochim. Biophys. Acta* 808:421–27

23. Brockmann, H., Lipinski, A. 1983. Bacteriochlorophyll *g*. A new bacteriochlorophyll from *Heliobacterium chlorum*. *Arch. Microbiol.* 136:17–19

24. Brunisholz, R. A., Jay, F., Suter, F., Zuber, H. 1985. The light-harvesting polypeptides of *Rhodopseudomonas viridis*. The complete amino acid sequences of B1015-α, B1015-β, and B1015-γ. *Biol. Chem. Hoppe-Seyler* 366:87–98

25. Brunisholz, R. A., Suter, F., Zuber, H. 1984. The light-harvesting polypeptides of *Rhodospirillum rubrum*. The amino-acid sequence of the second light-harvesting polypeptide B880-β (B-870β) of *Rhodospirillum rubrum* S1 and the carotenoidless mutant G-9$^+$. Aspects of the molecular structure of the two light-harvesting polypeptides B880-α (B870α) and B880-β (B870-β) and of the antenna complex B880 (B870) from *Rhodospirillum rubrum*. *Hoppe-Seyler's Z. Physiol. Chem.* 365:675–88

26. Brunisholz, R. A., Wiemken, V., Suter, F., Bachofen, R., Zuber, H. 1984. The light-harvesting polypeptides of *Rhodospirillum rubrum*. II. Localization of the amino terminal regions of the light-harvesting polypeptides B870-α and B870-β and the reaction centre subunits L at the cytoplasmic side of the photosynthetic membrane of *Rhodospirillum rubrum*. *Hoppe-Seyler's Z. Physiol. Chem.* 365:689–701

27. Brunisholz, R. A., Zuber, H., Valentine, J., Lindsay, J. G., Woolley, K. J., Cogdell, R. J. 1986. The membrane location of the B890-complex from *Rhodospirillum rubrum* and the effect of

carotenoid on the conformation of its two apoproteins exposed at the cytoplasmic surface. *Biochim. Biophys. Acta* 849:295–303

28. Bryant, D. A. 1986. The cyanobacterial photosynthetic apparatus: comparisons to those of higher plants and photosynthetic bacteria. *Physiological Ecology of Picoplankton, Can. Bull. Fish. Aquat. Sci.* In press

29. Burke, J. J., Ditto, C. L., Arntzen, C. J. 1978. Involvement of the light-harvesting complex in cation regulation of excitation energy distribution in chloroplasts. *Arch. Biochem. Biophys.* 187:252–63

30. Bylina, E. J., Ismail, S., Youvan, D. C. 1986. Site-specific mutagenesis of bacteriochlorophyll-binding sites affects biogenesis of the photosynthetic apparatus. In *Microbial Energy Transduction*, ed. D. C. Youvan, F. Daldal, pp. 63–70. New York: Cold Spring Harbor Lab. 181 pp.

31. Camm, E. L., Green, B. R. 1983. Isolation of PSII reaction centre and its relationship to the minor chlorophyll-protein complexes. *J. Cell Biochem.* 23:171–79

32. Camm, E. L., Green, B. R. 1983. Relationship between the two minor chlorophyll *a*-protein complexes and the photosystem II reaction centre. *Biochim. Biophys. Acta* 724:291–93

33. Campillo, A. J., Hyer, R. C., Monger, T. G., Parson, W. W., Shapiro, S. L. 1977. Light collection and harvesting processes in bacterial photosynthesis investigated on a picosecond time scale. *Proc. Natl. Acad. Sci. USA* 74:1997–2001

34. Chereskin, B. M., Clement-Metral, J. D., Gantt, E. 1985. Characterization of a purified photosystem II-phycobilisome particle preparation from *Porphyridium cruentum*. *Plant Physiol.* 77:626–29

35. Clement-Metral, J. D., Gantt, E., Redlinger, T. 1985. A photosystem II-phycobilisome preparation from the red alga *Porphyridium cruentum*: oxygen evolution, ultrastructure and polypeptide resolution. *Arch. Biochem. Biophys.* 238:10–17

36. Cogdell, R. J. 1985. Carotenoid-bacteriochlorophyll interactions. In *Antennas and Reaction Centers of Photosynthetic Bacteria: Structure, Interactions and Dynamics*, ed. M. E. Michel-Beyerle, 42:62–66. Berlin: Springer-Verlag. 367 pp.

37. Cogdell, R. J. 1986. Light-harvesting complexes in the purple photosynthetic bacteria. See Ref. 171, pp. 252–59

38. Cogdell, R. J., Woolley, K. J., Dawkins, D. J., Lindsay, J. G. 1986. The structure and function of bacterial antenna complexes. See Ref. 30, pp. 47–51

39. Coombs, J., Greenwood, A. D. 1976. Compartmentation of the photosynthetic apparatus, In *The Intact Chloroplast*, ed. J. Barber, 1:1–51. Amsterdam/New York: Elsevier

40. Cruden, D. L., Stanier, R. Y. 1970. The characterization of chlorobium vesicles and membranes isolated from green bacteria. *Arch. Microbiol.* 72:115–34

41. Daurat-Larroque, S. T., Brew, K., Fenna, R. E. 1986. The complete amino acid sequence of a bacteriochlorophyll *a*-protein from *Prosthecochloris aestuarii*. *J. Biol. Chem.* 261:3607–15

42. Davis, M. S., Forman, A., Hanson, L. K., Thornber, J. P., Fajer, J. 1979. Anion and cation radicals of bacteriochlorophyll and bacteriophaeophytin *b*. Their role in the primary charge separation of *Rhodopseudomonas viridis*. *J. Phys. Chem.* 83:3325–32

43. Debus, R. J., Feher, G., Okamura, M. Y. 1986. Iron-depleted reaction centers from *Rhodopseudomonas sphaeroides* R26.1: characterization and reconstitution with Fe^{2+}, Mn^{2+}, Co^+, Ni^{2+}, Cu^{2+}, and Zn^{2+}. *Biochemistry* 25:2276–87

44. Debus, R. J., Okamura, M. Y., Feher, G. 1985. LM complex of reaction centers from *Rhodopseudomonas sphaeroides* R-26: Characterization and reconstitution with H subunit. *Biochemistry* 24:2488–2500

45. Deisenhofer, J., Epp, O., Miki, K., Huber, R., Michel, H. 1984. X-ray structure analysis of a membrane protein complex. Electron density map at 3 Å resolution and a model of the chromophores of the photosynthetic reaction center from *Rhodopseudomonas viridis*. *J. Mol. Biol.* 180:385–98

46. Deisenhofer, J., Epp, O., Miki, K., Huber, R., Michel, H. 1985. Structure of the protein subunits in the photosynthetic reaction centre of *Rhodopseudomonas viridis* at 3 Å resolution. *Nature* 318:618–24

47. Deisenhofer, J., Michel, H., Huber, R. 1985. The structural basis of photosynthetic light reactions in bacteria. *Trends Biochem. Sci.* 10:243–48

48. Deng, X., Melis, A. 1986. Phosphorylation of the light-harvesting complex II in higher plant chloroplasts: Effect on photosystem II and photosystem I absorption cross section. *Photobiochem. Photobiophys.* In press

49. Döring, G., Renger, G., Vater, J., Witt, H. T. 1969. Properties of the photoactive chlorophyll-a_{II} in photosynthesis. Z. Naturforsch. Teil B 24:1139–43

50. Döring, G., Stiehl, H. H., Witt, H. T. 1967. A second chlorophyll reaction in the electron chain of photosynthesis. Registration by the repetitive excitation technique. Z. Naturforsch. Teil B 22:639–44

51. Dörnemann, D., Senger, H. 1986. The structure of chlorophyll RCI, a chromophore of the reaction center of photosystem I. Photochem. Photobiol. 43: 573–81

52. Drews, G. 1985. Structure and functional organization of light-harvesting complexes and photochemical reaction centers in membranes of phototrophic bacteria. Microbiol. Rev. 49:59–70

53. Drews, G., Oelze, J. 1981. Organization and differentiation of membranes of phototrophic bacteria. Adv. Microb. Physiol. 22:1–92

54. Dunahay, T. G., Staehelin, L. A. 1986. Isolation and characterization of a new minor chlorophyll a/b-protein complex (CP24) from spinach. Plant Physiol. 80:429–34

55. Dunsmuir, P. 1985. The petunia chlorophyll a/b binding protein genes: a comparison of Cab genes from different gene families. Nucleic Acids Res. 13:2503–2518

56. Engelhardt, H., Baumeister, W., Saxton, W. O. 1983. Electron microscopy of photosynthetic membranes containing bacteriochlorophyll b. Arch. Microbiol. 135:169–75

57. Evans, M. C. W. 1982. Iron-sulfur centers in photosynthetic electron transport. In Iron-Sulfur Proteins, ed. T. G. Spiro, 4:249–84. New York: Wiley

58. Evans, M. C. W., Diner, B. A., Nugent, J. H. A. 1981. Characteristics of the photosystem II reaction center. I. Electron acceptors. Biochim. Biophys. Acta 682:97–105

59. Feher, G., Okamura, M. Y. 1984. Structure and function of the reaction center from Rhodopseudomonas sphaeroides. Advances in Photosynthesis Research, ed. C. Sybesma, 2:155–64. The Hague: Nijhoff/Dr W Junk

60. Feick, R. G., Fuller, R. C. 1984. Topography of the photosynthetic apparatus of Chloroflexus aurantiacus. Biochemistry 23:3693–3700

61. Fork, D. C., Satoh, K. 1986. The control by state transitions of the distribution of excitation energy in photosynthesis. Ann. Rev. Plant Physiol. 37:335–61

62. Fuller, R. C., Sprague, S. G., Gest, H., Blankenship, R. E. 1985. A unique photosynthetic reaction center from Heliobacterium chlorum. FEBS Lett. 182:345–49

63. Gerola, P. D., Olson, J. M. 1986. A new bacteriochlorophyll a-protein complex associated with chlorosomes of green sulfur bacteria. Biochim. Biophys. Acta 848:69–76

64. Gest, H., Favinger, J. L. 1983. Heliobacterium chlorum, an anoxygenic brownish-green photosynthetic bacterium containing a "new" form of bacteriochlorophyll. Arch. Microbiol. 136:11–16

65. Ghirardi, M. L., McCauley, S. W., Melis, A. 1986. Photochemical apparatus organization in the thylakoid membrane of Hordeum vulgare wild type and chlorophyll b-less chlorina f2 mutant. Biochim. Biophys. Acta. 851:331–39

66. Giesbrecht, P., Drews, G. 1966. Über die Organisation und die makromolekulare Architektur der Thylakoide "lebender" Bakterien. Arch. Mikrobiol. 54: 297–330

67. Glazer, A. N. 1982. Phycobilisomes: structure and dynamics. Ann. Rev. Microbiol. 36:173–98

68. Glazer, A. N. 1984. Phycobilisome. A macromolecular complex optimized for light energy transfer. Biochim. Biophys. Acta 768:29–51

69. Glazer, A. N. 1985. Light harvesting by phycobilisomes. Ann. Rev. Biophys. Biophys. Chem. 14:47–77

70. Glazer, A. N., Lundell, D. J., Yamanaka, G., Williams, R. C. 1983. The structure of a "simple" phycobilisome. Ann. Microbiol. 134:159–80

71. Glick, R. E., McCauley, S. W., Gruissem, W., Melis, A. 1986. Light quality regulates expression of chloroplast genes and assembly of photosynthetic membrane complexes. Proc. Natl. Acad. Sci. USA 83:4287–91

72. Glick, R. E., McCauley, S. W., Melis, A. 1985. Effect of light quality on chloroplast-membrane organization and function in pea. Planta 164:487–94

73. Golecki, J., Oelze, J. 1980. Differences in the architecture of cytoplasmic and intracytoplasmic membranes of three chemotrophically and phototrophically grown species of Rhodospirillaceae. J. Bacteriol. 144:781–88

74. Griffith, M., Sistrom, W. R., Cohen-Bazire, G., Stanier, R. Y. 1955. Function of carotenoids in photosynthesis. Nature 176:1211–15

75. Guglielmi, G., Cohen-Bazire, G., Bryant, D. A. 1981. The structure of

Gloeobacter violaceus and its phycobilisomes. *Arch. Microbiol.* 129:181–89
76. Haehnel, W. 1984. Photosynthetic electron transport in higher plants. *Ann. Rev. Plant Physiol.* 35:659–93
77. Haworth, P., Kyle, D. J., Horton, P., Arntzen, C. J. 1982. Chloroplast membrane protein phosphorylation. *Photochem. Photobiol.* 36:743–48
78. Haworth, P., Melis, A. 1983. Phosphorylation of chloroplast thylakoid membrane proteins does not increase the absorption cross-section of photosystem I. *FEBS Lett.* 160:277–80
79. Haworth, P., Watson, J. L., Arntzen, C. J. 1983. The detection, isolation and characterization of a light-harvesting complex which is specifically associated with photosystem I. *Biochim. Biophys. Acta* 724:151–58
80. Heathcote, P., Williams-Smith, D. L., Sihra, C. K., Evans, M. C. W. 1978. The role of the membrane-bound iron-sulfur centers A and B in the photosystem I reaction center of spinach chloroplasts. *Biochim. Biophys. Acta* 503:333–42
81. Hill, R., Bendall, R. 1960. Function of the two cytochrome components in chloroplasts. A working hypothesis. *Nature* 186:136–37
82. Hiyama, T., Ke, B. 1972. Difference spectra and extinction coefficients of P700. *Biochim. Biophys. Acta* 267:160–71
83. Holo, H., Broch-Due, M., Ormerod, J. G. 1985. Glycolipids and the structure of chlorosomes in green bacteria. *Arch. Microbiol.* 143:94–99
84. Holten, D., Windsor, M. W., Parson, W. W., Thornber, J. P. 1978. Primary photochemical processes in isolated reaction centers of *Rhodopseudomonas viridis*. *Biochim. Biophys. Acta* 501:112–26
85. Horton, P., Black, M. T. 1982. On the nature of the fluorescence decrease due to phosphorylation of chloroplast membrane proteins. *Biochim. Biophys. Acta* 680:22–27
86. Hurt, E. C., Hauska, G. 1984. Purification of membrane-bound cytochromes and a photoactive P840 protein complex of the green sulfur bacterium *Chlorobium limicola f. thiosulfatophilum*. *FEBS Lett.* 168:149–54
87. Ikegami, I., Katoh, S. 1973. Studies on chlorophyll fluorescence in chloroplasts. II. Effect of ferricyanide on the induction of fluorescence in the presence of DCMU. *Plant Cell Physiol.* 14:829–36
87a. Ikegami, I., Ke, B. 1984. A 160-kDa photosystem I reaction-center complex.

Low temperature fluorescence spectroscopy. *Biochim. Biophys. Acta* 764:80–85
88. Imhoff, J. F., Trüper, H. G., Pfennig, N. 1984. Rearrangement of the species and genera of the phototrophic "purple nonsulfur bacteria". *Int. J. Syst. Bacteriol.* 34:340–43
89. Jacob, J. S., Miller, K. R. 1983. Structure of a bacterial photosynthetic membrane: isolation, polypeptide composition, and selective proteolysis. *Arch. Biochem. Biophys.* 223:283–90
90. Jay, F. A., Lambillotte, M. 1985. Isolation of oriented vesicles from thylakoids of *Rhodopseudomonas viridis* and topographical studies. *Eur. J. Cell Biol.* 37:7–13
91. Jay, F., Lambillotte, M., Mühlethaler, K. 1983. Localization of *Rhodopseudomonas viridis* reaction centre and light harvesting proteins using ferritin-antibody labelling. *Eur. J. Cell Biol.* 30:1–8
92. Jay, F., Lambillotte, M., Stark, W., Mühlethaler, K. 1984. The preparation and characterization of native photoreceptor units from the thylakoids of *Rhodopseudomonas viridis*. *EMBO J.* 3:773–76
93. Jay, F. A., Lambillotte, M., Wyss, F. 1985. Immune electron microscopy with monoclonal antibodies against *Rhodopseudomonas viridis* thylakoid polypeptides. *Eur. J. Cell Biol.* 37:14–20
93a. Jennings, R. C., Zucchelli, G. 1985. The influence of membrane stacking on light absorption by chloroplasts. *Photobiochem. Photobiophys.* 9:215–21
94. Klimov, V. V., Dolan, E., Shaw, E. R., Ke, B. 1980. Interaction between the intermediary electron acceptor (pheophytin) and a possible plastoquinone-iron complex in photosystem II reaction centers. *Proc. Natl. Acad. Sci. USA* 77:7227–31
95. Klimov, V. V., Klevanik, A. V., Shuvalov, V. A. 1977. Reduction of pheophytin in the primary light reaction of photosystem II. *FEBS Lett.* 82:183–86
96. Knaff, D. B., Malkin, R. 1976. Iron-sulfur proteins of the green photosynthetic bacterium *Chlorobium*. *Biochim. Biophys. Acta* 325:94–101
97. Kramer, H. J. M., Westerhuis, W. H. J., Amesz, J. 1985. Low-temperature spectroscopy of intact algae. *Physiol. Vég.* 23:535–43
98. Kühlbrandt, W. 1984. Three-dimensional structure of the light-harvesting chlorophyll a/b protein complex. *Nature* 307:478–80

99. Kura-Hotta, M., Satoh, K., Katoh, S. 1986. Functional linkage between phycobilisome oxygen-evolving photosystem II preparations isolated from the thermophilic cyanobacterium *Synechococcus* sp. *Arch. Biochem. Biophys.* 249:1–7

100. Kyle, D. J., Haworth, P., Arntzen, C. J. 1982. Thylakoid membrane protein phosphorylation leads to a decrease in connectivity between photosystem II reaction centers. *Biochim. Biophys. Acta* 680:336–42

101. Kyle, D. J., Kuang, T. Y., Watson, J. L., Arntzen, C. J. 1984. Movement of a subpopulation of the light-harvesting complex (LHC II) from grana to stroma lamellae as a consequence of its phosphorylation. *Biochim. Biophys. Acta* 765:89–96

102. Kyle, D. J., Staehelin, L. A., Arntzen, C. J. 1983. Lateral mobility of the light-harvesting complex in chloroplast membranes controls excitation energy distribution in higher plants. *Arch. Biochem. Biophys.* 222:527–41

103. Lagoutte, B., Setif, P., Duranton, J. 1984. Tentative identification of the apoproteins of iron-sulfur centers of photosystem I. *FEBS Lett.* 174:24–29

104. Lam, E., Baltimore, B., Ortiz, W., Chollar, S., Melis, A., Malkin, R. 1983. Characterization of a resolved oxygen-evolving photosystem II preparation from spinach thylakoids. *Biochim. Biophys. Acta* 724:201–11

105. Lam, E., Ortiz, W., Malkin, R. 1984. Chlorophyll *a/b* proteins of photosystem I. *FEBS Lett.* 168:10–14

106. Lam, E., Ortiz, W., Mayfield, S., Malkin, R. 1984. Isolation and characterization of a light-harvesting chlorophyll *a/b* protein complex associated with PSI. *Plant Physiol.* 74:650–55

107. Larsson, U. K., Andersson, B. 1985. Different degrees of phosphorylation and lateral mobility of two polypeptides belonging to the light-harvesting complex of photosystem II. *Biochim. Biophys. Acta* 809:396–402

108. Larsson, U. K., Ögren, E., Öquist, G., Andersson, B. 1987. Electron transport and fluorescence studies on the functional interaction between phospho-LHC II and PSI in isolated stroma lamellae vesicles. *Photobiochem. Photobiophys.* In press

109. Ley, A. C., Mauzerall, D. C. 1982. Absolute absorption cross-sections for photosystem II and the minimum quantum requirement for photosynthesis in *Chlorella vulgaris*. *Biochim. Biophys. Acta* 680:95–106

110. Li, J. 1985. Light-harvesting chlorophyll *a/b*-proteins: Three-dimensional structure of a reconstituted membrane lattice in negative stain. *Proc. Natl. Acad. Sci. USA* 82:386–90

111. Ludwig, F. R., Jay, F. A. 1985. Reversible chemical cross-linking of the light-harvesting polypeptides of *Rhodopseudomonas viridis*. *Eur. J. Biochem.* 151:83–87

112. Lundell, D. J., Glazer, A. N., Melis, A., Malkin, R. 1985. Characterization of a cyanobacterial photosystem I complex. *J. Biol. Chem.* 260:646–54

113. Malkin, R. 1982. Photosystem I. *Ann. Rev. Plant Physiol.* 33:455–79

114. Manodori, A., Alhadeff, M., Glazer, A. N., Melis, A. 1984. Photochemical apparatus organization in *Synechococcus* 6301 *(Anacystis nidulans)*. Effect of phycobilisome mutation. *Arch. Microbiol.* 139:117–23

115. Manodori, A., Melis, A. 1985. Phycobilisome-photosystem II association in *Synechococcus* 6301. *FEBS Lett.* 181:79–82

116. Manodori, A., Melis, A. 1986. Cyanobacterial acclimation to photosystem I or photosystem II light. *Plant Physiol.* 82:185–89

117. Mansfield, R. W., Evans, M. C. W. 1985. Optical difference spectrum of the electron acceptor A_0 in photosystem I. *FEBS Lett.* 190:237–41

118. Matoo, A. K., Hoffman-Falk, H., Marder, J. B., Edelman, M. 1984. Regulation of protein metabolism: Coupling of photosynthetic electron-transport to *in vivo* degradation of the rapidly metabolized 32 kDa protein of the chloroplast membranes. *Proc. Natl. Acad. Sci. USA* 81:1380–84

119. Matoo, A. K., Marder, J. B., Gaba, V., Edelman, M. 1986. Control of 32 kDa thylakoid protein degradation as a consequence of herbicide binding to its receptor. In *Regulation of Chloroplast Differentiation*, ed. G. Akoyunoglou, H. Senger, pp. 607–613. New York: Liss

120. Matthews, B. W., Fenna, R. E., Bolognesi, M. C., Schmid, M. F., Olson, J. M. 1979. Structure of a bacteriochlorophyll *a*-protein from the green photosynthetic bacterium *Prosthecochloris aestuarii*. *J. Mol. Biol.* 131:259–85

121. Melis, A. 1984. Light regulation of photosynthetic membrane structure, organization and function. *J. Cell. Biochem.* 24:271–85

122. Melis, A. 1985. Functional properties of photosystem II_β in spinach chloroplasts. *Biochim. Biophys. Acta* 808:334–42

123. Melis, A., Anderson, J. M. 1983. Structural and functional organization of the photosystems in spinach chloroplasts. Antenna size, relative electron-transport capacity and chlorophyll composition. *Biochim. Biophys. Acta* 724: 473–84

124. Melis, A., Brown, J. S. 1980. Stoichiometry of system I and system II reaction centers and of plastoquinone in different photosynthetic membranes. *Proc. Natl. Acad. Sci. USA* 77:4712–16

125. Melis, A., Harvey, G. W. 1981. Regulation of photosystem stoichiometry, chlorophyll *a* and chlorophyll *b* content and relation to chloroplast ultrastructure. *Biochim. Biophys. Acta* 637:138–45

126. Melis, A., Manodori, A., Glick, R. E., Ghirardi, M. L., McCauley, S. W., Neale, P. J. 1985. The mechanism of photosynthetic membrane adaptation to environmental stress conditions: a hypothesis on the role of electron-transport capacity and of ATP/NADPH pool in the regulation of thylakoid membrane organization and function. *Physiol. Vég.* 23:757–65

127. Melis, A., Spangfort, M., Andersson, B. 1987. Light-absorption and electron-transport balance between photosystem II and photosystem I in spinach chloroplasts. *Photochem. Photobiol.* 45:129–36

128. Meyer, R., Snozzi, M., Bachofen, R. 1981. Freeze-fracture studies of reaction centers from *Rhodospirillum rubrum* in chromatophores and liposomes. *Arch. Microbiol.* 130:125–28

129. Michel, H., Deisenhofer, J. 1986. X-ray diffraction studies on a crystalline bacterial photosynthetic reaction center: a progress report and conclusions on the structure of photosystem II reaction centers. See Ref. 171, pp. 371–81

130. Michel, H., Weyer, K. A., Gruenberg, H., Dunger, I., Oesterhelt, D., Lottspeich, F. 1986. The 'light' and 'medium' subunits of the photosynthetic reaction centre from *Rhodopseudomonas viridis:* isolation of the genes, nucleotide and amino acid sequence. *EMBO J.* 5:1149–58

131. Michel, H., Weyer, K. A., Gruenberg, H., Lottspeich, F. 1985. The "heavy" subunit of the photosynthetic reaction centre from *Rhodopseudomonas viridis:* isolation of the gene, nucleotide and amino acid sequence. *EMBO J.* 4:1667–72

132. Miller, K. R., Jacob, J. S. 1983. Two dimensional crystals formed from photosynthetic reaction centers. *J. Cell Biol.* 97:1266–70

133. Miller, K. R., Jacob, J. S. 1985. The *Rhodopseudomonas viridis* photosynthetic membrane: arrangement *in situ*. *Arch. Microbiol.* 142:333–39

134. Monger, T. G., Parson, W. W. 1977. Singlet-triplet fusion in *Rhodopseudomonas sphaeroides* chromatophores. A probe of the organization of the photosynthetic apparatus. *Biochim. Biophys. Acta* 460:393–407

135. Mullet, J. E. 1983. The amino acid sequence of the polypeptide segment which regulates membrane adhesion (grana stacking) in chloroplasts. *J. Biol. Chem.* 258:9941–48

136. Mullet, J. E., Arntzen, C. J. 1981. Identification of a 32–34 kDa polypeptide as a herbicide receptor protein in photosystem II. *Biochim. Biophys. Acta* 635:236–48

137. Mullet, J. E., Burke, J. J., Arntzen, C. J. 1980. Chlorophyll-proteins of photosystem I. *Plant Physiol.* 65:814–22

138. Mullet, J. E., Burke, J. J., Arntzen, C. J. 1980. A developmental study of photosystem I peripheral chlorophyll proteins. *Plant Physiol.* 65:823–27

139. Myers, J., Graham, J. R., Wang, R. T. K. 1980. Light harvesting in *Anacystis nidulans* studied in pigment mutants. *Plant Physiol.* 66: 1144–49

140. Nakatani, H. Y., Ke, B., Dolan, E., Arntzen, C. J. 1984. Identity of the photosystem II reaction center polypeptide. *Biochim. Biophys. Acta* 765: 347–52

141. Nelson, N., Notsani, B. E. 1977. Function and organization of individual polypeptides in chloroplast photosystem I reaction center. In *Bioenergetics of Membranes,* ed. L. Packer, G. Papageorgiou, A. Trebst, pp. 233–44. Amsterdam/New York: Oxford Press

142. Norris, J. R., Katz, J. J. 1978. Oxidized bacteriochlorophyll as a photoproduct. See Ref. 1a, pp. 397–418

143. Oelze, J., Golecki, J. R. 1975. Properties of reaction center depleted membranes of *Rhodospirillum rubrum*. *Arch. Microbiol.* 102:59–64

144. Okamura, M. Y., Feher, G., Nelson, N. 1982. Reaction Centers. In *Photosynthesis: Energy Conversion by Plants and Bacteria,* ed. Govindjee, 1:195–272. New York: Academic

145. Ortiz, W., Lam, E., Ghirardi, M. L., Malkin, R. 1984. Antenna function of a chlorophyll *a/b* protein complex of photosystem I. *Biochim. Biophys. Acta* 766:505–9

146. Park, R. B., Sane, P. V. 1971. Distribution of function and structure in chloroplast lamellae. *Ann. Rev. Plant Physiol.* 22:395–430

147. Parson, W. W. 1982. Photosynthetic

bacterial reaction centers: Interactions among the bacteriochlorophylls and bacteriophaeophytins. *Ann. Rev. Biophys. Bioeng.* 11:57–80

148. Parson, W. W., Woodbury, N. W. T., Becker, M., Kirmaier, C., Holten, D. 1985. Kinetics and mechanisms of initial electron-transfer reactions in *Rhodopseudomonas sphaeroides* reaction centers. See Ref. 36, pp. 278–85

149. Peters, J., Welte, W., Drews, G. 1984. Topographical relationships of polypeptides in the photosynthetic membrane of *Rhodopseudomonas viridis* investigated by reversible chemical cross-linking. *FEBS Lett.* 171:167–70

150. Petrouleas, V., Diner, B. A. 1986. Identification of Q_{400}, a high-potential electron acceptor of photosystem II, with the iron of the quinone-iron acceptor complex. *Biochim. Biophys. Acta* 849:264–75

151. Pfennig, N. 1978. General physiology and ecology of photosynthetic bacteria. See Ref. 1a, 3–18

152. Pfennig, N., Trüper, H. G. 1983. Taxonomy of phototrophic green and purple bacteria: A review. *Ann. Microbiol. (B)* 134:9–20

153. Pierson, B. K., Castenholz, R. W. 1978. Photosynthetic apparatus and cell membranes of the green bacteria. See Ref. 1a, pp. 179–97

154. Pierson, B. K., Giovannoni, S. J., Stahl, D. A., Castenholz, R. W. 1985. *Heliothrix oregonensis*, gen. nov., sp. nov., a phototrophic filamentous gliding bacterium containing bacteriochlorophyll *a*. *Arch. Microbiol.* 142:164–67

155. Pierson, B. K., Thornber, J. P. 1983. Isolation and spectral characterization of photochemical reaction centers from the thermophilic green bacterium *Chloroflexus aurantiacus* strain J-10-fl. *Proc. Natl. Acad. Sci. USA* 80:80–84

156. Pierson, B. K., Thornber, J. P., Seftor, R. E. B. 1983. Partial purification, subunit structure, and thermal stability of the photochemical reaction center of the thermophilic green bacterium *Chloroflexus aurantiacus*. *Biochim. Biophys. Acta* 723:322–26

157. Prince, R. C., Gest, H., Blankenship, R. E. 1985. Thermodynamic properties of the photochemical reaction center of *Heliobacterium chlorum*. *Biochim. Biophys. Acta* 810:377–84

158. Rippka, R., Waterbury, J. B., Cohen-Bazire, G. 1974. A cyanobacterium which lacks thylakoids. *Arch. Microbiol.* 100:419–36

159. Rutherford, A. W., Zimmermann, J. I. 1984. A new EPR signal attributed to the primary plastosemiquinone acceptor of PSII. *Biochim. Biophys. Acta* 767:168–75

160. Sakurai, H., San Pietro, A. 1985. Association of Fe-S center(s) with the large subunit(s) of photosystem I particles. *J. Biochem.* 98:69–76

161. Schmidt, K. 1980. A comparative study on the composition of chlorosomes and cytoplasmic membranes from *Chloroflexus aurantiacus* strain OK-70-fl and *Chlorobium limicola f. thiosulfatophilum* strain 6230. *Arch. Microbiol.* 124:21–31

162. Schoeder, H. U., Lockau, W. 1986. Phylloquinone copurifies with the large subunit of PSI. *FEBS Lett.* 199:23–27

163. Shopes, R. J., Wraight, C. A. 1983. Primary and secondary electron acceptors in *Rps. viridis*. *Biophys. J.* 41:40a

164. Shuvalov, V. A., Dolan, E., Ke, B. 1979. Spectral and kinetic evidence for two early electron acceptors in photosystem I. *Proc. Natl. Acad. Sci. USA* 76:770–73

165. Shuvalov, V. A., Ke, B., Dolan, E. 1979. Kinetic and spectral properties of the intermediary electron acceptor A_1 in photosystem I. *FEBS Lett.* 100:5–8

166. Sistrom, W. R. 1978. Control of antenna pigments components. See Ref. 1a, pp. 841–48

167. Smith, K. M., Kehres, L. A. 1983. Aggregation of the bacteriochlorophylls *c*,*d*, and *e*. Models for the antenna chlorophylls of green and brown photosynthetic bacteria. *J. Am. Chem. Soc.* 105:1357–89

168. Sprague, S. G., Varga, A. R. 1986. Membrane architecture of anoxygenic photosynthetic bacteria. See Ref. 171, pp. 603–19

169. Stackebrandt, E., Woese, C. R. 1981. The evolution of prokaryotes. In *Molecular and Cellular Aspects of Microbial Evolution. Soc. Gen. Microb. Symp. 32*, ed. M. H. Carlisle, I. F. Collins, B. E. B. Moseley, pp. 1–31. Cambridge: Cambridge Univ. 368 pp.

170. Staehelin, L. A. 1986. Chloroplast structure and supramolecular organization of photosynthetic membranes. See Ref. 171, pp. 1–83

171. Staehelin, L. A., Arntzen, C. J., eds. 1986. *Encyclopedia of Plant Physiology, Photosynthesis III. Photosynthetic Membranes and Light Harvesting Systems* (NS), vol. 19. Berlin: Springer-Verlag. 802 pp.

172. Staehelin, L. A., Golecki, J. R., Drews, G. 1980. Supramolecular organization of chlorosomes (Chlorobium vesicles) and of their membrane attachment sites

in *Chlorobium limicola*. *Biochim. Biophys. Acta* 589:30–45

173. Staehelin, L. A., Golecki, J. R., Fuller, R. C., Drews, G. 1978. Visualization of the supramolecular architecture of chlorosomes (chlorobium type vesicles) in freeze-fractured cells of *Chloroflexus aurantiacus*. *Arch. Microbiol.* 119:269–77

174. Stark, W., Kühlbrandt, W., Wildhaber, I., Wehrli, E., Mühlethaler, K. 1984. The structure of the photoreceptor unit of *Rhodopseudomonas viridis*. *EMBO J.* 3:777–83

175. Steinback, K. E., Bose, S., Kyle, D. J. 1982. Phosphorylation of the light-harvesting chlorophyll-protein regulates excitation energy distribution between photosystem II and photosystem I. *Arch. Biochem. Biophys.* 216:356–61

176. Stiehl, H. H., Witt, H. T. 1968. Die kurzzeitigen ultravioletten Differenz-spektren bei der Photosynthese. *Z. Naturforsch. Teil B* 23:220–24

177. Tadros, M. H., Suter, F., Drews, G., Zuber, H. 1983. The complete amino acid sequence of the large bacteriochlorophyll binding polypeptide from light-harvesting complex II (B800–850) of *Rhodopseudomonas capsulata*. *Eur. J. Biochem.* 129:533–36

178. Tadros, M. H., Suter, F., Seydewitz, H. H., Witt, J., Zuber, H., Drews, G. 1984. Isolation and complete amino acid sequence of the small polypeptide from light-harvesting pigment protein complex I (B870) of *Rhodopseudomonas capsulata*. *Eur. J. Biochem.* 138:209–12

179. Takahashi, Y., Hirota, K., Katoh, S. 1985. Multiple forms of P700-chlorophyll *a*-protein complexes from *Synechococcus* sp.: The iron, quinone and carotenoid contents. *Photosynth. Res.* 6:183–92

180. Telfer, A., Bottin, H., Barber, J., Mathis, P. 1984. The effect of magnesium and phosphorylation of light-harvesting Chl *a/b*-protein on the yield of P700 photooxidation in pea chloroplasts. *Biochim. Biophys. Acta* 764:324–30

181. Thielen, A. P. G. M., van Gorkom, H. J. 1981. Quantum efficiency and antenna size of photosystems II_α, II_β and I in tobacco chloroplasts. *Biochim. Biophys. Acta* 635:111–20

182. Thornber, J. P., Cogdell, R. J., Seftor, R. E. B., Webster, G. D. 1980. Further studies on the composition and spectral properties of the photochemical reaction center of bacteriochlorophyll *b*-containing bacteria. *Biochim. Biophys. Acta* 593:60–75

183. Trebst, A. 1986. The topology of the plastoquinone and herbicide binding peptides of photosystem II in thylakoid membrane. *Z. Naturforsch. Teil C* 41:240–45

184. Trebst, A., Depka, B. 1985. The architecture of photosystem II in plant photosynthesis. Which peptide subunits carry the reaction center of PSII? See Ref. 36, 42:216–24

185. Tronrud, D. E., Schmid, M. F., Matthews, B. W. 1986. Structure and X-ray amino acid sequence of a bacteriochlorophyll *a* protein from *Prosthecochloris aestuarii* refined at 1.9 Å resolution. *J. Mol. Biol.* 188:443–54

186. Trüper, H. G., Pfennig, N. 1981. Characterization and identification of the anoxygenic phototrophic bacteria. In *The Prokaryotes*, ed. M. P. Starr, H. Stolp, H. G. Trüper, A. Balows, H. G. Schlegel, 1:299–312. Berlin: Springer-Verlag. 1102 pp.

187. Ueki, T., Katoka, M., Mitsui, T. 1976. Structural order in chromatophore membranes of *Rhodospirillum rubrum*. *Nature* 262:809–10

188. van Gorkom, H. J. 1974. Identification of the reduced primary electron acceptor of photosystem II as a bound semiquinone anion. *Biochim. Biophys. Acta* 347:439–42

189. van Grondelle, R., Hunter, C. N., Bakker, J. G. C., Kramer, H. J. M. 1983. Size and structure of antenna complexes of photosynthetic bacteria as studied by singlet-singlet quenching of the bacteriochlorophyll fluorescence yield. *Biochim. Biophys. Acta* 723:30–36

190. Vasmel, H., van Dorssen, R. J., de Vos, G. J., Amesz, J. 1986. Pigment organization and energy transfer in the green photosynthetic bacterium *Chloroflexus aurantiacus*. I. The cytoplasmic membrane. *Photosynth. Res.* 7:281–94

191. Vierling, E., Alberte, R. S. 1983. P700 chlorophyll *a*-protein. Purification, characterization, and antibody preparation. *Plant Physiol.* 72:625–33

192. Vos, M., van Grondelle, R., van der Kooij, F. W., van de Poll, D., Amesz, J., Duysens, L. M. N. 1986. Singlet-singlet annihilation at low temperatures in the antenna of purple bacteria. *Biochim. Biophys. Acta* 850:501–12

193. Vredenberg, W. J., Duysens, L. M. N. 1963. Transfer of energy from bacteriochlorophyll to a reaction centre during bacterial photosynthesis. *Nature* 197:355–57

194. Wakim, B., Golecki, J. R., Oelze, J. 1978. The unusual mode of altering the cellular membrane content by *Rhodos-*

pirillum tenue. FEMS Microbiol. Lett. 4:199–201
195. Wang, R. T., Myers, J. 1974. On the State 1-State 2 phenomenon in photosynthesis. *Biochim. Biophys. Acta* 347: 134–40
196. Watanabe, T., Kobayashi, M., Hongu, A., Nakazato, M., Hiyama, T., Murata, N. 1985. Evidence that chlorophyll *a*' dimer constitutes the photochemical reaction centre 1 (P700) in photosynthetic apparatus. *FEBS Lett.* 191:252–56
197. Wechsler, T., Brunisholz, R., Suter, F., Fuller, R. C., Zuber, H. 1985. The complete amino acid sequence of a bacteriochlorophyll *a* binding polypeptide isolated from the cytoplasmic membrane of the green photosynthetic bacterium *Chloroflexus aurantiacus. FEBS Lett.* 191:34–38
198. Wechsler, T., Suter, F., Fuller, R. C., Zuber, H. 1985. The complete amino acid sequence of the bacteriochlorophyll *c* binding polypeptide from chlorosomes of the green photosynthetic bacterium *Chloroflexus aurantiacus. FEBS Lett.* 181:173–78
199. Wettern, M. 1986. Localization of 32 kDa chloroplast protein pools in thylakoids: significance in atrazine binding. *Plant Sci.* 43:173–77
199a. Whitmarsh, J., Ort, D. R. 1984. Stoichiometries of electron transport complexes in spinach chloroplasts. *Arch. Biochem. Biophys.* 231:378–89
200. Wiemken, V., Bachofen, R. 1984. Probing the smallest functional unit of the reaction center of *Rhodospirillum rubrum* G-9 with proteinases. *FEBS Lett.* 166:155–59
201. Williams, J. C., Steiner, L. A., Feher, G., Simon, M. I. 1984. Primary structure of the L subunit of the reaction center from *Rhodopseudomonas sphaeroides. Proc. Natl. Acad. Sci. USA* 81:7303–7
202. Williams, J. C., Steiner, L. A., Ogden, R. C., Simon, M. I., Feher, G. 1983. Primary structure of the M subunit of the reaction center from *Rhodopseudomonas sphaeroides. Proc. Natl. Acad. Sci. USA* 80:6505–9
203. Woodbury, N. W., Becker, M., Middendorf, D., Parson, W. W. 1985. Picosecond kinetics of the initial photochemical electron-transfer reaction in bacterial photosynthesis reaction centers. *Biochemistry* 24:7516–21

204. Wraight, C. A. 1982. Current attitudes in photosynthetic research. See Ref. 144, pp. 17–61
205. Wraight, C. A. 1985. Modulation of herbicide-binding by the redox state of Q_{400}, an endogeneous component of photosystem II. *Biochim. Biophys. Acta* 809:320–30
206. Yamagishi, A., Katoh, S. 1983. Two chlorophyll-binding subunits of the photosystem II reaction center isolated from the thermophilic cyanobacterium *Synechococcus* sp. *Arch. Biochem. Biophys.* 225:836–46
207. Yamagishi, A., Katoh, S. 1984. A photoactive photosystem II reaction-center complex lacking a chlorophyll-binding 40 kilodalton subunit from the thermophilic cyanobacterium *Synechococcus* sp. *Biochim. Biophys. Acta* 765:118–24
208. Yamagishi, A., Katoh, S. 1985. Further characterization of the photosystem II reaction center complex preparations from the thermophilic cyanobacterium *Synechococcus* sp. *Biochim. Biophys. Acta* 807:74–80
209. Youvan, D. C., Bylina, E. J., Alberti, M., Begusch, H., Hearst, J. E. 1984. Nucleotide and deduced polypeptide sequences of the photosynthetic reaction-center, B870 antenna, and flanking polypeptides from *R. capsulata. Cell* 27:949–57
210. Youvan, D. C., Ismail, S. 1985. Light-harvesting II (B800-850 complexes) structural genes from *Rhodopseudomonas capsulata. Proc. Natl. Acad. Sci. USA* 82:58–62
211. Zinth, W., Kaiser, W., Michel, H. 1983. Efficient photochemical activity and strong dichroism of single crystals of reaction centers from *Rhodopseudomonas viridis. Biochim. Biophys. Acta* 723:128–31
212. Zuber, H. 1985. Structure and function of light-harvesting complexes and their polypeptides. *Photochem. Photobiol.* 42:821–44
213. Zuber, H., Sidler, W., Füglistaller, P., Brunisholz, R., Theiler, R. 1985. Structural studies on the light-harvesting polypeptides from cyanobacteria and bacteria. In *Molecular Biology of the Photosynthetic Apparatus,* ed. K. E. Steinback, S. Bonitz, C. J. Arntzen, L. Bogorad, pp. 183–95. New York: Cold Spring Harbor Lab. 437 pp.

Ann. Rev. Plant Physiol. 1987. 38:47–72

SOME ASPECTS OF CALCIUM-DEPENDENT REGULATION IN PLANT METABOLISM

H. Kauss

Department of Biology, University of Kaiserslautern, Postfach 3049, D-6750 Kaiserslautern, West Germany

CONTENTS

INTRODUCTION.. 47
AN INDIRECT VERSUS A DIRECT ACTION OF Ca^{2+} 48
MAINTENANCE AND FUNCTION OF Ca^{2+} GRADIENTS............................ 50
    Cytoplasmic Ca^{2+} Concentration.. 50
    Active Ca^{2+} Transport at ER, Plasma Membrane, and Tonoplast 51
    Mitochondria... 53
    Chloroplasts.. 54
CALLOSE SYNTHESIS AND 1,3-β-GLUCAN SYNTHASE.......................... 55
    Callose Formation by Enforced Ca^{2+} Influx.. 55
    Ca^{2+} Dependence of the Plasma Membrane–Located 1,3-β-Glucan Synthase 58
    Some Open Questions and Implications.. 62
Ca^{2+}-DEPENDENT GENERATION OF A REGULATIVE PROTEINASE.............. 64
    Proteinase Generation by Ca^{2+} in Cell Homogenates 65
    Contribution of Ca^{2+} to Cell Volume Regulation 68

INTRODUCTION

The last comprehensive contribution in this series on the function of Ca^{2+} in plants emphasized the involvement of this ion in development (33). Plant development represents the integral in space and time of numerous metabolic reactions. The present article aims, therefore, to complement its predecessor

47

0066-4294/87/0601-0047$02.00

by reviewing how Ca^{2+} might operate at the level of the underlying biochemical reactions. The fascination of the idea that Ca^{2+} also represents a signal substance in plant cells is reflected by the fact that several of the other contributions in this volume also touch on certain Ca^{2+}-regulated enzymes, reaction sequences, and related physiological responses (see the subject index). In order to avoid duplication as far as possible and foreseeable, several important target enzymes for Ca^{2+} are mentioned only marginally here.

In the last sections I present two less well-known aspects of Ca^{2+}-dependent regulation in plants, which have emerged in the last few years. These and others are used to illustrate recent trends in our understanding of the function of Ca^{2+} in plants. For a while, plant physiologists automatically linked effects caused by Ca^{2+} to calmodulin, because the unique features of this calciprotein (solubility, low molecular weight, change in charge after binding of Ca^{2+}, affinity to certain drugs, and conservative evolution) enabled rapid experimental progress. [Calmodulin and calmodulin-mediated processes in plants appear to have been sufficiently treated in recent reviews (21, 23, 33, and the citations therein). They are mentioned here only in regard to one special system (see the final section).] It has become evident, however, first for animal cells and now in plants, that target enzymes for Ca^{2+} are not necessarily also calmodulin-dependent. One likely example from plants, the plasma membrane–located $1,3$-β-glucan synthase, is discussed here in detail. Two others—the Ca^{2+}/phospholipid-dependent protein kinase and the Ca^{2+}-binding subunit of the quinate : NAD^+ oxidoreductase from dark-grown carrot cells—are discussed by A.M. Boudet and R. Ranjeva elsewhere in this volume.

Plant physiologists are at present so fascinated by the role of Ca^{2+} as a second messenger in plant cells that they often tend to neglect the complexity of Ca^{2+}-mediated effects. One aspect of this problem is discussed below; others are related to the integration of Ca^{2+}-triggered reactions into the physiological network. As in animal systems (33), this will most likely be elucidated once the function of cAMP and cGMP in plant systems can be experimentally shown. The occurrence of these signal substances in higher plants appears now to be established (70).

AN INDIRECT VERSUS A DIRECT ACTION OF Ca^{2+}

The concept of "stimulus-response coupling" by Ca^{2+} envisages that a physiological stimulus leads to changes in the flux rates of Ca^{2+} between external or internal stocks and the cytoplasm. The resulting change in concentration of cytoplasmic free Ca^{2+} represents then the signal able to regulate the activity of respective target enzymes, and is thus expressed as a physiological effect. A rigorous proof of this chain of events would require that the cytoplasmic

[Ca^{2+}] or at least its relative change be determined. As outlined below, this measurement is not yet possible. The next best approach to the problem has been to change the external [Ca^{2+}], either by adding $CaCl_2$ over the naturally present level or by removing it with EGTA. With these procedures responses at the level of physiological effects have been observed in numerous cases (33) and are at present often taken as evidence for a role of Ca^{2+} in stimulus-response coupling as outlined above. One or two decades ago a tendency prevailed to explain such effects by the stabilization of membranes (3, 35).

Right-side-out plasma membrane vesicles contain considerable amounts of surface-bound Ca^{2+} (56), and this appears to reflect the in situ circumstance (14). The potent Ca^{2+}-buffering capacity of the cell-wall matrix (19) makes it difficult to estimate the external concentration necessary to saturate in situ the binding ability of the external membrane surface, although under constant conditions the free [Ca^{2+}] at the membrane surface might be in the same order of magnitude as the external [Ca^{2+}]. Nevertheless, various physiological effects [e.g. α-amylase secretion in barley aleurone (36, 37) and root viability (29, 30, 33)] require mM external concentrations of Ca^{2+} for saturation. This appears to be far above a Ca^{2+} concentration necessary to serve simply as a stock for holding a gradient to the low cytoplasmic [Ca^{2+}]. In addition, certain other divalent ions (Sr^{2+}, Ba^{2+}, sometimes Mg^{2+}) can partly replace Ca^{2+}, indicating that the effects are related to a more general quality of the ions and might not be due to those ions entering the cell to serve as a rather specific second messenger.

These two features suggest that the external Ca^{2+} plays a dual role (see also 33). One component of its action may indeed be indirect and result from a conditioning or stabilization of either the phospholipid phase or proteins integral to the membrane. It appears to be superimposed on the second component, namely to serve as a Ca^{2+} pool for the internal messenger role. This latter component of Ca^{2+} function is more properly reflected by physiological effects requiring only μM concentrations of external Ca^{2+}, e.g. callose deposition (54; see also below) or red light–induced germination of *Onoclea* spores (90). With these two examples conditions can also be found that appear to indicate superimposed secondary Ca^{2+} effects: For callose deposition in the presence of mM concentrations of divalent ions (= nutrient solution), higher amounts of the respective elicitor chitosan (54) or digitonin (H. Kauss, unpublished) are needed. Similarly, germination and early outgrowth of *Onoclea* spores occurs not only at 1 mM Ca^{2+} but also at 1 mM Sr^{2+} or partly at 1 mM Ba^{2+} (90).

On removal of some surface-bound Ca^{2+} by washing with water, exposition to low or high temperature, low pH, or high concontration of NaCl, various responses such as K^+ leakage, membrane depolarization, and concomitant Ca^{2+} influx can be provoked (14, 29, 30, 79). Although in plant cells the primary event in the membrane and the causal relationship among these three parameters are not clear, changes in any of them will surely have impacts on various metabolic reactions and will thus influence physiological effects. The "stabilization" component of external Ca^{2+} obviously also controls the "message" component represented by the Ca^{2+} ion actually moving into the cytoplasm (29, 30, 79). Related to these effects is the observation that removal of wall/surface-located Ca^{2+} with EGTA renders the membrane more vulnerable to perturbing substances such as polycations (54) and amphipathic substances capable of inducing callose formation and other related reactions (see the section below on callose).

The suggested dual role of Ca^{2+} can also be demonstrated for certain animal cells. In the immunoglobulin E–mediated degranulation of mast cells—a classic example of the involvement of external Ca^{2+} in exocytosis (11, 72)—the entry of Ca^{2+} appears not to be an obligatory event in stimulus-response coupling (61). The authors concluded that an unbinding of Ca^{2+} from the plasma membrane causes long-term effects, possibly damaging membrane components indirectly involved in stimulus-secretion coupling. It should be kept in mind that the complex role of Ca^{2+} in exocytosis, although known in numerous systems for a long time (11), remains incompletely understood at the biochemical level even in cells where an entry of Ca^{2+} is well established—e.g. in neuro-transmitter release from electrically excitable neurons (18). This situation may change in the near future: Some of the important molecules involved have recently been identified, and the participation of protein kinase C has been demonstrated (for citations see 31).

MAINTENANCE AND FUNCTION OF Ca^{2+} GRADIENTS

Cytoplasmic Ca^{2+} Concentration

Investigators generally agree, based on circumstantial evidence, that in higher plants, as in animal cells, the free $[Ca^{2+}]$ in the cytoplasm is held rather low (33). Absolute values cannot be given because no general method for their measurement is available. Direct determination using microinjection of the photoprotein aequorin has been possible only for characean cells (91). For *Chara corallina* under nonstimulated conditions a mean value slightly above 0.1 μM was calculated assuming plausible concentrations of other cytoplasmic constituents (see Figure 1). This value was transiently raised to about 7 μM on depolarization of the cell by nonphysiological electrical stimulation, a condition under which cytoplasmic streaming fully stopped. The values given should not be confused with the total Ca^{2+} which in *Chara* is 3–12 mM in the vacuole and 2–8 mM in the cytoplasm (91), indicating that most of the cellular Ca^{2+} is bound to membrane surfaces and to organic and inorganic anions, as well as locked into various membranous compartments.

There are manifold reasons why it is difficult to measure $[Ca^{2+}]$ directly in the cytoplasm of higher plant cells. On the one hand, an attempted microinjection of indicators into the cytoplasm appears often to end up in the vacuole. The ester quin2/AM appears to permeate plant cells only in special cases such as the cell wall–free endosperm of *Haemanthus* where individual cells were monitored with image processing to show a transient local increase of free Ca^{2+} at the mitotic spindle poles (52). In general, quin2/AM appears either not to permeate the plant plasma membrane or to be already hydrolyzed outside the cell (27, 92). The problems may be solved in the near future with a new generation of permeant indicators; first positive results indicate that indeed cytoplasmic $[Ca^{2+}]$ may be below 1 μM (10, 80). Alternatively, fluorescent nonpermeant probes might be brought into protoplasts by electroporation; results with protoplasts from mung bean root tips again show that $[Ca^{2+}]$ is below 1 μM, but they also demonstrate that calibration and interpretation of the measured values is difficult (27).

Active Ca^{2+} Transport at ER, Plasma Membrane, and Tonoplast

The Ca^{2+} gradients arising from the differences in concentration and from membrane potential must be maintained by transport systems (33, 67, 81) that allow active transport by coupling them to the cellular energy metabolism. On a long-term scale, Ca^{2+} entering the cell has mainly either to be pumped out again into the apoplastic space or stored in the vacuole (64). Cytoplasmic organelles such as mitochondria and chloroplasts appear to provide only limited storage capacity and may function, therefore, for peak loads and for short times only, or take up Ca^{2+} for other reasons (see below). This suggests that the major membranes involved in long-term Ca^{2+} transport are the plasma membrane and the tonoplast (Figure 1). The ER is intermediate in this respect, depending on the developmental state of the cell. In cells exhibiting active exocytosis [e.g. tip-growing pollen tubes; see citations in (77)] this process may be important in a final extrusion of the Ca^{2+} accumulated in the ER.

There are many reports on the use of subcellular fractions from plant cells for the in vitro study of active Ca^{2+} accumulation into membrane vesicles. The relevant "pumps" have been studied initially by following the accumulation of Ca^{2+} into microsomes, without providing much evidence on the origin of the membranes (21, 23, 76). Such studies have shown that two principal types of mechanisms exist in higher plants.

One is a Ca^{2+}/Mg^{2+}-ATPase that is at least partially dependent on calmodulin and represents a "primary pump" similar to that known from many animal plasma membranes (23, 67, 81). There is good evidence now that this vanadate-sensitive system is located at the ER of higher plants (8, 9, 82). Its affinity for Ca^{2+} appears to be relatively high; half-saturation is reached below 1 μM (8, 9), which suggests that the ER may help to ensure the low cytoplasmic $[Ca^{2+}]$ of unstimulated cells. Suggestions that the Ca^{2+}-ATPase is also present at the plasma membrane (21, 23) appear to be less sustained by experimental in vitro results. This is mainly because only the fraction of microsomes that represents resealed and tight inside-out plasma membrane vesicles is responsible for the respective ATP-driven Ca^{2+} accumulation in vitro. This fraction appears to be usually rather small; most of the plasma membrane is recovered as sealed outside-out vesicles [e.g. presumably 80–90% in suspension-cultured soybean cells (45); for arguments see also (60)]. Although the direct experimental demonstration of a Ca^{2+}-ATPase operating as a Ca^{2+} pump in the plasma membrane of higher plants is not yet convincing, the assumption that it might exist is sustained by the demonstration of a similar enzyme in plasma membranes of cyanobacteria (62). Such a transport Ca^{2+}/Mg^{2+}-ATPase in plasma membranes from higher plants might–owing to its regulation by calmodulin—also be a site where aluminum exerts its toxicity on plants (32, 84).

The other established mechanism for active Ca^{2+} transport is a "secondary pump" that makes use of H^+ gradients as a driving force and that was characterized as a Ca^{2+}/H^+ antiport using protonophores (76, 82, and citations therein). Demonstration of this system is facilitated by the presence of BSA during isolation of the membrane fractions. Density gradient centrifugation suggested that at least one of the locations is the tonoplast (82). This was confirmed using vesicles derived from isolated vacuoles (6, 9). Some properties of the in vitro system are remarkable. Oxalate facilitates Ca^{2+} uptake over a longer period; the apparent half-saturation for external Ca^{2+} is far above 10 μM and depends on the internal pH (6, 9, 82). If the low affinity for Ca^{2+} applies also to the in vivo situation, then the tonoplast preferentially has to cope with high cytoplasmic concentrations of Ca^{2+}. The H^+ gradients necessary to provide the energy for the Ca^{2+} accumulation may be generated by the tonoplast-located H^+-ATPase (88). Many animal cells carry in addition to their Ca^{2+}/Mg^{2+}-ATPAse also a Ca^{2+}/Na^+-antiport system responsible in part for the active Ca^{2+} transport over the plasma membrane (81). In analogy to these observations one could speculate that a Ca^{2+}/H^+ antiport in higher plant cells might also operate at the plasma membrane, fueled by the respective outward-directed H^+-ATPase (88) or the pyridine nucleotide oxidation system located there (4). Such a Ca^{2+}/H^+ antiport has been shown with plasma membrane vesicles from *Neurospora* (87) and is also suggested by experiments with tonoplast-free *Chara* cells, which can extrude Ca^{2+} without using the energy from ATP hydrolysis (63; see also the contribution of Tazawa et al in this volume).

The biochemical characteristics of the above active Ca^{2+}-transport systems are being rapidly explored. At the physiological level it is already well documented that lateral or polar Ca^{2+} transport is under the control of various physiological stimuli such as hormones, gravity, or light. The rather complicated relationships cannot be discussed here in detail; several relevant reviews are available (24, 33, 74, 75, 89). These physiological data strongly suggest that both the channels responsible for an entry of Ca^{2+} following the gradient (see the section below on callose) and the fueled active Ca^{2+} pumps are regulated. Some of the observed physiological effects might be of a long-term developmental nature, but others are fast enough to suggest a direct connection between stimulus perception and presumed Ca^{2+} flux—e.g. in certain phytochrome-mediated effects (33) and gravitropism (74, 75). How such a coupling is effected at the biochemical level remains to be solved experimentally within the next few years. A first successful in vitro system may be represented by a tonoplast-enriched subcellular fraction from corn mesocotyl in which the rates of ATP-dependent Ca^{2+} accumulation are influenced by the presence of auxins (J. Gross, personal communication).

Mitochondria

Mitochondria isolated from various plants have been used to demonstrate Ca^{2+} accumulation in vitro. Uptake requires the addition of succinate or malate as respiratory substrates and/or of exogenous ATP and thus appears to be energy dependent (66, 67, 93). The simultaneous presence of phosphate is essential. Although the exact nature of the uptake carrier remains unclear, a Ca^{2+}/phosphate cotransport (66) and the possibility that part of the Ca^{2+} taken up is precipitated within the mitochondria must be considered. To what extent do mitochondria in vivo contribute to the maintainance of cytoplasmic $[Ca^{2+}]$? The interpretation of the in vitro experimental results is controversial. A major problem is that considerable uptake in vitro is observed only at concentrations above 10 μM; all measurements were therefore most likely performed in a concentration range attained in vivo only under extreme conditions, if at all.

The problem is reminiscent of the situation with vertebrate mitochondria (12). They take up Ca^{2+} by a uniporter, simply by electrophoresis in terms of energy supply. Release of Ca^{2+} appears to occur by independent mechanisms—a prominent one operates by Ca^{2+}/Na^{+} exchange. Under near-physiological conditions both directions of Ca^{2+} movement are balanced: Below about 1 μM external Ca^{2+}, a net efflux occurs; above this range a net uptake prevails. In vitro experiments with $[Ca^{2+}]$ at 10 μM or higher are regarded as unphysiological because long-term Ca^{2+} uptake under these conditions of "massive loading" leads to damage of the mitochondria, in part caused directly by precipitation of calcium phosphate. It has therefore been suggested that in vertebrates the main physiological role of Ca^{2+} transport by mitochondria is to regulate not cytoplasmic $[Ca^{2+}]$ but the $[Ca^{2+}]$ within the mitochondrial matrix in order to control oxidative metabolism (12, 20).

The Ca^{2+}-dependent enzymes reported from plant mitochondria, namely the NADH dehydrogenase (66, 67) and NAD kinase (21–23), are external enzymes and thus likely to be under the control of fluctuations of cytoplasmic Ca^{2+}. Accordingly the metabolic parameters of plant mitochondria are not greatly affected by Ca^{2+} uptake. In addition, the Ca^{2+}-uptake rates calculated on a protein basis are low in plants when compared to animal mitochondria (66). The lack of convincing Ca^{2+} effects on metabolic characteristics of plant mitochondria may be partly related to technical difficulties. The presumed "physiological" $[Ca^{2+}]$ surrounding the mitochondria in undisturbed cells might be far below 1 μM, and cell homogenization exposes the organelles to tremendous amounts of Ca^{2+}, mainly coming from the vacuolar pool. One wonders why no attempts were made to maintain the organelles under low-Ca^{2+} conditions throughout the whole isolation procedure. In such a preparation more pronounced changes of metabolic parameters might be induced on addition of Ca^{2+}. We must clearly await future experimental progress before we can finally evaluate the Ca^{2+} accumulation by plant mitochondria. Based on the available data it appears premature to adopt for plant cells the view that Ca^{2+} transport is mainly a regulator of mitochondrial metabolism.

Chloroplasts

Intact illuminated chloroplasts can take up considerable amounts of Ca^{2+} (7, 58, 66, 68). The idea that this may contribute primarily to maintenance of a low cytoplasmic $[Ca^{2+}]$ appears scarcely to have developed, presumably because light may play a role only for green tissues and only part of the time. Ca^{2+} uptake appears to be mediated by an uniport-type carrier (59) and indirectly driven by the photosynthetic electron transport, the driving force being most likely the membrane-potential component of the proton motive force (58).

Several potential target enzymes for Ca^{2+} have been demonstrated in fractionated chloroplasts, namely a Ca^{2+}/calmodulin-dependent NAD kinase (69) and a Ca^{2+}/phospholipid-dependent protein kinase (85) at the envelope. Both these enzymes are, however, also present in chloroplasts as Ca^{2+}-independent forms. In addition, sugar phosphatases appear to be inhibited (15), and the specificity of their activation by reduced thioredoxin is altered by Ca^{2+} (83). Electron flow at photosystem II of higher plants (15) and of cyanobacteria (7, 65) requires the presence of bound Ca^{2+}. The interplay of these various sites and types of Ca^{2+} action in chloroplasts is both far from clear and beyond the scope of this article. It may be mentioned, however, that changes in the level of stromal nicotinamide coenzymes have been reported upon Ca^{2+} influx into isolated chloroplasts (58, 68). The Ca^{2+} dependence of these enzymes has been cited merely to sustain the opinion (67) that Ca^{2+} uptake by chloroplasts is of significance mainly for regulation of their own metabolism. In this respect again, as discussed above for mitochondria, it appears amazing that precautions are not usually taken to prevent heavy Ca^{2+} loading during chloroplast isolation and that studies on Ca^{2+} uptake and its influence on metabolic parameters are performed at a rather high experimental $[Ca^{2+}]$.

CALLOSE SYNTHESIS AND 1,3-β-GLUCAN SYNTHASE

Callose Formation by Enforced Ca^{2+} Influx

A number of recent investigations have sought to elucidate metabolic changes considered to be of importance in plant resistance to pathogens. A convenient experimental system comprises suspension-cultured soybean cells, which can be stimulated by chitosan (i.e. fragments of deacetylated chitin) to produce the phytoalexin glyceollin. This induction also resulted in changes of alkali-soluble cell-wall phenolic compounds and increased the content of 1,3-linked glucan in the cell walls from less than 1 to about 10% (54, 55). The chitosan-treated cells stained with decolorized aniline blue; staining was distributed either cap-like or in patches irregularly over the cell surface. This cytochemical procedure is based on the formation of a specific complex between a fluorochrome present in the commercial stain and 1,3-β-glucan chains ("callose") exhibiting a linear arrangement long enough to allow an ordered open helical conformation (25). Although some other linkages or sugars may be present in the polymers stained, the term *callose* is, for pragmatic reasons, used to denote 1,3-β-glucan throughout this chapter. Similarly, *elicitor* is used for substances capable of inducing callose deposition. The staining procedure was adapted to a quantitative fluorometric assay (54) that allowed study of the physiology of callose synthesis in more detail.

The deposition of callose starts about 10 min after addition of chitosan to the cells, increases in rate for about 30 min, and continues at a constant rate for several hours. In contrast to phytoalexin formation, it is therefore rapid enough to be followed under nonsterile conditions, in a buffer in which a low $[Ca^{2+}]$ can easily be measured and controlled. An external concentration of 5–10 μM is sufficient (54); routinely 20–30 μM concentrations are used to allow a more accurate determination of changes of $[Ca^{2+}]$ by means of commercial Ca^{2+} electrodes. In addition to chitosan, other polycations (poly-L-Lys, poly-L-Orn) can elicit callose synthesis, as can certain amphipathic compounds [Polymyxin B, Echinocandin B, acylcarnitine, digitonin; (45, 54); Figure 1]. These substances alone or in combination may be regarded as analogous to nonspecific toxins produced from fungal hyphae, which might make contact with a cell of an infected plant.

When used in doses low enough to allow cells to survive but high enough to cause a limited electrolyte leakage, all the above-mentioned substances elicit callose deposition. This indicates that a certain perturbation of membrane permeability is a prerequisite for callose formation. If only one elicitor is considered, and its doses are increased within a certain range, the amount of callose produced increases in parallel with electrolyte leakage. There is no quantitative correlation, however, between the degree of leakage and the extent of callose formation if different elicitors are compared (45, 54). Similarly, the combination of 1 μM Echinocandin B with chitosan (45) even leads to a doubled callose formation with only a tiny further increase in leakage. These observations favor the hypothesis that the primary damage to the membrane and the stimulation of callose deposition are only indirectly

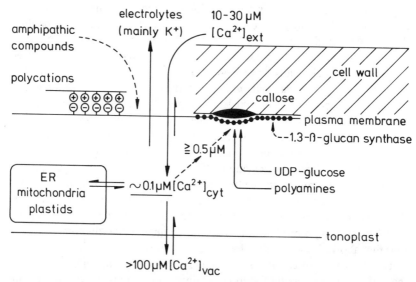

Figure 1   Induction of callose deposition onto cell walls of suspension-cultured soybean cells. Polycations or certain amphipathic compounds at low concentrations perturb the membrane permeability to allow a limited leakage of electrolytes, K^+ representing the major cation. It is suggested that this is also accompanied by transient membrane depolarization and an increased influx of Ca^{2+}. The resulting elevated cytoplasmic $[Ca^{2+}]$ can directly activate the 1,3-β-D-glucan synthase, which appears to be vectorially located in the plasma membrane and which in vitro is rendered more active and more sensitive to Ca^{2+} by polyamines. The events depicted side by side should be imagined to occur in the same surface region to explain the often localized callose deposition, whose restriction to the affected area may be promoted by the potent Ca^{2+}-buffering capacity of the cytoplasm.

linked. A similar conclusion can be drawn from the observation that there is no clear correlation in the time course of callose formation and electrolyte leakage. Depending on the type and concentration of the elicitor used, callose formation occurs with a lag time of less than 5–30 min; and the bulk of electrolyte leakage can precede callose deposition (54) or occur more or less simultaneously.

Callose formation is fully inhibited when EGTA is added shortly after chitosan; it is also inhibited immediately if the EGTA is added at a time when callose synthesis is proceeding at a constant rate. If the EGTA is titrated back with $CaCl_2$ to the initial free $[Ca^{2+}]$, callose synthesis is restored, but only in part. Addition of EGTA has no significant influence on electrolyte leakage in control cells, whereas the leakage is greatly enhanced if the complexing agent is applied in the presence of the elicitors (54). The membrane is apparently rendered more vulnerable to damage by removing bound Ca^{2+} from its surface. The effects were less drastic when the external $[Ca^{2+}]$ was greatly lowered not by EGTA but by the presence of ion-exchange beads in the cell suspension (54). Under these conditions, chitosan-induced callose synthesis was also diminished, indicating again that some external Ca^{2+} is necessary. The various elicitors used are chemically diverse and are therefore likely to have different initial modes of action on membranes. The primary unknown events are then indirectly coupled to the enzymes responsible for callose synthesis by the common signal "increased Ca^{2+} influx" (41, 54).

In the first experiments (54) mainly polycations were used to induce callose formation; more recently (45) and currently (Th. Waldmann, H. Kauss, unpublished results) amphipathic compounds are preferred because, unlike chitosan (94), they do not displace Ca^{2+} and other ionically bound substances from the wall. In addition, when such substances are used, callose is more evenly distributed through the total population of cells; it also appears over the entire cell surface as a haze of small dots (45). In partially plasmolyzed cells, these dots appear not only at the wall but also all over the plasma membrane (Figure 1 does not fully reflect this distribution). These observations indicate that the soybean cells can react equally all over their plasma membrane surfaces.

Using digitonin or chitosan as an elicitor, we have determined that the major ion contributing to the electrolytes released is K^+. A considerable part (5–15%) of the cellular K^+ exits within the first half-hour after start of the induction whereas the efflux is diminished again at a time when callose synthesis proceeds at constant rates (Th. Waldmann and H. Kauss, unpublished). Although nothing is known yet about the respective counter-ions, the massive amounts of K^+ lost suggest that the possibility of at least transient changes in the membrane potential as well as in the osmotic pressure should be evaluated; both parameters might effect metabolic regulation. Poly-L-Lys under conditions similar to those under which it induces callose synthesis in soybean cells (54) was also shown to cause efflux of a limited pool of K^+ and concomitant depolarization of tobacco cell membranes (78).

When control soybean cells are suspended in a buffer (54) at 20–30 μM external $[Ca^{2+}]$, a net uptake of Ca^{2+} can be followed using an ion-sensitive electrode. Based on total cell water content, the Ca^{2+} entering within 3 hr is equivalent to more than 100 μM. The only compartment able to accommodate such high concentrations appears to be the vacuole (Figure 1). The Ca^{2+} uptake increases significantly in the presence of 6–15 μM digitonin, and the concomitant callose synthesis is greatly inhibited in a dose-dependent manner by La^{3+} and nifedipine, (Th. Waldmann and H. Kauss, unpublished). These drugs act as "Ca^{2+}-channel blockers" in animal cells and also interfere with certain Ca^{2+}-dependent physiological processes in plants (33, 77). Their action suggests, therefore, that the digitonin-enforced Ca^{2+} influx may occur through Ca^{2+}-specific channels, rather than through nonspecific pores from which other electrolytes would simultaneously leak. Both types of inhibitors, however, have particular limitations. La^{3+} is considered not to enter cells (33), although a rigorous proof for this conclusion is not yet available for the soybean cells. If some La^{3+} reaches the cytoplasmic side of the plasma membrane it could directly inhibit the 1,3-β-glucan synthase, an enzyme that is most likely responsible for callose formation and that is also inhibited in vitro by La^{3+} (44). Similarly, in digitonin-opened microsomes, nifedipine at high concentrations also slightly inhibits this enzyme (Th. Waldmann and H. Kauss, unpublished).

One could speculate that the hypothetical Ca^{2+}-specific transport sites, suggested from the above results, might be opened directly (11, 33) by the presumed membrane depolarization occurring due to K^+ efflux. Such a simple causal relationship appears unlikely, however, considering that no clear correlation is found between degree or time course of electrolyte leakage and callose formation. This lack of correlation suggests that either some of the elicitors additionally may have a direct influence on Ca^{2+} channels or that yet unknown chemical events caused by the membrane perturbation are superimposed. Some circumstantial evidence suggests that phospholipid degradation products could play a role in this respect (45). This speculative interpretation is complicated, in addition, by the possibility that the target enzyme for Ca^{2+}, the 1,3-β-glucan synthase, may also be influenced directly by some of the elicitors and/or by phospholipid degradation products (see below).

Ca^{2+} Dependence of the Plasma Membrane–Located 1,3-β-Glucan Synthase

The suggestion that a local increase in cytoplasmic $[Ca^{2+}]$ might be one of the parameters that trigger callose formation gained plausibility with the recognition that the 1,3-β-glucan synthase is a new and unexpected target for Ca^{2+}. This enzyme is a well-known plasma-membrane marker (called GS II in this context) that has fooled numerous investigators interested in cellulose biosynthesis in vitro. (For details and a possible role of undisturbed membrane potential in regulating cellulose synthesis, and suggestions on the possible relationship between synthesis of callose and cellulose, see the contribution of D. P. Delmer in this volume). The 1,3-β-glucan synthase exhibits high activity in conventionally prepared homogenates from any plant tissue examined, although it is obviously not operating in most undisturbed cells. Washed microsomes from suspension-cultured soybean cells (41, 48) exhibit 1,3-β-glucan synthase activity that is strictly dependent on Ca^{2+} in the μM range. In the absence of Mg^{2+}, the requirement is almost absolute, with up to 40-fold activation (41). At mM concentrations, Mg^{2+} alone also activates slightly; thus in the presence of Mg^{2+} the activation by Ca^{2+} appears as only about 10-fold (48). A stimulation of the 1,3-β-glucan synthase by Ca^{2+} has more recently been confirmed by several other groups (5, 17, 67a). It was previously overlooked because traces of endogenous Ca^{2+} present in homogenates or assay buffers prepared without EGTA are sufficient almost to saturate the Ca^{2+} requirement; further additions of Ca^{2+} therefore cause only slight effects.

The activation by Ca^{2+} is readily reversible. Neither activation nor inactivation in washed microsomes requires the addition of any nucleotide. Therefore, the effects appear to result not from a covalent modification (e.g. phosphorylation/dephosphorylation) but from a direct allosteric interaction of Ca^{2+} with the enzyme. This situation is in contrast to a report that the formation of so-called β-glucans in homogenates from corn coleoptiles can be enhanced with Ca^{2+} both directly and by a presumed phosphorylation process (71).

When it became clear that the 1,3-β-glucan synthase in microsomes from soybean cells is Ca^{2+} dependent, the question arose whether this activation is mediated by calmodulin. The enzyme also becomes active in the absence of Ca^{2+} on limited proteolysis with trypsin in the presence of digitonin (48), a property most evident when assayed in the presence of Mg^{2+} (44). The Ca^{2+}-stimulated native enzyme as well as the trypsinized preparation were both inhibited to about the same extent by calmidazolium and trifluoperazine (48). Since these drugs bind to calmodulin only in the presence of Ca^{2+} and since relatively high concentrations were required, it was concluded that their effect resulted from a nonspecific interaction. Thus activation of the 1,3-β-glucan synthase by Ca^{2+} is not mediated by calmodulin but probably involves another calciprotein, a built-in Ca^{2+}-binding site, or phospholipids.

It is of interest that high concentrations of the above cited drugs, which are widely regarded as "calmodulin-inhibitors," can cause a nonspecific inhibition of well-established Ca^{2+} target enzymes, *without* interfering with calmodulin. Examples from animal sources are the Ca^{2+}-dependent but calmodulin-independent protein kinase C, as well as the Ca^{2+}/calmodulin-dependent Ca^{2+}-transport ATPase and phosphodiesterase (citations in 41). Inhibition of these enzymes appears to result from the phospholipid-binding potency of the amphipathic drugs. It appears possible, therefore, that other membrane-bound enzymes might also be affected. The effects caused by these inhibitors on physiological parameters in plant tissues should, therefore, be interpreted with great caution before an involvement of calmodulin or Ca^{2+} in the respective process is considered.

In addition to calmidazolium and trifluoperazine, Polymyxin B also inhibits the 1,3-β-glucan synthase, and the effects were taken as a first indication of a phospholipid requirement of the membrane-bound enzyme (48). The assumption was supported by the more recent observation that various other substances known to interfere with phospholipids can also influence the 1,3-β-glucan synthase activity (45). This enzyme has routinely to be assayed in the presence of digitonin because 80–90% of its activity in soybean microsomes is "latent." The enzyme is obviously present in microsomes mainly in outside-out plasma membrane vesicles which have to be opened to allow access of the substrate UDP-glucose and presumably also of the activator Ca^{2+} to the cytoplasmic side of the membrane. In the presence of saturating concentrations of digitonin, certain polyunsaturated free fatty acids, lysophosphatidylcholine, acylcarnitine, Echinocandin B, and platelet-activating factor (PAF) showed in principle similar effects, although some quantitative differences occurred (45). At low concentrations they all significantly stimulated the 1,3-β-glucan synthase over the activity attained with digitonin alone, whereas at more elevated concentrations they caused strong inhibition. The inhibition was similarly observed when the enzyme was activated by Ca^{2+} or by trypsinization, suggesting that the effect is not due to an interference of the substances with the Ca^{2+} necessary for activation (45).

Both the stimulation at low and the inhibition at higher concentrations indicate that the amphipathic compounds act by replacement of endogenous boundary lipids necessary for optimal activity of the enzyme and, therefore, reflect phospholipid requirement. These findings may be relevant for the induction of callose synthesis: Phospholipid degradation may take place in membranes perturbed by the elicitors, and stimulation or inhibition of the 1,3-β-glucan synthase by the respective products might be superimposed on the regulation by Ca^{2+}.

In addition to Ca^{2+} (and possibly phospholipid degradation products) still another parameter was found to be important for the in vitro activity of 1,3-β-glucan synthase. This enzyme is usually assayed at rather high (0.5–1.0 mM) concentrations of UDP-glucose since it exhibits sigmoidal dependence of its activity on the substrate concentration with unfavorable low activity below about 50 μM. Although Ca^{2+} was found to be essential for the activity of 1,3-β-glucan synthase, a low apparent substrate affinity was still observed even at saturating concentrations of Ca^{2+} and in the presence of the beneficial compounds cellobiose and glycerol (42, 44). The enzyme was alternatively activated, in the absence of Ca^{2+} (i.e. in the presence of EGTA), by polycationic substances such as poly-L-Lys, poly-L-Arg, poly-L-Orn, chitosan, histone, or ruthenium red. Under these conditions a better apparent substrate affinity was observed (44), and it was suggested on the basis of inhibitor experiments with La^{3+} that these polycationic activators might interact with the 1,3-β-glucan synthase at a site different from the Ca^{2+}-binding site. It remains unclear whether this direct influence of these exogenous polycations with the 1,3-β-glucan synthase may play a role during the in vivo induction of callose synthesis, in addition to enforced Ca^{2+} influx (see preceding subsection).

The natural polyamines spermine and spermidine caused relatively little in vitro activation of the 1,3-β-glucan synthase as long as EGTA was added to complex the 2–5 μM endogenous Ca^{2+} (44). If, however, Ca^{2+} was also present at this concentration in the assay in addition to the polyamines, the enzyme exhibited a considerable increase of its apparent substrate affinity and V_{max}. In addition, in the presence of 200 μM spermine the Ca^{2+} concentration for half-saturation of the 1,3-β-glucan synthase (determined at 33 μM UDP-glucose) is lowered from about 4 to about 0.6 μM (46). Polyamines thus appear to render the enzyme more sensitive towards the signal represented by Ca^{2+} (Figure 1). The activation by Ca^{2+} and by the polyamines or poly-L-Orn was found to be synergistic, indicating that both types of activator exert their effects at the same enzyme (46, 47). Polyamines are present in all eukaryotic cells in considerable amounts and have been suggested to play a regulatory role in plants (86) and animals (73), although little is known regarding their biochemical function and the cytoplasmic concentrations are not yet established. The 1,3-β-glucan synthase represents a possible target enzyme as its activation already becomes evident at the unusually low concentration of 4 μM spermine (46). On a quantitative basis the presumably artificial activator poly-L-Orn is, however, more effective than spermine (46, 47). It appears worthwhile to consider whether similar endogenous polycations occur and play a role in regulation of the 1,3-β-glucan synthase.

All the above experiments were performed using microsome preparations from suspension-cultured soybean cells. Similar results have also been observed with suspension cultures of parsley and tobacco, as well as tissues from various other plants (H. Kauss et al, unpublished results). With microsomes the classification of the enzyme studied as the plasma membrane–located 1,3-β-glucan synthase was based mainly on its requirement for high substrate concentrations and on the tentative identification of the reaction products by degradation (45, 47, 48, 54). It was, however, recently confirmed for soybean cells that the activation by Ca^{2+} and polycations is indeed a property of the plasma membrane–located 1,3-β-glucan synthase (so-called GS II). These experiments have been carried out with density gradient centrifugation and with fractions enriched in outside-out plasma membrane vesicles by two-phase partitioning in a dextran polyethyleneglycol system. The labeled product synthesized from UDP-[^{14}C]glucose by the latter purified membrane preparation in the presence of Ca^{2+} and spermine was characterized by gas chromatography of the permethylated derivatives and was found to contain at least 99% 1,3-glucan; no evidence for 1,4-linked glucose was obtained (J. Fink, W. Jeblick, W. Blaschek, H. Kauss, in preparation). A similar product composition was reported for the Ca^{2+}-activated 1,3-β-glucan synthase in microsomes from various other plants (17, 67a).

The 1,3-β-glucan synthase was also solubilized from the enriched soybean plasma membranes by digitonin and was affinity-chromatographed on a spermine-Sepharose column (47). The resulting soluble preparation retained in part its synergistic activation by Ca^{2+} and spermine or poly-L-Orn. This indicates that these two types of activators indeed interact directly with the enzyme or enzyme complex rather than exerting their function indirectly by a stabilization of the membrane. Taken together, the results provide convincing evidence that the plasma membrane–located 1,3-β-glucan synthase has the regulatory properties necessary to function as a target enzyme for Ca^{2+}, which is likely to enter the cell as a signal indicating cell damage.

Some Open Questions and Implications

Discussion of the induction of callose synthesis in soybean cells in the two preceding subsections and its depiction in Figure 1 have been restricted to the involvement of the 1,3-β-glucan synthase located in the plasma membrane. This may be an oversimplification, because several mechanisms for callose deposition appear possible. The above enzyme is presumably involved in the formation of "wound" callose or "plasma membrane" callose occurring in conjunction with traumatic events causing membrane perturbation—e.g. the closing of plasmodesmata and sieve pores, cell wall strengthening after tissue injury, or the formation of papillae at sites of attempted fungal penetration.

It may be of interest that deposition of callose, induced as suggested above by enforced Ca^{2+} influx, does not necessarily require chemical treatment or an actual injury of the cell. It may also be caused by physical stress without obvious cell damage [e.g. by bending of root hairs (2), pressing in of the cell surface (1, 28), and sonication (16)]. This implies, hypothetically, an opening of Ca^{2+} channels, presumably by stretching or condensing the plasma membrane (see Figure 1). It has been considered that these parameters are also involved in the sensing of pressure gradients during regulation of ion trans-port–mediated turgor in giant algae (95) and volume regulation in *Poter-ioochromonas* (see the last section, below). They are likely to play a role also in the opening of ion channels responsive to pressure, shown recently using the patch-clamp method (25a).

There is no convincing evidence that secretory vesicles are involved in the above cases. However, callose might also be deposited from Golgi-derived secretory vesicles, for instance, as a transient first wall material during cell plate formation, a process considered to be indirectly also under the influence of Ca^{2+} (33). Different mechanisms of 1,3-β-glucan deposition may even operate in the same cell, separated in space and time. One can imagine, for example, that in pollen tubes, vesicle-mediate secretion might constantly deliver 1,3-β-glucans to the elongating part of the wall behind the tip. Material rich in callose is also produced at a distance behind the tip to form "plugs"; such plugs are believed to be deposited at intervals, cutting off the spent grain and tube from the growing tip which contains the cytoplasm and organelles (31a). Deposit of these callose plugs may be mediated directly from the plasma membrane.

Another illustration of the different mechanisms for callose formation that may exist is the liverwort *Riella*. Every nonmeristematic cell of its unistratose thallus has a certain surface region specialized for vesicle-mediated exocy-tosis (57) where callose deposition can be induced with 100 μM chlorotet-racycline or 100 μM verapamil in a light-dependent manner (28). Stimulation by additional mM Ca^{2+} indicates that this ion is somehow involved in the secretion of callose at predetermined positions. The overall process appears complex, since it is also inhibited by the ionophores A 23187, valiomycin, and monensin. In contrast, "wound"-callose deposition anywhere on the entire cell surface of *Riella* can be induced by pressing a sharp edge onto it. Nifedipine (100 μM) strongly increases only the formation of callose in cell plates of meristematic cells. The relatively high concentrations of the drugs necessary for the above effects suggest that induction may imply a nonspecific perturbation of membrane permeability, in a way similar to that of amphipath-ic compounds in soybean cells. The example illustrates the different sensitiv-ity of the various mechanisms of callose deposition and also the difficulties likely to be encountered during future attempts to discriminate experimentally between the underlying biochemical events. As mentioned at the end of the second section, the general role of Ca^{2+} in exocytosis is not yet understood at the biochemical level. This has to be considered with regard to a possible participation of Ca^{2+} in cases where callose deposition may occur by vesicle-mediated secretion.

Addition of chitosan (55) or digitonin (H. Kauss et al, unpublished) to suspension-cultured soybean cells not only elicits callose synthesis but also induces some other metabolic changes such as formation of the phytoalexin glyceollin and some precursors. Callose formation is a rapid process suggested to be, at least in part, due to direct activation of the 1,3-β-glucan synthase by Ca^{2+} (see above). Consistent with this suggestion is the observation that the overall activity of this enzyme did not increase 2 hr after the addition of chitosan, conditions under which callose synthesis proceeds at a high constant rate (54). Callose formation, therefore, is unlikely to involve de novo enzyme synthesis. In contrast, induction of phytoalexin synthesis, which is a late event in the syndrome of defense reactions, is regulated in a number of systems (including soybean cells) by de novo synthesis of some enzymes (13). Chitosan-induced callose formation proceeds in a nutrient medium for at least 7 hr (54). This indicates that the cytoplasmic [Ca^{2+}] has remained elevated to a certain degree during this time. It is tempting to speculate therefore that elevated [Ca^{2+}] might also trigger the onset of a de novo synthesis of enzymes for phytoalexin formation, as well as for other enzymes—e.g. chitinase and 1,3-β-glucanase, reported to be induced by certain elicitors (citations in 41). However, treatment of soybean cells for 2 hr with a crude fungal glucan able to induce glyceollin synthesis did not lead to callose formation (54). This observation does not rule out the above hypothesis, because de novo synthesis of enzymes for glyceollin formation may already be induced below a certain threshold of [Ca^{2+}] whereas the massive event of callose formation might require a more drastic increase (or additional parameters such as phospholipid degradation products or membrane depolarization).

As mentioned above in the section on the maintenance and function of Ca^{2+} gradients, the perception of geotropic and thigmotropic stimuli may imply the alteration of cellular Ca^{2+} fluxes. Although the causal relationships are far from conclusive, it is of interest that an early manifestation of the stimuli is either a direct increase in, or the development of a predisposition for enhanced callose synthesis (34, 74). Induction of this process may require an elevated cytoplasmic [Ca^{2+}] (see Figure 1). Callose synthesis in the course of stimulus transduction may, therefore, represent a hint that indeed a transient increase in intracellular [Ca^{2+}] has occurred. It would be of interest in regard to this suggestion to determine whether callose formation also occurs in conjunction with other physiological events thought to involve Ca^{2+} and membrane depolarization—e.g. during peroxidase regulation (26).

Ca^{2+}-DEPENDENT GENERATION OF A REGULATIVE PROTEINASE

Studies on the osmoregulation in *Poterioochromonas malhamensis* have uncovered a possible new function of Ca^{2+} in the metabolic regulation of plant cells. Readers interested in more physiological and biochemical details of the system may consult a recent review (43) and the literature cited there. The overall process is discussed here only to the extent necessary to understand the role Ca^{2+} may play as a signal transmitter. Such a role is strongly suggested by results obtained with cell homogenates, whereas attempts to show that changes in cytoplasmic $[Ca^{2+}]$ in vivo really contribute to osmoregulation of *Poterioochromonas* are not yet satisfying).

Proteinase Generation by Ca^{2+} in Cell Homogenates

Poterioochromonas is a wall-less naked golden-brown alga. An increase in the concentration of external solutes causes the cells to shrink within 1–2 min; the water lost is regained in the following 1–2 hr owing to an internal formation of solutes, mainly isofloridoside (IF = α-galactosyl-1→1-glycerol). The high concentrations of this substance are accumulated in a reaction sequence fairly similar to sucrose synthesis in higher plants, namely by transfer of a galactosyl residue from UDP galactose to position 1 of glycerol-3-phosphate by the enzyme IFP-synthase. This is followed by dephosphorylation of IFP (= isofloridoside phosphate) to form IF (43). Chase experiments showed that during and shortly after shrinkage a prominent part of the regulation of the IF pathway may occur at the IFP-synthase level, and several observations suggested that this might involve covalent modification of the enzyme. The overall IF metabolism includes several other regulated steps, as the substance shows a rapid turnover in cells that have regained volume. A mobilization of the reserve 1,3-β-glucan chrysolaminarin has also to be considered. Nevertheless, most of the experiments performed were focussed on the rapid activation of the IFP synthase, because this might provide a promising answer to the question how the stimulus set by the decrease in cell volume or surface area is rapidly transduced to the biochemical reactions.

Because chase experiments suggested that the IFP synthase exhibits low activity in standard-volume cells, the experimental procedures were optimized to prepare crude homogenates from these cells with a IFP synthase activity as low as possible, a strategy that might appear unusual from a biochemical view. The aim was to achieve an in vitro system in which the physiological event could be imitated, namely to raise the activity by experimental manipulations, and in which differences between cells of standard volume and those shrunken for a few minutes would also be preserved. Several precautions during cell homogenization had to be strictly obeyed (43). If any of these was missed, the activation system described below was artificially triggered during and shortly after cell homogenization, resulting in fully active IFP synthase. The most important manipulation in this respect was that EGTA or EDTA had to be added to the homogenization buffer.

Under these conditions the IFP-synthase activity was low in homogenates from standard cells and was the higher the more the cell volume was previously diminished. It also had disappeared again a few minutes after the shrunken cells were artificially reswollen by adding water. These results indicated that the physiological stimulus had been manifested at the level of the IFP synthase.

The IFP-synthase enzyme is present in the homogenates in a soluble inactive form. It was observed that the homogenates in addition contain an auxiliary enzyme capable of activating the inactive IFP-synthase and that this ability also reflected the physiological situation (increased activation rate in homogenates from cells shrunken for 2 min, decreased again a few minutes after cell reswelling). Circumstantial evidence suggested that this activating enzyme may be a proteinase (51).

In homogenates from standard volume cells, prepared in the presence of EGTA or EDTA, the physiological stimulus could be mimicked in vitro by titration of the complexing agent with $CaCl_2$ to give a free $[Ca^{2+}]$ of about 1 μM or higher (38). Under these conditions the IFP synthase gained activity in a reaction depending on time, temperature, and pH; obviously an activating enzyme was initiated. Most important, the Ca^{2+} had to be present for one minute only. If it was lowered after that time again by the complexing agents, the events caused by Ca^{2+} were irreversibly set. It was concluded, therefore, that the activating enzyme, presumed to be a proteinase, was not directly stimulated by Ca^{2+} but was the output of a Ca^{2+}-dependent process (38) that requires the presence of membranes and results in the release of the activating enzyme (50).

The observation (40) that this Ca^{2+}-triggered presumed sequence of membrane-located reactions exhibited higher activity when the cells were homogenized in the presence of fluoride or molybdate may help to explain the biochemical events occurring at the membrane. These substances are phosphatase inhibitors; their effects may, therefore, indicate the participation of a phosphorylated constituent whose dephosphorylation during cell homogenization is slowed down.

An enzyme capable of activating the inactive IFP synthase was purified from a membrane pellet of *Poterioochromonas* cells by hydrophobic and affinity chromatography on immobilized fetuin. It was shown to be a proteinase, apparently a quite specific one, since no artificial chromogenic substrate has yet been found. The insulin B chain can serve as an alternative substrate which appears to be split on either side of its Leu[15] [(53); H. Ahmad, H. Kauss, unpublished]. As predicted, the proteinase does not require Ca^{2+}, appears to have a serine in its active site, and is inhibited by chymostatin and antipain (53). These inhibitors also allowed us to establish that the activating enzyme generated in vivo on cell shrinkage or on addition of Ca^{2+} to the homogenate is indeed identical to or similar to the above-mentioned purified proteinase (53).

However, it is not yet clear what happens to the inactive IFP synthase on activation. This latter enzyme is present in the cells in various oligomeric forms, identified by monoclonal antibodies on native PAGE. There is some indication that activation occurs simply by proteolytic nicking of the 67-kD monomer to result in still adhering 52- + 15-kD polypeptides (G. Brunner, H. Kauss, unpublished). It is also known that the IFP synthase once activated is subject to rapid further degradation in crude cell homogenates, an observation that may explain both the presence of cross-reactive polypeptides observed and the disappearance of the active enzyme in vivo after artificial swelling of previously shrunken cells.

Is the Ca^{2+}-dependent generation of the proteinase in crude cell homogenates mediated by calmodulin? The process is indeed enhanced up to 3-fold upon addition of calmodulin (40, 50). The calmodulin from both *Poterioochromonas* and bovine brain exhibits similar activity. However, relatively high concentrations are needed, and the effects require elevated $[Ca^{2+}]$. This could be due to several factors: Either the presumed Ca^{2+}/calmodulin-dependent process is nearly saturated by endogenous calmodulin and/or its localization in the membranes does not readily allow access of exogenous calmodulin. Alternatively, an unknown Ca^{2+}-sensing calciprotein is responsible and is only mimicked partly by added calmodulin.

Attempts to show with inhibitors that possibly membrane-bound endogenous calmodulin is responsible for the Ca^{2+}-mediated generation of the proteinase have not yet been conclusive. Trifluoperazine, fluphenazine, and chloropromazine showed activation without Ca^{2+}, and this was obviously due to their detergent-like properties (39). Calmidazolium caused significant inhibition in the presence of Ca^{2+} but only in a narrow concentration range (40). This could, nevertheless, be taken as positive evidence for a possible involvement of calmodulin, thus sustaining the conclusions drawn from the experiments with exogenous calmodulin. These results should be regarded with caution. At slightly higher concentrations, calmidazolium also acts the way some detergents and the above-mentioned other drugs do, namely increasing the generation of the regulative proteinase in a Ca^{2+}-independent way and thereby possibly causing nonspecific responses (see also the section on 1,3-β-glucan synthase, above).

As indicated above, the reaction sequence presumed to be located in or at membranes and triggered by Ca^{2+} to generate the proteinase can also be brought to action in the absence of Ca^{2+} by some detergents and, at elevated concentrations, by the detergent-like action of some drugs regarded to be specific for Ca^{2+}/calmodulin (39). The reactions also start "spontaneously"— i.e. without free Ca^{2+}—when the crude homogenate is preincubated at pH 6.2

for prolonged periods. The exact times for this latter process are subject to some variation between individual batches of algae (43). This indicates that the system might be triggered not only by Ca^{2+} but also by other unknown factors. The "spontaneous" as well as the Ca^{2+}-stimulated operation can be inhibited (43) by some substances known to interfere with phospholipids (polymyxin B, doxorubicin, palmitoyl-DL-carnitine). This again indicates that the unknown reaction sequence may take place in membranes and also that its operation may be started by alterations in the chemical composition or packing density of phospholipids. Whether this is of physiological significance for the perception of the stimulus set by shrinkage has yet to be elaborated.

Contribution of Ca^{2+} to Cell Volume Regulation

The above-discussed in vitro results suggested that an increase in cytoplasmic $[Ca^{2+}]$ caused by changes of the cellular Ca^{2+} compartmentation or by the fluxes of Ca^{2+} over the plasma membrane might also be one of the in vivo signals able to transmit the stimulus "cell shrinkage" to the described biochemical events. However, rather indirect and partly contradictory evidence is available with regard to this question.

Treatment of *Poterioochromonas* cells with the ionophore A 23187 increased the influx of $^{45}Ca^{2+}$ (49). This substance was also able to mimick in standard-volume cells some effects on the IFP-synthase activation system that are known to be triggered by cell shrinkage (38). The solvent DMSO, however, could act similarly to some extent, suggesting that the results can only be taken as a weak indication that an increased Ca^{2+} influx caused the observed effects. In addition, when the cells are suspended in a buffer containing EGTA the shrinkage-induced increase in IF production is not diminished (49). This could be interpreted to mean that if Ca^{2+} is necessary for the stimulus coupling it must come from internal pools. Such a suggestion agrees with the observation that in the simultaneous presence of EGTA and the ionophore A 23187 the shrinkage-induced IF formation slowed down considerably, presumably because the Ca^{2+} coming from internal stocks now flowed out of the cells and was complexed outside. It should be kept in mind that complexing of surface-bound Ca^{2+} and simultaneous application of a lipophilic agent may also have caused nonspecific effects (see the section on indirect action of Ca^{2+}, above).

An influence of cell shrinkage on the rearrangement of cellular Ca^{2+} pools is also evident from the following observation. When the cells were preincubated with chlorotetracycline (15–30 μM) a yellow fluorescence was seen at one or a few large spots, presumably representing membrane-lined compartments containing Ca^{2+} at high concentrations. The fluorescence of such a cell suspension in buffer containing EGTA was measured in a conventional spectrofluorometer and found to increase rapidly on shrinkage (49). Such measurements, however, are hampered by the possibility that the light-scattering characteristics of the algal cells and/or the polarity of internal membranes may change on osmotically induced shrinkage. Unfortunately, a directed determination of cytoplasmic Ca^{2+} with quin2/AM has not yet been possible. When preincubated *Poterioochromonas* cells were observed under the fluorescence microscope they did not show any fluorescence, whereas controls with mouse myeloma cells were positive. Obviously quin2/AM is not taken up by the wall-less algal cells (U. Rausch, H. Kauss, unpublished) and the questions raised must await new methods.

ACKNOWLEDGMENTS

Financial support by the DFG over many years has enabled the research in my laboratory. I wish also to thank H. Ahmad, G. Brunner, A. Clark, D. P. Delmer, J. Gross, R. Grotha, B. G. Pickard, and U. Rausch for critically discussing and carefully correcting parts of the manuscript.

Literature Cited

1. Aist, J. R. 1976. Papillae and related wound plugs of plant cells. *Ann. Rev. Phytopathol.* 14:145–63
2. Aist, J. R. 1977. Mechanically induced wall appositions of plant cells can prevent penetration by a parasitic fungus. *Science* 197:568–71
3. Bangerth, F. 1979. Calcium-related physiological disorders of plants. *Ann. Rev. Phytopathol.* 17:97–122
4. Barr, R., Sandelius, A. S., Crane, F. L., Morré, D. J. 1985. Oxidation of reduced pyridine nucleotides by plasma membranes of soybean hypocotyls. *Biochem. Biophys. Res. Commun.* 131: 943–48
5. Blaschek, W., Semler, U., Franz, G. 1985. The influence of potential inhibitors on the in vivo and in vitro cell-wall β-glucan biosynthesis in tobacco cells. *J. Plant Physiol.* 120: 457–70
6. Blumwald, E., Poole, R. J. 1986. Kinetics of Ca^{2+}/H^+ antiport in isolated tonoplast vesicles from storage tissue of *Beta vulgaris* L. *Plant Physiol.* 80:727–31
7. Brand, J. J., Becker, D. W. 1984. Evidence for direct roles of calcium in photosynthesis. *J. Bioenerg. Biomembranes* 16:239–49
8. Buckhout, T. J. 1984. Characterization of Ca^{2+} transport in purified endoplasmic reticulum membrane vesicles from *Lepidium sativum* L. roots. *Plant Physiol.* 76:962–67

9. Bush, D. R., Sze, H. 1986. Calcium transport in tonoplast and endoplasmic reticulum vesicles isolated from cultured carrot cells. *Plant Physiol.* 80:549–55
10. Bush, D. S., Jones, R. L. 1985. Calcium uptake and exchange in barley aleurone layers and protoplasts during Ca^{2+}-stimulated amylase secretion. In *Molecular and Cellular Aspects of Calcium in Plant Development*, ed. A. J. Trewavas, pp. 335–38. New York/London: Plenum. 452 pp.
11. Campbell, A. K. 1983. *Intracellular Calcium: Its Universal Role as Regulator.* New York: Wiley. 556 pp.
12. Carafoli, E. 1982. The transport of calcium across the inner membrane of mitochondria. In *Membrane Transport of Calcium,* ed. E. Carafoli, pp. 109–39. London/New York: Academic. 266 pp.
13. Chappell, J., Hahlbrock, K. 1984. Transcription of plant defense genes in response to UV light or fungal elicitor. *Nature* 311:76–78
14. Cramer, G. R., Läuchli, A., Polito, V. 1985. Displacement of Ca^{2+} by Na^+ from the plasmalemma of root cells. A primary response to salt stress? *Plant Physiol.* 79:207–11
15. Crane, F. L., Barr, R. 1985. Function of calcium-calmodulin in chloroplasts. See Ref. 10, pp. 269–76
16. Currier, H. B., Webster, D. H. 1964. Callose formation and subsequent disappearance: Studies in ultrasound

stimulation. *Plant Physiol.* 39:843–47

17. Delmer, D. P., Cooper, G., Alexander, D., Cooper, J., Hayashi, T., et al. 1985. New approaches to the study of cellulose biosynthesis. *J. Cell Sci. Suppl.* 2:33–50

18. De Lorenzo, R. J. 1985. Calcium and calmodulin control of neurotransmitter synthesis and release. In *Calcium and Cell Physiology*, ed. D. Marmé, pp. 265–84. Berlin: Springer-Verlag. 390 pp.

19. Demarty, M., Morvan, C., Thellier, M. 1984. Calcium and cell wall. *Plant, Cell, Environ.* 7:441–48

20. Denton, R. M., Mc Cormack, J. G. 1985. Physiological role of Ca^{2+} transport by mitochondria. *Nature* 315:635

21. Dieter, P. 1984. Calmodulin and calmodulin-mediated processes in plants. *Plant, Cell, Environ.* 7:371–80

22. Dieter, P., Marmé, D. 1984. A Ca^{2+}, calmodulin-dependent NAD kinase from corn is located in the outer mitochondrial membrane. *J. Biol. Chem.* 259:184–89

23. Dieter, P., Salimath, B. P., Marmé, D. 1984. The role of calcium and calmodulin in higher plants. *Ann. Proc. Phytochem. Soc. Eur.* 23:213–29

24. Evans, M. L. 1985. The action of auxin on plant cell elongation. *CRC Crit. Rev. Plant Sci.* 2:317–65

25. Evans, N. A., Hoyne, P. A. 1984. Characteristics and specificity of the interaction of a fluorochrome from aniline blue (sirofluor) with polysaccharides. *Carbohydr. Polym.* 4:215–30

25a. Falke, L., Edwards, K. L., Misler, S., Pickard, B. P. 1986. A mechanotransductive ion channel in patches from cultured tobacco cell plasmalemma. *Plant Physiol. Suppl.* 80:40

26. Gaspar, T., Penel, C., Castillo, F. J., Greppin, H. 1985. A two-step control of basic and acidic peroxidases and its significance for growth and development. *Physiol. Plant.* 64:418–23

27. Gilroy, S., Hughes, W. A., Trewavas, A. J. 1986. The measurement of intracellular calcium levels in protoplasts from higher plant cells. *FEBS Lett.* 199:217–21

28. Grotha, R. 1986. Tetracycline, verapamil and nifedipine induce callose deposition at specific cell sites in *Riella helicophylla. Planta.* In press

29. Hanson, J. B. 1984. The functions of calcium in plant nutrition. *Adv. Plant Nutr.* 1:149–208

30. Hanson, J. B., Rincon, M., Rogers, S. A. 1986. Control on calcium influx in corn root cells. See Ref. 10, pp. 253–60

31. Harris, K. M., Kongsamut, S., Miller, R. J. 1986. Protein kinase C mediated regulation of calcium channels in PC-12 pheochromocytoma cells. *Biochem. Biophys. Res. Commun.* 134:1298–1305

31a. Harris, P. J., Anderson, M. A., Bacic, A., Clarke, A. E. 1984. Cell-cell recognition in plants with special reference to the pollen-stigma interaction. *Oxford Surveys Plant Mol. Cell Biol.* 1:161–203

32. Haug, A., Weis, C. 1986. Aluminum-induced changes in calmodulin. See Ref. 10, pp. 19–26

33. Hepler, P. K., Wayne, R. O. 1985. Calcium and plant development. *Ann. Rev. Plant Physiol.* 36:397–439

34. Jaffe, M. J., Hubermann, M., Johnson, J., Telewski, F. W. 1985. Thigmomorphogenesis: the induction of callose formation and ethylene evolution by mechanical perturbation in bean stems. *Physiol. Plant.* 64:271–79

35. Jones, R. G. W., Lunt, O. R. 1967. The function of calcium in plants. *Bot. Rev.* 33:407–26

36. Jones, R. L., Carbonell, J. 1984. Regulation of the synthesis of barley aleurone α-amylase by gibberellic acid and calcium ions. *Plant Physiol.* 76:213–18

37. Jones, R. L., Deikman, J., Melroy, D. 1986. Role of Ca^{2+} in the regulation of α-amylase synthesis and secretion in barley aleurone. See Ref. 10, pp. 49–56

38. Kauss, H. 1981. Sensing of volume changes by *Poterioochromonas* involves a Ca^{2+}-regulated system which controls activation of isofloridoside-phosphate synthase. *Plant Physiol.* 68:420–24

39. Kauss, H. 1982. Volume regulation: Activation of a membrane-associated cryptic enzyme system by detergent-like action of phenothiazine drugs. *Plant Sci. Lett.* 26:103–9

40. Kauss, H. 1983. Volume regulation in *Poterioochromonas:* Involvement of calmodulin in the Ca^{2+}-stimulated activation of isofloridoside-phosphate synthase. *Plant Physiol.* 71:169–72

41. Kauss, H. 1985. Callose biosynthesis as a Ca^{2+}-regulated process and possible relations to the induction of other metabolic changes. *J. Cell Sci. Suppl.* 2:89–103

42. Kauss, H. 1986. Ca^{2+}-dependence of callose synthesis and the role of polyamines in the activation of 1,3-β-glucan synthase by Ca^{2+}. See Ref. 10, pp. 131–37

43. Kauss, H. 1986. A membrane-derived proteinase capable of activating a galactosyl-transferase involved in volume regulation of *Poterioochromonas.* In *Plant Proteolytic Enzymes II*, ed. M. J. Dalling, pp. 91–102. Boca Raton:CRC Press

44. Kauss, H., Jeblick, W. 1985. Activation by polyamines, polycations, and Ruthenium Red of the Ca²⁺-dependent glucan synthase from soybean cells. *FEBS Lett.* 185:226–30

45. Kauss, H., Jeblick, W. 1986. Influence of free fatty acids, lysophosphatidylcholine, platelet activating factor, acylcarnitine, and Echinocandin B on 1,3-β-D-glucan synthase and callose synthesis. *Plant Physiol.* 80:7–13

46. Kauss, H., Jeblick, W. 1986. Synergistic activation of 1,3-β-glucan synthase by Ca²⁺ and polyamines. *Plant Sci.* 43:103–7

47. Kauss, H., Jeblick, W. 1986. Solubilization, affinity chromatography and Ca²⁺/polyamine activation of the plasma membrane located 1,3-β-D-glucan synthase. *Plant Sci.* In press

48. Kauss, H., Köhle, H., Jeblick, W. 1983. Proteolytic activation and stimulation by Ca²⁺ of the glucan synthase from soybean cells. *FEBS Lett.* 158:84–88

49. Kauss, H., Rausch, U. 1984. Compartmentation of Ca²⁺ and its possible role in volume regulation of *Poterioochromonas.* In *Compartments in Algal Cells and Their Interaction,* ed. W. Wiessner, D. Robinson, R. C. Starr, pp. 147–56. Heidelberg: Springer-Verlag

50. Kauss, H., Thomson, K. S. 1982. Biochemistry of volume control in *Poterioochromonas.* In *Plasmalemma and Tonoplast: Their Function in the Plant Cell,* ed. D. Marmé, E. Marrè, R. Hertel, pp. 255–62. New York: Elsevier

51. Kauss, H., Thomson, K. S., Tetour, M., Jeblick, W. 1978. Proteolytic activation of galactosyl transferase involved in osmotic regulation. *Plant Physiol.* 61:35–37

52. Keith, C. H., Ratan, R., Maxfield, F. R., Bayer, A., Shelanski, M. L. 1985. Local cytoplasmic calcium gradients in living mitotic cells. *Nature* 316:848–50

53. Köhle, D., Kauss, H. 1984. Purification of a membrane-derived proteinase capable of activating a galactosyltransferase involved in volume regulation. *Biochim. Biophys. Acta* 799:59–67

54. Köhle, H., Jeblick, W., Poten, F., Blaschek, W., Kauss, H. 1985. Chitosan-elicited callose synthesis in soybean cells as a Ca²⁺-dependent process. *Plant Physiol.* 77:544–51

55. Köhle, H., Young, D. H., Kauss, H. 1983. Physiological changes in suspension-cultured soybean cells elicitied by treatment with chitosan. *Plant Sci. Lett.* 33:221–30

56. Körner, L. E., Kjellbom, P., Larsson, C., Møller, I. M. 1985. Surface properties of right side-out plasma membrane vesicles isolated from barley roots and leaves. *Plant Physiol.* 79:72–79

57. Kramer, A., Lehmann, H. 1986. Feinstrukturelle Untersuchungen zur Entwicklung des 'metachromatischen Körpers' in Zellen des Lebermoses *Riella helicophylla* (Bory et Mont.) Mont. *Ber. Dtsch. Bot. Ges.* 99:111–21

58. Kreimer, G., Melkonian, M., Holtum, J. A. M., Latzko, E. 1985. Characterization of calcium fluxes across the envelope of intact spinach chloroplasts. *Planta* 166:515–23

59. Kreimer, G., Melkonian, M., Latzko, E. 1985. An electrogenic uniport mediates light-dependent Ca²⁺-influx into intact spinach chloroplasts. *FEBS Lett.* 180:253–58

60. Larsson, C., Kjellbom, P., Widell, S., Lundborg, T. 1984. Sidedness of plant plasma membrane vesicles purified by partitioning in aqueous two-phase systems. *FEBS Lett.* 171:271–76

61. Lindau, M., Fernandez, J. M. 1986. IgE-mediated degranulation of mast cells does not require opening of ion channels. *Nature* 319:150–53

62. Lockau, W., Pfeffer, S. 1983. ATP-dependent calcium transport in membrane vesicles of the cyanobacterium, *Anabaena variabilis. Biochim. Biophys. Acta* 733:124–32

63. Lühring, H., Tazawa, M. 1985. Effect of cytoplasmic Ca²⁺ on the membrane potential and membrane resistance of *Chara* plasmalemma. *Plant Cell Physiol.* 26:635–46

64. Macklon, A. E. S. 1984. Calcium fluxes at plasmalemma and tonoplast. *Plant, Cell, Environ.* 7:407–13

65. Mohanty, P., Brand, J. J., Fork, D. C. 1985. Calcium depletion alters energy transfer and prevents state changes in intact *Anacystis. Photosynth. Res.* 6:349–61

66. Moore, A. L., Åkerman, K. E. O. 1984. Calcium and plant organelles. *Plant, Cell, Environ.* 7:423–29

67. Moore, A. L., Proudlove, M. O., Åkerman, K. E. O. 1986. The role of intracellular organelles in the regulation of cytosolic calcium levels. See Ref. 10, pp. 277–84

67a. Morrow, D. L., Lucas, W. T. 1986. (1→3)-β-D-glucan synthase from sugar beet. *Plant Physiol* 81:171–76

68. Muto, S., Miyachi, S. 1985. Roles of calmodulin dependent and independent NAD kinases in regulation of nicotinamide coenzyme levels of green plant cells. See Ref. 10, pp. 107–14

69. Muto, S., Shimogawara, K. 1985. Calcium and phospholipid-dependent phosphorylation of ribulose-1,5-bisphosphate carboxylase/oxygenase small subunit by a chloroplast envelope-bound protein kinase in situ. *FEBS Lett.* 193:88–92

70. Newton, R. P., Kingston, E. E., Evans, D. E., Younis, L. Y., Brown, E. G. 1984. Occurrence of guanosine 3',5'-cyclic monophosphate (cyclic GMP) and associated enzyme systems in *Phaseolus vulgaris. Phytochemistry* 13:1367–72

71. Paliyath, G., Poovaiah, B. W. 1985. Promotion of β-glucan synthase activity in corn microsomal membranes by calcium and calmodulin. *Plant Physiol. Suppl.* 77:4

72. Peachell, P. T., Pearce, F. L. 1985. Calcium regulation of histamine secretion from mast cells. See Ref. 18, pp. 311–27

73. Pegg, A. E. 1986. Recent advances in the biochemistry of polyamines in eukaryotes. *Biochem. J.* 234:249–62

74. Pickard, B. G. 1985. Early events in geotropism of seedling shoots. *Ann. Rev. Plant Physiol.* 36:55–75

75. Pickard, B. G. 1985. Roles of hormones, protons and calcium in geotropism. *Encycl. Plant Physiol. (NS)* 11:193–281

76. Rasi-Caldogno, F., De Michelis, M. I., Pugliarello, M. C. 1982. Active transport of Ca^{2+} in membrane vesicles from pea. Evidence for a H^+/Ca^{2+} antiport. *Biochim. Biophys. Acta* 693:287–95

77. Reiss, H.-D., Herth, W. 1985. Nifedipine-sensitive calcium channels are involved in polar growth of lily pollen tubes. *J. Cell Sci.* 76:247–54

78. Reuveni, M., Lerner, H. R., Poljakoff-Mayber, A. 1985. Changes in membrane potential as a demonstration of selective pore formation in the plasmalemma by poly-L-lysine treatment. *Plant Physiol.* 79:406–10

79. Rincon, M., Hanson, J. B. 1985. Voltage-regulated Ca^{2+} channels in corn roots. *Plant Physiol. Suppl.* 77:20

80. Saunders, M. J. 1985. Cytokinin activates and redistributes plasma membrane ion channels creating a zone of high free Ca^{2+} that predicts the site of cell division. *J. Cell Biol.* 101 (Suppl. 5.2):4a

81. Schatzmann, H. J. 1985. Calcium extrusion across the plasma membrane by the calcium-pump and the $Ca^{2+}-Na^+$ exchange system. See Ref. 18, pp. 18–52

82. Schumaker, K. S., Sze, H. 1985. A Ca^{2+}/H^+ antiport system driven by the proton electrochemical gradient of a tonoplast H^+-ATPase from oat roots. *Plant Physiol.* 79:1111–17

83. Schürmann, P., Roux, J., Salvi, L. 1985. Modification of thioredoxin specificity of chloroplast fructose-1,6-bisphosphatase by substrate and Ca^{2+}. *Physiol. Veg.* 23:813–18

84. Siegel, N., Haug, A. 1983. Calmodulin-dependent formation of membrane potential in barley root plasma membrane vesicles: a biochemical model of aluminum toxicity in plants. *Physiol. Plant.* 59:285–91

85. Simon, P., Bonzon, M., Greppin, H., Marmé, D. 1984. Subchloroplastic localization of NAD kinase activity: evidence for a Ca^{2+}, calmodulin-dependent activity at the envelope and for a Ca^{2+}, calmodulin-independent activity in the stroma of pea chloroplasts. *FEBS Lett.* 167:332–38

86. Smith, T. A. 1985. Polyamines. *Ann. Rev. Plant Physiol.* 36:117–43

87. Stroobant, P., Scarborough, G. A. 1979. Active transport of calcium in *Neurospora* plasma membrane vesicles. *Proc. Natl. Acad. Sci. USA* 76:3102–6

88. Sze, H. 1985. H^+-translocating ATPases: advances using membrane vesicles. *Ann. Rev. Plant Physiol.* 36:175–208

89. Taiz, L. 1984. Plant cell expansion: regulation of cell wall mechanical properties. *Ann. Rev. Plant Physiol.* 35:585–657

90. Wayne, R., Hepler, P. K. 1984. The role of calcium ions in phytochrome-mediated germination of spores on *Onoclea sensibilis* L. *Planta* 160:12–20

91. Williamson, R. E., Ashley, C. C. 1982. Free Ca^{2+} and cytoplasmic streaming in the alga *Chara. Nature* 296:647–51

92. Wolniak, S. M., Bart, K. M. 1985. The buffering of calcium with quin2 reversibly forestalls anaphase onset in stamen hair cells of *Tradescantia. Eur. J. Cell Biol.* 39:33–40

93. Yamaya, T., Oaks, A., Matsumoto, H. 1984. Stimulation of mitochondrial calcium uptake by light during growth of corn shoots. *Plant Physiol.* 75:773–77

94. Young, D. H., Kauss, H. 1983. Release of calcium from suspension-cultured *Glycine max* cells by chitosan, other polycations, and polyamines in relation to effects on membrane permeability. *Plant Physiol.* 73:698–702

95. Zimmermann, U. 1978. Physics of turgor- and osmoregulation. *Ann. Rev. Plant Physiol.* 29:121–48

Ann. Rev. Plant. Physiol. 1987. 38:73–93

PHOSPHORYLATION OF PROTEINS IN PLANTS: Regulatory Effects and Potential Involvement in Stimulus/Response Coupling

R. Ranjeva and A. M. Boudet

Centre de Physiologie Végétale de l'Université Paul Sabatier, U.A., CNRS No. 241, 118 route de Narbonne, F-31062 Toulouse, France

CONTENTS

INTRODUCTION 73
CASCADE SYSTEMS AS REGULATORY DEVICES 74
PHOSPHORYLATION OF PROTEINS IN PLANT ORGANELLES 75
  Nuclei 75
  Plastids and Mitochondria 77
  Microsomal Membranes 79
PHOSPHORYLATION OF ENZYMES 81
  Regulation of Enzyme Activities 81
  Modifications of Regulatory Properties 84
PROTEIN PHOSPHORYLATION AND STIMULUS-RESPONSE COUPLING 85
REFLECTIONS AND PROSPECTS 87
  Receptor Phosphorylation 87
  Transmembrane Signalling 88
GENERAL CONCLUSIONS 89

INTRODUCTION

In animal cell biology, approximately 5% of the research deals in some manner with protein phosphorylation (18). Knowledge in this area has progressed to the point that chapters are devoted not only to general considerations (16, 50) but also to specific aspects such as tyrosine protein kinases (44) or serine and threonine protein kinases (51).

73

0066-4294/87/0601-0073$02.00

In the Annual Review of Plant Physiology, protein phosphorylation was discussed more than ten years ago by Trewavas (96). However, at that time the majority of examples quoted were from animal systems, whereas only a few were from plants.

During the last decade, considerable data have been accumulated especially after the discovery of calcium-regulated protein kinases in plant extracts (38, 77). This finding opened new frontiers by connecting the concept of cascade reactions with that of a second messenger; specific aspects of the latter are discussed by Kauss in this volume (48).

In this chapter we review significant findings on protein phosphorylation in plants. Although the amount of data concerning protein kinases is growing, no specific description of these enzymes will be made. We rather emphasize the functional significance of protein phosphorylation and its possible role in stimulus-response coupling.

CASCADE SYSTEMS AS REGULATORY DEVICES

Phosphorylation is one of the most studied posttranslational modifications affecting proteins. Basically, the system involves a minimum of three proteins and two reactions:

$$\text{protein} + n\text{ATP} \rightarrow \text{protein-Pn} + \text{ADP}, \qquad\qquad 1.$$

$$\text{(native)} \qquad\qquad \text{(modified)}$$

$$\text{protein-Pn} + H_2O \rightarrow n\text{Pi} + \text{protein}. \qquad\qquad 2.$$

Reaction 1 is catalyzed by protein kinase(s), and Reaction 2 (the opposite) by phosphoprotein phosphatase(s). Such a system, which involves a reversible shift between modified and unmodified substrate protein forms catalyzed by converter enzymes, is defined as a cascade system (16).

Theoretical considerations and extensive analysis of kinetics have shown that cascade systems are very efficient in regulating and coordinating reticulated networks of metabolic pathways. They combine the advantages of signal amplification and the flexibility of regulation by involving different components that act in additive, opposite, or synergistic ways (89).

If the substrate is an enzyme involved in a second cascade cycle, then all the regulatory advantages afforded by the system are multiplied by several orders of magnitude (89). Such a situation is illustrated by the coordinated glycogen phosphorylase-glycogen synthase system (18). In all cases, cascade systems are able to sense the fluctuations in the concentrations of substrates and regulatory molecules and therefore act as an integrative network. It has been definitely shown that these systems are a fundamental part of cellular regulatory devices in animal systems and a mechanism by which neural and hormonal stimuli may be amplified (18).

PHOSPHORYLATION OF PROTEINS IN PLANT ORGANELLES

Nuclei

The intensity and complexity of protein phosphorylation in the plant nucleus seem to be equivalent to that found in animal nuclei (97). Proteins are easily phosphorylated in isolated nuclei, and the phosphorylations seem to be catalyzed by endogenous protein kinase (28, 63). However, the physiological significance of these reversible phosphorylations is not clear in plants and most of the time is inferred from hypotheses or results obtained with other organisms.

Significant work in this area has been carried out by Trewavas and co-workers (15, 98); they have characterized a large number of phosphorylated proteins in plant nuclei and have attempted to correlate the phosphorylation patterns with different physiological events. These workers have provided evidence for changes in chromatin phosphorylation patterns during barley germination and after abscisic acid treatment of *Lemna minor* (15), thus demonstrating the diverse occurrence of the phosphorylation process during development. Melanson & Trewavas (59) have emphasized the relation between cell division and nuclear protein phosphorylation in artichoke tuber. Thus, enhanced phosphorylation of histone H1 occurred during the G2/M phase as in animal systems (92), and the reported data are consistent with the notion that additional histone H1 phosphorylation takes place during the mitotic phase of the plant cell cycle.

An important function of phosphorylated nuclear proteins could be their involvement in regulation of gene expression. This was suggested some years ago (49) for non-histone proteins in animal cells and was substantiated by further experimental evidence (52). However, although there are many reports on the phosphorylation of specific nucleus-associated proteins, there are only a few examples of a direct correlation with increased gene expression in eukaryotes. The phosphorylation of nuclear proteins has been studied in crown gall tumors of higher plants, which synthesize large amounts of RNA during their intense growth phase. Nuclei from transformed cell cultures of *Nicotiana tabaccum* (Ti plasmid) did not exhibit enhanced protein kinase activity (2), as is known to be the case, upon transformation of animal tissues. In contrast, in potato tubers transformed by *Agrobacterium tumefaciens*, chromatin-bound protein kinases were activated and chromatin protein phosphorylation was enhanced concomitantly with the dramatic increase in

RNA synthesis in growing tumors (46). A causative relationship between increases in RNA synthesis and protein phosphorylation has yet to be demonstrated.

Other attempts have been made to determine the possible relationship between gene activity and protein phosphorylation. Soybean hypocotyls are induced to synthesize large amounts of rRNA by addition of dichlorophenoxyacetic acid (2,4-D), but Murray & Key (64) were not able to decide whether the changes in phosphorylation of nuclear proteins were the cause or the result of the induced RNA synthesis. Chromatin isolated from GA3-treated pea embryos showed a higher degree of phosphorylation in vitro when compared with control plants (104). This increase in phosphorylation was correlated with increased transcriptional activity in vitro and seemed related to a greater accessibility of the protein substrates in the nuclei. These promising data have not been further developed.

Recently, Roux and co-workers (23, 24) have obtained potentially important results concerning the regulation of nuclear protein phosphorylation. Highly purified nuclei containing both their outer and inner membranes were isolated from etiolated pea plumules. The addition of micromolar concentrations of Ca^{2+} to the nuclear preparation increased the level of phosphorylation in several nuclear proteins, and low concentrations of the calmodulin antagonists inhibited the calcium-induced response. In the presence of Ca^{2+}, red light induced the phosphorylation of at least three proteins while far-red light abolished this effect. The photoreversibility of this response by red and far-red light indicates that phytochrome is likely to be the photoreceptor. The red light response was inhibited both by calcium chelation and by calmodulin antagonists. Because phytochrome is known to regulate gene expression in pea, these results raise the possibility that nuclear protein phosphorylation may be an important step between the photoactivation of phytochrome and the subsequent gene expression. Such an experimental system may be valuable for understanding the role of nuclear protein phosphorylation in the response of plants to external stimuli.

In conclusion, although no specific functions can be clearly ascribed to phosphorylated plant nuclear proteins, the following points must be emphasized:

• Phosphorylation of plant nuclear proteins is associated with active metabolism.
• Despite the large number of polypeptides concerned, the phosphorylation of nuclear proteins does not seem to be a random phenomenon. Comparative experiments on normal and transformed potato tuber nurse tissue show that in this system the phosphorylation pattern of chromosomal proteins is highly and reproducibly stage specific (46). This has also been shown for

resting versus wounded and control versus hormone-treated cells (87). Thus, nuclear protein phosphorylation appears to be a tissue- and developmental-stage-specific process.

Future experiments should provide unambiguous demonstration of the role of protein phosphorylation in plant nuclei. They should preferentially use strictly synchronized cells to study the possible effects of phosphorylated proteins in the structural maintenance of chromatin. Furthermore, in vitro transcription systems would be appropriate to probe the question as to whether chromatin phosphoproteins direct specific gene activation.

Plastids and Mitochondria

There is a paucity of data concerning protein phosphorylation in plant mitochondria. However, these organelles contain protein kinase(s) that can phosphorylate different endogenous polypeptides in an ATP-dependent manner (21). The kinase, which is Ca^{2+}/calmodulin and cAMP independent, resembles protein kinase C from mammalian systems.

In contrast, a large number of experiments have shown that plastid proteins can be extensively phosphorylated. Proteins from amyloplasts isolated from sycamore cells can be reversibly phosphorylated, some of them in a calcium/calmodulin-dependent manner (58). These results, along with data on tobacco and carrot cells, suggest a possible regulation of starch metabolism in nonphotosynthetic cells through reversible protein phosphorylation (8). However, the identification of the phosphorylated proteins is a prerequisite to the development of a serious hypothesis concerning the significance of the observed phosphorylation processes. Most studies of phosphorylation of plastid proteins focus on the chloroplastic proteins from different structural subcompartments, i.e. envelope, stroma and thylakoids. Except in the case of thylakoids, the functional significance of these reversible posttranslational phosphorylation reactions is not known.

Protein phosphorylation is a widespread phenomenon within chloroplast membranes (54, 90). Up to ten phosphorylated proteins can be detected in the envelope membrane fraction from pea. Most of the labelled proteins are localized in the outer envelope, as is the protein kinase activity (90). This preferential location suggests a possible role of the phosphorylated substrates in regulating functional interactions between chloroplast and the other compartments.

Stromal components also undergo phosphorylation. In spinach more than fifteen polypeptides of stromal origin, including both the large and the small subunit of rubisco, were phosphorylated (30). The thylakoid protein kinase played no part in the phosphorylation of stromal proteins. These are catalyzed by soluble protein kinase(s) that is(are) insensitive to light. The phosphorylation level of stromal proteins was enhanced with increased CO_2 fixation in isolated chloroplasts (30). Phosphorylated ribosomal proteins, protein kinase, and phosphoprotein phosphatase activities have been also detected in the stroma of spinach chloroplasts (35a; R. Mache, personal communication).

Soll & Buchanan (91) characterized a protein kinase, on the outer membrane of spinach chloroplasts, that phosphorylates the mature form of the small subunit of rubisco (a stromal protein). The kinase is insensitive to light, cAMP, thioredoxin, and plastoquinone. The crude kinase was slightly activated by micromolar concentrations of Ca^{2+} but was independent of exogenous calmodulin (65). If the envelope was lipid-depleted or treated by various phospholipases, then the small subunit (SSU) of rubisco was no longer phosphorylated. The phosphorylation was restored by adding back the phospholipid fraction (65). Full activation was obtained when both calcium and phospholipids were added.

Because of the special location of this protein kinase, there was speculation that it could be involved in the ATP-dependent transport of proteins into the chloroplast (91). Additional experiments have not yet provided conclusive evidence to support this possibility (90).

Several thylakoid proteins are easily phosphorylated either in vivo or in vitro; among them is a 25 kilodalton polypeptide that was unambiguously identified as a component of the light-harvesting chlorophyll (LHCP) *a/b*-binding complex. An abundant literature summarized in recent reviews (1, 4) supports the concept of the involvement of reversible phosphorylation in the regulation of light energy distribution between the two photosystems in higher plants and algae.

The main components of the system are as follows. The phosphorylation of LHCP is catalyzed by a light-dependent protein kinase. The light effect is mediated by plastoquinone, the redox state of which depends on the light-driven electron transport chain. Thus, the protein kinase is activated when the pool of plastoquinone becomes reduced (43). The phosphorylated LHCP is reversibly hydrolyzed by a light-insensitive phosphoprotein phosphatase, which is bound to the thylakoid membrane (3). The phosphorylation cycle of LHCP induces changes in energy transfer between photosystem (PS) I and photosystem (PS) II (5). When LHCP is phosphorylated, light energy is transferred to PS I; conversely, the distribution of excitation energy is in favor of PS II when LHCP is dephosphorylated. Such a system controlled by the redox state of plastoquinone is assumed to correct the imbalance of energy transfer between the two photosystems and to increase the efficiency of noncyclic electron transport (10).

In vivo and in vitro kinetics studies in higher plants and algae show the involvement of this process in State I/State II transitions induced by changing light conditions (41, 94). This overall mechanism would maintain the quantum yield of photosynthesis at a high and constant level over regions of the spectrum in which excitation of the two photosystems would otherwise be unequal.

The functional alterations of the photosystems are accompanied by structural modifications within the chloroplastic membranes. It is thought that when phosphorylation occurs a mobile part of LHCP migrates from appressed thylakoid regions (rich in PS II) to non-appressed regions (rich in PS I). The reverse movements would occur when LHCP is dephosphorylated. Details on this regulation model are provided (see 10). Phosphorylated and nonphosphorylated forms of LHCP have been recently separated by two-dimensional gel electrophoresis (83). Molecular heterogeneity among the molecules of LHCP has been suggested, since only some of them can be phosphorylated.

The phenomenon of reversible phosphorylation of polypeptides associated with PS II could be a mechanism for protection against photoinhibition caused by strong light or other stressful conditions that induce a light-dependent inactivation of photosynthetic electron transport. This has been recently discussed in the case of rice seedlings, where chilling temperatures that cause photoinhibition result in the inhibition of thylakoid protein phosphorylation (60) in vitro as well as in vivo.

In addition to the regulation of energy distribution between the two photosystems, phosphorylation of thylakoid proteins seems to induce other functional modifications—particularly a decrease in electron capacity of PS II. One of the most significant effects seems to be located on the acceptor side, because phosphorylation induces a decrease of the electron transfer from Q_A to Q_B. The effects of phosphorylation on both the capacity and absorption cross section of PS II are additive and represent a powerful feedback response to the over-reduction of the plastoquinone, the accumulation of which is assumed to induce the destruction of the 32 kilodalton Q_B protein (53). Different protein kinases from thylakoid membranes have been purified (19, 57), and some of them may phosphorylate LHCP (20). These enzymes are currently under investigation as researchers attempt to isolate the corresponding genes.

In brief, the above system illustrates the powerful regulation device represented by reversible protein phosphorylation at the level of specific events of plant functioning. It also underlines the flexibility and complexity of regulation mechanisms involving reversible protein phosphorylation.

Microsomal Membranes

During the last four years increasing attention has been devoted to the phosphorylation of proteins bound either to internal or border membranes (plasmalemma, tonoplast) of higher plant cells. Phosphorylation of membrane-bound proteins may be crucial in regulating the structure and function of membrane components, and consequently the function of the membrane itself. These rapid and reversible posttranslational changes could then play an important role, as they do in animal cells, in the transduction of different stimuli.

For technical reasons, most studies have been performed with crude membrane preparations (microsomes). Ralph et al (74) were the first to demonstrate that membrane proteins from plant tissues were phosphorylated by membrane-associated protein kinase(s). This observation was also made a few years later (31) on membrane fraction from rhizome tissues of Jerusalem artichoke. Similarly, Hetherington & Trewavas (38–40) demonstrated that membrane fractions from pea were able to phosphorylate and dephosphorylate their constitutive proteins. After these initial demonstrations, other groups have reported the phosphorylation of membrane-located proteins in various plants (62, 68, 71, 85, 101, 102). Plasmalemma- and tonoplast-bound proteins appear to be phosphorylated.

In corn coleoptiles, the majority of phosphorylated membrane proteins are glycoproteins (12). Some of the phosphorylated proteins are also iodinated after iodination of protoplasts and their location on the plasmalemma is strongly suggested. In endosperm cells of *Lolium multiflorum* the phosphorylation of proteins located in a well-defined plasma membrane fraction has also been demonstrated (71). Yet, as far as well-identified membranes are concerned, one of the most significant results concerns the characterization of phosphorylated proteins on the tonoplast of sycamore cells. The in vitro phosphorylation/dephosphorylation of these proteins was mediated by tonoplast-bound kinases and phosphatases (95).

Both Ca^{2+}-dependent and -independent phosphorylation of membrane proteins has been reported (38, 68, 85). In tonoplast from *Acer* cells the phosphorylation of proteins is highly calcium-calmodulin dependent (95), whereas in plasma membrane-enriched fractions of *Lolium* few polypeptides were subject to Ca^{2+}-dependent phosphorylation (71). Membranes from pea buds, and particularly enriched plasma membrane fractions, contained a protein kinase that is activated 5- to 15-fold by micromolar levels of Ca^{2+} (40). In some cases (85), calmodulin enhanced the effect of Ca^{2+} ions; in other (73), calmodulin antagonists inhibited the phosphorylation of several proteins despite the fact that calmodulin does not exert any effect by itself. Such results are likely the consequence of the presence of endogenous calmodulin in the isolated membranes.

An important problem is to identify the biochemical function of the membrane-bound phosphorylated polypeptides. At the present time the only functionally characterized membrane substrate seems to be the microsomal H^+ ATPase (106). Obviously, changes in the phosphorylation level of membranes may induce dramatic changes in electrostatic charges and subsequently in permeability or in functions of transport systems or membrane-bound enzymes. In this way the involvement of Ca^{2+} in the phosphorylation of membrane proteins can provide a basis for understanding the molecular mechanisms by which Ca^{2+} modulates various physiological processes in plants.

PHOSPHORYLATION OF ENZYMES

Only a few plant enzymes are known to undergo reversible phosphorylation. The discovery of this process has been largely circumstantial and stems from the observation that some crude enzyme preparations are activated or inhibited when stored at room temperature.

For example, pyruvate dehydrogenase behaves as a latent enzyme that gradually appears (79), whereas quinate NAD^+ oxidoreductase activity disappears as a function of time (81). In both cases, the time-dependent activation/inhibition is stimulated by Mg^{2+}, a cofactor for protein phosphatase and inhibited by fluoride, a phosphatase inhibitor.

The interpretation of this functional analysis is based only on changes in catalytic activities and is limited because phosphorylation may affect enzyme activities in indirect ways, for example, by altering the solubility or the sensitivity of an enzyme to low molecular weight effectors.

Another experimental procedure has been developed that involves the loading of cells or organelles with labelled phosphate (either as H_3PO_4 or $\gamma^{32}P$- ATP) and the subsequent identification of labelled substrate-enzymes by immunoprecipitation and/or PAGE analysis. These two different methods have been adopted and have produced the data described here.

Regulation of Enzyme Activities

PYRUVATE DEHYDROGENASE   The pyruvate dehydrogenase multienzyme complex (PDC) catalyzes the irreversible oxidative decarboxylation of pyruvate according to the reaction:

$$Pyruvate + CoA + NAD^+ \rightarrow Acetyl\text{-}CoA + CO_2 + NADH + H^+.$$

The complex occupies a crossroad position in metabolism because it provides a link between glycolysis and the Krebs cycle, but it is also located at a branching point leading to the synthesis of fatty acids and of several amino acids. Because of its key position, PDC was a good candidate for involvement in sophisticated regulatory processes, as has already been illustrated in animal systems (80).

Reports, issued mainly from Randall's laboratory, have progressively elucidated some aspects of PDC control in plant mitochondria and plastids (76, 79, 84).

The very first set of data established that PDC was deactivated in an ATP-dependent manner (84) concomitantly with the incorporation of $\gamma^{32}P$ into a purified preparation of mitochondrial complex from different plant tissues (79). The incubation of intact mitochondria with ATP resulted in the inactivation of the enzyme that was reactivated on supplementing the medium with Mg^{2+} or that was maintained in an inactivated state if NaF was added (79). The loss of ^{32}P label from the Mg^{2+}-treated complex occurred concomitantly with the activation of PDC, which is precipitated by specific antibodies (76). The addition of pyruvate, which prevents the deactivation of PDC, also prevents the labelling of the immunoprecipitated protein. PDC is the first plant enzyme that has been shown to be regulated by reversible phosphorylation, and the overall regulatory process is very similar to that found in animal systems.

QUINATE : NAD$^+$ OXIDOREDUCTASE   Plants may accumulate large amounts of the alicyclic acids shikimate and quinate (QA). The latter may represent up to 10% of the dry weight in the green leaves (9). Even if it is not directly involved in the shikimate pathway that leads to aromatic compounds (32), QA may be considered as a reservoir of alicyclic acid that may be reversibly injected into the main pathway. The enzyme catalyzing the reaction is quinate : NAD$^+$ oxidoreductase (QORase):

$$QA + NAD^+ \rightarrow DHQ + NADH,H^+.$$

QORase from carrot cells, a soluble 42 kilodalton protein, has been shown to be activated on phosphorylation and inactivated upon dephosphorylation (81, 82). These covalent modifications affect serine residues (32) and are under the control of cellular protein kinase(s) phosphatase(s).

QORase is only phosphorylated in the presence of physiological concentrations of calcium and calmodulin (CAM) (78), and the activating protein-kinase may bind to immobilized calmodulin (32). The calcium dependence probably operates in vivo, since the manipulation of the cellular calcium with ionophore leads to the inhibition of the phosphorylation of different proteins and the correlative decrease in QORase activity (33). The original activity may be recovered in vitro upon addition of Ca^{2+} and ATP-Mg.

QORase is, therefore, regulated through a two-cycle cascade system involving reversible phosphorylation and calcium-CAM dependent protein kinase. It may be assumed that minute changes in the calcium content may lead to the activation or the inhibition of the overall process. The kinetic properties of QORase are changed as a function of its phosphorylation degree; namely, dephosphorylation reduces the rate of conversion of dehydroquinate into QA but has less effect on the opposite reaction (82). Therefore, the phosphorylation state may be a way to control the direction of the reaction, i.e. the injection into or the removal from a storage pool of QA.

The above-mentioned data have been obtained with light-grown cells, and another degree of complexity was revealed on transfer to dark. Thus, the QORase behaves as a dimeric protein containing two different subunits (34). In addition to the changes in the molecular structure, the enzyme becomes directly activatable by high calcium concentration even when purified to homogeneity (35). It has been clearly established that the additional subunit acts as a calcium-binding protein that senses the changes in concentration of the cation but that also protects the catalytic subunit from dephosphorylation (35).

PYRUVATE, Pi DIKINASE   Pyruvate, Pi dikinase (PPDK), catalyzes a double transphosphorylation, giving rise to phosphoenolpyruvate (PEP), the primary CO_2 acceptor in C_4 and CAM plants according to the following reaction:

Pyruvate + Pi + ATP $\rightarrow$ PEP + AMP + PPi.

During catalysis, the active and unmodified PPDK is transiently phosphory-lated on a histidine residue. The phosphointermediate then becomes a sub-strate for a converter enzyme that inactivates PPDK by an ADP-dependent process (14). The use of different ^{32}P-labelled ADP has shown that only the phosphate located in the β position on the adenylate is specifically transferred onto a threonine residue of the enzyme during inactivation of PPDK. This work is elegantly reviewed in (13).

PPDK is converted back to its active state by a phosphorolytic split of the threonine phosphate ester, thus giving rise to active dephosphoenzyme and PPi provided that Pi is added to the medium. Pi-dependent removal of Pi from PPDK is catalyzed by the same converter enzyme as the ADP-dependent inactivation. Data obtained by Foyer (29, 30), while establishing that PPDK is definitely phosphorylatable, failed to correlate the degree of phosphoryla-tion with the enzyme activity.

The PPDK system is unique in different ways:

1. The phosphate donor for the phosphorylation process is ADP and not ATP as is generally the case.
2. The activation process involves the phosphorolytic cleavage to yield PPi.
3. The activation-deactivation is catalyzed by the same regulatory protein, which appears, therefore, to be a bifunctional enzyme that contains two active sites catalyzing opposing reactions.

PPDK activity is high in light-grown but almost absent in dark-grown tissues, and it is assumed that the above-mentioned complex regulatory system operates to control this light-dependent activation in vivo. According to Burnell & Hatch (13), light leads to an increase in ATP concentrations and, thereby, to a relative decrease in ADP, giving rise to an active enzyme. On the contrary, when light intensity decreases, the relative ADP concentration increases. Since ADP is both a substrate for the inactivating enzyme and a potent inhibitor of phosphorolysis, the population of inactive enzyme becomes proportionately higher.

PROTON-PUMPING SYSTEM   In corn roots, conditions that favor membrane protein phosphorylation such as cold shock cause a more than 50% loss of uncoupler-sensitive ATPase activity (36). The stimulating work of Zocchi (106) provided experimental evidence suggesting that H^+ ATPase activity is controlled through reversible phosphorylation even though phosphate binding to the purified enzyme has not been characterized.

The phosphorylation of a microsomal fraction carried out by a Ca^{2+}/ calmodulin-stimulated process decreases its H^+ ATPase activity (106). Interestingly, the inhibitory effect of phosphorylation is greater on the NO_3^--sensitive H^+ ATPase activity (tonoplast located) than on the vanadate-sensitive activity. Restoration of H^+ ATPase activity was achieved by allowing the phosphorylated membranes to dephosphorylate in the presence of endogeneous phosphoprotein phosphatase. These results have been recently extended to the plasma membrane ATPase of different plants and fungi (93). In the case of plant cells the indirect Ca^{2+} regulation of proton-pumping ATPase could represent a component in the interplay between cytoplasmic pH and pCa^{2+}.

Modifications of Regulatory Properties

PEP CARBOXYLASE   PEP carboxylase (PEPC) catalyzes the first step in the fixation of CO_2 for C_4 and CAM plants. In the latter case, the enzyme allows the plants to accumulate malate during the night; the malate is used as a CO_2 source for rubisco and the Calvin cycle during the day while the stomata are closed (11). From a functional point of view, PEPC extracted from CAM plants during the light period is inhibited by malate (Ki $\simeq$ 0.3 mM) (66). but activated by glucose-6-phosphate, whereas PEPC from dark-grown plants is insensitive to these effectors (malate Ki $\simeq$ 3mM) (66). During light/dark transitions, there are no changes either in enzyme-specific activity or in the amount of enzyme that can be determined immunologically (11). Therefore, the shifts in sensitivity may be the consequence of posttranslational modifications of native PEPC that are controlled by light.

Accordingly, if PEPC was immunoprecipitated from extracts of leaves prelabelled with ^{32}P, only the enzyme from the night extract contained ^{32}P (11, 66). It has been pointed out that, with the notable exception of PEPC, the general distribution of labelling in dark or light extracts was similar. Therefore, the difference in PEPC label is not due to a general effect of light-dark transitions on the phosphorylation degree of the bulk of proteins.

These data, obtained from *Bryophyllum fedtschenkoi* (66), have been extended to other CAM plants (11). Thus, prelabelling with $^{32}\gamma$P ATP yielded radioactive PEPC during the night but never during the day. More convincingly, the light-dependent increase in the sensitivity to malate may be mimicked in vitro by incubating the phosphorylated (dark) form with exogenous phosphatase. Therefore, reversible phosphorylation may explain the observed changes in regulatory properties of PEPC and more generally of plant enzymes.

PROTEIN PHOSPHORYLATION AND STIMULUS/RESPONSE COUPLING

In animal systems, it has been clearly demonstrated that phosphorylation of proteins is an important component in the integration of external and internal stimuli (18). From a very general point of view, the coupling of the stimulus to the response involves discrete steps (67):

1. perception of the stimulus by membrane-bound receptors;
2. activation of membrane activities: opening of channels (56), production of "second messengers" by hydrolysis of membrane components, namely phosphoinositides (6, 42), or stimulation of enzymes such as adenylate cyclase (55);
3. an increase in the concentration of second messengers (inositol polyphosphate and diacylglycerol, cAMP, calcium);
4. activation of second messenger–dependent enzymes, especially protein kinases;
5. phosphorylation of proteins and amplification of the stimulus;
6. dephosphorylation and return to a resting situation for turning off the stimulus.

Whereas such a general scheme is implicitly accepted as applicable to plant systems, there are at the present time no firm experimental data that demonstrate this assumption. However, it has been shown in plants that light and growth substances that exert pleiotropic effects are good candidates as stimuli and that calcium-dependent phosphorylation of proteins serves as an amplification device. Therefore, most discussion focuses on these aspects, since it is now generally accepted that cAMP, despite its presence in plants, has no effect on protein phosphorylation (85).

In the case of phosphorylation of plant proteins, light may regulate the process in a more or less direct way. Previously mentioned examples have shown that the process may be mediated by energy charge (13) or reducing power (43) rather than by a classical second messenger.

In addition to these situations where the substrates and enzymes are known, light plays a more general role because of its potential to control the movements of calcium. Thus, light stimulates the calcium influx into chloroplasts and membrane vesicles but may conversely facilitate release of Ca^{2+} from internal stores (37). Calcium has also been shown to be part of the transduction system of light (phytochrome) effects (22, 24). As a consequence, Ca^{2+}-dependent processes, and especially protein phosphorylation, may also be light dependent (23, 37). Thus the pleiotropic effects of light may in part be explained by the activation of soluble or membrane-bound protein kinases. Data from Yanshing et al (105) and Quail et al (72) indicate that phytochrome itself is subject to reversible phosphorylation. Such a direct link opens new perspectives in the integration of disparate data:

- various membrane receptors, for example insulin receptor, are auto-phosphorylatable (47);
- movements of Ca^{2+} via Ca^{2+}-channel may be modulated by phosphorylation (56);
- phytochrome may associate-dissociate to membranes and is auto-phosphorylatable (P. M. Quail, personal communication).

Phytochrome is a key component of the overall system, and its role as either kinase and/or channel needs to be examined.

As in the case of light, the role of growth substances in protein phosphorylation has also been examined. Auxin (88) and GA (105) regulate the process under in vitro conditions; cytokinins also affect the in vivo pattern (75).

A kinase that specifically phosphorylates a cytokinin-binding protein has been isolated from wheat germ (69, 70). It is tempting, therefore, to draw a parallel with animal hormone or neurotransmitter receptors, even though the consequences of the phosphorylation and the actual role of the cytokinin-binding protein remain obscure (69).

Additional data on growth-substance effects stem from the work of Morré (62). Thus, in vitro addition of the synthetic auxin 2,4-D to membranes leads to the increased phosphorylation of two specific polypeptides (99), whereas the weak auxin analog 2,3-D is not stimulatory. The auxin effects are significantly inhibited by calcium, and the process is observed for a wide range of calcium and auxin concentrations (62, 99).

Though their significance is not completely understood, such results demonstrate some antagonism between Ca^{2+} and 2,4-D. The situation becomes more complex if purified plasma membrane is used, since it appears that 2,4-D and Ca^{2+} independently stimulate protein phosphorylation (99). Furthermore, 2,4-D increases the phosphatidylinositol turnover (61) and the release of divalent ions from membranes, including Ca^{2+}.

Thus, at least with the plasma membrane systems, it seems that 2,4-D provokes different reactions that are consistent with the coupling of hormone binding with changes in membrane activities. The next step would be to consider the topographical distribution of the different components of the system.

Other compounds known to influence a large number of physiological processes in plants are polyamines (100). These substances vary in amounts in response to various stimuli including light, stress, and hormones. Accordingly, they may be considered as messengers and/or growth regulators, and they have been tested as possible protein phosphorylation effectors (25, 100). The data obtained show that spermine promotes the phosphorylation of various membrane and soluble proteins and that most of these proteins are different from those phosphorylated in the presence of calcium. In addition to Ca^{2+} it has been shown that spermine stimulates the phosphorylation of several nuclear proteins in pea (25). Spermine, and to a lesser extent spermidine, were also shown to promote the in vitro phosphorylation of several membrane proteins in corn coleoptiles (100).

Ca^{2+}-dependent protein phosphorylation changes are observed during growth and development. Thus, a general decrease for both soluble- and membrane-bound proteins occurs in tomatoes during ripening (73) and in apples during senescence (68). However, it is not clear if the recorded variations reflect changes in substrate composition or enzyme activities or/and specificities. Therefore, a comparison of the properties of protein kinases at different stages of development of a given plant must be made. Such a study would allow one to identify the limiting factors (enzymes or substrates?).

REFLECTIONS AND PROSPECTS

The disparate data obtained suggest that plants may use protein phosphorylation as an effective device for responding to endogenous stimuli and to changes in the environment. The number of phosphorylated proteins is large, and increased knowledge of these endogenous substrates and their physiological role is required.

Receptor Phosphorylation

An elegant experimental procedure for membrane-bound protein kinase characterization has been devised by Blowers et al (7). Taking advantage of the observation that auxin-binding protein can be solubilized from membranes (103), the authors have extended this approach to other proteins. Thus, after solubilization, the proteins were electrophoresed and blotted onto nitrocellulose membrane (7). Under these conditions, the transferred protein kinases are able to carry out autophosphorylation, particularly in a Ca^{2+}-dependent manner. Such a procedure would be useful for determining kinase activity and the ability of the same protein to bind regulatory compounds such as growth substances or ion channel probes. Systematic screening would allow the identification of many phosphorylatable bands, apparently free of biological (enzymatic) activity.

Protein kinases normally phosphorylate serine and threonine residues of protein substrates. However, in response to oncogenes or hormones the tyrosine residues, some of which are located on the hormone receptor itself, are also phosphorylated (44). The gene product of the *Rous sarcoma* virus SRC oncogene (pp60V SRC), for example, is a protein kinase that phosphorylates tyrosine residues; therefore, it is tempting to identify tyrosine-phosphorylating activity in transformed crown gall tissues (26). Three phosphoaminoacids have been characterized in these transformed issues: phosphoserine, phosphothreonine, and phosphotyrosine in the ratio 89.4 : 8.5 : 2.1. The level of phosphotyrosine is significantly higher than what is normally found in animal cells. However, the comparison with normal cells did not reveal significant differences. The lack of effect of transformation is in contrast to animal cells where transformation results in a 10-fold increase in tyrosine phosphorylation in the virally infected cells.

Transmembrane Signalling

Some of the most promising new data in the field of phosphorylation of proteins in plants comes from Marmé and co-workers (86) and Elliott & Skinner (27). They have established that an enzyme resembling protein kinase C probably occurs in plants. Thus, the activity of the protein kinase depends upon phosphatidylserine and Ca^{2+} at micromolar concentrations.

In animals, extracellular signals that activate cellular functions and proliferation through interaction with membrane receptors provoke the breakdown of inositol phospholipids and the formation of inositol trisphosphate (IP_3) and diacylglycerol (DAG) (6). IP_3 is thought to be responsible for the transient release of Ca^{2+} from internal stores (6), thus leading to a short-term response, whereas DAG stimulates protein kinase C and is thought to induce a sustained response. Protein kinase C occurs as a soluble or a membrane-bound enzyme, depending upon the physiological conditions (67). A shuttle system between cytosol and membrane would allow the phosphorylation of either soluble or membrane-bound proteins and therefore would allow protein kinase C to participate in transmembrane signalling (67). In this way, preliminary data suggest that phytochrome effects may be mediated through protein kinase C (22).

GENERAL CONCLUSIONS

Accumulation of knowledge about protein phosphorylation in plants lags significantly behind the imposing progress made in animal systems (for example, several genes coding for specific protein kinases have been identified).

At the present time, it is clear that the reversible phosphorylation of proteins is an integral part of the basic regulatory mechanisms of all eukaryotic cells as is, for example, allostery. Until now, animal models have been considered as reference systems. However, it is reasonable to conclude that plants may use the phosphorylation/dephosphorylation process as a regulatory device at least as frequently as animals do. Such an assumption may be inferred from the fact that the overall process is very rapid and therefore suitable for organisms that have to adapt themselves to continuing changes in their environment. Particularly relevant is the example of photosynthesis, which involves the phosphorylation of LHCP. Phosphorylation is also involved in controlling the activity or regulatory properties of the few substrate enzymes that have been identified. Circadian rhythmicity of enzyme activities may be also due to correlative changes, in the phosphorylation degree of the enzyme (45).

In general, phosphoproteins in prokaryotes are few, and protein kinase activity is low. Protein phosphorylation seems to be more common, with a parallel increase in complexity of eukaryotic organisms. A multicellular organism clearly requires a means of cell-cell signaling. In this regard, it is interesting to note that protein kinases are associated with the plasma membrane in plants and thus are in a position to transduce signals from outside the cell. The identification of a larger number of phosphorylated substrates is now a prerequisite for further progress in the understanding of protein phosphorylation in message transduction and amplification.

ACKNOWLEDGMENTS

Our sincere thanks to colleagues for making available some of their unpublished data. We are indebted to Stanley Roux (Austin, Texas) for critical discussions and to Danielle Lefebvre for handling the manuscript.

Literature Cited

1. Allen, J. F. 1983. Protein phosphorylation. Carburetor of photosynthesis? *Trends Biochem. Sci.* 8:369–73
2. Arfmann, H. A., Willmitzer, L. 1982. Endogenous protein kinase activity of tobacco nuclei. Comparison of transformed, non-transformed cell cultures and the intact plant of *Nicotiana tabacum*. *Plant Sci. Lett.* 2:31–38
3. Bennett, J. 1980. Chloroplast phosphoproteins. Evidence for a thylakoid bound phosphoprotein phosphatase. *Eur. J. Biochem.* 104:85–89
4. Bennett, J. 1984. Chloroplast protein phosphorylation and the regulation of photosynthesis. *Physiol. Plant.* 60:583–90
5. Bennett, J., Steinback, K. E., Arntzen,

C. J. 1980. Chloroplast phosphoproteins: regulation of activation energy transfer by phosphorylation of thylakoid membrane polypeptides. *Proc. Natl. Acad. Sci. USA* 80:5253–57
6. Berridge, M. J. 1984. Inositol triphosphate and diacylglycerol as second messenger. *Biochem. J.* 220:345–60
7. Blowers, D. P., Hetherington, A., Trewavas, A. 1985. Isolation of plasmamembrane-bound calcium/calmodulin-regulated protein kinase from pea using western blotting. *Planta* 166:208–15
8. Böcher, M., Erdmann, H., Heim, S., Wylegalla, C. 1985. Protein kinase activity of *in vitro* cultured plant cells in relation to growth and starch metabolism. *J. Plant Physiol.* 119:209–18
9. Boudet, A. 1973. Les acides quinique et shikimique chez les Angiospermes arborescentes. *Phytochemistry* 12:363–70
10. Briantais, J. M., Vernotte, C., Krause, G. H., Weis, E. 1986. Chlorophyll a fluorescence of higher plants: chloroplasts and leaves. In *Light Emission by Plants and Bacteria,* ed. Govindjee, D. C. Fork, J. Anesz, pp. 540–83. New York: Academic
11. Brulfert, J., Vidal, J., Le Maréchal, P., Gadal, P., Queiroz, O. 1986. Phosphorylation-dephosphorylation process as a probable mechanism for the diurnal regulatory changes of phosphoenolpyruvate carboxylase in CAM plants. *Biochem. Biophys. Res. Commun.* 136:151–59
12. Brummer, B., Parish, R. W. 1983. Identification of specific proteins and glycoproteins associated with membrane fractions isolated from *Zea mays* L. coleoptiles. *Planta* 157:446–53
13. Burnell, J. N., Hatch, M. D. 1985. Light-dark modulation of leaf pyruvate, Pi dikinase. *Trends Biochem. Sci.* 115:288–91
14. Burnell, J. N., Hatch, M. D. 1983. Dark-light regulation of pyruvate, Pi dikinase in C$_4$ plants: evidence that the same protein catalyzes activation and inactivation. *Biochem. Biophys. Res. Commun.* 111:288–93
15. Chapman, K. S. R., Trewavas, A., Van Loon, L. C. 1975. Regulation of the phosphorylation of chromatin-associated proteins in *Lemna* and *Hordeum*. *Plant Physiol.* 55:293–96
16. Chock, P. B., Rhee, S. G., Stadtman, E. R. 1980. Interconvertible enzyme cascades in cellular regulation. *Ann. Rev. Biochem.* 49:813–43
17. Deleted in proof
18. Cohen, P. 1982. The role of protein phosphorylation in neural and hormonal control of cellular activity. *Nature* 296: 613–20
19. Coughlan, S. J., Hind, G. 1986. Purification of a protein kinase from spinach thylakoids. *Plant Physiol. Suppl.* 80:65
20. Coughlan, S. J., Hind, G. 1986. Protein kinases of the thylakoid membranes. *J. Biol. Chem.* In press
21. Danko, J. S., Markwell, J. P. 1985. Protein phosphorylation in plant mitochondria. *Plant Physiol.* 79:311–14
22. Das, R., Sopory, S. K. 1985. Evidence of regulation of calcium uptake by phytochrome in maize protoplasts. *Biochem. Biophys. Res. Commun.* 128: 1455–62
23. Datta, N., Chen, Y. R., Roux, S. J. 1985. Phytochrome and calcium stimulation of protein phosphorylation in isolated pea nuclei. *Biochem. Biophys. Res. Commun.* 128:1403–8
24. Datta, N., Chen, Y. R., Roux, S. J. 1986. Modulation of enzyme activities in isolated pea nuclei by phytochrome, calcium and calmodulin. In *Molecular and Cellular Aspects of Calcium in Plant Development*, ed. A. J. Trewavas, pp. 115–22. New York: Plenum
25. Datta, N., Hardison, L. K., Roux, S. J. 1986. Polyamine stimulation of protein phosphorylation in isolated pea nuclei. *Plant Physiol. Suppl.* 80:15
26. Elliott, D. C., Geytenbeek, M. 1985. Identification of products of protein phosphorylation in T37-transformed cells and comparison with normal cells. *Biochim. Biophys. Acta* 845:317–23
27. Elliott, D. C., Skinner, J. D. 1986. Calcium-dependent, phospholipid-activated protein kinase in plants. *Phytochemistry* 25:39–44
28. Erdmann, H., Bocher, M., Wagner, K. G. 1982. Two protein kinases from nuclei of cultured tobacco cells with properties similar to the cyclic nucleotide-independent enzymes (N I and N II) from animal tissue. *FEBS Lett.* 137: 245–48
29. Foyer, Ch. 1984. Phosphorylation of a stromal enzyme protein in maize *(Zea mays)* mesophyll chloroplasts. *Biochem. J.* 222:247–53
30. Foyer, Ch. 1985. Stromal-protein phosphorylation in spinach *(Spinacia oleracea)* chloroplasts. *Biochem. J.* 231:97–104
31. Giannattasio, M., Tucci, G. F., Carratu, G., Ponzi, G. 1979. Membrane-bound protein kinase activity in Jerusalem artichoke rhizome tissues. *Plant Sci. Lett.* 15:373–85

32. Graziana, A. 1985. *Regulation de la voie préaromatique chez les végétaux supérieurs.* Thèse Doct. Sci., Univ. Paul Sabatier, Toulouse, France. 214 pp.

33. Graziana, A., Ranjeva, R., Boudet, A. M. 1983. Provoked changes in cellular calcium controlled protein phosphorylation and activity of quinate: NAD⁺ oxidoreductase in carrot cells. *FEBS Lett.* 156:325–28

34. Graziana, A., Ranjeva, R., Salimath, B. P., Boudet, A. M. 1983. The reversible association of quinate: NAD⁺ oxidoreductase from carrot cells with a putative regulatory subunit depends on light conditions. *FEBS Lett.* 163:306–10

35. Graziana, A., Dillenschneider, M., Ranjeva, R. 1984. A calcium binding protein is a regulatory subunit of quinate: NAD⁺ oxidoreductase from dark grown carrot cells. *Biochem. Biophys. Res. Commun.* 125:774–83

35a. Guitton, C., Dorne, A. M., Mache, R. 1984. In organello and in vitro phosphorylation of chloroplast ribosomal proteins. *Biochem. Biophys. Res. Commun.* 121:297–303

36. Hanson, J. B., Rincon, M., Rogers, S. A. 1986. Controls on calcium influx in corn root cells. *Phytochemistry* 25:253–60

37. Hepler, P. K., Wayne, R. O. 1985. Calcium and plant development. *Ann. Rev. Plant Physiol.* 36:397–439

38. Hetherington, A., Trewavas, A. 1982. Calcium-dependent protein kinase in pea shoot membranes. *FEBS Lett.* 145:67–71

39. Hetherington, A., Trewavas, A. 1984. The regulation of membrane-bound protein kinases by phospholipid and calcium. *Ann. Proc. Phytochem. Soc. Eur.* 24:181–97

40. Hetherington, A., Trewavas, A. 1984. Activation of a pea membrane protein kinase by calcium ions. *Planta* 161:409–17

41. Hodges, M., Barber, J. 1983. Photosynthetic adaptation of pea plants grown at different light intensities: state 1–state 2 transitions and associated chlorophyll fluorescence changes. *Planta* 157:166–73

42. Hokin, L. E. 1985. Receptors and phosphoinositide-generated second messengers. *Ann. Rev. Biochem.* 54:205–35

43. Horton, P., Allen, J. F., Black, M. T., Bennett, J. 1981. Regulation of phosphorylation of chloroplast membrane polypeptides by the redox state of plastoquinone. *FEBS Lett.* 125:193–96

44. Hunter, T., Cooper, J. A. 1985. Protein-tyrosine kinases. *Ann. Rev. Biochem.* 54:897–930

45. Jerebzoff, S., Jerebzoff-Quintin, S. 1984. Cyclic activity of L-asparaginase through reversible phosphorylation in *Leptosphaeria michotii*. *FEBS Lett.* 171:67–71

46. Kahl, G., Schafer, W. 1984. Phosphorylation of chromosomal proteins changes during the development of crown gall tumors. *Plant Cell Physiol.* 25:1187–96

47. Kasuga, M., Zick, Y., Blithe, D. L., Crettaz, M., Kahn, C. R. 1982. Tyrosine-specific protein kinase activity is associated with purified insulin receptor. *Nature* 298:667–69

48. Kauss, H. 1987. Some aspects of calcium-dependent regulation in plant metabolism. *Ann. Rev. Plant Physiol.* 38:47–72

49. Kleinsmith, L. J., Stein, J., Stein, G. 1975. Direct evidence for a functional relationship between nonhistone chromosomal protein phosphorylation and gene transcription. In *Chromosomal Proteins and Their Role in the Regulation of Gene Expression*, ed. G. S. Stein, L. J. Kleinsmith, pp. 59–80. New York: Academic

50. Krebs, E. G., Beavo, J. A. 1979. Phosphorylation-dephosphorylation of enzymes. *Ann. Rev. Biochem.* 48:923–59

51. Krebs, E. G., Edelman, A. M., Blumenthal, D. K. 1987. Protein serine/threonine kinases. *Ann. Rev. Biochem.* 56: In press

52. Kuehn, G. D., Affolter, H. U., Atmar, V. J., Seebeck, T., Gubler, U., Braun, R. 1979. Polyamine-mediated phosphorylation of a nucleolar protein from *Physarum polycephalum* that stimulates rRNA synthesis. *Proc. Natl. Acad. Sci. USA* 76:2541–45

53. Kyle, D. J., Ohad, I., Arntzen, C. J. 1984. Membrane protein damage and repair: selective loss of a quinone-protein function in chloroplast membranes. *Proc. Natl. Acad. Sci. USA* 81:4070–74

54. Laing, W. A., Christeller, J. T. 1984. Chloroplast phosphoproteins: distribution of phosphoproteins within spinach chloroplasts. *Plant Sci. Lett.* 36:99–104

55. Lefkowitz, R. J., Stadel, J. M., Caron, M. G. 1983. Adenylate cyclase-coupled beta-adrenergic receptors: Structure and mechanisms of activation and desensitization. *Ann. Rev. Biochem.* 52:159–86

56. Levitan, I. B. 1985. Phosphorylation of ion channels. *J. Membr. Biol.* 87:177–90

57. Lin, Z. F., Lucero, H. A., Racker, E. 1982. Protein kinases from spinach chloroplasts. I. Purification and identification of two distinct protein kinases. *J. Biol. Chem.* 257:12153–56
58. Macherel, D., Viale, A., Akazawa, T. 1986. Protein phosphorylation in amyloplasts isolated from suspension-cultured cells of sycamore (*Acer pseudoplatanus* L.). *Plant Physiol.* 80:1041–44
59. Melanson, D., Trewavas, A. J. 1982. Changes in tissue protein pattern in relation to auxin induction of DNA synthesis. *Plant Cell Environ.* 5:53–64
60. Moll, B. A., Steinback, K. E. 1986. Chilling sensitivity in *Oriza sativa:* the role of protein phosphorylation in protection against photoinhibition. *Plant Physiol.* 80:420–23
61. Morré, D. J., Gripshover, B., Monroe, A., Morré, J. T. 1984. Phosphatidylinositol turnover in isolated soybean membranes stimulated by the synthetic growth hormone 2,4-dichlorophenoxyacetic acid. *J. Biol. Chem.* 259: 15364–68
62. Morré, D. J., Morré, J. T., Varnold, R. L. 1984. Phosphorylation of membrane located proteins of soybean *in vitro* and response to auxin. *Plant Physiol.* 75:265–68
63. Murray, M. G., Guilfoyle, T. J., Key, J. L. 1978. Isolation and preliminary characterization of a casein kinase from cauliflower nuclei. *Plant Physiol.* 62: 434–37
64. Murray, M. G., Key, J. L. 1978. 2,4-Dichlorophenoxyacetic acid-enhanced phosphorylation of soybean nuclear proteins. *Plant Physiol.* 61:190–98
65. Muto, S., Shimogawara, K. 1985. Calcium and phospholipid dependent phosphorylation of ribulose-1,5-biphosphate carboxylase/oxygenase small subunit by a chloroplast envelope-bound protein kinase *in situ.* *FEBS Lett.* 193:88–92
66. Nimmo, G. A., Nimmo, H. G., Fewson, C. A., Wilkins, M. B. 1984. Diurnal changes in the properties of phosphoenolpyruvate carboxylase in *Bryophyllum* leaves: a possible covalent modification. *FEBS Lett.* 178:199–203
67. Nishizuka, Y. 1984. The role of protein kinase C in cell surface signal transduction and tumour promotion. *Nature* 308:693–98
68. Paliyath, G., Poovaiah, B. W. 1985. Calcium and calmodulin-promoted phosphorylation of membrane proteins during senescence in apples. *Plant Cell Physiol.* 26:977–86
69. Polya, G. M., Davies, J. R. 1983. Reso-

lution and properties of a protein kinase catalyzing the phosphorylation of a wheat-germ cytokinin-binding protein. *Plant Physiol.* 71:482–88
70. Polya, G. M., Davies, J. R., Micucci, V. 1983. Properties of a calmodulin activated Ca^{2+}-dependent protein kinase from wheat germ. *Biochim. Biophys. Acta* 761:1–12
71. Polya, G. M., Schibeci, A., Micucci, V. 1984. Phosphorylation of membrane proteins from cultured *Lolium multiflorum* (ryegrass) endosperm cells. *Plant Sci. Lett.* 36:51–57
72. Quail, P. H., Briggs, W. R., Pratt, L. H. 1978. *in vivo* phosphorylation of phytochrome. *Carnegie Inst. Washington Yearb.* 77:342–44
73. Raghothama, K. G., Veluthambi, K., Poovaiah, B. W. 1985. Stage-specific changes in calcium-regulated protein phosphorylation in developing tomato fruits. *Plant Cell Physiol.* 26:1565–72
74. Ralph, R. K., Bullivant, S., Wojcik, S. J. 1976. Effects of kinetin on phosphorylation of leaf membrane proteins. *Biochim. Biophys. Acta* 421:319–27
75. Ralph, R. K., McCombs, P. J. A., Tener, G., Wojcik, S. J. 1972. Evidence for modification of protein phosphorylation by cytokinins. *Biochem. J.* 130:901–11
76. Randall, D. D., Williams, M., Rapp, B. 1981. Phosphorylation dephosphorylation of pyruvate dehydrogenase complex from pea leaf mitochondria. *Arch. Biochem. Biophys.* 207:437–44
77. Ranjeva, R., Graziana, A., Ranty, B., Cavalié, G., Boudet, A. M. 1984. Phosphorylation of proteins in plants: a step in the integration of extra and intracellular stimuli? *Physiol. Vég.* 22:365–76
78. Ranjeva, R., Refeno, G., Boudet, A. M., Marmé, D. 1983. Activation of plant quinate: NAD^+ 3-oxidoreductase by calcium and calmodulin. *Proc. Natl. Acad. Sci. USA* 80:5222–24
79. Rao, P. K., Randall, D. D. 1977. Regulation of plant pyruvate dehydrogenase complex by phosphorylation. *Plant Physiol.* 60:34–39
80. Reed, L. J. 1981. Regulation of mammalian pyruvate dehydrogenase complex by a phosphorylation-dephosphorylation cycle. *Curr. Topics Cell. Regul.* 18:95–106
81. Refeno, G., Ranjeva, R., Boudet, A. M. 1982. Modulation of quinate:NAD^+ oxidoreductase through reversible phosphorylation in carrot cell-suspensions. *Planta* 154:193–98
82. Refeno, G., Ranjeva, R., Delvare, S.,

Boudet, A. M. 1982. Functional properties of protein kinase(s) and phosphatase(s) converting quinate:NAD$^+$ oxidoreductase into active and deactivated forms in carrot cell suspension cultures. *Plant Cell Physiol.* 7:1137–44

83. Remy, R., Ambard-Bretteville, F., Dubertret, C. 1985. Separation of phosphorylated from non-phosphorylated LHCP polypeptides by two-dimensional electrophoresis. *FEBS Lett.* 188:43–47

84. Rubin, P. M., Randall, D. D. 1977. Regulation of plant pyruvate dehydrogenase complex by phosphorylation. *Plant Physiol.* 60:34–39

85. Salimath, B. P., Marmé, D. 1983. Protein phosphorylation and its regulation by calcium and calmodulin in membrane fractions from zucchini hypocotyls. *Planta* 158:560–68

86. Schäfer, A., Bygrave, F., Matzenauer, S., Marmé, D. 1985. Identification of a calcium and phospholipid dependent protein kinase in plant tissue. *FEBS Lett.* 187:25–28

87. Schafer, W., Kahl, G. 1982. Phosphorylation of chromosomal proteins in resting and wounded potato tuber tissues. *Plant Cell Physiol.* 23:137–46

88. Schafer, W., Kahl, G. 1981. Auxin-induced changes in chromosomal protein phosphorylation in wounded potato tuber parenchyma. *Plant Mol. Biol.* 6:5–17

89. Shacter, E., Chock, P. B., Stadtman, E. R. 1984. Regulation through phosphorylation/dephosphorylation cascade systems. *J. Biol. Chem.* 259:12252–59

90. Soll, J. 1985. Phosphoproteins and protein-kinase activity in isolated envelopes of pea (*Pisum sativum* L.) chloroplasts. *Planta* 166:394–400

91. Soll, J., Buchanan, B. B. 1983. Phosphorylation of chloroplast ribulose bisphosphate carboxylase/oxygenase small subunit by an envelope-bound protein kinase *in situ*. *J. Biol. Chem.* 258:6686–89

92. Stratton, B. R., Trewavas, A. J. 1981. Phosphorylation of histone H1 during the cell cycle of artichoke. *Plant Cell Environ.* 4:419–26

93. Sussman, M. R. 1985. Protein kinase mediated phosphorylation of the 100,000 molecular weight plasma membrane proton ATPase in plants and fungi. *Plant Physiol.* 77(Suppl. 4):87

94. Telfer, A., Allen, J. F., Barber, J., Bennett, J. 1983. Thylakoid protein phosphorylation during state 1/state 2 transitions in osmotically shocked pea chloroplasts. *Biochim. Biophys. Acta* 722:176–81

95. Teulières, C., Alibert, G., Ranjeva, R. 1985. Reversible phosphorylation of tonoplast proteins involves tonoplast-bound calcium-calmodulin dependent protein kinase(s) and protein phosphatase(s). *Plant Cell Rep.* 4:199–201

96. Trewavas, A. 1976. Post–translational modification of proteins by phosphorylation. *Ann. Rev. Plant Physiol.* 27:349–74

97. Trewavas, A. 1979. Nuclear phosphoproteins in germinating cereal embryos and their relationship to the control of mRNA synthesis and the onset of cell division. In *Recent Advances in the Biochemistry of Cereals*, ed. D. L. Laidman, R. G. Wyn Jones, pp. 175–208. New York: Academic

98. Van Loon, L. C., Trewavas, A., Chapman, K. S. R. 1975. Phosphorylation of chromatin-associated proteins in *Lemna* and *Hordeum*. *Plant Physiol.* 55:288–92

99. Varnold, R. L., Morré, D. J. 1985. Phosphorylation of membrane-located proteins of soybean hypocotyl: inhibition by calcium in the presence of 2-4 dichlorophenoxyacetic acid. *Bot. Gaz.* 146:315–19

100. Veluthambi, K., Poovaiah, B. W. 1984. Polyamine-stimulated phosphorylation of proteins from corn (*Zea mays* L.) coleoptiles. *Biochem. Biophys. Res. Commun.* 122:1374–80

101. Veluthambi, K., Poovaiah, B. W. 1984. Calcium promoted protein phosphorylation in plants. *Science* 223:167–69

102. Veluthambi, K., Poovaiah, B. W. 1984. Calcium- and calmodulin-regulated phosphorylation of soluble and membrane proteins from corn coleoptiles. *Plant Physiol.* 76:359–65

103. Venis, M. A. 1977. Solubilization and partial purification of auxin binding sites of corn membranes. *Nature* 266:268–69

104. Wielgat, B., Kleczkowski, K. 1981. Gibberellic acid enhanced phosphorylation of pea chromatin proteins. *Plant Sci. Lett.* 21:381–88

105. Yanshing, W., Cheng, H. C., Walsh, D. A., Lagarias, G. C. 1986. *in vitro* phosphorylation of phytochrome as a probe of light induced conformational changes. *J. Cell Biochem. B* 10:41

106. Zocchi, G. 1985. Phosphorylation/dephosphorylation of membrane proteins controls the microsomal proton-translocating ATPase activity of corn *Zea mays* roots. *Plant Sci.* 40:153–59

Ann. Rev. Plant Physiol. 1987. 38:95–117

MEMBRANE CONTROL IN THE CHARACEAE

Masashi Tazawa, Teruo Shimmen, and Tetsuro Mimura

Department of Botany, University of Tokyo, Hongo, Tokyo 113, Japan

CONTENTS

INTRODUCTION 95
OSMOREGULATION 96
    Osmotic Pressure 96
    Turgor 98
MEMBRANE EXCITATION 101
    Plasmalemma 101
    Tonoplast 103
INTERACTION BETWEEN PLASMALEMMA AND CHLOROPLASTS
    AND MITOCHONDRIA 104
    Electrogenic H^+ ATPase 104
    HCO_3^- and OH^- Transport 106
    K^+ Channel Activation 107
CONTROL OF CELL MOTILITY 108
CONCLUDING REMARKS 110

INTRODUCTION

Of all plant cells, internodal cells of the Characeae have been used most frequently in the study of membrane physiology. In 1930 Umrath (152) inserted a glass microelectrode into *Nitella* cells and recorded action potentials. This was earlier than the first intracellular recording of the action potential in animal cells: frog muscle fibers by Nastuk & Hodgkin (88) and myelinated nerve fibers by Tasaki (134). Using *Nitella*, Osterhout & Hill (99) succeeded in demonstrating that the conduction of action potentials is mediated by a local current. This observation provided the earliest evidence for the

0066-4294/87/0601-0095$02.00

so-called local-circuit theory for the conduction of impulses, which was later established for both myelinated nerve fiber (133) and nonmyelinated fibers (42). Like most plant cells, characean cells have a plasmalemma and a tonoplast. The plasmalemma regulates the intracellular environment, maintaining physiological activity constant for changes in external conditions. To achieve homeostasis of internal conditions the plasmalemma must sense not only external signals but also changes in the chemical condition of the cytoplasm. Thus the functions of the plasmalemma are closely related to those of the cytoplasm. The vacuolar membrane, the tonoplast, also regulates cytoplasmic conditions, utilizing the large central vacuole as a buffer compartment (cf 10).

To investigate the interrelationship of membranes and cytoplasm or the mechanism of cytoplasmic homeostasis in relation to membrane functions, it is useful to modify the composition of the vacuole and cytoplasm and measure the responses of the plasmalemma and the tonoplast. Internodal cells of the Characeae are cylindrical. Their large size makes the surgical modification of cells possible (cf 138).

The vacuolar perfusion of plant cells was first developed by Blinks (9) using *Halicystis*. Vacuolar perfusion of characean cells was first attempted by Kamiya & Kuroda (46), and the technique was later improved by Tazawa (137). Prolonged survival of *Nitella* cells after replacement of the vacuolar sap with artificial solutions (Figure 1: vacuole-perfused cell) enabled us to study various functions of the tonoplast. Perfusion of the cytoplasmic space is possible if the tonoplast is removed. Williamson (155) and Tazawa et al (139) succeeded in preparing tonoplast-free cells (Figure 1), thereby facilitating the study of plasmalemma functions through direct modification of the cytoplasmic composition. On the other hand, the plasmalemma permeabilized cell model (Figure 1) (116, 117) provides an excellent experimental system for studying tonoplast functions in relation to cytoplasmic conditions.

Here we confine discussion to membrane (plasmalemma and tonoplast) functions that either control or are controlled by the physiological activities.

OSMOREGULATION

Osmoregulation is an essential function of plant cells that enables plants to grow and survive under a wide range of external osmotic pressures and salinities. Topics relating to osmoregulation have therefore often been dealt with in this *Annual Review* (26, 32, 37, 83, 157).

Osmotic Pressure

The simplest way of demonstrating the capacity of plant cells to regulate cellular osmotic pressure $(\pi_i)^1$ is to modify it artificially. This was elegantly

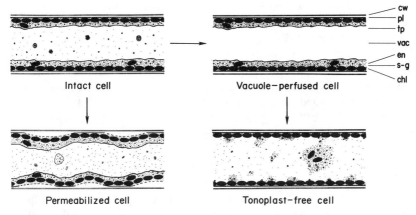

Figure 1 Schematic longitudinal section of corticated characean internodal cells. Vacuole-perfused cell: the vacuole is perfused with a medium containing Ca^{2+} which keeps the tonoplast intact (137). Tonoplast-free cell: the vacuole is perfused with a medium containing ethyleneglycol-bis-(β-aminoethyl ether)N,N'-tetraacetic acid (EGTA). The tonoplast disintegrates and some endoplasm is detached from the cortical gel and floats in the central space originally occupied by the vacuole (139). Permeabilized cell: the cell is treated with a slightly hypertonic medium containing EGTA at low temperature. The plasmalemma detaches from the cell wall owing to plasmolysis and is permeabilized (117). cw: cell wall; pl: plasmalemma; chl: chloroplast layer; s-g: sol-gel interface where actin bundles are present; en: streaming endoplasm; tp: tonoplast; vac: vacuole.

performed using the transcellular osmosis method in *Nitella* (47). Two cell fragments were isolated from the same internode, one having lower (L) and the other higher (H) osmotic pressure. Both L- and H-fragments recovered the original π_i within a few days by absorbing or releasing mainly K^+ and Cl^- (141, 142). Since L- and H-cells in artificial pond water have either lower or higher turgor pressure than normal cells, π_i regulation may be equivalent to turgor regulation. But when the turgor of H-cells was reduced to almost zero by adding sucrose to artificial pond water, they still reduced π_i and were plasmolyzed. Plasmolysis was not observed when the external osmotic pressure (π_o) was lowered to the natural π_i (47). *Nitella* cells clearly regulate not the turgor pressure but the osmotic pressure.

The osmotic pressure of intact *Nitella* cells incubated in an almost isotonic medium (1 mM KCl + 250 mM sucrose) did not differ from that of cells incubated in 1 mM KCl (136). The same is true for *Chara* in that when turgor was decreased by the addition of sorbitol (108) or raffinose (6), π_i did not increase. Thus freshwater Characeae do not appear to regulate turgor.

Since π_i regulation in *Nitella* is achieved mainly via K^+, the question arises of whether cells regulate π_i or $[K^+]_i$. Vacuolar perfusion with isotonic media

[1]Subscripts: i = inside; o = outside; v = vacuole; c = cytoplasm.

revealed that cells retained K^+ even when its concentration differed greatly from the normal level (80 mM). Under isotonic conditions, variation of $[Na^+]_v$ (0–70 mM) or $[Ca^{2+}]_v$ (0–40 mM) did not affect $[K^+]_v$ (80 nM). All these facts show that *Nitella* cells are relatively insensitive to the ion species and ionic concentrations in the vacuole as long as the osmotic pressure is maintained at the normal level (87). However when the vacuolar ionic composition deviated too much from the normal one (K^+ 68–92 mM, Na^+ 12–45 mM) (60), cells could not maintain the composition even when the sap was isotonic to normal cell sap. Cells lost K^+ when the vacuole contained 160 mM KCl (87). Replacement of cell sap with an isotonic medium containing 150 mM NaCl + 10 mM $CaCl_2$ caused rapid release of a large amount of Na^+ accompanied by uptake of a small amount of K^+, thus resulting in a marked decrease in π_i (142). Clearly the mechanism of π_i-regulation can only function normally when the ionic composition of the vacuole is controlled within certain limits.

What is the physiological significance of π_i regulation? Since the water permeability of the tonoplast is extremely high (63), the cytoplasm is always in osmotic equilibrium with the vacuole—i.e. π_c is equal to π_v. When π_v is increased or decreased by transcellular osmosis, π_c immediately follows π_v. This causes changes in the cytoplasmic volume, as is the case in wall-less unicellular organisms. Thus π_i regulation in walled cells may be similar to volume regulation in wall-less cells. Wall-less marine algae regulate cell volume by synthesizing organic solutes (49), whereas *Nitella* regulate cytoplasmic volume using inorganic ions (141).

Solutes in the cytoplasm are either condensed or diluted in response to changes in π_v. It is probable that a change in the activity of a particular ion, either K^+ or Cl^-, directly affects electrogenesis or ion channels in the plasmalemma and/or the tonoplast. In tonoplast-free *Chara* cells the plasmalemma was depolarized by 40 mV when $[K^+]_c$ was increased from 65 mM to 100 mM and by 70–80 mV when $[Cl^-]_c$ was increased from 29 mM to 48 mM or more (114). Cl^- influx is also sensitive to $[Cl^-]_c$ (106).

Turgor

Only euryhaline brackish Characeae can regulate turgor. *Lamprothamnium papulosum* can maintain turgor pressure almost constant for changes in (π_o) between 550 and 1350 mOsm (7). Another species, *L. succinctum,* can also regulate its turgor for π_o ranging from zero to twofold diluted artificial sea water (½ ASW, $\pi_o = 0.54$ Osm) (95). The main solutes in the vacuole, K^+ and Cl^-, increase in parallel with π_o (7, 96). Hypotonic regulation (regulation induced by hypotonic treatment) was complete within 1–2 hr whereas hypertonic regulation (regulation induced by hypertonic treatment) lasted one to several days (8, 95). Hypotonic treatment induced rapid membrane depolarization and a drastic increase in membrane conductance (96). Both K^+ and Cl^- are assumed to be released passively (44, 96). Hypertonic treatment caused slow membrane hyperpolarization without a significant change in membrane conductance (96, 100). Under these circumstances net K^+ influx is down its electrochemical gradient but Cl^- is actively absorbed.

The error signal, the difference between the elevated turgor and the reference turgor, may be transduced by the plasmalemma to other signals controlling ion movements. Although we are ignorant of the details of this transduction process, changes in membrane thickness due to compression or stretching may directly influence ion transport via ion pumps and channels (18). Gutknecht et al (33) criticized this hypothesis on the grounds that a rise in absolute hydrostatic pressure, which compresses both plasmalemma and tonoplast, did not affect turgor regulation in marine algae. They proposed an alternative model in which anisotropic changes in appression, tension, or stretch applied to the plasmalemma cause the redistribution of membrane components to minimize strain.

In *L. succinctum* the presence of several millimolar external Ca^{2+} is essential for the induction of hypotonic turgor regulation (97). Membrane depolarization took place at high (3.9 mM) and low (0.01 mM) $[Ca^{2+}]_o$. Inhibition of the electrogenic H^+ pump (7) may be responsible for depolarization, since the depolarized level is close to the K^+-equilibrium potential (= -100 mV). In contrast to membrane depolarization, a drastic increase in membrane conductance occurred only at high $[Ca^{2+}]_o$. Thus membrane depolarization is not directly coupled with increased membrane conductance. The calcium content of the cytoplasm increased significantly upon hypotonic treatment when $[Ca^{2+}]_o$ was 3.9 mM. This increase was not observed at low $[Ca^{2+}]_o$ (0.01 mM) (Y. Okazaki and M. Tazawa, submitted to *Plant Cell Environ.*). Cytoplasmic streaming, which is very sensitive to free Ca^{2+} (35, 148, 156), almost stopped at the peak of membrane conductance and then recovered paralleling the decrease in membrane conductance (98). Hence a transient rise in $[Ca^{2+}]_c$ may induce increased membrane permeability to K^+ and Cl^-. Since the Ca^{2+} channel antagonist nifedipine (25) completely abolished hypotonic turgor regulation (98a), Ca^{2+} is assumed to enter the cell via the Ca^{2+} channel. Thus an abrupt increase in turgor has two effects on the membrane, membrane depolarization and Ca^{2+}-channel activation. There are two possible explanations for Ca^{2+}-channel activation. One is direct control of the Ca^{2+} channel by turgor—i.e. the Ca^{2+} channel itself is the pressure transducer. The second possibility is membrane depolarization–induced Ca^{2+}-channel activation. Voltage-dependent Ca^{2+}-channel activation is well known in the plasma membranes of animal cells (34).

The Ca^{2+} channel in *Lamprothamnium* is assumed to be opened by a sudden increase in turgor even at very low Ca^{2+} concentrations (0.01 mM). There was a lag of about 1 min before the abrupt rise in membrane conductance took place after hypotonic treatment; but an instantaneous rise in membrane conductance was observed when cells pretreated for 30 min with hypotonic medium containing 0.01 mM $[Ca^{2+}]_o$ were transferred to hypotonic medium containing 3.9 mM $[Ca^{2+}]_o$ (97).

During hypotonic regulation it is probable that increased $[Ca^{2+}]_c$ activates the Cl^- channel. Ca^{2+}-dependent Cl^--channel activation has been reported in the plasma membranes of various animal cells (cf 74). A similar mechanism was also proposed for plasmalemma and tonoplast excitation in characean cells (52, 73) and the water mold *Blastocladiella* (15). This hypothesis was tested by changing the cytoplasmic free Ca^{2+} concentration in tonoplast-free *Lamprothamnium* cells. In fact, however, membrane conductance was not affected by raising $[Ca^{2+}]_c$ to 10^{-5} M, which is 10^2 times higher than the normal level (Y. Okazaki, M. Tazawa, unpublished). Some soluble cytoplasmic factor linking the Ca^{2+} signal with Cl^--channel activation may exist that is lost or dispersed into the large vacuolar space after disintegration of the tonoplast. Recently, Shiina & Tazawa (111a) obtained results suggesting that protein phosphorylation/dephosphorylation controls membrane excitability in *Nitellopsis*. It is tempting to assume that Ca^{2+}-dependent protein phosphorylation/dephosphorylation is involved in Cl^--channel activation, the final step in the chain of reactions comprised in turgor regulation.

The next problem is to explain how a turgor increase sensed by the plasmalemma can cause large effluxes of K^+ and Cl^- across the tonoplast. During the initial phase of hypotonic turgor regulation, large ion fluxes (1 nmol cm^{-2} sec^{-1}), occur not only across the plasmalemma but also across the tonoplast in *L. succinctum*. There are two possible explanations for these high tonoplast effluxes. First, a rapid loss of ions across the plasmalemma might cause a drastic decrease in the concentrations of K^+ and Cl^- in the cytoplasm. This would break the ionic equilibrium between the cytoplasm and the vacuole, and induce movements of K^+ and Cl^-. Since the tonoplast potential is near zero (96), net effluxes of ions from the vacuole to the cytoplasm would occur down concentration gradients as suggested also in *L. papulosum* (100). The second possibility is that a transient rise in $[Ca^{2+}]_c$ might in some way stimulate K^+ and Cl^- channels not only in the plasmalemma but also in the tonoplast, the main barrier to ion movement in *Lamprothamnium* (44). In the water mold *Blastocladiella*, the action current under voltage-clamp conditions consists of an initial rapid Ca^{2+} current and a delayed Cl^- current (15). A sudden rise in turgor caused by decreasing π_o evoked action potentials involving Cl^- efflux. This mechanism is also assumed to operate for turgor regulation.

In hypertonic turgor regulation, marked membrane hyperpolarization preceded an increase in π_i (96). Hyperpolarization is assumed to be caused by activation of the ATP-dependent electrogenic H^+ pump (7). Membrane hyperpolarization increases the electrochemical potential gradient for passive K^+ influx, and activation of the H^+ extruding pump produces an electrochemical potential gradient for H^+ which could be used for $2H^+/Cl^-$ symport (107, 127, 130).

MEMBRANE EXCITATION

Action currents in plants were first recorded by Burdon-Sanderson in 1873 (12) on the trap lobes of *Dionaea*. Membrane excitation coupled with leaf movement evoked the interest of plant physiologists and has been studied intensively (cf 123). In charophytes, membrane excitation is coupled with intracellular cytoplasmic movement. The ionic processes occurring in the plasmalemma during excitation provide the basis for membrane control of cell motility (see below).

Characean cells generate action potentials in response to various stimuli. The strength-duration (threshold-voltage-duration) relation is hyperbolic, indicating that Weiss's experimental formula holds (29, 124). The action potential is transmitted along the cell (122, 124), and electrical impedance decreases during the action potential (16).

Plasmalemma

In charean cells, both plasmalemma and tonoplast generate action potentials (24). Beilby & Coster (2, 3), analyzed the excitation process for the *Chara* plasmalemma using Hodgkin-Huxley type equations. Under voltage-clamp conditions they found two temporally separated inward currents responsible for membrane depolarization in addition to the leak current. Hirono & Mitsui (38), using endoplasm-rich cells, analyzed the membrane current using a single inward current and an outward K^+ current.

One of the ionic mechanisms responsible for the plasmalemma excitation involves Cl^-. Cl^- efflux increases upon membrane excitation (30, 92). Increasing external $[Cl^-]_o$ suppressed the inward current under voltage-clamp conditions (58). The Cl^- current under voltage-clamp conditions decreased after 6–10 hr of Cl^- starvation, presumably owing to decreased $[Cl^-]_c$ (1). Besides Cl^-, Ca^{2+} is also assumed to carry inward current, since the peak of the action potential (39) and the inward current (21, 22) are dependent on $[Ca^{2+}]_o$.

In tonoplast-free *Chara* cells, action potentials were generated even when $[Cl^-]_i$ was drastically reduced; and the peak value of the action potential was not dependent upon $[Cl^-]_o$ (0.3–30 mM) (114). In *Nitella* no Cl^- efflux was detected upon excitation, although K^+ efflux did occur (131). Action potentials could be elicited without a measurable increase in Cl^- efflux in tonoplast-free cells (54). However, Hope & Findlay (40) reported that the Ca^{2+}

influx was insufficient to explain the inward current. On the other hand, Hayama et al (35) found in intact *Chara* cells a large increase in Ca^{2+} influx amounting to 300 nmol m^{-2} s^{-1} upon excitation. This is approximately equal to the K^+ efflux found in tonoplast-free cells (54). Thus it is likely that the action current is carried either by Ca^{2+} or by both Cl^- and Ca^{2+} depending on the physiological conditions. In tonoplast-free cells the mechanism of Cl^--channel activation seems to be impaired. Some soluble activating factor may be lost after disintegration of the tonoplast. In this case, then, we can study mechanisms of Ca^{2+}-channel activation using tonoplast-free cells without interference of Cl^- channels.

The involvement of Cl^- and Ca^{2+} currents in a single action potential has been suggested (2, 73). In *Chara*, Beilby & Coster (2) assumed that activation of the Cl^- channel precedes that of the Ca^{2+} channel. On the other hand, Lunevsky et al (73) suggested that an increase in cytoplasmic Ca^{2+} activates the Cl^- channel in *Nitellopsis*. In tonoplast-free *Nitellopsis* cells, increasing $[Ca^{2+}]_i$ depolarized the plasmalemma significantly and increased the membrane conductance a great deal without an increase in Cl^- efflux (80). It is possible that the elevated $[Ca^{2+}]_i$ increases K^+ conductance of the plasmalemma. On the other hand, in perfused *Chara* cells the membrane potential was not significantly affected by elevating $[Ca^{2+}]_i$ to 2.5×10^{-5} M (71). Furthermore, action potentials were observed even at 10^{-4} M $[Ca^{2+}]_i$ (72). The discrepancy may be attributed to the difference of materials.

The K^+ efflux that constitutes the repolarizing current increases upon excitation (92, 130, 131). Involvement of a K^+ channel in the repolarization process was suggested by the prolongation of the *Nitella* action potential by tetraethylammonium (TEA) (5, 118). The after-hyperpolarization, which is due to activation of the K^+ channel, was observed in *Nitella axilliformis* (118). In *Chara*, however, TEA did not affect the inward current under voltage-clamp condition (2) and induced only a slight prolongation of the action potential (118).

Action potentials in characean cells can also be explained using the two-stable-state hypothesis of the membrane developed by Tasaki (135). This hypothesis maintains that in the resting state negative charges on the outer surface of the membrane are occupied by divalent cations, whereas in the excited state they are occupied by monovalent cations. Oda (91) found two stable states in *Chara* in media containing high K^+ concentrations. The presence of two stable plasmalemma states was demonstrated more clearly in tonoplast-free *Chara* cells (112). In a medium containing 4 mM K^+, transition of the membrane between the two states could easily be controlled by an electrical current. The membrane state can also be controlled by external ions without electrical stimulation—i.e. in high-Ca^{2+} medium the membrane takes the resting state while in high-K^+ medium it resembles the excited state.

Mg^{2+} and ATP are essential not only for fuelling the membrane H^+-ATPase but also for maintaining membrane excitability (113). Although removal of either Mg^{2+} or ATP instantaneously inhibited the electrogenic pump, membrane excitability was sustained for some time (143). Even when the electrogenic pump in intact cells was inhibited by anoxia (79) or by triphenyltin chloride (59), membrane excitability was still maintained. Thus Mg-ATP does not affect membrane excitability by inhibiting the electrogenic H^+ pump. At present we cannot give any definite answer concerning the role of Mg-ATP in membrane excitation. Some cytoplasmic factor may exist that requires Mg-ATP and is lost with time after disintegration of the tonoplast. In view of the fact that in animal cells the activities of ion channels are controlled by protein phosphorylation (65), it is assumed that removal of ATP from *Chara* cells may also cause protein dephosphorylation that leads to inhibition of membrane excitability. However, excitability was lost even in the presence of sufficient Mg-ATP, when the protein phosphatase inhibitor α-naphthyl-phosphate was perfused into tonoplast-free *Nitellopsis* cells. Perfusion of the phosphatase inhibitor with a protein kinase inhibitor restored membrane excitability (111a). Thus protein phosphorylation seems to inhibit membrane excitability, while protein dephosphorylation seems to stimulate it. Therefore in the control of membrane excitation, Mg-ATP's role is in stimulating protein dephosphorylation. Mg-ATP-dependent protein phosphatase, which is actually a complex of protein phosphatase-1 and inhibitor-2 (101), is broadly distributed in animal tissues (31, 75) and in *Neurospora* (146). If a similar protein phosphatase were also present in *Chara* cytoplasm, depletion of Mg-ATP would reduce phosphatase activity and result in protein phosphorylation, depressing membrane excitability. The washing out of some factor controlling ion-channel activity in internally perfused squid giant axons and snail neurons has been suggested (cf 65). The importance of Mg-ATP and EGTA was suggested in relation to Ca^{2+} buffering near the axolemma (14), which is essential for maintaining Ca^{2+}-channel activation (13).

Tonoplast

Both plasmalemma and tonoplast depolarize on excitation (23, 24). It is supposed that the tonoplast action potential is caused by an increase in the Cl^- permeability of the tonoplast, since the tonoplast membrane potential shifts toward E_{Cl} (17, 23, 24). When the Cl^- concentration in the vacuole ($[Cl^-]_v$) of *Nitella* was drastically reduced by vacuolar perfusion (i.e. E_{Cl} becoming vacuole negative), the direction of the tonoplast action potential was reversed (55). The tonoplast action potential can only be generated when the plasmalemma is excited (23, 55). There should therefore be a mechanism coupling plasmalemma excitation and tonoplast excitation. Ca^{2+} is assumed to be the coupling factor (55), since $[Ca^{2+}]_c$ increases upon excitation (156).

Replacement of external Ca^{2+} with Ba^{2+}, Mg^{2+}, or Mn^{2+} did not inhibit the generation of the plasmalemma action potential but did inhibit the generation of the tonoplast action potential and the cessation of cytoplasmic streaming (52) caused by an increase in cytoplasmic Ca^{2+} (156). Furthermore, injection of Ca^{2+} into the cytoplasm of *Chara* induced a transient change in the tonoplast potential similar to the action potential (52).

Endoplasmic drops isolated from *Nitella* internodes (48) also exhibit two stable membrane states (64). The surface membrane, which originates from the tonoplast (105), is depolarized in solutions of high ionic strength with a high monovalent/divalent cation ratio. The membrane shifts to the polarized resting state on replacing the solution with a medium of low ionic strength with a low monovalent/divalent cation ratio (43). When the membrane in the resting state is stimulated electrically, action potentials are elicited that are of much shorter duration than normal plasmalemma or tonoplast action potentials (23, 24). This is indicative of modification of the tonoplast in vitro, since the membrane in situ can not generate action potentials independent of the plasmalemma action potential. The resting membrane displayed a lower refractive index and higher surface tension than the depolarized membrane (43, 151). Thus the two membrane states differ not only in ionic permeability but also in structure, as reflected by changes in the surface tension and refractive index of the membrane.

INTERACTION BETWEEN PLASMALEMMA AND CHLOROPLASTS AND MITOCHONDRIA

Some transport functions of the plasmalemma are strongly regulated by intracellular metabolism. In this section, we focus on plasmalemma characteristics controlled by photosynthesis and respiration.

Electrogenic H^+-ATPase

The light-induced potential change in Characeae was first reported by Brown in 1938 (11). It is dependent on photosynthesis (86, 89). Using metabolic inhibitors and changing the external pH, Spanswick (128, 129) and Saito & Senda (102, 103) attempted to explain the light-induced potential change as an activation of the electrogenic H^+-pump (61) caused by photosynthesis. This concept was more directly demonstrated in tonoplast-free *Chara* cells, in which depletion of ATP completely abolished the light response (53). It should be added that Mg-ATP-dependent electrogenesis (113) can be explained in terms of Mg-ATP-dependent H^+ efflux (132).

No structural connection is known between the plasmalemma and chloroplasts. Some chemical signal that controls the activity of the H^+-ATPase may be released from chloroplasts. First, light-induced changes in substrate (H^+ and ATP) levels for the H^+-ATPase should be examined as candidates for the control signal. Cytoplasmic pH (pH_c), which is 7.3–7.5 in the dark, increases upon illumination by 0.2–0.4 pH units (76, 125, 154). On the other hand, it was found using tonoplast-free cells that the optimal pH for ATP-dependent electrogenesis was about 6.5–7.0 (28, 145). Therefore, alkalization of the cytoplasm caused by light would tend to reduce H^+-ATPase activity. Furthermore, the light-induced potential change occurred even when cells were perfused with media of high pH-buffering capacity. Recently, Felle & Bertl (20) suggested for rhizoids of *Riccia* that changes in pH_c measured with a pH-sensitive microelectrode were correlated with transient changes in membrane potential induced by light. We discuss this below in relation to HCO_3^- and OH^- transport, which are activated by photosynthesis.

It is difficult to accept the contention that light stimulates the electrogenic H^+-ATPase by elevating the cytosolic ATP level, since no significant difference in the ATP levels during light and dark has been detected (51). In tonoplast-free cells, however, the intracellular levels of adenine nucleotides changed significantly upon illumination; ATP increased, while ADP and AMP decreased (81). These changes in adenine nucleotide levels were large enough to explain the light-induced membrane hyperpolarization in terms of enzyme kinetics [K_m for ATP 0.1 mM (78); K_i for both ADP and AMP 0.4 mM (79)]. When cells were perfused with an ATP-regenerating system (pyruvate kinase and phosphoenolpyruvate), the membrane remained in the hyperpolarized state even in the dark and did not show further hyperpolarization upon illumination (81). Thus the photosynthesis-dependent H^+-ATPase activation in tonoplast-free cells can be accounted for in terms of changes in intracellular adenine nucleotide levels.

To explain the difference between intact and tonoplast-free cells, we assume that adenine nucleotide levels near the plasmalemma are different from those in the bulk cytoplasm (81). In characean cells, the chloroplast layer is located very close to the plasmalemma. Sometimes the gap between its outer surface and the plasmalemma is less than 0.1 μm (27). Furthermore, the rate of diffusion of solutes in the gel layer is expected to be lower than in the flowing sol layer owing to lack of mechanical agitation. Under these conditions, adenine nucleotide levels measured for the whole cytoplasm may not be representative of levels near the plasmalemma. Unfortunately, we have no means of detecting local adenine nucleotide levels. A real solution to this problem will only come when this technical difficulty is solved.

Brown (11) also reported that the membrane potential of *Chara* changed upon anoxic treatment. In *Nitella*, Kitasato (61) using uncouplers and Mimura et al (77) using electron transport inhibitors showed that part of the membrane potential is dependent on respiration under dark conditions. In the case of intact characean cells, Mimura et al (79) demonstrated that at least 60% of the membrane depolarization induced by anoxia can be explained by changes in cytoplasmic adenine nucleotide levels.

In conclusion, control of the electrogenic H^+-ATPase by energy metabolism in tonoplast-free cells occurs through changes in cytoplasmic adenine nucleotide levels. Recently, inhibition of the plasmalemma H^+-ATPase (158) in corn roots by protein phosphorylation was suggested. In perfused *Nitellopsis* cells the membrane potential was depolarized by the inhibition of protein phosphatase. Depolarization was reversed by inhibition of protein kinase (111a). In intact characean cells, not only H^+-ATPase substrates (ATP, H^+) but also other factors such as protein phosphorylation may be involved in controlling H^+-ATPase activity.

HCO_3^- and OH^- Transport

Charophytes utilize HCO_3^- for carbon assimilation in alkaline solutions. Many species of Characeae form alternating acid and alkaline bands along the internodal cell axis. Acid regions utilize HCO_3^-, while alkaline regions extrude OH^- produced in the cytoplasm when HCO_3^- is utilized for CO_2 fixation. Since Spear et al (130) demonstrated this banding phenomenon using pH dyes, Lucas and his coworkers have intensively investigated the mechanisms of HCO_3^- transport and band formation by measuring the cell-surface pH and the electric currents flowing between acid and alkaline bands (67, 68). The outward current, which coincides with the HCO_3^- influx, flows out of the acid region; the inward current, which coincides with the OH^- efflux, flows into the alkaline region. Ogata (93) measured the surface potential distribution using a scanning water-film electrode and found that the surface potential in alkaline regions is more negative than in acid regions.

As for the relationship between photosynthesis and HCO_3^- transport, the following problems remain unsolved. First, how does the cell utilize the external HCO_3^-? One plausible explanation is that HCO_3^- is utilized after conversion to CO_2 occurring in the unstirred layer at the acid bands. An alternative explanation is that HCO_3^- itself is taken up into the cytoplasm, either by uniport or by symport with H^+, utilizing the H^+ gradient produced by the H^+-ATPase. Unfortunately we cannot decide which model is correct. Using a vibrating electrode, Lucas (66) measured very large currents in both acid and alkaline regions. Sometimes these reached about 300 mA/m^2, which is more than ten times larger than the H^+ current (about 20 mA/m^2) found in *Nitellopsis* cells in darkness (132). Let us assume a parallel circuit model for the plasmalemma containing diffusion-channel and electrogenic-pump components. Adopting a diffusional conductance of 1 S/m^2 and an electromotive force for the diffusion channel of -100 mV (70, 85), and electric current amounting to 200–250 mA/m^2 would result in a membrane potential of -300 to -350 mV. Such a potential was found transiently in *Chara* when the light was turned off (66). From an electrophysiological point of view, it is not unreasonable to assume that the outward current in the acid region of *Chara* cells is carried by H$^+$ transported by the electrogenic H$^+$-ATPase. The H$^+$ current would then play an important role in HCO$_3^-$ utilization.

The current in the alkaline region may be carried only by OH$^-$ efflux. The mechanism by which such a large electrogenic current is generated remains obscure.

When the light is turned off in the presence of HCO$_3^-$, the inward current in alkaline regions diminished earlier than the outward current in acid regions (66). If OH$^-$ efflux occurs via a uniport system, the membrane would be depolarized and an earlier decay in OH$^-$ efflux would result in membrane hyperpolarization. In fact, a large transient hyperpolarization was observed on turning off the light in the presence of HCO$_3^-$ (66). The next problem to be solved is how the kinetics of transport in acid (H$^+$ and HCO$_3^-$) and alkaline (OH$^-$) regions are regulated. Felle & Bertl (20) showed that a transient membrane hyperpolarization followed transient cytoplasmic acidification in *Riccia*. The transient hyperpolarization in *Chara*, although much larger than the hyperpolarization occurring in *Riccia*, may be partially explained in terms of changes in pH$_c$. Measurement of local pH$_c$ changes are keenly awaited.

The formation of acid and alkaline bands raises the question of how functional differentiation develops in single cells. Inhibition of cytoplasmic streaming with cytochalasin B disturbed band formation (69). Using centrifuged cells, Lucas & Shimmen (70) demonstrated the involvement of large organelles in the band formation.

K$^+$-Channel Activation

The plasmalemma of characean cells is hyperpolarized by light even when ATP is depleted using hexokinase and glucose (50). The reaction becomes enhanced when cells are stained with neutral red. Since hyperpolarization occurs quickly, with a half time of about 10 sec, it is termed the rapid light-induced hyperpolarization (rapid LIH). This LIH, which was discovered in ATP-depleted tonoplast-free cells, was found also in intact cells having high levels of ATP whose membrane potential was depolarized by the application of outward current. The membrane potential in LIH approaches the K$^+$-equilibrium potential. The amplitude of LIH is therefore dependent on how far the membrane potential in darkness is away from the K$^+$-equilibrium potential (115). The LIH is inhibited by DCMU (50) and by the K$^+$-channel blockers tetraethylammonium and nonyltriethylammonium (144). Thus it seems likely that the plasmalemma is equipped with a K^+ channel dependent on photosynthesis.

Recently Mimura & Tazawa (82) found that the rapid LIH could be induced even under continuous rapid perfusion of the cell. This argues against the view that some diffusible chemical signal is transferred from chloroplasts to K^+ channels in the plasmalemma. On the other hand, we found that increasing the ionic strength of the perfusion medium suppressed the LIH. The surface-charge density of isolated intact chloroplasts increases upon illumination and is dependent on external ionic concentration (109). Based on these results, we propose that light-dependent changes in the surface-charge density of chloroplasts affect the charge distribution at the inner surface of the plasmalemma, which in turn affects K^+ channels. Since neutral red had no effect on oxygen evolution, the pigment, which is lipophilic, may interact with the hydrophobic region of the membrane or with K^+ channel proteins and alter the membrane structure giving rise to susceptibility to changes in the electrical surface charge of chloroplasts. The voltage dependence of K^+ conductance is well known in characean cells (62). The K^+ channel activated by light in the presence of neutral red may be the same one that is activated voltage dependently, since both are inhibited by tetraethylammonium.

CONTROL OF CELL MOTILITY

Rotational streaming in characean cells was reviewed in detail by Kamiya (45) on a structural, physical, and chemical basis. Here we deal with the membrane control of cytoplasmic streaming. Cytoplasmic streaming stops the moment action potentials are generated. This excitation-cessation coupling, first found by Hörmann (41), is a typical example of the membrane control of cell function and is comparable with the excitation-contraction coupling observed in muscle cells. Vacuolar perfusion studies demonstrated that the cause of cessation of streaming upon membrane excitation is not gelation of the streaming endoplasm but rather loss of motive force (140). Sometimes chloroplasts are detached from the gel layer with actin bundles and rotate in the flowing endoplasm. On membrane excitation, the rotation of such chloroplasts also ceases. Some delay was observed before chloroplasts located far from the plasmalemma stopped rotating (36), suggesting that some chemical signal is generated at the plasmalemma that propagates through the endoplasm. In cytoplasm-rich cells prepared by centrifugation, streaming also stopped on plasmalemma excitation, showing that the presence of the vacuole is not necessary for excitation-induced cessation (35).

Changes in $[Ca^{2+}]_c$ were detected by measuring light emission from the photoprotein aequorin injected into the cytoplasm (156). The $[Ca^{2+}]_c$, which is normally 0.1–1 μM, increased to 10 μM or more upon membrane excitation of *Chara* and *Nitella*. Ca^{2+} injected iontophoretically into *Nitella* cytoplasm reversibly inhibited streaming (56). Ca^{2+} was also iontophoretically injected into endoplasmic droplets isolated from *Chara* (36). Ca^{2+} reversibly inhibited the rotation of chloroplasts in these droplets. Thus, Ca^{2+} is supposed to be the signal causing excitation-induced streaming cessation. Ca^{2+} influx increases upon membrane excitation (35). The increase in $[Ca^{2+}]_c$ estimated from this influx, assuming that incoming Ca^{2+} distributes homogeneously in the cytoplasm without binding to cytoplasmic constituents, is more than 4μM for $[Ca^{2+}]_o$ higher than 0.1 mM (35). This Ca^{2+} concentration is sufficient to stop cytoplasmic streaming (147, 156).

In tonoplast-free cells, Kikuyama & Tazawa (57) measured the Ca^{2+} influx using aequorin light emission. The light emission consisted of an initial rapid component and a subsequent slow component. The first component, which was found also in intact cells (156), corresponded to the initial spike of the action potential and the slow component to the prolonged plateau depolarization. It was assumed that the initial light emission corresponds to the release of Ca^{2+} from some internal Ca^{2+} store, since light emission decreased upon repeated excitation. The slow component was assumed to represent influx of Ca^{2+} from the external medium, since it was dependent on $[Ca^{2+}]_o$. Since the initial light emission was also dependent on $[Ca^{2+}]_o$, it may represent the real Ca^{2+} spike. This assumption is further supported by the fact that a large net $^{45}Ca^{2+}$ influx was found on excitation of intact *Chara* cells (35). The Ca^{2+} influx estimated from the initial peak of light emission for a $[Ca^{2+}]_o$ of 0.1–1 mM, was 12–38 pmol/cm^2/impulse, which is almost equal to the Ca^{2+} influx found by Hayama et al (35) using ^{45}Ca (5–50 pmol/cm^2/impulse).

The relationship between $[Ca^{2+}]_c$ and streaming rate was first studied quantitatively using tonoplast-free cells. When the streaming rate was measured 10 min after the introduction of Ca^{2+} into tonoplast-free cells, streaming was partially inhibited by 1 μM Ca^{2+} (35). When it was measured 3 min after Ca^{2+} application, it was not inhibited at all at 10^{-4} M and was inhibited only partially at 5×10^{-4} M Ca^{2+} (149). In view of the fact that the $[Ca^{2+}]_c$ in intact cells is only 6.7–43 μM when streaming is inhibited by excitation (156), the Ca^{2+} sensitivity of streaming is clearly impaired after removal of the tonoplast. Since the endoplasm disperses into the space originally occupied by the vacuole after disintegration of the tonoplast, some Ca^{2+}-sensitizing component may be lost.

The relationship between $[Ca^{2+}]_c$ and streaming rate was studied using the plasmalemma-permeabilized cell model, prepared by inducing plasmolysis in an ice-cooled EGTA medium (117). In this model the flowing endoplasm is sandwiched between the tonoplast and the cell wall (Figure 1). Diffusible substances such as ATP and Ca^{2+} are applied to the cytoplasm from the outside. Streaming was almost completely inhibited by 1 μM Ca^{2+} (148).

Control of the streaming by physiologically reasonable Ca^{2+} concentrations made the Ca^{2+} hypothesis plausible.

To study which component, actin or myosin, is responsible for Ca^{2+} sensitivity, reconstitution experiments were carried out in which plastic beads coated with foreign myosins or heavy meromyosin were introduced into tonoplast-free characean cells, causing movement of the beads along actin bundles (110, 111, 119, 120). The sliding of skeletal muscle myosin along characean actin bundles is insensitive to Ca^{2+} (120, 121, 153). However, it became Ca^{2+} sensitive when native skeletal muscle tropomyosin, the Ca^{2+}-sensitizing component of the actin-myosin interaction (19), was incorporated into *Chara* actin bundles (120, 121). Sliding did not occur in the absence of Ca^{2+} and was activated by Ca^{2+}. The absence of Ca^{2+} sensitivity in characean actin bundles suggests that Ca^{2+} control of excitation-induced cessation of cytoplasmic streaming is mediated through myosin only.

The involvement of calmodulin in excitation-cessation coupling in characean cells has also been studied. Calmodulin antagonists applied to the exterior of intact cells did not affect excitation-cessation coupling (4, 147). When the antagonists were applied to plasmalemma-permeabilized cells, they did not affect Ca^{2+}-induced cessation but rather inhibited the recovery from Ca^{2+} inhibition (147). ATP-r-S, which irreversibly thiophosphorylates protein under the action of protein kinase, irreversibly inhibited recovery from Ca^{2+} inhibition (150). Y. Tominaga, R. Wayne, H. Y. L. Tung, and M. Tazawa (to be submitted to *Protoplasma*) applied protein phosphatase and its inhibitors either to tonoplast-free cells or permeabilized cells and found that protein phosphatase inhibitors inhibited the streaming even in the absence of Ca^{2+}. On the other hand, protein phosphatase prompted the recovery of streaming even in the presence of Ca^{2+}. Based on these results, they have proposed a phosphorylation-dephosphorylation cycle of putative myosin in the Ca^{2+}-induced cessation of cytoplasmic streaming and its subsequent recovery.

One phenomenon suggests not the membrane control of cytoplasmic streaming, but the reverse. Nishizaki (90) found oscillations in the membrane potential. Since the period of these oscillations decreased with increased cell length and with decreased rate of cytoplasmic streaming, it was concluded that the oscillation was caused by the rotation of some factor in the flowing endoplasm (94). This factor should be distributed heterogeneously in the endoplasm and may affect ion channels and/or the H^+ pump in the plasmalemma or tonoplast.

CONCLUDING REMARKS

For want of space we have omitted another important aspect of membrane control—homeostasis of ions and amino acids. However, several important results merit brief mention. Using intracellular perfusion, Sanders (106, 107) found that Cl^- influx was very sensitive to changes in cytoplasmic pH and Cl^- concentration. The mechanism of cytoplasmic pH regulation in plant cells has been thoroughly reviewed by Smith & Raven (126) in terms of biophysical and biochemical pH-stats and intracellular buffer capacity. Using vacuolar perfusion, Moriyasu et al (84, 85) demonstrated that the vacuolar pH in *Chara* is maintained constant by the balance between the passive H^+ efflux and the active H^+ influx driven by the tonoplast H^+-ATPase. Using plasmalemma-permeabilized cells of *Nitella,* T. Shimmen and E. MacRobbie (unpublished) have demonstrated two types of proton pumps in the tonoplast, one fuelled by Mg-ATP and the other by Mg-pyrophosphate. Since the latter H^+ pump can carry H^+ to the same extent as the Mg-ATP-dependent H^+ pump, its significant contribution to the regulation of cytoplasmic and vacuolar pH is expected. Presence of the proton-translocating pyrophosphatase in the tonoplast has also been demonstrated in higher plants (99a, 154a). Various amino acids were loaded into the vacuole by vacuolar perfusion and their transport and metabolism were studied in *Chara* by Sakano & Tazawa (104). Two interesting facts emerged; the tonoplast transport of alanine is strongly dependent on metabolism of this amino acid in the cytoplasm, and the total cytoplasmic concentration of free amino acids is maintained constant even under high vacuolar loading of amino acids.

In recent studies, it has been proved that Ca^{2+} plays a central role in the membrane control of various physiological functions in characean cells. Ca^{2+}-channel activation caused by a sudden increase in turgor or by electrical stimulation results in an increase in the cytoplasmic Ca^{2+} level, which is normally maintained strictly at about 0.1 μM (156). This increase in Ca^{2+} triggers a sequence of events leading to turgor regulation via activation of K^+ and Cl^- channels in the plasmalemma and tonoplast or to the cessation of cytoplasmic streaming. Calmodulin, which is present at 400 μg/ml cell homogenate (147), seems to participate with Ca^{2+} in protein dephosphorylation (150). Thus protein kinases and protein phosphatases emerge as cytoplasmic factors that may transduce the Ca^{2+} signal affecting various cellular functions. It is well known that in animal cells phosphorylation modulates the opening of ion channels (65). In view of the fact that the H^+-ATPase activity (158) and the membrane excitation (111a) are affected by protein kinase, protein phosphatase, and their inhibitors, it is possible that protein phosphorylation and/or dephosphorylation are involved in membrane control in plant cells. Further investigations along these lines are keenly awaited.

ACKNOWLEDGMENTS

We thank Drs. J. Dainty, A. B. Hope, N. Kamiya, and W. J. Lucas for critical reading of the manuscript; Ms. N. Ebihara for extraordinary help with word processing; Mr. M. Reeves for his help with the English text.

Literature Cited

1. Beilby, M. J. 1981. Excitation-revealed changes in cytoplasmic Cl⁻ concentration in "Cl⁻ starved" *Chara* cells. *J. Membr. Biol.* 62:207–18
2. Beilby, M. J., Coster, H. G. L. 1979. The action potential in *Chara corallina*. II. Two activation-inactivation transients in voltage clamps of the plasmalemma. *Aust. J. Plant Physiol.* 6:323–35
3. Beilby, M. J., Coster, H. G. L. 1979. The action potential in *Chara corallina*. III. The Hodgkin-Huxley parameters for the plasmalemma. *Aust. J. Plant Physiol.* 6:337–53
4. Beilby, M. J., MacRobbie, E. A. C. 1984. Is calmodulin involved in electrophysiology of *Chara corallina J. Exp. Bot.* 35:568–80
5. Belton, P., Van Netten, C. 1971. The effects of pharmacological agents on the electrical responses of cells of *Nitella flexilis. Can. J. Physiol. Pharmacol.* 49:824–32
6. Bisson, M. A., Bartholomew, D. 1984. Osmoregulation or turgor regulation in *Chara? Plant Physiol.* 74:252–55
7. Bisson, M. A., Kirst, G. O. 1980. *Lamprothamnium*, a euryhaline charophyte. I. Osmotic relations and membrane potential at steady state. *J. Exp. Bot.* 31:1223–35
8. Bisson, M. A., Kirst, G. O. 1980. *Lamprothamnium*, a euryhaline charophyte. II. Time course of turgor regulation. *J. Exp. Bot.* 31:1237–44
9. Blinks, L. R. 1935. Protoplasmic potentials in *Halycystis*. IV. Vacuolar perfusion with artificial sap and sea water. *J. Gen. Physiol.* 18:409–20
10. Boller, T., Wiemken, A. 1986. Dynamics of vacuolar compartmentation. *Ann. Rev. Plant Physiol.* 37:137–64
11. Brown, S. O. 1938. Relation between light and the electric polarity of *Chara*. *Plant Physiol.* 13:713–36
12. Burdon-Sanderson, J. 1873. Note on the electrical phenomena which accompany irritation of the leaf of *Dionaea muscipula*. *Proc. R. Soc. London* 21:495–96
13. Byerly, L., Hagiwara, S. 1982. Calcium currents in internally perfused nerve cell bodies of *Limnea stagnalis. J. Physiol.* 322:503–28
14. Byerly, L., Moody, W. J. 1984. Intracellular calcium ions and calcium currents in perfused neurones of the snail, *Lymnaea stagnalis. J. Physiol.* 352:637–52
15. Caldwell, J. H., Brunt, J. V., Harold, F. M. 1986. Calcium-dependent anion channel in the water mold. *Blastocladiella emersonii. J. Membr. Biol.* 86:85–97
16. Cole, K. S., Curtis, H. J. 1938. Electrical impedance of *Nitella* during activity. *J. Gen. Physiol.* 22:37–64
17. Coster, H. G. L. 1966. Chloride in cells of *Chara australis. Aust. J. Biol. Sci.* 19:545–54
18. Coster, H. G. L., Steudle, E., Zimmermann, U. 1977. Turgor pressure sensing in plant cell membranes. *Plant Physiol.* 58:636–43
19. Ebashi, S. 1976. Excitation-contraction coupling. *Ann. Rev. Physiol.* 38:293–313
20. Felle, H., Bertl, A. 1986. Light-induced cytoplasmic pH changes and their interrelation to the activity of the electrogenic proton pump in *Riccia fluitans. Biochim. Biophys. Acta* 848:176–82
21. Findlay, G. P. 1961. Voltage-clamp experiments with *Nitella. Nature* 191:812–14
22. Findlay, G. P. 1962. Calcium ions and the action potential in *Nitella. Aust. J. Biol. Sci.* 15:69–82
23. Findlay, G. P. 1970. Membrane electrical behavior in *Nitellopsis obtusa. Aust. J. Biol. Sci.* 23:1033–45
24. Findlay, G. P., Hope, A. B. 1964. Ionic relations of cells of *Chara australis*. VII. The separate electrical characteristics of the plasmalemma and tonoplast. *Aust. J. Biol. Sci.* 17:67–77
25. Fleckenstein, A., Tritthart, H., Döring, H.-J., Byon, K. Y. 1972. BAY a 1040-ein hochaktiver Ca⁺⁺-antagonistischer Inhibitor der elektro-mechanischen Koppelungsprozesse im Warmblüter-myokard. *Arzneim.-Forsch.* 22:22–33
26. Flowers, T. J., Troke, P. F., Yeo, A. R. 1977. The mechanism of salt tolerance in halophytes. *Ann. Rev. Plant Physiol.* 28:89–121
27. Franceschi, V. R., Lucas, W. J. 1980. Structure and possible function(s) of charasomes: complex plasmalemma-cell wall elaborations present in some characean species. *Protoplasma* 104:253–71
28. Fujii, S., Shimmen, T., Tazawa, M. 1979. Effect of intracellular pH on the light-induced potential change and electrogenic activity in tonoplast-free cells of *Chara australis. Plant Cell Physiol.* 20:1315–28
29. Fujita, M., Mizuguchi, K. 1956. Excitation in *Nitella*, especially in relation to

electric stimulation. *Cytologia* 21:135–45

30. Gaffey, C. T., Mullins, L. J. 1958. Ion fluxes during the action potential in *Chara*. *J. Physiol.* 144:505–24

31. Goris, J., Dopere, F., Vandenheede, J. R., Merlevede, W. 1980. Regulation of liver phosphorylase phosphatase. ATP-Mg-mediated activation of the partially purified dog-liver enzyme. *FEBS Lett.* 117:117–21

32. Greenway, H., Munns, R. 1980. Mechanisms of salt tolerance in nonhalophytes. *Ann. Rev. Plant Physiol.* 31:149–90

33. Gutknecht, J., Hastings, D. S., Bisson, M. A. 1978. Ion transport and turgor pressure regulation in giant algal cells. In *Membrane Transport in Biology*, ed. G. Giebisch, D. C. Tosteson, H. H. Ussing. Berlin/New York: Springer-Verlag

34. Hagiwara, S., Byerly, L. 1981. Calcium channel. *Ann. Rev. Neurosci.* 4:69–125

35. Hayama, T., Shimmen, T., Tazawa, M. 1979. Participation of Ca^{2+} in cessation of cytoplasmic streaming induced by membrane excitation in Characeae internodal cells. *Protoplasma* 99:305–21

36. Hayama, T., Tazawa, M. 1980. Ca^{2+} reversibly inhibits active rotation of chloroplasts in isolated cytoplasmic droplets of *Chara*. *Protoplasma* 102:1–9

37. Hellebust, J. A. 1976. Osmoregulation. *Ann. Rev. Plant Physiol.* 27:485–505

38. Hirono, C., Mitsui, T. 1981. Time course of activation in plasmalemma of *Nitella axilliformis*. In *Nerve Membrane-Biochemistry and Function of Channel Proteins*, ed. G. Matsumoto, M. Kotani, pp. 135–49. Tokyo: Univ. Press Tokyo

39. Hope, A. B. 1961. Ionic relations of cells of *Chara australis*. V. The action potential. *Aust. J. Biol. Sci.* 15:69–82

40. Hope, A. B., Findlay, G. P. 1964. The action potential in *Chara*. *Plant Cell Physiol.* 5:377–79

41. Hörmann, G. 1898. Studien über die Protoplasmaströmung bei den Characeen. Jena: Gustav Fischer Verlag

42. Huxley, A. F., Stämpfli, R. 1949. Evidence for saltatory conduction in peripheral myelinated nerve fibres. *J. Physiol.* 108:315–39

43. Inoue, I., Ueda, T., Kobatake, Y. 1973. Structure of excitable membranes formed on the surface of protoplasmic drops isolated from *Nitella*. I. Conformation of surface membrane determined from the refractive index and from enzyme actions. *Biochim. Biophys. Acta* 298:653–63

44. Jefferies, R. L., Reid, R. J. 1984. *Lamprothamnium*, a euryhaline charophyte. III. Ionic fluxes and compartmental analysis at steady state. *J. Exp. Bot.* 35:912–24

45. Kamiya, N. 1981. Physical and chemical basis of cytoplasmic streaming. *Ann. Rev. Plant Physiol.* 32:205–36

46. Kamiya, N., Kuroda, K. 1955. Some experiments on cell amputation. *20th Ann. Meet. Bot. Soc. Japan*, pp. 46–47 (In Japanese)

47. Kamiya, N., Kuroda, K. 1956. Artificial modification of the osmotic pressure of the plant cell. *Protoplasma* 46:423–36

48. Kamiya, N., Kuroda, K. 1957. Cell operation in *Nitella*. I. Cell amputation and effusion of the endoplasm. *Proc. Jpn. Acad.* 33:149–52

49. Kauss, H. 1978. Osmotic regulation in algae. *Prog. Phytochem.* 5:1–27

50. Kawamura, G., Tazawa, M. 1980. Rapid light-induced potential change in *Chara* cells stained with neutral red in the absence of internal Mg-ATP. *Plant Cell Physiol.* 21:547–59

51. Keifer, D. W., Spanswick, R. M. 1979. Correlation of adenosine triphosphate levels in *Chara corallina* with the activity of the electrogenic pump. *Plant Physiol.* 64:165–68

52. Kikuyama, M. 1986. Tonoplast action potential of Characeae. *Plant Cell Physiol.* In press

53. Kikuyama, M., Hayama, T., Fujii, S., Tazawa, M. 1979. Relationship between light-induced potential change and internal ATP concentration in tonoplast-free *Chara* cells. *Plant Cell Physiol.* 20:993–1002

54. Kikuyama, M., Oda, K., Shimmen, T., Hayama, T., Tazawa, M. 1984. Potassium and chloride effluxes during excitation of Characeae cells. *Plant Cell Physiol.* 25:965–74

55. Kikuyama, M., Tazawa, M. 1976. Tonoplast action potential in *Nitella* in relation to vacuolar chloride concentration. *J. Membr. Biol.* 29:95–110

56. Kikuyama, M., Tazawa, M. 1982. Ca^{2+} ion reversibly inhibits the cytoplasmic streaming of *Nitella*. *Protoplasma* 113:241–43

57. Kikuyama, M., Tazawa, M. 1983. Transient increase of intracellular Ca^{2+} during excitation of tonoplast-free *Chara* cells. *Protoplasma* 117:62–67

58. Kishimoto, U. 1962. Current voltage relation in *Nitella*. *Biol. Bull.* 121:370–71

59. Kishimoto, U., Kami-ike, U., Takeuchi, U. 1980. The role of elec-

trogenic pump in *Chara corallina*. *J. Membr. Biol.* 55:149–56
60. Kishimoto, U., Tazawa, M. 1965. Ionic composition and electric response of *Lamprothamnium succinctum*. *Plant Cell Physiol.* 6:529–36
61. Kitasato, H. 1968. The influence of H⁺ on the membrane potential and ion fluxes of *Nitella*. *J. Gen. Physiol.* 52:60–87
62. Kitasato, H. 1973. K permeability of *Nitella clavata* in the depolarized state. *J. Gen. Physiol.* 62:535–49
63. Kiyosawa, K., Tazawa, M. 1977. Hydraulic conductivity of tonoplast-free *Chara* cells. *J. Membr. Biol.* 37:157–66
64. Kobatake, Y., Inoue, I., Ueda, T. 1975. Physical chemistry of excitable membranes. *Adv. Biophys.* 7:43–89
65. Levitan, I. B. 1985. Phosphorylation of ion channels. *J. Membr. Biol.* 87:177–90
66. Lucas, W. J. 1982. Mechanism of acquisition of exogenous bicarbonate by internodal cells of *Chara corallina*. *Planta* 156:181–92
67. Lucas, W. J. 1983. Photosynthetic assimilation of exogenous HCO₃⁻ by aquatic plants. *Ann. Rev. Plant Physiol.* 34:71–104
68. Lucas, W. J. 1985. Bicarbonate utilization by *Chara*: A reanalysis. In *Inorganic Carbon Uptake by Aquatic Photosynthetic Organisms*, ed. W. J. Lucas, J. A. Berry, pp. 229–54. Rockville, MD: Am. Soc. Plant Physiol.
69. Lucas, W. J., Dainty, J. 1977. Spatial distribution of functional OH⁻ carriers along a characean internodal cell: Determined by the effect of cytochalasin-B on H₂CO₃⁻ assimilation. *J. Membr. Biol.* 32:75–92
70. Lucas, W. J., Shimmen, T. 1981. Intracellular perfusion and cell centrifugation studies on plasmalemma transport processes in *Chara corallina*. *J. Membr. Biol.* 58:227–247
71. Lühring, H., Tazawa, M. 1985. Effect of cytoplasmic Ca²⁺ on the membrane potential and membrane resistance of *Chara* plasmalemma. *Plant Cell Physiol.* 26:635–46
72. Lühring, H., Tazawa, M. 1985. Cytoplasmic Ca²⁺ on the excitability of *Chara* plasmalemma. *Plant Cell Physiol.* 26:769–74
73. Lunevsky, V. Z., Zherelova, O. M., Vostrikov, I. Y., Berestovsky, G. N. 1983. Excitation of *Characeae* cell membranes as a result of activation of calcium and chloride channels. *J. Membr. Biol.* 72:43–58
74. Mayer, M. L. 1985. A calcium-

activated chloride current generates the after-depolarization of rat sensory neurones in culture. *J. Physiol.* 364:217–39
75. Merlevede, W., Riley, G. A. 1966. The activation and inactivation of phosphorylase phosphatase from bovine adrenal cortex. *J. Biol. Chem.* 241:3517–24
76. Mimura, T., Kirino, Y. 1984. Changes in cytoplasmic pH measured by ³¹P-NMR in cells of *Nitellopsis obtusa*. *Plant Cell Physiol.* 25:813–20
77. Mimura, T., Shimmen, T., Tazawa, M. 1982. Respiration-dependent membrane hyperpolarization in tonoplast-free cells of *Nitella axilliformis*. *Plant Cell Physiol.* 23:1419–25
78. Mimura, T., Shimmen, T., Tazawa, M. 1983. Dependence of the membrane potential on intracellular ATP concentration in tonoplast-free cells of *Nitellopsis obtusa*. *Planta* 157:97–104
79. Mimura, T., Shimmen, T., Tazawa, M. 1984. Adenine-nucleotide levels and metabolism-dependent membrane potential in cells of *Nitellopsis obtusa* Groves. *Planta* 162:77–84
80. Mimura, T., Tazawa, M. 1983. Effect of intracellular Ca²⁺ on membrane potential and membrane resistance in tonoplast-free cells of *Nitellopsis obtusa*. *Protoplasma* 118:49–55
81. Mimura, T., Tazawa, M. 1986. Light-induced membrane hyperpolarization and adenine nucleotide levels in perfused characean cells. *Plant Cell Physiol.* 27:319–30
82. Mimura, T., Tazawa, M. 1986. Analysis of rapid light-induced potential change in cells of *Chara corallina*. *Plant Cell Physiol.* 27:895–902
83. Morgan, J. M. 1984. Osmoregulation and water stress in higher plants. *Ann. Rev. Plant Physiol.* 35:299–319
84. Moriyasu, Y., Shimmen, T., Tazawa, M. 1984. Vacuolar pH regulation in *Chara australis*. *Cell Struct. Funct.* 9:225–34
85. Moriyasu, Y., Shimmen, T., Tazawa, M. 1984. Electric characteristics of the vacuolar membrane of *Chara* in relation to pHᵥ regulation. *Cell Struct. Funct.* 9:235–46
86. Nagai, R., Tazawa, M. 1962. Changes in resting potential and ion absorption induced by light in a single plant cell. *Plant Cell Physiol.* 3:323–39
87. Nakagawa, S., Kataoka, H., Tazawa, M. 1974. Osmotic and ionic regulation in *Nitella*. *Plant Cell Physiol.* 15:457–68
88. Nastuk, W. L., Hodgkin, A. L. 1950.

The electrical activity of single muscle fibers. *J. Cell. Comp. Physiol.* 35:39–73

89. Nishizaki, Y. 1968. Light-induced changes of bioelectric potential in *Chara. Plant Cell Physiol.* 9:377–87

90. Nishizaki, Y. 1968. Rhythmic changes in the resting potential of a single plant cell. *Plant Cell Physiol.* 9:613–16

91. Oda, K. 1962. Polarized and depolarized states of the membrane in *Chara braunii*, with special reference to the transition between the two states. *Sci. Rep. Tohoku Univ. Ser. 4* 28:1–16

92. Oda, K. 1976. Simultaneous recording of potassium and chloride effluxes during an action potential in *Chara corallina. Plant Cell Physiol.* 17:1085–88

93. Ogata, K. 1983. The water-film electrode: A new device for measuring the characean electro-potential and -conductance distributions along the length of the internode. *Plant Cell Physiol.* 24:695–703

94. Ogata, K., Kishimoto, U. 1976. Rhythmic change of membrane potential and cyclosis of *Nitella* internode. *Plant Cell Physiol.* 17:201–8

95. Okazaki, Y., Shimmen, T., Tazawa, M. 1984. Turgor regulation in a brackish charophyte, *Lamprothamnium succinctum.* I. Artificial modification of intracellular osmotic pressure. *Plant Cell Physiol.* 25:565–71

96. Okazaki, Y., Shimmen, T., Tazawa, M. 1984. Turgor regulation in a brackish charophyte, *Lamprothamnium succinctum.* II. Changes in K^+, Na^+ and Cl^- concentrations, membrane potential and membrane resistance during turgor regulation. *Plant Cell Physiol.* 25:573–81

97. Okazaki, Y., Tazawa, M. 1986. Involvement of calcium ion in turgor regulation upon hypotonic treatment in *Lamprothamnium succinctum. Plant Cell Environ.* 9:185–90

98. Okazaki, Y., Tazawa, M. 1986. Effect of calcium ion on cytoplasmic streaming during turgor regulation in a brackish water charophyte *Lamprothamnium. Plant Cell Environ.* 9:491–94

98a. Okazaki, Y., Tazawa, M. 1986. Ca^{2+} antagonist nifedipine inhibits turgor regulation upon hypotonic treatment in internodal cells of *Lamprothamnium. Protoplasma.* 134:65–66

99. Osterhout, W. J. V., Hill, S. E. 1930. Salt bridges and negative variations. *J. Gen. Physiol.* 13:547–52

99a. Rea, P. A., Poole, R. J. 1985. Proton-translocating inorganic pyrophophatase in red beet (*Beta vulgaris* L) tonoplast vesicles. *Plant Physiol.* 77:46–52

100. Reid, R. J., Jefferies, R. L., Pitman, M. G. 1984. Lamprothamnium, a euryhaline charophyte. IV. Membrane potential, ionic fluxes and metabolic activity during turgor adjustment. *J. Exp. Bot.* 35:925–37

101. Resink, T. J., Hemmings, B. A., Tung, H. Y. L., Cohen, P. 1983. Characterisation of a reconstituted Mg-ATP-dependent protein phosphatase. *Eur. J. Biochem.* 133:455–61

102. Saito, K., Senda, M. 1973. The light-dependent effect of external pH on the membrane potential of *Nitella. Plant Cell Physiol.* 14:147–56

103. Saito, K., Senda, M. 1973. The effect of external pH on the membrane potential of *Nitella* and its linkage to metabolism. *Plant Cell Physiol.* 14:1045–52

104. Sakano, K., Tazawa, M. 1985. Metabolic conversion of amino acids loaded in the vacuole of *Chara australis* internodal cells. *Plant Physiol.* 78:673–77

105. Sakano, K., Tazawa, M. 1986. Tonoplast origin of the envelope membrane of cytoplasmic droplets prepared from *Chara* internodal cells. *Protoplasma* 131:247–49

106. Sanders, D. 1980. Control of Cl influx in *Chara* by cytoplasmic Cl^- concentration. *J. Membr. Biol.* 52:51–60

107. Sanders, D. 1980. The mechanism of Cl^- transport at the plasma membrane of *Chara corallina.* I. Cotransport with H^+. *J. Membr. Biol.* 53:129–41

108. Sanders, D. 1981. Physiological control of chloride transport in *Chara corallina.* I. Effects on low temperature, cell turgor pressure, and anions. *Plant Physiol.* 67:1113–18

109. Schapendonk, H. C. M., Hemrika-Wagner, A. M., Theuvenet, A. P., Sang, H. W. W. F., Vredenberg, W. J., Kraayenhof, R. 1980. Energy-dependent changes of the electrokinetic properties of chloroplasts. *Biochemistry* 19:1922–27

110. Sheetz, M. P., Spudich, J. A. 1983. Movement of myosin-coated fluorescent beads on actin cables *in vitro. Nature* 303:31–35

111. Sheetz, M. P., Spudich, J. A. 1983. Movement of myosin-coated structures on actin cables. *Cell Motility* 3:485–89

111a. Shiina, T., Tazawa, M., 1986. Regulation of membrane excitation by protein phosphorylation in *Nitellopsis obtusa. Protoplasma* 134:60–61

112. Shimmen, T., Kikuyama, M., Tazawa, M. 1976. Demonstration of two stable potential states of plasmalemma of

116   TAZAWA, SHIMMEN & MIMURA

Chara without tonoplast. *J. Membr. Biol.* 30:249–70

113. Shimmen, T., Tazawa, M. 1977. Control of membrane potential and excitability of *Chara* cells with ATP and Mg^{2+}. *J. Membr. Biol.* 37:167–92

114. Shimmen, T., Tazawa, M. 1980. Intracellular chloride and potassium ions in relation to excitability of *Chara* membrane. *J. Membr. Biol.* 55:223–32

115. Shimmen, T., Tazawa, M. 1981. Demonstration of voltage dependence of light-induced potential change in *Chara*. *Plant Cell Physiol.* 22:807–18

116. Shimmen, T., Tazawa, M. 1982. Cytoplasmic streaming in the cell model of *Nitella*. *Protoplasma* 112:101–6

117. Shimmen, T., Tazawa, M. 1983. Control of cytoplasmic streaming by ATP, Mg^{2+} and cytochalasin B in permeabilized Characeae cell. *Protoplasma* 115:18–24

118. Shimmen, T., Tazawa, M. 1983. Activation of K^+-channel in membrane excitation of *Nitella axilliformis*. *Plant Cell Physiol.* 24:1511–24

119. Shimmen, T., Yano, M. 1984. Active sliding movement of latex beads coated with skeletal muscle myosin on *Chara* actin bundles. *Protoplasma* 121:132–37

120. Shimmen, T., Yano, M. 1985. Ca^{2+} regulation of myosin sliding along *Chara* actin bundles mediated by native tropomyosin. *Proc. Jpn. Acad.* 61(B): 86–89

121. Shimmen, T., Yano, M. 1986. Regulation of myosin sliding along *Chara* actin bundles by native skeletal muscle tropomyosin. *Protoplasma* 132:129–36

122. Sibaoka, T. 1958. Conduction of action potential in the plant cell. *Trans. Bose Res. Inst.* 22:43–56

123. Sibaoka, T. 1966. Action potentials in plant organs. *20th Symp. Soc. Exp. Biol.* 20:49–74

124. Sibaoka, T., Oda, K. 1956. Shock stoppage of the protoplasmic streaming in relation to the action potential in *Chara*. *Sci. Rep. Tohoku Univ. Ser. 4* 22:157–66

125. Smith, F. A. 1984. Regulation of the cytoplasmic pH of *Chara corallina*: Response to changes in external pH. *J. Exp. Bot.* 35:43–50

126. Smith, F. A., Raven, J. A. 1979. Intracellular pH and its regulation. *Ann. Rev. Plant Physiol.* 30:289–311

127. Smith, F. A., Walker, N. A. 1976. Chloride transport in *Chara corallina* and the electrochemical potential difference for hydrogen ions. *J. Exp. Bot.* 27:451–59

128. Spanswick, R. M. 1972. Evidence for an electrogenic ion pump in *Nitella translucens*. I. The effects of pH, K^+, Na^+, light and temperature on the membrane potential and resistance. *Biochim. Biophys. Acta* 288:73–89

129. Spanswick, R. M. 1974. Evidence for an electrogenic ion pump in *Nitella translucens*. II. Control of the light-stimulated component of the membrane potential. *Biochim. Biophys. Acta* 332: 387–98

130. Spear, D. G., Barr, J. K., Barr, C. E. 1969. Localization of hydrogen ion and chloride ion fluxes in *Nitella*. *J. Gen. Physiol.* 54:397–414

131. Spyropoulos, C. S., Tasaki, I., Hayward, G. 1961. Fractionation of tracer effluxes during action potential. *Science* 133:2064–65

132. Takeshige, K., Shimmen, T., Tazawa, M. 1986. Quantitative analysis of ATP-dependent H^+ efflux and pump current driven by an electrogenic pump in *Nitellopsis obtusa*. *Plant Cell Physiol.* 27:337–48

133. Tasaki, I. 1939. The electro-saltatory transmission of the nerve impulse and the effect of narcosis upon the nerve fiber. *Am. J. Physiol.* 127:211–27

134. Tasaki, I. 1952. Properties of myelinated fibers in frog sciatic nerve and in spinal cord as examined with microelectrodes. *Jpn. J. Physiol.* 3:73–94

135. Tasaki, I. 1968. *Nerve Excitation.* Springfield, IL: Thomas

136. Tazawa, M. 1961. Weitere Untersuchungen zur Osmoregulation der *Nitella*-Zelle. *Protoplasma* 53:227–58

137. Tazawa, M. 1964. Studies on *Nitella* having artificial cell sap. I. Replacement of the cell sap with artificial solutions. *Plant Cell Physiol.* 5:33–43

138. Tazawa, M. 1980. Cytoplasmic streaming and membrane phenomena in cells of Characeae. In *Handbook of Phycological Methods. Developmental and Cytological Methods*, ed. E. Gantt, pp. 179–93. London/New York/New Rochelle/Melbourne/Sydney: Cambridge Univ. Press

139. Tazawa, M., Kikuyama, M., Shimmen, T. 1976. Electric characteristics and cytoplasmic streaming of Characeae cells lacking tonoplast. *Cell Struct. Funct.* 1:165–76

140. Tazawa, M., Kishimoto, U. 1968. Cessation of cytoplasmic streaming of *Chara* internodes during action potential. *Plant Cell Physiol.* 9:361–68

141. Tazawa, M., Nagai, R. 1960. Die Mitwirkung von Ionen bei der Osmoregulation der *Nitella*zelle. *Plant Cell Physiol.* 1:255–67

142. Tazawa, M., Nagai, R. 1966. Studies on osmoregulation of *Nitella* internode with modified cell saps. *Z. Pflanzenphysiol.* 54:333–44

143. Tazawa, M., Shimmen, T. 1980. Action potential in Characeae: Some characteristics revealed by internal perfusion studies. In *Plant Membrane Transport: Current Conceptual Issues,* ed. R. M. Spanswick, W. J. Lucas, J. Dainty, pp. 349–62. Amsterdam: Elsevier/North-Holland Biomed. Press

144. Tazawa, M., Shimmen, T. 1980. Demonstration of the K^+ channel in the plasmalemma of tonoplast-free cells of *Chara australis. Plant Cell Physiol.* 21:1535–40

145. Tazawa, M., Shimmen, T. 1982. Artificial control of cytoplasmic pH and its bearing on cytoplasmic streaming, electrogenesis and excitability of Characeae cells. *Bot. Mag.* 95:147–54

146. Tellez-Ionen, M. T., Torres, H. N. 1973. Regulation of glycogen phosphorylase *a* phosphatase in *Neurospora crassa. Biochim. Biophys. Acta* 297: 399–412

147. Tominaga, Y., Muto, S., Shimmen, T., Tazawa, M. 1985. Calmodulin and Ca^{2+}-controlled cytoplasmic streaming in characean cells. *Cell Struct. Funct.* 10:315–25

148. Tominaga, Y., Shimmen, T., Tazawa, M. 1983. Control of cytoplasmic streaming by extracellular Ca^{2+} in permeabilized *Nitella* cells. *Protoplasma* 116:75–77

149. Tominaga, Y., Tazawa, M. 1981. Reversible inhibition of cytoplasmic streaming by intracellular Ca^{2+} in tonoplast-free cells of *Chara australis. Protoplasma* 109:103–11

150. Tominaga, Y., Tazawa, M. 1986. Mechanism of Ca^{2+}-control of cytoplasmic streaming in Characeae. In *Molecu-*

lar and Cellular Aspects of Calcium in Plant Development, ed. A. J. Trewavas, pp. 399–400. New York/London: Plenum

151. Ueda, T., Inoue, I., Kobatake, Y. 1973. Studies of excitable membrane formed on the surface of protoplasmic drops isolated from *Nitella.* II. Tension at the surface of protoplasmic drops. *Biochim. Biophys. Acta* 318:326–34

152. Umrath, K. 1930. Untersuchungen über Plasma und Plasma-strömung an Characeen. IV. Potentialmessungen an *Nitella mucronata* mit besonderer Berücksichtigung der Erregungserscheinungen. *Protoplasma* 9:576–97

153. Vale, R. D., Szent-Gyorgyi, A. G., Sheetz, M. P. 1984. Movement of scallop myosin on *Nitella* actin filaments: Regulation by calcium. *Proc. Natl. Acad. Sci. USA* 81:6775–78

154. Walker, N. A., Smith, F. A. 1975. Intracellular pH in *Chara corallina* measured by DMO distribution. *Plant Sci. Lett.* 4:125–32

154a. Wang, Y., Leigh, R. A., Kaestner, K. H., Sze, H. 1986. Electrogenic H^+-pumping pyrophosphatase in tonoplast vesicles of oat roots. *Plant Physiol.* 81:497–502

155. Williamson, R. E. 1975. Cytoplasmic streaming in *Chara:* A cell model activated by ATP and inhibited by cytochalasin–B. *J. Cell. Sci.* 17:655–68

156. Williamson, R. E., Ashley, C. C. 1982. Free Ca^{2+} and cytoplasmic streaming in the alga *Chara. Nature* 296:647–51

157. Zimmermann, U. 1978. Physics of turgor- and osmoregulation. *Ann. Rev. Plant Physiol.* 29:121–48

158. Zocchi, G. 1985. Phosphorylation-dephosphorylation of membrane proteins controls the microsomal H^+-ATPase activity of corn roots. *Plant Sci.* 40:153–59

Ann. Rev. Plant Physiol. 1987. 38:119–39

THE PLANT CYTOSKELETON: The Impact of Fluorescence Microscopy

Clive W. Lloyd

Department of Cell Biology, John Innes Institute, Colney Lane, Norwich, NR4 7UH, United Kingdom

CONTENTS

INTRODUCTION 119
THE INTERPHASE MICROTUBULE ARRAY 120
   Microtubule Nucleation 120
   Microtubule Organization 121
   Proteins Co-Localizing with the Cortical Array 126
THE PREPROPHASE BAND 128
   Proteins Co-Localizing with the PPB 130
THE MITOTIC SPINDLE 130
   Proteins Co-Localizing with the Spindle 130
THE PHRAGMOPLAST 131
   Proteins Co-Localizing with the Phragmoplast 132
THE RELATIONSHIP BETWEEN THE PPB ZONE AND THE
   PHRAGMOPLAST 132
TIP-GROWING CELLS 133
ORGANELLE MOVEMENT 134
PROSPECTS 135

INTRODUCTION

Two major advances have been made since the plant cytoskeleton was last considered in this series (24). Our understanding of the biochemistry of plant microtubules (MTs) (reviewed in 8) is growing rapidly, and fluorescence microscopy, which is the subject of the present review, is beginning to change our conception of the plant cytoskeleton.

Werner Franke and his colleagues (18) first demonstrated that the mitotic

0066-4294/87/0601-0119$02.00

apparatus of wall-less endosperm cells could be stained with antitubulin antibodies; my colleagues and I showed (39, 40) that staining of the cortical cytoskeleton in walled plant cells is practicable, and Wick and her collaborators (80) introduced the technique of staining fixed meristematic cells with antibodies. Recently, the range of plant cells studied by fluorescence microscopy has greatly increased, thus allowing plant cells to be seen—literally as well as metaphorically—in a new light.

The first phase of plant cytoskeletal research exploited the power of electron microscopy (EM); now to this two-dimensional closely focused image, fluorescence microscopy adds the ability to see whole cells in three dimensions. Because many cells at different stages of the cell cycle can be screened simultaneously, fluorescence microscopy also injects a temporal element, and we are more conscious of cytoskeletal dynamics. Of course, immunofluorescence is not an "improvement"; it is a different technique with its own advantages and limitations. But now that a variety of techniques is available (including freeze substitution, immunogold, high voltage EM, dry cleaving), the composite view of the plant cell will undoubtedly become fuller and more satisfying.

This review follows the stages of the microtubule cycle. The way in which each microtubule array gives way to its successor is interesting not only for its own sake but because the precise workings of this cycle—and the interactions of microtubules with other cytoskeletal elements—will undoubtedly illuminate the higher-order issue of cellular morphogenesis. This duality is well illustrated by the way in which the preprophase band of microtubules succeeds the interphase, cortical array: Not only do we wish to know whether the latter is a new array or a metamorphosis of the former, but the solution may tell us more about directional cell expansion and the determination of the division plane. The role of microtubules in cellular morphogenesis has been covered elsewhere (see 35) and forms the background for the present review.

THE INTERPHASE MICROTUBULE ARRAY

Microtubule Nucleation

Microtubules are believed to form in vivo by directed self-assembly—that is, by end addition of tubulin polypeptides upon microtubules seeded at microtubule nucleation sites. These sites specify where (and perhaps when) microtubule polymerization is initiated and therefore have a pivotal role in controlling the microtubule cycle. In cases where such sites dictate the form that the array will subsequently take, they are referred to as microtubule organizing centers (MTOCs). A familiar example is the highly structured centriole, or basal body, that forms a template for the outgrowing flagellar microtubules. No such organized sites have been seen in higher plants. Loci where microtubules converge have been observed in the electron microscope (19, 25), but whether or not these are nucleation sites is open to question. In principle, it is possible that the cortical interphase microtubules re-grow from the cortex after their disappearance during division. But there is now increasing evidence for the idea that interphase microtubules are polymerized from sites at the nuclear envelope. During the early development of microspores in *Lilium* there is no well-formed cellulosic wall, and it has been shown with antitubulin that microtubules radiate from the nuclear envelope to the plasma membrane (11, 63). Later, the mature microspore will have a cortical cytoskeleton and a cellulosic wall, but at this earlier stage microtubules can only be discerned as spokes radiating from the nucleus. *Haemanthus* endosperm is similar in not having a regular cellulosic wall. Antitubulin antibodies labelled by the colloidal gold method also reveal that microtubules radiate from the nuclear envelope in the absence of a cortical microtubular cytoskeleton (1, 9, 58). In plant cells with a regular cellulosic wall, there is also evidence that microtubules radiate from the nuclear envelope (2). Using immunofluorescence, Wick & Duniec (78) have demonstrated the existence of microtubules that radiate between nucleus and cortex in onion root tip cells.

The major obstacle to assigning the nuclear envelope as the site of microtubule nucleation in higher plants has been the lack of a stain for microtubule nucleating material. But recently a human autoimmune serum has been identified that recognizes the amorphous microtubule nucleating component of the animal cell's centrosome. At a 1984 symposium (36) and then in greater detail (4), results were presented on the application of this serum to meristematic plant cells. It was shown that the serum cross-reacts with sites around the onion cell's nucleus. Immediately following cytokinesis, often before phragmoplast microtubules have completely disappeared, cytoplasmic microtubules radiate from the nucleus, representing a distinct stage before the cortical array is formed (4, 36, 78). At this early and probably brief phase of the cycle, presumptive microtubule nucleation sites are stained around the nucleus and not at the cortex. At no stage of the microtubule cycle (4) could we detect labelling at the cortex. As anticipated, the broad spindle poles were labelled during mitosis; approaching cytokinesis the fluorescent perinuclear material became redistributed towards the side of the sister nuclei proximal to the phragmoplast. The picture is therefore one of nucleating material that helps initiate the spindle, the phragmoplast, and the interphase array—but all from sites capable of regrouping around the nucleus. This picture has been confirmed by Wick (77).

Microtubule Organization

Because microtubules radiate from the nucleus before forming the cortical array, it would appear that the acts of microtubule initiation and organization are separate events. For this reason, the peri-nuclear sites are nucleation sites (MTNS), not organizing centers. Organization seems to occur after the radial microtubules have contacted the plasma membrane, and it may well be that the latter phase depends upon microtubule-microtubule and microtubule-plasma membrane interactions.

The behavior of the mature array is reviewed first in order to form a background for subsequent discussion of how the array is established after the radial MTs have contacted the cortex.

THE BEHAVIOR OF HELICAL CORTICAL ARRAYS   Compared to electron microscopy, immunofluorescence has diminished the scale of observation, and it would now appear that the interphase microtubule array in uniformly expanding cells can be explained as an integral system of variable helices. First discussed in review (34), the dynamic helical model proposed that long microtubules, capable of sliding relative to neighbors and to the plasma membrane, would inevitably form helices of variable pitch, rather than hoops. The advantage of the helix is that this conformation can be "unwound," converting flat-pitched helices to steeply-pitched arrays and thereby accounting for the various orientations in which cortical microtubules have been observed. Immunofluorescence of nonelongating onion root hairs confirmed that microtubules formed left-handed– or right-handed–helices along the length of the cell (33, 41). Helical arrays have subsequently been seen in *Allium* root hairs, *Urtica* root hairs (73), cortical cells of *Raphanus* (72), and in cotton fibers (22, 61).

To confirm the idea that helically wound cortical microtubules could change their pitch by moving relative to the plasma membrane, Roberts et al (56) tested the effect of ethylene on mung bean and pea cortical and epidermal cells. These tissues have been extensively used in studying wall texture; it is known that successive wall lamellae contain helices of cellulose microfibrils of changing pitch and that the rhythm of wall deposition can be shifted by ethylene. In turn, ethylene is known to affect microtubules in these cells; it converts mainly transverse microtubules and innermost wall microfibrils to a longitudinal orientation concomitant with inhibition of cell elongation and encouragement of lateral expansion. By immunofluorescence (56), ethylene was shown to switch microtubule orientation from transverse to axial via an intermediate stage where the cytoskeleton consisted of multistart 45° helices of single-handedness.

For animal tissue culture cells the emerging concept is of the dynamic instability of microtubules: some growing upon nucleation sites while neighbors are catastrophically shrinking (45). It is not yet known whether this applies to the plasma-membrane–associated (i.e. better stabilized) cortical microtubules of plants, but in no cell were microtubules seen to be grossly depolymerized during the ethylene-induced realignments. Accepting that there might be some statistical turnover of microtubules, it appears that the whole cortical array can unwind against the plasma membrane. This is analogous to the stretching of a bed spring (34, 38, 56), although there must also be sliding between adjacent microtubules and/or the breakage of cross-links. The helical model emphasizes the point that the cortical cytoskeleton behaves as a whole and provides a unitary template that could influence the alignment of cellulose deposition along an entire cell face.

Some of the implications of a dynamic helical model for wall texture are discussed more fully elsewhere (34). However, although the gradual unwinding or compression of a helix could account for the deposition of wall fibrils at angles from 0 to 90° to the cell's long axis, it does not yet account for the way in which wall lamellae of one helical sign can be sandwiched between lamellae of opposite helical sign. Some of the possibilities are the following: (*a*) the entire MT array depolymerizes, then repolymerizes into an opposite helical sign; (*b*) the array countertwists itself without depolymerizing; (*c*) some wall layers (i.e. hemicellulose) contribute to the wall pattern but are not influenced by MTs; and (*d*) cell expansion subsequently unwinds some wall layers, changing their pitch and perhaps even their sign. Admittedly, working back from complex wall patterns to the behavior of the underlying MTs seems to require that a microtubular template undergo some mind-boggling contortions. But it should not be forgotten that the very existence of microtubular helices, let alone their capacity for realignment, was unsuspected not too long ago, and so the further gaps in our knowledge may be filled by only minor imaginative leaps.

That cortical arrays can be helical and can realign is now supported by further evidence. Immunofluorescence of the in vitro differentiation of *Zinnia* mesophyll cells into tracheary elements confirms earlier EM studies showing that evenly distributed cortical MTs bunch together as differentiation proceeds (16, 57) and that this precedes the formation of the cellulosic bands of secondary thickening. The bands may then be lignified in the absence of microtubules (57), but it is clear that the bunching of microtubules comes first in this sequence. Before the *Zinnia* mesophyll cells redifferentiate in culture, the evenly distributed microtubules are axial; but after a few days in vitro the MTs become wound around the transverse axis. It is in the transverse axis that bands of secondary thickening normally form. However, if the microtubule-stabilizing agent, taxol, is administered before the MTs have undergone reorientation, the secondary thickening occurs longitudinally and not transversely (15). This has been interpreted as supporting the dynamic helical model (38). Even though taxol (which has the dual action of driving more tubulin into MTs and of bundling MTs) inhibits the 90° reorientation, it does not prevent the local bunching of MTs.

In *Zinnia* suspension cells, the MTs can be randomly arranged in the minority isodiametric cells but they are transverse in the majority elongate cells (17). Secondary thickening occurs as webs in the former but as bands in the latter, corresponding to the pattern of MTs. Amiprophos-methyl (APM) depolymerizes microtubules and causes *Zinnia* cells to swell, so that when they recover, the balance of isodiametric to elongate cells is altered in favor of the former. In the isodiametric cells, the recovering MT array is random and upon differentiation microtubules bunch together to underlie web-like rather than band-like secondary thickenings. These studies differ from most microtubule/microfibril parallelism studies in that there is a sudden and dramatic, rather than a subtle, shift in wall patterning. They clearly demonstrate that cellulose deposition follows in the wake of a dynamic cytoplasmic template.

Similarly, MTs originally distributed evenly along the plasma membrane of *Cobaea* seed hairs (55) become grouped into bands of 10–18. These form a helix that underlies the helical cellulosic thickening of the seed hairs. One of the original proposals for the helical model was that MTs would be long, relative to the cell's circumference, and evidence was provided showing that the average length of MTs on negatively stained fragments of carrot protoplasts was 11 μm with some MTs double that length (34). Dry cleaving *Cobaea* hairs (55) shows that the MTs are between 20 and 30 μm long. It is still possible that such techniques select against a population of smaller, less stable MTs. But, together with immunofluorescence evidence that suboptimal buffering of root hairs causes fragmentation of interphase microtubules (41), there are grounds for believing that conventionally embedded material does not reflect the length of the longer microtubules. Cotton fibers contain spectacular microtubular helices (61). In developing fibers the cortical MTs wind transversely (flat-pitched helices) until, at about 15–18 days postanthesis, their pitch increases so that they are seen as "spirals" (45° helices). This change of pitch is coordinated over the entire length of the fiber and matches the Calcofluor staining (cellulose) pattern of the wall. In older hairs the cellulose microfibrils, in lamellae, regularly alternate their helical sign (gyre) at reversal points along the fiber. Seagull (61) has now shown that the MTs underlying these reversals undergo a complementary reversal—the MTs in one gyre flowing smoothly, via an arc, into the neighboring gyre.

Because cellulose may be deposited outside the plasma membrane, guided by a dynamic cytoplasmic template, it is important to discuss the behavior of cellulose-synthesizing complexes. Herth (29) proposed for chitin biosynthesis—and the argument also holds for cellulose—that mobile membrane synthases are driven along the plane of the plasma membrane by the crystallization of nascent polyglucan chains. The underlying membrane-associated microtubules are thought to provide lanes within which synthases move. For the *Zinnia* differentiation system it has been hypothesized (57) that the bunching of microtubules would also cause the synthases to be swept into corresponding bands. Schneider & Herth (59) confirm that membrane particle rosettes (presumptive synthases) are restricted to areas underlying wall thickenings in developing xylem elements of maize roots. This collection of reports provides one of the clearest images of how the movement and patterning of cortical microtubules can influence wall patterns via intramembranous synthases.

The linkages between microtubules and between microtubules and the plasma membrane almost certainly affect the stability of the interphase array. It has also been hypothesized (34, 35) that maximization of cross-bridging would tighten the helical array (producing "transverse" microtubules), whereas minimal interaction should favor other conformations. Unfortunately, little is known of these cross-bridges, but rapid freezing (32) spectacularly emphasizes their regularity and extensiveness.

Information on the stability and behavior of microtubules (and indirect information of their bridges) can be drawn from studies using gibberellic acid (GA). In dividing suspension cultures of *Vicia hajastana,* Simmonds and her colleagues (67, 68) have shown by immunofluorescence that whereas cortical microtubules are variably transverse when cells are grown in 2,4-D they are more strictly transverse when they elongate in the presence of gibberellic acid. These whole-cell images correspond to much of the foregoing EM evidence on the effect of GA on microtubules (see 34).

Mita & Shibaoka's study (44) provides clear evidence of the effects of gibberellic acid on both microtubule alignment and stability. The basal parts of young onion plants become bulbous under long day conditions. This is believed to involve the disorganization of microtubules, for the effect can be mimicked by the depolymerizing agents, colchicine and cremart. However, GA suppresses the swelling caused by these agents, and it also stabilizes microtubules to the disrupting effects of low temperature. In GA-treated cells almost all microtubules were running strictly transverse to the cell axis, whereas there was a greater range of angular dispersion in controls. Like GA, taxol stabilizes plant microtubules to the cold (46) and enhances lateral association of microtubules. Increased stability and parallelism of microtubules in GA-treated cells suggests increased lateral association between microtubules. If the dynamic helical model is to have any utility, then it must accommodate agents of physiological change, such as GA and ethylene, which would alter the orientation of microtubules within the integral interphase array by maximizing or minimizing the degree of interaction between microtubules.

ESTABLISHMENT OF THE CORTICAL ARRAY   Having discussed some of the influences on the behavior of the mature interphase array, it is now appropriate to consider the formation of the immature array following its nucleation. There are two separate hypotheses concerning the way in which the cortical array becomes organized (35): microtubules could reassociate with membrane linkage sites inherited from the previous array or they could renegotiate novel conformations according to prevailing conditions of the kind discussed above. The latter approach places more emphasis on the self-determining capacity of plant microtubules, for which there is some evidence.

Bajer & Mole-Bajer (1) note in *Haemanthus* endosperm cells that microtubules emerge from the nucleus and whorl around the cortex. In cytoplasts (enucleate cell fragments), small whorls form and their rapid formation is thought to be a function of the intrinsic properties of plant microtubules. Without a regular cellulosic wall, a membrane-associated cortical array is not seen, but cytoplasts nevertheless demonstrate the autonomous ability of microtubules to form circumferential arrays.

Hogetsu's (30) study on recovery of microtubule arrays in the alga *Closterium ehrenbergii* after chilling also throws light upon the process of microtubule organization. There are two cortical arrays in this alga: a ring of microtubules around the premitotic nucleus, and transverse, more evenly distributed tubules along the expanding semi-cell during interphase. Subjected to chilling in ice water, the wall microtubules first fragmented before becoming completely disorganized. Upon rewarming, wall MTs formed star-like arrays (within 5 min), then random arrangements, and after 30 min ordered, transverse arrays were seen. In the early stages of reformation the MT ring could be seen among the randomly arranged wall MTs. This observation suggests that the band may well have a cortical "template" for reformation but that wall MTs do not; they are initially nucleated and only later do they sort themselves out into organized transverse arrays. A further point of interest is that such transverse conformations only form in young, expanding semi-cells and not in the old, nonexpanding semi-cells in which the random arrangement gradually disappeared.

Other workers (69) have also pointed out that transverse MT arrays exist predominantly in young, expanding epidermal cells of *Vigna angularis*. Increasing the osmolality of the medium surrounding epidermal cells produces an ethylene-like effect in causing flat-pitched MT helices to unwind towards steeply pitched arrays that accompany and probably precede lateral expansion (56). It is possible that some factors that affect cell expansion might also affect the quality of linkage between microtubules, thereby influencing the direction of cell expansion.

Proteins Co-Localizing with the Cortical Array

Antibodies to the calcium-dependent regulator protein, calmodulin, do not appear to stain the cortical cytoskeleton even when any weak fluorescence should be magnified by viewing cells end-on (79).

When the cortex is exposed, *en face*, by dry cleaving (28, 71, 73), or is sectioned following freeze-substitution (32, 70) or is viewed in the high voltage electron microscope (28), fine filaments have been seen running along and between cortical microtubules. Their identity is unknown but there are two candidates: F-actin or some nonactomyosin filament. Fluorescent analogs of the F-actin-binding phallotoxins from *Amanita phalloides* have been used over the past few years to demonstrate the presence of an extensive network of actin cables in interphase plant cells. Phallacidin coupled to a fluorescent dye stains actin cables in *Chara* (3) and in conifer root parenchymatous cells (53). Other fluorescent phallotoxins have subsequently established the presence of this network in a range of cell types (6, 50–52).

Focusing at the cortex of double-stained onion root tip cells (6) revealed only microtubules labelled with fluorescein-conjugated antibodies. Rhodaminyl-lysine phallotoxin (RLP)-stained F-actin cables could only be seen by refocusing deeper within the cell where the cables run at 90° to the transverse, cortical MTs. It is possible, though, that single thin filaments may not be stabilized by fixation conditions allowing preservation of microtubules. In support of this, we (72a) have used two methods, which avoid aldehyde fixation, to demonstrate that extremely fine phalloidin-stained elements occur in the cell cortex in patterns reminiscent of MT helices. Detergent extraction and electrically induced permeabilization (electroporation) of carrot suspension cells, *Tradescantia* stamen hairs, and onion epidermis reveal the presence of aligned filaments at the cortex in the presence of RLP.

Indirect immunofluorescence using antiactin antibodies has not yet been as successful as phallotoxins, but Menzel & Schliwa (42, 43) have reported that a broadly cross-reactive monoclonal antibody originally raised against chicken gizzard actin does stain F-actin in *Bryopsis*. In the cortical cytoplasm of this green alga microtubules form wavy longitudinal bundles, and this staining pattern is largely superimposable upon that produced by the antiactin antibody. After treatment with the antimicrotubule agent amiprophos-methyl, microtubules reform in *Bryopsis* in large bundles that are organized in tandem with actin. Given these indications of microtubule/F-actin interaction, it is possible that the single filaments observed in the cortex of higher plant cells by various EM techniques are indeed actin.

Another possibility is that the cortical thin filaments are related to the third cytoskeletal system of animal cells—the intermediate filaments. Extracting carrot protoplasts with detergent revealed bundles of 7-nm fibrils that were associated with the nucleus but broke down into finer, single fibrils among the cortical microtubules (54). In several respects these fibrils were unlike actin; in particular, their gel migration characteristics and optical diffraction pattern were different. Further work (7) confirmed that the fibrillar bundles consisted of polypeptides with molecular weights between 50 and 68 kilodaltons. Using a monoclonal antibody to an epitope (intermediate filament antigen, IFA) detected in all classes of intermediate filament across a broad phylogenetic range, it has been shown (7) that four of the polypeptides cross-react with this monoclonal antibody. It is of particular interest that anti-IFA stained onion root tip cells; it stained all four microtubule arrays and stained the cortex in a transverse linear-punctate manner. This disjointed pattern is not consistent with the continuous pattern of staining produced by antitubulin antibodies. Indeed, by isolating plant tubulin it was shown that anti-IFA does not immunoblot plant tubulin; neither do antitubilins immunoblot the fibrillar bundles. It would seem therefore that an antigen, unrelated to tubulin but related to intermediate filaments, co-localizes with cortical microtubules. This antigen could represent the thin filament or it could take the form of a fine, more pervasive cortical meshwork. At the moment, we can only guess at the function of possible ancillary filaments.

Lim et al (32a) recently reported that a monoclonal antibody to troponin T recognized epitopes on the cortical microtubules of onion cells. These antigens were described as microtubule-associated proteins, but (see 72a) it is conceivable that they could be part of an actomyosin system co-distributing with plant microtubules.

THE PREPROPHASE BAND

In their report on the immunofluorescence microscopy of *Allium cepa* L. meristematic cells, Wick and her colleagues (80) described the staining of the preprophase band (PPB) of microtubules with antitubulin antibodies. In addition to the band MTs, fluorescence was associated with the nuclear envelope. A later paper (78) described microtubules that traversed nucleus and cortex. Linkage between the nucleus and the cortical PPB agrees with these authors' observations that the PPB segregates with the nucleus in fragmented cells.

In squashes of root tips there is a high proportion of cells containing a preprophase band, indicating that this is not a transient stage. Fluorescent dyes that stain chromatin indicate that the band is present in cells with chromatin in various stages of condensation: from "speckled" to "brain coral" to fully condensed chromosomes (C. W. Lloyd, personal observations). Indeed, the PPB and spindle arrays can coexist briefly (78). Broad as well as double bands seem to give way to tighter bands seen when chromatin is most condensed (78). This raises questions about the formation of the band and its duration during the cycle. According to the definition of prophase as preparation for mitosis, Roberts and his colleagues (57) recommend that "prophase" should replace the term "preprophase." Protoplasts of *Vicia hajastana* suspension cells contain circumferential bands of microtubules, and because most of them are associated with condensing chromatin Simmonds has also termed these "prophase bands" (65).

The fact that these bands are now seen in suspension cells that form disorganized callus (65, 67, 68) shows that they are not invariable markers of tissue organization. Bands are also found in nonembryogenic carrot suspensions (J. H. Doonan and C. W. Lloyd, unpublished). Recalling Hogetsu's observations (30) on the stable microtubule ring (cf PPB) in *Closterium,* it may well be that the band is a region of microtubule stability. In tissues the band may be influenced or stabilized by supracellular forces (e.g. ionic gradients, strains), but the fact that bands (albeit, sometimes broad) form in suspension cells demonstrates that the hypothetical stabilizing forces are at work within single cells and protoplasts.

Using the herbicide chloroisopropylphenylcarbamate (CIPC), it has been shown (5) that tripolar spindles are formed and then are succeeded by three-limbed phragmoplasts. All PPBs were planar and appeared normal and therefore could not have predicted the tortuous three-dimensional paths of centrifugal outgrowth shown by tri- and multi-limbed, CIPC-treated phragmoplasts. There is not an obligatory relationship between the former PPB site and the direction of outgrowth of the phragmoplast, although there is evidence (20, 23) that the cortical site does have some attractive function during the final stage of phragmoplast growth.

In filamentous protonemata of the moss *Physcomitrella patens* (12), immunofluorescence reveals that a phragmoplast of typical appearance develops within a cell whose plane of division is precisely controlled by light and gravity, yet no preprophase band occurs at this stage of development. Again, no imprinting by a PPB is necessary for cross-wall formation. However, cells of leafy side shoots that grow not by tip growth but by uniform cell expansion *do* contain PPBs. The interphase microtubules in tip-growing cells are longitudinal; they persist at the tip during division, are mainly endoplasmic, and have little connection with the cortex.

Cells of leafy shoots of this moss contain cortical, helical arrays that "disappear" during division, and we (13a) have suggested that the band might only form where there are cortical MTs. That is, membrane-associated interphase MTs "bunch up" to form bands much in the way that broad PPBs appear to "bunch up" to form tighter bands. Because at present there is no way to selectively tag plant MTs, it cannot be said whether the band forms exclusively from old MTs or whether it is entirely composed of nascent MTs or of a mixture of old and new microtubules. However, since presumptive MT nucleation sites were not detected within the PPB (4), and since broad bands appear to give way to tighter bands, even a totally reconstructed PPB would appear to involve some realignment at the cortex following nucleation at the nuclear surface.

Proteins Co-Localizing with the PPB

The concentration of MTs in a PPB represents an opportunity to magnify the signal from any MT-associated protein, and it would certainly be important if any such protein could be shown to mark the PPB zone until the phragmoplast grew out to that zone.

Calmodulin does not appear to be concentrated in the PPB (79) nor does myosin (49a). However, the PPB can be stained with phalloidin if aldehyde fixation is avoided (72a). Troponin T may also be a component of the band (32a). The intermediate filament antigen (IFA) is also concentrated in the PPB of onion root tip cells (7), and fuzzy radial spokes can be seen to pass between nucleus and cortex.

Clearly, the number of cytoskeletal antigens detected in the division zone is increasing.

THE MITOTIC SPINDLE

Mitosis in plants is likely to be similar in broad outline to the process occurring in other eukaryotes, and space does not permit a detailed review of this large topic.

Immunofluorescence studies have shown that presumptive MTNS (4) begin to group at opposing spindle poles at prophase and occur along the broad poles during metaphase.

Approaching division, the PPB MTs have depolymerized and nuclear-envelope–associated anti-tubulin fluorescence becomes more prominent (78, 80). During prophase, microtubules progress from pole to pole, forming a cage around the nucleus. After nuclear envelope breakdown, bundles of cone-shaped kinetochore microtubules abut the chromosomes at the metaphase plate. This picture has, however, been challenged (31). It is suggested that the prophase spindle is not a precursor for the metaphase spindle but that the former breaks down and the latter results from a new polymerization. However, the amount of background fluorescence, not seen in other anti-tubulin studies (5, 78), somewhat undermines the strength of this conclusion.

During anaphase, the half spindles move apart and the kinetochore bundles shorten (5, 78), and at the anaphase/telophase transition, phragmoplast microtubules appear in the mid-zone.

Proteins Co-Localizing with the Spindle

Calcium fluxes are thought to play a part in controlling mitosis. The calcium-binding regulatory protein may mediate these effects by sensitizing microtubules to the cation—hence the interest in localizing calmodulin during the different stages of mitosis. Two groups have localized this protein during mitosis in plants. Vantard and co-workers (75) find that in *Haemanthus* endosperm cells calmodulin can be detected at polar microtubule converging centers and with kinetochore bundle MTs. Wick and co-workers (79) also find calmodulin at the polar regions and co-localized with the spindle in pea and onion cells. In the former study (75), the calmodulin-staining pattern was similar to that for membrane-bound calcium (as detected by chlorotetracycline, CTC). This endorses an earlier study (82) in which cones of CTC fluorescence, coincident with kinetochore bundles in metaphase, disappeared at anaphase.

Fluorescent phallotoxin (6) did not stain the spindles of onion meristematic cells that had been fixed for 45 min with formaldehyde. However, Seagull et al (62), by fixing suspension cells in a phosphate rather than a PIPES-based buffer, have found that metaphase and anaphase spindles can be stained with rhodamine-conjugated phallacidin. They found that interphase networks of microfilaments disappear by the end of prophase. But by using detergent or electroporation to introduce labelled phalloidin into unfixed carrot cells we (72a) find that not only do spindles stain but a network of cytoplasmic filaments impinges upon the nucleus and is present throughout cell division.

Monoclonal antibodies to troponin T stain the mitotic spindle of onion root tip cells (32a). Anti-IFA also stains the spindle of onion root tip cells (7).

THE PHRAGMOPLAST

The phragmoplast is first seen as two opposing bundles of microtubules at the spindle equator at anaphase/telophase (5, 78). This double ring expands centrifugally—depositing the cell plate at its midline—until it meets the mother cell walls.

Staining with antiserum that recognizes MT nucleation sites (4) shows that these sites redistribute from opposing poles at prophase to those faces of the reconstituting nuclei that are proximal to the early phragmoplast. This agrees with the hook-decoration studies of Euteneuer and her colleagues (14). Exogenous tubulin can be added to microtubules under special conditions, producing C-shaped hooks that indicate the intrinsic polarity of the microtubules. Their study of the *Haemanthus* phragmoplast indicated that the growing ("plus") ends of the microtubules were the interdigitating ends of the two half-sets distal to the nucleus. Nongrowing "minus" ends are generally associated with nucleating sites, and so it would now seem that microtubules nucleated from opposing nuclear faces interdigitate at their plus ends.

Somewhat unusual but perhaps revealing phragmoplasts are formed in chromosome-free cytoplasts derived from *Haemanthus* endosperm (1). Where cones of cytoplasmic microtubules appose one another, they deposit a cell plate. Similarly, in some nucleated cells recovering from cold treatment, microtubules radiating from the nucleus have been observed to encircle the cell. When these bundles of peripheral microtubules split into clockwise and counter-clockwise groups, they form a miniature phragmoplast at the site where the two groups collide. This occurs in cells with interphase chromatin configurations and further illustrates the dynamic and self-reorganizing nature of plant microtubules.

Proteins Co-Localizing with the Phragmoplast

Apart from tubulin, other antigens have now been localized in the phragmoplast by fluorescence microscopy. Clayton & Lloyd (5) used rhodaminyl lysine phallotoxin (RLP) to demonstrate the occurrence of F-actin in the phragmoplast, and it was suggested that F-actin propelled to the midline vesicles that were involved in cell plate deposition. Rhodamine-labelled phalloidin has now also been used on *Tradescantia* stamen hair cells, with results confirming that F-actin probably exists in the phragmoplast (26).

Phallacidin-stained fibers have been seen radiating from the outer edge of a phragmoplast in suspension cells (62). However, a more extensive phalloidin-labelled network exists in the cytoplasm of carrot cells undergoing cytokinesis (72a). This network occurs in the cortex but also impinges upon the phragmoplast as it did upon the mitotic spindle.

Antitroponin T stains the onion cell's phragmoplast (32a) as does a monoclonal antibody to myosin heavy chain (49a).

Antibodies to calmodulin indicate that it, too, is a component of the phragmoplast (79).

The monoclonal antibody to a "universal" intermediate filament antigen also stains the phragmoplast in onion meristematic cells (7).

THE RELATIONSHIP BETWEEN THE PPB ZONE AND THE PHRAGMOPLAST

Although the PPB may not be obligatory it is, nevertheless, generally predictive of the division plane. There is great interest, therefore, in identifying molecules that could imprint the cortical zone before mitosis in order to explain the attractive force of the zone for the outgrowing phragmoplast.

The presence of microtubules, alone, could explain the imprinting if they somehow change the zone. Packard & Stack (49) proposed that the band caused a local cellulosic thickening of the wall (cf xylem). Galatis and his colleagues (21) confirm that this occurs during guard mother cell formation in some Leguminosae species in which the cell plate accurately meets the middle of the thickening. Even so, this does not explain why the phragmoplast should expand centrifugally out to that zone.

The problem is well illustrated by large vacuolate cells in which the nucleus is suspended across the vacuole by a thin raft of cytoplasm—the phragmosome. In *Nautilocalyx* epidermal cells (76) a cortical band of microtubules marks the zone at which the phragmosome unites with the cortex at pre-prophase. The nucleus originally resides in the cytoplasm along a side wall but migrates into the phragmosome where it divides and along which the phragmoplast develops. The phragmosome is a coalescence of transvacuolar cytoplasmic strands and, although the PPB disappears, it is almost certainly true that the skeletal elements anchoring the nucleus within the phragmosomal plane must persist throughout mitosis.

The only elements so far detected within phragmosomes by EM are micro-tubules and uncharacterized bundles of "microfilaments" (2, 22). Even if antigens that mark the perimeter of the division plane until cytokinesis are eventually detected, the nature of phragmosomal division should encourage the search for the complementary radial elements that, in principle, seem necessary to bridge nucleus and cortex. The network of cytoplasmic RLP-staining elements recently detected in carrot suspension cells (72a) remains in the cortex throughout the cell cycle while part of the network impinges upon both spindle and phragmoplast. These early indications suggest that F-actin could represent the radial part of the division cytoskeleton, but further work is necessary to confirm this.

TIP-GROWING CELLS

In tip growth, precursors are directed to a confined growing point in contrast to intercalary growth where materials are inserted more generally along the cell surface. Consistent with this pattern, membrane particle rosettes (pre-sumptive cellulose-synthesizing particles) are concentrated at the tip of *Funaria* filaments—each rosette presumed to be inserted by a Golgi vesicle (see 60). In the related moss *Physcomitrella patens* (12), it has been demon-strated by immunofluorescence that more-or-less axial microtubules course through the cytoplasm; they are not extensively associated with the plasma membrane but focus at the cell's tip. Unlike meristematic cells of higher plants, these interphase MTs do not depolymerize during cell division.

Cortical microtubules have now been detected within the apical dome of onion (33), radish, *Nigella,* and Brussels sprout (41), and they appear to be sensitive to the nature of the osmoticum. In these studies microtubules were shown to form 45° helices in onion and *Nigella* root hairs and to end at a region at the tip of the apical dome. In radish and Brussels sprout root hairs, the cortical MTs were net axial. In their study of root hairs, Traas and his co-workers (73) found that *Urtica* hairs contained helical arrays, whereas hairs from other plants contained net axial MTs. In pollen tubes, too, the cortical microtubules occur in net axial or S-helical conformations (10). In tip growth, therefore, it seems that microtubules are generally parallel rather than normal to the direction of cell extension.

In *Vicia hirsuta* root hairs, the cortical MTs are also net axial, except in this case a second set of endoplasmic MTs can be seen (C. W. Lloyd, K. J. Pearce, D. Rawlings, R. W. Ridge, and P. J. Shaw, unpublished). These progress from the nucleus in bundles of 2–10, fan out within the apical dome, and appear to be continuous with the cortical microtubules upon the dome. If the nucleus is the site of MT nucleation, then this would help explain the often-recorded observation that the nucleus migrates in register with the tip. Axial bundles of F-actin also enter the apical dome of *Vicia hirsuta* root hairs as has been established by Parthasarthy's group (51, 52) for other root hairs and pollen tubes.

ORGANELLE MOVEMENT

Latex beads (as model organelles) coated with myosin migrate along *Nitella* actin filaments if calcium and the regulatory light chains of myosin are present (74). Similar experiments with actin filaments in *Chara* demonstrate that Mg·ATP is required for active movement (64). Consistent with the involvement of actomyosin in plant intracellular movement, Witztum & Parthasarathy (81) report that the clustering and banding of chloroplasts in *Egeria, Elodea,* and *Hydrilla* leaves is inhibited by cytochalasin B.

Giant cells of the marine alga *Bryopsis* contain in their cortical cytoplasm bundles of longitudinal MTs that extensively but not exclusively co-distribute with actin fibers (42, 43). Amiprophos-methyl (APM) and cytochalasin D both inhibited movement of chloroplasts parallel to the bundles: The former affected actin bundles as well as MTs, whereas the latter did not affect MTs. These authors have concluded that both actin filaments and MTs are required for chloroplast movement.

In protonemata of the moss *Physcomitrella patens* (13), the nucleus in subapical cells is connected to the apex-proximal cross wall by a bundle of microtubules. The nucleus migrates along this bundle (which becomes shorter and thicker) in preparation for the division at that apex-proximal site, which gives rise to a side branch. Depolymerization of those microtubules with the herbicide cremart prevents nuclear migration, but the herbicide's effect can be overcome by adding the microtubule-stabilizing agent taxol. This implies an obligatory requirement for MTs in nuclear migration.

In tip-growing root hairs (C. W. Lloyd, K. J. Pearce, D. Rawlings, R. W. Ridge, and P. J. Shaw, unpublished) the nucleus migrates in register with the apex. Both F-actin and MTs are associated with the nucleus, and endoplasmic MTs progress from the nucleus to the tip. Treatment with cremart (or APM or CIPC) inhibits tip growth and causes the nucleus to migrate toward the base.

Cytochalasin D treatment arrests tip growth and nuclear movement, but when cytochalasin is added to cremart-treated hairs the uncoupled basipetal migration of the nucleus is prevented. This suggests that the more-or-less constant nucleus-to-tip distance is measured by MTs but that under MT-depolymerizing conditions, F-actin propels the nucleus to the base.

This involvement of microtubules in organellar movement, and the interaction in some instances with actin filaments, holds out the hope that similar kinds of interplay will be found in the migration and positioning of the nucleus prior to mitosis.

PROSPECTS

Immunofluorescence microscopy is changing the way in which we think about plant cells. Plant cells contain two or more extensive, interacting, cytoskeletal networks that are clearly involved in dynamic aspects of cell behavior. There are probably few, if any, plant cells that are not suitable subjects for immunofluorescence, and so the limiting step to progress is likely to be in preparing antibodies to a wider range of antigens. Most work is performed with antibodies raised against nonplant proteins, depending upon antigenic conservation, but there are likely to be features of the plant cytoskeleton that are peculiar to plants and that will necessitate biochemical characterization. Genes for plant cytoskeletal proteins (tubulin, actin) are being sequenced and cloned. This is exciting, but will not obviate the need to know how major and accessory cytoskeletal proteins interact, how they are affected by other regulatory proteins and ions, and how plant growth substances influence the behavior of cytoskeletal arrays. Purification of plant microtubules by taxol (47) has been an important advance, although other methods will be required to show how plant microtubules de- and re-polymerise in vitro. However, now that plant tubulin biochemistry is underway it should not be long before antibodies are raised against microtubule-associated proteins.

As with most nonmuscle cells, the biochemical characterization of actomyosin is proceeding slowly in plants. Fluorescent phallotoxins are useful probes that are small enough to enter cells that have only been minimally pretreated. Further ultrastructural studies will require antibodies for the characterization of components of the actomyosin system, but there are grounds for suspecting that the system is especially sensitive to fixation. Microinjection of directly-labelled antibodies into living cells may be necessary to overcome this.

Research into the plant cytoskeleton will not only underpin our theoretical understanding of the cell cycle, but it is likely to have an impact upon all areas of applied biology in which dividing and expanding cells are used. There is a growing appreciation of how the cytoskeletal cycle impinges upon the regeneration of protoplasts into cells (27); how aberrant microtubule organization results in genetic abnormalities in protoplasts (66); how some herbicides affect microtubule-based functions (e.g. 48); and the way in which exogenous growth substances affect cell division and expansion in tissue culture cells (37, 67).

In a relatively short time we can expect to see papers emerging on the molecular biology of the plant cytoskeleton. But within the area of this review we can anticipate the isolation of proteins other than tubulin and actin as immunogens, and the characterization of systems that will allow a better appreciation of the three-dimensional complexity and the dynamic nature of the cytoskeleton.

Literature Cited

1. Bajer, A. S., Mole-Bajer, J. 1986. Reorganization of microtubules in endosperm cells and cell fragments of the higher plant *Haemanthus in vivo*. *J. Cell Biol.* 102:263–81
2. Bakhuizen, R., van Spronsen, P. C., Sluiman-den Hertog, F. A. J., Venverloo, C. J., Goosen-de Roo, L. 1985. Nuclear envelope-radiating microtubules in plant cells during interphase-mitosis transition. *Protoplasma* 128:43–51
3. Barak, L. S., Yocum, R. R., Nothnagel, E. A., Webb, W. B. 1980. Fluorescence staining of the actin cytoskeleton in living cells with 7-nitrobenz-2-oxa-1,3-diazole phallacidin. *Proc. Natl. Acad. Sci. USA* 77:980–84
4. Clayton, L., Black, C. M., Lloyd, C. W. 1985. Microtubule nucleating sites in higher plant cells identified by an auto-antibody against pericentriolar material. *J. Cell Biol.* 101:319–24
5. Clayton, L., Lloyd, C. W. 1984. The relationship between the division plane and spindle geometry in *Allium* cells treated with CIPC and griseofulvin: an anti-tubulin study. *Eur. J. Cell Biol.* 34:248–53
6. Clayton, L., Lloyd, C. W. 1984. Actin organization during the cell cycle in meristematic plant cells. Actin is present in the cytokinetic phragmoplast. *Exp. Cell Res.* 156:231–38
7. Dawson, P. J., Hulme, J. S., Lloyd, C. W. 1985. Monoclonal antibody to intermediate filament antigen cross-reacts with higher plant cells. *J. Cell Biol.* 100:1793–98
8. Dawson, P. J., Lloyd, C. W. 1987. A comparison of plant and animal tubulins. In *The Biochemistry of Plants: A Comprehensive Treatise*, Vol. 9: *Metabo-*

lism, ed. D. D. Davies. London/New York: Academic
9. de Mey, J., Lambert, A. M., Bajer, A. S., Moeremans, M., de Brabander, M. 1982. Visualization of microtubules in interphase and mitotic plant cells of *Haemanthus* endosperm with the immuno-gold staining (IGS) method. *Proc. Natl. Acad. Sci. USA* 79:1898–1902
10. Derksen, J., Pierson, E. S., Traas, J. A. 1985. Microtubules in vegetative and generative cells of pollen tubes. *Eur. J. Cell Biol.* 38:142–48
11. Dickinson, H. G., Sheldon, J. M. 1984. A radial system of microtubules extending between the nuclear envelope and the plasma membrane during early male haplophase in flowering plants. *Planta* 161:86–90
12. Doonan, J. H., Cove, D. J., Lloyd, C. W. 1985. Immunofluorescence microscopy of microtubules in intact lineages of the moss, *Physcomitrella patens*. I. Normal and CIPC-treated tip cells. *J. Cell Sci.* 75:13–147
13. Doonan, J. H., Jenkins, G. I., Cove, D. J., Lloyd, C. W. 1986. Microtubules connect the migrating nucleus to the prospective division site during side branch formation in the moss *Physcomitrella patens*. *Eur. J. Cell Biol.* 41:157–64
13a. Doonan, J. H., Cove, D. J., Corke, F. M. K., Lloyd, C. W. 1987. The preprophase band of microtubules absent from tip-growing moss filaments arises in leafy shoots during transition to intercalary growth. In *Cell Motility and the Cytoskeleton*. In press
14. Euteneuer, U., McIntosh, J. R. 1980. Polarity of midbody and phragmoplast microtubules. *J. Cell Biol.* 87:509–15
15. Falconer, M. M., Seagull, R. W. 1985.

Xylogenesis in tissue culture: Taxol effects on microtubule reorientation and lateral association in differentiating cells. *Protoplasma* 128:157–66

16. Falconer, M. M., Seagull, R. W. 1985. Immunofluorescent and calcofluor white staining of developing tracheary elements in *Zinnia elegans* L. suspension cultures. *Protoplasma* 125:190–98

17. Falconer, M. M., Seagull, R. W. 1986. Xylogenesis in tissue culture II: Microtubules, cell shape and secondary wall patterns. *Protoplasma* 133:140–48

18. Franke, W. W., Seib, E., Osborn, M., Weber, K., Herth, W., Falk, H. 1977. Tubulin-containing structures in the anastral mitotic apparatus of endosperm cells of the plant *Leucojum aestivum* as revealed by immunofluorescence microscopy. *Cytobiologie* 15:24–48

19. Galatis, B., Apostolakos, P., Katsaros, C. 1983. Microtubules and their organizing centres in differentiating guard cells of *Adiantum capillus veneris*. *Protoplasma* 115:176–92

20. Galatis, B., Apostolakos, P., Katsaros, C. 1984. Positional inconsistency between preprophase microtubule band and final cell plate arrangement during triangular subsidiary cell and atypical hair cell formation in two Triticum species. *Can. J. Bot.* 62:343–59

21. Galatis, B., Apostolakos, P., Katsaros, C., Loukari, A. 1982. Pre-prophase microtubule band and local wall thickening in guard mother cells of some Leguminosae. *Ann. Bot.* 59:779–91

22. Goosen-de Roo, L., Bakhuizen, R., van Spronsen, P., Libbenga, K. R. 1984. The presence of extended phragmosomes containing cytoskeletal elements in fusiform cambial cells of *Fraxinus excelsior* L. *Protoplasma* 122:145–52

23. Gunning, B. E. S. 1982. The cytokinetic apparatus: Its development and spatial regulation. In *The Cytoskeleton in Plant Growth and Development,* ed. C. W. Lloyd, pp. 229–92. London: Academic

24. Gunning, B. E. S., Hardham, A. R. 1982. Microtubules. *Ann. Rev. Plant Physiol.* 33:651–98

25. Gunning, B. E. S., Hardham, A. R., Hughes, J. E. 1978. Evidence for initiation of microtubules in discrete regions of the cell cortex in *Azolla* root-tip cells, and an hypothesis on the development of cortical arrays of microtubules. *Planta* 143:161–79

26. Gunning, B. E. S., Wick, S. M. 1985. Preprophase bands, phragmoplasts, and spatial control of cytokinesis. *J. Cell Sci. Suppl.* 2:157–79

27. Hahne, G., Hoffman, F. 1984. Di-

methylsulfoxide can initiate cell divisions of arrested callus protoplasts by promoting microtubule assembly. *Proc. Natl. Acad. Sci. USA* 81:5449–53

28. Hawes, C. 1985. Conventional and high voltage electron microscopy of the cytoskeleton and cytoplasmic matrix of carrot (*Daucus carota* L.) cells grown in suspension culture. *Eur. J. Cell Biol.* 38:201–10

29. Herth, W. 1980. Calcofluor White and Congo Red inhibit chitin microfibril assembly of *Poterioochromonas:* Evidence for a gap between polymerization and microfibril formation. *J. Cell Biol.* 87:442–50

30. Hogetsu, T. 1986. Re-formation of microtubules in *Closterium ehrenbergii* Meneghini after cold-induced depolymerization. *Planta* 167:437–43

31. Kubiak, J., de Brabander, M., de Mey, J., Tarkowska, J. A. 1986. Origin of the mitotic spindle in onion root cells. *Protoplasma* 130:51–56

32. Lancelle, S. A., Callaham, D. A., Hepler, P. K. 1986. A method for rapid freeze fixation of plant cells. *Protoplasma* 131:153–65

32a. Lim, S-S., Hering, G. E., Borisy, G. G. 1986. Widespread occurrence of antitroponin T crossreactive components in non-muscle cells. *J. Cell Sci.* 85:1–19

33. Lloyd, C. W. 1983. Helical microtubular arrays in onion root hairs. *Nature* 305:311–13

34. Lloyd, C. W. 1984. Toward a dynamic helical model for the influence of microtubules on wall patterns in plants. *Int. Rev. Cytol.* 86:1–51

35. Lloyd, C. W. 1986. Microtubules and the cellular morphogenesis of plants. In *Developmental Order: A Comprehensive Treatise,* Vol. 2: *The Cellular Basis of Morphogenesis,* ed. L. W. Browder, pp. 31–57. New York/London: Plenum

36. Lloyd, C. W., Clayton, L., Dawson, P. J., Doonan, J. H., Hulme, J. S., Roberts, I. N., Wells, B. 1985. The cytoskeleton underlying cross walls and side walls in plants; molecules and macromolecular assemblies. *J. Cell Sci. Suppl.* 2:143–55

37. Lloyd, C. W., Lowe, S. B., Peace, G. W. 1980. The mode of action of 2,4-D in counteracting the elongation of carrot cells grown in culture. *J. Cell Sci.* 45:257–68

38. Lloyd, C. W., Seagull, R. W. 1985. A new spring for plant cell biology: microtubules as dynamic helices. *Trends Biochem. Sci.* 10:476–78

39. Lloyd, C. W., Slabas, A. R., Powell, A. J., Lowe, S. B. 1980. Microtubules,

138   LLOYD

protoplasts and plant cell shape. An immunofluorescent study. *Planta* 147:500–
6
40. Lloyd, C. W., Slabas, A. R., Powell,
A. J., MacDonald, G., Badley, R. A.
1979. Cytoplasmic microtubules of
higher plant cells visualised with anti-
tubulin antibodies. *Nature* 279:239–41
41. Lloyd, C. W., Wells, B. 1985. Microtubules are at the tips of root hairs and
form helical patterns corresponding to
inner wall fibrils. *J. Cell Sci.* 75:225–38
42. Menzel, D., Schliwa, M. 1986. Motility
in the siphonous green alga Bryopsis. I.
Spatial organization of the cytoskeleton
and organelle movements. *Eur. J. Cell
Biol.* 40:275–85
43. Menzel, D., Schliwa, M. 1986. Motility
in the siphonous green alga Bryopsis. II.
Chloroplast movement requires organized arrays of both microtubules and
actin filaments. *Eur. J. Cell Biol.* 40:
286–95
44. Mita, T., Shibaoka, H. 1984. Gibberellin stabilizes microtubules in onion
leaf sheath cells. *Protoplasma* 119:100–
9
45. Mitchison, T. J., Kirshner, M. 1984.
Dynamic instability of microtubule
growth. *Nature* 312:237–42
46. Mole-Bajer, J., Bajer, A. S. 1983. Action of taxol on mitosis-modification of
microtubule arrangements and function
of the mitotic spindle in Haemanthus endosperm. *J. Cell Biol.* 96:527–40
47. Morejohn, L., Fosket, D. E. 1982.
Higher plant tubulin identified by self-
assembly into microtubules *in vitro*. *Nature* 297:426–28
48. Morejohn, L. C., Fosket, D. E. 1984.
Inhibition of plant microtubule polymerization *in vitro* by the phosphoric amide
herbicide amiprophosmethyl. *Science*
224:874–76
49. Packard, M. J., Stack, S. M. 1976. The
preprophase band: Possible involvement
in the formation of the cell wall. *J. Cell
Sci.* 22:403–11
49a. Parke, J., Miller, C., Amderton, B. H.
1986. Higher plant myosin heavy-chain
identified using a monoclonal antibody.
Eur. J. Cell Biol. 41:9–13
50. Parthasarathy, M. V. 1985. F-actin
architecture in coleoptile epidermal
cells. *Eur. J. Cell Biol.* 39:1–12
51. Parthasarathy, M. V., Perdue, T. D.,
Witztum, A., Albernaz, J. 1985. Actin
network as a normal component of the
cytoskeleton in many vascular plant
cells. *Am. J. Bot.* 72:1318–23
52. Perdue, T. D., Parthasarathy, M. V.
1985. *In situ* localization of F-actin in
pollen tubes. *Eur. J. Cell Biol.* 39:13–20

53. Pesacreta, T. C., Carley, W. W., Webb,
W. W., Parthasarathy, M. V. 1982. F-
actin in conifer roots. *Eur. J. Cell Biol.*
39:21–26
54. Powell, A. J., Peace, G. W., Slabas, A.
R., Lloyd, C. W. 1982. The detergent-
resistant cytoskeleton of higher plant
protoplasts contains nucleus-associated
fibrillar bundles in addition to microtubules. *J. Cell Sci.* 56:319–35
55. Quader, H., Deichgraber, G., Schnepf,
E. 1986. The cytoskeleton in Cobaea
seed hairs: patterning during cell wall
differentiation. *Planta* 168:1–10
56. Roberts, I. N., Lloyd, C. W., Roberts,
K. 1985. Ethylene-induced microtubule
reorientations: mediation by helical
arrays. *Planta* 164:439–47
57. Roberts, K., Burgess, J., Roberts, I. N.,
Linstead, P. 1985. Microtubule rearrangements during plant growth and development: An immunofluorescence
study. In *Botanical Microscopy, 1985,*
ed. A. W. Robards, pp. 105–27. New
York: Oxford Univ. Press
58. Schmitt, A.-C., Vantard, M., de Mey,
J., Lambert, A.-M. 1983. Aster-like
microtubule centres establish spindle
polarity during interphase-mitosis transition in higher plant cells. *Plant Cell Rep.*
2:285–88
59. Schneider, B., Herth, W. 1986. Distribution of plasma membrane rosettes
and kinetics of cellulose formation in
xylem development of higher plants.
Protoplasma 131:142–52
60. Schnepf, E., Witte, O., Rudolph, U.,
Deichgraber, G., Reiss, H.-D. 1985.
Tip cell growth and the frequency distribution of particle rosettes in the plasma membrane: experimental studies in
Funaria protonema cells. *Protoplasma*
127:222–29
61. Seagull, R. W. 1986. Changes in microtubule organization and wall microfibril
orientation during in vitro cotton fiber
development: an immunofluorescent
study. *Can. J. Bot.* 64:1373–81
62. Seagull, R. W., Falconer, M. M.,
Weerdenburg, C. A. 1986. Microfilaments: Dynamic arrays in higher plant
cells. *J. Cell Biol.* In press
63. Sheldon, J. M., Dickinson, H. G. 1986.
Pollen wall formation in *Lilium:* The
effect of chaotropic agents, and the
organization of the microtubular cytoskeleton during pattern development.
Planta 168:11–23
64. Shimmen, T., Yano, M. 1984. Active
sliding movement of latex beads coated
with skeletal muscle myosin on *Chara*
actin bundles. *Protoplasma* 121:132–37
65. Simmonds, D. H. 1986. Prophase bands

of microtubules occur in protoplast cultures of *Vicia hajastana* Grossh. *Planta* 167:469–72

66. Simmonds, D. H., Setterfield, G. 1986. Aberrant microtubule organization can result in genetic abnormalities in protoplast cultures of *Vicia hajastana* Grossh. *Planta* 176:460–68

67. Simmonds, D., Setterfield, G., Brown, D. L. 1983. Organization of microtubules in dividing and elongating cells of *Vicia hajastana* Grossh. in suspension culture. *Eur. J. Cell Biol.* 32:59–66

68. Simmonds, D., Setterfield, G., Tanchak, M., Brown, D. L. 1982. Microtubule organization in cultured plant cells. *Proc. 5th Int. Congr. Plant Tissue Cell Culture, Tokyo,* pp. 31–34

69. Takeda, K., Shibaoka, H. 1981. Changes in microfibril arrangement on the inner surface of the epidermal cell walls in the epicotyl of *Vigna angularis* Ohwi et Ohashi during cell growth. *Planta* 151:385–92

70. Tiwari, S. C., Wick, S. M., Williamson, R. E., Gunning, B. E. S. 1984. Cytoskeleton and integration of cellular function in cells of higher plants. *J. Cell Biol.* 99 (1):S63–S69

71. Traas, J. 1984. Visualization of the membrane bound cytoskeleton and coated pits of plant cells by means of dry cleaving. *Protoplasma* 119:212–18

72. Traas, J. A., Braat, P., Derksen, J. 1984. Changes in microtubule arrays during the differentiation of cortical root cells of *Raphanus sativus. Eur. J. Cell Biol.* 34:229–38

72a. Traas, J. A., Doonan, J. H., Rawlins, D., Shaw, P. J., Watts, J. W., Lloyd, C. W. 1986. An actin network is present in the cytoplasm throughout the division cycle of carrot cells: Actin co-distributes with the four microtubule arrays. Submitted for publication

73. Traas, J. A., Braat, P., Emons, A. M. C., Meekes, H., Derksen, J. 1985. Microtubules in root hairs. *J. Cell Sci.* 76:303–20

74. Vale, R. D., Szent-Gyorgyi, A. G.,

Sheetz, M. P. 1984. Movement of scallop myosin on *Nitella* actin filaments: Regulation by calcium. *Proc. Natl. Acad. Sci. USA* 81:6775–78

75. Vantard, M., Lambert, A.-M., de Mey, J., Picquot, P., van Eldik, L. J. 1985. Characterization and immunocytochemical distribution of calmodulin in higher plant endosperm cells: Localization in the mitotic apparatus. *J. Cell Biol.* 101:488–99

76. Venverloo, C. J., Hovenkamp, P. H., Weeda, A. J., Libbenga, A. K. 1980. Cell division in Nautilocalyx explants. I. Phragmosome, preprophase band and plane of division. *Z. Pflanzenphysiol.* 100:161–74

77. Wick, S. M. 1985. Immunofluorescence microscopy of tubulin and microtubule arrays in plant cells. III. Transition between mitotic/cytokinetic and interphase microtubule arrays. *Cell Biol. Int. Reps.* 9:357–71

78. Wick, S. M., Duniec, J. 1984. Immunofluorescence microscopy of tubulin and microtubule arrays in plant cells. II. Transition between the preprophase band and the mitotic spindle. *Protoplasma* 122:45–55

79. Wick, S. M., Muto, S., Duniec, J. 1985. Double immuno-fluorescence labelling of calmodulin and tubulin in dividing plant cells. *Protoplasma* 126: 198–206

80. Wick, S. M., Seagull, R. W., Osborn, M., Weber, K., Gunning, B. E. S. 1981. Immunofluorescence microscopy of organized microtubule arrays in structurally stabilized meristematic plant cells. *J. Cell Biol.* 89:605–90

81. Witztum, A., Parthasarathy, M. V. 1985. Role of actin in chloroplast clustering and bending in leaves of *Egeria, Elodea* and *Hydrilla. Eur. J. Cell Biol.* 39:21–26

82. Wolniak, S. M., Hepler, P. K., Jackson, W. T. 1980. Detection of the membrane-calcium distribution during mitosis in *Haemanthus* endosperm with chlorotetracycline. *J. Cell Biol.* 87:23–32

Ann. Rev. Plant Physiol. 1987. 38:141–53

GENETICS OF WHEAT STORAGE PROTEINS AND THE EFFECT OF ALLELIC VARIATION ON BREAD-MAKING QUALITY

Peter I. Payne

Plant Breeding Institute, Maris Lane, Trumpington, Cambridge CB2 2LQ, United Kingdom

CONTENTS

Introduction 141
Gliadin and Glutenin 142
HMW Subunit Genes (Glu-1) 143
Genes for LMW Subunits, ω- and γ-Gliadins (Gli-1) 146
Genes for α- and β-Gliadins (Gli-2) 148
Minor Gene Loci 149
Allelic Variation 150
Exploitation of Variation in Plant Breeding 150

Introduction

In most years the world's largest crop for human consumption is wheat, followed closely by maize and rice (27). The protein content of wheat grains, which usually varies from 9 to 15% of the dry weight, is small when compared with leguminous seeds (1). Nevertheless, more wheat protein is eaten by man than protein from any other source (27). The largest tissue in the grain is the endosperm, and it contains the majority of the protein, mostly in the form of prolamins (43). Unfortunately, the amino-acid composition of the prolamin proteins is not ideally suited for human nutrition because these proteins contain large amounts of glutamine and proline and correspondingly inadequate levels of the essential amino acids, particularly lysine (1). How-

141

ever, the prolamins play a fundamental role in the processing of wheat into its many food products (53). When flour is mixed with water to form a dough, it becomes both extensible and elastic—properties conferred primarily by its protein (53). Furthermore, the balance between elasticity and extensibility, though genetically controlled, can differ greatly between varieties and has a major influence on the type of food product that can be prepared. Thus, in North America and Western Europe, wheat varieties that produce strongly elastic doughs with some extensibility are used to make bread, whereas those giving highly extensible doughs are used primarily to make biscuits (cookies). In other parts of the world, wheat with properties intermediate between these two extremes is used to make chapati, the staple of much of Pakistan and India, and noodles in China and Japan.

Future breeding of varieties that are optimized for processing into specific foods will be greatly aided by an understanding of the genetic basis for varietal differences in viscoelasticity. In this chapter, I review our current knowledge of the classical genetics of the two major prolamin groups, gliadin and glutenin, and incorporate relevant results from recent developments in molecular biology.

Gliadin and Glutenin

Traditionally, two storage protein groups have been recognized in the endosperm: gliadin and glutenin (53). They are synthesized on the endoplasmic reticulum in the developing endosperm and are deposited in protein bodies (43). Originally distinguished by their relative solubilities in aqueous alcohols (53), they are now more usually defined by their molecular size in dissociating solvents (14, 28). Gliadin molecules are small, $\sim$ 35,000 mol wt, and they do not have a disulfide-bonded subunit structure (53). When fractionated by gel electrophoresis at low pH, they separate into four groups, α, β, γ, and ω (53). Two-dimensional gel electrophoresis suggests that a single variety contains about 45 gliadin components (54). Glutenins are large, heterogeneous molecules built up from some 19 different subunits that are connected by disulfide bonds (53). The glutenin subunits fall into two unequal groups, the predominant low-molecular-weight (LMW) subunits (16) and the high-molecular-weight (HMW) subunits (31).

The last few years has seen rapid advances in our knowledge of wheat proteins, mainly through the application of various techniques in molecular biology. It is now clear that glutenin and gliadin are two groups of proteins that belong to the same storage-protein family, the prolamins, as opposed to the globulin storage proteins found principally in leguminous seeds (43). The biochemistry and molecular biology of wheat prolamins have been described in an extensive review (43), and research in this area will only be covered here when it impinges directly on the classical genetics of glutenin subunit and gliadin genes.

Bread wheat is a hexaploid species, containing three different but related genomes of 7 chromosome pairs named arbitrarily A, B, and D. The cytogenetics of wheat chromosomes has been studied extensively for several decades, and each of the 21 chromosome pairs has been given a chromosome number (1 to 7) followed by the genome assignment (A, B, or D) (21). Spontaneous, partial, or whole chromosome deletions are common in wheat, and because the genetic information is effectively triplicated they are not usually lethal. These chromosome mutants have been exploited by cytogeneticists in the development of various kinds of genetic stocks (41). Those most commonly used to determine the chromosome location of prolamin genes have been the nullisomic (N), tetrasomic (T) lines of variety Chinese Spring (CS) (41a). Thus CS N1AT1D does not possess any 1A chromosomes but has twice the normal dosage of 1D chromosomes. The absence of storage proteins in this line compared to euploid CS indicates that their controlling genes occur on chromosome 1A. The chromosome arm location of these genes can then be determined by analyzing ditelocentric (DT) lines, CS DT 1AL (short arms of 1A missing), and CS DT 1AS (long arms missing). The results from this type of analysis have been reviewed (30, 32) and they show the following.

1. Genes controlling the HMW subunits of glutenin occur on the long arms of chromosomes 1A, 1B, and 1D.
2. Genes controlling the LMW subunits of glutenin, ω-gliadins, and γ-gliadins occur on the short arms of the same set of chromosomes.
3. Genes for the α- and β-gliadins occur on the short arms of chromosomes 6A, 6B, and 6D.

Analysis of varieties and another type of genetic stock, intervarietal chromosome substitution lines, have shown that wheat prolamins are extremely polymorphic, owing to the mutation of their genes into many allelic forms (9, 18, 31, 49).

These approaches do not usually allow one to distinguish between the location of structural and regulatory genes, hence the use of the term controlling genes. However, recent work employing Southern blotting (11, 52), in which storage-protein cDNA sequences have been hybridized to restriction enzyme digests of wheat DNA from different Chinese Spring NT lines, has shown unambiguously that "structural gene" can replace "controlling gene" in the above conclusions.

HMW Subunit Genes (Glu-1)

The position of these gene loci on the long arms of the group 1 chromosomes has been determined by both recombination and physical mapping. Recombination (R) between the genes and their respective centromeres is similar for all three chromosomes and very low (R = 9%) (35). The loci that contain the HMW glutenin subunit genes are collectively called *Glu-1* and individually *Glu-A1, Glu-B1,* and *Glu-D1* for chromosomes 1A, 1B, and 1D respectively (35). Recently, two chromosome mutants have been detected, one lacking half of the long arm of chromosome 1B and the other lacking a similar segment from 1D (P. I. Payne, J. Seekings, and L. M. Holt, unpublished results). The former lacks *Glu-B1* and the latter *Glu-D1,* indicating that physically the *Glu-1* loci occur on the distal half of the chromosome arm (Figure 1). There are several other examples of discrepancies between physical and recombination mapping, and they are thought to be caused by the presence of large segments of chromatin on either side of the centromere that either recombine infrequently or not at all.

Varieties of wheat each contain between 3 and 5 major HMW glutenin subunits, 2 coded by genes at *Glu-D1,* 1 or 2 by *Glu-B1,* and 1 or none by *Glu-A1* (22, 31). On the basis of electrophoretic mobility and isoelectric

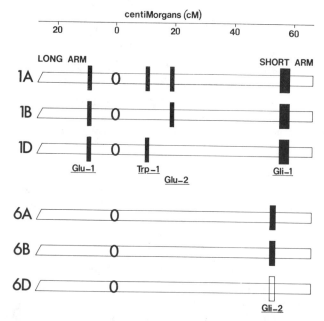

Figure 1   Chromosomal position of the genes coding for storage proteins of wheat endosperm (6, 17, 23, 35, 40, 45). The distance between *Glu-1* and *Gli-1* has been normalized to 66 cM (35). Because recombination frequency is affected by environmental factors and by the genetic relatedness of the parents of crosses, the values given should not be regarded as absolute. However, the relative distances between loci are much more reliable. *Glu-2* refers to the locus on chromosome 1B. Another locus, described as *Gld-B6* (10), occurs close to *Glu-2* on chromosome 1B or is identical with it. The locus at the equivalent position on 1A is called *Gld-2-1A* (48). The relative thicknesses of the vertical bars reflect approximately the number of genes at each locus (11, 12). The position of *Gli-2* on chromosome 6D, depicted by an open rectangle, has not been studied. Genes at Trp-1 code for triplet band subunits.

focusing, the subunits coded at each locus were subdivided into "x" and "y" types, with the single 1A-encoded subunit being of the x type (31). The molecular basis for the subdivision was later strengthened by the finding that 1Ax, 1Bx, and 1Dx subunits gave electrophoretic fingerprint patterns after partial chymotrypsin and complete V8 protease digestions that were more similar to each other than to those given by 1By and 1Dy subunits (10a, 34). Another difference is that the x type subunits have only about half the cysteine content of y type subunits.

More recently, the complete nucleotide sequences of three genes coding for HMW glutenin subunits have been published, and all show close homologies (8, 50, 51). By comparing them with the published, N-terminal amino acid sequences of HMW glutenin subunits (42), one of the genes was shown to correspond to subunit 12, a 1Dy subunit (51), and another (50) was interpreted by Harberd et al. (12) to correspond to subunit 2, a 1Dx subunit. The third gene, derived from chromosome 1A, contains a stop codon in the middle of the coding sequence (8) and so probably does not become translated into a HMW subunit. Its nucleotide sequence is more closely related to the 1Dy than to the 1Dx subunit sequence and so probably specifies a 1Ay subunit. 1Ay subunits are not synthesized in any variety of bread wheat (31), but they are found in a restricted number of diploid relatives that contain A-type genomes (J. G. Waines and P. I. Payne, unpublished data). Such a finding of a nonfunctional gene (pseudogene) was not unexpected, since earlier work had demonstrated that Chinese Spring contains two gene classes on chromosome 1A that code for HMW glutenin subunits (52), although neither are synthesized in the endosperm (31).

The *Glu-1* loci contain few genes. Quantitative Southern hybridization experiments involving cDNA molecules that specify HMW glutenin subunit sequences and restriction endonuclease digestion fragments of wheat DNA suggest that there are between 2 and 6 genes per locus (8, 52). More detailed analyses using three different restriction endonucleases and four different varieties strongly suggest only two genes per *Glu-1* locus, i.e. single genes for each x and each y subunit (12). Such a hypothesis would explain most easily the spontaneous and random deletions of single HMW glutenin subunits that occur in populations of wheat landraces gathered worldwide from ancient agriculture (30). A spontaneous reduction in the intensity of a single band, as might happen if one of a two or three gene family was inactivated by a change in the coding sequence, was not detected in this study. By contrast a genotype from Israel, TAA 36, was shown to overproduce a 1Bx subunit, relative to its other HMW subunits of glutenin (9). This may have arisen through duplication of the 1Bx genes or through a mutation in the regulatory region of the gene, causing it to be transcribed more efficiently. The additional, minor HMW glutenin subunits detected by one- and two-dimensional electrophoresis, such as the 1Bz subunits (13) and the *Glt-B3* subunits (9), are most likely to be posttranslationally modified forms of major subunits, in this case the 1By subunits. Similarly the minor *Glt-A2* subunits (9) are likely to be derived from the major 1Ax subunits, 1 and 2*, and are not the products of separate genes. The two genes at each *Glu-1* locus are tightly linked (9, 22, 31), although two recombinants were detected in the analysis of 1883 progeny (34), giving a recombination frequency of approximately 0.11%.

The accumulated data on the *Glu-1* loci indicate that long before the formation of bread-wheat, the ancestral HMW glutenin subunit gene duplicated and probably became physically separated. No further gene amplification occurred, but the two genes diversified into many allelic forms to code for the present-day x and y subunits by mechanisms such as unequal crossing over and point mutations.

Genes for LMW Subunits, ω- and γ-Gliadins (Gli-1)

It has been known for many years that the genes coding for ω- and γ-gliadins are tightly clustered at a single, major locus on each of the short arms of the group 1 chromosomes (24, 32, 49). These loci were initially described as *Gld-1A*, *Gld-1B*, and *Gld-1D* (49) but were changed to *Gli-A1*, *Gli-B1*, and *Gli-D1* (35) to conform with the Recommended International Rules for Gene Symbols in Wheat (3). Because recombination is extremely rare among genes at each locus, much more so than the recombination between x and y genes at *Glu-1*, the concept of gliadin heredity blocks was popularized by Sozinov and colleagues (49). To the author's knowledge, there have been no publications that have convincingly demonstrated a recombination between ω- and γ-gliadins from the many thousands of segregating progeny that have been analyzed by electrophoresis. More recent studies have shown that a third family of genes occurs at each of the *Gli-1* loci that code for LMW glutenin subunits (16, 36).

The *Gli-1* loci are either very weakly or not significantly linked to *Glu-1* loci on the three group 1 chromosomes (R = 30–50%) (23, 35). Because recombination frequency between *Glu-1* (35) and the centromere is small, it follows that recombination between the centromere and *Gli-1* on the centromere is great. This was shown directly by telocentric mapping of *Gli-B1* (R = 42%) (40). The *Gli-B1* locus has also been shown to be physically located towards the end of the short arm of chromosome 1B. A chromosome mutant that has completely lost the satellite of this chromosome, which normally accounts for the terminal third of the short arm, was shown to lack *Gli-B1* (29). *Gli-B1* probably lies close to the distal end of the satellite because recombination between the ribosomal RNA genes *(Nor-B1)* at the nucleolar organizing region and *Gli-B1* is appreciable (R = 25%) (47). The results to date suggest that the *Gli-1* loci occur at the same relative positions on the three homoeologous chromosomes, indicating that no major rearrangements have taken place on the group 1 chromosomes in the recent evolution of wheat from its diploid progenitors.

During the last few years, the biochemistry and molecular biology of the three protein groups coded at *Gli-1* have been studied in great detail. An endosperm-specific cDNA clone, pTag 544, was shown to specifically hybridize to DNA fragments derived from the group 1 chromosomes (4). Hybrid-release translation of proteins in vitro and more recent sequencing studies on the clone have shown that its DNA corresponds to part of a LMW glutenin subunit gene (V. Colot and D. Bartels, unpublished results). Quantitative Southern hybridizations to restriction enzyme digested shoot DNA indicate that these genes are present in 10–15 copies per haploid genome (11), that is some 3–5 genes per *Gli-1* locus. Clones of γ-gliadins were also identified, and quantitative hybridization results indicated about the same number of genes per locus (11). Clones of ω-gliadin gene sequences have not been studied so far, but if it is assumed that their genes occur in similar numbers to those coding for γ-gliadins and LMW glutenin subunits, then each *Gli-1* locus will contain about 9–15 genes. This is greater than the number of major proteins that are expressed from each of these loci. For instance, Chinese Spring contains eight major proteins coded by *Gli-B1:* three ω-gliadins, three γ-gliadins, and two LMW glutenin subunits (36). It is not known whether the minor protein components coded at this locus are the products of different genes and are expressed (or conserved) in small amounts or whether they are simply posttranslational modifications of major proteins. It is likely, though, from corresponding studies of *Glu-1* genes that there are one or more pseudogenes at each *Gli-1* locus. If it is assumed that the above estimate of gene number per *Gli-1* locus is reliable and that minor proteins are not coded by individual genes, then it is probable that some of the dominant proteins, and the γ-gliadins in particular, are the products of two or more different genes. Different, allelic γ-gliadins can be expressed in quite different amounts (compare GLD 1A 6 and GLD 1A 2, Figure 2 of Ref. 49), but the same γ-gliadin is always present in the same relative amounts in different varieties. One explanation of this is that different alleles contain different numbers of active genes.

The primary structures of the three sets of proteins coded at *Gli-1* are currently being studied intensively by N-terminal amino-acid sequencing and by nucleotide sequencing of cDNA and genomic clones (4, 26). The work has already demonstrated that the proteins share sufficient sequence homology to show that their encoding genes have evolved from a common ancestor. It is also likely that the γ-gliadins and the LMW glutenin subunits are more similar to each other than either is to the ω-gliadins. The genes at the *Gli-1* loci probably occur as three family groups, although some intermingling of genes may have occurred by rare recombinations during more recent evolution. The order of these gene families on the group 1 chromosomes is not known.

Genes for α- and β-Gliadins (Gli-2)

Amino acid sequencing of α- and β-gliadins and nucleotide sequencing of their genes have shown that they have closely related primary structures and belong to a common protein group (2, 39), in spite of their distinctive electrophoretic properties at low pH (53). The *Gli-2* loci therefore probably only contain one gene family, as opposed to the three gene families at each *Gli-1* locus. *Gli-B2* occurs on the satellite segment of the short arm of chromosome 6B, about 22 recombination units from the ribosomal RNA genes at the nucleolus organizer (6). By telocentric mapping, recombination between *Gli-A2* and the centromere was 35% (L. M. Holt and P. I. Payne, unpublished data). Both *Gli-A2* and *Gli-B2* probably occur at similar positions on their respective chromosomes, towards their distal ends. There are no recombination data available for *Gli-D2*, but a similar, relative position is anticipated. Like *Gli-1* but unlike *Glu-1*, occasional recombination within a *Gli-2* locus has not been reported.

Southern blot hybridizations using α-/β-gliadin gene sequences as a probe have given conflicting values for gene family sizes at *Gli-2*. One estimate was 9–12 genes (11), and the other more than 33 genes (26). Because the former authors used similar hybridization procedures to obtain realistic values for the number of genes at *Glu-1* (52), an estimate of gene number in low double figures seems the most likely. The gliadin proteins of Chinese Spring have been fractionated by at least three different, two-dimensional electrophoretic procedures (20, 32, 54), and the number of principal components was shown to be 5, 5, and 10 on chromosome 6A; 6, 6, and 5 on chromosome 6B; and 5, 5, and 6 on chromosome 6D. Each major protein is thus probably coded by one or two genes. It is also probable that each *Gli-2* locus contains one or more pseudogenes.

Nucleotide sequencing has shown close homologies between genes at the *Gli-1* and *Gli-2* loci, for instance between γ- and α-/β-gliadins (26), and there can be no doubt that they arose from the duplication and divergence of a common ancestral gene. The most obvious hypothesis to account for the spatial separation of *Gli-1* and *Gli-2* is an ancient interchromosomal translocation that occurred in an ancestral line prior to the evolution of the A-, B-, and D-genome diploid species (41b). But which chromosome, 1 or 6, carried the ancestral storage-protein genes? A comparative study of species in the genera *Triticum, Aegilops, Secale,* and *Hordeum* convincingly shows that it is chromosome 1. All *Triticum* and *Aegilops* species that were analyzed contain prolamin genes on homoeologous chromosomes 1 and 6 only, as does *Secale montanum,* a wild rye (44). Cultivated rye, *S. cereale,* on the other hand, contains the genes on homoeologous chromosomes 1 and 2 (44). *Hordeum* species are different again, with prolamin genes occurring only on chromosome 1H in cultivated barley (5) and on 1 and 7 in a wild relative, *H. chilense* (33). It is likely that prolamin genes on homoeologous chromosomes 2 and 7 occurred through translocations.

Minor Gene Loci

Convincing evidence that prolamins are coded at additional, minor loci was presented by Sobko (48) and by Galili & Feldman (10). In the former study, a minor component whose electrophoretic mobility was within the range of ω-gliadins was shown to be coded at a separate locus *Gld-2-1A*, on the short arm of chromosome 1A, 31 recombination units from *Gli-A1*. Because *Gli-A1* is close to the end of the chromosome, *Gld-2-1A* was placed proximal to this locus. Galili & Feldman (9) analyzed 109 varieties by SDS-PAGE and showed that variety Thatcher contained a rare but distinctive protein of similar mobility to the ω-gliadins. Its controlling genes were shown to occur on the short arm of chromosome 1B at *Gld-B6*, which lies in between *Glu-B1* and *Gli-B1* and 26 recombination units from the latter (10). These results suggest that *Gld-2-1A* and *Gld-B6* may be homoeoallelic. More recently, the genes coding for the D group of LMW glutenin subunits have been mapped on chromosomes 1B and 1D (17). The genes on 1B were also shown to be intermediate between *Glu-B1* and *Gli-B1*, and the locus was appropriately called *Glu-B2*. Because this chromosome map position was similar to that of *Gld-B6*, crosses are being made to see if the two loci are identical or separate.

When the genes on chromosome 1D that code for the D group of LMW glutenin subunits were mapped, they were shown to be unbreakably linked to *Gli-D1* and were not found at a separate locus (38). The different map positions of these genes is therefore the first case, in wheat, of apparently homoeologous storage proteins being coded by genes at different positions on their respective chromosomes. It was suggested that the ancestral location of the genes coding for the D group of LMW glutenin subunits was *Gli-1* but that a chromosome rearrangement of the ancestral 1B (and probably 1A) chromosomes had caused them to become separated and positioned closer to the centromere.

Recently discovered is another group of proteins, the so-called triplet bands that are controlled by genes on chromosomes 1A and 1D (46). Judging from their biochemical properties, it is unlikely that the triplet bands are prolamins; yet they are located in protein bodies like storage proteins. It has been suggested that they may belong to the globulin storage proteins, which are found more commonly in leguminous seeds. Their genes occur close to the centromeres on the short arms of 1A and 1D, about 40.1 and 36.5 recombination units from *Gli-A1* and *Gli-D1*, respectively (45).

Allelic Variation

Although there are only nine major gene loci in bread wheat that contain prolamin genes, they exhibit extensive allelic variation. Thus, in a sample of 300 wheat varieties, 3 alleles of *Glu-A1* were identified, 11 of *Glu-B1*, and 5 of *Glu-D1* (37). Similarly after analyzing an undisclosed number of varieties, 15, 18, and 8 alleles were described for the three *Gli-1* loci and 13, 11, and 10 for the three *Gli-2* loci (25). The total number of permutations of prolamin storage proteins for a genotype is thus astronomical, about 500 million. Even this is an underestimate, for bread-wheat landraces, grown for centuries in the primitive agriculture of many parts of the world, contain several novel prolamins (30). Many of the prolamin variants are easily distinguished by aluminium lactate-PAGE at pH 3.1; this technique is routinely used in many wheat-growing countries of the world to identify varieties and to estimate the relative proportions of varieties in grain mixtures (7, 19).

The molecular basis for allelic variation in prolamin genes has been studied in greatest detail for the HMW glutenin subunits (12). They have molecular weights of $\sim$ 70,000, and they contain three principal domains (8, 50, 51): short, hydrophilic stretches at their C- and N-termini and a long hydrophobic region composed of repeating sequences. Restriction enzyme digestions of DNA from various intervarietal chromosome substitution lines show a positive correlation between electrophoretic mobilities of the intact subunits by SDS-PAGE and the length of the central repetitive domain. Allelic variation in this set of proteins is thus primarily the result of differences in molecular weight, and these have no doubt arisen in evolution by such events as unequal crossing over. However, two-dimensional electrophoresis of gliadins from different varieties (20, 32, 54) suggests that these proteins can vary in both molecular size and charge, rather than molecular size alone.

Exploitation of Variation in Plant Breeding

Apart from varietal identification, variation in prolamin composition is beginning to be exploited by wheat breeders in the development of new varieties with improved bread-making qualities. In most wheat-growing areas of the world, glutenin elasticity is the major limiting factor in the conversion of varietal flours into bread. Glutenin can range from being highly extensible, like the gliadins, to being highly elastic, and this property is primarily genetically controlled. The reason for this variation at the molecular level is not clear. It is known for glutenins that molecular weight is directly related to elasticity (15), and it has been hypothesized that these two variables are governed by the specific composition of subunits that make up the glutenin of a specific variety (30). Analyses of random lines from many crosses have shown that the different alleles at the *Glu-1* and *Gli-1* loci have contrasting associations between dough elasticity and good bread-making quality (4a, 30,

49). Furthermore, the effects of combining "good-quality" alleles coded at different loci in the same genotype has at least partially additive effects upon bread-making quality (30). Thus, a new, improved strategy for wheat breeders is to cross varieties that have complementary, good-quality prolamins and to screen embryoless half grains at later generations by electrophoresis. A few of the progeny will have the best combination of prolamins possible for the cross, and their bread-making qualities will be better than those of their parents. The remaining embryo-parts of these selected grains can then be grown and evaluated for other characters in later generations.

ACKNOWLEDGMENT

I am grateful to my colleague, Dr. R. D. Thompson, for helpful discussions.

Literature Cited

1. Altschul, A. M. 1965. *Proteins: Their Chemistry and Politics.* London: Chapman & Hall. 337 pp.
2. Anderson, O. D., Litts, J. C., Gautier, M.-F., Greene, F. C. 1984. Nucleic acid sequence and chromosome assignment of a wheat storage protein gene. *Nucleic Acids Res.* 12:8129–44
3. Anonymous. 1968. Recommendations on gene symbols and stock preservation for wheat and its relatives. *Proc. 3rd Int. Wheat Genet. Symp.*, pp. 466–67. London: Butterworths
4. Bartels, D., Thompson, R. D. 1983. The characterisation of cDNA clones coding for wheat storage proteins. *Nucleic Acids Res.* 11:2961–77
4a. Branlard, G. 1983. Correlation between gliadin bands. *Theor. Appl. Genet.* 64:163–68
5. Doll, H., Brown, A. H. D. 1979. Hordein variation in wild *(Hordeum spontaneum)* and cultivated barley. *Can. J. Genet. Cytol.* 21:391–404
6. Dvořák, J., Chen, K.-C. 1983. Distribution of nonstructural variation between wheat cultivars along chromosome arm 6Bp: evidence from the linkage map and physical map of the arm. *Genetics* 106:325–33
7. Ellis, J. W. S. 1984. The cereal grain trade in the United Kingdom: the problem of cereal variety. *Philos. Trans. R. Soc. London Ser. B* 304:395–407
8. Forde, J., Malpica, J.-M., Halford, N. G., Shewry, P. R., Anderson, O. D., et al. 1985. The nucleotide sequence of a HMW glutenin subunit gene located on chromosome 1A of wheat *(Triticum aestivum). Nucleic Acids Res.* 13:6817–32
9. Galili, G., Feldman, M. 1983. Genetic control of endosperm proteins in wheat. 2. Variation in high molecular weight glutenin and gliadin subunits of *Triticum aestivum. Theor. Appl. Genet.* 66:77–86
10. Galili, G., Feldman, M. 1984. Mapping of glutenin and gliadin genes located on chromosome 1B of common wheat. *Mol. Gen. Genet.* 193:293–98
10a. Galili, G., Feldman, M. 1985. Structural homology of endosperm high molecular weight glutenin subunits of common wheat *(Triticum aestivum). Theor. Appl. Genet.* 70:634–42
11. Harberd, N. P., Bartels, D., Thompson, R. D. 1985. Analysis of the gliadin multigene loci in bread using nullisomic-tetrasomic lines. *Mol. Gen. Genet.* 198:234–42
12. Harberd, N. P., Bartels, D., Thompson, R. D. 1986. DNA restriction fragment variation in the gene family encoding HMW glutenin subunits of wheat. *Biochem. Genet.* 24:579–96
13. Holt, L. M., Astin, R., Payne, P. I. 1981. Structural and genetical studies on the high-molecular-weight subunits of wheat glutenin. Part 2. Relative isoelectric points determined by two-dimensional fractionation in polyacrylamide gels. *Theor. Appl. Genet.* 60:237–43
14. Huebner, F. R. 1970. Comparative studies on glutenins from different classes of wheat. *J. Agric. Food Chem.* 18:256–59
15. Huebner, F. R., Wall, J. S. 1976. Fractionation and quantitative differences of glutenin from wheat varieties varying in baking quality. *Cereal Chem.* 53:258–69
16. Jackson, E. A., Holt, L. M., Payne, P. I. 1983. Characterization of high

molecular–weight gliadin and low-molecular-weight glutenin subunits of wheat endosperm by two-dimensional electrophoresis and the chromosomal localization of their controlling genes. *Theor. Appl. Genet.* 66:29–37

17. Jackson, E. A., Holt, L. M., Payne, P. I. 1985. *Glu-B2,* a storage protein locus controlling the D-group of LMW glutenin subunits in bread wheat *(Triticum aestivum). Genet. Res.* 46:11–17

18. Kasarda, D. D., Bernadin, J. E., Qualset, C. O. 1976. Relationships of gliadin protein components to chromosomes in hexaploid wheats *(Triticum aestivum* L). *Proc. Natl. Acad. Sci. USA* 73:3646–50

19. Khan, K., McDonald, C. E., Banasik, O. J. 1983. Polyacrylamide gel electrophoresis of gliadin proteins for wheat variety identification—procedural modifications and observations. *Cereal Chem.* 60:178–81

20. Lafiandra, D., Kasarda, D. D., Morris, R. 1984. Chromosomal assignment of genes coding for the wheat gliadin protein components of the cultivars "Cheyenne" and "Chinese Spring" by two-dimensional (two-pH) electrophoresis. *Theor. Appl. Genet.* 68:531–39

21. Law, C. N. 1981. Chromosome manipulation in wheat. In *Chromosomes Today,* ed. M. D. Bennett, M. Bobrow, G. M. Hewitt, 7:194–205. London: Allen & Unwin

22. Lawrence, G. J., Shepherd, K. W. 1980. Variation in glutenin protein subunits of wheat. *Aust. J. Biol. Sci.* 33:221–33

23. Lawrence, G. J., Shepherd, K. W. 1981. Inheritance of glutenin protein subunits of wheat. *Theor. Appl. Genet.* 60:333–37

24. Mecham, D. K., Kasarda, D. D., Qualset, C. O. 1978. Genetic aspects of wheat gliadin proteins. *Biochem. Genet.* 16:831–53

25. Metakovsky, E. V., Novoselskaya, A. Yu., Kopus, M. M., Sobko, T. A., Sozinov, A. A. 1984. Blocks of gliadin components in winter wheat detected by one-dimensional polyacrylamide gel electrophoresis. *Theor. Appl. Genet.* 67:559–68

26. Okita, T. W., Cheesbrough, V., Reeves, C. D. 1985. Evolution and heterogeneity of the α/β-type sequences and γ-type gliadin DNA sequences. *J. Biol. Chem.* 260:8203–13

27. Payne, P. I. 1983. Breeding for protein quantity and protein quality in seed crops. In *Seed Proteins,* ed. J. Daussant, J. Mosse, J. Vaughan, pp. 223–53. Lon-

don/New York: Academic. 335 pp.

28. Payne, P. I., Corfield, K. G. 1979. Subunit composition of wheat glutenin proteins, isolated by gel filtration in a dissociating medium. *Planta* 145:83–88

29. Payne, P. I., Holt, L. M., Hutchingson, J., Bennett, M. D. 1984. Development and characterisation of a line of bread wheat, *Triticum aestivum,* which lacks the short-arm satellite of chromosome 1B and the *Gli-B1* locus. *Theor. Appl. Genet.* 68:327–34

30. Payne, P. I., Holt, L. M., Jackson, E. A., Law, C. N. 1984. Wheat storage proteins: their genetics and their potential for manipulation by plant breeding. *Philos. Trans. R. Soc. London Ser. B* 304:359–71

31. Payne, P. I., Holt, L. M., Law, C. N. 1981. Structural and genetical studies on the high-molecular-weight subunits of wheat glutenin. Part I: Alleic variation in subunits amongst varieties of wheat. *Theor. Appl. Genet.* 60:229–36

32. Payne, P. I., Holt, L. M., Lawrence, G. J., Law, C. N. 1982. The genetics of gliadin and glutenin, the major storage proteins of the wheat endosperm. *Qual. Plant. Plant Foods Hum. Nutr.* 31:229–41

33. Payne, P. I., Holt, L. M., Reader, S. M., Miller, T. E. 1986. Chromosomal location of genes coding for endosperm proteins of *Hordeum chilense,* determined by two-dimensional electrophoresis of wheat—*H. chilense* chromosome addition lines. *Biochem. Genet.* In press

34. Payne, P. I., Holt, L. M., Thompson, R. D., Bartels, D., Harberd, N. P., et al. 1983. *Proc. 6th Int. Wheat Genet. Symp.,* pp. 827–34. Kyoto: Univ. Press

35. Payne, P. I., Holt, L. M., Worland, A. J., Law, C. N. 1982. Structural and genetical studies on the high-molecular-weight subunits of wheat glutenin. Part 3. Telocentric mapping of the subunit genes on the long arms of the homoeologous group 1 chromosomes. *Theor. Appl. Genet.* 63:129–38

36. Payne, P. I., Jackson, E. A., Holt, L. M., Law, C. N. 1984. Genetic linkage between endosperm storage protein genes on each of the short arms of chromosomes 1A and 1B in wheat. *Theor. Appl. Genet.* 67:235–43

37. Payne, P. I., Lawrence, G. J. 1983. Catalogue of alleles for the complex gene loci, *Glu-A1, Glu-B1, Glu-D1* which code for high-molecular-weight subunits of glutenin in hexaploid wheat. *Cereal Res. Commun.* 11:29–35

38. Payne, P. I., Roberts, M. S., Holt, L. M. 1986. Location of genes controlling

the D group of LMW glutenin subunits on chromosome 1D of bread wheat. *Genet. Res.* 47:175–79

39. Rafalski, J. A., Scheets, K., Metzler, M., Peterson, D. M., Hedgcoth, C., Soll, D. G. 1984. Developmentally regulated plant genes: the nucleotide sequence of a wheat gliadin genomic clone. *EMBO J.* 3:1409–1415

40. Rybalka, A. I., Sozinov, A. A. 1979. Mapping of the locus of *Gld-1B,* which controls the biosynthesis of reserve proteins in soft wheats. *Tsitol. Genet.* 13:276–82

41. Sears, E. R. 1954. The aneuploids of common wheat. *Univ. Mo. Agric. Exp. Stn., Res. Bull. 572.* Columbia. 59 pp.

41a. Sears, E. R. 1966. Nullisomic-tetrasomic combinations in hexaploid wheat. In *Chromosome Manipulation and Plant Genetics,* ed. R. Riley, K. R. Lewis, pp. 29–45. Edinburgh: Oliver & Boyd

41b. Shepherd, K. W., Jennings, A. C. 1970. Genetic control of rye endosperm proteins. *Experientia* 27:98–99

42. Shewry, P. R., Field, J. M., Faulks, A. J., Parmar, S., Miflin, B. J., et al. 1984. The purification and N-terminal amino-acid sequence analysis of the high molecular weight glutenin polypeptides of wheat. *Biochim. Biophys. Acta* 788:23–34

43. Shewry, P. R., Miflin, B. J. 1985. Seed storage proteins of economically important cereals. *Adv. Cereal Sci. Technol.* 7:1–84

44. Shewry, P. R., Parmar, S., Miller, T. E. 1985. Chromosomal location of the structural genes for the M_r 75,000 γ-secalins in *Secale montanum* Guss: evidence for a translocation involving chromosomes 2R and 6R in cultivated rye *(Secale cereale). Heredity* 54:381–83

45. Singh, N. K., Shepherd, K. W. 1984. A new approach to study the variation and genetic control of disulphide-linked endosperm proteins in wheat and rye.

Proc. 2nd Int. Workshop Wheat Gluten Proteins, Wageningen, pp. 129–36

46. Singh, N. K., Shepherd, K. W. 1985. The structure and genetic control of a new class of disulphide-linked proteins in wheat endosperm. *Theor. Appl. Genet.* 71:79–92

47. Snape, J. W., Flavell, R. B., O'Dell, M., Hughes, W. G., Payne, P. I. 1985. Intrachromosomal mapping of the nucleolar organiser region relative to three marker loci on chromosome 1B of wheat *(Triticum aestivum). Theor. Appl. Genet.* 69:263–70

48. Sobko, T. I. 1984. Identification of a new locus which controls the synthesis of alcohol-soluble endosperm proteins in soft winter wheat. *J. Agric. Sci. Kiev* 7:78–80

49. Sozinov, A. A., Poperelya, F. A. 1982. Polymorphism of prolamins and variability of grain quality. *Qual. Plant. Plant Foods Hum. Nutr.* 31:243–49

50. Sugiyama, T., Rafalski, A., Peterson, D., Soll, D. 1985. A wheat HMW glutenin subunit gene reveals a highly repeated structure. *Nucleic Acids Res.* 13:8729–37

51. Thompson, R. D., Bartels, D., Harberd, N. P. 1985. Nucleotide sequence of a gene from chromosome 1D of wheat encoding a HMW-glutenin subunit. *Nucleic Acids Res.* 13:6833–46

52. Thompson, R. D., Bartels, D., Harberd, N. P., Flavell, R. B. 1983. Characterisation of the multigene family coding for HMW glutenin subunits in wheat using cDNA clones. *Theor. Appl. Genet.* 67:87–96

53. Wall, J. S. 1979. The role of wheat proteins in determining baking quality. In *Recent Advances in the Biochemistry of Cereals,* ed. D. L. Laidman, R. G. Wyn Jones, pp. 275–311. London/New York: Academic

54. Wrigley, C. W., Shepherd, K. W. 1973. Electrofocusing of grain proteins from wheat genotypes. *Ann. NY Acad. Sci.* 209:154–62

FRUIT RIPENING

C. J. Brady

Plant Physiology Unit, CSIRO Division of Food Research and School of Biological
Science, Macquarie University, North Ryde, New South Wales, Australia 2109

CONTENTS

1. INTRODUCTION 155
2. RIPENING IS DEVELOPMENTALLY REGULATED............................. 156
   2.1 Ripening as Senescence 156
   2.2 Control by Growth Regulators.. 157
   2.3 Gene Expression in Ripening Fruit 160
   2.4 Ripening Mutants.. 163
3. MEMBRANES AND ORGANIZATIONAL RESISTANCE........................... 164
4. OXIDATIVE REACTIONS 166
   4.1 The Respiratory Climacteric.. 166
   4.2 The Climacteric and Protein Synthesis....................................... 166
   4.3 Other Oxidative Reactions.. 169
5. THE SOFTENING OF FRUITS... 169
   5.1 Chemistry and Enzymology of Softening..................................... 169
   5.2 A Role for Calcium 170
   5.3 Softening as a Regulator of Ripening 171
6. CONCLUDING REMARKS 171

1. INTRODUCTION

Fruit ripening has been reviewed a number of times in this series. The
physiology of ripening was treated briefly in Coombe's review on fruit
development (39), and the subject was reviewed thoroughly by Sacher (128)
in 1973. Yang & Hoffman's 1984 paper (158) on the synthesis and action of
ethylene is pertinent, and they discussed several aspects of the physiology and
biochemistry of ripening. A number of reviews dealing generally with ripening or with the biochemistry and/or molecular biology of particular fruits have
appeared during the last few years (51, 52, 55, 58, 59, 64, 74, 88, 93, 121,
135).

0066-4294/87/0601-0155$02.00

Studies of the mechanisms that regulate ripening can be traced to the 1920s when the dominant theory emphasized "organizational resistance." This theory, developed in detail by Blackman & Parija in 1928 (18), suggested that the ripening events, which were perceived as largely catabolic in nature, were the consequence of a breakdown in the resistances that kept cellular compartments contained. This theory held sway, in a largely unmodified form, for 40 years or more (128). As evidence accumulated that protein and perhaps RNA synthesis played vital roles in the induction of ripening, an alternative theory emerged: ripening was presented in positive terms as a process of tissue differentiation. The techniques of modern molecular biologists have enabled this theory to be examined in a precise way, and evidence for the direct genomic control of the ripening of climacteric fruits is now accumulating rapidly. Nonetheless, there is evidence that cellular compartments are modified through ripening, and recent evidence suggests that lipid oxidation and/or phase changes within membranes contribute to changing metabolite distribution within cells as ripening proceeds. How novel transcriptional and/or translational events and changes in metabolite partitioning interact and contribute interdependently or independently to ripening is a matter of conjecture. It may well be that both theories are more or less correct and both mechanisms are involved but to differing extents in different fruits. Fruit ripening appears to be a well-regulated, genetically determined event, but coming as it does at the end of ontogeny in a fleshy, energy-rich tissue that is destined for dispersal and disposal, it would be surprising if the mechanisms of control were the same for all species. The reviewer naturally seeks generalities and uniformity of pattern. The reader should be warned that the emphasis given to a few conclusions based on a few species that today may seem to point to firm guiding principles may tomorrow seem tangential or particular as more species and more aspects are examined.

2. RIPENING IS DEVELOPMENTALLY REGULATED

2.1 Ripening as Senescence

The distinction between ripening and senescence has never been finely drawn. Watada et al (154) have defined ripening as changes that "occur from the latter stages of growth and development through the early stages of senescence and result in characteristic aesthetic and/or food quality." They define senescence as processes that follow physiological or horticultural maturity and "lead to death of tissue." These are sensible, pragmatic definitions, although it is possible to challenge the concept that each of the processes that contribute to the senescence syndrome leads directly to tissue death. The dismantling of the chloroplast photosystem apparatus is a prominent feature of the senescence of leaves and many fruits, but it is not lethal. It is perhaps more accurate to suggest that the senescence processes increase the probability of death, for example, by dehydration or microbial invasion, for there is little evidence that senescence includes programmed death (21). There is no doubt that the intrusion of the ripening syndrome into tissue development hastens the onset of senescence and therefore the probability of injury and death. This is clearly seen in the comparison of the nonripening tomato mutants and normal tomatoes (127). The nonripening mutants, which are nonclimacteric, eventually senesce, lose chloroplast components and cell wall hemicelluloses (60), and become more susceptible to disease. The normal fruit undergo these same changes much earlier as part of ripening. It follows that senescence and ripening have common features; however, ripening includes processes that are not part of the senescence syndrome. Pigment accumulation and the cell wall changes that result in the softening of fruit are common ripening changes that are not usually involved in the senescence processes.

In the climacteric fruits, ripening is characterized by ethylene production, which is apparently autocatalytic (35). McMurchie et al (95) introduced the concept of two systems for ethylene production: system 1 is the ethylene-production system common to climacteric and nonclimacteric fruits and operating in climacteric fruits until ripening commences. In these mature fruits, as also in some flowers, exposure to ethylene induces a large increase in the ethylene-forming system (68, 95), and this is seen as the induction of "system 2" ethylene. Solomos & Laties (136) suggested that the autocatalytic induction may be consequential to changes in "organizational resistance" resulting in the lifting of previously imposed controls. More recent evidence (158) indicates substantial increases in the activities of the regulatory enzymes; it has not yet been shown definitively that the increases in activities result from the synthesis of more enzyme, but cycloheximide does inhibit the response (83). The induction of system 2 ethylene results in massive increases in ethylene production by the tissues, and ripening and then senescence follow. Climacteric fruits are now distinguished as those in which system 2 ethylene operates. The concept of two systems for the control of ethylene synthesis has been a useful one and should be better defined. In time, we will know if separate genes contribute to the two systems or if two levels of regulation apply.

2.2 Control by Growth Regulators

There is indisputable evidence that ethylene is involved in the induction of ripening in fruits (158). A main concern of the commercial postharvest horticultural industry is to limit the exposure of harvested fruit to ethylene. An aspect of fruit ripening that attracts molecular biologists is the prospect that ethylene induces a mature, nongrowing tissue to rapidly differentiate into a new state—to switch from nonripening to ripening. Because the tissue is already mature and the form and structure of cells decided, it is thought that the molecular steps involved when the "switch" is thrown may be relatively few; the system therefore has appeal as one in which to study the interaction between the signalling chemical, ethylene, and the genome (58, 59).

A complication is that the tissues are exposed to low levels of ethylene (produced by system 1 ethylene) throughout development. Exposure to larger concentrations of ethylene produces, in fruit tissues, a response that varies between species and with fruit maturity (15). In the more sensitive species, for example cantaloupe (92) or banana (26), ripening is immediately induced, but the more immature the fruit, the higher the concentration of ethylene that is required. In the less sensitive species, for example tomatoes or apples, ethylene treatments reduce the time before ripening occurs (91, 119). Since ripening is not immediately induced in tomatoes by ethylene treatments, system 2 ethylene synthesis is not activated. Ethylene synthesis is normally limited by the supply of the immediate precursor, 1-aminocyclopropane-1-carboxylic acid (ACC). During ripening, as system 2 ethylene is activated, ACC synthase activity and the content of ACC rise sharply (32, 68). Immature tomatoes exposed to exogenous ethylene show no increase in ACC synthase activity, but they do show a difference in their regulation: the ethylene-forming enzyme (EFE) is induced when ACC synthase is not. In cucumbers, EFE is constant during development and ripening, while ACC synthase and system 2 ethylene synthesis increase with ripening (139). Maturing apples gradually develop the capacity to produce EFE in response to applied ethylene (33).

A role for ACC synthase in regulating ethylene synthesis and ripening is demonstrated by the inhibition of the induction of ripening when L-2-amino-L-alkoxy-trans-3-butenoic acid (aminovinylglycine, AVG) is administered to preclimacteric pears (106, 126) or apples (32, 36). The inhibition of the induction of ripening by AVG is overcome by subsequent exposures to ethylene. In D'anjou pears, which require low temperature storage prior to normal ripening, Blankenship & Richardson (19) found that EFE increased during cold pretreatment. In this fruit, as in tomatoes, system 2 EFE develops prior to system 2 ACC synthase. In nectarines (28), EFE normally increases during ripening, but has, at all maturities, a capacity in excess of that of ACC synthase. In this fruit exogenous ethylene decreases EFE activity.

Bufler & Bangerth (7, 34) applied ethylene or propylene to apples stored under hypobaric conditions. Fruits stored hypobarically for short periods retained the capacity to generate system 2 ACC synthase; those stored for longer periods lost this capacity. Apparently neither immature tomatoes nor immature or overmature apples are able to produce system 2 ethylene or ripen in response to applied ethylene. Little is known of the internal factors that regulate the response of fruit to ethylene; possibilities include the concentration of other growth regulators and the integration of earlier responses to growth regulators. Thus the response to ethylene may depend upon the set of genes that are transcribed or have been transcribed when the exposure to ethylene occurs.

Some fruits, e.g. avocados (14), do not ripen while attached to the tree and gradually increase their sensitivity to ethylene with time after harvest. This has raised the concept of a ripening inhibitor or juvenility factor derived from the tree. In avocados, as with immature apples or tomatoes, high concentrations of ethylene somehow reduce the resistance of the fruit to ethylene and reduce the time before system 2 ethylene operates. Whether one sees this as a titration of the resistance of the tissue to ethylene or as the need for a slow ethylene-provoked developmental program is currently a matter of choice.

There have been many attempts to define roles for the other well-known plant growth regulators in the regulation of ripening or as resistance factors. This subject was thoroughly reviewed by McGlasson et al (93), and more recent studies have provided no general clarification. Abscisic acid commonly rises late in development or during ripening (77, 159), and treatment with abscisic acid may advance ripening (76, 86, 98, 151, 159). Abscisic acid frequently promotes the senescence of green tissues (155), and abscisic acid increases in osmotically or salt-stressed tissues (156). Salt treatments of plants advance ripening, at least in tomatoes (61, 96), so there is a presumptive case for a role for abscisic acid in some ripening reactions. However, the case has not been substantiated or finely drawn, and Tsay et al (143), found no consistent relationship between the abscisic acid content of fruits and their ripening or senescence.

The hypothesis that endogenous auxins play roles as ripening inhibitors (46) is attractive. Evaluation of the hypothesis is complicated because auxins both increase ethylene production and modify tissue response to ethylene (158). In apples, Mousdale & Knee (103) found a peak in indole-3-acetic acid concentration as system 2 ethylene was initiated, but they were unable to conclude that the two were related.

Research in this area has been hampered by the difficulty of equating responses to externally applied regulators with endogenous control by the growth regulator, by the interactions of the different regulators, and by analytical difficulties when attempts were made to monitor endogenous concentrations. The latter difficulties are receding, but a recognition of the importance of distribution within and between cells and of the desirability of monitoring receptors (142) as well as growth regulators limits studies in this area.

It is possible that the influence of ethylene receptors may be monitored, albeit indirectly, by the apparent blocking of these receptors by anionic silver

(148, 149). Silver inhibits the ripening of banana (129) and tomato (66, 129) fruit tissues at nontoxic concentrations, and the inhibition is not reversed by externally supplied ethylene. In tomato fruit discs, silver inhibits further ripening if it is applied to partly ripened tissue (G. A. Tucker and C. J. Brady, unpublished); this indicates that ethylene action is involved through the ripening period and is not limited to an initial triggering of ripening. High concentrations of externally added calcium inhibit the induction of ripening in a number of fruits (45). In avocados, infiltration with 0.4 M calcium prevented normal ripening (44). Treatment with ethylene overcame the calcium effect. The effect of calcium was to prevent system 2 ethylene synthesis without modifying the capacity to respond to the higher concentrations of ethylene that are produced by system 2 ethylene.

2.3 Gene Expression in Ripening Fruit

The concept of ripening as a differentiation event was initially stimulated by an apparent increase in synthetic activity early in the climacteric rise. For example, in avocados, there were large increments in the incorporation of precursors into protein (122) and nucleic acid (123). The inhibition of protein or nucleic acid synthesis (128) prevented normal ripening, and this was taken as evidence that novel translational and/or transcriptional events occurred during the early stages of ripening. The evidence was far from satisfactory. To pulse label or to introduce inhibitors, tissue slices were used so that interactions between the effects of slicing (132) and those of ripening were always involved. There was no clear evidence of the synthesis of specific proteins that had a role in ripening. It was suggested that malic enzyme was newly synthesized in the early climacteric of pears (49), but the evidence failed to distinguish net synthesis from turnover. In bananas, there was evidence that most of the increment in protein synthesis early in the climacteric rise resulted in an increase in turnover and the replacement of preexisting species of protein (25).

More recently, evidence for a redirection of protein synthesis during ripening has emerged. Increases in the activities of β-1:4 glucanhydrolase (cellulase) in avocados (37) and in invertase (72) and polygalacturonase in tomatoes (22, 144, 145) result from increases in the amount of enzyme protein measured immunologically. The contents of these hydrolytic enzymes increase following a change in their rates of synthesis and/or degradation and not as a result of the activation of a precursor molecule synthesized during earlier development. Either the synthesis of these proteins increases dramatically when ripening commences or else a previously unstable translation product becomes stable. Because the enzymes normally function in the cell wall (cellulase, polygalacturonase, invertase) or in the vacuole (invertase), stability could conceivably be the consequence of a developing mechanism for transporting or processing the primary translation products.

In avocados, Christoffersen et al (38) showed that, with the advent of ripening, there was a change in the population of translatable messenger RNA molecules. In vitro translation of poly (A^+)-RNA populations showed three products that were not apparent in translations of RNA from nonripening fruit. In an extension of this work, Tucker & Laties (147) identified, by immunoprecipitation, a 53,000 dalton product of the in vitro translation system as the precursor of cellulase. Ripening-specific cDNAs were located in a cDNA library constructed from the poly (A^+)-RNA of ripe avocado fruit. A cDNA specific for cellulase was identified by hybrid selection, in vitro translation, and product immunoprecipitation (37). By hybridization, the messenger RNA for the cellulase precursor was shown to increase at least 50-fold during ripening (37).

Bennett & Christoffersen (13) determined the sequence of a full length cDNA and the N-terminal sequence of the mature cellulase protein and suggested that the primary translation product is a 54,000 dalton polypeptide. In vivo removal of a signal peptide during membrane passage yields a processed polypeptide of 52,800 daltons; glycosylation produces a membrane-associated form of 56,500 daltons that is trimmed to the mature protein of 54,200 daltons, which is presumably located in the cell wall region.

This work on the ripening avocado provides direct evidence of a qualitative change in protein synthesis during ripening. A result of the change is an increase during ripening of a gene product which presumably plays a role in the softening of the fruit (62). The experiments do not involve slicing and the consequential interactions with wounding (132). The evidence establishes a change in the steady state level of messenger RNA after ripening is initiated, and the cDNA probes allow an approach to the gene(s) and their promoter sequences. It is not yet established that the transcription of the cellulase gene is activated when ripening initiates, but clearly the control of cellulase synthesis is at a pre-translational level.

Similar evidence has arisen from studies of the tomato fruit. Within 48 hours of the initiation of ripening—indicated by system 2 ethylene synthesis—changes in the population of messenger RNAs were demonstrated (57, 137). A few messenger RNAs decreased in relative intensity, and six to eight increased in relative intensity. The screening of cDNA libraries prepared from the poly(A^+) RNA of ripening fruit confirmed the presence of a number of messenger RNAs that are unique to, or greatly enhanced in, ripening fruit (56, 87, 134). Mansson et al (87) found that about 1% of the homologues of the cDNAs they prepared from the poly(A^+) RNA of ripe fruit were most prominent in mature, nonripening fruit, about 2% were most prominent in ripening fruit, while the remainder were equally prominent in unripe and ripe fruit. Piechulla et al (112) detected transcripts for a range of chloroplast proteins in immature green, tomato fruits; transcripts of only two of these were present in ripe fruit. They confirmed earlier observations that ripe fruit contain chloroplast ribosomal RNA (11, 137) and that the plastids of ripe tomato fruit appear to remain active in protein synthesis (11).

Polygalacturonase protein increases in tomatoes during ripening (24), and in soft cultivars it becomes a major protein in the ripe fruit (22). The large increment in net synthesis suggests that the polygalacturonase messenger RNA may become very prominent during ripening, and a number of groups have sought to characterize its messenger RNA and thence its gene (55). The enzyme exists in three molecular forms in extracts of ripe tomatoes (1), but there may be only one active polypeptide, the other forms representing differential glycosylation and the presence of an inactive subunit (1, 101, 117). The most prominent enzyme, with a single subunit, has been assigned various molecular weights: 43,000 (1), 45,000 (41), 46,000 (57), and 47,500 (101) daltons. Grierson and co-workers described the precipitation, with polygalacturonase-specific antisera, of an in vitro synthesized polypeptide of 48,000 daltons (57). They also observed that a cDNA to a ripening-enhanced poly(A$^+$) RNA species hybrid selected a messenger RNA that coded for a polypeptide of 48,000 daltons, which was precipitated by the same serum (134). The putative polygalacturonase messenger RNA was of relatively low abundance. Sato et al (130, 131) immunoprecipitated, with an antiserum to polygalacturonase, a polypeptide of 54,000 daltons from translations of the RNA of ripe tomatoes. By their estimate, somewhat less than 1% of the translatable messenger RNA in ripe tomatoes coded for polygalacturonase. There are other reports of a 54,000 dalton precursor to tomato polygalacturonase (16, 17, 41, 100). Products in the 54,000 to 55,000 dalton region are prominent among the polypeptides synthesized in vitro from the RNA from ripe tomato fruit, whether the translation is by wheat germ (137) or reticulocyte lysate translation system (57, 134); translates of RNA from green fruit do not have prominent products of this size. DellaPenna et al (41) prepared a cDNA library from the poly(A$^+$) RNA of ripe tomato fruit and inserted the library into an expression vector. Polygalacturonase cDNA clones were identified by immunological screening. A 2.1-kb messenger RNA for polygalacturonase was identified and was shown to code for a 54,000 dalton polypeptide. The messenger RNA was not detected in immature green fruit, was first detected as ripening commenced, and greatly increased in abundance as the fruit colored. The evidence is now substantial that the primary transcript for tomato polygalacturonase is a polypeptide of about 54,000 daltons and that it includes an unusually long signal peptide. Sequence studies have confirmed the identity of cDNA probes (W. R. Hiatt and R. E. Sheey; D. DellaPenna and A. B. Bennett; D. Grierson, unpublished) and will aid identification of the processing steps between the primary translation product and the mature enzymes.

2.4 Ripening Mutants

Early, mid, and late season varieties are widely recognized among horticultural crops. Their occurrence implies that there is a genetic component regulating the initiation of ripening, but it does not establish a specific genetic control of the onset of ripening. "Earliness" may be one aspect of a pleiotropic effect perhaps exercised through growth regulator or receptor concentrations.

Several spontaneous mutants of tomatoes appear to be specifically "ripening" mutants. Two of these, the ripening inhibitor or *rin* mutation on chromosome 5 and the nonripening or *nor* mutation on chromosome 10, are recessive mutations that result in nonclimacteric fruits (141). These fruits fail to produce system 2 ethylene and fail to undergo normal ripening. When exposed to high concentrations of ethylene under specific conditions *rin* fruit develop a moderate amount of lycopene but do not soften appreciably (48, 97, 141); *nor* fruit produce some lycopene during senescence, and there is a slight rise in ethylene production. With salt treatment, *nor* fruit, at least in some genetic backgrounds, may develop near-normal levels of carotenes and some polygalacturonase activity (3). Both *rin* and *nor* are sensitive to ethylene in that, in the presence of added ethylene, the respiration rate rises (141); in *rin* plants, ethylene treatment induces an increase in EFE (83), but the full system 2 apparatus is not induced and there is no massive increase in ACC synthase. The two mutants lack the capacity to induce system 2 ethylene synthesis, and they also lack the capacity to respond to ethylene exposure in terms of normal ripening. A search for mutants that produce system 2 ethylene but do not ripen in response to added ethylene, or ripen in the presence of ethylene without producing system 2 ethylene, may be worthwhile. The development of molecular probes for ripening and system 2 ethylene (presumably ACC synthase) would substantially aid such a search. *rin* and *nor* fail to develop the capacity to ripen and, in this sense, are developmental mutants. It follows that ripening is a stage of development specifically controlled by a small gene set.

In addition to *rin* and *nor* which are nonclimacteric, there are a number of mutants in which all ripening parameters occur slowly and continue over a long period. "Never ripe" (Nr) is a dominant allele on chromosome 9 (141), "green ripe" (Gr) is a partly dominant allele (73), and "alcobaca" is a recessive allele that is allelic with (84) or positioned close to (105) *nor* on chromosome 10. The slow-ripening mutants have a depressed and extended climacteric, color slowly and sometimes incompletely, and produce small amounts of polygalacturonase, which catalyzes a slow softening of the fruits. Exposure to increased levels of ethylene does not hasten the ripening process.

The nonripening mutants, although recessives, influence the ripening of fruits in heterozygous plants. The heterozygotes ripen more slowly than the normal parent, and polygalacturonase production is much reduced (24, 65). In the heterozygote, the *nor* gene generally has a greater effect than does *rin*, but the influence of the mutant genes is greatly influenced by the genetic background.

Tigchelaar et al (141) recognized the value of the ripening mutants as an approach to the "conditioning events" that allow ripening to proceed. They suggested that the synthesis of polygalacturonase may be the critical event in the induction of ripening and that its action had a catalytic role in the induction of intracellular ripening. This seems unlikely in that the initiation of polygalacturonase synthesis is a relatively late event in tomato ripening (22, 41, 138) and there is some evidence that *nor* (3) and *rin* (146) are able to synthesize polygalacturonase in some circumstances or some tissues. The slow-ripening mutants synthesize polygalacturonase. The presence and the intactness of the polygalacturonase gene(s) will undoubtedly be evaluated directly now that molecular probes for the gene are available. Presently, the physiological and biochemical evidence suggests that the mutations are in a range of genes that regulate the expression of the ripening syndrome, rather than in genes that code for a catalyst of one or more of the recognized ripening events.

Ripening mutants have been recognized in other species. A slow-ripening nectarine genotype (29) fails to produce normal levels of ethylene. When stimulated by added ethylene or propylene, endogenous ethylene synthesis is increased and ripening advanced (27). M. E. Patterson has recognized a nonclimacteric apple sport (personal communication), and it seems likely that, with the realization of the value of slow or nonripening genotypes, other ripening mutants will be described.

3. MEMBRANES AND ORGANIZATIONAL RESISTANCE

There is now firm evidence that the synthesis of polygalacturonase in tomatoes (23) and cellulase in avocados occurs during the climacteric and continues in ripe fruit (4, 37). A functional ribosomal system is required, and the recovery of polysomes from ripe fruit is consistent with this finding (137, 147). There are strict ionic requirements for polysome stability and function (157), and these requirements are apparently met within the cytoplasm of the ripe fruits. This implies that the plasmalemma and tonoplast maintain their functions at least as long as the net synthesis—and presumably the processing and export—of the wall-hydrolyzing enzymes occurs. One must conclude that there is no breakdown in permeability barriers even in quite ripe fruit.

Studies of the ultrastructure of ripening avocados (113) and tomatoes (40) have led to the same conclusion. Although some changes occurred in the structure of mitochondria, the endoplasmic reticulum, and the plasmalemma during ripening, these changes were associated (113) with metabolic events, in particular with the export of the wall-hydrolyzing enzymes. Other ultrastructural studies of fruits have suggested a disturbance of membrane function in senescent (post-climacteric) fruit cells, but there are questions as to how accurately structures were maintained in the ripe tissues in which wall degradation (12, 115) was advanced. The avocado study (113) noted that a phase change in the plasmalemma bilayer may occur in ripening fruit, but the evidence suggested that the change occurred at the climacteric peak and was transitory. This would distinguish any such event from the phase disturbances that occur in senescent cotyledon (94, 114) and floral tissues (20) and that have been ascribed to ethylene action. Rather similar changes were described in the microviscosity of "microsomal" membranes of senescing apples (108).

A number of analyses of fruit lipids have been reported, and some of these were coupled with observations on the permeability characteristics of liposomes prepared from the lipids or lipid subfractions. Losses of membrane lipids from tomato (75), of plastid membrane lipids (53) and of phospholipids (85) from apple, and an increase in phosphatidylcholine (9) in ripening apples and of unsaturation in the membrane lipids of ripening bananas (153) have all been described. In senescing leaf and flower tissues, changes in fluidity have been related to increases in neutral lipids—sterols, free fatty acids or hydroperoxides—unaccompanied by changes in the phospholipid classes or the acyl species (8, 42). An increase in lipid peroxides during ripening has been observed in tomatoes (47), and in hydrogen peroxide in pears (30). Analytical and liposome studies suffer from averaging effects as the constituents of several membranes are mixed. They may provide evidence that the potential for leakage changes in one or more membranes, but they cannot be definitive because they make no allowance for the capacity of the cells to cope with increased leakiness using active carriers or other means. The changes that have been observed, as in bananas (153) and apples (9), emphasize the dynamic nature of the tissues and the capacity for change that exists. A capacity for change may also mean a capacity for repair, an aspect that has been emphasized by Romani (125).

Much of the evidence for a disruption of tissue compartments during ripening involves the use of tissue slices (128). Methods using slices necessarily involve an interaction between the initial condition of the tissue and the response to slicing. The interaction may depend upon cell wall properties, on the osmotic gradient between tissue and bathing medium (133), or on the rate of the metabolic responses to slicing. Most researchers who have measured uptake or efflux from fruit slices have found some change associated with ripening, although sometimes the changes are slight. In tomatoes, there is evidence for a small redistribution of potassium between cell compartments (152), for an increase in hydraulic conductivity (116), for a decrease in the capacity for amino acid uptake (5), for an increase in amino acid efflux (6), and for a decrease in cytoplasmic volume (6). None of the changes suggest a catastrophic leakage between compartments [although if cytoplasmic potassium concentration is actually as low as 50 mM (152), the effect would be considerable (157)], and some of the changes become significant only late in ripening, which suggests an association with senescence rather than with ripening per se.

The sum of evidence is that fruit cells do not suffer a massive breakdown in cellular compartmentation during ripening but that there may well be a tendency towards a greater leakiness of membranes. The evidence for the latter is cumulative and borrows heavily on that pertaining to petals and cotyledons. No summation of ambiguous evidence, no matter how consistent in its direction, can satisfactorily establish a conclusion, and evidence from analyses, from liposomes, and from tissue slices all have some ambiguity. In situ analyses by NMR or other means may eventually prove steady state changes in distribution; changes in fluxes may be harder to prove.

4. OXIDATIVE REACTIONS

4.1 The Respiratory Climacteric

Much of the fascination in studies of fruit ripening lies in attempting to explain the respiratory climacteric and its relation to other ripening events. There have been periods when the climacteric rise was thought to be directly related to the anabolic needs of ripening (122) and others when it was considered the prime reflection of the decline in organizational resistance (18, 136). Both periods seem to have passed, but no entirely satisfactory explanation of the climacteric has emerged. The subject has been reviewed recently (121, 135), as were the pathways for electron flow (80), and no extended treatment is required here.

The increase in respiration appears (a) to be a consequence of the increase in endogenous ethylene, (b) to result in increased ATP levels and perhaps in an increase in energy charge in the cells (135), and (c) to be associated with an increase in fructose-2,6-bisphosphate concentration and a subsequent increase in glycolytic flux (A. B. Bennett, unpublished). The contributions of the cytochrome pathway, the cyanide-resistant alternative pathway, and the extent of by-passing of site 1 phosphorylation (54, 99) remain matters for debate. Since ATP concentrations increase through the climacteric period, the increase in respiration provides a supply of chemical energy in excess of the demands of the tissue.

4.2 The Climacteric Rise and Protein Synthesis

Energy balance and energy use in nongrowing tissues—the so-called maintenance energy requirement—are often calculated, but in terms of cell biology are poorly understood. The main sinks for maintenance energy are thought to be the preservation of ion gradients and protein turnover (111). When nonclimacteric fruits, or climacteric fruits that lack the maturity to ripen, are exposed to ethylene, the respiration rate increases. When ethylene is withdrawn, the respiration rate falls. There is a similar response to ethylene in several nonfruit tissues (135). Clearly this respiratory rise is a response to ethylene and not a demand for energy with which to synthesize ripening-specific enzymes. Nonetheless, associated with the increased respiration, there is an increase in protein synthesis (25, 147; J. Speirs, C. J. Brady, E. Lee, unpublished), a large component of which is protein turnover. Thus, when ethylene stimulates respiration, one of the two main sinks of maintenance energy—protein turnover—increases. It is entirely possible that the energy demand for the maintenance of ion gradients also increases; this would occur if ethylene perturbates membrane function and increases passive diffusivity (see Section 3).

If one looks broadly at respiration in nongrowing plant tissues, it is not at all unusual to observe that the respiration rate is regulated by substrate supply (79). This is so, even in tissues amply filled with carbohydrate reserves. Presumably, compartmentation within cells is critical to this regulation. It is postulated that the cyanide-resistant pathway operates when electron flow exceeds the acceptor-regulated cytochrome pathway (79) and, in particular, when substrate supply is excessive. In fruit tissues, it may be that ethylene treatment increases substrate supply—be it by an effect on glycolysis, on tonoplast fluxes, or otherwise—with a resultant increase in respiration rate and in chemical energy. Of the postulated major sinks for maintenance energy, protein turnover may be the most flexible, so that an increase in protein synthesis is observed as a consequence of ethylene action. There is other evidence in plants that protein turnover is related to the respiration rate (120).

An increase in protein synthesis early in the climacteric period (43, 128, 147) has been measured in many fruits. Because the increase appears to be tied to the respiration increase (25), alternative and opposite interpretations are available. The initial effect of ethylene may be to increase respiration and with it available energy; if the rates of the initiation reactions are controlled by energy supply, perhaps via the energy charge, an increase in polysomes would result. On the other hand, if an early effect of ethylene treatment is to increase the supply of messenger RNA available for translation, then the demand for energy would increase and a rise in respiration rate would follow. A choice between the opposites is not presently possible. We know neither the ATP yield from the respiratory increase nor the absolute rates of protein synthesis involved. Clearly the second alternative demands a tight coupling of the energy use in protein synthesis with the energy yield from respiration. The first alternative may show this too, but it is not a prerequisite of the model.

In avocados, Tucker & Laties (147) delineated two regions of the climacteric period that are associated with changes in protein synthesis. The initial part of the rise—in this case provoked by external ethylene—was associated with a threefold increase in polysomal RNA and poly(A$^+$) RNA recoveries per unit fresh weight and presumably a threefold increase in protein turnover. The second phase was associated with a decline in polysome recovery and with the ripening-specific enzyme cellulase representing an increasing proportion of the protein synthesized. Subsequent studies have established that the synthesis of cellulase is regulated at a step preceding translation (Section 2.3), but the mechanisms leading to the increases in translation of the generic or "housekeeping" messengers have not been sought. In ripe avocado fruit, the ripening-specific messenger RNAs account for only a small part of the total (about 1.4% for cellulase), and this seems also to be the case in tomatoes (55).

Evidence for an increase and then a decline in protein synthesis also comes from estimates of the incorporation of precursors into protein in tissue slices (5, 25, 122, 128) and from estimates of the recovery of polysomes from intact fruits (43, 147). The evidence with slices is very consistent in several species, but there are problems of interpretation associated with changes in endogenous dilution of exogenous precursors, with amino acid recycling, and with interaction with the response to slicing. Measuring polysome content (137, 147) avoids these complications but is subject to the assumption that recoveries are constant as the chemistry of the fruits changes during ripening. Total RNA analyses to support the polysome estimates are needed. In tomatoes, there is an increase in the proportion of ribosomes recovered as polysomes when the climacteric commences (137), and an apparent decline in ribosome recovery late in ripening. The results suggest an increase in the rate of protein synthesis during the climacteric rise, followed by a decline later in ripening; again, RNA analyses are required to consolidate the conclusion.

A large amount of circumstantial evidence links the ethylene-provoked rise in respiration to an increase in protein turnover. Such a situation fits the description of the struggle for homeostasis offered by Romani (125), but it is well to remember that other differentiation signals, for example phytochrome activation (102), provoke a large generic effect on protein synthesis as well as a specific effect involving some transcripts. Changes in cell structure and function may be more easily accommodated if the general rate of replacement of macromolecules is increased. The two phases detected in avocados (147) may occur commonly, but they do not imply an absolute segregation, in time, of the generic and specific responses. In tomatoes, there are early and late responding changes in messenger RNAs (55), and the early changes may be within the first "generic" increase in protein synthesis.

4.3 Other Oxidative Reactions

Extension of postharvest life by storage in low concentrations of oxygen cannot be easily explained by effects on either the cyanide-sensitive or -resistant respiratory pathways (135) and focuses attention on other oxidative events with a lower affinity for oxygen. The inhibition of ripening by auxin (46, 150) and the peroxidative destruction of indole-3-acetic acid (78) have been posed as important regulators of ripening, but unambiguous evidence in favor of this thesis has been difficult to obtain. Attempts have been made to relate peroxidase-catalyzed reactions to ethylene synthesis (47, 50), and high concentrations of free radical scavengers have been shown to inhibit ethylene production in fruit tissues (2); however, whether or not free radicals have a direct role in ethylene biosynthesis remains doubtful (158). Evidence that superoxide radicals are involved in reactions whose products perturb membrane function in flowers has been mentioned previously (Section 3; 42, 90) and may be relevant to fruit ripening, because the ethylene-provoked respiratory climacteric in the floral tissues has an obvious parallel in the climacteric fruits. Lesham (81) has proposed that cytokinins, which generally are anti-senescence agents, function as free radical scavengers, implying that free radicals play a critical role in inducing senescence. Chloroplast de-differentiation and chlorophyll destruction are part of the ripening syndrome in many fruits, and Matile (89) has suggested that peroxidase or lipoxygenase are involved in chlorophyll catabolism.

5. THE SOFTENING OF FRUITS

The softening process is an integral part of the ripening of almost all fruits. It has immense commercial importance because the postharvest life of the fruit is to a large extent limited by increasing softness, which brings with it an increase in physical damage during handling and an increase in disease susceptibility. To the physiologist and biochemist the softening reactions are significant because they involve cell wall changes that have no parallel in leaf senescence and that appear to occur only in the fruit and perhaps in abscission zones. It is no accident that cell wall hydrolyzing enzymes have featured prominently in recent studies of the molecular biology of ripening.

5.1 Chemistry and Enzymology of Softening

The chemical changes involved in softening and the enzymes contributing to the changes have been reviewed recently (10, 70) and will be briefly mentioned here. Understanding of the changes that occur during ripening is limited because knowledge of the structures of the walls in mature fruit and of the enzymes that modify the walls is very limited. Happily, both deficiencies are being addressed and more detailed studies undertaken.

Softening is normally accompanied by an increase in the concentration of soluble pectic polysaccharide (10, 70). The increase in soluble uronic acid residues is often correlated with an increase in the polyuronide hydrolyzing enzymes, especially endopolygalacturonase [poly(1,4-α-D-galacturonide)glycanohydrolase, EC3.2.1.15]. In tomatoes there is evidence of a decrease in the average size of the soluble polyuronide chains late in ripening (69); although hemicelluloses are also normally degraded during ripening, degradation of the polyuronides is not dependent on changes in hemicellulose components (69). In tomatoes, too, there is a reasonable correlation between the amount of polygalacturonase, the rate of softening, and pectin degradation (22–24, 65), although information on the pectins remains sparse. There is also a correlation in avocados (4, 115), pears (12), and tomatoes (40) between the appearance of polygalacturonase activity in ripening fruit and a loss of electron density in the middle lamellar region of cell walls. In a number of other fruits there is a correlation in time between the appearance of endopolygalacturonase activity and the initiation of softening (70, 104, 109, 110, 124). An increase in pectin solubility is not adequate evidence of endopolygalacturonase activity, as this may occur by the action of other enzymes or by a change in organic acids, other chelating agents, or pH in the fruit cell walls. In apples (10) and strawberries (71), pectin solubility increases but endopolygalacturonase activity is not detected (but see 82 for apples).

Cellulase activity is almost always found in fruit, and the level of activity often increases greatly during ripening (10, 70). A precise role for cellulase, its contribution to softening, or even its substrate in the walls have yet to be defined. A number of other glycosidases occur in walls and some, like xylanase in papaya walls (109), increase during ripening, which suggests, but does not establish, that they play a role in softening. In tomatoes, the nonripening mutants *rin* and *nor* (70, 140) have been useful in distinguishing changes in the walls that are associated specifically with ripening from those that occur in the absence of ripening and may be considered as part of the senescence syndrome.

5.2 A Role for Calcium

High concentrations of calcium are known to inhibit and sometimes prevent fruit ripening and to lead to firmer fruit (45). While calcium undoubtedly has subtle effects on cell membranes, on membrane proteins (107, 118), and as a second messenger (63), the high concentrations of calcium required to delay ripening or inhibit senescence suggest a gross effect, perhaps in the wall. When one considers the potential activity of the polygalacturonase and pectin methyl esterase in fruit tissues, the attack on wall polyuronide appears to be very limited and, in tomatoes, there is evidence that calcium limits wall hydrolysis (23, 31). Claims that calcium is "solubilized" or otherwise redistributed in ripening fruit have been made, but these claims are based on inadequate techniques (23), and there is a need to evaluate calcium distribution between the vacuole and the wall by using sound fixation and analytical methods.

5.3 Softening as a Regulator of Ripening

Although it is unlikely that wall hydrolysis has an initiating role in the ripening of climacteric fruits (141), genotypes that soften rapidly generally ripen rapidly, and soft fruits appear to senesce rapidly. The interrelationships between wall and intracellular events have been little studied. Ripening is often seen as a series of independent events, perhaps coordinated by system 2 ethylene but otherwise operating in parallel. According to this view, the expression of system 2 ethylene results, among other things, in the induction of the synthesis of wall-hydrolyzing enzymes, which in turn cause the wall to soften; the synthesis and export from the cytoplasm of the wall-hydrolyzing enzymes are the full extent of the interaction between cytoplasm and wall. There is, however, the possibility that the cytoplasm is influenced by a change in the chemistry and the physical state of the wall. Wall hydrolysis appears to decrease the ability of the walls to bind a range of proteins (67), so one consequence of softening may be the release, for interaction with the plasmalemma, of proteins previously immobilized on the wall. A second possibility lies in the release of carbohydrate oligomers from the wall. Such oligomers may function as elicitors (127) and may stimulate ethylene production and other stress responses. Is it possible that they have an intensifying role on the latter stages of ripening? Invasion by microorganisms often advances or hastens ripening. Ethylene production undoubtedly contributes to this process, but an effect of the products or consequences of wall breakdown have not been excluded. Changes in wall properties will modify water relations in fruit tissues, and this will contribute to solute redistribution between tissue compartments.

6. CONCLUDING REMARKS

Throughout most studies of the physiology of fruit ripening, researchers have been hampered and frustrated by the bulk of these organs. Bulk has limited the applicability of many of the methods that elsewhere have aided the exploration of intermediary metabolism. There is, of course, some use in knowing what the potential of a cell is—which enzymes, what potential activities?—but in a bulky, nongrowing tissue tuned down to "maintenance only" reactions, which enzymes and what potential is not answer enough. The literature is crammed with reports of changes in the amount of this or that enzyme as maturity or storage environment changed, but always there was equivocation as to the meaning of the change. Substrate concentrations and enzyme environment were always unknown. To measure flux rates, the bulk was overcome by using tissue slices, but it is nonsense to extrapolate measurements from a rapidly repairing slice to a bulky organ in "maintenance only" gear.

Recent years have seen some lifting of the frustration imposed by tissue bulk. Indeed the fact that the bulky tissues are generally rather homogenous in cell type may prove an advantage in the use of physical analytical methods, such as NMR and X-ray probes. It is anticipated that detailed and focused analysis will yield firm information on aspects such as vacuolar, cytoplasmic, and free space pH and ion concentrations and changes that occur over time in major metabolites. Similarly, carefully applied polysome analysis will yield information on protein synthesis in the intact, bulky tissues, and the wise use of immunological and nucleotide probes will show changes in the synthesis and location of particular macromolecules. The next review on ripening in this series will have fewer "suggests" or "is consistent with" or "implies" and more positive statements than is presently possible.

This is both an opportune and difficult time to review fruit ripening. Opportune because we stand on the edge of rapid progress and it is useful to contemplate directions; difficult because progress is now so rapid that surmise will be overtaken by knowledge as the review pupates in press. The time is near when the number of genes for avocado cellulase and tomato polygalacturonase will be known, when the promoter sequences for ripening-related genes will be revealed, and when the consequences of having more, or fewer, or even none of these genes can be assessed. The time may be near, too, when the terms "system 1" and "system 2" ethylene—remarkably useful terms for a decade and a half—can be replaced by a description in molecular terms of what underlies the concept. Now we can be certain that ripening is genetically and developmentally regulated. Soon we may know a little of what those words imply.

Literature Cited

1. Ali, Z. M., Brady, C. J. 1982. Purification and characterisation of the polygalacturonases of tomato fruits. *Aust. J. Plant Physiol.* 9:171–78
2. Apelbaum, A., Wang, S., Burgoon, A. C., Baker, J., Lieberman, M. 1981. Inhibition of the conversion of ACC to ethylene by structural analogs, inhibitors of electron transfer, uncouplers of oxidative phosphorylation and free radical scavengers. *Plant Physiol.* 67:74–79
3. Arad, S., Mizrahi, Y. 1983. Stress in-

duced ripening of the nonripening tomato, *Lycopersicon esculentum* mutant *nor. Physiol. Plant* 59:213–17
4. Awad, M., Young, R. E. 1979. Postharvest variation in cellulase, polygalacturonase and pectin methyl esterase in avocado (*Persea americana* Mill. cv. Fuerte) fruits in relation to respiration and ethylene production. *Plant Physiol.* 64:306–8
5. Baker, J. E., Anderson, J. D., Hruschka, W. R. 1985. Protein synthesis in

tomato fruit pericarp tissue during ripening. Characteristics of amino acid incorporation. *J. Plant Physiol.* 120:167–79

6. Baker, J. E., Saftner, R. A. 1984. Compartmentation of amino acids in tomato fruit pericarp tissue. See Ref. 52, pp. 317–27

7. Bangerth, F. 1984. Changes in sensitivity for ethylene during storage of apple and banana fruits under hypobaric conditions. *Sci. Hortic.* 24:151–64

8. Barber, R. F., Thompson, J. E. 1983. Neutral lipids rigidify unsaturated acyl chains in senescing membranes. *J. Exp. Bot.* 34:268–76

9. Bartley, I. M. 1984. Lipid metabolism in ripening apples. *Phytochemistry* 24:2857–59

10. Bartley, I. M., Knee, M. 1982. The chemistry of textural changes in fruit during storage. *Food Chem.* 9:47–58

11. Bathgate, B., Purton, M., Grierson, D., Goodenough, P. W. 1985. Plastid changes during the conversion of chloroplasts to chromoplasts in ripening tomatoes. *Planta* 165:197–204

12. Ben-Aire, R., Kislev, N., Frenkel, C. 1979. Ultrastructural changes in the cell walls of ripening apple and pear fruit. *Plant Physiol.* 64:197–202

13. Bennett, A. B., Christoffersen, R. E. 1986. Synthesis and processing of cellulase from ripening avocado fruit. *Plant Physiol.* 81:830–35

14. Biale, J. B., Young, R. E. 1971. The avocado pear. In *The Biochemistry of Fruit and Their Products,* ed. A. C. Hulme, 2:1–63. London: Academic. 788 pp.

15. Biale, J. B., Young, R. E. 1981. Respiration and ripening in fruits—retrospect and prospect. See Ref. 51, pp. 1–39

16. Biggs, M. S., Tieman, D. M., Handa, A. K. 1985. Ripening associated gene expression in tomato fruit pericarp. *HortScience* 20:538 (Abstr.)

17. Biggs, M. S., Harriman, R. W., Handa, A. K. 1986. Changes in gene expression during tomato fruit ripening. *Plant Physiol.* 81:395–403

18. Blackman, F. F., Parija, P. 1928. Analytical studies in plant respiration. I. The respiration of a population of senescent ripening apples. *Proc. R. Soc. London Ser.* 103:412–45

19. Blankenship, S. M., Richardson, D. G. 1985. Development of ethylene biosynthesis and ethylene-induced ripening in cultivar D'anjou pears *(Pyrus communis)* during the cold requirement for ripening. *J. Am. Soc. Hortic. Sci.* 110:520–23

20. Borochov, A., Halevy, A. H., Shinitzky, M. 1982. Senescence and the fluidity of rose petal membranes: relationship to phospholipid metabolism. *Plant Physiol.* 69:296–99

21. Brady, C. J. 1973. Changes accompanying growth and senescence and effect of physiological stress. In *Chemistry and Biochemistry of Herbage,* ed. G. W. Butler, R. W. Bailey, 2:317–51. London: Academic. 455 pp.

22. Brady, C. J., MacAlpine, G., McGlasson, W. B., Ueda, Y. 1982. Polygalacturonase in tomato fruits and the induction of ripening. *Aust. J. Plant Physiol.* 9:171–78

23. Brady, C. J., McGlasson, W. B., Pearson, J. A., Meldrum, S. K., Kopeliovitch, E. 1985. Interactions between the amount and molecular forms of polygalacturonase, calcium and firmness in tomato fruit. *J. Am. Soc. Hortic. Sci.* 110:254–58

24. Brady, C. J., Meldrum, S. K., McGlasson, W. B., Ali, Z. M. 1983. Differential accumulation of the molecular forms of polygalacturonase in tomato mutants. *J. Food Biochem.* 7:7–14

25. Brady, C. J., O'Connell, P. B. H. 1976. On the significance of increased protein synthesis in ripening banana fruits. *Aust. J. Plant Physiol.* 3:301–10

26. Brady, C. J., O'Connell, P. B. H., Smydzuk, J., Wade, N. L. 1970. Permeability, sugar accumulation and respiration rate in ripening banana fruits. *Aust. J. Biol. Sci.* 23:1143–50

27. Brecht, J. K., Kader, A. A. 1984. Ethylene production by fruit of some slow-ripening nectarine *Prunus persica* genotypes. *J. Am. Soc. Hortic. Sci.* 109:763–67

28. Brecht, J. K., Kader, A. A. 1984. Regulation of ethylene production by ripening nectarine fruit *Prunus persica* as influenced by ethylene and low temperature. *J. Am. Soc. Hortic. Sci.* 109:869–72

29. Brecht, J. K., Kader, A. A., Ramming, D. W. 1984. Description and postharvest physiology of some slow-ripening nectarine genotypes. *J. Am. Soc. Hortic. Sci.* 109:596–600

30. Brennan, T., Frenkel, C. 1977. Involvement of hydrogen peroxide in the regulation of senescence in pears. *Plant Physiol.* 59:411–16

31. Buescher, R. W., Hobson, G. E. 1982. Role of calcium and chelating agents in regulating the degradation of tomato fruit tissue by polygalacturonase. *J. Food Biochem.* 6:147–60

32. Bufler, G. 1984. Ethylene-enhanced 1-

aminocyclopropane-1-carboxylic acid synthase activity in ripening apples. *Plant Physiol.* 75:192–95

33. Bufler, G. 1986. Ethylene-promoted conversion of 1-aminocyclopropane-1-carboxylic acid to ethylene in peel of apple at various stages of fruit development. *Plant Physiol.* 80:539–43

34. Bufler, G., Bangerth, F. 1983. Effects of propylene and oxygen on the ethylene producing system of apples (*Malus sylvestris* cultivar Golden Delicious). *Plant Physiol.* 58:486–92

35. Burg, S. P., Burg, E. A. 1967. Molecular requirements for the biological activity of ethylene. *Plant Physiol.* 42:144–52

36. Child, R. D., Williams, A. A., Hoad, G. V., Baines, C. R. 1984. The effects of aminoethoxyvinylglycine on maturity and postharvest changes in Cox's Orange Pippin Apples. *J. Sci. Food Agric.* 35:773–81

37. Christoffersen, R. E., Tucker, M. L., Laties, G. G. 1984. Cellulase gene expression in ripening avocado, *Persea americana* cultivar Hass, fruit; the accumulation of messenger RNA and protein as demonstrated by complementary DNA hybridization and immunodetection. *Plant Mol. Biol.* 3:385–92

38. Christoffersen, R. E., Warm, E., Laties, G. G. 1982. Gene expression during fruit ripening in avocado. *Planta* 155:52–57

39. Coombe, B. G. 1976. The development of fleshy fruits. *Ann. Rev. Plant Physiol.* 27:207–28

40. Crookes, P. R., Grierson, D. 1983. Ultrastructure of tomato fruit ripening and the role of polygalacturonase isoenzymes in cell wall degradation. *Plant Physiol.* 72:1088–93

41. DellaPenna, D., Alexander, D. G., Bennett, A. B. 1986. Molecular cloning of tomato fruit polygalacturonase: analysis of polygalacturonase mRNA levels during ripening. *Proc. Natl. Acad. Sci. USA* 83:6420–24

42. Dhindsa, R. J., Dhindsa, P. P., Thorpe, T. A. 1981. Leaf senescence: correlated with increased levels of membrane permeability and lipid peroxidation and decreased levels of superoxide dismutase and catalase. *J. Exp. Bot.* 32:93–101

43. Drouet, A., Hartmann, C. 1982. Polyribosomes from pear fruit. II. Changes occurring in pulp tissues during ripening and senescence. *Plant Physiol.* 69:885–87

44. Eaks, I. L. 1985. Effect of calcium on

ripening, respiratory rate, ethylene production and quality of avocado fruit. *J. Am. Soc. Hortic. Sci.* 110:145–48

45. Ferguson, I. B. 1984. Calcium in plant senescence and fruit ripening. *Plant Cell Environ.* 7:477–89

46. Frenkel, C., Dyck, R. 1973. Auxin inhibition of ripening in Bartlett pears. *Plant Physiol.* 51:6–9

47. Frenkel, C., Eskin, M. 1977. Ethylene evolution as related to changes in hydroperoxides in ripening tomato fruits. *HortScience* 12:552–53

48. Frenkel, C., Garrison, S. A. 1976. Initiation of lycopene synthesis in the tomato mutant *rin* as influenced by oxygen and ethylene interactions. *HortScience* 11:20–21

49. Frenkel, C., Klein, I., Dilley, D. R. 1968. Protein synthesis in relation to ripening of pome fruits. *Plant Physiol.* 43:1146–53

50. Frenkel, C., Mukai, M. K. 1984. Possible role of fruit cell wall oxidative activity in ethylene evolution. See Ref. 52, pp. 303–16

51. Friend, J., Rhodes, M. J. C. 1981. *Recent Advances in the Biochemistry of Fruit and Vegetables.* London: Academic. 275 pp.

52. Fuchs, Y., Chalutz, E., eds. 1984. *Ethylene: Biochemical, Physiological and Applied Aspects.* The Hague: Martinus Nijhoff/Dr. W. Junk. 348 pp.

53. Galliard, T. 1968. Aspects of lipid metabolism in higher plants. II. The identification and quantitative analysis of lipids from the pulp of pre– and post–climacteric apples. *Phytochemistry* 7:1915–22

54. Goodenough, P. W., Prosser, I. M., Young, K. 1985. NADP-linked malic enzyme and malate metabolism in ageing tomato fruits. *Phytochemistry* 24:1157–62

55. Grierson, D. 1985. Gene expression in ripening tomato fruit. *CRC Crit. Rev. Plant Sci.* 3:113–32

56. Grierson, D., Slater, A., Maunders, M., Crookes, P., Tucker, G. A., et al. 1985. Regulation of the expression of tomato fruit ripening genes: the involvement of ethylene. In *Ethylene and Plant Development,* ed. J. Roberts, G. A. Tucker, pp. 147–61. Oxford: Butterworths. 416 pp.

57. Grierson, D., Slater, A., Speirs, J., Tucker, G. A. 1985. The appearance of polygalacturonase mRNA in tomatoes—one of a series of changes in gene expression during development and ripening. *Planta* 163:263–71

58. Grierson, D., Tucker, G. A., Robertson, N. G. 1981. The molecular biology of ripening. See Ref. 51, pp. 147–58
59. Grierson, D., Tucker, G. A., Robertson, N. G. 1981. The regulation of gene expression during the ripening of tomato fruits. In *Quality in Stored and Processed Vegetables and Fruit*, ed. P. W. Goodenough, R. K. Atkin, pp. 179–91. London: Academic. 398 pp.
60. Gross, K. C., Wallner, S. J. 1979. Degradation of cell wall polysaccharides during tomato fruit ripening. *Plant Physiol.* 63:117–20
61. Harman, J., Hobson, G. E., Grimbly, P. E. 1984. The effect of salt treatment on the ripening of slow and 'non-ripening' tomato mutants. *Plant Physiol.* 75S:57 (Abstr.)
62. Hatfield, R., Nevins, D. J. 1986. Characterization of the hydrolytic activity of avocado cellulase. *Plant Cell Physiol.* 27:541–42
63. Hepler, P. K., Wayne, R. O. 1985. Calcium and plant development. *Ann. Rev. Plant Physiol.* 36:397–439
64. Hobson, G. E. 1979. What factors are involved in the onset of ripening in climacteric fruit? *Curr. Adv. Plant Sci.* 37:1–11
65. Hobson, G. E. 1980. Effect of the introduction of non-ripening mutant genes on the composition and enzyme content of tomato fruit. *J. Sci. Food Agric.* 31:578–84
66. Hobson, G. E., Nichols, R., Davies, J. N., Atkey, P. T. 1984. The inhibition of tomato fruit ripening by silver. *J. Plant Physiol.* 116:21–29
67. Hobson, G. E., Richardson, C., Gillham, D. J. 1983. Release of protein from normal and mutant tomato cell walls. *Plant Physiol.* 71:635–38
68. Hoffman, N. E., Yang, S.-F. 1980. Changes in 1-amino-cyclopropane-1-carboxylic acid content in ripening fruits in relation to their ethylene production rates. *J. Am. Soc. Hortic. Sci.* 105:492–95
69. Huber, D. J. 1983. Polyuronide degradation and hemicellulose modifications in ripening tomato fruit. *J. Am. Soc. Hortic. Sci.* 108:405–9
70. Huber, D. J. 1983. The role of cell wall hydrolases in fruit softening. *Hortic. Rev.* 5:169–219
71. Huber, D. J. 1984. Strawberry *(Fragaria ananassa)* fruit softening, the potential roles of polyuronides and hemicelluloses. *J. Food Sci.* 49:1310–15
72. Iki, K., Sekiguchi, K., Kurata, K., Tada, T., Nakagawa, H., et al. 1978. Immunological properties of β-fructo-furanosidase from ripening fruit. *Phytochemistry* 17:311–12
73. Jarret, R. L., Tigchelaar, E. C., Handa, A. K. 1984. Ripening behaviour of the green-ripe tomato, *Lycopersicon esculentum* mutant. *J. Am. Soc. Hortic. Sci.* 109:712–17
74. Jerie, P. H., Hall, M. A., Zeroni, M. 1978. Aspects of the role of ethylene in fruit ripening. *Acta Hortic.* 80:325–32
75. Kalra, S. K., Brooks, J. L. 1973. Lipids of ripening tomato fruit and its mitochondrial fraction. *Phytochemistry* 12:487–92
76. Khudairi, A. K. 1972. The ripening of tomatoes. *Am. Sci.* 60:696–707
77. Kitamura, T., Itamura, H., Fukushima, T. 1983. Ripening changes in respiration ethylene emanation and abscisic acid content of plum *(Prunus salicina)* fruit. *J. Jpn. Soc. Hortic. Sci.* 52:325–31 (In Japanese)
78. Kokkinakis, D. M., Brooks, J. L. 1979. Hydrogen-peroxide mediated oxidation of indole-3-acetic acid by tomato peroxidase and molecular oxygen. *Plant Physiol.* 64:220–23
79. Lambers, H., Szaniawski, R. K., de Visser, R. 1983. Respiration for growth, maintenance and ion uptake. An evaluation of concepts, methods, values and their significance. *Physiol. Plant* 45:351–56
80. Laties, G. G. 1982. The cyanide-resistant, alternative path in higher plant respiration. *Ann. Rev. Plant Physiol.* 33:519–55
81. Lesham, Y. Y. 1984. Interaction of cytokinins with lipid-associated oxy free radicals during senescence: a prospective mode of cytokinin action. *Can. J. Bot.* 62:2943–49
82. Liang, Y. F., Ao, L. D., Wang, M. X., Yi, W. J. 1982. The role of polygalacturonase in the ripening of apple fruits. *Acta Bot. Sin.* 24:143–46
83. Liu, Y., Hoffman, N. E., Yang, S.-F. 1985. Promotion by ethylene of the capability to convert 1-aminocyclopropane-1-carboxylic acid to ethylene in preclimacteric tomato and cantaloupe fruits. *Plant Physiol.* 77:407–11
84. Lobo, M., Bassett, M. J., Hannah, L. C. 1984. Inheritance and characterization of the fruit ripening mutation in cultivar 'alcobaca' tomato, *Lycopersicon esculentum.* *J. Am. Soc. Hortic. Sci.* 109:741–45
85. Lurie, S., Ben-Aire, R. 1983. Microsomal membrane changes during the ripening of apple fruit. *Plant Physiol.* 73:636–38

86. Majmudar, G., Modi, V. V., Palejwala, V. A. 1981. Effect of plant growth regulators on mango ripening. *Indian J. Exp. Biol.* 19:885–86
87. Mansson, P.-E., Hsu, D., Stalker, D. 1985. Characterization of fruit specific cDNAs from tomato. *Mol. Gen. Genet.* 200:356–61
88. Marriott, J. 1980. Bananas—physiology and biochemistry of storage and ripening for optimum quality. *CRC Crit. Rev. Food Sci. Nutr.* 13:41–88
89. Matile, P. 1980. Catabolism of chlorophyll: involvement of peroxidase? *Z. Pflanzenphysiol.* 99:475–78
90. Mayak, S., Legge, R. L., Thompson, J. E. 1983. Superoxide production by microsomal membranes from senescing carnation flowers: an effect on membrane fluidity. *Phytochemistry* 22:1375–80
91. McGlasson, W. B., Dostal, H. C., Tigchelaar, E. C. 1975. Comparison of propylene-induced responses of immature fruit of normal and mutant tomatoes. *Plant Physiol.* 85:218–22
92. McGlasson, W. B., Pratt, H. K. 1964. Effects of ethylene on cantaloupe fruits harvested at various ages. *Plant Physiol.* 39:120–27
93. McGlasson, W. B., Wade, N. L., Adato, I. 1978. Phytohormones and stress phenomena. In *Phytohormones and Related Compounds—A Comprehensive Treatise*, ed. D. S. Letham, P. B. Goodwin, T. J. V. Higgins, 2:447–93. Amsterdam: Elsevier/North Holland Biomed. 648 pp.
94. McKersie, B. D., Thompson, J. E. 1977. Lipid crystallization in senescent membranes from cotyledons. *Plant Physiol.* 59:803–7
95. McMurchie, E. J., McGlasson, W. B., Eaks, I. L. 1972. Treatment of fruit with propylene gives information about the biogenesis of ethylene. *Nature* 237:235–36
96. Mizrahi, Y. 1982. Effect of salinity on tomato fruit ripening. *Plant Physiol.* 69:966–70
97. Mizrahi, Y., Dostal, H. C., Cherry, J. H. 1975. Ethylene induced ripening in attached *rin* fruits, a non-ripening mutant of tomato. *HortScience* 10:414–15
98. Mizrahi, Y., Dostal, H. C., McGlasson, W. B., Cherry, J. H. 1975. Effects of abscisic acid and benzyladenine on fruits of normal and *rin* mutant tomatoes. *Plant Physiol.* 56:544–46
99. Moreau, F., Romani, R. 1982. Malate oxidation and cyanide insensitive respiration in avocado, *Persea america-*

na, mitochondria during the climacteric cycle. *Plant Physiol.* 70:1385–90
100. Morgens, P. H., Pyle, J. B., Hershberger, W. L., Dunn, L. J., Abeles, F. B., Callahan, A. M. 1985. Molecular events in tomato fruit ripening. *1st Int. Congr. Plant Mol. Biology, Savannah*, p. 98 (Abstr.)
101. Moshrefi, M., Luh, B. S. 1983. Carbohydrate composition and electrophoretic properties of tomato polygalacturonase isoenzymes. *Eur. J. Biochem.* 135:511–14
102. Mösinger, E., Schopfer, P. 1983. Polysome assembly and RNA synthesis during phytochrome-mediated photomorphogenesis in mustard cotyledons. *Planta* 158:501–11
103. Mousdale, D. A., Knee, M. 1982. Indole-3-acetic acid and ethylene levels in ripening apple fruits. *J. Exp. Bot.* 32:753–58
104. Mowlah, G., Itoo, S. 1983. Changes in pectic components, ascorbic acid, pectic enzymes and cellulase activity in ripening and stored guava (*Psidium guajava* L.). *J. Jpn. Soc. Food Sci. Technol.* 30:454–61
105. Mutschler, M. A. 1984. Inheritance and linkage of the 'Alcobaca' ripening mutant in tomato. *J. Am. Soc. Hortic. Sci.* 109:500–3
106. Ness, P. J., Romani, R. J. 1980. Effects of aminoethoxyvinylglycine and counter effects of ethylene on ripening of 'Bartlett' pear fruits. *Plant Physiol.* 65:372–76
107. Paliyath, G., Poovaiah, B. W. 1985. Calcium- and calmodulin-promoted phosphorylation of membrane proteins during senescence in apples. *Plant Cell Physiol.* 26:977–86
108. Paliyath, G., Poovaiah, B. W., Munske, G. R., Magnuson, J. A. 1984. Membrane fluidity in senescing apples: effects of temperature and calcium. *Plant Cell Physiol.* 25:1083–87
109. Paull, R. E., Chen, N. J. 1983. Postharvest variation in cell wall degrading enzymes of papaya (*Carica papaya* L.) during fruit ripening. *Plant Physiol.* 72:382–85
110. Paull, R. E., Deputy, J., Chen, N. J. 1983. Changes in organic acids, sugars and head space volatiles during fruit ripening of soursop (*Annona muricata* L.). *J. Am. Soc. Hortic. Sci.* 108:931–34
111. Penning de Vries, F. W. T. 1975. The cost of maintenance processes in plant cells. *Ann. Bot.* 39:77–92
112. Piechulla, B., Imlay, K. R. C., Gruissem, W. 1985. Plastid gene expression

during fruit ripening in tomato. *Plant Mol. Biol.* 5:373–84

113. Platt-Aloia, K. A., Thomson, W. W. 1981. Ultrastructure of the mesocarp of mature avocado fruit and changes associated with ripening. *Ann. Bot.* 48:451–65

114. Platt-Aloia, K. A., Thomson, W. W. 1985. Freeze-fracture evidence of gel-phase lipid in membranes of senescing cowpea cotyledons. *Planta* 163:360–69

115. Platt-Aloia, K. A., Thomson, W. W., Young, R. E. 1980. Ultrastructural changes in the walls of ripening avocados: transmission, scanning and freeze fracture microscopy. *Bot. Gaz.* 141: 366–73

116. Poovaiah, B. W., Mizrahi, Y., Dostal, H. C., Cherry, J. H., Leopold, A. C. 1975. Water permeability during tomato fruit development in normal and *rin* non-ripening mutant. *Plant Physiol.* 56:813–15

117. Pressey, R. 1984. Purification and characterization of tomato polygalacturonase converter. *Eur. J. Biochem.* 144: 217–21

118. Raghothama, K. G., Veluthambi, K., Poovaiah, B. W. 1985. Regulation of protein phosphorylation by calcium and polyamines in developing tomato fruits. *Plant Physiol.* 77S:4 (Abstr.)

119. Rai, R. M., Tewari, J. D., Pant, N. 1983. Physiological effect of growth regulators on ripening of cultivar early—Shanbury apples. *Prog. Hortic.* 15:276–82

120. Rasi-Caldogno, F., De Michelis, M. I. 1978. Correlation between oxygen availability, energy charge and protein synthesis in squash cotyledons isolated from germinating seed. *Plant Physiol.* 61:85–88

121. Rhodes, M. J. C. 1980. Respiration and senescence of plant organs. In *The Biochemistry of Plants: A Comprehensive Treatise*, ed. P. K. Stumpf, E. E. Conn, 2:419–62. London: Academic. 693 pp.

122. Richmond, A., Biale, J. B. 1966. Protein and nucleic acid metabolism in fruits. I. Studies of amino acid incorporation during the climacteric rise of avocado. *Plant Physiol.* 41:1247–53

123. Richmond, A., Biale, J. B. 1967. Protein and nucleic acid metabolism in fruits. II. RNA synthesis during the respiratory rise of the avocado. *Biochim. Biophys. Acta* 138:625–27

124. Roe, B., Bruemmer, J. H. 1981. Changes in pectic substances and enzymes during ripening and storage of

'Keitt' mangoes. *J. Food Sci.* 46:186–89

125. Romani, R. 1984. Respiration, ethylene, senescence and homeostasis in an integrated view of postharvest life. *Can. J. Bot.* 62:2950–55

126. Romani, R., Labavitch, J., Yamashita, T., Hess, B., Rae, H. 1983. Pre-harvest aminoethoxyvinyl glycine treatment of cultivar Bartlett pear fruits: effects on ripening color change and volatiles. *J. Am. Soc. Hortic. Sci.* 108:1046–49

127. Ryan, C. A., Bishop, P., Pearce, G., Darvill, A. G., McNeil, M., Albersheim, P. 1981. A sycamore cell wall polysaccharide and a chemically related tomato leaf polysaccharide possess similar proteinase inhibitor-inducing activities. *Plant Physiol.* 68:616–18

128. Sacher, J. A. 1973. Senescence and postharvest physiology. *Ann. Rev. Plant Physiol.* 24:197–224

129. Saltveit, M. E., Bradford, K. J., Dilley, D. R. 1978. Silver ion inhibits ethylene synthesis and action in ripening fruits. *J. Am. Soc. Hortic. Sci.* 103:472–75

130. Sato, T., Kusaba, S., Nakagawa, H., Ogura, N. 1984. Cell-free synthesis of a putative precursor of polygalacturonase in tomato fruits. *Plant Cell Physiol.* 25:1069–71

131. Sato, T., Kusaba, S., Nakagawa, H., Ogura, N. 1985. Polygalacturonase mRNA of tomato: size and content in ripe fruit. *Plant Cell Physiol.* 26:211–14

132. Schuster, A. M., Davies, E. 1983. RNA and protein metabolism in pea epicotyls. I. The aging process. *Plant Physiol.* 73:809–16

133. Simon, E. W. 1977. Leakage from fruit cells in water. *J. Exp. Bot.* 28:1147–52

134. Slater, A., Maunders, M. J., Edwards, K., Schuch, W., Grierson, D. 1985. Isolation and characterization of cDNA clones for tomato polygalacturonase and other ripening-related proteins. *Plant Mol. Biol.* 5:137–47

135. Solomos, T. 1983. Respiration and energy metabolism in senescing plant tissues. In *Post-Harvest Physiology and Crop Improvement*, ed. M. Lieberman, pp. 61–98. New York: Plenum 572 pp.

136. Solomos, T., Laties, G. G. 1973. Cellular organization and fruit ripening. *Nature* 245:390–92

137. Speirs, J., Brady, C. J., Grierson, D., Lee, E. 1984. Changes in ribosome organization and messenger RNA abundance in ripening tomato fruits. *Aust. J. Plant Physiol.* 11:225–34

138. Su, L-Y., McKeon, T., Grierson, D., Cantwell, M., Yang, S.-F. 1984. Development of 1-aminocyclopropane-1-

carboxylic acid synthase and polygalacturonase activities during the maturation and ripening of tomato fruit. *HortScience* 19:576–78

139. Terai, H., Mizuno, S. 1985. Changes in 1-aminocyclopropane-1-carboxylic acid and ethylene forming enzyme activity in growing and ripening fruits of tomato and cucumber. *J. Jpn. Soc. Hortic. Sci.* 53:467–73

140. Themmer, A. P. N., Tucker, G. A., Grierson, D. 1982. The degradation of isolated tomato walls by purified polygalacturonase in vitro. *Plant Physiol.* 60:122–24

141. Tigchelaar, E. C., McGlasson, W. B., Buescher, R. W. 1978. Genetic regulation of tomato fruit ripening. *HortScience* 13:508–13

142. Trewavas, A. 1981. How do plant growth substances work? *Plant Cell Environ.* 4:203–28

143. Tsay, L. M., Mizuno, S., Kozukue, N. 1984. Changes in respiration, ethylene evolution and abscisic acid content during ripening and senescence of fruits picked at young and mature stage. *J. Jpn. Soc. Hortic. Sci.* 52:458–63

144. Tucker, G. A., Grierson, D. 1982. Synthesis of polygalacturonase during tomato fruit ripening. *Planta* 155:64–67

145. Tucker, G. A., Robertson, N. G., Grierson, D. 1980. Changes in polygalacturonase isoenzymes during the 'ripening' of normal and mutant tomato fruit. *Eur. J. Biochem.* 112:119–24

146. Tucker, G. A., Schindler, C. B., Roberts, J. A. 1984. Flower abscision in mutant tomato, *Lycopersicon,* plants. *Planta* 160:164–67

147. Tucker, M. L., Laties, G. G. 1984. Interrelationship of gene expression, polysome prevalence and respiration during ripening of ethylene and/or cyanide treated avocado fruit, *Persea americana* cultivar Hass. *Plant Physiol.* 74: 307–15

148. Veen, H. 1983. Silver thiosulphate: an experimental tool in plant science. *Sci. Hortic.* 20:211–24

149. Veen, H. 1985. Antagonistic effect of silver thiosulphate or 2,5-norbornadiene on 1-aminocyclopropane-1-carboxylic

acid stimulated growth of pistils in carnation buds. *Physiol. Plant* 65:2–8

150. Vendrell, M. 1969. Reversion of senescence: effects of 2,4-dichlorophenoxyacetic acid and indoleacetic acid on respiration, ethylene production and ripening of banana fruit slices. *Aust. J. Biol. Sci.* 22:601–10

151. Vendrell, M. 1985. Effect of abscisic acid and ethephon on several parameters of ripening in banana fruit tissue. *Plant Sci.* 40:19–24

152. Vickery, R. S., Bruinsma, J. 1973. Compartments and permeability for potassium in developing fruits of tomato (*Lycopersicon esculentum* Mill.). *J. Exp. Bot.* 24:1261–70

153. Wade, N. L., Bishop, D. G. 1978. Changes in the lipid composition of ripening banana fruits and evidence for an associated increase in cell membrane permeability. *Biochim. Biophys. Acta* 529:454–64

154. Watada, A. E., Herner, R. C., Kader, A. A., Romani, R. J., Staby, G. L. 1984. Terminology for the description of developmental stages of horticultural crops. *HortScience* 19:20–21

155. Woolhouse, H. W., Batt, T. 1976. The nature and regulation of senescence in plastids. In *Perspectives in Experimental Biology, Botany,* ed. N. Sunderland, 2:163–75. Oxford: Pergamon 523 pp.

156. Wright, S. T. C., Hiron, R. W. 1970. The accumulation of abscisic acid in plants during wilting and under other stress conditions. In *Plant Growth Substances,* ed. D. Carr, pp. 291–98. Berlin: Springer-Verlag. 837 pp.

157. Wyn Jones, R. G., Brady, C. J., Speirs, J. 1979. Ionic and osmotic relations in plant cells. In *Recent Advances in the Biochemistry of Cereals,* ed. D. L. Laidman, R. G. Wyn Jones, pp. 63–103. London: Academic. 391 pp.

158. Yang, S.-F., Hoffman, N. E. 1984. Ethylene biosynthesis and its regulation in higher plants. *Ann. Rev. Plant Physiol.* 35:155–89

159. Zhang, W., Lu, Z.-S. 1983. Relationship between abscisic acid and apricot (*Prunus armeniacae*) fruit ripening. *Acta Bot. Sin.* 25:537–43

Ann. Rev. Plant Physiol. 1987. 38:179–204

DIFFERENTIATION OF VASCULAR TISSUES

Roni Aloni

Department of Botany, Tel Aviv University, Tel Aviv 69978, Israel

CONTENTS

INTRODUCTION 179
STRUCTURE AND FUNCTION OF VASCULAR TISSUES 180
VASCULAR MERISTEMS 182
THE ROLE OF CELL DIVISION IN VASCULAR DIFFERENTIATION 184
CONTROL OF CONDUIT SIZE AND DENSITY 186
    General Patterns of Vascular Elements 186
    The Problem of Conduit Size Control 187
    Control of Vascular Adaptation–a New Hypothesis 188
CONTROL OF PHLOEM AND XYLEM DIFFERENTIATION 190
    Induction of Vascular Tissues by Leaves and by Auxin 190
    The Role of Roots and Cytokinin in Vascular Differentiation 191
    Effects of Pressure and Ethylene on Vascular Tissues 192
    The Relation Between Phloem and Xylem Differentiation 192
CONTROL OF FIBER DIFFERENTIATION 194
    The Role of Leaves, Auxin, and Gibberellin in Fiber Induction 194
    The Role of Roots and Cytokinin in Fiber Differentiation 196
CONCLUDING REMARKS 197

INTRODUCTION

The study of vascular differentiation, from the level of the individual cell to that of the entire plant, fascinates imaginative students. The complexity of vascular differentiation raises endless questions about control mechanisms, some of which are discussed here. Since vascular differentiation was last reviewed in this series by Shininger in 1979 (138), the subject has expanded at an impressive rate; many publications have appeared, including excellent

0066-4294/87/0601-0179$02.00

reviews by Sachs (125), Savidge & Wareing (133), and Jacobs (85). A monograph on vascular differentiation and plant growth regulators is in preparation (120).

I provide a summary of the structure and function of vascular tissues for those unfamiliar with the subject and as a preamble to a discussion of the control of vascular differentiation. In the hope that further research will yield a better understanding of vascular development, attention is given to the vascular meristems and the role of cell division in the process of differentiation.

I focus on three major topics in vascular differentiation and the recent advances made in each. The first is the control of cell dimensions in vascular systems, wherein the diameter of the vessels is the important parameter for evaluating the ascent of sap and the adaptation of plants to their environment. Here I put forward a new hypothesis on the control of vascular adaptation which I hope will stimulate research to test its applicability in different plants. The second topic is the hormonal control of phloem and xylem differentiation. Here I weigh the evidence for and against the hypothesis that sugar is a factor controlling the relations between phloem and xylem. The third topic is growth regulators and their role in the control of fiber differentiation. A study of cell differentiation in the vascular system has enormous economic importance in terms of increasing fiber yield in industrial plants and improving wood productivity and quality.

STRUCTURE AND FUNCTION OF VASCULAR TISSUES

The vascular tissues of the plant connect the leaves and other parts of the shoot with the roots and thus enable efficient long-distance transport between the organs. The vascular system is composed of two kinds of conducting tissues: the *phloem,* through which organic materials are transported; and the *xylem,* which is the conduit for water and soil-derived nutrients. Vascular development in the plant is an open type of differentiation process persisting throughout the growth period from apical and lateral meristems (52). New vascular tissues develop continuously in dynamic relationship to one another. This continuous development of new vascular tissues enables regeneration of the plant and its adaptation to interruptions and changes in the environment.

In the plant kingdom, the earliest examples of an organized vascular system are encountered among the brown algae, which possess phloem with no xylem (e.g. 53). The water-conducting system evolved much later; its adaptation and refinement became mandatory for the development of plants on land, because the lignified conducting and supporting cells in the xylem provide the long-distance water conduction and the mechanical support needed in a terrestrial habitat.

The vascular tissues are complex, being comprised of several types of cells. The conducting cells in the phloem are the *sieve cells* in gymnosperms and the highly specialized *sieve elements* in angiosperms (e.g. see 53, 62). During their maturation they undergo a selective autophagy, and the plasmodesmata in their end walls are enlarged to form sieve-area pores. This results in continuity of the cytoplasm of contiguous sieve-tube members along the sieve tubes (54). At maturity the sieve elements lack nuclei. For further details on phloem structure and function, the reader is directed to several reviews (e.g. 38, 56, 57, 78).

In the xylem the conducting elements are the tracheary cells. These function, in the conduction of water, as nonliving cells after autolysis of their cytoplasm. Tracheary cells are characterized by secondary wall thickenings that enable them to retain their shape when dead, despite the pressure of the surrounding cells. Two functional conduits are recognizable in the xylem: the *tracheid,* which is a nonperforated long cell, with bordered pits; and the *vessel,* which is a long continuous tube made up of numerous vessel elements connected end-to-end by perforation plates and limited in length by imperforate walls at both extremities.

It is important to point out that the vessels and not vessel elements are the operating conduction units of the xylem in angiosperms. Therefore, their dimensions are important parameters in the study of long-distance water transport, xylem adaptations, or pathology of the xylem. Thus, for instance, increase in the vessel diameter markedly increases the efficiency of water conduction, owing to decrease in the resistance to flow, whereas increase in both vessel diameter and length decreases safety (in terms of cavitation) of water conduction (166).

Vessels do not end randomly in the young stems. In fact, the frequency of vessel endings in the young stem of *Vitis, Populus, Olea* (129), and *Medicago* (154) is significantly higher in the nodes. Consequently, the node regions are considered "safety zones" because gaseous emboli and fungal spores fail to pass through the endings (129). An extreme example of the nodes as safe zones is found in the palm *Rhapis excelsa*. This species has wide metaxylem vessels extending through the stem that must remain functional for many years. Because the palm stem lacks a cambium, new regenerative vessels will not differentiate in case of damage. The leaves, which are the disposable organs, also have wide metaxylem vessels. At the node the vessel system of a leaf connects to the vessels of the stem via narrow tracheids. The tracheids in the node region protect the vessels of the stem from cavitation when a leaf drops off (170). Vascular tissues of angiosperm and gymnosperm trees are highly compartmented. In a sense, trees are multiple plants built up of long cones of tissues in a concentric pattern. Each growth season usually results in a separated cone. This compartmentation pattern within trees helps them and other perennial plants to protect themselves against injuries (135). Wounding induces regenerative vessels that are short and narrow; they are therefore safer than normal vessels and have an adaptive value in case of repeated injury (19).

In large shrubs and diffuse-porous trees the longest vessels are about 1 m long (168). In ring-porous species the largest vessels formed in the earlywood reach the length of the stem itself (75, 168). A maximum vessel length of 18 m was measured in *Fraxinus* (75). Vessel length distribution analysis however, has shown that generally the xylem comprises many more short vessels than long ones (168). Lianas have very large (up to 8 m or more) and wide (up to 500 μm) vessels (59, 168). The large vessels in lianas remain conductive for two to several years, whereas the wide earlywood vessels in the ring-porous species conduct water for only one year (59, 166).

Measurements in the ring-porous species *Ulmus americana* have shown that over 90% of the water flows through the outermost ring, mainly through its wide earlywood vessels; in the inner rings, only the narrow vessels of the latewood are functional (51). Lianas were found to be the most efficient in terms of water conductivity compared to other woody species (59).

The safer conduit (vis-à-vis freezing and gas-induced cavitation or embolism) in the xylem is the tracheid (166), though there are recent findings that contradict this view (147). This relatively hardy xylem element explains why conifers, which are dependent upon tracheids for water transport, are such effective competitors in cold and dry habitats (59).

Fibers are long and narrow cells possessing thick secondary walls that are usually heavily lignified. They are found in both the phloem and the xylem. Primary fibers arise from short initials that greatly elongate in the course of their maturation, mainly by intrusive growth at their apical and basal ends. In *Boehmeria nivea* the fiber initials in the primary phloem are 10 μm long and may grow to 55 cm (1). These fibers are present through several internodes in the mature stem. Development of fibers in *B. nivea* is gradual and may take months. In *Coleus*, however, the development of primary phloem fibers requires only a few weeks (4) and in *Pisum* it is a matter of days (70).

VASCULAR MERISTEMS

In the plant body the vascular tissues are formed from embryonic tissues, called vascular meristems, whose cells retain the ability to divide and continually multiply. This ensures flexibility and adaptation of the vascular system to constant changes within and outside the plant. Apart from that, vascular elements are formed also by re-differentiation of parenchyma cells. This occurs where lateral roots interconnect the vascular system of the main axis and also around wounds (e.g. 14).

Two meristematic stages are distinguished: the procambium and the cambium. The *procambium* is the apical meristem that produces primary phloem and primary xylem in the embryo and in the young portions of shoots and roots. For a review of procambium development, see Shininger (138). The *cambium* is a lateral meristem found in gymnosperms and dicotyledons. It develops in the older parts of the plant axis, where it produces the secondary phloem and secondary xylem. There is a vast literature on the cambium, and the reader is referred to selected reviews (e.g. 62, 98, 113, 125, 133).

Larson (97) proposed to recognize also the metacambium, as an intermediate stage between the procambium and cambium. However, most students of vascular meristems proclaim the difficulties in determining the transition between the procambium and the cambium and emphasize that the transition between the meristems is a gradual process with no clear demarcation. Accordingly, the procambium and the cambium are usually regarded today as a continuum representing two developmental stages of a single meristem (e.g. see 98).

The cambium stage is first detected in the cotyledonary node of seedlings (127) and in the young portions of the stem within the nodal regions (29). The appearance of the cambium in the nodal regions is ascribed to higher levels of auxin in these regions (29, 127). This ascription is in accordance with results of experimental wounding of young stems whose meristems were in the procambial stage. Such wounding induces a cambial stage alongside the wound owing to a concentration of the stimuli for vascular differentiation within a narrow area around the injury, leaving procambium both above and below the wound (63).

The vascular meristems are polar in both the radial and longitudinal directions. The radial polarity, which is determined in a very early stage of stem development (140), and also the longitudinal polarity are putatively a consequence of radial and longitudinal streams of polar inductive stimuli operating in these directions rather than the result of relatively static gradients of auxin and sucrose in the transverse direction, as proposed by Warren Wilson (150). Findings from partial girdling of trunks of *Ailanthus,* achieved by Carmi et al (34), suggest that the ray initials in the cambium differentiate along channels of stimulus moving between the differentiating phloem and the differentiating xylem and that their pattern and spacing are controlled not in the cambium but in the differentiating vascular tissues. Although the ray-inducing stimulus remains to be identified, a difference has been detected in the rate of auxin transport within the rays in the centrifugal versus the centripetal direction (165). This difference in the auxin transport within the rays may contribute to the control of the radial polarity of the vascular meristems.

The longitudinal polarity of the vascular meristems is attributed to the polar flow of auxin from leaves to roots (41, 117, 131, 133, 134). In the cambium region, auxin apparently moves polarly in waves or oscillations of transport rates (155, 161, 162). The movement of a controlling stimulus within the cambium in wave pattern might explain the occurrence of morphogenetic waves in structure and activity of the cambium and in its derivative cells (77, 142). Both auxin and gibberellic acid were found to control cambium activity (90, 149, 162).

Convincing evidence that endogenous factors within the meristematic cells control cambium activity comes from the tropical tree *Avicennia*. As a tropical tree, it does not have a dormant period but rather functions continuously, producing successive anomalous cambia that form alternating rings of xylem and phloem (72, 163). Gill (72) found that two to six rings can be formed during one year and that the number of rings is correlated to the diameter of the branch. Zamski (164) has shown that there is no simple concordance between the initiation of the successive cambia and the sequence of leaf and lateral branch formation. Both these investigators suggest that an endogenous factor controls the autonomous activity of the cambium in *Avicennia*.

The focus of any study of meristem differentiation should be on (*a*) the nature and physiology of the inductive stimuli and the influence of the development of the plant organs and changes in the environment of the plant on their levels and movement in the meristem; and (*b*) the endogenous factors in meristematic cells that control their activity, autonomic rhythms, and sensitivity to the inducing stimuli.

THE ROLE OF CELL DIVISION IN VASCULAR DIFFERENTIATION

In intact plants the phloem and xylem elements differentiate from new cells that are continuously produced in the procambium or cambium by the process of cell division. Additionally, DNA synthesis is often observed during the process of differentiation, and this results in endopolyploidy and the production of multinucleate stages of the differentiating vascular elements. The association of vascular differentiation with DNA synthesis and cell division raises questions as to whether cell division prior to differentiation is a prerequisite for gene reprogramming, and whether endoduplication of DNA is needed to control the process of differentiation.

In the phloem of angiosperms, cell division occurs during the process of differentiation of the sieve element and its companion cell. Both these cells are ontogenetically related because they develop from the same meristematic cell (53). The meristematic parent cell divides once or several times. Usually, the larger daughter cell will become the sieve element and one or more of the other daughter cells will develop to companion cells. The questions raised above have not yet been resolved in studies of the phloem.

Fibers also show nuclear division activity during their development. The nucleus of the fiber cell divides several times in the course of fiber development (88, 116). Because these nuclear divisions are not usually followed by cytokinesis, they result in multinucleate fibers. Occasionally, cytokinesis occurs by formation of a few septa at the later stages of fiber differentiation (88, 116). Nuclear divisions in *Agava* may be accompanied by chromosome multiplication (42). There is evidence that fiber initials in cultured hypocotyl segments of *Helianthus* need cytokinin at the stage of nuclear division and will not differentiate into fibers in the absence of the hormone (7). This finding seems to suggest the need for more than one operating nucleus in the process of fiber differentiation to enable efficient control on the intrusive growth at the two ends of a very long cell.

Most of the evidence for and against cell division as a prerequisite for differentiation was accumulated on tracheary elements. There is evidence that parenchyma cells can differentiate into tracheary elements without DNA synthesis or cell division (e.g. 68, 69, 146). Usually, dramatic alterations occur in the structure and content of the nucleus during the early stages of tracheary element differentiation. The nuclear volume substantially increases and is accompanied by endopolyploidy (81, 92, 101). Endoduplication may raise the DNA contents to levels of 4C (50), 8C, 16C (99, 115, 145) or even 64C (101). In *Marsilea* the differentiating tracheary elements become multi-nucleate during their development (101). The regulatory role of DNA content level in differentiating fibers or tracheary elements has yet to be confirmed.

Evidence in support of the hypothesis that cell division must precede differentiation comes from the studies of Fosket (66) with explants of *Coleus* stems and of Shininger (137) with explants of pea roots. These investigators have shown that inhibition of DNA synthesis with 5-fluorodeoxyuridine (FUdR) abrogated DNA synthesis or cell division in the explants and totally prevented tracheary element formation. Malawer & Phillips (103), with the aid of tritiated (H^3) thymidine applied to Jerusalem artichoke explants, have demonstrated that all the differentiating tracheary cells incorporate the label, owing to DNA synthesis prior to differentiation.

On the other hand, the counter hypothesis that DNA synthesis is not essential for differentiation and that tracheary elements can differentiate "directly" without cell division is clearly supported by studies with isolated mesophyll cells of *Zinnia elegans* (e.g. 68, 69). Fukuda & Komamine (69) have further shown that the molecular events responsible for gene reprogramming following hormonal stimulation for xylogenesis occur during the G_1 phase in the cell cycle (the period before DNA synthesis) and can proceed in the *Zinnia* system independently of the progression of the cell cycle.

Following the model proposed for milk protein formation (148), Dodds (49) has suggested that the G_1 phase may be subdivided into two phases by a "critical event" for the initiation of xylogenesis. Accordingly, cells like those in the *Zinnia* system, which differentiate "directly" to tracheary elements, are in the "early" stage of G_1 phase and are able to pass through the "critical event(s)" episode of G_1 and differentiate without needing to undergo DNA synthesis. Those cells that require division prior to differentiation are arrested at the "later" stage of the G_1 phase and consequently have to go through the cell cycle (DNA synthesis and mitosis) in order to surpass the "critical event" and differentiate into tracheary elements. However, the nature of the "critical event(s)" is still unknown and there is need for further research to confirm Dodds' suggestion. Results reported by Phillips & Arnott (114) do not support Dodds' hypothesis. A useful summary on the general subject of cell division cycle in plants has been recently published (30a).

CONTROL OF CONDUIT SIZE AND DENSITY

General Patterns of Vascular Elements

Nehemiah Grew (1641–1712) was the first to note that vessels in the roots are generally wider than those in the trunk (21). Sanio (130) subsequently discovered the general increase in the size of tracheids in the secondary xylem of *Pinus sylvestris,* in the stem and branches—an increase proceeding outwards from the inner growth ring through a number of annual growth rings until a constant size is attained. He also found that tracheids are smaller in branches and bigger in the trunk. These findings of Sanio were later confirmed in *Sequoia sempervirens* by Bailey (25), who clearly demonstrated a continuous increase in the length and width of secondary tracheids proceeding from branches to trunk and down into the roots. Such a general increment in tracheid size from leaves to roots was subsequently reported by other investigators (e.g. 31, 45) in additional species. Interestingly, the increase in tracheid diameter from leaves to roots is positively correlated with an increase in the duration of tracheid expansion in the same direction (45, 127).

A similar polar pattern of gradual increase in conduit size from leaves to roots was found also in vessels. Thus a basipetal increase in vessel diameter was reported along the leaves (35), while a continuous increase in vessel diameter and vessel length was demonstrated from twigs to branches, down along the stem and into the roots of *Acer rubrum* trees (169). Vessel diameter continued to increase in the long conducting lateral roots with increasing distance from the stem (61, 112, 139). This basic pattern was observed in dicotyledons as well as in monocotyledons (31, 144, 166). The basipetal increase in vessel diameter is associated with a basipetal decrease in vessel density (i.e. number of vessels per transverse-sectional area). Thus, vessel density is greater in branches, where the vessels are small, than in roots where they are larger (31, 32, 65).

Another well-known gradient in vascular element size occurs in the radial direction in each annual growth ring; here there are wide tracheids or vessels that differentiate at the beginning of the growing season (earlywood) and narrow elements that appear at the end of the season (latewood) (87, 167).

The Problem of Conduit Size Control

There are contradictory hypotheses regarding the control of vascular element size. The first, called the hypothesis of tracheid diameter regulation (96), maintains that gradients of tracheid diameter throughout the stem are "positively" regulated by parallel gradients of auxin. This hypothesis, which was based on photoperiodic studies of growth ring formation in *Pinus*, ascribes the formation of wide tracheids (like those in the earlywood) to high levels of auxin associated with shoot extension and leaf development, and the formation of narrow tracheids (as in latewood) to the low levels of auxin associated with cessation of shoot growth (95, 96).

Experimental results obtained with the ring-porous species *Robinia pseudacacia* by Digby & Wareing (48) indeed show a linear increase in vessel diameter with increasing auxin concentration from 1 mg/l to 1000 mg/l (with the gibberellic acid at 100 mg/l in all cases). However, this increase in vessel diameter could perhaps be ascribed to the disparity in rates between gibberellic acid and auxin.

The hypothesis that proposes a positive correlation between auxin concentration and tracheid diameter in *Pinus* (96) and vessel diameter in *Robinia* (48) is contrary to what one would expect from the overall pattern of vascular element size. Thus the smallest vascular elements differentiate near the leaves, where the highest auxin levels are expected, while the largest elements are formed in the roots, at the greatest distance from the auxin sources. To resolve this apparent contradiction and to better explain the general increase in conduit size and the decrease in conduit density from leaves to roots, a six-point hypothesis has been proposed as follows (18): (*i*) Basipetal polar flow of auxin from leaves to roots establishes a gradient of decreasing auxin concentration in this direction. (*ii*) Local structural or physiological obstruction of auxin flow results in a local increase in auxin concentration. (*iii*) The distance from the source of auxin to the differentiating cells controls the amount of auxin flowing through the differentiating cells at a given time, thus determining the cells' position in the gradient. (*iv*) The rate of conduit differentiation is positively correlated with the amount of auxin that the differentiating cells receive; consequently, the duration of the differentiation process increases from leaves to roots. (*v*) The final size of a conduit is determined by the rate of cell differentiation. Because cell expansion ceases after the secondary wall is deposited, rapid differentiation results in narrow vascular elements while slow differentiation permits more cell expansion and therefore results in wide tracheary elements. Hence, decreasing auxin concentration from leaves to roots leads to an increase in conduit size in this direction. (*vi*) Conduit density is controlled by, and positively correlated with, the auxin concentration; consequently, vessel density decreases from leaves to roots.

Experiments with *Phaseolus* seedlings (18), *Acer* stems (19), and *Pinus* seedlings (127) support this six-point hypothesis and show that the rates of vessel and tracheid differentiation decrease with increasing distance from the auxin source. The rates of vessel formation in *Phaseolus* (18) and of tracheid differentiation in *Picea* (45) were found to be constant at any given distance from the auxin source.

Auxin applied to decapitated stems of *Phaseolus* (18) and *Pinus* species (76, 131) induced both the differentiation of numerous tracheary elements immediately below the site of application and a progressive decrease in the number of elements with increasing distance below the auxin source. The auxin induced a substantial gradient of increasing vessel diameter and decreasing vessel density from the auxin source towards the roots (18, 19). Thus, high auxin concentration yielded numerous small tracheary elements, whereas low auxin concentration resulted in slow differentiation and, therefore, in fewer and larger elements (18).

The Control of Vascular Adaptation—a New Hypothesis

Vascular plants grow in different environments, ranging from deserts to rain forests and from alpine and arctic regions to the tropics. Comparative anatomical studies (e.g. 22, 24) reveal similarities in structure of the vascular system in plants grown in extreme habitats versus ones grown in mesomorphic environments. Desert (33, 64), arctic, and alpine shrubs (31) as well as artificially induced or naturally occurring dwarf trees (23) show a high density of very narrow vessels. Such vascular systems are typical of extreme habitats and are deemed adaptive safety mechanisms against drought and freezing. Conversely, forest trees and lianas, which are typical in the tropics and rain forests, have vessels of very wide diameter (31, 151), which affords maximal efficiency of water conduction (51, 59, 166) and is considered to be an adaptation to mesic conditions. The influence of various environmental factors on the structure of xylem was recently reviewed by Creber & Chaloner (37), but the mechanisms whereby the vascular systems of plants are influenced by or adapted to their environment have yet to be elucidated. I suggest that the environment controls the plant's vascular system through its control of the plant's development, height, and shape. To explain how the ecological conditions control the size and frequency of vessels and fibers in plants, I propose the following tripartite vascular adaptation hypothesis:

1. Curtailment of growth factor in the plant's immediate environment

limits the final size of the plant and results in small and suppressed shoots, whereas favorable conditions that do not curtail plant development allow the plant to attain its appropriate shape and maximal height.

2. The duration of the growth period affects the rate of plant development. In extreme and limiting habitats the active growth period is relatively short and results in small plants, whereas stable and moderately comfortable conditions like those found in the humid tropics allow growth activity throughout the year, thereby enabling more growth and consequently large and well-developed plants.

3. The height of the plant and the degree of its branching determine gradients of auxin along the plant's axis. An increase in the plant's length and a diminution of its branching enhances the gradients of auxin from the young leaves (the provenance of auxin) to the lower parts of the stem. In small shrubs, which are typical of extremely cold, dry, and saline habitats, as well as in grazing areas or in locations with insufficient soil for the roots, the distances from the young leaves to the roots are very short and no substantial gradient of auxin can be formed. Therefore, the levels of auxin along these small plants are relatively high and result in the differentiation of numerous very small vessels in the greatest densities [as predicted by the six-point hypothesis (18)] and also in the production of small fibers with thick secondary walls, as stipulated by Aloni (5). Conversely, in the large trees and in the long lianas, the very great distances from the young leaves to the roots enable a substantial decrease in auxin levels in the lower parts of the stem and in the roots; this leads to the differentiation of very wide vessels in low density, along with large fibers having relatively thin secondary walls.

Experiments on *Hibiscus cannabinus* support this adaptation hypothesis. In these experiments, the *Hibiscus* plants were subjected to stress conditions that resulted in small and retarded plants. When the retarded plants were compared with well-developed control plants grown under favorable conditions, it was found that the vascular system in the stems of the suppressed plants had typically narrow vascular elements with relatively thick secondary walls as opposed to the much larger vascular elements in the control plants (R. Aloni and T. Plotkin, unpublished). Additional experiments conducted on the ring-porous trees *Robinia pseudacacia* and *Melia azedarach* have shown that the wide earlywood vessels that form in the spring are induced and differentiate, before or at the early stages of leaf development, by very low auxin levels which occur along the cambium at the beginning of the growth season. Application of moderate or high auxin concentrations via the terminal bud to disbudded trees in the spring, before the buds break, inhibits the formation of wide vessels and results in narrow vessels in the earlywood (R. Aloni, in preparation). In the liana *Luffa cylindrica*, the long stem contains very wide vessels that are arranged in the mature internodes at a distance from the young leaves. When the leaves are excised and are replaced by auxin, high concentrations of the auxin result in many narrow vessels in the older internodes, whereas low auxin concentrations yield a few wide vessels (11).

In comparative anatomical studies it is important to ascertain precisely the distance of each studied wood specimen from the young leaves, because the size and pattern of the vessels and fibers in the twigs, branches, or trunks are variable within the same plant. Clearly there is no sense in comparing a twig from a small shrub with the branch or trunk of a large tree. Furthermore, I propose that the familiar increase in the size of vascular elements in the developing shoot (e.g. tracheids, vessels, and fibers) with increasing age of a tree, such as can be traced in a transverse section from the inner growth ring towards the outer ones (25, 31, 65, 130), has to be attributed to the continuously increasing distance of the young growing leaves from the vascular tissues along the plant axis.

CONTROL OF PHLOEM AND XYLEM DIFFERENTIATION

Induction of Vascular Tissues by Leaves and by Auxin

In the spring, developing buds and young growing leaves stimulate cambium reactivation and the formation of phloem and xylem, which extend down from the developing buds towards the roots (e.g. see 117). The removal of young leaves from the stem reduces or even prevents vascular differentiation below the excised leaves (e.g. see 84). Leaves promote a roots-directed vascular regeneration (83) but have no influence on, or even slightly inhibit, vascular regeneration in the direction of the shoot tip (12). The stimulatory effect of the leaves on root-directed vascular differentiation can be demonstrated by the grafting of shoot apices with a few leaf primordia on callus, which results in the formation of vascular tissues below the graft in the callus tissue (153).

The pioneering study of Jacobs (83) clearly showed that auxin, indole-3-acetic acid (IAA), produced by the young growing leaves was the main factor in limiting and controlling xylem regeneration around the wound in Coleus stems. Auxin alone could, both qualitatively and quantitatively, supplant the effect of the leaves on vascular regeneration in Coleus (83, 143). The polar movement of auxin from the young leaves towards the roots through procambium, cambium, or parenchyma tissues triggers a complex sequence of changes that ultimately results in the formation of a vascular strand along the flow of auxin. Once developed, this vascular strand remains the preferable pathway of auxin transport, inasmuch as cells possessing the ability to transport auxin are associated with vascular tissues (82, 82a, 110). Consequently, new streams of auxin emanating from young developing leaves are directed towards the vascular strands. Sachs (123) has shown that a preexisting vascular strand that is not supplied with auxin (e.g. one descending from an old leaf) acts as a sink for any new stream of auxin. Therefore, a new strand will be formed towards the preexisting strand that has a low supply of auxin. On the other hand, a strand that is well supplied with auxin (e.g. one descending from a young leaf) prevents the expression of another source of auxin in its neighborhood and will not interlink with a new strand for as long as it is well supplied with auxin (123).

An additional factor in vascular control is the auxin-transport capacity of mature vascular tissues. Auxin from mature leaves moves rapidly in a nonpolar fashion in the sieve tubes (e.g. 73, 109). When the phloem below mature leaves is damaged, there is a quantitative increase in vascular differentiation, which promotes a replacement of long nonfunctional, damaged tissues (26). It is believed that this promoting effect of the wound results from additional auxin in the wound region that arrives from the mature phloem. This is an ancillary mechanism that enables leaves to regulate their supportive vascular system.

An interesting feature of vascular tissues are the circular vessels in the form of closed rings. These can be induced above transverse wounds (126) and occur naturally above constrictions in suppressed buds of *Ficus religiosa* (17). Because the circular vessels differentiate above constriction areas, where the polar flow of auxin is interrupted in the longitudinal direction from leaves to roots, their differentiation is attributed to the inductive effect of auxin flow in the closed rings of these locations (17, 126).

The role of auxin as the main regulatory factor in vascular differentiation is well established and has been demonstrated in different experimental systems (6, 18, 30, 43, 47, 71, 100, 104).

The Role of Roots and Cytokinin in Vascular Differentiation

The root need not be present to obtain vascular differentiation in stem tissues (e.g. see 27, 143). This is true also for vascular differentiation in tissue culture, which likewise occurs in the absence of roots (e.g. see 6, 7, 108). Roots do, however, have two major known functions in vascular differentiation, namely: (*a*) The root orients the pattern of vascular differentiation from the leaves towards the root tip by acting as a sink for the continuous flow of auxin deriving from the leaves (122); (*b*) the root apices are sources of inductive stimuli that promote vascular development. The major stimulus is cytokinin. Various cytokinins promote xylem differentiation (39, 40, 67, 107, 108). Zeatin "replaces" the effect of the root on phloem regeneration in stem internodes (80) and induces fiber differentiation in tissue culture (7). Kinetin "replaces" the effect of the root on the differentiation of fibers in stems of derooted seedlings (128).

Effects of Pressure and Ethylene on Vascular Tissues

Brown & Sax (28) were the first to demonstrate the need for mechanical pressure for normal development of secondary vascular tissues. Plant tissues synthesize ethylene in response to external pressures (e.g. see 160). The bending of shoots is known to induce the formation of reaction wood in the stressed shoots. Nelson & Hillis (111) have shown that when seedlings of *Eucalyptus gomphocephala* are placed in the horizontal position they produce higher amounts of ethylene in their upper halves, where the reaction wood was induced. Although there are indications that auxin (96) and ethylene (111, 132) are involved in the elicitation of reaction wood, the mechanism that controls its formation is not clearly understood.

Ethyelene is known to affect xylem differentiation, and the subject has been reviewed by Roberts (119) and Roberts & Miller (121). Miller & Roberts (106) have shown that the ethylene-releasing agent 2-chloroethylphosphonic acid (CEPA) or the ethylene precursor L-methionine promoted xylem differentiation in *Lactuca sativa* pith explants. Addition of ethylene inhibitor such as silver (105) to the culture medium inhibited lignification and xylem differentiation. The inhibition of silver was completely reversed by the addition of L-methionine to the medium (105). Miller, et al (105) suggested that ethylene may play a role in controlling lignification during xylogenesis by inducing wall-bound peroxidase activity.

Yamamoto & Kozlowski have recently found that application of CEPA to seedlings induces thicker bark in *Pinus halepensis* (158) and in *Ulmus americana* (157) as a result of an increase in phloem production and an increase in intercellular space. The CEPA induced more tracheid production in *Pinus* (158), and it increased the number of vessels and reduced their diameter in *Ulmus* (157). In both species the CEPA treatment increased the amount of ray tissue. Flooding of soil stimulates ethylene production and increases phloem and xylem differentiation in the stem of *Thuja orientalis* seedlings (159).

The Relation Between Phloem and Xylem Differentiation

Plant vascular systems are usually composed of phloem and xylem. In the intact plant, xylem does not differentiate in the absence of phloem, though phloem often develops in the absence of xylem. As already mentioned, the earliest organized vascular system is found in brown algae, and it consists of phloem with no xylem (e.g. 53). Along the stem of angiosperms, in addition to the colateral bundles (which consist of both phloem and xylem) there are also bundles of phloem with no adjunct xylem. In *Coleus* on each colateral bundle, there is usually a bundle of phloem (13, 30). Phloem anastomoses are lateral sieve tubes with no xylem that occur between the longitudinal bundles; they are encountered in many plant species, are common in summer conditions (16), and their differentiation is dependent on light intensity (15). However, high auxin concentration applied to decapitated *Luffa* stems induced xylem in the phloem anastomoses (9). A mycelium-like network of internal phloem with no xylem was found in the inner mesocarp of the lateral pod walls of the fruit of *Vigna unguiculata* (89). Mature needle leaves of *Pinus* perennially produce secondary phloem but no secondary xylem (58, 60). In callus grown in culture, sieve elements differentiate with no tracheary elements at low auxin levels (6).

In the young organs of intact plants, the phloem always differentiates before the xylem. For a review see Esau (52). This pattern of vascular development is also true in tissue culture conditions (6, 36) as well as in vascular regeneration around a wound (13, 84), where sieve element differentiation is detected a day or more before tracheary differentiation can be observed. Differentiation of secondary phloem may precede that of secondary xylem by several weeks (55).

In vivo and in vitro studies on the control of phloem and xylem differentiation have yielded contradictory results. Thus Jacobs (83) has shown, in the stems of *Coleus,* that auxin was the limiting and controlling factor for xylem differentiation around a wound. Subsequently, LaMotte & Jacobs (94) reported that auxin was also the limiting and controlling factor in phloem regeneration. Sucrose did not affect phloem regeneration in *Coleus* stems (80, 94). Thompson & Jacobs reaffirmed the earlier finding that auxin was the common controlling factor for phloem and xylem regeneration; in this respect it supplanted the role of the leaves both qualitatively and quantitatively. Phloem was observed at low auxin levels, and regeneration of both phloem and xylem was induced by high auxin concentrations (143).

On the other hand, in tissue culture of *Syringa,* Wetmore & Rier (153) have shown that in order to induce phloem and xylem differentiation there is need to apply a sugar together with the auxin. These authors failed to detect correlation between auxin concentration and the quantity of vascular tissue, such as had been reported for *Coleus* stems; with auxin concentration kept constant, low sucrose levels (1.5–2.5%) induced strong xylem differentiation with little or no phloem, whereas differentiation of phloem with little or no xylem was obtained with higher sucrose levels (3–4%); intermediate sucrose concentrations (2.5–3.5%) favored the formation of phloem and xylem. Subsequent experiments with fern prothalli confirmed that at low sugar concentrations (1.5–3%) xylem was formed, while at higher concentrations (4.5–5%) phloem differentiated (152). However, the later results of Rier & Beslow (118) with callus of *Parthenocissus* tend to contradict the earlier findings in tissue culture, because the number of xylem elements was found to be directly proportional to the sucrose concentration in the medium, at least up to 8%. These latter results were confirmed quantitatively in excised *Coleus* internodes (27) and in cultured tuber tissue of *Helianthus* (108).

Jeffs & Northcote (86) reported that maltose, trehalose, and sucrose, but not various other sugars, induced nodules in *Phaseolus* that contained both phloem and xylem. They suggested that these three α-glucosyl disaccharides exert a specific effect on vascular differentiation in addition to their value as a carbon source. Minocha & Halperin (108) reported a similar effect of these disaccharides on the differentiation of xylem elements in callus of *Helianthus;* they also added glucose, a monosaccharide, to this category. By comparing growth to differentiation they found that the effect of the three disaccharides was not specific for vascular differentiation. Wright & Northcote (156) found that in callus of sycamore trees the differentiation of both phloem and xylem occurred when any sugar promoting good growth was used. They concluded that sucrose is important only as a carbon source and that any other sugar that is a sufficiently good carbon source will promote vascular differentiation.

Aloni (6) reported that low auxin levels induced sieve elements but not tracheary elements in tissue cultures of *Syringa, Daucus,* and *Glycine.* High auxin levels resulted in the differentiation of both phloem and xylem. Changes in sucrose concentration, while the auxin concentration was kept constant, did not exert a specific effect on either sieve element differentiation or the ratio between phloem and xylem. Sucrose did, however, affect the quantity of callose on the sieve plates: an increase in the former resulted in an increase in the latter. Callose serves as an indicator of sieve elements in tissue culture; thus when low sugar concentrations were reported to induce xylem but not phloem in callus (152, 153), it is possible that the callose-poor sieve elements were not detected. In summary, it appears that low levels of auxin, which is the limiting and controlling factor for both phloem and xylem differentiation in plant as well as in tissue culture, induce phloem but not xylem differentiation. The latter only takes place at high auxin levels. It follows, therefore, that xylem does not differentiate in the absence of phloem and always accompanies the pattern of the phloem.

There is evidence that gibberellic acid (GA_3) promotes phloem differentiation (44, 60). On the other hand, GA_3 applied to the storage root of carrot significantly reduced the amount of secondary phloem production and decreased the phloem/xylem ratio for parenchyma cells (103a). Digby & Wareing (48) reported that the relative levels of applied auxin and gibberellic acid were important in determining whether mainly xylem or phloem tissue was produced in stems of *Popolus robusta.* High IAA/low GA_3 concentrations favored xylem formation, whereas low IAA/high GA_3 levels favored phloem production.

CONTROL OF FIBER DIFFERENTIATION

The Role of Leaves, Auxin, and Gibberellin in Fiber Induction

The differentiation of fibers concurrently with the conducting elements in the vascular bundles (53, 62) raises the question as to whether there is a common limiting or controlling factor for fibers and vascular elements. To answer this question we must first trace the stimuli that induce fiber differentiation, and this is usually done by excision experiments. In the lower internods of *Pisum sativum* there are special strands of primary phloem fibers only (10). The differentiation of these special strands depends on stimuli originating in the very young leaf primordia; early removal of these primordia prevents formation of the fiber strands (124). In wounding experiments in *Coleus* stems, the removal of leaves distal or proximal to the wound has shown that the signals for fiber differentiation flow in a strictly polar fashion from leaves to roots (2, 4, 70). An open wound, in which the overlying and underlying tissues are separated with parafilm, prevents fiber differentiation in the tissue directly below the wound and causes more and faster primary phloem fiber differentiation in the tissues above and lateral to it (2). Excision experiments in *Coleus* have shown that the young (4) and mature (2, 4) leaves are the sources of signals for fiber differentiation; when the leaves are excised, no fiber differentiation occurs in the internodes below. Young leaves yield shorter fibers than do mature leaves, suggesting that more than one stimulus is involved in the induction process (4). The signals for fiber differentiation travel polarly from leaves to roots and induce fibers along their pathway in the stem (2, 4) and in the root (70).

Aloni (5) has shown that the role of the leaves in the differentiation of primary phloem fibers in *Coleus* can be replaced by the exogenous application of combined indole-3-acetic acid and gibberellic acid. Both these growth regulators have been found to be limiting and controlling factors in fiber differentiation. IAA, when administered alone, causes the differentiation of only a few phloem fibers, whereas GA_3 by itself does not exert any effect on fiber differentiation, but when both hormones are applied together they induce a considerable number of fibers. This synergistic effect demonstrates the usefulness of the combined administration of auxin and gibberellin as a perfect substitute, both qualitatively and quantitatively, for the aforementioned role of leaves in *Coleus* (5).

The effect of the hormones, however, diminishes considerably with increasing distance from the source of induction (5); this finding is compatible with the decrease in the number of fibers and the increase in their size along the plant axis from leaves to root apices (10, 46). Despite this, the ring width increased down the stem, presumably as a result of the increased duration of fiber production and increased cell size (48a). The observed correlation is in accord with the six-point hypothesis (18).

When various proportions of both hormones are applied within lanolin pastes to decapitated and excised stems, we find that high levels of IAA stimulate rapid differentiation of fibers with thick secondary walls, while high levels of GA_3 result in long fibers with thin secondary walls (5). The induction of long fibers by GA_3 is compatible with the findings of earlier studies on the effect of GA_3 applied in the form of spray to the leaves of intact plants (20, 141).

The combined presence of both growth regulators is also a requirement for secondary fibers in the xylem of *Populus,* where the applied GA_3 effected elongation of the xylem fibers but only in the presence of IAA (48). In *Phaseolus* a combination of IAA and GA_3 induced xylem made up almost entirely of secondary fibers (79). However, in *Xanthium,* the combination of auxin and gibberellin did not effect the differentiation of secondary xylem fibers (136).

The experiments with IAA and GA_3 lead to the conclusion that the differentiation of fibers along with the conducting elements in the vascular bundles results from their common dependence on the polar movement of auxin. Auxin has been shown to be the limiting factor for the differentiation of both vessel elements (18, 83, 123) and sieve elements (6, 94). The study of fiber differentiation shows that IAA is also a limiting factor in this process. The difference, however, between the differentiation of the conducting cells and that of fibers is that the latter process necessitates an additional stimulus, namely, gibberellin.

Field experiments on a wide variety of plants utilized as fiber sources for industry have shown that weekly spraying of the leaves with a mixture of naphthalene acetic acid and gibberellic acid increases the crop of fibers by 50 to 100%. The process also improves the quality of the fibers by yielding fibers of greater length. The highest increment in number of fibers occurs in the young and fast growing parts of the plant (8).

A novel type of differentiation, in which parenchyma cells between the longitudinal vascular strands re-differentiate to form regenerative phloem fibers, was induced around wounds inflicted on young internodes of *Coleus* (3). Just as vessels and sieve elements are commonly induced today from parenchyma cells in tissue culture (6, 153), so also the discovery that parenchyma cells can be induced to re-differentiate to form fibers makes possible the study of fiber differentiation in *in vitro* systems as well (7).

The Role of Roots and Cytokinin in Fiber Differentiation

Aloni (7) has suggested that the correlation between development of the plant body and differentiation of its supportive tissues arises from a common dependence on the same shoot/root feedback control signals. Indeed, experiments with cultured hypocotyl segments (7) and seedlings of *Helianthus annuus* (128) support this view and indicate that fiber differentiation in the secondary xylem is dependent on the inductive signal originating in root apices, namely, on cytokinin. Cytokinin, however, induces fiber differentiation in the explants only in the presence of IAA plus GA_3 (7). Cytokinin has been found to be both a limiting and controlling factor in the early stages of fiber differentiation, when many nuclear divisions take place in the young fiber cells. No fiber differentiation occurs in the absence of zeatin or kinetin, and at low cytokinin levels there is positive correlation between cytokinin concentration in the medium and the number of fibers in the explants (7). Kinetin has been found to increase the length of secondary xylem fibers in *Adhatoda* (102).

Experiments with intact *Helianthus* seedlings have shown that a brief exposure to kinetin, when the latter is applied to the roots of the seedlings, exerts a promoting effect on fiber differentiation in the stem, which, however, is detectable only a few days after removal of the kinetin from the medium (128). This delayed promoting effect can be interpreted as the influence of kinetin on early stages of fiber differentiation, probably affecting the rate of cell divisions in the vascular cambium. Positive correlation has been found between the kinetin supplied to the growth medium solution and the rate of fiber formation within and between the vascular bundles in the hypocotyl of young *Helianthus* plants. Reducing the root-originated cytokinin supply, either by removal of root apices or by lowering the transpiration rate, diminished the number of newly formed secondary xylem fibers in the stem, but this decrease was markedly reversed in the presence of kinetin (128). Apart from its direct effect on fiber differentiation (7), cytokinin also controls fiber differentiation indirectly, inasmuch as it influences the development of the leaves, which are the source of IAA and GA$_3$ (5).

CONCLUDING REMARKS

Auxin seems to be the major agent-signal involved in the control of all aspects of plant vascular differentiation. Further investigations on auxin transport mechanisms are needed to improve our understanding of control mechanisms of vascular differentiation. Apart from the two main pathways of auxin, namely polar movement in wave pattern (161) via the vascular meristems and nonpolar transport through the sieve tubes (109), there is evidence also for additional variegation in both transport pathways and retention sites during the growth season (91). It has been suggested that a signal that moves in wave form may also convey morphogenetic and positional information (74). This suggestion merits further investigation. The cellular complex of plant vascular tissues is probably controlled by fluxes of signals of auxin and other growth regulators. These fluctuations during plant development and throughout the season deserve to be thoroughly studied. The new vascular adaptation hypothesis propounded herein should be tested on a variety of species and under different experimental conditions in order to elucidate the possible role of external environmental factors in plant vascular differentiation.

Space limitations do not permit discussion of certain aspects of vascular differentiation at either the tissue or subcellular levels. Future progress in this field, however, will undoubtedly be predicated on the molecular approach (e.g. 93), and this, coupled with the holistic approach, should yield a more complete picture of our subject.

Literature Cited

1. Aldaba, V. C. 1927. The structure and development of the cell wall in plants. I. Bast fibers of *Boehmeria* and *Linum*. *Am. J. Bot.* 14:16–24
2. Aloni, R. 1976. Polarity of induction and pattern of primary phloem fiber differentiation in *Coleus*. *Am. J. Bot.* 63:877–89
3. Aloni, R. 1976. Regeneration of phloem fibres round a wound: a new experimental system for studying the physiology of fibre differentiation. *Ann. Bot.* 40:395–97
4. Aloni, R. 1978. Source of induction and sites of primary phloem fibre differentiation in *Coleus blumei*. *Ann. Bot.* 42:1261–69
5. Aloni, R. 1979. Role of auxin and gibberellin in differentiation of primary phloem fibers. *Plant Physiol.* 63:609–14
6. Aloni, R. 1980. Role of auxin and sucrose in the differentiation of sieve and tracheary elements in plant tissue cultures. *Planta* 150:255–63
7. Aloni, R. 1982. Role of cytokinin in differentiation of secondary xylem fibers. *Plant Physiol.* 70:1631–33
8. Aloni, R. 1985. Plant growth method and composition. *USA Patent No. 4,507,144*
9. Aloni, R. 1987. The induction of vascular tissues by auxin. In *Plant Hormones and Their Role in Plant Growth and Development*, ed. P. J. Davies. *Agric. Biotechnol. Ser.* The Hague: Nijhoff/Junk. In press
10. Aloni, R., Gad, A. E. 1982. Anatomy of the primary phloem fiber system in *Pisum sativum*. *Am. J. Bot.* 69:979–84
11. Aloni, R., Indig, E. F. 1985. The control of vessel diameter in the stem of *Luffa*. *Am. J. Bot.* 72:806 (Abstr.)
12. Aloni, R., Jacobs, W. P. 1977. Polarity of tracheary regeneration in young internodes of *Coleus* (Labiatae). *Am. J. Bot.* 64:395–403
13. Aloni, R., Jacobs, W. P. 1977. The time course of sieve tube and vessel regeneration and their relation to phloem anastomoses in mature internodes of *Coleus*. *Am. J. Bot.* 64:615–21

14. Aloni, R., Plotkin, T. 1985. Wound induced and naturally occurring regenerative differentiation of xylem in *Zea mays* L. *Planta* 163:126–32
15. Aloni, R., Raviv, A., Wolf, A. 1986. Effect of light intensity on the differentiation of phloem anastomoses in cucumber seedlings. *Am. J. Bot.* 73:618 (Abstr.)
16. Aloni, R., Sachs, T. 1973. The three-dimensional structure of primary phloem systems. *Planta* 113:345–53
17. Aloni, R., Wolf, A. 1984. Suppressed buds embedded in the bark across the bole and the occurrence of their circular vessels in *Ficus religiosa*. *Am. J. Bot.* 71:1060–66
18. Aloni, R., Zimmermann, M. H. 1983. The control of vessel size and density along the plant axis—a new hypothesis. *Differentiation* 24:203–8
19. Aloni, R., Zimmermann, M. H. 1984. Length, width and pattern of regenerative vessels along strips of vascular tissue. *Bot. Gaz.* 145:50–54
20. Atal, C. K. 1961. Effect of gibberellin on the fibers of hemp. *Econ. Bot.* 15:133–39
21. Baas, P. 1982. Systematic, phylogenetic, and ecological wood anatomy. In *New Perspectives in Wood Anatomy*, ed. P. Baas, pp. 23–58. The Hague/Boston/London: Nijhoff/Junk
22. Baas, P., Carlquist, S. 1985. A comparison of the ecological wood anatomy of the floras of southern California and Israel. *IAWA Bull.* (NS) 6:349–53
23. Baas, P., Chenglee, L., Xinyling, Z., Keming, C., Yuefen, D. 1984. Some effects of dwarf growth on wood structure. *IAWA Bull.* (NS) 5:45–63
24. Baas, P., Werker, E., Fahn, A. 1983. Some ecological trends in vessel characters. *IAWA Bull.* (NS) 4:141–59
25. Bailey, I. W. 1958. The structure of tracheids in relation to the movement of liquids, suspensions and undissolved gases. In *The Physiology of Forest Trees*, ed. K. V. Thimann, pp. 71–82. New York: Ronald
26. Benayoun, J., Aloni, R., Sachs, T.

1975. Regeneration around wounds and the control of vascular differentiation. *Ann. Bot.* 39:447–54

27. Beslow, D. T., Rier, J. P. 1969. Sucrose concentration and xylem regeneration in *Coleus* internodes *in vitro*. *Plant Cell Physiol.* 10:69–77

28. Brown, C. L., Sax, K. 1962. The influence of pressure on the differentiation of secondary tissues. *Am. J. Bot.* 49: 683–91

29. Bruck, D. K., Paolillo, D. J. Jr. 1984. Anatomy of nodes vs. internodes in *Coleus:* the nodal cambium. *Am. J. Bot.* 71:142–50

30. Bruck, D. K., Paolillo, D. J. Jr. 1984. Replacement of leaf primordia with IAA in the induction of vascular differentiation in the stem of *Coleus. New Phytol.* 96:353–70

30a. Bryant, J. A., Francis, D., ed. 1985. *The Cell Division Cycle in Plants.* Cambridge/London/New York: Cambridge Univ. Press

31. Carlquist, S. 1975. *Ecological Strategies of Xylem Evolution.* Berkeley: Univ. Calif. Press

32. Carlquist, S. 1976. Wood anatomy of Roridulaceae: ecological and phylogenetic implications. *Am. J. Bot.* 63:1003–8

33. Carlquist, S., Hoekman, D. A. 1985. Ecological wood anatomy of the woody southern California flora. *IAWA Bull.* (NS) 6:319–47

34. Carmi, A., Sachs, T., Fahn, A. 1972. The relation of ray spacing to cambial growth. *New Phytol.* 71:349–53

35. Colbert, J. T., Evert, R. F. 1982. Leaf vasculature in sugar cane (*Saccharum officinarum* L.). *Planta* 156:136–51

36. Comer, A. E., Leonardo, L. 1981. Phloem differentiation in *Coleus* pith explants. *Physiol. Plant.* 51:130–32

37. Creber, G. T., Chaloner, W. G. 1984. Influence of environmental factors on the wood structure of living and fossil trees. *Bot. Rev.* 50:357–448

38. Cronshaw, J. 1981. Phloem structure and function. *Ann. Rev. Plant Physiol.* 32:465–84

39. Dalessandro, G. 1973. Interaction of auxin, cytokinin, and gibberellin on cell division and xylem differentiation in cultured explants of Jerusalem artichoke. *Plant Cell Physiol.* 14:1167–76

40. Dalessandro, G., Roberts, L. W. 1971. Induction of xylogenesis in pith parenchyma explants of *Lactuca. Am. J. Bot.* 58:378–85

41. Dann, I. R., Jerie, P. H., Chalmers, D. J. 1985. Short-term changes in cambial growth and endogenous IAA concentrations in relation to phloem girdling

of peach, *Prunus persica. Aust. J. Plant Physiol.* 12:395–402

42. Datta, P. C. 1971. Karyological anatomy of fibre development in the leaves of *Agava americana* L. var. *marginata alba* Trel. *Ann. Bot.* 35:421–27

43. DeGroote, D. K., Larson, P. R. 1984. Correlations between net auxin and secondary xylem development in young *Populus deltoides. Physiol. Plant.* 60: 459–66

44. DeMaggio, A. E. 1966. Phloem differentiation: induced stimulation by gibberellic acid. *Science* 152:370–72

45. Denne, M. P. 1972. A comparison of root and shoot-wood development in conifer seedlings. *Ann. Bot.* 36:579–87

46. Denne, M. P., Whitbread, V. 1978. Variation of fibre length within trees of *Fraxinus excelsior. Can. J. Forest Res.* 8:253–60

47. Denne, M. P., Wilson, J. E. 1977. Some quantitative effects of indoleacetic acid on the wood production and tracheid dimensions of *Picea. Planta* 134:223–28

48. Digby, J., Wareing, P. F. 1966. The effect of applied growth hormones on cambial division and the differentiation of the cambial derivatives. *Ann. Bot.* 30:539–48

48a. Dodd, R. S. 1985. Within-tree variation in wood production and wood quality in sycamore (*Acer pseudoplatanus*): its relation to crown characteristics. *Can. J. For. Res.* 15:56–65

49. Dodds, J. H. 1981. Relationship of the cell cycle to xylem cell differentiation: a new model. *Plant Cell Environ.* 4:145–46

50. Dodds, J. H., Phillips, R. 1977. DNA and histone content of immature tracheary elements from cultured artichoke explants. *Planta* 135:213–16

51. Ellmore, G. S., Ewers, F. W. 1985. Hydraulic conductivity in trunk xylem of elm, *Ulmus americana. IAWA Bull.* (NS) 6:303–7

52. Esau, K. 1965. *Vascular Differentiation in Plants.* New York/Chicago/London: Holt, Rinehart & Winston

53. Esau, K. 1969. In *Encyclopedia of Plant Anatomy. The Phloem,* ed. W. Zimmermann, P. Ozenda, H. D. Wulff, Vol. 5, Pt. 2. Berlin: Gebruder Borntraeger

54. Esau, K., Thorsch, J. 1985. Sieve plate pores and plasmodesmata, the communication channels of the symplast: ultrastructural aspects and developmental relations. *Am. J. Bot.* 72:1641–53

55. Evert, R. F. 1963. The cambium and seasonal development of the phloem in *Pyrus malus. Am. J. Bot.* 50:149–59

56. Evert, R. F. 1977. Phloem structure and histochemistry. *Ann. Rev. Plant Physiol.* 28:199–222

57. Evert, R. F. 1984. Comparative structure of phloem. In *Contemporary Problems in Plant Anatomy*, ed. R. A. White, W. C. Dickison, pp. 145–234. New York: Academic

58. Ewers, F. W. 1982. Secondary growth in needle leaves of *Pinus longaeva* (bristlecone pine) and other conifers: quantitative data. *Am. J. Bot.* 69:1552–59

59. Ewers, F. W. 1985. Xylem structure and water conduction in conifer trees, dicot trees and lianas. *IAWA Bull.* (NS) 6:309–17

60. Ewers, F. W., Aloni, R. 1985. Effects of applied auxin and gibberellin on phloem and xylem production in needle leaves of *Pinus*. *Bot. Gaz.* 146:466–71

61. Fahn, A. 1964. Some anatomical adaptations of desert plants. *Phytomorphology* 14:93–102

62. Fahn, A. 1982. *Plant Anatomy.* Oxford: Pergamon. 3rd ed.

63. Fahn, A., Ben-Sasson, R., Sachs, T. 1972. The relation between the procambium and the cambium. In *Research Trends in Plant Anatomy*, ed. A. K. M. Ghouse, pp. 161–70. New Delhi: Tata McGraw-Hill

64. Fahn, A., Werker, E., Baas, P. 1986. *Wood Anatomy and Identification of Trees and Shrubs from Israel and Adjacent Regions.* Jerusalem: Israel Acad. Sci.

65. Fegel, A. C. 1941. Comparative anatomy and varying physical properties of trunk, branch and root wood in certain northeastern trees. *Bull. NY State Coll. For. Syracuse Univ.*, Vol. 14, No. 2b. *Tech. Publ.* 55:1–20

66. Fosket, D. E. 1970. The time course of xylem differentiation and its relation to DNA synthesis in cultured *Coleus* segments. *Plant Physiol.* 46:64–68

67. Fosket, D. E., Torrey, J. G. 1969. Hormonal control of cell proliferation and xylem differentiation in cultured tissues of *Glycine max* var. *Biloxi*. *Plant Physiol.* 44:871–80

68. Fukuda, H., Komamine, A. 1980. Direct evidence for cytodifferentiation to tracheary elements without intervening mitosis in a culture of single cells isolated from the mesophyll of *Zinnia elegans*. *Plant Physiol.* 65:61–64

69. Fukuda, H., Komamine, A. 1981. Relationship between tracheary element differentiation and the cell cycle in single cells isolated from the mesophyll of

Zinnia elegans. *Physiol. Plant.* 52:423–30

70. Gad, A. E., Aloni, R. 1984. Time course and polarity of primary phloem fibre differentiation in *Pisum sativum*. *Ann. Bot.* 54:423–28

71. Gersani, M., Sachs, T. 1984. Polarity reorientation in beans expressed by vascular differentiation and polar auxin transport. *Differentiation* 25:205–8

72. Gill, A. M. 1971. Endogenous control of growth-ring development in *Avicennia*. *For. Sci.* 17:462–65

73. Goldsmith, M. H. M., Cataldo, A. D., Karn, J., Brenneman, T., Trip, P. 1974. The rapid non-polar transport of auxin in the phloem of intact *Coleus* plants. *Planta* 116:301–17

74. Goodwin, B. C., Cohen, M. H. 1969. A phase-shift model for the spatial and temporal organization of developing system. *J. Theor. Biol.* 25:49–107

75. Greenidge, K. N. H. 1952. An approach to the study of vessel length in hardwood species. *Am. J. Bot.* 39:570–74

76. Hejnowicz, A., Tomaszewski, M. 1969. Growth regulators and wood formation in *Pinus silvestris*. *Physiol. Plant.* 22:984–92

77. Hejnowicz, Z. 1973. Morphogenetic waves in cambia of trees. *Plant Sci. Lett.* 1:359–66

78. Hendrix, J. E. 1983. Phloem function: an integrated view. *What's New Plant Physiol.* 14:45–48

79. Hess, T., Sachs, T. 1972. The influence of a mature leaf on xylem differentiation. *New Phytol.* 71:903–14

80. Houck, D. F., LaMotte, C. E. 1977. Primary phloem regeneration without concomitant xylem regeneration: its hormone control in *Coleus*. *Am. J. Bot.* 64:799–809

81. Innocenti, A. M., Avanzi, S. 1971. Some cytological aspects of the differentiation of metaxylem in the root of *Allium cepa*. *Caryologia* 24:283–92

82. Jacobs, M., Gilbert, S. F. 1983. Basal localization of the presumptive auxin transport carrier in pea stem cells. *Science* 220:1297–300

82a. Jacobs, M., Short, T. W. 1986. Further characterization of the presumptive auxin transport carrier using monoclonal antibodies. In *Plant Growth Substances*, ed. M. Bopp, pp. 218–26. Berlin/Heidelberg/New York: Springer Verlag

83. Jacobs, W. P. 1952. The role of auxin in differentiation of xylem around a wound. *Am. J. Bot.* 39:301–9

84. Jacobs, W. P. 1970. Regeneration and

differentiation of sieve tube elements. *Int. Rev. Cytol.* 28:239–73
85. Jacobs, W. P. 1984. Functions of hormones at tissue level of organization. In *Encyclopedia of Plant Physiology: Hormonal Regulation of Development II. The Functions of Hormones from the Level of the Cell to the Whole Plant* (NS), ed. T. K. Scott, 10:149–71. Berlin/Heidelberg/New York: Springer-Verlag
86. Jeffs, R. A., Northcote, D. H. 1967. The influence of indol-3yl acetic acid and sugar on the pattern of induced differentiation in plant tissue culture. *J. Cell Sci.* 2:77–88
87. Kramer, P. J., Kozlowski, T. T. 1979. *Physiology of Woody Plants.* Orlando/San Francisco/New York: Academic
88. Kundu, B. C., Sen, S. 1960. Origin and development of fibres in ramie *(Boehmeria nivea* Gaud.). *Proc. Natl. Inst. Sci. India B* 26(Suppl.):190–98
89. Kuo, J., Pate, J. S. 1985. Unusual network of internal phloem in the pod mesocarp of cowpea [*Vigna unguiculata* (L.) Walp. (Fabaceae)]. *Ann. Bot.* 55:635–47
90. Lachaud, S. 1983. Xylogénèse chez les Dicotylédones arborescentes. IV. Influence des bourgeons, de l'acide β-indolyl acétique et de l'acide gibberellique sur la réactivation cambiale et la xylogénèse dans les jeunes tiges de Hêtre. *Can. J. Bot.* 61:1768–74
91. Lachaud, S., Bonnemain, J. L. 1984. Seasonal variations in the polar transport pathways and retention sites of [^{3}H] indole-3-acetic acid in young branches of *Fagus sylvatica* L. *Planta* 161:207–15
92. Lai, V., Srivastava, L. M. 1976. Nuclear changes during differentiation of xylem vessel elements. *Cytobiologie* 12:220–43
93. Lamb, C. J. 1983. Molecular approaches to the study of cell differentiation and development in higher plants: the biochemistry of xylem and phloem production. In *Biochemistry of Cellular Regulation,* ed. M. E. Buckingham, pp. 145–78. Boca Raton, FL: CRC
94. LaMotte, C. E., Jacobs, W. P. 1963. A role of auxin in phloem regeneration in *Coleus* internodes. *Dev. Biol.* 8:80–98
95. Larson, P. R. 1964. Some indirect effects of environment on wood formation. In *Formation of Wood in Forest Trees,* ed. M. H. Zimmermann, pp. 345–65. New York: Academic
96. Larson, P. R. 1969. Wood formation and the concept of wood quality. *Sch. For. Bull.* 74. New Haven: Yale Univ

97. Larson, P. R. 1976. Procambium vs. cambium and protoxylem vs. metaxylem in *Populus deltoides* seedlings. *Am. J. Bot.* 63:1332–48
98. Larson, P. R. 1982. The concept of cambium. In *New Perspectives in Wood Anatomy,* pp. 88–121. The Hague/Boston/London: Nijhoff/Junk
99. Libbenga, K. R., Torrey, J. G. 1973. Hormone-induced endo-reduplication prior to mitosis in cultured pea root cortex. *Am. J. Bot.* 60:293–99
100. Liskova, D. 1985. Regulation of tracheid differentiation by various auxins in spruce *(Picea excelsa)* tissue culture. *Biologia* 40:431–36
101. List, A. 1963. Some observations on DNA content and cell and nuclear volume growth in the developing xylem of certain higher plants. *Am. J. Bot.* 50:320–29
102. Maity, C., Nag, P., Datta, P. C. 1976. *In vitro* culture of the wood of *Adhatoda vasica. Phyton* 17:247–53
103. Malawer, C. L., Phillips, R. 1979. The cell cycle in relation to induced xylem differentiation: tritiated thymidine incorporation in cultured tuber explants of *Helianthus tuberosus* L. *Plant Sci. Lett.* 15:47–55
103a. McKee, J. M. T., Morris, G. E. L. 1986. Effects of gibberellic acid and chlormequat on the proportion of phloem and xylem parenchyma in the storage root of carrot (*Daucus carrota* L.). *Plant Growth Regul.* 4:203–11
104. Meicenheimer, R. D., Larson, P. R. 1985. Exogenous auxin and N-1-naphthylphthalamic acid effects on *Populus deltoides* xylogenesis. *J. Exp. Bot.* 36:320–29
105. Miller, A. R., Crawford, D. L., Roberts, L. W. 1985. Lignification and xylogenesis in *Lactuca* pith explants cultured *in vitro* in the presence of auxin and cytokinin: a role for endogenous ethylene. *J. Exp. Bot.* 36:110–18
106. Miller, A. R., Roberts, L. W. 1984. Ethylene biosynthesis and xylogenesis in *Lactuca* pith explants cultured *in vitro* in the presence of auxin and cytokinin: the effect of ethylene precursors and inhibitors. *J. Exp. Bot.* 35:691–98
107. Minocha, S. C. 1984. The role of benzyladenin in the differentiation of tracheary elements in Jerusalem artichoke tuber explants cultured *in vitro. J. Exp. Bot.* 35:1003–15
108. Minocha, S. C., Halperin, W. 1974. Hormones and metabolites which control tracheid differentiation, with or without concomitant effects on growth

in cultured tuber tissue of *Helianthus tuberosus* L. *Planta* 116:319–31

109. Morris, D. A., Kadir, G. O. 1972. Pathways of auxin transport in the intact pea seedling (*Pisum sativum* L.) *Planta* 107:171–82

110. Morris, D. A., Thomas, A. G. 1978. A microautoradiographic study of auxin transport in the stem of intact pea seedlings (*Pisum sativum* L.). *J. Exp. Bot.* 29:147–58

111. Nelson, N. D., Hillis, W. E. 1978. Ethylene and tension wood formation in *Eucalyptus gomphocephala*. *Wood Sci. Tech.* 12:309–15

112. Patel, R. N. 1965. A comparison of the anatomy of the secondary xylem in roots and stems. *Holzforschung* 19:72–79

113. Philipson, W. R., Ward, J. M., Butterfield, B. G. 1971. *The Vascular Cambium, Its Development and Activity.* London: Chapman & Hall

114. Phillips, R., Arnott, S. M. 1983. Studies on induced tracheary element differentiation in cultured tissues of tubers of the Jerusalem artichoke, *Helianthus tuberosus*. *Histochem. J.* 15:427–36

115. Phillips, R., Torrey, J. G. 1974. DNA levels in differentiating tracheary elements. *Dev. Biol.* 39:322–25

116. Pizzolato, T. D., Heimsch, C. 1975. Ontogeny of the protophloem fibers and secondary xylem fibers within the stem of *Coleus*. I. A light microscope study. *Can J. Bot.* 53:1658–71

117. Reinders-Gouwentak, C. A. 1965. Physiology of the cambium and other secondary meristems of the shoot. In *Handbuch der Pflanzenphysiologie*, Vol. XV/1, ed. W. Ruhland, pp. 1077–105. Berlin/Göttingen/Heidelberg: Springer-Verlag

118. Rier, J. P., Beslow, D. T. 1967. Sucrose concentration and the differentiation of xylem in callus. *Bot. Gaz.* 128:73–77

119. Roberts, L. W. 1976. *Cytodifferentiation in Plants, Xylogenesis as a Model System.* Cambridge/London/New York: Cambridge Univ.

120. Roberts, L. W., Gahan, P. B., Aloni, R. 1987. In *Springer Series in Wood Science. Vascular Differentiation and Plant Growth Regulators* ed. T. E. Timell. Berlin/Heidelberg/New York: Springer-Verlag. In preparation

121. Roberts, L. W., Miller, A. R. 1982. Ethylene and xylem differentiation. *What's New Plant Physiol.* 13:13–16

122. Sachs, T. 1968. The role of the root in the induction of xylem differentiation in pea. *Ann. Bot.* 32:391–99

123. Sachs, T. 1969. Polarity and the induction of organized vascular tissues. *Ann. Bot.* 33:263–75

124. Sachs, T. 1972. The induction of fibre differentiation in peas. *Ann. Bot.* 36:189–97

125. Sachs, T. 1981. The control of the patterned differentiation of vascular tissues. *Adv. Bot. Res.* 9:152–255

126. Sachs, T., Cohen, D. 1982. Circular vessels and the control of vascular differentiation in plants. *Differentiation* 21:22–26

127. Saks, Y., Aloni, R. 1985. Polar gradients of tracheid number and diameter during primary and secondary xylem development in young seedlings of *Pinus pinea* L. *Ann. Bot.* 56:771–78

128. Saks, Y., Feigenbaum, P., Aloni, R. 1984. Regulatory effect of cytokinin on secondary xylem fiber formation in an *in vivo* system. *Plant Physiol.* 76:638–42

129. Salleo, S., LoGullo, M. A., Siracusano, L. 1984. Distribution of vessel ends in stems of some diffuse- and ring-porous trees: the nodal regions as 'safety zones' of the water conducting system. *Ann. Bot.* 54:543–52

130. Sanio, K. 1872. Über die Grösse der Holzzellen bei der gemeinen Kiefer (*Pinus sylvestris*). *Jahrb. Wiss. Bot.* 8:401–20

131. Savidge, R. A. 1983. The role of plant hormones in higher plant cellular differentiation. II. Experiments with the vascular cambium and sclereid and tracheid differentiation in the pine, *Pinus contorta*. *Histochem. J.* 15:447–66

132. Savidge, R. A., Mutumba, M. C., Heald, J. K., Wareing, P. F. 1983. Gas chromatography - mass spectroscopy identification of 1-aminocyclopropane-1-carboxylic acid in compression wood vascular cambium of *Pinus contorta* Dougl. *Plant Physiol.* 71:434–36

133. Savidge, R. A., Wareing, P. F. 1981. Plant-growth regulators and the differentiation of vascular elements. In *Xylem Cell Development*, ed. J. R. Barnett, pp. 192–235. Kent: Castle House

134. Savidge, R. A., Wareing, P. F. 1984. Seasonal cambial activity and xylem development in *Pinus contorta* in relation to endogenous IAA and (s)-abscisic acid levels. *Can. J. For. Res.* 14:676–82

135. Shigo, A. L. 1984. Compartmentalization: a conceptual framework for understanding how trees grow and defend themselves. *Ann. Rev. Phytopathol.* 22:189–214

136. Shininger, T. L. 1971. The regulation of

cambial division and secondary xylem differentiation in *Xanthium* by auxin and gibberellin. *Plant Physiol.* 47:417–22

137. Shininger, T. L. 1975. Is DNA synthesis required for the induction of differentiation in quiescent root cortical parenchyma? *Dev. Biol.* 45:137–50

138. Shininger, T. L. 1979. The control of vascular development. *Ann. Rev. Plant Physiol.* 30:313–37

139. Sieber, M. 1985. Anatomical structure of roots of two species of *Khaya* in Ghana. In *Xylorama, Trends in Wood Research*, ed. L. J. Kučera, pp. 176–83. Basel: Birkhäuser

140. Siebers, A. M. 1971. Initiation of radial polarity in the interfascicular cambium of *Ricinus communis* L. *Acta Bot. Neerl.* 20:211–20

141. Stant, M. Y. 1963. The effect of gibberellic acid on cell width and cell wall of some phloem fibres. *Ann. Bot.* 27:185–96

142. Stieber, J. 1985. Wave nature and a theory of cambial activity. *Can. J. Bot.* 63:1942–50

143. Thompson, N. P., Jacobs, W. P. 1966. Polarity of IAA effect on sieve-tube and xylem regeneration in *Coleus* and tomato stems. *Plant Physiol.* 41:673–82

144. Tomlinson, P. B., Zimmermann, M. H. 1967. The 'wood' of monocotyledons. *IAWA Bull.* 2:4–24

145. Torrey, J. G., Fosket, D. E. 1970. Cell division in relation to cytodifferentiation in cultured pea root segments. *Am. J. Bot.* 57:1072–80

146. Turgeon, R. 1975. Differentiation of wound vessel members without DNA synthesis, mitosis or cell division. *Nature* 257:806–8

147. Tyree, M. T., Dixon, M. A. 1986. Water stress induced cavitation and embolism in some woody plants. *Physiol. Plant.* 66:397–405

148. Vonderhaar, B. K., Topper, Y. J. 1974. Role of the cell cycle in hormone-dependent differentiation. *J. Cell Biol.* 63:707–12

149. Wareing, P. F., Hanney, C. E. A., Digby, J. 1964. The role of endogenous hormones in cambial activity and xylem differentiation. In *The Formation of Wood in Forest Trees*, ed. M. H. Zimmermann, pp. 323–44. New York: Academic

150. Warren Wilson, J. 1978. The position of regenerating cambia: auxin/sucrose ratio and the gradient induction hypothesis. *Proc. R. Soc. London Ser. B* 203:153–76

151. Welle, B. J. H. ter. 1985. Differences in wood anatomy of lianas and trees. *IAWA Bull.* (NS) 6:70 (Abstr.)

152. Wetmore, R. H., DeMaggio, A. E., Rier, J. P. 1964. Contemporary outlook on the differentiation of vascular tissue. *Phytomorphology* 14:203–17

153. Wetmore, R. H., Rier, J. P. 1963. Experimental induction of vascular tissue in callus of angiosperms. *Am. J. Bot.* 50:418–30

154. Wiebe, H. H., Greer, R. L., Van Alfen, N. K. 1984. Frequency and grouping of vessel endings in alfalfa (*Medicago sativa*) shoots. *New Phytol.* 97:583–90

155. Wodzicki, T. J., Knegt, E., Wodzicki, A. B., Bruinsma, J. 1984. Is indolyl-3-acetic acid involved in the wave-like pattern of auxin efflux from *Pinus sylvestris* stem segments? *Physiol. Plant.* 61:209–13

156. Wright, K., Northcote, D. H. 1972. Induced root differentiation in sycamore callus. *J. Cell Sci.* 2:319–37

157. Yamamoto, F., Angeles, G., Kozlowski, T. T. 1987. Effect of ethrel on stem anatomy of *Ulmus americana* seedlings. *IAWA Bull.* (NS). 8: In press

158. Yamamoto, F., Kozlowski, T. T. 1987. Effect of ethrel on growth and stem anatomy of *Pinus halepensis* seedlings. *IAWA Bull.* (NS). 8: In press

159. Yamamoto, F., Kozlowski, T. T. 1987. Effect of flooding of soil on growth, stem anatomy and ethylene production of *Thuja orientalis* seedlings. *IAWA Bull.* (NS). 8: In press

160. Yang, S. F., Hoffman, N. E. 1984. Ethylene biosynthesis and its regulation in higher plants. *Ann. Rev. Plant Physiol.* 35:155–89

161. Zajaczkowski, S., Wodzicki, T. J., Romberger, J. A. 1984. Auxin waves and plant morphogenesis. See Ref. 85, pp. 244–62

162. Zakrzewski, J. 1983. Hormonal control of cambial activity and vessel differentiation in *Quercus robur*. *Physiol. Plant.* 57:537–42

163. Zamski, E. 1979. The mode of secondary growth and the three-dimensional structure of the phloem in *Avicennia*. *Bot. Gaz.* 140:67–76

164. Zamski, E. 1981. Does successive cambia differentiation in *Avicennia* depend on leaf and branch initiation? *Isr. J. Bot.* 30:57–64

165. Zamski, E., Wareing, P. F. 1974. Vertical and radial movement of auxin in young sycamore plants. *New Phytol.* 73:61–69

166. Zimmermann, M. H. 1983. In *Springer Series in Wood Science. Xylem Structure*

and the Ascent of Sap, ed. T. E. Timell. Berlin/Heidelberg/New York: Springer–Verlag

167. Zimmermann, M. H., Brown, C. L. 1971. Trees Structure and Function. New York/Heidelberg, Berlin: Springer–Verlag

168. Zimmermann, M. H., Jeje, A. A. 1981. Vessel-length distribution in stems of some American woody plants. Can. J. Bot. 59:1882–92

169. Zimmermann, M. H., Potter, D. 1982. Vessel-length distribution in branches, stem and roots of Acer rubrum L. IAWA Bull. (NS) 3:103–9

170. Zimmermann, M. H., Sperry, J. S. 1983. Anatomy of the palm Rhapis excelsa. 9: Xylem structure of the leaf insertion. J. Arnold Arbor. 64:599–609

SOME MOLECULAR ASPECTS OF PLANT PEROXIDASE BIOSYNTHETIC STUDIES

R. B. van Huystee

Department of Plant Sciences, University of Western Ontario, London, Ontario N6A 5B7, Canada

CONTENTS

INTRODUCTION 205
ISOZYMES AND THEIR IMPLICATION 206
MOLECULAR STRUCTURE 207
BIOSYNTHESIS AND ENZYME ACTIVITY 207
EXTRACTION, PURIFICATION, AND ANTIBODY ELICITATION 208
IN VITRO SYNTHESIS OF APOPROTEIN MOIETY 209
HEME SYNTHESIS 209
GLYCOSIDIC PROSTHETIC GROUP(S) AND TRANSPORT 211
INTRA- AND EXTRACELLULAR PROTEINS AND ISOZYMES 212
FUNCTION 213
IMMUNOLOGICAL STUDIES 214
SUMMARY 214

INTRODUCTION

The early history of research on peroxidase is most capably reviewed by Saunders et al (67). The first color reaction of biological material with guaiacum as substrate was noted in 1809, but the term peroxidase was used for the first time nearly a century later for an enzyme isolated from horseradish. In fact, peroxidase is widely distributed in the plant kingdom and also

205

0066-4294/87/0601-0205$02.00

occurs in the animal kingdom. Many of the spectral aspects of this hemoprotein and its reaction products have been described in the above-mentioned text
and elsewhere (e.g. 23). Different types of peroxidases were recognized, and
an ever-increasing number of functions were ascribed to them, particularly in
plant physiology (31). Therefore, while the basic chemical reactions of
peroxidase (E.C.1.11.1.7) are well established, namely that of the peroxidatic
cycle involving H_2O_2 and a large array of hydrogen donors (23), the questions
of how its many isozymes function (31) and how its action in differentiation is
controlled (68) remain largely unanswered. I do not review all publications on
these topics here. A comprehensive tabulation has appeared recently (31).
Instead, I offer a critical review of certain aspects of peroxidase in terms of
structure and biosynthesis.

ISOZYMES AND THEIR IMPLICATION

Isozymes are generally evaluated by zymogram patterns after gel electrophoresis. Because zymograms are carried out on native proteins, they
allow for polymer and other complex formations. Isozymes may form from
several different causes (49). For example, modification of a glycosidic
moiety (27) may be considered a bonafide posttranslational alteration, even
though it would be advisable not to base too much emphasis on the Fergusson
analysis for determinations of glycosylation (81, 82). However, possible
modifications of the protein chain as a result of interaction with phenolics in
the formation of isozymes (76, 87) are unsuitable candidates for the term
isozyme. Reaction products between protein and phenolics are a bane for
most researchers (1).

These arguments notwithstanding, many interesting studies with zymograms on the genetics of iszoymes have been done, specifically with petunia
(99). At least three specific peroxidase (PRXa, -b, and -c) isozymes have
been identified in petunia and their genes mapped to chromosomes III, I, and
IV, respectively (83). Other forms may be derived from genes on chromosome VII and/or by posttranslational modification from these gene products.
Isozyme patterns are often used to confirm the derivation of materials. For
example, isozyme patterns have been used to examine progeny after regeneration of tobacco plants from "hairy root" formation via *Agrobacterium rhizogenes* (6).

Despite these informative investigations, more detailed molecular studies
are now called for in order to determine more precisely the reasons for the
variations in the isozymes at the level of intra- and interspecies relatedness
(12, 98).

MOLECULAR STRUCTURE

Several investigators undertook to isolate peroxidase isozymes from horseradish roots (57, 74) and studied their physical properties (40). Similar studies have been completed for peroxidases from rice (54) and turnip (50).

A complete amino acid sequence was carried out on the horseradish peroxidase c (HRPc) (97), which was then compared with turnip peroxidase 7 (51). The molecular weights of peptide chain vary from 33890 to 31060, and there are 8 vs 1 glycosidic side chains for HRPc and turnip 7, respectively. Their amino acid sequence is only 49% identical. This 51% divergence between these two peroxidases suggests major alterations during evolution (51).

The next step in structural studies would be to produce crystalline proteins, as was done for yeast cytochrome c peroxidase (100). These protein crystals could then be used for diffraction studies to construct the 3-D conformation of the enzyme (60). Unfortunately this next step has not yet been possible for plant peroxidases (10), but alignment of amino acid sequences and subsequent computer comparison of plant and yeast peroxidase structures has allowed certain conclusions: Overall, the structures of these three peroxidases resemble each other; differences occur in certain surface areas (98). Peptide sequence comparison between yeast cytochrome c peroxidase and HRPc or turnip 7 indicates 18% and 16% identity, respectively. Incidentally, the yeast peroxidase is nonglycosylated.

Variations in molecular structure in a specific protein in different organisms are well known—e.g. cytochrome c (28). The use of monoclonal antibodies has greatly aided in the identification of species-specific variations (71). The significant advantages of these immunological studies of structural variations are discussed further below. It suffices to mention here that the use of detailed immunological assays has led to a review of structural homology in other proteins—e.g. tubulin—that were considered antigenically similar for a wide variety of species (53).

BIOSYNTHESIS AND ENZYME ACTIVITY

Frequently a mere increase of specific peroxidase activity has been equated with synthesis of peroxidase despite the reports that many factors may influence enzyme activity during extraction (31). In addition, moieties associated with the peptide chain may also influence the activity of the enzyme—e.g. the heme moiety (15). Some early studies on peroxidase biosynthesis employed labelling of newly synthesized proteins, particularly with deuterated water (3, 75). It was acknowledged at the time that the heavy deuterium label may have been incorporated into the peptide chain and/or the glycosidic side chain. Therefore, experiments using ^{14}C-labelled amino acids are more appropriate for protein biosynthesis (59). Alternatively, inhibitors such as cycloheximide (96) or actinomycin D (32) were used to measure peroxidase synthesis or the half-life of its mRNA. These studies were carried out without the benefit of the molecular techniques now available for detailed analysis. One of the most frequently used approaches nowadays is through immunological assays, which identify the molecule regardless of its enzymatic activity (4). The elicitation of specific antibodies requires a pure protein fraction. The extraction and purification of peroxidase have proven to be demanding.

EXTRACTION, PURIFICATION, AND ANTIBODY ELICITATION

The hemoprotein nature of peroxidase provides a means to measure purification by determining the absorption at 403 and 280 nm at each step of purification (48). The increasing ratio (called RZ) reflects the extent of purification of heme protein from contaminating proteins. The RZ for various purified peroxidase isozymes may vary (57). The variation in RZ does not depend so much on the heme moiety (A_{403} nm) as on the aromatic amino acid composition (A_{280} nm) of the peptide chain (100). Extraction of HRPc rendered 200 mg protein per 50 kg tissue at a purity of RZ = 3.0 (10). Preparations of equal purity are now commercially available. However, because protein surface areas show the greatest variation among peroxidases (98), the antibodies raised against HRP may not react appropriately with peroxidase of other plant sources used as an antigen. Moreover, because horseradish is the richest source of peroxidase (61), the purification of peroxidase may be a greater problem for other plant sources. Therefore, if a mode of harvesting other than extraction from tissue is feasible, it should be considered.

Peroxidase occurs in cell walls (47), where it participates in lignin synthesis (33, 34). This peroxidase must also occur in the intercellular space and consequently in the medium of suspension cells (46). A culture of peanut suspension cells was found to be an enriched source of peanut peroxidase (93). The cationic isozyme constitutes three fourths of the peroxidase activity in the medium (48), and following sequential acetone and ammonium sulfate precipitation and resuspension it is readily purified by chromatography on carboxy-methyl (CM) cellulose to an RZ > 3.0 (95). Antibodies raised against this pure fraction (molecular mass 40 kd) will precipitate a single radioactive protein (77). An anionic peroxidase, which is slightly more massive (2 kd) and has a different peptide sequence (78), is found in the flow-through volume of CM cellulose columns. Antibodies raised against the cationic isozyme did not cross-react with the anionic peanut peroxidase (95) or with the petunia anionic peroxidase (84). Nor did they cross-react with HRP (95) or with peroxidase from *Trametes versicolor* (45). These results are in agreement with detailed sequence analysis of three isolated peroxidases, namely that protein surface areas are the most variable parts of the peroxidase (98). Fortunately, the antibodies did react with the extract of leaves from the peanut plant (16a). This suggests that no major alterations in the peptide chain of peroxidase had occurred during peanut cell culture. Additional support for genetic stability in peanut cells used in culture came through the finding that chromosomal aberrations were found in fewer than 1% of peanut cells in a suspension subcultured for seven years (39).

IN VITRO SYNTHESIS OF APOPROTEIN MOIETY

The biosynthesis of the peptide chain on polysomes isolated from peanut cells was studied by indirect immunoprecipitation (77). Further in vitro studies using isolated poly(A)-rich RNA from these cells indicated that at day 4 of subculturing on fresh media, peroxidase synthesis represents 2% of total protein synthesis (78). Similar and even higher values were found at day 8 of culturing via in vivo studies (94). Conversely, in peanut leaves the synthesis was at best 10% (i.e. 0.2% of total protein synthesis) of that in cultured cells (16a). The enhanced peroxidase synthesis in cultured peanut cells has not yet been explained. The greatly increased extracellular space of the cultured cells in suspension medium over that of the extracellular space in the plant cell wall in situ may produce a physiological sink, particularly because biweekly transfers of the cells into fresh medium barely allows for equilibrium conditions to be reached (41). The proper means for analysis of the enhanced peroxidase synthesis is to construct a peroxidase cDNA and use this as a hybridization probe for the occurrence, synthesis, and half-life of mRNAs for peroxidase (92). Ten-fold increases of enzyme have been reported from other projects involving cultured cells (7, 65). This increased synthesis of enzyme in, and the biweekly harvesting of, the medium may be beneficial for studies using other extracellular proteins that are normally not abundant and are therefore difficult to study (90).

HEME SYNTHESIS

Synthesis of peroxidase protein leads to a nonfunctional enzyme because peroxidase activity is only reestablished by association of the apoprotein with the heme moiety to form the holoenzyme (15). Cultured plant cells are ideal for studies of heme biosynthesis because they are frequently achlorophyllous;

the major porphyrin metabolism in plants leading to chlorophyll synthesis does not interfere with the analysis (91). The simultaneous biosynthesis of the protein and heme moieties required to form peroxidase may be measured in cultured cells by adding ^{3}H-leucine and ^{14}C-amino-levulinic acid to the medium. The synthesis of doubly labelled peroxidase can be demonstrated in extracts of these cells (86). Assays for the initial metabolic steps in the porphyrin pathway show that amino levulinic dehydratase activity in cultured cells equals that in the green plant (41), but the synthesis of amino levulinic acid in cells is only 10% of that in green tissue (88). Considering that chlorophyll synthesis in green tissue demands nearly all the porphyrin synthesized in the plant, the 10% of amino levulinic acid remaining in cultured cells is ample for heme synthesis destined for peroxidase (91). Double-labelling experiments (see above) indicate a slower rate for ^{14}C than for ^{3}H incorporation in the hemoprotein, which suggests that the pool of the heme moiety is larger than that of the protein moiety in cultured cells (88).

Based on analogy to results obtained for heme synthesis from animal studies (35), it was assumed originally that the heme was synthesized by the reaction of glycine with succinyl-CoA (59). Because of the difficulties in isolating plant peroxidase referred to above, no direct evidence for plant heme synthesis was then available. However, in the medium of a peanut cell suspension, peroxidase is the only heme protein secreted (14); it constitutes 16% of all proteins in the medium (16b). It also comprises 15% of all cell-wall proteins of petunia (85). Following immunoprecipitation, peroxidase may be cleaved into a heme and an apoprotein. When the peanut cells were incubated in medium containing either [^{14}C]glycine or [^{14}C]glutamic acid and the radioactivity in the separated moieties was subsequently measured, apoprotein and heme showed a nearly equal incorporation of each amino acid into the protein moiety. Conversely, only glutamic acid was incorporated to a significant extent in the heme moiety. Therefore, glutamic acid not glycine is the precursor for the heme moiety of plant peroxidase (16b).

A mass ion spectrum of the cleaved heme moiety gave an M_r of 616, suggesting that the heme moiety is a protoheme (19). Further, the results from incubating the cells with [^{14}C]amino levulinic acid and measuring radioactivity in the subsequently isolated amyloplasts and mitochondria strongly suggest that the synthesis of heme occurs in the mitochondria (19) as it does in animal cells.

A final aspect of the involvement of the heme moiety with peroxidase synthesis is its potential role as a regulator of the synthesis of the apoprotein synthesis (91). Biosynthesis of the protein moiety of another heme protein, catalase, is regulated by heme concentration (37). It has yet to be determined whether this is also true for peroxidase. If it were true, the mitochondria would play a second role in determining peroxidase protein synthesis, because mitochondria are the site of heme moiety synthesis.

GLYCOSIDIC PROSTHETIC GROUP(S) AND TRANSPORT

The second prosthetic group on the peptide chain is (are) the glycosidic side chain(s). All major HRP isozymes contain 15–17% of their molecular weight as carbohydrates; a minor isozyme contains only 7% carbohydrates (57). Less clearly identified in terms of composition are the glycosidic moieties of peroxidase from pea (22), flax (27), petunia (38), and potato (25). In all reported cases, including the peroxidase from cultured peanut cells (86), some peroxidase binds to Concanavalin A-Sepharose. Whether those that do not bind (those that appear in the flow-through fraction of this affinity chromatography) are nonglycosylated peptides (22) is uncertain. Isozymes from other than HRP may also have different glycosidic side chains (38). Concanavalin A has a broad specificity for some but not all sugars (72). At least six glycosidic prosthetic groups of HRP show affinity to Concanavalin A. The size of their N-linked (asparagine-linked) chains ranges from 1600 to 3000 with mannose and glucosamine as their predominant sugars (21). Not surprisingly, incubating slices of horseradish root with $[U\text{-}^{14}C]$mannose resulted in the incorporation of the labelled sugar into all three isozymes of HRP. However, when incubated for 48 hr some of the mannose was transformed into fucose, arabinose, and xylose as well as into some amino acids (44). Moreover, double-labelling experiments employing $[^3H]$leucine and $[^{14}C]$mannose with and without cycloheximide suggested that the peptide chain may be synthesized well in advance of the addition of the glycosidic prosthetic group. Delayed glycosylation may explain the continued release of peroxidase from cultured peanut cells for 2 hr in the presence of cycloheximide (96) but not in the presence of antimycin A (94). Because a considerable pool of heme exists in these cells (87, 88), the control of peroxidase release may well rest with the glycosylation process (89). Preceding glycosylation, the peptide is synthesized by membrane-bound polysomes (78). This and the glycosylation process may proceed rapidly, because within an hour of incubation of peanut cells with $[^{35}S]$methionine at least some labelled peroxidase is found in the suspension medium (94). The transport of proteins through the Golgi apparatus in cultured carrot cells requires 20 min (36).

Peroxidase could be extracted from cells by using either a low-ionic- or a high-ionic-strength buffer. This initially suggested the presence of at least two classes of isozyme (41). However, once the antibodies against cationic peroxidase were available, it appeared that at least part of each extracted peroxidase fraction could be precipitated (94). Because cross-reactivity was largely excluded (see above), it is possible that the two peroxidase fractions extracted are two phases of the development of the glycoprotein. Furthermore, because any change in peroxidase release in vivo is most closely related to the amount of newly synthetized peroxidase extracted by high-ionic rather than low-ionic buffer (94), it could lead to the conclusion that the former extract contained the peroxidase being glycosylated. Immunoprecipitation of cellular peroxidase reveals multiple dense protein bands, suggestive of increasing glycosylation (62).

The site of glycosylation of proteins is presumed to be the Golgi apparatus (26, 66, 80). Therefore, organelles (i.e. nuclei, amyloplasts, mitochondria, and microsomes) from cultured peanut cells were isolated and purified to a reasonable homogeneity. Then the organelles were each extracted sequentially with the low- and high-ionic-strength buffer and the peroxidase immunoprecipitated. The amounts of peroxidase in the microsomal pellet were at least ten times greater than in the other three groups of organelles (20). Because the microsomal pellet without further detailed fractionation contains the Golgi apparatus (36), the high-ionic buffer extract may contain peroxidase in the process of being glycosylated. Additional proof for this assumption was derived from studies with the inhibitor monensin. Monensin has been shown to influence the kinetics of flux through the Golgi apparatus in carrot cells (55). Moreover, it induced swelling of the cisternae of the Golgi apparatus (8). Similar swelling was observed in the cisternae of the Golgi apparatus of monensin-treated peanut cells (14). In addition, the peroxidase recovered from the high-ionic buffer extract following monensin treatment of peanut cells was 150% of control (93). This information agrees with the findings that the pathway of protein release proceeds via the Golgi apparatus (26). Further probing with other specific inhibitors such as tunicamycin (62) may be used for studies of details of the glycosylation process for peroxidase (73). Detailed analysis of the oligosaccharide synthesis could be a complex investigation (69).

INTRA- AND EXTRACELLULAR PROTEINS AND ISOZYMES

The release of peroxidase by peanut cells is inhibited by uncouplers of oxidative phosphorylation (94). Release of peroxidase by plant cells is also regulated by calcium (79). Intracellularly, the calcium effect has been noted to determine whether peroxidase occurs in a soluble form or bound to membranes (58). That calcium is normally an integral part of peroxidase molecule has been established for HRPc (98). The transport of a specific peroxidase isozyme across membranes has been attributed to the presence of two calcium molecules per mole of protein in HRPc (13). In addition, calcium has been found to be required for maintenance of protein structure near the heme moiety of HRPc (56). Whether this is the case for peroxidase from all sources remains to be determined. However, the cationic peanut peroxidase, which greatly resembles HRPc in its amino acid composition (95), is the major extracellular peroxidase (48) and also contains two molecules per mole peroxidase (R. B. Van Huystee, unpublished data).

The anionic peanut peroxidase is found in much lower amounts in the spent medium of the suspension culture (48) and therefore has been studied less. Because the shape and action of the two isozymes do not differ significantly (17) it is difficult to find an obvious reason for the low concentration of the anionic isozyme in the medium. Perhaps because the peptide sequences of the two proteins differ (78) the two genes behave analoguously to those for amylase, where either hormonal action (5) or cell development (43) regulates the differential expression. A delay in expression of the anionic isozyme gene precludes its release and accumulation in the medium. The glycosylation of the protein may also play a role, because inhibition of glycosylation of the cationic peanut peroxidase isozyme prevents release (62).

With amylase there are glycosylated forms in rice (52) and nonglycosylated forms in mung bean (81). Therefore there is no definitive result for amylase as to whether glycosylation determines protein transport. Another factor in amylase secretion is calcium (12a). Does the difference in calcium content between the two isozymes really play a role in transport, as contended for HRPc (13)? Regardless of the factor(s) that determine transport, one of the basic tenets of isozyme forms is that they may occur in different sites of the cell (49). Even though the concept of compartmentation of isozymes is well established (70), few if any reasons are advanced for their transport to a specific compartment. In general it is accepted that as in the case of peroxidase they have a specific function to play in the particular compartment (9).

FUNCTION

Peroxidase appears to influence a great array of physiological processes (31). However, even the use of purified isozymes in assays for substrate specificity did not determine a definitive function for either cationic or anionic peroxidase (17). This may appear to contradict earlier findings (84). However, in order to determine specificities accurately the purity of all samples should be nearly identical (93). Kinetic rate assays have shown a great similarity between purified enzymes from various plant sources (42).

General consensus holds that a major function of the extracellular peroxidase concerns lignin synthesis, regardless whether in connection with cell differentiation or with host defenses against pathogenic invasions (31). The latter may include peroxidase action in suberization of cells (25). Incidentally, a peroxidase isolated from a white rot fungus can dissociate lignin again (64). Elegant studies of cell differentiation have shown a parallel relationship between the increase of peroxidase activity and that of the lignin content of the tracheids formed (30). Moreover, detailed studies have shown that peroxidase specifically polymerizes feruloyl radicals in the formation of cross-links in the plant cell wall (29). However, it remains to trace the various events in precise detail, specifically in chronological order. The most appealing approach for such work is to employ immunological assays. The study of the suberization process would be a good example (24).

IMMUNOLOGICAL STUDIES

As detailed above, immunoprecipitation of a specific product from among all other in vivo or in vitro synthesized proteins is the most accurate means of measuring peroxidase synthesis. However, cross-reactivity between a polyclonal serum and proteins other than peroxidase frequently occurs. Even attempts to obtain monospecific antibodies through immunoaffinity (18) have not prevented all nonspecific cross-reactivity. Therefore, the use of monoclonal antibodies is strongly recommended (93). Monoclonal antibodies have been successfully used in studies with other heme proteins—e.g. cytochrome P_{450} isozymes (63). Moreover, because monoclonal antibodies recognize only a specific epitope [a sequence of amino acids or carbohydrates (2)] their use may be equally well suited for comparisons of sequences in two related proteins or glycoproteins. Once the epitope specificity of the monoclonal antibody has been established, it may also be used in biosynthetic studies, particularly as it relates to glycosylation. The specific path of the glycoprotein may be followed through the Golgi apparatus and be clearly demonstrated through immuno-gold-labelled antibodies (11).

SUMMARY

Even within one plant family, plant peroxidases vary greatly in amino acid sequence and in number and composition of glycosidic side chains (98). Particularly the variation in the amino acid complement on the surface of the protein may pose problems in immunological reactions if antibodies raised against peroxidase from one plant source are used in studies with peroxidases from another. If the involvement of peroxidase in plant cell development is to be clarified, many investigations will probably employ immunoassays and eventually molecular cloning.

In terms of the biosynthesis of the molecule, three components are recognized. The apoprotein moiety is synthesized predominantly on the rough endoplasmic reticulum. The heme moiety is synthesized in mitochondria, utilizing glutamic acid not glycine as its precursor. Finally, the glycosidic prosthetic group is added as the polypeptide proceeds through the Golgi apparatus. Where or when in this chain of biosynthetic events the calcium is added is unknown as is the timing of heme addition. The main question that has not yet been addressed is the process of peroxidase induction. This will require a peroxidase cDNA to probe for mRNA transcription and translation (92). However, to isolate mRNA for peroxidase, highly selective antibodies will be required to screen expression libraries. To accomplish this cloning project as well as to study the biosynthetic pathway in situ, monoclonal antibodies against specific epitopes should be raised.

ACKNOWLEDGEMENT

I am grateful to the NSERC for a grant in support of the research not yet published and to I. van Huystee for editorial help.

Literature Cited

1. Anderson, J. W. 1968. Extraction of enzymes and subcellular organelles from plant tissues. *Phytochemistry* 7:1973–88
2. Anderson, M. A., Sandrin, M. S., Clark, A. E. 1984. A high proportion of hybridomas raised to a plant extract secrete antibody to arabinose or galactose. *Plant Physiol.* 75:1013–16
3. Anstine, W., Jacobsen, J. V., Scandalios, J. G., Varner, J. E. 1970. Deuterium oxide as a density label of peroxidases in germinating barley embryos. *Plant Physiol.* 45:148–52
4. Apel, K., Kloppstech, K. 1978. The plastid membranes of barley *(Hordeum vulgare)*. Light induced appearance of mRNA coding for the apoprotein of the light harvesting chlorophyll a/b protein. *Eur. J. Biochem.* 85:581–88
5. Baulcombe, D. C., Buffard, D. 1983. Gibberellic acid regulated expression of α-amylase and six other genes in wheat aleurone layers. *Planta* 157:493–501
6. Benvenuto, E., Ancora, G., Spano, L., Costantino, P. 1983. Morphogenesis an isoperoxidase characterization in tobacco "hairy root" regenerants. *Z. Pflanzenphysiol.* 110:239–45
7. Berlin, J. 1984. Plant cell cultures: A future source of natural products. *Endeavour* (NS) 8:5–8
8. Boss, W. F., Morré, D. J., Mollenhauer, H. H. 1984. Monensin-induced swelling of Golgi apparatus cisternae mediated by a proton gradient. *Eur. J. Cell. Biol.* 34:1–8
9. Boyer, N., Desbiez, M. O., Hofinger, M., Gaspar, Th. 1983. Effect of lithium on thigmomorphogenesis in *Bryonia di-*

ocia ethylene production and sensitivity. *Plant Physiol.* 72:522–24
10. Braithwaite, A. 1976. Unit cell dimensions of crystalline horseradish peroxidase. *J. Mol. Biol.* 106:229–30
11. Brewin, N. J., Robertson, J. G., Wood, E. A., Wells, B., Larkins, A. P., et al. 1985. Monoclonal antibodies to antigens in the peribacteroid membrane from *Rhizobium*-induced root nodules of pea cross-react with plasma membranes and Golgi bodies. *EMBO J.* 4: 605–11
12. Cairns, E., van Huystee, R. B., Cairns, W. L. 1980. Peanut and horseradish peroxidase isoenzymes. Intraspecies and interspecies immunological relatedness. *Physiol. Plant.* 49:78–82
12a. Carbonell, J., Jones, R. L. 1984. A comparison of the effects of Ca^{2+} and gibberellic acid on enzyme synthesis and secretion in barley aleurone. *Physiol. Plant.* 63:345–50
13. Chan, K. Y., Bunt, A. H., Haschke, R. H. 1981. *In vitro* retrograde neuritic transport of horseradish–peroxidase isoenzymes by sympathetic neurons. *Neuroscience* 6:59–69
14. Chibbar, R. N. 1984. *Biosynthesis of peroxidase in peanut cell suspension culture.* PhD thesis. Univ. West. Ontario, London, Ontario. 191 pp.
15. Chibbar, R. N., Cella, R., van Huystee, R. B. 1984. The heme moiety in peanut peroxidase. *Can. J. Biochem.* 62:1046–50
16a. Chibbar, R. N., van Huystee, R. B. 1983. Immunological relatedness of peanut peroxidase from cells in suspension

cultures to that in peanut plants. *Z. Pflanzenphysiol.* 109:191–96

16b. Chibbar, R. N., van Huystee, R. B. 1983. Glutamic acid is the haem precursor for peroxidase synthesized by peanut cells in suspension culture. *Phytochemistry* 22:1721–23

17. Chibbar, R. N., van Huystee, R. B. 1984. Characterization of peroxidase in plant cells. *Plant Physiol.* 75:956–58

18. Chibbar, R. N., van Huystee, R. B. 1984. Immunoaffinity studies on cationic peanut peroxidase fraction. *J. Plant Physiol.* 116:365–73

19. Chibbar, R. N., van Huystee, R. B. 1986. Site of haem synthesis in cultured peanut cells. *Phytochemistry* 25:585–87

20. Chibbar, R. N., van Huystee, R. B. 1986. Immunochemical localization of peroxidase in cultured peanut cells. *J. Plant Physiol.* 123:477–86

21. Clarke, J., Shannon, L. M. 1976. The isolation and characterization of the glycopeptides from horseradish peroxidase isoenzyme c. *Biochim. Biophys. Acta* 427:428–42

22. Darbyshire, B. 1973. The glycoprotein nature of IAA oxidase/peroxidase fractions and their development in pea roots. *Physiol. Plant.* 29:293–97

23. Dunford, H. B., Stillman, J. S. 1976. On the function and mechanism of action of peroxidase. *Coord. Chem. Rev.* 19:187–251

24. Espelie, K. E., Franceschi, V. R., Kolattukudy, P. E. 1986. Immunocytochemical localization and time course of appearance of an anionic peroxidase associated with suberization in wound-healing potato tuber tissue. *Plant Physiol.* 81:487–92

25. Espelie, K. E., Kolattukudy, P. E. 1985. Purification and characterization of an abscisic acid-inducible anionic peroxidase associated with suberization in potato. *Arch. Biochem. Biophys.* 240:539–45

26. Fernandez, D. E., Staehelin, L. A. 1985. Structural organization of ultra rapidly frozen barley aleurone cells actively involved in protein secretion. *Planta* 165:455–68

27. Fieldes, M. A., Tyson, H. 1984. Possible post-translational modification, and its genetic control, in flax genotroph isozymes. *Biochem. Genet.* 22:99–114

28. Fitch, W. M., Margoliash, E. 1968. The construction of phylogenetic trees. II. How well do they reflect past history? *Brookhaven Symp. Biol.* 21:217–42

29. Fry, S. C. 1986. Cross-linking of matrix polymers in the growing cell walls of angiosperms. *Ann. Rev. Plant Physiol.* 37:165–86

30. Fukuda, H., Komamine, A. 1982. Lignin synthesis and its related enzymes as markers of tracheary-element differentiation in single cells isolated from the mesophyll of *Zinnia elegans. Planta* 155:423–30

31. Gaspar, Th., Penel, C., Thorpe, T., Greppin, H. 1982. *Peroxidases 1970–1980*. A survey of their biochemical and physiological roles in higher plants. Univ. Genève. 324 pp.

32. Gayler, K. R., Glasziou, K. T. 1968. Plant enzyme synthesis: mRNA for peroxidase in sugar-cane stem tissue. *Phytochemistry* 7:1247–51

33. Goldberg, R., Catesson, A.-M., Czaninski, Y. 1983. Some properties of syringaldazine oxidase, a peroxidase specifically involved in the lignification process. *Z. Pflanzenphysiol.* 110:267–79

34. Goldberg, R., Lé, T., Catesson, A.-M. 1985. Localization and properties of cell wall enzyme activities related to the final stages of lignin biosynthesis. *J. Exp. Bot.* 36:503–10

35. Granick, S., Beale, S. Y. 1978. Hemes, chlorophylls and related compounds. Biosynthesis and metabolic regulation. *Adv. Enzymol.* 46:33–203

36. Gripshover, B., Morré, D. J., Boss, W. F. 1984. Fractionation of suspension cultures of wild carrot and kinetics of membrane labeling. *Protoplasma* 123:213–20

37. Hamilton, B., Hofbauer, R., Ruis, H. 1982. Translational control of catalase synthesis by hemin in yeast *Saccharomyces cerevisiae. Proc. Natl. Acad. Sci. USA* 79:7609–13

38. Hendriks, T., vandenBerg, B. M., Schram, A. W. 1985. Cellular location of peroxidase isoenzymes in leaf tissue of petunia and their affinity for Concanavalin A-sepharose. *Planta* 164:89–95

39. Inoue, M., van Huystee, R. B. 1984. Effects of caffeine treatment alone and in combination with gamma exposure on cultured peanut cells. *Can. J. Bot.* 62:1890–95

40. Kay, E., Shannon, L. M., Lew, J. Y. 1967. Peroxidase isozymes from horseradish roots. II. Catalytic properties. *J. Biol. Chem.* 242:2470–73

41. Kossatz, V. C., van Huystee, R. B. 1976. The specific activities of peroxidase and amino levulinic acid dehydratase during the growth cycle of peanut suspension culture. *Can. J. Bot.* 54:2089–94

42. Lambert, A.-M., Dunford, H. B., van Huystee, R. B., Lobarzewski, J. 1985. Spectral and kinetic properties of a

cationic peroxidase secreted by cultured peanut cells. *Can. J. Biochem.* 63:1086–92

43. Lazarus, C. M., Baulcombe, D. C., Martienssen, R. A. 1985. α–Amylase genes of wheat are two multigene families which are differentially expressed. *Plant Mol. Biol.* 5:13–24

44. Lew, J. Y., Shannon, L. M. 1973. Incorporation of carbohydrate residues into peroxidase isoenzymes in horseradish roots. *Plant Physiol.* 52:462–65

45. Lobarzewski, J., van Huystee, R. B. 1982. Purification of cationic peanut peroxidases by immunoaffinity chromatography. *Plant Sci. Lett.* 26:39–49

46. Mäder, M. 1976. Die Lokalisation der Peroxidase—Isoenzymgruppe G_1 in der Zellwand von Tabak-Geweben. *Planta* 131:11–15

47. Mäder, M., Meyer, Y., Bopp, M. 1975. Lokalisation der Peroxidase—Isoenzyme in den Protoplasten und Zellwänden von *Nicotiana tabacum* L. *Planta* 122:259–68

48. Maldonado, B. A., van Huystee, R. B. 1980. Isolation of a cationic peroxidase from cultured peanut cells. *Can. J. Bot.* 58:2280–84

49. Markert, C. L. 1977. Isozymes: the development of a concept and its application. In *Isozymes: Current Topics in Biological and Medical Research*, ed. M. C. Ratazzi, J. G. Scandalios, G. S. Whitt, 1:1–17. New York: Liss

50. Mazza, G., Charles, C., Bouchet, M., Ricard, J., Reynaud, J. 1968. Isolement, purification et propriétés physico-chimiques des peroxidases de navet. *Biochim. Biophys. Acta* 167:89–98

51. Mazza, G., Welinder, K. G. 1980. Covalent structure of turnip peroxidase 7. Cyanogen bromide fragments, complete structure and comparison to horseradish peroxidase c. *Eur. J. Biochem.* 108:481–89

52. Miyata, S., Akazawa, T. 1983. Biosynthesis of rice seed α-amylase: proteolytic processing and glycosylation of precursor polypeptides by microsomes. *J. Cell. Biol.* 96:802–6

53. Morejohn, L. C., Bureau, T. E., Tocchi, L. P., Fosket, D. H. 1984. Tubulins from different higher plant species are immunologically non identical and bind colchicine differently. *Proc. Natl. Acad. Sci. USA* 81:1440–44

54. Morita, Y., Ida, S. 1968. Studies on respiratory enzymes in rice kernel. 1. Isolation and purification of cytochrome c and peroxidase 556 from rice embryo. *Agric. Biol. Chem.* 32:441–47

55. Morré, D. J., Boss, W. F., Grimes, H.,

Mollenhauer, H. H. 1983. Kinetics of Golgi apparatus membrane flux following monensin treatment of embryogenic carrot cells. *Eur. J. Cell Biol.* 30:25–32

56. Ogawa, S., Shiro, Y., Marishima, I. 1979. Calcium binding by horseradish peroxidase c and the heme environmental structure. *Biochem. Biophys. Res. Commun.* 90:674–78

57. Paul, K.-G., Stigbrand, T. 1970. Four isoperoxidases from horseradish roots. *Acta Chem. Scand.* 24:3607–17

58. Penel, C. E., Greppin, H. 1979. Effect of calcium on subcellular distribution of peroxidase. *Phytochemistry* 18:29–33

59. Penon, P., Cecchini, J.-P., Miassod, R., Ricard, J., Teissere, M., Pinna, M. H. 1970. Peroxidases associated with lentil root ribosomes. *Phytochemistry* 9:73–86

60. Poulos, T. L., Freer, S. T., Alden, R. A., Edwards, S. L., Skogland, U. et al. 1980. The crystalline structure of cytochrome c peroxidase. *J. Biol. Chem.* 255:575–80

61. Ramshaw, J. A. M. 1982. Structures of plant proteins. In *Encyclopedia of Plant Physiology: Nucleic Acids and Proteins in Plants* I. *Structure, Biochemistry and Physiology of Proteins*, ed. D. Boulter, B. Parthier, 14A:237–39. Berlin/Heidelberg/New York: Springer-Verlag

62. Ravi, K., Hu, C., Reddi, P. S., van Huystee, R. B. 1986. Effect of tunicamycin on peroxidase release by cultured peanut suspension cells. *J. Exp. Bot.* 37(184):1708–15

63. Reik, L. M., Levin, W., Ryan, D. E., Maines, S. L., Thomas, P. E. 1985. Monoclonal antibodies distinguish among isozymes of the cytochrome P-450b subfamily. *Arch. Biochem. Biophys.* 242:365–82

64. Renganathan, V., Miki, K., Gold, M. H. 1986. Role of molecular oxygen in lignin peroxidase reactions. *Arch. Biochem. Biophys.* 246:155–61

65. Robbins, M. P., Bolwell, G. P., Dixon, R. A. 1985. Metabolic changes in elicitor-treated bean cells. Selectivity in enzyme induction in relation to phytoalexin accumulation. *Eur. J. Biochem.* 148:563–69

66. Robinson, D. G. 1985. Site of glycosylation. In *Plant Membranes*, p. 163. New York/Toronto: Wiley Interscience. 331 pp.

67. Saunders, B. C., Holmes-Siedle, A. G., Stark, B. P. 1964. *Peroxidase. The Properties and Uses of a Versatile Enzyme and Some Related Catalysts.* London: Butterworths. 271 pp.

68. Scandalios, J. G. 1974. Isozymes in de-

velocity and differentiation. *Ann. Rev. Plant Physiol.* 25:225–58
69. Schachter, J. 1986. Biosynthetic controls that determine the branching and microheterogeneity of protein bound oligosaccharides. *Biochem. Cell Biol.* 64:163–81
70. Schnarrenberger, C., Herbert, M., Kruger, I. 1983. Intracellular compartmentation of isozymes of sugar phosphate metabolism in green leaves. *Isozymes: Curr. Top. Biol. Med. Res.*, 8:23–51 Liss.
71. Schneider, H. A. W., Liedgens, W. 1981. An evolutionary tree based on monoclonal antibody-recognized surface features of a plastid enzyme (5-aminolevulinate dehydratase). *Z. Naturforsch. Teil C* 36:44–50
72. Scouten, W. H. 1981. Purification of glycoproteins by way of immobilized lectins. In *Affinity Chromatography. Bioselective Adsorption on Inert Matrices. Chemical Analysis*, 59:148–55. New York/Toronto: Wiley Interscience. 348 pp.
73. Sevier, E. D., Shannon, L. M. 1977. Plant glycoprotein biosynthesis. Uridine diphosphate N-acetyl glucosaminyl transferase from horseradish root. *Biochim. Biophys. Acta* 497:578–85
74. Shannon, L. M., Kay, E., Lew, J. Y. 1966. Peroxidase isozymes from horseradish roots. 1. Isolation and physical properties. *J. Biol. Chem.* 241:2166–72
75. Siegel, B. Z., Galston, A. W. 1966. Biosynthesis of deuterated isoperoxidase in rye plants grown in D_2O. *Proc. Natl. Acad. Sci.* 56:1040–42
76. Srivastava, O. P., van Huystee, R. B. 1977. Interactions among phenolics and peroxidase isozymes. *Bot. Gaz.* 138:457–64
77. Stephan, D., van Huystee, R. B. 1980. Peroxidase biosynthesis as part of protein synthesis of cultured peanut cells. *Can. J. Biochem.* 58:715–19
78. Stephan, D., van Huystee, R. B. 1981. Some aspects of peroxidase synthesis by cultured peanut cell. *Z. Pflanzenphysiol.* 101:313–21
79. Sticher, L., Penel, C., Greppin, H. 1981. Calcium requirement for the secretion of peroxidase by plant cell suspension. *J. Cell Sci.* 48:345–55
80. Strous, G. J., Van Kerkhof, P., Willemsen, A., Slot, J. W., Geuze, H. J. 1985. Effect of monensin on the metabolism, localization, and biosynthesis of N- and O-linked oligosaccharides of galactosyl transferase. *Eur. J. Cell Biol.* 36:256–62
81. Tomura, H., Koshiba, T. 1985. Biosynthesis of α amylase in *Vigna*

mungo cotyledon. *Plant Physiol.* 79:939–42
82. Tyson, H., Fieldes, M. A. 1982. Molecular weight and net charge of peroxidase isozymes in F_1 hybrids between L and S flax genotrophs. *Biochem. Genet.* 20:919–28
83. vandenBerg, B. M. 1984. *On the regulation of peroxidase gene expression in Petunia*. PhD thesis. Univ. Amsterdam, Holland. 83 pp.
84. vandenBerg, B. M., Chibbar, R. N., van Huystee, R. B. 1983. A comparative study of a cationic peroxidase from peanut and an anionic peroxidase from petunia. *Plant Cell Rep.* 2:304–7
85. vandenBerg, B. M., van Huystee, R. B. 1984. Rapid isolation of plant peroxidase. Purification of peroxidase *a* from petunia. *Physiol. Plant.* 60:299–304
86. van Huystee, R. B. 1976. A study of peroxidase synthesis by means of double labelling and affinity chromatography. *Can. J. Bot.* 54:876–80
87. van Huystee, R. B. 1977. Relationship of the heme moiety of peroxidase and the free heme pool in cultured peanut cells. *Z. Pflanzenphysiol.* 84:427–33
88. van Huystee, R. B. 1977. Porphyrin and peroxidase synthesis in cultured peanut cells. *Can. J. Bot.* 55:1340–44
89. van Huystee, R. B. 1978. Peroxidase synthesis in and membranes from cultured peanut cells. *Phytochemistry* 17:191–93
90. van Huystee, R. B. 1985. Plant cells, their development and secondary metabolites. In *Biotechnology: Applications and Research*, ed. P. N. Cheremisinoff, R. P. Ouellette, pp. 215–20. Lancaster, Penn: Technomic. 699 pp.
91. van Huystee, R. B., Cairns, W. L. 1980. Appraisal of studies on induction of peroxidase and associated porphyrin metabolism. *Bot. Rev.* 46:429–46
92. van Huystee, R. B., Cairns, W. L. 1982. Progress and prospects in the use of peroxidase to study cell development. *Phytochemistry* 21:1843–47
93. van Huystee, R. B., Chibbar, R. N. 1986. Peanut peroxidase—a model system for peroxidase analysis. *Curr. Top. Biol. Med. Res.* 13:155–79
94. van Huystee, R. B., Lobarzewski, J. 1982. An immunological study of peroxidase release by cultured peanut cells. *Plant Sci. Lett.* 27:59–67
95. van Huystee, R. B., Maldonado, B. 1982. Some physico-chemical properties of a major cationic peroxidase from cultured peanut cells. *Physiol. Plant.* 54:88–92

96. van Huystee, R. B., Turcon, G. 1973. Rapid release of peroxidase by peanut cells in suspension culture. *Can. J. Bot.* 51:1169–75

97. Welinder, K. G. 1979. Amino acid sequence studies of horseradish peroxidase. Amino and carboxyl termini, cyanogen bromide and tryptic fragments, a complete sequence, and some structural characteristics of horseradish peroxidase c. *Eur. J. Biochem.* 96:483–502

98. Welinder, K. G. 1985. Plant peroxidases. Their primary, secondary and tertiary structures, and relation to cytochrome c peroxidase. *Eur. J. Biochem.* 151:497–504

99. Wysman, H. J. W. 1983. Petunia. In *Isozymes in Plant Genetics and Breeding,* Part B, pp. 229–52, ed. S. D. Tanksley, T. J. Orton. Amsterdam: Elsevier. 472 pp.

100. Yonetani, T. 1967. Studies on cytochrome c peroxidase. X. Crystalline apo- and reconstituted holoenzymes. *J. Biol. Chem.* 242:5008–13

Ann. Rev. Plant Physiol. 1987. 38:221–57

REGULATION OF GENE EXPRESSION IN HIGHER PLANTS

Cris Kuhlemeier, Pamela J. Green, and Nam-Hai Chua

Laboratory of Plant Molecular Biology, The Rockefeller University, 1230 York Avenue, New York, New York 10021–6399

CONTENTS

INTRODUCTION 221
GENE TRANSFER SYSTEMS 222
LIGHT-REGULATED GENE EXPRESSION 224
    The Complex Nature of the Light Response 225
    Cis-Acting Elements for Light-Regulated Transcription 227
    Trans-Acting Elements for Transcription of Light-Regulated Genes 230
    Signal Transduction 233
TISSUE-SPECIFIC EXPRESSION 234
    Tissue-Specific Expression of Light-Regulated Genes 234
    Seed Storage Protein Genes 235
    Hormonal Regulation of Seed Protein Gene Expression 237
AUXIN-REGULATED GENES 238
HEAT SHOCK GENES 239
PLANT-MICROBE INTERACTIONS 242
    Nodulin Genes 242
    Defense Genes 243
CONSTITUTIVE GENES 244
    Cellular Genes 244
    Ribosomal RNA Genes 244
    Viral Genes 244
    Agrobacterial Genes 245
CONCLUDING REMARKS 246

INTRODUCTION

Many plant genes are expressed in a highly regulated manner. Gene products may be present only in certain cell types, at specific stages of development or

0066-4294/87/0601-0221$02.00

only following the application of distinct environmental stimuli. A specific gene can be turned on by very different inducers, or a single stimulus may have totally different effects on different genes. Gene cloning coupled with efficient systems for gene transfer has made it increasingly possible to unravel the complexity of plant gene regulation. Initial gene transfer experiments primarily made use of undifferentiated calli. With these systems it was demonstrated that foreign genes could be introduced into a plant genome and faithfully expressed: that is, the mRNAs had the cognate 5' and 3' termini, were spliced correctly, and in some cases gave rise to protein products. Gene regulation could be studied to a limited extent. Further development of gene transfer technologies allowed for the introduction of genes into cells that retained the capacity to differentiate. From these cells, fertile plants could be regenerated that contained a foreign gene in all cells. Such transgenic plants provide a much better background for studying gene regulation in its full complexity.

In this chapter we discuss recent developments in the analysis of gene regulation in higher plants. Rather than providing an exhaustive overview of all genes studied, we focus on a few that have been characterized in detail. We concentrate our discussion on transcriptional regulation because the most recent developments are in this area. Accordingly, we discuss in detail whatever is known about the *cis*-regulatory DNA sequences and the *trans*-acting factors that bind to them.

GENE TRANSFER SYSTEMS

The study of gene expression and gene regulation has been greatly advanced by the ability to reintroduce defined DNA segments into plant cells. Plant gene transfer systems differ from animal systems in several ways, and it is important to emphasize these differences. Most importantly, a typical plant cell is totipotent, that is, it can give rise to a complete, differentiated organism (225). This has two consequences. First, it is not necessary to transform a germ line cell to obtain a normal fully developed plant containing a transferred gene in all its cells. Any somatic cell is in principle a suitable primary target for DNA uptake. Second, isolated cells do not necessarily retain a differentiated state determined by the tissue of origin, as is the case in animal cell culture. Thus, leaf mesophyll cells, which are the cells in which *rbcS* genes are most highly expressed, rapidly turn off the expression of these genes after protoplast isolation (58, 223). This means that rapid transient assay systems are probably only feasible for certain household genes, which are expressed regardless of the differentiated state of the cell. The purpose of this section is to catalog the plant gene transfer systems used thus far, with little emphasis on the mechanisms involved.

Agrobacterium tumefaciens is the gene transfer intermediate most frequently used for higher plants. The bacterium is naturally capable of transferring a piece of DNA into the genomes of most dicotyledonous plants (21, 63, 87, 149). The expression of genes located on this transferred DNA (T-DNA) inside the plant causes a perturbation of the endogenous hormone balance and leads to tumor formation. Foreign genes inserted into the T-DNA are cotransferred and integrated into the plant genome. It has been shown that for DNA transfer only the T-DNA borders and some flanking sequences need to be present in *cis* (160, 227). By deleting T-DNA encoded genes and adding selectable markers, so-called "disarmed" agrobacteria were obtained, which allow gene transfer without disturbing the endogenous hormone balance (20, 90, 91, 113, 233). In this way one can generate transgenic plants that differ from the wild type only by the presence of the transferred gene to be studied.

The natural host-range of agrobacteria is limited to dicots, and reports of *Agrobacterium*-mediated gene transfer into monocots are few. None of these reports involves the commercially important cereals. Two members of the Liliaceae and Amaryllidaceae were infected with *A. tumefaciens*, and nopaline synthase activity was demonstrated (88). Similarly, *Asparagus* stem sections were challenged with *Agrobacterium,* and again opine production was evident (83). In none of these cases, however, has either integration of foreign DNA into the host genome or sexual transmission into the progeny been demonstrated.

Direct transfer methods are becoming increasingly popular, and such methods may be instrumental in the stable introduction of foreign DNA into monocots. PEG- and Ca^{2+}-mediated uptake of naked DNA was shown to lead to stable integration of DNA into tobacco protoplasts from which normal plants could subsequently be regenerated (79, 155, 161). Using direct transfer methods, the expression of foreign genes was observed in the monocots *Triticum monococcum* (126, 230), *Oryza sativa* (217), and *Lolium multiflorum* (162). Electroporation has been successful for the transformation of carrot and monocot cells (64), and with this system the only case of physical linkage between the donor DNA and the host genome in monocot cells has been reported (65). No regeneration of transformed monocot cells into normal plants has been obtained so far.

Although transient assay systems may not be generally applicable in plants, great success has been obtained with some systems. The efficiency of *Agrobacterium*-mediated leaf disc transformation has been optimized to such an extent that expression of T-DNA gene products could be assayed five days after infection (92, 93). Naked DNA uptake techniques were used to study the effect of anti-sense RNA on the expression of cotransferred sense DNA in carrot protoplasts (49). The activity of the promoter of the maize *shrunken* gene was examined using a transient assay in *Triticum* protoplasts (230). It will be interesting to see whether this promoter will respond to anaerobic stress in the *Triticum* cells, as it does in maize (199). In transient expression systems, the DNA is taken up by the cell without integration into the host genome. The advantage is that the expression of a transferred gene will not depend on particular neighboring sequences in the recipient DNA, and thus there will be no position effects. On the other hand, the copy number of the introduced DNA is uncontrolled, which may lead to discrepancies with results obtained in situations where a gene is stably present in a single copy. It is evident that transient assay systems are valuable because of the speed with which results can be obtained and that they will be very useful, particularly in conjunction with the so much slower regeneration of transgenic plants.

LIGHT-REGULATED GENE EXPRESSION

Light-induced changes provide the basis for much of plant development. Perhaps this is best illustrated when seedlings previously grown in darkness are exposed to light. The leaves respond by beginning to produce chlorophyll and by converting precursor plastids known as etioplasts into photosyntheti-cally active chloroplasts. Simultaneously, the increased synthesis of nuclear-encoded proteins occurs. Photocontrol of gene expression was first suggested by these events and other observations documenting light-modulated enzyme activities (for reviews see 116 and 211). Transcriptional regulation became a major focal point after it was found that in vitro translation of poly(A) RNA from light- and dark-grown plants yielded different sets of polypeptides (4, 210). Quantitative mRNA hybridization (38, 72, 84, 148, 195) and nuclear runoff transcription experiments (19, 67, 68, 138, 190) confirmed that the transcription of many plant genes involved in photosynthesis is controlled by light (50a, 211).

The first step in all photoresponses is the absorption of light by a photore-ceptor. Phytochrome, a red light–absorbing pigment, is the most highly characterized photoreceptor in higher plants (99, 164, 165). It can exist in two stable spectral forms, Pr and Pfr. The inactive form of phytochrome (Pr) is converted to the active form (Pfr) when it absorbs red light. The conversion is photoreversible, i.e. Pfr converts to Pr with far-red light. Thus, the induction of transcription with a pulse of red light that is reversed with far-red light is characteristic of a phytochrome response. Prominent light-induced genes controlled by phytochrome include those encoding the small subunit *(rbcS)* of ribulose-1,5-bisphosphate carboxylase (176, 206, 209) and the chlorophyll a/b binding (Cab) protein of the light-harvesting chlorophyll-protein complex (5, 201, 206). Phytochrome negatively regulates the expression of its own gene(s) (35, 36, 154) as well as the gene(s) for protochlorophyllide reductase

(6, 138) in some plants (134). In addition to phytochrome, one or more blue/UV photoreceptors play an important role in plant gene expression as has been demonstrated with studies on the genetic control of flavonoid biosynthesis (for reviews see 26, 178, 183).

After a photoreceptor absorbs the effective light, little is known about how the signal is transduced to bring about changes in gene expression. One approach to this problem (albeit distant) has been to characterize the complex nature of a plant's response to light at the molecular level. This has been done most definitively for responses involving RNA abundance, but light effects have also been reported for posttranscriptional processes such as mRNA degradation, translation, and protein turnover (15, 17, 194). The second major direction of current work involves analysis of the genes that are transcriptionally regulated by light; this is important irrespective of whether the light signal is transduced through a long transduction chain or is conveyed more directly. In either event, it is likely that transcriptional responses are eventually mediated by sequence-specific DNA-binding proteins. This is implied by the finding that regulatory DNA sequences exist upstream of light-regulated genes (60, 193) and also by work in animal systems on transcriptional responses to environmental and cell-specific factors (48). The identification of DNA-protein interactions involved in light regulation is an important step towards understanding the pathways of signal transduction and their interconnections.

The Complex Nature of the Light Response

Photoregulated plant genes often respond to the same light stimulus in different ways. This was clearly illustrated in a recent study by Kaufman et al (106) that charted the kinetics of accumulation of 11 phytochrome-regulated (105) transcripts in pea seedlings following a red light pulse. Four patterns of accumulation were evident: six transcripts accumulated in a linear fashion; two increased rapidly and leveled off, two exhibited long lag phases, and for one, the red pulse appeared to increase the rate of accumulation. Transcription in the *cab* and *rbcS* families follows the first pattern while ferredoxin transcripts follow the second (see 177). The transduction pathways leading to rapid increases in transcript abundance following light stimuli may be shorter or inherently different from those requiring longer lag times. The kinetics of far-red reversibility, i.e. escape times, also differ among the transcripts in the aforementioned study. In one case the induction becomes irreversible well before RNA accumulation begins, while in another, complete reversibility still can occur after induction reaches maximal levels. Again, a different chain of events likely links the two responses to phytochrome. More specific conclusions await the separation of mRNA stability and transcriptional effects.

Response diversity can also be observed among members of the same gene family with respect to temporal as well as overall expression levels (38, 44, 61, 216). For example, in etiolated tissues the mRNAs for two members of the pea *rbcS* gene family (*rbcS-3A* and *rbcS-3C*) accumulate to high levels following 24 h in white light while two others (*rbcS-E9* and *rbcS-8.0*) are barely detectable after the same treatment (61). These differences are likely to be dependent on the developmental stage of leaves, because all the rbcS transcripts accumulate with similar kinetics when dark-adapted mature leaves are exposed to light. In mature green leaves *rbcS-3A, rbcS-3C,* and *rbcS-8.0* account for about 40%, 34%, and 19% of the total *rbcS* transcript, respectively. Together, *rbcS-E9* and another pea *rbcS* gene, SS3.6 (84), contribute about 7% (61).

The spectral sensitivity of gene activation is an additional parameter that can be related to leaf development. Many groups (e.g. 16, 59, 176, 206) have demonstrated that *rbcS* transcription in etiolated tissues is induced by a pulse of red light in a far-red reversible manner. Thus, in immature tissues light-regulation of *rbcS* appears to only involve phytochrome. In contrast, Fluhr & Chua (59) have demonstrated that a red flash has no effect on *rbcS* transcript abundance in dark-adapted, mature pea leaves. When continuous illumination with red or blue light at equal fluence was performed, blue light was much more effective. The blue light effect was obtained both in light grown mature pea and transgenic petunia leaves, and in each case it was reversible by far-red light (59). These data suggest that in mature leaves maximal induction may require concerted action of phytochrome and a blue photoreceptor. Similarly, Oelmüller & Mohr (152) have shown that cooperation between blue/UV photoreceptors and phytochrome is important for maximal anthocyanin synthesis in milo seedlings. In tobacco suspension culture cells blue light has been shown to induce *rbcS* and *cab* gene expression, although in this case long-term illumination was used (167).

Fluence is another aspect of illumination that influences gene expression differentially. Responses to red light can be grouped into two categories: those triggered by low fluence (LF) and those triggered by very low fluence (VLF) illumination (107). The LF response is the classical far-red reversible phytochrome response. VLF responses appear to require much less Pfr for saturation—even less than that produced by far-red light alone. *rbcS* and ferredoxin transcripts (106, 107) in pea seedlings increase, and pro-tochlorophyllide reductase (PCR) transcript in barley (27) decrease in abundance only in the LF range. Yet, in pea and barley, *cab* (27, 107) responds to LF and VLF components. VLF red light is also a very effective attenuator of phytochrome transcription in *Avena* seedlings (35).

Clearly the light responses of photoregulated genes differ at many levels. However, this does not necessarily imply the operation of numerous independent transduction pathways. In nearly all cases phytochrome is assumed to play a pivotal role. The diversity could be accomplished by cooperation with other photoreceptors or by additional transduction chain components. Even with a common signal transduction pathway, differential gene expression could be dictated by the affinity of a critical transcription factor for target DNA sequences.

Cis-Acting Elements for Light-Regulated Transcription

Identifying the DNA sequences mediating light-regulated transcription is an essential step in understanding the molecular biology of light induction. The advent of Ti vector systems for plant cell transformation has provided the means by which normal and mutant gene constructs can be evaluated in vivo. Following the demonstration by Broglie et al (30) that the pea *rbcS-E9* gene could be stably introduced into petunia cells and faithfully transcribed in a light-inducible manner, the characterization of important *cis*-acting elements rapidly ensued. Chimeric genes consisting of an *rbcS* promoter fused to bacterial coding sequences were then used to show that approximately 1 kb of 5' flanking sequence was sufficient for light-induced expression. In these experiments promoters derived from *rbcS-E9* (136), the *rbcS*-SS3.6 gene (84), and a soybean *rbcS* gene (52) were analyzed in petunia, tobacco, and soybean calli, respectively.

Much of the more detailed work assessing the function of *cis*-acting elements has been performed using transgenic plants. This assay system offers several advantages over transformed calli for the study of light induction: (*a*) The response of plants (59, 60) to illumination is faster than in calli (136, 207), which undergo developmental changes during their slow induction period. (*b*) Transgenic seedlings (60, 145) or immature green leaves (59) can be used for phytochrome experiments, whereas the expression of *rbcS* in calli is insensitive to a red light pulse (F. Nagy, unpublished results). (*c*) Transgenic plants provide a normal hormonal environment with normal chloroplast development. (*d*) Transgenic plants are fertile and can be crossed with other plants as desired. (*e*) Organ, tissue, and cell-specific expression can be evaluated in transgenic plants in contrast to calli.

For these reasons three pea *rbcS* genes (*rbcS-E9, -3A,* and *-3C*) were introduced into tobacco or petunia using "disarmed" vectors that allow normal plant regeneration. Quantitative S_1 analysis for each of the genes showed that phytochrome regulation was fully recapitulated in the transgenic plants (59, 145, 146). The sensitivity of the S_1 assay allowed even subtle differences among the transcripts previously observed in pea to be faithfully reflected in transgenic tobacco or petunia. For example, in pea the far-red reversibility of expression for *rbcS-3A* is more complete than for *rbcS-3C* (59). This difference is also seen in transgenic plants. Moreover, expression of the *rbcS* genes was tissue-specific, reflecting the natural distribution of *rbcS* transcripts in pea plants (38, 61). In the case of *rbcS-E9*, the longest fragment used in the transgenic plant experiments contained 1052 bp of 5' upstream sequences. However, deleting all but 352 bp upstream of the transcription start still allowed regulated expressions at normal levels (146). This result is inconsistent with the low-level expression of the same deletion in nonmorphogenic calli (136), thus emphasizing the fact that gene regulation in calli does not always reflect the situation in normal plants. In other experiments, a wheat *cab* gene was introduced into tobacco and shown to be light-regulated and expressed in a tissue-specific manner (118). This finding indicates that dicots are capable of recognizing the regulatory signals of the monocot *cab* gene and that analysis of *cis* elements in transgenic plants is possible. However, it should be pointed out that the ability of dicots to express monocot genes is not universal. Expression of a wheat *rbcS* gene could not be detected upon transfer to tobacco plants (110). Because transcript level is determined not only by promoter strength but also by accurate processing of the pre-mRNA, Keith & Chua (110) examined both of these parameters for the wheat *rbcS* gene. Transcripts could be detected in transgenic tobacco following replacement of the wheat promoter with the CaMV 35S promoter. However, splicing of the wheat *rbcS* pre-mRNA occurred inefficiently, in contrast to the pea *rbcS-E9* transcript (110). That monocot introns may be processed inefficiently (although accurately) in transgenic dicots is supported by the case of a maize *adh1* intron (intron 6), which is also spliced with greatly reduced efficiency in tobacco plants (110). In light of these findings, it is not surprising that introns from the human growth hormone gene are not spliced in tobacco or sunflower tumors (9) or in transgenic tobacco plants (J. T. Odell et al, unpublished). The aforementioned wheat *cab* gene does not contain any introns (117), and thus the presence of a functional promoter is sufficient for its expression in dicots (118).

To further characterize *cis* elements in the 5' flanking region of the *cab* and *rbcS* genes, a number of deletion derivatives and chimeric genes have been constructed and analyzed in transgenic petunia and tobacco. Experiments with the intact *rbcS-3A* gene (59) indicated that 410 bp of upstream sequence allowed proper expression. Fluhr et al (60) isolated the *rbcS-3A* sequences from −410 to +15 (relative to the transcription start) and fused them to a chloramphenicol acetyl transferase (CAT) reporter gene (60). The expression of this chimeric gene in transgenic plants confirmed that sequences sufficient for light-dark regulation and tissue specificity were contained in the 410 bp upstream region. These same regulatory properties are also exhibited by a chimeric promoter construct consisting of the −330 to −50 region of *rbcS-3A* fused to a truncated version of the CaMV 35S promoter and the CAT gene

(60). The −330 to −50 region is transcriptionally active in either orientation and thus exhibits enhancer-like activity (70). Moreover, the −330 to −50 sequences are sufficient for phytochrome-regulated expression. An even smaller fragment (−317 to −82) from the *rbcS-E9* upstream region directs the same regulatory responses (60). In experiments using the calli assay system, Timko et al (207) identified a 900-bp enhancer-like element from the upstream region of SS3.6 (from −973 to −90). This fragment was shown to function in two orientations, but only 5' to the reporter gene neomycin phosphotransferase II (*nptII*), and it has not been tested for tissue specificity or phytochrome response. The phytochrome response was checked in transgenic tobacco seedlings using an 850-bp fragment from the SS3.6 gene, but this region only allowed induction with white light (193). A red flash was ineffective. It is possible that the *nptII* assays used in these experiments were not sufficiently sensitive to detect the red light induction of the SS3.6 transcript. Alternatively, the lack of response might be attributed solely to the construct or to the illumination conditions, or it might reflect a feature of SS3.6 regulation that differs from the other pea *rbcS* genes.

In recent attempts to narrow down the *rbcS* sequences required for regulation, we have identified a 140-bp sequence (from −189 to −50) that can mediate light induction of a bacterial CAT reporter gene (C. Kuhlemeier et al, unpublished results). Using an upstream enhancer element from the CaMV 35S promoter, we have been able to further localize light regulatory elements (LRE) to the 58 bp between −168 and −110. The activity of the 58-bp fragment cannot be observed without additional enhancement. This effect is probably due to the loss of a quantitative positive element. The −168 to −110 region contains two sequence "boxes," conserved among members of the pea rbcS gene family, which have been implicated in regulation in in vitro experiments (see the next section). These boxes are reiterated further upstream (approximately between −250 and −210), indicating that the region between −330 and −50 may contain more than one LRE. Consistent with this contention is the fact that when the conserved boxes are deleted from the −150 region the remaining sequences containing the second set of boxes are sufficient for tissue specificity and light induction. The LREs present upstream of −50 are independent of the TATA box region, which was shown in earlier *rbcS-E9* experiments (136) to have light regulatory activity in calli. Transgenic plant experiments indicate that sequences downstream of −100 of *rbcS-3A* also confer light inducibility (C. Kuhlemeier et al, unpublished results). These experiments strongly support the existence of multiple LREs upstream of the *rbcS-3A* and *rbcS-E9* genes. The function of these elements alone and in concert will be an important focal point of future work.

Considerable progress has also been made delineating the regulatory sequences in the *cab* upstream region. Analysis of the wheat *cab* chimeric gene constructed by Nagy et al (145) attributed tissue specificity and phytochrome-mediated light regulation to *cis* elements between -1816 and $+31$. In transgenic plants this construct exhibited the VLF response, which is characteristic of *cab* transcription. From a set of deletion derivatives made from an intact *cab* gene, the 5' border of one LRE was positioned at -180. In addition, *cab* sequences from -354 to -90 contribute bi-directional enhancement, tissue-specificity, and phytochrome control to the 35S TATA box fused to the CAT gene (F. Nagy et al, unpublished). The 266-bp fragment can exert control from the 3' end of the gene and therefore can be considered an enhancer. These results indicate that an LRE is located, at least in part, between -180 and -90. We are in the process of testing the regulatory capabilities of this short sequence and other possible LREs in the *cab* upstream region.

Similar work has been carried out on a pea *cab* gene by Simpson et al (191). Their experiments localized phytochrome regulation to a 2.5-kb upstream fragment (193). Tissue specificity and white light regulation could still be conferred when the 5' flanking sequence was shortened to 400 bp, although a quantitative element was lost (191, 193). These authors identified the sequence between -347 and -100 as sufficient for light-regulated and tissue-specific expression of an *nptII* reporter gene (192).

Recently, Kaulen et al (109) reported the UV light-induced expression of a chimeric chalcone synthase-*nptII* gene in transgenic tobacco plants and teratoma calli. *Cis* elements residing in a 1.2-kb chalcone synthase promoter fragment were sufficient for UV-dependent induction of *nptII* activity and *nptII* RNA in transgenic plants. Experiments with calli have implied that important quantitative elements are present between -1200 and -357 as well as between -357 and -39. However, the calli results have not yet been confirmed in transgenic plants and, in light of the inconsistent results concerning *rbcS-E9* in calli and plants, should be interpreted with caution. Light-dependent expression studies were not performed with the calli transformed with the deletion mutants, perhaps because of technical limitations such as those discussed earlier.

Trans-Acting Elements for Transcription of Light-Regulated Genes

One of the reasons for localizing the *cis*-acting elements mediating light regulation is to use them as a tool to identify *trans*-acting transcription factors. We have initiated such experiments in order to characterize nuclear protein factors that specifically interact with the *rbcS-3A* upstream region (from -330 to -50). A number of laboratories have recently reported using gel retardation assays to characterize regulatory factors in prokaryotes, yeast, and animal cells (for a review see 82). Using this approach we have obtained binding of the *rbcS-3A* upstream DNA fragment to one or more protein factors present in nuclei from mature pea leaves. The binding activity is insensitive to large amounts of nonspecific competitor DNA and has little or no affinity for the constitutive 35S promoter (between -300 and $+8$).

In DNA foot-printing experiments with pea nuclear extracts, two main sites are protected from DNase I digestion. The site in the -140 region contains the conserved sequence boxes (discussed earlier) that were previously suggested to play a regulatory role (60). The -140 site (see Table 1) is highly homologous to the second site in the -220 region, and each contains a GT motif (38) similar to the SV40 core enhancer (229). Seventy-three base pairs separate the invariant double Gs in the two motifs, indicating they are on the same face of the DNA helix. In gel retardation assays, binding to a -170 to -50 fragment containing one of the GT motifs could be competed with by excess amounts of the same unlabeled fragment or the fragment from -330 to -170 containing the other GT motif. Strikingly, competition was ineffective with the following: (*a*) a mutant -170 to -50 fragment containing a 12-bp substitution of the GT motif; (*b*) a mutant deleted for the GT motif and flanking conserved boxes, and (*c*) the 35S promoter (from -300 to $+8$), which contains three SV40 enhancer core sequences. The inability of the 35S fragment to compete effectively indicates that sequences outside the core are involved in conferring specificity to the elements. The 6-bp sequence TAATAT, which overlaps the *rbcS-3A* core enhancer motifs (see Table 1) in both factor-binding sites, may fulfill such a role.

The homology of the binding site sequences and competition experiments indicate that the same factor interacts more than once in the upstream fragment. There is much precedence for multiple factor interactions mediating transcription in animal cells. The upstream region of herpes virus thymidine kinase promoter appears to require the interaction of the cellular factor Sp1 at two sites and the CCAAT-binding protein CTF at another site for optimal expression (103). As reviewed by Dynan & Tjian (48), multiple Sp1 interactions have been observed for a number of genes. Another specific example is the *hsp70* gene of *Drosophila,* which has four heat shock transcription factor (HSTF) binding sites (212). Experiments now in progress should reveal whether the factor binding to the *rbcS* gene is influenced by light or tissue type.

It has been suggested that in addition to nuclear regulatory factors, at least one plastid-derived factor contributes to light-regulated transcription of specific nuclear genes (10, 193). Elimination of functional chloroplasts as a result of mutation or herbicide treatment precludes light induction of *cab* transcription (10, 132, 193). Whether this effect is direct or indirect, it certainly warrants further investigation. As a step towards studying light-regulated gene expression in vitro, Ernst & Oesterhelt (51) have looked at the

Table 1 Comparison of the GT sequence motif in the −140 region of *rbcS* genes in higher plants

Gene	GT-motif															References
Pea *rbcS-E9*	G	T	G	T	G	G	T	T	A	A	A	T	A	T	G	38
Pea *rbcS-3A*	G	T	G	T	G	G	T	T	A	A	A	T	A	T	G	61
Pea *rbcS-3C*	G	T	G	T	G	G	T	T	A	A	A	T	A	T	G	61
Pea *SS3.6*	G	T	G	T	G	G	T	T	A	A	A	T	A	T	G	84
Pea *rbcS-8.0*	G	T	G	T	G	G	T	T	C	A	A	T	A	T	G	208
Soybean SRS1	G	T	G	T	G	G	C	C	T	A	A	T	A	T	G	18
Tobacco Ntss23	G	T	G	T	G	G	A	T	A	T	A	T	A	A	G	133
N. plumbaginifolia rbcS-8B	G	T	G	T	G	G	A	T	A	T	A	T	A	A	A	163
Petunia SSU8	G	T	G	T	G	G	A	T	A	T	A	T	A	A	A	44, 216
Petunia SSU11	A	T	G	T	G	G	C	C	A	C	A	T	A	A	T	216
Petunia SSU611	A	T	G	T	G	G	C	C	A	C	A	T	A	A	C	44
Petunia SSU491	A	T	G	T	G	G	C	C	A	T	A	T	C	A	T	44
Petunia SSU112	A	T	G	T	G	G	C	C	A	T	A	T	A	A	T	44
Petunia SSU911	A	T	G	T	G	G	G	C	A	T	A	T	A	A	G	44
Consensus	G / A[a]	T	G	T	G	G	T / C[b]	T / C[b]	A	A / T	A	T	A	T / A	G[c]	

	1	2	3	4	5	6	7	8	9	10	11	12	13	14	15
A	5	—	—	—	—	—	3	—	12	5	13	—	13	8	2
G	9	—	14	—	14	14	1	—	—	—	—	—	—	—	7
C	—	—	—	—	—	—	5	6	1	3	1	1	1	—	1
T	—	14	—	14	—	—	5	8	1	6	—	13	—	6	3

[a] The As in this position are only found in petunia genes.
[b] The Cs in this position are only found in petunia and soybean genes.
[c] The G in this position is not conserved in petunia genes.

transcriptional effect of adding purified phytochrome to nuclei. The increase in overall runoff transcription they observe with nuclei from dark grown seedlings is difficult to interpret. One reason is that the isolated nuclei are likely to only elongate rather than initiate transcripts in response to exogenously added phytochrome (Pfr). Thus the results are not easily assimilated with phytochrome effects on initiation. It will be particularly important to monitor runoff transcription of a specific mRNA with this method in order to best assess its usefulness.

Signal Transduction

Variations in light-response kinetics, fluence requirements, tissue specificity, and developmental associations among responses argue for multiple pathways for signal transduction in higher plants. Yet how different must the pathways be to accommodate the observed diversity? A small number of branched rather than independent pathways would seem most economical. For example, very rapid photoresponses such as the opening of stomata (98), chloroplast rotation (186), or leaf closure (26) are unlikely to involve transcription. However, such responses might result from an early reaction in a transduction chain that leads eventually to the slower changes in transcription.

The conservation of the GT motif among all *rbcS* genes examined so far (see Table 1) may also be taken as evidence for at least a partially common pathway. Similar to the heat shock and Sp1 transcription factors in animal cells, DNA-binding activities in pea nuclei may interact with redundant regulatory elements conserved among a group of genes. Light-dependent modulation of transcription factors could be accomplished at several levels including (*a*) *de novo* factor synthesis, (*b*) posttranslational factor modification, (*c*) factor interaction or binding site blockage, (*d*) factor compartmentalization, and (*e*) the affinities of the factor for different sequences. Clearly, not all mechanisms require the binding activity to be absent in nuclei from nonexpressing cells.

At present, it is not possible to say how close the DNA-protein interactions are to the initial perception of light by the photoreceptor. Conceivably, the photoreceptor could bind directly to DNA sequences to activate or repress transcription. The native form of phytochrome has been isolated (224); however, there has as yet been no report of a phytochrome-DNA interaction. As Schafer & Briggs have pointed out (177), the low estimated number of Pfr molecules per cell argues for an amplification step between Pfr and transcriptional activation. Such an amplification step could involve the red light-induced, far-red reversible changes in phosphorylation noted by Datta et al (43), as detailed in the following model (172). Pfr could first increase cytoplasmic Ca^{2+} concentration, perhaps via a direct membrance interaction. This would activate Ca^{2+} binding modulator proteins, e.g. calmodulin, that could in turn activate cellular enzymes such as kinases. The kinases could then phosphorylate transcription factors specific for light-regulated genes and thereby induce or repress transcription. Other possible models might include rapidly induced gene products, ubiquitination of proteins, or second messengers other than Ca^{2+}.

TISSUE-SPECIFIC EXPRESSION

Tissue-Specific Expression of Light-Regulated Genes

The photosynthetic role of many light-regulated genes explains why they are expressed in an organ-specific manner. In higher plants such patterns and their molecular basis have been studied in the most detail for the rbcS and cab gene families. The rbcS gene products are most abundant in leaves (predominantly in mesophyll cells and guard cells) (130, 193, 234), but they are observed in other tissues as well (38). In pea, pericarps contain about 40% of the leaf transcript level, petals and seeds about 10%, and stems 5% (61). Little or no rbcS transcript can be found in roots. Similarly, petunia sepals contain about 11% of the leaf rbcS transcript level, petals 3%, stems 2%, and roots 0.2% (44); cab transcription also occurs at high levels in leaves (117). In barley, this has been correlated with increased DNaseI sensitivity of the cab gene in leaf chromatin compared to endosperm where transcription is low (200).

Another level of complexity is evident when the contributions of the individual gene family members are separated. This has been studied in detail for the rbcS genes from petunia and pea (44, 61). In both cases, the transcript levels from different rbcS genes vary more than fivefold in a given organ. The relative transcript ratios can also differ from organ to organ. For example, in pea the contribution of the rbcS-E9, SS3.6, and rbcS-8.0, relative to rbcS-3A, is less in petals and in seeds than in leaves and pericarps (61). DNA sequence differences upstream of the genes likely contribute to this differential expression.

With the advent of the transgenic plants, the analysis of the cis elements responsible for tissue-specific expression of light regulated genes has progressed rapidly. At present it is not known how the pathways leading to tissue-specific and light-induced transcription are related. Several short DNA sequences confer both modes of expression. These include −189 to −50 of rbcS-3A (C. Kuhlemeier et al, unpublished results), −317 to −82 of rbcS-E9 (144), −354 to −90 of a wheat cab gene (144), and −347 to −100 of a pea cab gene (192). The latter sequence was recently reported to silence the expression of reporter genes in the roots of transgenic plants (192), thus implicating the involvement of negative regulatory elements in cab transcription. It will be important to determine how such silencers modulate the enhancer-like elements that are present in the *cab* upstream regions (144, 145, 192).

Seed Storage Protein Genes

Seeds contain a relatively small number of highly abundant proteins that serve as a nitrogen reserve to sustain the first heterotrophic phase of development of the germinating seedling (85). Several classes of storage proteins are generally recognized, based on their physicochemical properties. In legumes and most other dicots, the major storage proteins are the salt-soluble globulins. The alcohol-soluble prolamins and acid- or base-soluble glutelins are the predominant seed proteins in monocots. The abundance of these proteins and of their mRNAs have made it possible to study the precise developmental and tissue-specific regulation of their expression (46, 47).

In pea, the major protein classes, vicilins and legumins, are synthesized as precursors in the cytoplasm of cells from the axis and cotyledons of the embryo and are targeted into protein bodies via cotranslational cleavage of a hydrophobic signal sequence (23, 74). Vicilins and legumins are encoded by multigene families, and both cDNA and genomic sequences have been obtained (7, 11, 127–129). Interestingly, the mRNAs for different vicilins and legumins show differences in the timing and pattern of accumulation (85). For instance, one vicilin mRNA appears 12 days after flowering (DAF) and reaches a peak around 22 DAF, whereas another vicilin mRNA does not appear before 17 DAF and reaches a maximum at 25 DAF.

How is the very specific expression of these and other seed protein genes (1, 8, 34) brought about? Obviously, regulatory *cis*-acting elements present in the genes or their flanking regions are likely to be important determinants. Sequence comparisons between legumin 5' flanking regions have led to the identification of short (< 100 bp) consensus sequences with a putative regulatory role (128). The functional significance of such sequences needs to be established by mutational analysis in conjunction with gene transfer methods. Several groups have initiated this type of experiments. Murai et al (141) used a nondisarmed Ti-vector to introduce a gene coding for bean phaseolin into sunflower tumors. The gene was expressed to give correctly sized mRNA, and immunoreactive phaseolin polypeptides were detected. Subsequent experiments using a partially disarmed Ti-vector system allowed the analysis of phaseolin gene products in regenerated tobacco plants (184). Phaseolin mRNA and protein products could be detected at high levels in seeds but not in leaves, thereby demonstrating correct organ-specific expression. Whereas in bean the phaseolin is found exclusively in the protein bodies of the embryo, tobacco seed storage proteins are present both in embryo and endosperm. Interestingly, the transferred bean gene directed the synthesis of phaseolin only in the tobacco embryo. This suggests that either the transcriptional and translational machinery of the tobacco cells is capable of discriminating between bean and tobacco regulatory sequences or, alternatively, that the gene may be expressed in the tobacco embryo endosperm but the gene products may fail to accumulate in this tissue. The bean-specific expression pattern is retained in tobacco not only with respect to tissue but also to temporal expression. The phaseolin proteins start to accumulate 15 days after pollination as in bean, whereas the synthesis of endogenous tobacco storage proteins begins after only 9 days.

Similar experiments were performed with a soybean gene coding for the alpha' subunit of beta-conglycinin, which was studied in transgenic petunia plants (12). In this case, the soybean gene was found to be expressed predominantly in the seeds of petunia, and the timing of the expression followed the pattern of the endogenous petunia proteins. A 5' deletion analysis indicated that 257 nucleotides of upstream sequence were sufficient for high-level regulated expression. A −159 deletion retained the regulatory properties, but was expressed at a low level. Deletion to −69 bp resulted in a complete loss of activity (33a).

A 17.1-kb soybean DNA fragment containing a seed lectin gene together with at least four nonseed protein genes was transferred to transgenic tobacco plants (153). The genes in this cluster were expressed in a manner similar to that in soybean, i.e. the lectin gene products accumulated in seeds, and the other genes were expressed in tobacco leaves, stems, and roots. These results suggest not only that the *cis*-acting elements determining the regulation of each of the genes in this cluster are recognized by the tobacco *trans*-acting factors but also that *cis*-acting elements directing expression of one gene do not affect expression of neighboring genes.

In monocots, the endosperm is the tissue containing most of the storage proteins. In maize the zeins, members of the class of alcohol-soluble proteins called prolamins, have been characterized in great detail (62, 80). Zeins are encoded by a number of multigene families with 5 to 25 members each. The zein proteins are synthesized between 10 and 50 days after pollination, with a peak around 35 days that can differ depending on the protein species. The accumulation and decline of these proteins is reflected at the mRNA level. Expression of a zein genomic clone has been studied in transformed sunflower tissue using an oncogenic vector (71, 131). In all cases transcription was shown to start at the authentic site, but the level of expression was relatively low (1% of the level in maize endosperm) and no protein was made. The low level of expression may not be surprising for two reasons. First, not all monocot genes are well expressed in dicots, as mentioned earlier. Although a wheat *cab* gene is expressed at high levels in transgenic tobacco, a wheat *rbcS* gene is not expressed at all, and there is no evidence as yet to suggest what will be the rule and what will be the exception. Second, zeins are expressed in a highly tissue-specific fashion, and sunflower tumor cells may not be a suitable tissue for the expression of these genes. A zein gene introduced in transgenic tobacco was not expressed in the seeds or in any other tissues (A. Matzke, personal communication). Demonstration of transcriptional activity of monocot genes in the seeds of transgenic dicots may be further complicated by the fact that the endosperm, which is the major storage tissue in monocots, is only a minor component of dicotyledonous seeds (74). Perhaps genes coding for monocot storage proteins that are expressed in both endosperm and embryo, such as globulins, may be better expressed in transgenic dicots (85, 168).

The previous paragraphs have outlined how seed storage protein genes are expressed in a very highly regulated manner. In a limited number of cases it has been shown by gene transfer experiments that all the information needed for regulated expression is contained within the genes and the flanking stretches of DNA. More detailed experiments are needed to further delineate these *cis*-acting sequences. The next step will be to ask what are the exogenous and endogenous factors that either directly or indirectly determine the regulation of these genes? Preliminary studies with a wheat gliadin gene have identified *trans*-acting factors that bind to putative regulatory elements. When a wheat nuclear extract was incubated with a fragment from the upstream region of this gliadin gene, specific binding to a conserved sequence was detected. Interestingly, this binding activity was only present in nuclear extracts derived from endosperm and not from other tissues (T. Okita, personal communication).

Hormonal Regulation of Seed Protein Expression

In higher plants the process of embryogenesis is usually preceded by a period of dormancy. This quiescent period can be bypassed by excising the embryos and culturing them under appropriate conditions, which suggests that the surrounding tissues are involved in the maintenance of dormancy (40, 41, 214). A number of studies have shown that the growth regulator abscisic acid (ABA) prevents precocious germination (54, 55). ABA is known to induce the synthesis of specific proteins, among which are certain seed storage proteins (66, 139, 140). The inducing effects of ABA on the expression of beta conglycinin genes are specific for the alpha' subunit; the genes for the other subunits are not affected (25). This provides an interesting example of differential regulation of related genes. Other ABA-inducible proteins, such as lectins and agglutinins, may be involved in maintaining dormancy, in protecting seeds against desiccation, and in preventing the deleterious effects of hydrolytic enzymes (86, 139, 140).

Whereas ABA is involved in dormancy, the plant hormone gibberellic acid (GA) is produced upon the onset of germination; this compound induces the synthesis of a number of enzymes involved in the breakdown of seed storage products. The best-studied example of such a protein is alpha-amylase. Genes coding for alpha-amylase isozymes from barley were induced upon the addition of GA, and this induction was reversible by ABA. Nuclear runoff experiments demonstrated that the hormones affected transcription initiation (101). Genes coding for different isozymes were not expressed coordinately (96, 121, 170). In barley, gene-specific probes were used to show that a gene coding for a low pI type alpha-amylase was expressed at relatively high basal levels, whereas the expression of a gene encoding a high pI enzyme depended entirely upon addition of GA (170).

For alpha-amylase, GA stimulates and ABA reduces the synthesis, but these hormones have opposite effects on the level of alpha-amylase/subtilisin inhibitor (139, 140). Assuming that regulation of the amylase inhibitor is also controlled at the level of transcription, a parallel with light-regulated genes comes to mind. Light induces the expression of *rbcS* and *cab* but has a negative effect on expression of phytochrome and protochlorophyllide reductase. Again, it is unknown where and how the transduction chain branches so that one stimulus produces opposite effects on different genes.

Receptors for ABA have been identified in *Vicia faba* guard cells, which respond to ABA (and water stress) by rapidly closing the stomata (89). This is an event that is unlikely to involve gene expression but rather is a direct physiological phenomenon. However, the characterization of the receptors will facilitate the study of signal transduction. In this respect, it may be worth noting that there is a certain chemical similarity between the plant hormones GA and ABA and the animal steroid hormones (89, 104). The animal glucocorticoid receptor stimulates transcription by binding to regulatory regions of glucocorticoid-responsive genes (232). If the chemical similarity mentioned above has biological significance, then studying the binding of ABA-receptors to ABA responsive genes is an exciting prospect.

AUXIN-REGULATED GENES

Like other plant hormones, auxins have very pleiotropic effects. For example, one particular auxin, indole-3-acetic acid (IAA), plays a role in many processes such as maintenance of apical dominance, cell elongation, and xylem differentiation (213). In addition, the regeneration of shoots in vitro requires the presence of auxins and cytokinins (142). It is therefore likely that some developmentally regulated genes will respond to a combination of hormones. In the case of auxin-mediated hypocotyl elongation, it has been forcefully argued that auxin action has a very rapid component that does not involve gene expression as well as a long-term component that does involve the selective induction of genes (218). To what extent the two responses share a signal transduction chain is not known. A number of cDNAs coding for auxin-regulated genes has been isolated (77, 78, 203, 204, 235, 236). However with the exception of beta-1,3-glucanase (53, 135), no functions have yet been assigned to the encoded proteins. Detailed analysis of the regulation of these genes showed that their induction was specific for auxins, and no effects of other growth regulators or environmental stimuli could be detected (226). In contrast, the gene for beta-1,3-glucanase is induced by the absence of hormones or if either auxins or cytokinins are added separately. When both hormones are present, expression was reduced to a low basal level (135). It will be of interest to analyze the regulatory regions of this gene for cis-acting elements responsive to different hormones and to see how these elements would interact to result in this complex pattern of regulation.

Another interesting aspect of auxin-regulated gene expression is the substantial information that has been gathered over the years concerning auxin receptors and auxin-binding proteins, which may be components of the signal transduction chain leading to auxin-induced gene expression (100, 123–25, 173–175). In one instance, a direct effect of an auxin-binding protein on transcription has been observed. Van der Linde et al (219) isolated a low abundant soluble protein with a very high affinity for IAA ($K_a = 1.6 \times 10^8 \cdot M^{-1}$ at 25°C). Addition of partially purified preparations of this protein to isolated tobacco nuclei resulted in an IAA-dependent stimulation of overall runoff transcription. Obviously, one would like to know: (a) What are the effects of this factor on the expression of specific auxin-induced genes? (b) Is there any effect on the initiation of transcription, as isolated nuclei usually do not reinitiate RNA synthesis?

The cellular enzymes for auxin biosynthesis have not been characterized. However, the presence in the *Agrobacterium* T-DNA of two genes encoding enzymatic steps for IAA-synthesis from tryptophan makes possible the manipulation of auxin levels in vivo (205). Overproduction of these enzymes by fusing the bacterial genes to highly expressed cell-specific or developmentally regulated promoters offers exciting prospects for gaining insight not only into the complete chain of events that leads to auxin-regulated gene expression but, beyond that, into the way plant hormones affect development.

HEAT SHOCK GENES

Cells from a wide variety of organisms express a set of so-called heat shock proteins (hsp's) as a protective response to environmental stress (3, 111, 158, 179). This phenomenon has attracted much interest as a model system for the coordinate regulation of gene expression. In *Drosophila,* a number of *hsp* genes has been characterized in great detail and DNA sequence elements have been defined that confer heat inducibility on heterologous reporter genes

(157, 159). The exact sequence requirements, however, seem to vary depending on the assay system used. When a 14 nucleotide heat shock consensus sequence was positioned upstream of the TATA box of a nonheat-inducible reporter gene and the construct was introduced into monkey COS cells or *Xenopus* oocytes, transcription was fully inducible by heat shock (157, 159). Strikingly different results were obtained when *hsp* genes were reintroduced into flies by P-element mediated transduction; several hundred base pairs of upstream sequence containing several heat-shock elements (HSEs) were required for induction of the *hsp23* gene (45), whereas for *hsp26* and *hsp70* (45, 156), sequences including at least two HSEs were required. These discrepancies indicate that the results obtained in heterologous systems need to be interpreted with some caution.

The data available from plants strongly suggest that the structure of heat shock proteins, their function, and the regulation of their expression are highly conserved between plants and animals. In plants, heat-induced proteins have been detected in the 80–90, 65–75, and 15–30 kilodalton size classes, and there is strong homology with *Drosophila* proteins (111, 181). In many organisms, various stress conditions such as ethanol, arsenite, anoxia, and heavy metals will also elicit the synthesis of these proteins (3). In soybean plants, arsenite and to some extent $CdCl_2$ (76) are effective, in addition to heat shock (122). In maize mesocotyls, water stress, abscisic acid, and wounding mimic the heat shock response (81). Plant heat shock proteins are expressed in almost all cells. In maize, these proteins can be induced in all tissues examined except germinating pollen (37), and a role in thermotolerance has been proposed as in other organisms (122, 182).

Runoff experiments with nuclei isolated either from control or heat-stressed soybean plants have demonstrated that the heat shock induction of the mRNAs for the low molecular weight proteins is at least partly at the level of transcription initiation (181). Posttranscriptional mechanisms, however, including preferential translation of hs mRNAs at elevated temperature may also be involved. The characteristic HSEs shown to be important for regulated transcription in other organisms have also been detected in plant heat shock genes by sequence comparison (42, 180). To assess the biological relevance of these HSEs in plants, gene transfer experiments have been performed. A 457-bp upstream sequence derived from the *Drosophila hsp70* gene was fused to a bacterial reporter gene, and the resulting chimera was introduced into tobacco calli (198). The *Drosophila* sequence was capable of expressing the reporter gene in a heat-regulated fashion, and thus the *cis*-acting elements present on the *Drosophila* DNA must be correctly recognized by the plant's transacting factors. Two maize genomic clones with strong homology to the *Drosophila hsp70* gene have been isolated (169). A construct containing 1.1 kb of 5' flanking sequence was transferred to transgenic petunia plants, and the expression of this gene was found to be thermally induced with transcription initiating at the correct site.

Soybean genes coding for low molecular weight heat shock proteins were introduced into sunflower using a nondisarmed vector system. Gurley et al (76) studied the expression of the soybean *hsp17.5E* gene with 3.25 kb of upstream flanking sequence in tumors incited on sunflower hypocotyls. Upon heat shock the steady state mRNA reached a level comparable to that in soybean hypocotyls. However, in the tumor tissue there was a substantial basal level at low temperature, which was not observed in the soybean hypocotyls. A 5' deletion to −1175 bp further increased the basal level without affecting the induced level. A further deletion to −95 bp decreased both basal and induced mRNAs. This could mean that a negative element was located upstream of −1175 and that positive elements are located further downstream. Schoffl & Baumann (180) transferred a related gene, *hs6871*, containing > 1 kb of 5' noncoding sequence into similar sunflower tumors. In a control experiment, they noted that the level of expression of endogenous sunflower heat shock genes was dramatically decreased in tumors compared to hypocotyls, suggesting that the dedifferentiated state negatively influenced the *hsp* transcript level. Using a sensitive S1 mapping technique, they were able to demonstrate correctly regulated expression of the soybean gene in the sunflower tumor tissue, albeit at a low level. Clearly the heat shock genes are correctly transcribed in tumor tissues, but there are discrepancies with respect to the basal and induced transcript levels between the two groups of experiments. This may result from differences in the transformation and/or expression conditions. Schoffl et al (181) repeated their transformation experiments with a disarmed gene transfer system that enabled them to study the regulation of heat shock gene expression in transgenic plants. Comparing sunflower tumors and transgenic tobacco leaves, these authors found that the soybean *hs6871* gene transcripts accumulated to much higher levels upon induction in the transgenic plants.

We have recently constructed a chimeric gene in which two soybean HSEs were placed upstream of a truncated 35S promoter driving a bacterial reporter gene (G. Strittmatter and N-H. Chua, unpublished). In transgenic tobacco plants the expression of this chimeric gene is heat-shock dependent and correctly initiated. Transgenic plants offer the unique opportunity to study the full complexity of gene regulation, and we are interested in learning how, in molecular terms, different stimuli can affect the expression of a single gene. Towards this end, we constructed chimeric genes containing enhancer-like elements from both a soybean *hsp* gene and from a pea *rbcS* (light-regulated) gene. By determining transcript levels in dark and light, at low and high temperatures, we hope to obtain insight into the interaction between different regulatory *cis*-acting elements.

PLANT-MICROBE INTERACTIONS

Nodulin Genes

Nitrogen fixation, which takes place in a specialized organ, the nodule, is the result of a complex interaction leading to the finely tuned symbiosis between a plant and a compatible *Rhizobium* (22, 114, 137, 222). The bacterium encodes the basic enzymatic machinery for converting molecular nitrogen into ammonia plus a number of genes required for symbiosis. The plant provides the peculiar microenvironment that is necessary for carrying out nitrogen fixation and it contributes the enzymes that assimilate the reduced nitrogen.

When *Rhizobium* attaches to the plant root hair a signal is transmitted that changes the developmental program of the root cells and turns on a set of plant genes called nodulin genes. A few nodulin gene products have known functions, such as leghemoglobin (29), which serves as an oxygen scavenger, and glutamine synthase (69), xanthine dehydrogenase (215), and uricase (150), which are involved in ammonium metabolism. The functions of most other nodulins are as yet undefined.

Nodulin genes vary in the time window of expression during nodule differentiation; some are turned on early, others later (137). This raises the question of how such differential expression is determined. Are there multiple signals conveyed by the bacterium? Alternatively, is there a single signal that is differentially perceived by the root cells, which may have different effects on nodulin genes? Two avenues for studying this problem have been followed. One is the use of bacterial and plant mutants. Stages in the development of nodules can now be defined by mutants that cause an arrest at that stage. For instance, "very early" bacterial mutants will fail to attach to the plant root hair, and "late" mutants will form nodules but may not be capable of nitrogen fixation (114). The effect of each of these bacterial mutants on the expression of the plant nodulin genes, and thus the effects of each subsequent step in the increasingly intimate interaction, can now be studied. As an example of this approach, Govers et al (73) first carefully analyzed the time course of appearance of mRNAs for a number of pea nodulins and thereby defined early and late nodulin genes. Subsequently, they followed the expression of these genes following interaction with three *Rhizobium* mutants affected in the later stages of nodule development. One protein previously characterized as a late nodulin was absent in these mutants, whereas all other major nodulins were unaffected. It is not known how the bacterial mutation leads to this plant gene response.

Another related approach is to study nodulin gene expression in bacterium-free plant cell cultures under conditions simulating the symbiotic state. Thus, uricase was turned on in soybean callus tissue under semi-anaerobic conditions (119). Isolation of plant mutants has also been attempted (33, 112).

Obviously, this is a laborious task that can only be successful if mutations are generated at a very high rate or if effective selection is possible. The regulatory elements of the nodulin genes have only been explored in a few cases. A severe problem until very recently (see below) has been the inability to generate transgenic legume plants. Analysis of transcription of a nodulin gene in a HeLa cell extract revealed that the transcription start sites were similar to the ones used in vivo (231). It remains to be seen, however, to what extent this kind of heterologous in vitro system can be useful for the unraveling of the regulatory events that take place in vivo.

Whereas most commonly used and commercially interesting legumes resist efficient regeneration, a breakthrough has been achieved with the wild legume *Lotus cornatus*. Jensen et al (102) succeeded in transforming this species with a leghemoglobin-CAT chimeric construct. Roots could be infected with a strain of *Agrobacterium rhizogenes,* and subsequently transformed *Lotus* plants containing the hybrid gene were obtained. Upon infection with *Rhizobium loti,* nodules were formed that expressed the introduced gene in a fashion that was correct by all the criteria applied. Evidently, this approach opens the way for functionally defining the *cis*-regulatory elements in leghemoglobin genes (97) and in any nodulin gene in a manner similar to that used for other plant genes.

Defense Genes

Because plants are sessile, they must be able to defend themselves against a variety of environmental conditions or factors that are potentially harmful to their survival (13, 185). The plant responds by the induction of specific defense genes (220, 221). UV light has been shown to lead to the rapid induction of enzymes involved in the biosynthesis of flavonoids, which serve as a UV-screen to shield the plant. Similarly, challenging plant cells with compounds of fungal origin leads to the induction of enzymes involved in lignin biosynthesis (14, 39, 50, 120, 189, 197). By lignifying cells neighboring the affected cells, the plant may protect itself from further damage. The genes for one of the key enzymes in flavonoid and lignin production, chalcone synthase, have been studied in detail, as discussed in the section on light-regulated genes (109).

The enzyme chitinase most likely plays a role in protecting plant cells against invading fungi that contain chitin in their cell walls (31). Interestingly, this enzyme is induced by the plant growth regulator ethylene, and the induction has been shown to be at the level of steady state mRNA (31). In most of these regulatory systems no genomic clones have been isolated, but the eventual characterization of the *cis*- and *trans*-acting elements that mediate these responses is expected to yield very interesting insights into the mechanisms of plant defense.

CONSTITUTIVE GENES

Cellular Genes

A gene is classified as constitutive when its expression remains constant on a per cell basis under all conditions examined. This implies that there must be no variation of mRNA abundance and/or translational efficiency during development, in different organs or upon application of various endogenous or environmental stimuli. As the analysis of plant gene expression becomes more detailed, genes formerly thought to be constitutive may eventually be found to be regulated by a hitherto unidentified condition. In fact, it is hard to find a cellular plant gene that is truly constitutive. So far, the only candidate is a nuclear gene coding for the beta-subunit of the mitochondrial ATPase complex (24), which is not regulated by light and is expressed at the same level in all tissues examined (24; F. Nagy et al, unpublished results; S. Tingey et al, unpublished results).

Ribosomal RNA Genes

The genes coding for cytoplasmic ribosomal RNAs are clustered in tandem arrays at the nucleolar organizers, and they are transcribed by RNA polymerase I. Their expression is constitutive only in the sense that they are not or are hardly influenced by exogenous factors such as light. It has been estimated from nuclear runoff experiments that the transcription rate in the light is twofold higher than in the dark (67, 211). Interestingly, changes in methylation patterns have been observed in pea seedlings, depending on light conditions (108) and during development (228). The expression of rRNA genes also responds to the cell cycle: Transcription decreases during interphase and comes to a halt in metaphase.

In *Xenopus* a detailed structural and functional characterization of the rRNA transcriptional signals has been performed (166). Multiple enhancer-like elements with sequence homology to the rRNA promoter work additively to increase the expression of a particular rRNA gene. In interspecies crosses, genes with the most copies of an enhancer are expressed, and genes with fewer enhancer copies are suppressed, a phenomenon called nucleolar dominance. The structural organization of plant ribosomal RNA genes is reminiscent of that in *Xenopus* and other organisms, and it is likely that this structural homology reflects a functional similarity. For a more complete discussion of plant rRNA organization and expression and of nucleolar dominance, the reader is referred to two excellent reviews (56, 57).

Viral Genes

Four RNA transcripts encoded by cauliflower mosaic virus (CaMV), a double-stranded DNA virus, have been detected in infected turnip plants (75).

The two major transcripts that are commonly referred to as 19S and 35S are the most well-studied. A 0.4-kb DNA fragment containing the 19S promoter was placed upstream of the *nptII* gene; this chimeric construct was transferred into petunia cells via oncogenic and partially disarmed Ti-vectors, and correct expression was demonstrated (115). A *nptII* gene flanked by expression signals of the 5' and 3' regions of the 19S gene was introduced into tobacco plants by direct DNA uptake (161). We have made chimeric genes in which a 60-bp sequence from −100 to −40 bp relative to the transcription start site was fused to the 19S TATA box or a *rbcS* TATA box to drive the expression of the bacterial CAT gene. In transgenic plants, this 60-bp fragment was found to be capable of increasing the expression from its own and the heterologous promoter; thus it has the characteristics of an enhancer element (C. Kuhlemeier et al, unpublished results). The 35S promoter has been characterized in transformed calli and in transgenic plants. Equal concentrations of RNA were found in leaves, stems, roots, and petals (146), and no light regulation was observed (59). From 5' deletion analysis it became clear that elements upstream of the TATA box greatly enhance the level of transcription (151). Sequences with homology to the SV40 core enhancer have been implicated in this enhancement, but more refined analysis is needed to pinpoint their true biological significance.

Agrobacterial Genes

The T-DNA of the *Agrobacterium* Ti plasmid encodes a number of genes that are poorly expressed in the bacterium but are active after transfer to the plant. Sequence analysis has revealed that these bacterial genes contain eukaryotic expression signals (63). These genes are commonly categorized as constitutive, because their expression does not seem to be influenced by factors such as light and heat shock. Nor has tissue specificity of the expression been observed. It should be kept in mind, however, that an exhaustive search for putative regulatory stimuli affecting these genes has not been made, and it is conceivable that more detailed studies might elucidate exogenous or endogenous factors that do regulate the expression of these genes.

The gene coding for nopaline synthase *(nos)* has been extensively used for the construction of chimeric selectable genes (21, 63, 113, 233). From these constructs it has become clear that the coding region is dispensable for expression in plants and that the 5' flanking region is the major determinant for full expression. Shaw et al (188) constructed mutants for the *nos* gene with internal deletions in the upstream noncoding region. After introduction into *Kalanchöe* tumors, they monitored nopaline synthase enzymatic activity and concluded that the 88 bp upstream of the transcription initiation site that contain a putative CCAAT-box were sufficient for full level of nopaline synthase activity and thus for transcription. No further upstream sequence was required. An et al (2) fused sections of the *nos* flanking region to CAT and measured promoter strength by monitoring CAT enzymatic activity in tobacco calli. In contrast with the previous study, they identified an element with a 5' boundary between -130 and -101 that was required for high-level expression. It is not clear how these two sets of results can be reconciled. Somewhat surprisingly, in the latter study, deletion of the TATA-box by itself did not completely abolish CAT activity. In this case pseudopromoters might be responsible for the remaining transcriptional activity.

The expression of another T-DNA gene with unknown function, referred to as *780* gene, has been studied at the RNA level (32). A large number of mutated *780* genes were placed individually in Ti-DNA vectors containing an additional, wild-type *780* gene. Transcript levels were studied using a quantitative S1 nuclease protection assay with the wild-type *780* gene as an internal standard. The use of this direct method for assaying transcription is more advantageous than the previous methods that relied on an indirect way of quantitating transcription, namely by measuring enzymatic activity of the final gene product. The 5' deletions delineated an activator element located between -440 and -229 from the transcription start site. Internal deletions identified additional sequences, which possess a down-regulating activity that apparently was modulated by yet another adjacent element. Removal of the TATA-box decreased activity to less than 0.1%. Gel retardation experiments have provided evidence for the specific binding of *trans*-acting factors to the activator element (W. Gurley, personal communication).

CONCLUDING REMARKS

The study of gene regulation relies heavily on methods that test the effects of mutations on the expression of genes. For higher eukaryotes the most fruitful approach has been to induce specific mutations in vitro and to monitor expression following reintroduction of the altered genes into a recipient host. In this respect the field of plant research has experienced a spectacular breakthrough with the advent of *Agrobacterium*-mediated gene transfer. The totipotency of plant cells obviates the need for germ-line transformation. Consequently, transgenic plants can be generated in large numbers with relatively little effort, compared to transgenic mice or flies. The problems with monocot transformation have been overcome through the use of direct gene transfer methods for protoplasts. Regeneration of transformed monocot cells is the next major hurdle to be overcome.

A general limitation inherent in the analysis of individual transgenic organisms is that the transgene is subject to chromosomal position effects. The ability to direct the integration to a particular chromosomal site would be a great asset. In addition, integration of genes into autonomously replicating viral vectors would allow expression to be studied when genes are in a controlled environment (28, 94, 95, 171). Moreover, viral replicons provide an excellent system for studying homologous recombination (187, 196, 202). A better understanding of homologous recombination in plants could eventually lead to the opportunity to selectively alter or replace endogenous genes. This will be a powerful tool to study the function of specific genes.

For a number of plant genes, *cis*-acting DNA sequences that mediate regulated expression have been identified. Genes that respond to a certain stimulus in a coordinated fashion, such as heat-shock genes, have conserved consensus sequences. It is not yet known if genes such as the *rbcS* genes and the genes for phytochrome, which respond in opposite ways to the same light stimulus, have consensus elements with biological significance. The identification of functional *cis*-acting DNA elements is just beginning to lead to the characterization of *trans*-acting factors with which they interact. From the characterization of exact factor-binding sites, much will be learned about multiple interactions, coordinate regulation, and other mechanistic considerations such as cooperativity. The *cis* elements can also be used as a tool to purify the factors—an important step towards isolating the corresponding regulatory genes. For the study of both *cis*- and *trans*-acting elements, a major experimental approach that is lacking for plants is in vitro transcription. Not only would such a system allow one to study the function of (partially) purified transcription factors, but it is also a necessary prerequisite for the in vitro reconstitution of regulatory signal transduction chains.

Genetic approaches for identifying regulatory genes have been extensively applied in prokaryotes and lower eukaryotes, but their use has been scarce in higher, multicellular organisms. However, especially in *Nicotiana,* where excellent tissue culture including the generation of haploids is available (147), selection of mutants affected in *trans*-acting factors seems feasible. Fusing a *cis*-regulatory element with a deleterious gene and selecting for surviving cells after random mutagenesis is an approach that merits further attention. Unraveling the complexity of transcriptional responses to individual and multiple stimuli, as well as the further study of the transduction chain network, require several approaches. Undoubtedly, the use of transgenic organisms, which allows the study of genes in their normal context, is indispensable. In addition, the biochemical and genetic characterization of all components of the transduction chains is of major importance. Such a combined approach will bring about the most rapid progress towards understanding plant development in molecular terms.

ACKNOWLEDGMENTS

We thank our many colleagues for providing us with preprints and other unpublished results during the preparation of this manuscript. We also thank

Drs. Gloria Coruzzi and John Mundy for critically reading the manuscript and Wendy Roine for her skilled secretarial assistance. This work was supported by a grant from the Monsanto Company and by the Pew Trusts. C. K. was supported by a fellowship from The Netherlands Organization for the Advancement of Pure Research (Z.W.O.). This article was based on literature available before September 1, 1986.

Literature Cited

1. Allen, B. V., Nessler, R. D., Thomas, T. L. 1985. Developmental expression of sunflower 11S storage protein genes. *Plant Mol. Biol.* 5:165–73
2. An, G., Ebert, P. R., Yi, B.-Y., Choi, C.-H. 1986. Both TATA box and upstream regions are required for the nopaline synthase promoter activity in transformed tobacco cells. *Mol. Gen. Genet.* 203:245–50
3. Ananthan, J., Goldberg, A. L., Voellmy, R. 1986. Abnormal proteins serve as eukaryotic stress signals and trigger the activation of heat shock genes. *Science* 232:522–24
4. Apel, K., Kloppstech, K. 1978. The plastid membranes of barley *(Hordeum vulgare):* Light-induced appearance of mRNA coding for the apoprotein of the light-harvesting chlorophyll a/b-protein. *Eur. J. Biochem.* 85:581–88
5. Apel, K. 1979. Phytochrome-induced appearance of mRNA activity for the apoprotein of the light-harvesting chlorophyll a/b-protein of barley *(Hordeum vulgare). Eur. J. Biochem.* 97:183–88
6. Apel, K. 1981. The protochlorophyllide holochrome of barley *(Hordeum vulgare L.).* Phytochrome-induced decrease of translatable mRNA coding for the NADPH: protochlorophyllide oxidoreductase. *Eur. J. Biochem.* 120: 89–93
7. Argos, P., Narayana, S. V. L., Nielsen, N. C. 1985. Structural similarity between legumin and vicilin storage proteins from legumes. *EMBO J.* 4:1111–17
8. Aspart, L., Meyer, Y., Laroche, M., Penon, P. 1984. Developmental regulation of the synthesis of proteins encoded by stored mRNA in radish embryos. *Plant Physiol.* 76:664–73
9. Barta, A., Sommergruber, K., Thompson, D., Hartmuth, K., Matzke, M. A., Matzke, A. J. M. 1986. The expression of a nopaline synthase-human growth hormone chimaeric gene in transformed tobacco and sunflower callus tissue. *Plant Mol. Biol.* 6:347–57

10. Batschauer, A., Mosinger, E., Kreuz, K., Dorr, I., Apel, K. 1986. The implication of a plastid-derived factor in the transcriptional control of nuclear genes encoding the light-harvesting chlorophyll a/b protein. *J. Biochem.* 154:625–34
11. Baumlein, H., Wobus, U., Pustell, J., Kafatos, F. C. 1986. The legumin gene family: structure of a B–type gene of *Vicia faba* and a possible legumin gene specific regulatory element. *Nucleic Acids Res.* 14:2707–20
12. Beachy, R. N., Chen, Z.-L., Horsch, R. B., Rogers, S. G., Hoffmann, N. J., Fraley, R. T. 1985. Accumulation and assembly of soybean beta-conglycinin in seeds of transformed petunia plants. *EMBO J.* 4:3047–53
13. Bell, A. A. 1981. Biochemical mechanisms of disease resistance. *Ann. Rev. Plant Physiol.* 32:21–81
14. Bell, J. N., Ryder, T. B., Wingate, V. P. M., Bailey, J. A., Lamb, C. J. 1986. Differential accumulation of plant defense gene transcripts in a compatible and an incompatible plant-pathogen interaction. *Mol. Cell. Biol.* 6:1615–23
15. Bennett, J. 1981. Biosynthesis of the light-harvesting chlorophyll a/b protein; polypeptide turnover in darkness. *Eur. J. Biochem.* 118:61–70
16. Bennett, J., Jenkins, G. I., Hartley, M. R. 1984. Differential regulation of the accumulation of the light-harvesting chlorophyll a/b-complex and ribulose bisphosphate carboxylase/oxygenase in greening pea leaves. *J. Cell Biochem.* 25:1–13
17. Berry, J. O., Nikolau, B. J., Carr, J. P., Klessig, D. F. 1985. Transcriptional and post-transcriptional regulation of ribulose 1,5-bisphosphate carboxylase gene expression in light- and dark-grown amaranth cotyledons. *Mol. Cell. Biol.* 5:2238–46
18. Berry-Lowe, S. L., McKnight, T. D., Shah, D. M., Meagher, R. B. 1982. The nucleotide sequence, expression, and evolution of one member of a multigene

family encoding the small subunit of ribulose-1,5-bisphosphate carboxylase in soybean. *J. Mol. Appl. Genet.* 1:483–98

19. Berry-Lowe, S. L., Meagher, R. B. 1985. Transcriptional regulation of a gene encoding the small subunit of ribulose-1,5-bisphosphate carboxylase in soybean tissue is linked to the phytochrome response. *Mol. Cell. Biol.* 5:1910–17

20. Bevan, M. 1984. Binary *Agrobacterium* vectors for plant transformation. *Nucleic Acids Res.* 12:8711–21

21. Bevan, M. W., Chilton, M.-D. 1982. T-DNA of the *Agrobacterium* Ti- and Ri-plasmids. *Ann. Rev. Genet.* 16:357–84

22. Bisseling, T., Govers, F., Stiekema, W. 1984. The identification of proteins and their mRNAs involved in the establishment of an effective symbiosis. *Oxford Surv. Plant Mol. Cell Biol.* 1:53–83

23. Bollini, R., Chrispeels, J. J. 1978. Characterization and subcellular localization of vicilin and phyto-hemagglutinin, the two major reserve proteins of *Phaseolus vulgaris. Planta* 142:291–98

24. Boutry, M., Chua, N.-H. 1985. A nuclear gene encoding the beta subunit of the mitochondrial ATPase in *N. plumbaginifolia. EMBO J.* 4:2159–65

25. Bray, E., Beachy, R. N. 1985. Regulation by ABA of beta-conglycinin expression in cultured developing soybean cotyledons. *Plant Physiol.* 79:746–50

26. Briggs, W. R., Iino, M. 1983. Blue-light absorbing photoreceptors in plants. *Philos. Trans. R. Soc. London Ser. B* 303:347–59

27. Briggs, W. R., Mosinger, E., Batschauer, A., Apel, K., Schafer, E. 1986. Molecular events in photoregulated greening in barley leaves. In *Molecular Biology of Plant Growth Control*, ed. J. E. Fox, M. Jacobs. New York: Liss. In press

28. Brisson, N., Paszkowski, J., Penswick, J. R., Gronenborn, B., Potrykus, I., Hohn, T. 1984. Expression of a bacterial gene in plants by using a viral vector. *Nature* 310:511–14

29. Brisson, N., Verma, D. P. S. 1982. Soybean leghemoglobin gene family: normal, pseudo, and truncated genes. *Proc. Natl. Acad. Sci. USA* 79:4055–59

30. Broglie, R., Coruzzi, G., Fraley, R. T., Rogers, S. G., Horsch, R. B., et al. 1984. Light-regulated expression of a pea ribulose-1,5-bisphosphate carboxylase small subunit gene in transformed plant cells. *Science* 224:838–43

31. Broglie, K. E., Gaynor, J. J., Broglie, R. M. 1986. Ethylene regulated gene expression: molecular cloning of the genes encoding an endochitinase from *Phaseolus vulgaris. Proc. Natl. Acad. Sci. USA* 83:6820–24

32. Bruce, W. B., Gurley, W. B. 1986. Functional domains of a T-DNA promoter active in crown gall tumors. *Mol. Cell. Biol.* In press

33. Carroll, B. J., McNeil, D. L., Gresshoff, P. M. 1985. Isolation and properties of soybean [*Glycine max* (L.) Merr.] mutants that nodulate in the presence of high nitrate concentrations. *Proc. Natl. Acad. Sci. USA* 82:4162–66

33a. Chen, Z-L., Schuler, M., Beachy, R. N. 1986. Functional analysis of regulatory elements in a plant embryo-specific gene. *Proc. Natl. Acad. Sci. USA* 83: In press

34. Chrispeels, M. J., Vitale, A., Staswick, P. 1984. Gene expression and synthesis of phytohemagglutinin in the embryonic axes of developing *Phaseolus vulgaris* seeds. *Plant Physiol.* 76:791–96

35. Colbert, J. T., Hershey, H. P., Quail, P. H. 1983. Autoregulatory control of translatable phytochrome mRNA levels. *Proc. Natl. Acad. Sci. USA* 80:2248–52

36. Colbert, J. T., Hershey, H. P., Quail, P. H. 1985. Phytochrome regulation of phytochrome mRNA abundance. *Plant Mol. Biol.* 5:91–101

37. Cooper, P., Ho, T.-H. D., Hauptmann, R. M. 1984. Tissue specificity of the heat-shock response in maize. *Plant Physiol.* 75:431–41

38. Coruzzi, G., Broglie, R., Edwards, C., Chua, N.-H. 1984. Tissue-specific and light-regulated expression of a pea nuclear gene encoding the small subunit of ribulose-1,5-bisphosphate carboxylase. *EMBO J.* 3:1671–79

39. Cramer, C. L., Ryder, T. B., Bell, J. N., Lamb, C. J. 1984. Rapid switching of plant gene expression induced by fungal elicitor. *Science* 227:1240–43

40. Crouch, M. L., Sussex, I. M. 1981. Development and storage protein synthesis in *Brassica napa* L. embryos *in vivo* and *in vitro. Planta* 153:64–74

41. Crouch, M. L., Tenbarge, K., Simon, A., Finkelstein, R., Scofield, S., Solberg, L. 1985. Storage protein mRNA levels can be regulated by abscisic acid in *Brassica* embryos. In *Molecular Form and Function of the Plant Genome*, ed. L. van Vloten-Doting, G. S. P. Groot, T. C. Hall, 83A:555–66. London/New York: Plenum. 693 pp.

42. Czarnecka, E., Gurley, W. B., Nagao, R. T., Mosquera, L. A., Key, J. L.

1985. DNA sequence and transcript mapping of a soybean gene encoding a small heat shock protein. *Proc. Natl. Acad. Sci. USA* 82:3726–30

43. Datta, N., Chen, Y.-R., Roux, S. J. 1985. Phytochrome and calcium stimulation of protein phosphorylation in isolated pea nuclei. *Biochem. Biophys. Res. Commun.* 128:1403–8

44. Dean, C., van den Elzen, P., Tamaki, S., Dunsmuir, P., Bedbrook, J. 1985. Differential expression of the eight genes of the petunia ribulose bisphosphate carboxylase small subunit multigene family. *EMBO J.* 4:3055–61

45. Dudler, R., Travers, A. A. 1984. Upstream elements necessary for optimal function of the hsp70 promoter in transformed flies. *Cell* 38:391–98

46. Dure, L. III. 1985. Embryogenesis and gene expression during seed formation. *Oxford Surv. Plant Mol. Cell Biol.* 2:179–97

47. Dure, L. III, Greenway, S. C., Galau, G. A. 1981. Developmental biochemistry of cottonseed embryogenesis and germination: changing messenger ribonucleic acid populations as shown by *in vitro* and *in vivo* protein synthesis. *Biochemistry* 20:4162–68

48. Dynan, W. S., Tjian, R. 1985. Control of eukaryotic messenger RNA synthesis by sequence-specific DNA-binding proteins. *Nature* 316:774–78

49. Ecker, J. R., Davis, R. W. 1986. Inhibition of gene expression in plant cells by expression of antisense RNA. *Proc. Natl. Acad. Sci. USA* 83:5372–76

50. Edwards, K., Cramer, C. L., Bolwell, G. P., Dixon, R. A., Schuch, W., Lamb, C. J. 1985. Rapid transient induction of phenylalanine ammonia-lyase mRNA in elicitor-treated bean cells. *Proc. Natl. Acad. Sci. USA* 82:6731–35

50a. Ellis, R. J. 1986. Photoregulation of gene expression. *Biosci. Rep.* 6:127–36

51. Ernst, D., Oesterhelt, D. 1984. Purified phytochrome influences *in vitro* transcription in rye nuclei. *EMBO J.* 3:3075–78

52. Facciotti, D., O'Neal, J. K., Lee, S., Shewmaker, C. K. 1985. Light-inducible expression of a chimeric gene in soybean tissue transformed with *Agrobacterium. Biotechnology* 3:241–46

53. Felix, G., Meins, F. Jr. 1985. Purification, immunoassay and characterization of an abundant, cytokinin-regulated polypeptide in cultured tobacco tissues. *Planta* 164:423–28

54. Finkelstein, R. R., Crouch, M. L. 1986. Rapeseed embryo development in culture on high osmoticum is similar to that in seeds. *Plant Physiol.* 81:907–12

55. Finkelstein, R. R., Tenbarge, K. M., Shumway, J. E., Crouch, M. L. 1985. Role of ABA in maturation of rapeseed embryos. *Plant Physiol.* 798:630–36

56. Flavell, R. B. 1986. The structure and control of expression of ribosomal RNA genes. *Philos. Trans. R. Soc. London Ser. B.* In press

57. Flavell, R. B., O'Dell, M., Thompson, W. F., Vincentz, M., Sardana, R., Barker, R. F. 1986. The differential expression of ribosomal RNA genes. Submitted for publication

58. Fleck, J., Durr, A., Lett, M. C., Hirth, L. 1979. Changes in protein synthesis during the initial stage of life of tobacco protoplasts. *Planta* 145:279–85

59. Fluhr, R., Chua, N.-H. 1986. Developmental regulation of two genes encoding ribulose-bisphosphate carboxylase small subunit in pea and transgenic petunia plants: phytochrome responses and blue-light induction. *Proc. Natl. Acad. Sci. USA* 83:2358–62

60. Fluhr, R., Kuhlemeier, C., Nagy, F., Chua, N.-H. 1986. Organ-specific and light-induced expression of plant genes. *Science* 232:1106–12

61. Fluhr, R., Moses, P., Morelli, G., Coruzzi, G., Chua, N.-H. 1986. Expression dynamics of the pea rbcS multigene family and organ distribution of the transcripts. *EMBO J.* 5:2063–71

62. Forde, B. G., Heyworth, A., Pywell, J., Kreis, M. 1985. Nucleotide sequence of a B1 hordein gene and the identification of possible upstream regulatory elements in endosperm storage protein genes from barley, wheat and maize. *Nucleic Acids Res.* 13:7327–39

63. Fraley, R. T., Rogers, S. G., Horsch, R. B. 1986. Genetic transformation in higher plants. *CRC Crit. Rev. Plant Sci.* 4:1–46

64. Fromm, M., Taylor, L. P., Walbot, V. 1985. Expression of genes transferred into monocot and dicot plant cells by electroporation. *Proc. Natl. Acad. Sci. USA* 82:5824–28

65. Fromm, M. E., Taylor, L. P., Walbot, V. 1986. Stable transformation of maize after gene transfer by electroporation. *Nature* 319:791–93

66. Galau, G. A., Hughes, D. W., Dure, L. III. 1986. Abscisic acid induction of cloned cotton late embryogenesis-abundant *(Lea)* mRNAs. *Plant Mol. Biol.* 7:155–70

67. Gallagher, T. F., Ellis, R. J. 1982. Light-stimulated transcription of genes for two chloroplast polypeptides in iso-

lated pea leaf nuclei. *EMBO J.* 1:1493–98

68. Gallagher, T. F., Jenkins, G. I., Ellis, R. J. 1985. Rapid modulation of transcription of nuclear genes encoding chloroplast proteins by light. *FEBS Lett.* 186:241–45

69. Gebhardt, C., Oliver, J. E., Forde, B. G., Saarelainen, R., Miflin, B. J. 1986. Primary structure and differential expression of glutamine synthetase genes in nodules, roots and leaves of *Phaseolus vulgaris. EMBO J.* 5:1429–35

70. Gluzman, Y., Shenk, T., eds. 1984. *Enhancers and Eukaryotic Gene Expression.* New York: Cold Spring Harbor Lab.

71. Goldsbrough, P. B., Gelvin, S. B., Larkins, B. A. 1986. Expression of maize zein genes in transformed sunflower cells. *Mol. Gen. Genet.* 202:374–81

72. Gollmer, I., Apel, K. 1983. The phytochrome-controlled accumulation of mRNA sequences encoding the light-harvesting chlorophyll a/b-protein of barley (*Hordeum vulgare* L.). *Eur. J. Biochem.* 133:309–13

73. Govers, F., Gloudemans, T., Moerman, M., van Kammen, A., Bisseling, T. 1985. Expression of plant genes during the development of pea root nodules. *EMBO J.* 4:861–67

74. Greenwood, J. S., Chrispeels, M. J. 1985. Correct targeting of the bean storage protein phaseolin in the seeds of transformed tobacco. *Plant Physiol.* 79:65–71

75. Guilley, H., Dudley, R. K., Jonard, G., Balazs, E., Richards, K. E. 1982. Transcription of cauliflower mosaic virus DNA: detection of promoter sequences, and characterization of transcripts. *Cell* 30:763–73

76. Gurley, W. B., Czarnecka, E., Nagao, R. T., Key, J. L. 1986. Upstream sequences required for efficient expression of a soybean heat shock gene. *Mol. Cell. Biol.* 6:559–65

77. Hagen, G., Guilfoyle, T. J. 1985. Rapid induction of selective transcription by auxins. *Mol. Cell. Biol.* 5:1197–1203

78. Hagen, G., Kleinschmidt, A., Guilfoyle, T. 1984. Auxin-regulated gene expression in intact soybean hypocotyl and excised hypocotyl sections. *Planta* 162:147–53

79. Hain, R., Stabel, P., Czernilofsky, A. P., Steinbiss, H. H., Herrera-Estrella, L., Schell, J. 1985. Uptake, integration, expression and genetic transmission of a selectable chimaeric gene by plant protoplasts. *Mol. Gen. Genet.* 199:161–68

80. Heidecker, G., Messing, J. 1986. Structural analysis of plant genes. *Ann. Rev. Plant Physiol.* 37:439–66

81. Heikkila, J. J., Papp, J. E., Schultz, G. A., Bewley, J. D. 1984. Induction of heat shock protein messenger RNA in maize mesocotyls by water stress, abscisic acid, and wounding. *Plant Physiol.* 76:270–74

82. Hendrickson, W. 1985. Protein-DNA interactions studied by the gel electrophoresis-DNA binding assay. *Biotechniques* 3:198–207

83. Hernalsteens, J.-P., Thia-Toong, L., Schell, J., Van Montagu, M. 1984. An *Agrobacterium*-transformed cell culture from the monocot *Asparagus officinalis. EMBO J.* 3:3039–41

84. Herrera-Estrella, L., van den Broeck, G., Maenhaut, R., Van Montagu, M., Schell, J., et al. 1984. Light-inducible and chloroplast-associated expression of a chimaeric gene introduced into *Nicotiana tabacum* using a Ti plasmid vector. *Nature* 310:115–20

85. Higgins, T. J. V. 1984. Synthesis and regulation of major proteins in seeds. *Ann. Rev. Plant Physiol.* 35:191–221

86. Hoffman, L. M., Sengupta-Gopalan, C., Paaren, H. E. 1984. Structure of soybean Kunitz trypsin inhibitor mRNA determined from cDNA by using oligodeoxynucleotide primers. *Plant Mol. Biol.* 3:111–17

87. Hooykaas, P. J. J., Schilperoort, R. A. 1984. The molecular genetics of crown gall tumorigenesis. *Adv. Genet.* 22:210–83

88. Hooykaas-Van Slogteren, G. M. S., Hooykaas, P. J. J., Schilperoort, R. A. 1984. Expression of Ti plasmid genes in monocotyledonous plants infected with *Agrobacterium tumefaciens. Nature* 311:763–64

89. Hornberg, C., Weiler, E. W. 1984. High-affinity binding sites for abscisic acid on the plasmalemma of *Vicia faba* guard cells. *Nature* 310:321–24

90. Horsch, R. B., Fraley, R. T., Rogers, S. G., Sanders, P. R., Lloyd, A., Hoffmann, N. 1984. Inheritance of functional foreign genes in plants. *Science* 223:496–98

91. Horsch, R. B., Fry, J. E., Hoffmann, N. L., Eichholtz, D., Rogers, S. G., Fraley, R. T. 1985. A simple and general method for transferring genes into plants. *Science* 227:1229–321

92. Horsch, R. B., Klee, H. J. 1986. Rapid assay of foreign gene expression in leaf discs transformed by *Agrobacterium tumefaciens:* role of T-DNA borders in

the transfer process. *Proc. Natl. Acad. Sci. USA* 83:4428–32
93. Horsch, R. B., Klee, H. J., Stachel, S., Winans, S. C., Nester, E. W., et al. 1986. Analysis of *Agrobacterium tumefaciens* virulence mutans in leaf discs. *Proc. Natl. Acad. Sci. USA* 83:2571–75
94. Howell, S. H. 1982. Plant molecular vehicles: potential vectors for introducing foreign DNA into plants. *Ann. Rev. Plant Physiol.* 33:609–50
95. Howell, S. H., Walker, L. L., Dudley, R. K. 1980. Cloned cauliflower mosaic virus DNA infects turnips *(Brassica hapa)*. *Science.* 208:1265–67
96. Huang, J.-K., Swegle, M., Dandekar, A. M., Muthukrishnan, S. 1984. Expression and regulation of alpha-amylase gene family in barley aleurones. *J. Mol. Appl. Genet.* 2:579–88
97. Hyldig-Nielsen, J. J., Jensen, E. O., Paludan, K., Wiborg, O., Garrett, R., et al. 1982. The primary structures of two leghemoglobin genes from soybean. *Nucleic Acids Res.* 10:689–701
98. Iino, M., Ogawa, T., Zeiger, E. 1985. Kinetic properties of the blue-light response of stomata. *Proc. Natl. Acad. Sci. USA* 82:8019–23
99. Jabben, M., Holmes, M. G. 1983. Phytochrome in light-grown plants. In *Encyclopedia of Plant Physiology: Photomorphogenesis* (NS), ed. W. Shropshire, H. Mohr, 27:704–19. Berlin: Springer-Verlag. 832 pp.
100. Jacobs, M., Gilbert, S. F. 1983. Basal localization of the presumptive auxin transport carrier in pea stem cells. *Science* 220:1297–300
101. Jacobsen, J. V., Beach, L. R. 1985. Control of transcription of alpha-amylase and rRNA genes in barley aleurone protoplasts by gibberellin and abscisic acid. *Nature* 316:275–77
102. Jensen, J. S., Marcker, K. A. 1986. Nodule-specific expression of a chimaeric soybean leghaemoglobin gene in transgenic *Lotus corniculatus*. *Nature* 321:669–74
103. Jones, K. A., Yamamoto, K. R., Tjian, R. 1984. Two distinct transcription factors bind to the HSV thymidine kinase promoter *in vitro*. *Cell* 42:559–72
104. Jones, R. L., MacMillan, J. 1985. Gibberellins. In *Advanced Plant Physiology*, ed. M. B. Wilkins, pp. 21–52. London: Pitman
105. Kaufman, L. S., Briggs, W. R., Thompson, W. F. 1985. Phytochrome control of specific mRNA levels in developing pea buds. *Plant Physiol.* 78:388–93

106. Kaufman, L. S., Roberts, L. L., Briggs, W. R., Thompson, W. F. 1986. Phytochrome control of specific mRNA levels in developing pea buds: kinetics of accumulation, reciprocity, and escape kinetics of the low fluence response. *Plant Physiol.* In press
107. Kaufman, L. S., Thompson, W. F., Briggs, W. R. 1984. Different red light requirements for phytochrome-induced accumulation of *cab* RNA and *rbcS* RNA. *Science* 226:1447–49
108. Kaufman, L. S., Watson, J. C., Thompson, W. F. 1986. Light-regulated changes in DNaseI hypersensitive sites in the rRNA genes of *Pisum sativum*. *Proc. Natl. Acad. Sci. USA.* In press
109. Kaulen, H., Schell, J., Kreuzaler, F. 1986. Light-induced expression of the chimeric chalcone synthase-NPTII gene in tobacco cells. *EMBO J.* 5:1–8
110. Keith, B., Chua, N.-H. 1986. Monocot and dicot pre-mRNAs are processed with different efficiencies in transgenic tobacco. *EMBO J.* 5:2419–26
111. Key, J. L., Lin, C. Y., Chen, Y. M. 1981. Heat shock proteins of higher plants. *Proc. Natl. Acad. Sci. USA* 78:3526–30
112. Keyser, H. H., Bohlool, B. B., Hu, T. S., Weber, D. F. 1981. Fast-growing Rhizobia isolated from root nodules of soybean. *Science* 215:1631–32
113. Klee, H. J., Yanofsky, M. F., Nester, E. W. 1985. Vectors for transformation of higher plants. *Biotechnology* 3:637–42
114. Kondorosi, E., Kondorosi, A. 1986. Nodule induction on plant roots by *Rhizobium*. *Trends Biol. Sci.* 11:296–99
115. Koziel, M. G., Adams, T. L., Hazlet, M. A., Damm, D., Miller, J., et al. 1984. A cauliflower mosaic virus promoter directs expression of kanamycin resistance in morphogenic transformed plant cells. *J. Mol. Appl. Genet.* 2:549–62
116. Lamb, C. J., Lawton, M. A. 1983. Photocontrol of gene expression. In *Encyclopedia of Plant Physiology: Photomorphogenesis* (NS), ed. W. Shropshire, H. Mohr, 16A:213–43. Berlin: Springer-Verlag
117. Lamppa, G., Morelli, G., Chua, N.-H. 1985. Structure and developmental regulation of a wheat gene encoding the major chlorophyll a/b-binding polypeptide. *Mol. Cell. Biol.* 5:1370–78
118. Lamppa, G., Nagy, F., Chua, N.-H. 1985. Light-regulated and organ-specific expression of a wheat Cab gene in transgenic tobacco. *Nature* 316:750–52

119. Larsen, K., Jochimsen, B. U. 1986. Expression of nodule-specific uricase in soybean callus tissue is regulated by oxygen. *EMBO J.* 5:15–19

120. Lawton, M. A., Lamb, C. J. 1986. Transcriptional activation of plant defense genes by fungal elicitor, wounding and infection. *Mol. Cell. Biol.* Submitted for publication

121. Lazarus, C. M., Baulcombe, D. C., Martienssen, R. A. 1985. Alpha-amylase genes of wheat are two multigene families which are differentially expressed. *Plant Mol. Biol.* 5:13–24

122. Lin, C.-Y., Roberts, J. K., Key, J. L. 1984. Acquisition of thermotolerance in soybean seedlings. *Plant Physiol.* 74:152–60

123. Lobler, M., Klambt, D. 1985. Auxinbinding protein from coleoptile membranes of corn (*Zea mays* L.). I. Purification by immunological methods and characterization. *J. Biol. Chem.* 260:9848–53

124. Lobler, M., Klambt, D. 1985. Auxinbinding protein from coleoptile membranes of corn (*Zea mays* L.). II. Localization of a putative auxin receptor. *J. Biol. Chem.* 260:9854–59

125. Lomax, T. L., Mehlhorn, R. J., Briggs, W. R. 1985. Active auxin uptake by zucchini membrane vesicles: quantitation using ESR volume and delta pH determinations. *Proc. Natl. Acad. Sci. USA* 82:6541–45

126. Lorz, H., Baker, B., Schell, J. 1985. Gene transfer to cereal cells mediated by protoplast transformation. *Mol. Gen. Genet.* 199:178–82

127. Lycett, G. W., Croy, R. R. D., Shirsat, A. H., Boulter, D. 1984. The complete nucleotide sequence of a legumin gene from pea (*Pisum sativum* L.). *Nucleic Acids Res.* 12:4493–506

128. Lycett, G. W., Croy, R. R. D., Shirsat, A. H., Richards, D. M., Boulter, D. 1985. The 5'-flanking regions of three pea legumin genes: comparison of the DNA sequences. *Nucleic Acids Res.* 13:6733–43

129. Lycett, G. W., Delauney, A. J., Gatehouse, J. A., Gilroy, J., Croy, R. R. D., Boulter, D. 1983. The vicilin gene family of pea (*Pisum sativum* L.): a complete cDNA coding sequence for preprovicilin. *Nucleic Acids Res.* 11:2367–80

130. Madhavan, S., Smith, B. N. 1982. Localization of ribulose bisphosphate carboxylase in the guard cells by an indirect, immunofluorescence technique. *Plant Physiol.* 69:273–77

131. Matzke, M. A., Susani, M., Binns, A. N., Lewis, E. D., Rubenstein, I.,

132. Mayfield, S. P., Taylor, W. C. 1984. Carotenoid-deficient maize seedlings fail to accumulate light-harvesting chlorophyll a/b binding protein (LHCP) mRNA. *Eur. J. Biochem.* 144:79–84

133. Mazur, B. J., Chui, C.-F. 1985. Sequence of a genomic DNA clone for the small subunit oif ribulose bis-phosphate carboxylase-oxygenase from tobacco. *Nucleic Acids Res.* 7:2373–86

134. Meyer, G., Bliedung, H., Kloppstech, K. 1983. NADPH-protochlorophyllide oxidoreductase: reciprocal regulation in mono-and dicotyledonous plants. *Plant Cell Rep.* 2:26–29

135. Mohnen, D., Shinshi, H., Felix, G., Meins, F. Jr. 1985. Hormonal regulation of beta–1,3-glucanase messenger–RNA levels in cultured tobacco tissues. *EMBO J.* 4:1631–35

136. Morelli, G., Nagy, F., Fraley, R. T., Rogers, S. G., Chua, N.-H. 1985. A short conserved sequence is involved in the light-inducibility of a gene encoding ribulose 1,5-bisphosphate carboxylase small subunit of pea. *Nature* 315:200–4

137. Morrison, N. A., Bisseling, T., Verma, D. P. S. 1986. Development and differentiation of the root nodule: involvement of plant and bacterial genes. In *Developmental Biology: A Comprehensive Synthesis*, ed. L. W. Browder. New York: Plenum. In press

138. Mosinger, E., Batschauer, A., Schafer, E., Apel, K. 1985. Phytochrome control of in vitro transcription of specific genes in isolated nuclei from barley (*Hordeum vulagare*). *Eur. J. Biochem.* 147:137–42

139. Mundy, J. 1984. Hormonal regulation of alpha-amylase/subtilism inhibitor synthesis in germinating barley. *Carlsberg Res. Commun.* 49:439–44

140. Mundy, J., Hejgaard, J., Hansen, A., Hallgren, L., Jorgensen, K. G., Munck, L. 1986. Differential synthesis in vitro of barley aleurone and starchy endosperm proteins. *Plant Physiol.* 81:630–36

141. Murai, N., Sutton, D. W., Murray, M. G., Slighton, J. L., Merlo, D. J., et al. 1983. Phaseolin gene from bean is expressed after transfer to sunflower via tumor-inducing plasmid vectors. *Science* 222:476–82

142. Murashige, T., Skoog, F. 1962. A revised medium for rapid growth and bioassays with tobacco tissue cultures. *Physiol. Plant.* 15:473–97

143. Deleted in proof

Matzke, A. J. M. 1984. Transcription of a zein gene introduced into sunflower using a Ti-plasmid vector. *EMBO J.* 3:1525–31

144. Nagy, F., Fluhr, R., Kuhlemeier, C., Kay, S., Boutry, M., et al. 1986. Cis-acting elements for selected expression of two photosynthetic genes in transgenic plants. *Philos. Trans. R. Soc. London Ser. B* 314:493–500

145. Nagy, F., Kay, S. A., Boutry, M., Hsu, M.-Y., Chua, N.-H. 1986. Phytochrome-controlled expression of a wheat Cab gene in transgenic tobacco seedlings. *EMBO J.* 5:1119–24

146. Nagy, F., Morelli, G., Fraley, R. T., Rogers, S. G., Chua, N.-H. 1985. Photoregulated expression of a pea rbcS gene in leaves of transgenic plants. *EMBO J.* 4:3063–68

147. Negrutiu, I., Jacobs, M., Caboche, M. 1984. Advances in somatic cell genetics of higher plants—the protoplast approach in basic studies on mutagenesis and isolation of biochemical mutants. *Theor. Appl. Genet.* 67:289–304

148. Nelson, T., Harpster, M. H., Mayfield, S. P., Taylor, W. C. 1984. Light-regulated gene expression during maize leaf development. *J. Cell Biol.* 98:558–64

149. Nester, E. W., Gordon, M. P., Amasino, R. M., Yanofsky, M. F. 1984. Crown gall: a molecular and physiological analysis. *Ann. Rev. Plant Physiol.* 35:387–413

150. Nguyen, T., Zelechowska, M., Foster, V., Bergmann, H., Verma, D. P. S. 1985. Primary structure of the soybean nodulin-35 gene encoding uricase II localized in the peroxisomes of uninfected cells of nodules. *Proc. Natl. Acad. Sci. USA* 82:5040–44

151. Odell, J. T., Nagy, F., Chua, N.-H. 1985. Identification of DNA sequences required for activity of a plant promoter: the CaMV 35S promoter. *Nature* 313:810–12

152. Oelmüller, R., Mohr, H. 1985. Mode of coaction between blue-UV light and light absorbed by phytochrome in light-mediated anthocyanin formation in milo (*Sorghum vulgare* Pers.) seedling. *Proc. Natl. Acad. Sci. USA* 82:6124–28

153. Okamuro, J. K., Jofuku, K. D., Goldberg, R. B. 1986. The soybean seed lectin gene and flanking nonseed protein genes are developmentally regulated in transformed tobacco plants. *Proc. Natl. Acad. Sci. USA* In press

154. Otto, V., Schafer, E., Nagatani, A., Yamamoto, K. T., Furuya, M. 1984. Phytochrome control of its own synthesis in *Pisum sativum*. *Plant Cell Physiol.* 25:1579–84

155. Paszkowski, J., Shillito, R. D., Saul, M., Mandak, V., Hohn, T., et al. 1984.

Direct gene transfer to plants. *EMBO J.* 3:2717–22

156. Pauli, D., Spierer, A., Tissieres, A. 1986. Several hundred base pairs upstream of *Drosophila* hsp 23 and 26 genes are required for their heat induction in transformed flies. *EMBO J.* 5:755–61

157. Pelham, H. R. B. 1982. A regulatory upstream promoter element in the *Drosophila* Hsp70 heat-shock gene. *Cell* 30:517–28

158. Pelham, H. 1985. Activation of heat-shock genes in eukaryotes. *Trends Genet.* 1:31–35

159. Pelham, H. R. B., Bienz, M. 1982. A synthetic heat-shock promoter element confers heat-inducibility on the herpes simplex virus thymidine kinase gene. *EMBO J.* 1:1473–77

160. Peralta, E. G., Hellmiss, R., Ream, W. 1986. *Overdrive*, a T-DNA transmission enhancer on the *A. tumefaciens* tumour-inducing plasmid. *EMBO J.* 5:1137–42

161. Potrykus, I., Paszkowski, J., Saul, M. W., Petruska, J., Shillito, R. D. 1985. Molecular and general genetics of a hybrid foreign gene introduced into tobacco by direct gene transfer. *Mol. Gen. Genet.* 199:169–77

162. Potrykus, I., Saul, M. W., Petruska, J., Paszkowski, J., Shillito, R. D. 1985. Direct gene transfer to cells of a graminaceous monocot. *Mol. Gen. Genet.* 199:183–88

163. Poulsen, C., Fluhr, R., Kauffman, J. M., Boutry, M., Chua, N.-H. 1986. Characterization of an rbcS gene from *Nicotiana plumbaginifolia* and expression of an rbcS CAT chimeric gene in homologous and heterologous nuclear background. *Mol. Gen. Genet.* 205: 193–200

164. Pratt, L. H. 1979. Phytochrome: function and properties. *Photochem. Photobiol. Rev.* 4:59–124

165. Quail, P. H. 1983. Rapid action of phytochrome in photomorphogenesis. See Ref. 99, pp. 178–212

166. Reeder, R. H. 1984. Enhancers and ribosomal gene spacers. *Cell* 38:349–51

167. Richter, G., Wessel, K. 1985. Red light inhibits light-induced chloroplast development in cultured plant cells at the mRNA level. *Plant Mol. Biol.* 5:175–82

168. Robert, L. S., Adeli, K., Altosaar, I. 1985. Homology among 3S and 7S globulins from cereals and pea. *Plant Physiol.* 78:812–16

169. Rochester, D. E., Winer, J. A., Shah, D. M. 1986. The structure and expression of maize genes encoding the major

heat shock protein, hsp70. *EMBO J.* 5:451–58

170. Rogers, J. C. 1985. Two barley alpha-amylase gene families are regulated differently in aleurone cells. *J. Biol. Chem.* 280:3731–38

171. Rogers, S. G., Bisaro, D. M., Horsch, R. B., Fraley, R. T., Hoffmann, N. L., et al. 1986. Tomato golden mosaic virus A component DNA replicates autonomously in transgenic plants. *Cell* 45:593–600

172. Roux, S. J. 1984. Ca^{2+} and phytochrome action in plants. *Bioscience* 34:25–29

173. Rubery, P. H. 1981. Auxin receptors. *Ann. Rev. Plant Physiol.* 32:569–96

174. Sakai, S. 1985. Auxin-binding protein in etiolated mung bean seedlings: purification and properties of auxin-binding protein-II. *Plant Cell Physiol.* 26:185–92

175. Sakai, S., Hanagata, T. 1983. Purification of an auxin-binding protein from etiolated mung bean seedlings by affinity chromatography. *Plant Cell Physiol.* 24:685–93

176. Sasaki, Y., Sakihama, T., Kamikubo, T., Shinozaki, K. 1983. Phytochrome-mediated regulation of two mRNAs, encoded by nuclei and chloroplasts of ribulose-1,5-bisphosphate carboxylase/oxygenase. *Eur. J. Biochem.* 133:617–20

177. Schafer, E., Briggs, W. R. 1986. Photomorphogenesis from signal perception to gene expression. *Photobiochem. Photobiophys.* In press

178. Schafer, E., Haupt, W. 1983. Blue-light effects in phytochrome-mediated responses. See Ref. 99, pp. 723–44

179. Schlesinger, M. J., Ashburner, M., Tissieres, A. 1982. *Heat Shock From Bacteria to Man.* New York: Cold Spring Harbor Lab. 440 pp.

180. Schoffl, F., Baumann, G. 1985. Thermo-induced transcripts of a soybean heat shock gene after transfer into sunflower using a Ti plasmid vector. *EMBO J.* 4:1119–24

181. Schoffl, F., Baumann, G., Raschke, E., Bevan, M. 1986. The expression of heat shock genes in higher plants. *Philos. Trans. R. Soc. London Ser. B.* 314 453–68

182. Schoffl, F., Raschke, E., Nagao, R. T. 1984. The DNA sequence analysis of soybean heat-shock genes and identification of possible regulatory promoter elements. *EMBO J.* 3:2491–97

183. Senger, H., Briggs, W. R. 1981. The blue light receptor(s): primary reactions

and subsequent metabolic changes. *Photochem. Photobiol. Rev.* 6:1–38

184. Sengupta-Gopalan, C., Reichert, N. A., Barker, R. F., Hall, T. C., Kemp, J. D. 1985. Developmentally regulated expression of the bean beta-phaseolin gene in tobacco seed. *Proc. Natl. Acad. Sci. USA* 82:3320–24

185. Sequeira, L. 1983. Mechanisms of induced resistance in plants. *Ann. Rev. Microbiol.* 37:51–79

186. Serlin, B. S., Roux, S. J. 1984. Modulation of chloroplast movement in the green alga *Mougeotia* by the Ca^{2+} ionophore A23187 and by calmodulin antagonists. *Proc. Natl. Acad. Sci. USA* 81:6368–72

187. Shaul, Y., Laub, O., Walker, M. D., Rutter, W. J. 1985. Homologous recombination between a defective virus and a chromosomal sequence in mammalian cells. *Proc. Natl. Acad. Sci. USA* 82:3781–84

188. Shaw, C. H., Carter, G. H., Watson, M. D., Shaw, C. H. 1984. A functional map of the nopaline synthase promoter. *Nucleic Acids Res.* 12:7831–46

189. Showalter, A. M., Bell, J. N., Cramer, C. L., Bailey, J. A., Varner, J. E., Lamb, C. J. 1985. Accumulation of hydroxyproline-rich glycoprotein mRNAs in response to fungal elicitor and infection. *Proc. Natl. Acad. Sci. USA* 82:6551–55

190. Silverthorne, J., Tobin, E. M. 1984. Demonstration of transcriptional regulation of specific genes by phytochrome action. *Proc. Natl. Acad. Sci. USA* 81:1112–16

191. Simpson, J., Timko, M. J., Cashmore, A. R., Schell, J., Van Montagu, M., Herrera-Estrella, L. 1985. Light-inducible and tissue specific expression of a chimaeric gene under control of the 5' flanking sequence of a pea chlorophyll a/b binding protein gene. *EMBO J.* 4:2723–29

192. Simpson, J., Schell, J., Van Montagu, M., Herrera-Estrella, L. 1986. The light-inducible and tissue specific expression of a pea LHCP gene involves an upstream element combining enhancer and silencer-like properties. *Nature* 323:551–53

193. Simpson, J., Van Montagu, M., Herrera-Estrella, L. 1986. Photosynthesis-associated gene families: differences in response to tissue-specific and environmental factors. *Science* 233:34–38

194. Slovin, J. P., Tobin, E. M. 1982. Synthesis and turnover of the light-harvesting chlorophyll a/b-protein in

Lemna gibba grown with intermittent red light: possible translational control. *Planta* 154:465–72

195. Smith, S. M., Ellis, R. J. 1978. Light-stimulated accumulation of transcripts of nuclear and chloroplast genes for ribulose-bisphosphate carboxylase. *J. Mol. Appl. Genet.* 1:127–37

196. Smithies, O., Gregg, R. G., Boggs, S. S., Koralewski, M. A., Kucherlapati, R. S. 1985. Insertion of DNA sequences into the human chromosomal beta-globin locus by homologous recombination. *Nature* 317:230–34

197. Somssich, I. E., Schmelzer, E., Bollmann, J., Hahlbrock, K. 1986. Rapid activation by fungal elicitor of genes encoding "pathogenesis-related" proteins in cultured parsley cells. *Proc. Natl. Acad. Sci. USA* 83:2427–30

198. Spena, A., Hain, R., Ziervogel, U., Saedler, H., Schell, J. 1985. Construction of a heat-inducible gene for plants. Demonstration of heat-inducible activity of the *Drosophila melanogaster hsp 70* promoter in plants. *EMBO J.* 4:2739–43

199. Springer, B., Werr, W., Starlinger, P., Bennett, D. C., Zokolica, M., Freeling, M. 1986. The *Shrunken* gene on chromosome 9 of *Zea mays* L. is expressed in various plant tissues and encodes an anaerobic protein. In press

200. Steinmuller, K., Batschauer, A., Apel, K. 1986. Tissue-specific and light-dependent changes of chromatin organization in barley *(Hordeum vulgare)*. *Eur. J. Biochem.* 6: In press

201. Stiekema, W. J., Wimpee, C. F., Silverthorne, J., Tobin, E. M. 1983. Phytochrome control of the expression of two nuclear genes encoding chloroplast proteins in *Lemna gibba* L. G-3. *Plant Physiol.* 72:717–24

202. Subramani, S., Berg, P. 1983. Homologous and nonhomologous recombination in monkey cells. *Mol. Cell. Biol.* 3:1040–52

203. Theologis, A., Huynh, T. V., Davis, R. W. 1985. Rapid induction of specific mRNAs by auxin in pea epicotyl tissue. *J. Mol. Biol.* 183:53–68

204. Theologis, A., Ray, P. M. 1982. Early auxin-regulated polyadenylated mRNA sequences in pea stem tissue. *Proc. Natl. Acad. Sci. USA* 79:418–21

205. Thomashow, M. F., Hugly, S., Buchholz, W. G., Thomashow, L. S. 1986. Molecular basis for the auxin-independent phenotype of crown gall tumor tissues. *Science* 231:616–18

206. Thompson, W. F., Everett, M., Polans, N. O., Jorgensen, R. A., Palmer, J. D. 1983. Phytochrome control of RNA

levels in developing pea and mung-bean leaves. *Planta* 158:487–500

207. Timko, M. P., Kausch, A. P., Castresana, C., Fassler, J., Herrera-Estrella, L., et al. 1985. Light regulation of plant gene expression by an upstream enhancer-like element. *Nature* 318:579–82

208. Timko, M. P., Kausch, A. P., Hand, J. M., Cashmore, A. R., Herrera-Estrella, L., et al. 1985. Structure and expression of nuclear genes encoding polypeptides of the photosynthetic apparatus. In *Molecular Biology of the Photosynthetic Apparatus,* ed. K. E. Steinback, S. Bonitz, C. J. Arntzen, L. Bogorad, 198:381–96. New York: Cold Spring Harbor Lab.

209. Tobin, E. M. 1981. Phytochrome-mediated regulation of messenger RNAs for the small subunit of ribulose-1,5-bisphosphate carboxylase and the light-harvesting chlorophyll a/b-protein in *Lemna gibba. Plant Mol. Biol.* 1:35–51

210. Tobin, E. M., Klein, A. O. 1975. Isolation and translation of plant messenger RNA. *Plant Physiol.* 56:88–92

211. Tobin, E. M., Silverthorne, J. 1985. Light regulation of gene expression in higher plants. *Ann. Rev. Plant Physiol.* 36:569–93

212. Topol, J., Ruden, D. M., Parker, C. S. 1985. Sequences required for in vitro transcriptional activation of a *Drosophila* hsp70 gene. *Cell* 42:527–37

213. Trewavas, A. J., Cleland, R. E. 1983. Is plant development regulated by changes in the concentration of growth substances or by changes in the sensitivity to growth substances? *Trends Biochem. Sci.* 10:354–57

214. Triplett, B. A., Quatrano, R. S. 1982. Timing, localization, and control of wheat germ agglutinin synthesis in developing wheat embryos. *Dev. Biol.* 91:491–96

215. Triplett, E. W. 1985. Intercellular nodule localization and nodule specificity of xanthine dehydrogenase in soybean. *Plant Physiol.* 77:1004–9

216. Tumer, N. E., Clark, W. G., Tabor, G. J., Hironaka, C. M., Fraley, R. T., Shah, D. M. 1986. The genes encoding the small subunit of ribulose-1,5-bisphosphate carboxylase are expressed differentially in petunia leaves. *Nucleic Acids Res.* 14:3325–42

217. Uchimiya, H., Fushimi, T., Hashimoto, H., Harada, H., Syono, K., Sugawara, Y. 1986. Expression of a foreign gene in callus derived from DNA-treated protoplasts of rice *(Oryza sativa* L.). *Mol. Gen. Genet.* 204:204–7

218. Vanderhoef, L. N. 1980. Auxin-regulated cell enlargement: is there action at the level of gene expression. In *Genome Organization and Expression in Plants*, ed. C. J. Leaver, 29A:159–73. London/New York: Plenum. 607 pp.

219. van der Linde, P. C. G., Bouman, H. Mennes, A. M., Libbenga, K. R. 1984. A soluble auxin-binding protein from cultured tobacco tissues stimulates RNA synthesis *in vitro*. *Planta* 160:102–8

220. van Huijsduidnen, R. A. M. H., Cornelissen, B. J. C., van Loon, L. C., van Boom, J. H., Tromp, M., Bol, J. F. 1985. Virus-induced synthesis of messenger RNAs for precursors of pathogenesis-related proteins in tobacco. *EMBO J.* 4:2167–71

221. van Loon, L. C. 1985. Pathogenesis-related proteins. *Plant Mol. Biol.* 4:111–16

222. Verma, D. P. S., Fortin, M. G., Stanley, J., Mauro, V. P., Purohit, S., Morrison, N. 1986. Nodulins and nodulin genes of *Glycine max. Plant Mol. Biol.* 7:51–61

223. Vernet, T., Fleck, J., Durr, A., Fritsch, C., Pinck, M., Hirth, L. 1982. Expression of the gene coding for the small subunit of ribulose bisphosphate carboxylase during differentiation of tobacco plant protoplasts. *Eur. J. Biochem.* 126:489–95

224. Vierstra, R. D., Quail, P. H. 1982. Native phytochrome: inhibition of proteolysis yields a homogeneous monomer of 124 kilodaltons from *Avena. Proc. Natl. Acad. Sci. USA* 79:5272–76

225. Walbot, V. 1985. On the life strategies of plants and animals. *Trends Genet.* 6:166–69

226. Walker, J. C., Legocka, J., Edelman, L., Key, J. L. 1985. An analysis of growth regulatory interactions and gene expression during auxin-induced cell elongation using cloned complementary DNAs to auxin-responsive messenger RNAs. *Plant Physiol.* 77:848–50

227. Wang, K., Herrera-Estrella, L., Van Montagu, M., Zambryski, P. 1984. Right 25 bp terminus sequence of the nopaline T-DNA is essential for and determines direction of DNA transfer from *Agrobacterium* to the plant genome. *Cell* 38:455–62

228. Watson, J. C., Kaufman, L. S., Thompson, W. F. 1986. Developmental regulation of cytosine methylation in the nuclear ribosomal RNA genes of *Pisum sativum. J. Mol. Biol.* Submitted for publication

229. Weiher, H., Konig, M., Gruss, P. 1983. Multiple point mutations affecting the simian virus 40 enhancer. *Science* 219:626–31

230. Werr, W., Lorz, H. 1986. Transient gene expression in a gramineae cell line. *Mol. Gen. Genet.* 202:471–75

231. Wong, S. L., Verma, D. P. S. 1985. Promoter analysis of a soybean nuclear gene coding nor nodulin-23, a nodule-specific polypeptide involved in symbiosis with *Rhizobium. EMBO J.* 4:2431–38

232. Yamamoto, K. R. 1985. Steroid receptor regulated transcription of specific genes and gene networks. *Ann. Rev. Genet.* 19:209–52

233. Zambryski, P., Joos, H., Genetello, C., Leemans, J., Van Montagu, M., Schell, J. 1983. Ti-plasmid vector for the introduction of DNA into plant cells without alteration of their normal regeneration capacity. *EMBO J.* 2:2143–50

234. Zemel, E., Gepstein, S. 1985. Immunological evidence for the presence of ribulose bisphosphate carboxylase in guard cell chloroplasts. *Plant Physiol.* 78:586–90

235. Zurfluh, L. L., Guilfoyle, T. J. 1982. Auxin-induced changes in the population of translatable messenger RNA in elongating sections of soybean hypocotyl. *Plant Physiol.* 69:332–37

236. Zurfluh, L. L., Guilfoyle, T. J. 1982. Auxin- and ethylene-induced changes in the population of translatable messenger RNA in basal sections and intact soybean hypocotyl. *Plant Physiol.* 69:338–40

Ann. Rev. Plant Physiol. 1987. 38:259–90
Copyright © 1987 by Annual Reviews Inc. All rights reserved

CELLULOSE BIOSYNTHESIS

Deborah P. Delmer

ARCO Plant Cell Research Institute, 6560 Trinity Court, Dublin, California 94568

CONTENTS

INTRODUCTION 259
OVERVIEW OF CELLULOSE BIOSYNTHESIS 260
RELEVANCE OF CELLULOSE MICROFIBRIL STRUCTURE TO
    THE MECHANISM OF BIOSYNTHESIS 262
CELLULOSE SYNTHESIS IN BACTERIA AND FUNGI 264
CELLULOSE SYNTHESIS IN ALGAE AND HIGHER PLANTS 267
    Terminal Complexes: Are They Really Cellulose Synthases? 267
    The Role of Microtubules in Microfibril Orientation 271
    The Biochemistry of the Polymerization Process 273
THE RELATIONSHIP BETWEEN THE SYNTHESIS OF CELLULOSE
    AND CALLOSE 277
SPECIFICITY OF GLYCOSYLTRANSFERASES 279
IDENTIFICATION OF A RECEPTOR FOR DCB IN PLANTS 280
MODEL FOR THE REGULATION OF CELLULOSE AND CALLOSE
    SYNTHESIS 282

INTRODUCTION

About once every year, I call up some of my colleagues who study the structure of cellulose and request an update on new findings. The usual response is "We're still confused." Then they ask me what's new in biosynthesis, and I reply, "We're still confused too!" To those who don't follow closely the literature on cellulose structure and biosynthesis, this may seem to be an amazing conversation. For, while the chemical definition of cellulose as $(1\rightarrow4)$-β-D-glucan is correct, it is deceptively simple. One might wonder how, in this age of sophisticated scientific analyses, there can possibly be so much confusion about a polymer composed entirely of glucose residues joined

259

together in identical chemical linkages. This review attempts to clarify the nature of our confusion with respect to the biosynthesis of cellulose. Ambiguities in the structure of cellulose are discussed only as they relate to problems of biosynthesis, although it is becoming increasingly clear that understanding of structure will contribute to understanding of biosynthesis and vice versa.

Lest the reader turn away in despair after the first paragraph, I should also make clear that there have been some exciting new developments in this field in the past few years. Discovery of a unique form of regulation of cellulose synthesis in the bacterium *Acetobacter xylinum* has enabled workers to achieve high rates of in vitro synthesis of cellulose for the first time in any organism. Very clear demonstrations of organized protein complexes in the plasma membrane, found in association with the ends of microfibrils or at least associated with sites of active cellulose synthesis, lend increasing support to the notion that we can in fact observe by freeze-fracture techniques cellulose synthase complexes in higher plants as well as in algae. Development of techniques for *in situ* localization of microtubule networks has further strengthened the concept that the orientation of cellulose microfibrils is controlled by the orientation of the cortical microtubule network and that such a network is a very dynamic structure. A number of laboratories are increasingly intrigued by the notion that there is an intimate relationship between the biosynthesis of cellulose and of callose, $(1{\rightarrow}3)$-β-D-glucan, another glucan usually synthesized by plants as a wound response under conditions where cellulose synthesis is inhibited. And discovery of a receptor protein for 2,6-dichlorobenzonitrile (DCB), a specific inhibitor of cellulose synthesis, offers hope that at least one polypeptide involved in cellulose biosynthesis has been identified.

My intention is to concentrate on these recent developments in the field and to outline some possible future approaches to the study of this complex process. A previous review of mine in 1983 (30) provides a more detailed account of the older literature. Other reviews of the topic include those by Colvin (24), Maclachlan (95), Haigler (51), and a general review of cell wall biosynthesis by Delmer & Stone (36). The role of calcium in regulating the synthesis of callose is reviewed by Kauss (82) in this volume, as are some pertinent dynamic aspects of microtubules (92a). A book concerning cellulose chemistry and its applications has also recently been published (105).

OVERVIEW OF CELLULOSE BIOSYNTHESIS

Since the intention is to concentrate on recent developments in the field, this section of the review provides the reader with an overview of the field of cellulose biosynthesis as we understood it at the time of my last review of the topic in the early 1980s (30). The reader should consult that and other reviews for detailed references.

By 1983, there was general agreement that, in most algae and in all higher plants, the process of cellulose synthesis occurs at the plasma membrane. Freeze-fracture studies in algae such as *Oocystis,* which have large microfibrils of cellulose, revealed ordered linear arrays of proteins situated at the ends of microfibrils; these so-called terminal complexes were speculated to be the cellulose synthases. In other algae such as *Micrasterias,* the terminal complex structure assumed a different shape; in this case the complex consisted of a hexagonal array of from 3 to 175 rosette-shaped structures, each of which consisted of six particles organized also in hexagon shape within one rosette. On the outer (EF) fracture face, particles complementary to the central holes in the rosettes [which fractured on the inner (PF) face] were sometimes seen. Each rosette was proposed to be part of a cellulose synthase complex; it was not clear whether the complementary particles on the opposite fracture face were part of the original complex or not. Rosettes had been seen occasionally in plasma membranes from higher plants, and there had been one report of a "terminal globule" of protein attached to the end of a protruding microfibril; in another study on plant tissue, globules were reported on the outer face of the membrane and rosettes on the other face, similar to the situation in *Micrasterias.* At that time, the relationship of such rosettes and/or globules to cellulose synthesis was still somewhat controversial, particularly in higher plants.

The force generated by polymerization of the relatively rigid microfibrils was believed to drive the movement of the complexes within the fluid membrane. Microtubules were believed to guide somehow the movement of these complexes in cases where ordered deposition of microfibrils occurs.

UDP-glucose was known to be the substrate for cellulose synthesis in the bacterium *Acetobacter xylinum,* and indirect evidence indicated it is also the substrate in higher plants. There was no compelling evidence to support the involvement of lipid intermediates in the polymerization process. Although limited in vitro synthesis of cellulose was obtained with extracts from *A. xylinum,* when such synthesis from UDP-glucose was attempted with higher plant preparations, the major product synthesized was callose [$(1\rightarrow3)$-β-glucan]. Some limited synthesis of $(1\rightarrow4)$-β-glucan could be obtained, particularly when Mg^{2+} and low concentrations of substrate were used in the assay. However, much of this activity was not localized in the plasma membrane, where cellulose synthase activity was expected to reside, but was rather found associated with the Golgi fraction. This activity was speculated to be either a partially latent form of cellulose synthase en route to the plasma membrane or a part of an enzyme system for xyloglucan synthesis.

Reasons for the inability to achieve substantial rates of cellulose synthesis in vitro were not understood. The very high rates of in vitro synthesis of callose by an enzyme localized on the same membrane system and using the same substrate as cellulose synthase led to some speculation that the callose synthase may in fact also be the cellulose synthase, and that unknown factors change its specificity upon cell damage. Several reports suggested that callose might even be a precursor to cellulose, but the evidence to support this was not particularly compelling. Indirect evidence with plant systems and more direct evidence with *A. xylinum* indicated that the existence of a transmembrane electrical potential might be important for maintainence of cellulose synthase activity; it was not clear if loss of such a potential could be one factor responsible for lack of activity in vitro. Such was the state of our knowledge in the early part of this decade.

RELEVANCE OF CELLULOSE MICROFIBRIL STRUCTURE TO THE MECHANISM OF BIOSYNTHESIS

Haigler (51) has recently done an excellent job of reviewing the relationship of cellulose structure to mechanisms of biosynthesis. In this review I therefore only highlight the most important relationships. Cellulose, as synthesized in its natural state, does not consist of isolated molecules of linear $(1\rightarrow4)$-β-glucan chains. Such chains have a strong tendency to self-associate via intra- and interchain hydrogen bonding to form insoluble, and often highly crystalline, fibrils. In nature, these so-called microfibrils exist in varying diameters that depend upon the source and age of the tissue. The degree of polymerization (D.P.) of the individual glucan chains within microfibrils is also variable. In general, the D.P. of cellulose chains derived from the bacterium *Acetobacter xylinum* and from the primary walls of higher plants is relatively low (2,000–6000 residues/chain), while that from many of the algae and secondary walls of higher plants is much higher (>10,000) (30). We still have no clear idea how chain length is controlled during synthesis, but some ideas are beginning to emerge on how control of microfibril diameter may be achieved. Current controversial questions relating to microfibril structure and their relationship to biosynthetic mechanisms are summarized below.

1. Why do crystallite sizes, as determined by X-ray or electron diffraction techniques, vary for celluloses of different origin? (51) Does each single microfibril represent a single crystal, or are there substructures within larger microfibrils, such as the 3.5 nm "elementary fibrils" (30, 51). If so, does the elementary fibril represent the single crystal? This certainly does not seem to be the case for *Valonia,* where recent evidence suggests that there is a single crystal across the entire 30-nm width of these large microfibrils (12, 123, 124, 138). The issue is almost certainly relevant to the presumed involvement of multisubunit plasma-membrane-localized cellulose synthase complexes in the generation of microfibrils in nature. The characteristics of these putative

synthase complexes are considered in more detail in a later section, where I discuss the possibility that the unit size and type of synthase complexes present in any given organism may determine the crystallite size as well as the diameter of the microfibrils they produce.

2. Are the chains within a single microfibril parallel or antiparallel? Until recently, we have relied almost exclusively on X-ray crystallography for an answer to this question. The current view, based on such analyses, generally favors a parallel orientation for native cellulose (designated Cellulose I). The reader is referred to a recent review by French that summarizes data on the X-ray crystallography of cellulose (42). However, as pointed out by French (42), different assumptions are often used in generation of these model structures, and it is often difficult to exclude completely other models based on antiparallel structures. Independent support for a parallel model comes from two more recent studies using different analytical approaches. In the first of these, Hieta et al (71) showed that electron-dense staining of the reducing ends of chains was concentrated in only one end of microfibrils of the alga *Valonia,* indicating parallel alignment. In the second study, Chanzy & Henrissat (22) showed that a specific cellobiohydrolase that degrades only from the nonreducing end of the chains erodes only one end of *Valonia* microfibrils.

A parallel model is certainly simpler to envisage from the standpoint of biosynthesis. One need only be reminded of the complexities necessary for synthesis of the antiparallel strands of DNA to see the difficulty in imagining what mechanism might account for synthesis of antiparallel chains in cellulose. On the other hand, it is relatively easy to imagine an enzyme complex with multiple active sites employing an identical mechanism to polymerize a number of chains that simultaneously self-associate to form microfibrils.

Given that the parallel model for Cellulose I is correct, there still remain several unexplained findings. When Cellulose I is "mercerized" by swelling in alkali, it is converted to Cellulose II, a different crystal form that current X-ray crystallographic evidence indicates is a structure with antiparallel chains (42). How can such a drastic rearrangement of chains occur under conditions where the molecules do not even come into solution? Some models have been presented that could explain such a conversion within a single microfibril (133), but other possibilities should be considered as well. The obvious one is that Cellulose II is also parallel but of slightly different structure. This seems simple and appealing on the surface, but it does present problems from the standpoint of biosynthesis. Cellulose II is generally recognized as the most favored and thermodynamically stable structure, and it is considered that Cellulose I, although more unstable, is generated in nature as a result of the mechanism of simultaneous polymerization and crystallization. One must therefore ask why, if Cellulose II were also parallel, it would not be the preferred structure in nature? Another possibility is that, within an individual microfibril, all chains are parallel, but adjacent microfibrils may be antiparallel, and during swelling in alkali, preferential reassociation occurs between chains of adjacent microfibrils (88). Several studies may be relevant to this model. The first is the finding of Brown & Montezinos (14) that the putative cellulose synthases, observed by freeze-fracture of *Oocystis* plasma membranes, are often observed in pairs that appear to begin synthesis of microfibrils by movement in opposite directions. Revol & Goring (124) have also shown that, whereas the directionality of the c-axis of *Valonia* cellulose crystallites is unidirectional within a single microfibril, each lamella of the wall has crystallites with the c-axis existing in two directions.

3. The degree of crystallinity of native celluloses is variable, and even within microfibrils from one source there appear to be crystalline and para-crystalline regions. Atalla & Van der Hart (5) have deduced from analysis of spectra from solid-state carbon-13 NMR of various celluloses that there may be two different crystalline forms within a single source of cellulose, although the interpretation of these data is controversial (60). At present, we do not have any indication how these different crystalline or para-crystalline regions may be generated in vivo, but transient changes in the organization of synthase complexes might be proposed as one explanation. Other possibilities, at least for the para-crystalline regions, are that they represent regions where single glucan chains are terminated (51) or are simply formed by a statistically predictable change in the rotations of certain bonds, which leads to altered chain association (111).

CELLULOSE SYNTHESIS IN BACTERIA AND FUNGI

The bacterium *Acetobacter xylinum,* unlike fungi, algae, and higher plants, synthesizes cellulose, not as a cell wall component but as an extracellular secretion. The earlier view that synthesis occurred via enzymes secreted to the medium (24) was disproven by the studies of Brown's group (15), who clearly showed that *A. xylinum* possesses a linear row of synthetic sites along one of the long axes of the cell; fibrils secreted through pores at these sites (145) further self-associate to form microfibrils, all of which combine to form one large ribbon of cellulose [see Haigler (51) for a more detailed description of this process].

Because of the relative ease of obtaining *A. xylinum* cellulose in a purified form, this organism has attracted the attention of industry in recent years. It was observed that various compounds can associate with glucan molecules (e.g. fluorescent brighteners such as Calcofluor White, the dye Congo Red, and carboxymethylcellulose). When added to the medium during synthesis, these compounds intercalate between the nascent chains and prevent self-association, which indicates that altered forms of cellulose can be induced with *A. xylinum* (9, 51, 52). These observations are of interest from a theoretical as well as a practical standpoint. They indicate that polymerization and crystallization are tightly coupled events; furthermore, when crystallization is disrupted by these agents, the rate of glucan synthesis increases, indicating that crystallization may be a rate-limiting step in the process (9). More recently, Hayashi & Delmer showed that addition of purified pea xyloglucan to *A. xylinum* culture can also prevent the fasciation of microfibrils into the larger ribbon (54). This finding raises the intriguing possibility that the xyloglucan, which is secreted to primary cell walls of higher plants and is known to bind to cellulose there (55), may play a role in limiting the size of microfibrils during cell expansion.

The most impressive progress in unraveling the biochemistry of cellulose biosynthesis in the past five years has been achieved in studies with *A. xylinum*, largely by the efforts of Benziman's group in Israel. Up to 1980, it was possible to demonstrate synthesis of $(1{\rightarrow}4)$-β-glucan from UDP-glucose using membranes isolated from *A. xylinum*, but the rates of synthesis observed were far below those obtained in vivo (48, 139). The series of events that led to the achievement of high rates of in vitro synthesis merits some discussion, since it started almost by accident and ended with very clever detective work on the part of Benziman's group. When I visited Benziman's laboratory on sabbatical leave in 1981, I suggested that they study the effect of polyethylene glycol (PEG) on synthase activity. We had recently observed limited stimulation of $(1{\rightarrow}4)$-β-glucan synthesis by PEG with plant extracts (7, 31), but we had no idea how PEG worked and we found the effect to be somewhat variable. PEG did in fact stimulate the *A. xylinum* system, but rates of synthesis still did not approach those achieved in vivo. At about the same time, Cabib's group had reported stimulation of yeast $(1{\rightarrow}3)$-β-glucan synthase activity by GTP (107), and we were quite excited to find that GTP caused a very dramatic stimulation of cellulose synthase activity in *A. xylinum* (2). The peculiar fact, however, was that this stimulation was only observed with membranes prepared in PEG. Y. Aloni, the student pursuing these studies, then made the clever observation that there was present in PEG-membrane fractions a protein factor that could be washed from the membranes in buffer lacking PEG and that was necessary for GTP stimulation (2, 8). Subsequently, he showed that a similar stimulation could be obtained with GTP and the protein factor using synthase activity solubilized from the membranes with digitonin (3).

Further studies by Ross et al (129–131) have now clarified this unique system of regulation. The protein factor has been shown to be a specific guanyl cylase that converts, in a two-step reaction, two molecules of GTP to one molecule of cyclic diguanylic acid (see Figure 1 and Ref. 131). This latter compound is the true activator of the synthase and has a K_a of about 1 μM. There appears also to be additional regulation in the form of a membrane-bound enzyme system that degrades cyclic diguanylic acid, a system inhibited by Ca^{2+} (131). Recently, Lin et al (90) have observed the structure of the product synthesized by the solubilized, activated cellulose synthase, and they have shown that the cellulose produced is in the form of short fibrils about 1.7 nm in diameter. A weak electron diffraction pattern was observed, and subsequent work (91) has shown that, by cleaning the product in mild alkali, sharp diffraction patterns identical to the Cellulose I produced in vivo can be seen.

A. *xylinum* was the first system for which a lipid intermediate was proposed to function in the synthesis of a polysaccharide, in that case cellulose (86). However, in all of the recent studies on this enzyme system, no real evidence has been obtained to support the involvement of such an intermediate. The fact that high rates of synthesis, activated by cyclic dinucleotide, can be obtained with the solubilized enzyme system further supports lack of involvement of a lipid intermediate (3); in the cases where such lipid intermediates are involved in glycosylation of protein, it has been found necessary to add back the lipid acceptor to solubilized preparations in order to achieve synthesis (6). The previous reports of synthesis of a lipid-linked cellobiose in certain strains of A. *xylinum* (45) probably relate to the role of such a compound in the synthesis of a polymer resembling xanthan gum (27). The strain used by Benziman does not produce such a polymer, and it also does not seem able to synthesize lipid-linked cellobiose in vitro (M. Benziman, personal communication).

The fascinating discovery of regulation by a cyclic dinucleotide offered hope that a similar system regulating cellulose synthesis in higher plants would be found. That hope now seems to be in vain, for no one with whom I have consulted (M. Benziman, F. Meier, G. Franz, G. Maclachlan) has been able to obtain stimulation by GTP under any conditions. Furthermore, T. Hayashi and I have tried direct addition of the cyclic dinucleotide to plant membranes and have seen no effect whatsoever on synthase activity under

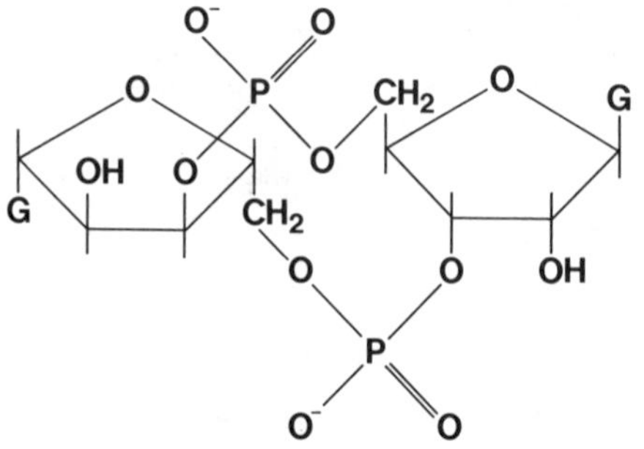

Figure 1   Structure of bis-(3'→5')-cyclic diguanylic acid. Ross et al (131) have shown that this compound is a highly specific activator of the cellulose synthase from *Acetobacter xylinum*. G = guanine.

conditions where we could observe no detectable degradation of the effector (unpublished). Although disappointed, perhaps we should not be surprised, since plants do not generally seem to use cyclic nucleotides as regulatory compounds. My personal opinion is that, while important in its own right, the discovery of cyclic dinucleotide regulation in *A. xylinum* will not be directly relevant to the mechanism of regulation in higher plants. The most important lesson for those of us working in plant systems is that the process can be exceedingly complex and may involve a combination of effectors and/or enzymes or regulatory proteins, all of which may need to be present under a specific set of conditions in order to obtain synthesis in vitro.

A few other genera of bacteria, including *Rhizobium* and *Agrobacterium*, also secrete some cellulose (30). We have been able to detect an apparent cellulose synthase activity in membrane preparations from *A. tumefaciens* (J. Cooper and D. Delmer, unpublished). Although we were unable to demonstrate any GTP-mediated regulation in test-tube assays, we could detect about a 4-fold stimulation of activity by addition of partially purified cyclic guanyl dinucleotide to an assay of digitonin-solubilized proteins separated by nondenaturing polyacrylamide gel electrophoresis (140). Although the system is not well-characterized, there is thus some indication that other bacteria also possess a regulatory system similar to that of *A. xylinum*.

Some in vitro synthesis of $(1\rightarrow4)$-β-glucan from UDP-glucose can be obtained using membranes derived from the cellulosic fungus *Saprolegnia* (40, 41, 47). This system falls somewhere between *A. xylinum*, yeast, and higher plants in its characteristics. Like higher plants (30, 117, 136), the membrane preparations synthesize largely $(1\rightarrow3)$-β-glucan at high substrate concentration, but $(1\rightarrow4)$-β-glucan synthesis can be enhanced in the presence of Mg^{2+} and low UDP-glucose concentration. Like yeast (107), glucan synthase activity is stimulated both by GTP and ATP, but unlike yeast and more like *A. xylinum* (2), GTP (and ATP) stimulates synthesis of $(1\rightarrow4)$-β-glucan [and inhibits synthesis of $(1\rightarrow3)$-β-glucan (40)]. These plasma-membrane-localized (47) glucan synthases in *Saprolegnia* certainly deserve more study, as they may help bridge the gap in our understanding of the differences and similarities in cellulose synthesis that exist between higher and lower forms of life.

CELLULOSE SYNTHESIS IN ALGAE AND HIGHER PLANTS

Terminal Complexes: Are They Really Cellulose Synthases?

As indicated in the overview, it is no doubt possible to observe (by freeze-fracture of plasma membranes of organisms that synthesize cellulose) organized protein complexes of large size. These so-called terminal complexes are of two types as shown in Figure 2. The linear complexes have, to date, been seen in the algae *Oocystis apiculata* (14), *Oocystis solitaria* (115, 127), *Glaucocystis nostochinearum* (144), *Valonia macrophysa* (78), *Boergesenia forbesii* (79), *Boodlea coacta* (99), and species of *Microdictyon* and *Chaetomorpha* (13), all of which are characterized by having microfibrils of relatively large diameters and high crystallinity. These complexes usually fracture with the EF (outer) face of the plasma membrane but sometimes fracture to the PF (inner) face, which indicates that they are probably transmembrane complexes (78). The complexes are often observed at the ends of the visible impressions of wall microfibrils. Conditions that lead to disruptions or perturbations in cellulose biosynthesis [addition of EDTA in *Oocystis* (100) or addition of the dye Congo Red or fluorescent brighteners in *Boergesenia* (79) or *Oocystis* (127)] lead to alterations or disappearance of the terminal complexes; restoration of normal synthesis is then accompanied by reappearance of normal complexes. The estimated number of subunits in these complexes is roughly similar to the estimated number of glucan chains in the microfibrils (14, 79). As for all static structures observed in electron microscopy, assignment of function is difficult, buy my personal opinion is that the evidence that these complexes are the synthases is very compelling.

Interpretation of observations in those organisms possessing the rosette and/or globule-type of complexes is more difficult. The clearest example to date is still that of the alga *Micrasterias* studied by Giddings et al (46), where very distinct hexagonal arrays of rosettes were seen on the PF (inner) fracture face of the plasma membrane. Because rosette structures fracture on this face, it is not usually possible to determine if their *in situ* position might be at the ends of microfibrils, since microfibril impressions appear as ridge-like impressions on the EF fracture face. However, in *Micrasterias,* microfibril impressions as grooves can sometimes be seen on the PF face as well, and the hexagonal arrays of rosettes were occasionally seen at the ends of such grooves. In a few cases, the authors were able to view complementary fracture faces and found on the EF face an array of particles that seemed to be complementary to the holes in the centers of the rosettes in the PF face. Given that these arrays of rosettes and particles [called globules by Mueller & Brown (103)] are seen at the ends of microfibrils and that the center-to-center spacing between rosette arrays matches the average center-to-center spacing between microfibrils, the overall conclusion from the study with *Micrasterias* does favor the idea that these structures are synthases. PF-face rosettes have now been reported in other algae including *Closterium* (73) and *Spirogyra* (67), in the moss *Funaria* (122) and the fern *Adiantum* (142), and in lower plants such as *Equisetum* (38), *Psilotum* (13), and *Gingko* (13). However, in these cases, the rosettes are largely but not always solitary; solitary globules on the EF face are also occasionally seen in these organisms.

The first report of (solitary) PF-face rosettes and EF-face globules in higher

A.

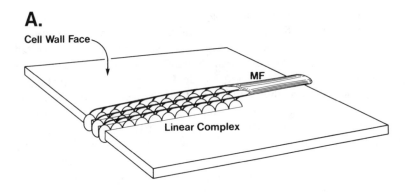

B.

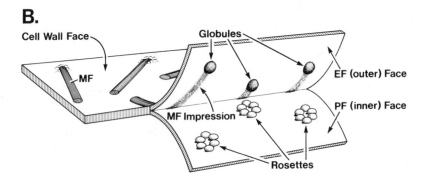

C.

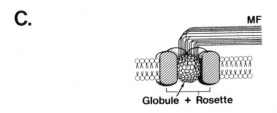

Figure 2 Stylized drawings of terminal complexes observed by freeze-fracture of plasma membranes of cellulosic algae and higher plants. (*A*) Linear terminal complexes found in some cellulosic algae. For simplicity the number of subunits drawn is less than that actually observed. (*B*) Rosettes and globules found in some cellulosic algae and in lower and higher plants. Linear complexes as in (*A*) usually fracture with the EF face but sometimes fracture with the PF face; rosettes fracture with the PF face and globules, with the EF face. In double-replica fracturing, globules are sometimes found in complementary location to rosettes, leading to the speculation that rosettes and globules may be part of the same complex in vivo as shown in (*C*). MF = microfibril.

plants by Mueller & Brown using *Avena* (103) appeared simultaneouly with that for *Micrasterias* (46). Subsequently, there have been several important additions to the literature using higher plants. Distinct rosettes were reported by Herth's group using *Vigna radiata* (68) and suspension-cultured cells of soybean (70). The latter report was of particular interest, since it pointed out the ephemeral nature of these structures; these workers found that simply centrifuging the cells at very low *g* forces was sufficient to lead to loss of ability to detect the rosettes. Such an observation may well be relevant to the apparent extreme lability of cellulose synthase activities in vitro. In my opinion, the most convincing evidence that rosettes, and perhaps also globules, may be parts of cellulose synthase complexes in plants comes from the recent report of Herth (69), who analyzed the distribution of such complexes by freeze-fracture of developing xylem vessel elements of cress *(Lepidium sativum)*. Using a double-replica technique, Herth observed abundant, but rather indistinct, terminal globules on the EF face and distinct rosettes on the PF face. Most important, both of these structures were restricted to the regions of secondary wall formation where the rate of cellulose synthesis is highest.

In sum, I am becoming more and more convinced that these structures are parts of the synthase complex. The relationship between the globules and the rosettes needs clarification. If both are part of the same complex, then this implies that the structure may contain nonidentical subunits, only one type of which may be involved in the actual catalysis of polymerization. One final note of caution should be added. Ordered arrays of plasma membrane particles have certainly been seen in organisms that do not produce cellulose (65 and references therein). The ultimate proof of function in cellulose biosynthesis for these structures can only come when cellulose synthases can be isolated in active form and purified and when antibodies can be raised that are shown to bind specifically to such complexes.

If one is willing to consider the hypothesis that these complexes are indeed synthases, then it is also interesting to consider whether the variable size and organization of such complexes may determine the diameters and degree of crystallinity of the microfibrils they produce. For higher plant celluloses, where elementary fibrils and variable crystallite sizes have often been reported, the synthetic complexes appear to be of the smaller "rosette" or "globule" type, whereas the larger, apparently single crystal (12, 123, 124, 138) microfibrils of algae like *Valonia* are apparently synthesized by much larger so-called "linear" synthetic complexes (78). Furthermore, the linear complexes of *Valonia* and *Boergesenia* are larger than those in *Oocystis,* and this difference correlates well with the difference in diameters of their microfibrils (79).

Based on current evidence, it would seem reasonable to propose that organisms that have rosette and/or globule structures as synthases will possess cellulose that may show the substructure characteristic of elementary fibrils, each of which is synthesized by a single complex. Overall microfibril size would then be determined by the degree of association of individual rosette/ globule complexes, which may be under genetic or developmental regulation. On the other hand, large, single-crystal microfibrils, with no evidence of elementary fibrils, should be found in those organisms possessing linear synthetic complexes. In the future, a study of *Microsterias,* an alga with large microfibrils, apparently synthesized by an aggregation of individual rosettes (46), would be a useful model to help test this hypothesis. The very powerful technique of high-resolution electron microscopy of ultrathin sections (138) would enable one to test whether these large microfibrils, similar in general appearance to *Valonia* microfibrils, would, upon closer examination, differ from *Valonia* in not being single crystals. In any case, a more thorough comparison of crystallite sizes in various organisms with terminal complex structure would be a useful addition to the literature.

The Role of Microtubules in Microfibril Orientation

In many of the algae and in the secondary walls of higher plants, microfibril deposition can be highly oriented. In many cases, the microfibrils are deposited parallel to each other in layers, with the axis of parallel orientation changing sharply from layer to layer. Another clear case of unique orientation occurs during the development of xylem vessels where complex spiral patterns of secondary wall thickening occur. Few people now dispute the concept that microtubules play some role in directing this orientation. Recent reviews of this topic include those by Robinson & Quader (128), Gunning & Hardham (49), Staehelin & Giddings (135), Heath & Seagull (59), and Lloyd (92, 92a). To summarize briefly from these reviews, electron microscopy has shown that microtubules usually (but not always) are oriented parallel to the direction of the microfibrils in the innermost, or most recently synthesized, part of the wall. Agents such as colchicine, which disrupt microtubules, can lead to loss of orientation of the newly synthesized microfibrils. No one has observed any direct connection between microtubules and anything resembling a synthase complex, but this may not be surprising since terminal complexes have not yet been observed in negatively stained sections. The movement of synthases is believed to be driven by the force generated during polymerization of the rigid microfibrils. This concept is supported by the video recordings of light microscopic observations by R. M. Brown, Jr.; these recordings show that the nonflagellated cells of *A. xylinum* move rapidly "backwards" in the medium during the synthesis of cellulose. Some current models favor the idea that the synthases are not directly attached to microtubules but rather are guided in a general direction by restriction of lateral movement within channels of oriented microtubules that are somehow anchored to the plasma membrane (13, 104, 135). It is surprising that no one has studied the effect of lowered temperatures on the extent and orientation of microfibril deposition; it would seem that movement of complexes might be severely restricted as the membrane undergoes temperature-dependent phase transitions. In any case, we are a long way from understanding the interaction, if any, between microtubules and synthase complexes.

A few recent reports offer substantial new evidence for the coincidence between microtubule and microfibril orientation. Unlike older reports that relied on two-dimensional sections, these recent studies employ the now highly developed technique of indirect immunofluorescence of permeabilized cells using antitubulin antibodies. With such a technique, it is possible to see the entire cortical microtubule network as an entity. In one study of pea stems by Roberts et al (126), the orientation of this network showed a pronounced shift upon addition of ethylene, which was mimicked by the subsequent shift in microfibril orientation. Similarly, in studies of xylem differentiation in cultured *Zinnia* mesophyll cells, Lloyd & Seagull (93) and Falconer & Seagull (39) observed the development of a cortical microtubule network pattern that was closely aligned with the subsequent spiraling pattern of secondary wall cellulose deposition. Stabilization of the microtubule network with taxol prior to induction of xylem formation prevented the subsequent development of the spiral pattern in both microtubules and the cell wall.

From these and other related studies, Lloyd has concluded that the cortical microtubule network is a dynamic array that winds around plant cells describing flat, medium, or steeply pitched helices (92–93). In addition to the unique cases of fibril orientation described above, such shifting helical patterns for the microtubule network may be reflected in the curious, but now well-documented, helicoidal patterns of microfibrils observed in many plant cell walls (106, 141). However, Neville & Levy (106) and the French group of Vian et al (see 141) have argued that these patterns arise by a self-assembly process following deposition.

Indeed, a number of recent studies on the deposition of helicoidal walls has led to a rather different concept of how microfibril orientation in such walls may be controlled (see 106, 141). This new concept proposes that such walls are assembled and oriented by a spontaneous self-assembly process that is characteristic of cholesteric liquid crystals and similar to the one proposed for the assembly of certain animal exoskeletons (11a). Such self-assembly requires rigid, elongated molecules possessing short and flexible side chains. Cellulose itself does not fulfill these criteria, but, if coated with matrix polysaccharides such as xyloglucan (54, 55) or glucuronoxylan (141), such a complex should have the appropriate structure to allow self-assembly into helicoidal arrays. The recent observations by Vian et al (141) that helicoidal texture in linden wood is found only in regions where glucuronoxylan is also present in the wall supports this concept. In such a model, a direct involvement of microtubules in guiding cellulose synthase complexes is discounted. Microtubules might instead play a role in the directed movement to the cell surface of secretory vesicles containing matrix polysaccharides and/or newly synthesized components of cellulose synthase complexes. In this regard, the recent observations of directed movement of organelles along microtubules in animals may be relevant (140a).

The Biochemistry of the Polymerization Process

Before proceeding further in this section, I believe it is important to discuss a technical problem that repeatedly affects all those who attempt to study the biochemistry of β-glucan synthesis—particularly for those who are beginning in this field. The problem, simply stated, is how to determine, easily and accurately, the linkage of glucosyl residues produced in vitro. For higher plants, this usually means distinguishing the production of $(1\rightarrow3)$-β-[^{14}C]glucan, $(1\rightarrow3, 1\rightarrow4)$-$\beta$-[^{14}C]glucan, and $(1\rightarrow4)$-β-[^{14}C]glucan from supplied UDP-[^{14}C]glucose. Unfortunately, at present I do not believe there is any single method of choice for these determinations. I have pointed out earlier (30) that relying on differences in solubility of the products is inadequate. Stone's group has used differential treatment of products from ryegrass membrane preparations with purified or partially purified glucanases including a $(1\rightarrow3)$glucan exohydrolase from *Euglena*, a $(1\rightarrow3, 1\rightarrow4)$-$\beta$-glucan endohydrolase from *Bacillus subtilis*, and a $(1\rightarrow4)$-β-glucan endohydrolase from *Streptomyces* (25, 63). The tedious part of this procedure is the rigorous purification and testing needed to ensure that no contaminating activities are present; once this is done, the technique seems reasonably reliable and rapid. However, workers should be aware; particularly for $(1\rightarrow4)$-β-glucan hydrolases, that even a pure enzyme may degrade $(1\rightarrow3)$-β-glucosyl linkages at a low rate. We have found this to be the case for the highly purified endohydrolase and cellobiohydrolase from *Trichoderma*. On the other hand, the purified $(1\rightarrow3)$-β-glucan hydrolases from *Euglena* and *Basidomycetes* seem not to hydrolyze $(1\rightarrow4)$-β-glucosyl residues. We have found that even a relatively crude preparation of *Mollusca* laminarinase marketed by Cal-Biochem will not degrade the cellulose produced in vitro by *A. xylinum* while very effectively degrading the $(1\rightarrow3)$-β-glucan synthesized in vitro by plant preparations; thus the enzyme has proven useful to us for preliminary linkage analyses (T. Hayashi and D. P. Delmer, unpublished). It would be much more useful, however, to have a single, specific $(1\rightarrow4)$-β-glucan hydrolase in order to detect low levels of $(1\rightarrow4)$-linkages in the presence of high levels of $(1\rightarrow3)$-β-glucan. Since enzyme treatments are so quick and easy to use, it is a pity that no reliable, pure enzymes are available commercially.

Methylation analysis still remains the method of choice for determination of absolute linkage. The method is tedious [although recently improved (53)], and special techniques are required to separate and collect the radioactive derivatives produced; these include collection of effuent from gas-liquid chromatography (GLC) columns (61, 80), or separation by thin-layer chromatography (TLC) (33, 125) or high-pressure liquid chromatography (HPLC) (64). The method may not be completely quantitative and will not distinguish if different linkages are in the same or different polymers, but it is so reliable that no worker should enter the field without learning this technique. Another useful technique is periodate oxidation, since (1→3)-linkages are resistant but (1→4)- linkages are susceptible to this treatment (61). Kauss' group has used a fluorimetric assay with aniline blue to detect callose production (87); if modified by use of the more specific fluoro-chrome Sirofluor (137), this might evolve into a highly specific assay for (1→3)-β-glucan.

Now we enter into the morass. First I shall consider the limited studies on biosynthesis done with algae. Although many algae have proven very useful for the cytological studies of terminal complexes, there have been very few studies on the biochemistry of cellulose synthesis using algae. The most widely quoted work with algae has been that of Hopp et al (75), who reported synthesis of glucolipids and glucoprotein from UDP-glucose using mem-branes isolated from the colorless alga *Prototheca zopfii*. Further addition of GDP-glucose resulted in transfer of radioactivity from the glucoprotein to an insoluble fraction that was termed cellulose, but which was only partially characterized. This work still remains difficult to interpret, since no further characterization of the final product has been reported. This group recently observed that (1→4)-linked glucose is present in the walls of the alga, but (1→3)-linked glucose and mannose (linkage not stated) were found as well (125). In another study (28), they showed that 2-deoxyglucose and coumarin did affect the synthesis of glucan polymer in vivo and also that the UDP- and GDP-derivatives of deoxyglucose inhibit the incorporation of glucose into the glucolipids, in a manner similar to that observed previously with coumarin (76).

All of these reports are intriguing, but they represent almost the only case where there is any indication of an involvement of lipid intermediates in cellulose synthesis; until the final proof that the glucan product produced is really cellulose, I remain somewhat skeptical. There is another recent report that tunicamycin inhibits cellulose synthesis in *Oocystis* (113); in this case the author speculated that some rapidly turning-over glycoprotein (rather than a lipid intermediate), possibly functioning as a primer, may be involved in the process. *Oocystis* is one of the few cases where continuous protein synthesis is required for cellulose synthesis (115); we have certainly found no such effect in plant systems or in *A. xylinum*. It is also worth noting that Jaffe & Leopold (81) have reported that 2-deoxyglucose inhibits callose synthesis in vivo in plants; whether this simply represents a disturbance in UDP-glucose pool levels or implies some specific requirement for lipid and/or protein glycosylation in vivo is not clear. Occasionally a paper implicates lipid intermediates in glucan synthesis in plants (e.g. 50), but, as recently pointed out by Camirand et al (20) and Stone (136), the evidence in sum does not support their involvement, at least in the case of the β-glucans which presently can be synthesized in vitro.

One additional observation with *Oocystis* merits notice; Quader & Robinson (114) have found that calcium ionophores or cryptates very effectively inhibit cellulose synthesis in vivo, suggesting that elevation of intracellular Ca^{2+} levels may be inhibitory to this process.

In higher plants, almost all attempts to obtain in vitro synthesis of $(1\rightarrow4)$-β-glucan have resulted in the synthesis of callose $[(1\rightarrow3)$-β-glucan] using radioactive UDP-glucose as a substrate with membrane preparations derived from higher plants (30). This statement is somewhat of an oversimplification, since limited synthesis of $(1\rightarrow4)$-β-glucan can be obtained in some systems, particularly at low ($<50 \mu$M) UDP-glucose concentrations in the presence of Mg^{2+} (see 4, 7, 10, 62–64, 80 for some recent examples). However, it is known from the detailed studies of Ray (117–119) and Hayashi & Matsuda (56, 57) that there is one, or perhaps two (56), Golgi-localized UDP-glucose: $(1\rightarrow4)$-β-glucan synthases found in most membrane preparations derived from dicot tissues engaged in synthesis of primary cell walls. At least one of these probably functions in the synthesis of xyloglucan; another possibility is that one such enzyme represents a partially latent form of cellulose synthase en route to the plasma membrane. In membrane preparations derived from cereals, such as the *Lolium* endosperm cells studied extensively by Stone's group, an additional complication arises because these membranes synthesize some $(1\rightarrow3)$-β-glucan, some $(1\rightarrow3, 1\rightarrow4)$-$\beta$-glucan, and some $(1\rightarrow4)$-β-glucan (62–64, 136).

It has not usually been possible to assign a plasma membrane location to any of these low-level $(1\rightarrow4)$-β-glucan synthase activities. In cotton fibers harvested after xyloglucan synthesis has almost ceased but when synthesis of cellulose is quite high in vivo, we usually find no in vitro synthesis of $(1\rightarrow4)$-β-glucan, even at low substrate concentration (33). There was one recent report of high rates of synthesis of $(1\rightarrow4)$-β-glucan using mung bean preparations (18), but these data do not appear to be reproducible (19, 33).

Interpretation of these depressing results is difficult. Some possible explanations might be as follows.

1. UDP-glucose is not really a precursor for cellulose.
2. The enzyme is very labile once cells are damaged.
3. The enzyme requires a special activator and/or regulatory protein that is/are lost to the soluble fraction during membrane isolation.
4. Any combination of the above.

The first possibility is difficult to disprove completely without evidence of plasma membrane–localized activity. However, in my past review, I outlined all of the indirect evidence which indicates that UDP-glucose is indeed the most likely precursor (30); as far as I know, the only recent evidence bearing on this point is a recent study by Inouhe et al (77) which involves a clarification of the mechanism of inhibition of cellulose synthesis in vivo by galactose in *Avena* coleoptiles. These workers showed that galactose effectively inhibits the UDP-glucose pyrophosphorylase of this tissue, thus providing more indirect evidence that UDP-glucose is the precursor to cellulose.

Likewise, it is difficult to determine if the problem lies in an inherent instability of the enzyme complex. However, several recent observations may be relevant. The first is that the rosette and globule structures seen in freeze-fracture preparations seem to be labile and can only be observed by omission of cryoprotectants, very rapid freezing techniques, and little or no handling of tissue prior to freezing (70). The second is a recent report by Jacob & Northcote (80) that freezing and thawing of membranes from celery petioles resulted in loss of ability to synthesize $(1{\rightarrow}4)$-β-glucan but not $(1{\rightarrow}3)$-β-glucan. However, the presence of Ca^{2+} and high concentrations of sucrose in their assays are more favorable conditions for $(1{\rightarrow}3)$-β-glucan synthases (58, 82); furthermore, the localization of the $(1{\rightarrow}4)$-β-glucan synthase activity was not determined. Since the Golgi-localized activity tends to be more labile than the callose synthase (117), Jacob & Northcote could have been studying the susceptibility of this enzyme to freeze-thawing. Cotton fiber membranes lacking any substantial in vitro activity for xyloglucan synthesis usually make only $(1{\rightarrow}3)$-β-glucan from UDP-glucose whether the membranes are prepared and assayed fresh or after freeze-thawing (D.P. Delmer, unpublished).

At present, I am most intrigued by the possibility that a critical factor or factors are dissociated from the enzyme in the course of isolation. In the concluding section of this review, I shall develop a model for the structure and regulation of a plant cellulose synthase complex. This model is unproven and is presented at this time mainly to stimulate new ways of thinking about this enzyme. The model is based upon three different types of observations: (*a*) the repeated observations of high activities for callose synthase in vitro in membranes derived from cells in which callose synthesis is usually latent in the unperturbed state in vivo; (*b*) the known mechanism of regulation of lactose synthase in mammals by a polypeptide called lactalbumin; and (*c*) our recent demonstration of a soluble polypeptide in cotton fibers that has the characteristics of a receptor for DCB, a specific inhibitor of cellulose synthesis. These three points are discussed individually and then tied together in the model.

THE RELATIONSHIP BETWEEN THE SYNTHESIS OF CELLULOSE AND CALLOSE

It now seems fairly well-accepted that there must be two UDP-glucose: β-glucan synthase activities on the plasma membrane of plant cells. One of these, the cellulose synthase, is active in intact cells; the other, the callose synthase, is generally latent in intact cells and becomes rapidly activated in response to a variety of perturbations [see Kauss (82), this volume, and Delmer & Stone (36) for discussion of the regulation of callose synthesis by Ca^{2+} in response to cell damage]. Studies with cotton fibers by Carpita & Delmer (21) provide one example of this inverse type of regulation. Cutting of fibers resulted in loss of capacity for cellulose synthesis concomitant with enhancement of callose synthesis; cutting under conditions that favored rapid resealing of the fibers, i.e. in polyethylene glycol, caused a partial reversal of this effect. Thus, synthesis of these two glucan polymers seems to show opposing modes of regulation, and the two glucans are rarely synthesized in the same cell at the same time.

In some cases callose and cellulose are both deposited in the same cell type without apparent external perturbation, although perhaps not simultaneously, nor in precisely the same location. One example is in the development of the cell plate, where callose is deposited, at least transiently, prior to the deposition of other wall components such as cellulose (43). Another is in pollen tubes where callose is probably located in the primary wall and cellulose in the secondary wall (116). Callose is also transiently deposited at the onset of secondary wall synthesis in the cotton fiber (16, 96, 132). In the first two examples, the deposition of cellulose and callose may be distinct in time; in the case of cotton fibers, there is some overlap in the two processes, although the rate of secondary wall cellulose synthesis does not really reach its maximum until the rate of callose synthesis declines (96).

In the past few years, a great deal more has become known about the properties of callose synthase from higher plants. It has been known for some time that β-linked disaccharides activate this enzyme (58, 63, 102). The major new finding is that this enzyme also shows marked activation by micromolar levels of Ca^{2+} (33, 35, 58, 82, 83, 85, 87, 102, 140). This finding came as a surprise, since this enzyme has been assayed for years without addition of Ca^{2+}. Indeed, an almost generally accepted definition of this enzyme has been as Glucan Synthetase II, a plasma membrane activity detected at high UDP-glucose concentrations in the absence of divalent cations (117).

Recently, Hayashi et al (58) clarified the effector requirements of this enzyme in a way that reconciled most of the older reports of diverse cation and disaccharide requirements as well as the variable values reported for the K_m for UDP-glucose. To summarize this study, the enzyme shows a complex interaction with Ca^{2+}, Mg^{2+}, β-glucoside, and substrate. Stimulation of activity by micromolar concentrations of Ca^{2+} and millimolar concentrations of β-glucoside is highest at low (<100 μM) substrate concentrations. These effectors act by both raising the V_{max} and enhancing the affinity for UDP-glucose. In the absence or in very limiting concentrations of these effectors, V versus S plots shift from hyperbolic to sigmoidal, and the K_m for UDP-glucose is elevated from about 0.2 mM to >1 mM. Thus, fairly good activity can be obtained in the absence of these effectors, but only at high substrate concentrations. Mg^{2+} has several interesting effects on the enzyme; in addition to enhancing the affinity for Ca^{2+}, Mg^{2+} appears to cause marked changes in the structure of the enzyme and on the macromolecular structure of the glucan product. The digitonin-solubilized enzyme assumes a much more rapidly sedimenting form in glycerol gradients run in the presence of Mg^{2+}; product synthesized in the presence of Mg^{2+}, in addition to Ca^{2+} and β-glucoside, is highly aggregated and alkali-insoluble, as opposed to product synthesized in the absence of Mg^{2+}, which was observed by EM to be dispersed, fibrillar material that is completely alkali-soluble. Since both products were shown to have identical primary structures [linear ($1{\rightarrow}3$)-β-glucans of $\sim$ 100,000 M_r], the effect of Mg^{2+} on product was thought to occur as a result of a change in enzyme conformation that allowed the glucan chains synthesized from the multiple active sites on the enzyme to align and interact more extensively during synthesis.

Other recent studies on callose synthases have shown a phospholipid requirement for a Triton-extracted form of the red beet enzyme (97), activation of the soybean enzyme by organic polycations (84), and conversion to a Ca^{2+}-independent form of the soybean enzyme by proteolysis (85). At present, we do not know what compound in vivo fulfills the requirement for β-glucoside. Heat-stable activating factors have been found in the supernatants of several plant extracts (18, 23; D. P. Delmer, unpublished), and one of these may represent the endogenous β-glucoside.

Several reports now indicate that callose synthases, as solublized in digitonin, are of high molecular weight (37, 58, 134, 140). Hayashi et al (58) observed numerous dougnut-shaped structures of about 20–30 nm in diameter in partially purified preparations of callose synthase and occasionally observed fibrils protruding from such structures after addition of substrate and effectors. If these structures were indeed callose synthases, then they resemble in size the terminal complexes seen in membranes of higher plants.

Now we come to the crucial question. Could the callose synthase and the cellulose synthase be the same enzyme as suggested several times previously (29, 80)? There is, of course, no definitive answer to this question at this time. The evidence suggests that they share a common intracellular localization and substrate, are possibly similar in size, and that one enzyme tends to be active at times when the other is inactive. The interesting effect of Mg^{2+} in apparently inducing a larger aggregate size for the enzyme (58) is reminiscent of the report by Montezinos & Brown that Mg^{2+} may play some role in maintaining the integrity of the terminal complexes in *Oocystis* (100).

SPECIFICITY OF GLYCOSYLTRANSFERASES

Is there any precedent for a glycosyltransferase that can transfer a sugar to form either of two different linkages? Watkins (143) recently reviewed the history of glycosyltransferases and pointed out that the concept of one enzyme: one linkage is certainly upheld in nearly all cases. There are, however, several interesting exceptions. A highly purified α-fucosyltransferase from human milk reportedly retains the capacity to transfer fucose either in (1→3)- or (1→4)-linkage (112). Stone (136) has proposed that the mixed-linked (1→3, 1→4)-β-glucans of cereals may be synthesized by a single enzyme; this model, analogous with the one developed earlier by Pattee (108), proposes that the stereochemical association of the binding subsites of the synthase with the terminal linkage existing on the growing nascent β-glucan chain establishes the identity of the next linkage formed. This is not much different from the recent proposal of Jacob & Northcote (80), in which the orientation of the primer or nascent glucan chain determines whether the plasma membrane glucan synthase in plants transfers to the 3-hydroxyl or the 4-hydroxyl of the terminal acceptor sugar. This explanation does not seem entirely satisfactory, since simply permeabilizing cells to UDP-glucose by addition of agents like dimethylsulfoxide (21) or chitosan (87) results in synthesis of callose and not cellulose. Another model for such regulation proposes that the presence of Ca^{2+} bound to the enzyme promotes synthesis of callose, and its absence favors cellulose synthesis. This model may also be too simple, since we have not been able to change the linkage of the product produced by the washing and incubation of membranes in EGTA (D. P. Delmer, unpublished); the possibility of very tightly-bound Ca^{2+} on the enzyme has not been excluded, however.

No direct evidence exists to support or refute either the Stone (136) or the Jacob & Northcote (80) models, but there is one well-known case where modulation of acceptor specificity is known to occur. This is the case of lactose synthesis in mammals (for a review, see 72, 89). Mammalian cells possess a Golgi-localized galactosyltransferase, the normal function of which is to transfer galactose from UDP-galactose in (1→4)-β-linkage to terminal

N-acetylglucosamine residues on proteins passing through the Golgi. During lactation in the mammary gland, changes in hormone levels lead to induction of a 14-kd acidic polypeptide called lactalbumin. When this polypeptide forms a specific complex with the galactosyltransferase, the acceptor specificity of the enzyme is changed. The enzyme now prefers to transfer galactose to free glucose, resulting in the synthesis of lactose. The linkage formed in each case is still $(1\rightarrow4)$-β, but the association of the enzyme with lactalbumin apparently prevents binding of the protein-N-acetylglucosamine acceptor and instead favors binding to glucose (89).

IDENTIFICATION OF A RECEPTOR FOR DCB IN PLANTS

The herbicide 2,6-dichlorobenzonitrile (DCB) is now recognized as an effective and apparently specific inhibitor of cellulose synthesis in algae and higher plants (36). Applied in vivo at micromolar concentrations, DCB inhibits cellulose synthesis with little or no short-term effects on synthesis of noncellulosic polysaccharides (11, 74, 101, 110), nuclear division or DNA synthesis (44, 98), protein synthesis (44), respiration (101), or the in vivo labelling patterns of UDP-glucose, phospholipids, or nuceloside mono-, di-, or triphosphates (D. P. Delmer, unpublished). It does inhibit cell wall regeneration and cytokinesis in protoplasts, presumably because these events require cellulose synthesis (98). DCB does not inhibit synthesis of the large amounts of $(1\rightarrow3)$-β-glucan or small amounts of $(1\rightarrow4)$-β-glucan synthesized in vitro by plant membrane preparations (11, 33).

Since DCB appears to be so effective and specific in its mode of action, we reasoned that it might interact specifically with some polypeptide involved in cellulose synthesis. We therefore synthesized a photoreactive analog of DCB, 2,6-dichlorophenylazide (DCPA), for use as an affinity-labelling reagent to detect such a polypeptide (26). Under UV illumination, DCPA will decompose to a reactive nitrene, which should react and form a covalent link with any adjacent molecule. Preliminary experiments (33) indicated that DCPA is an effective inhibitor of cellulose synthesis in vivo, and [^{3}H]-DCPA reacts specifically with an 18-kd polypeptide upon UV illumination of crude cotton fiber extracts. No labelling is observed if UV illumination is omitted, and labelling is substantially reduced in the presence of unlabelled DCB. Our initial hope that we might be labelling the catalytic polypeptide of the cellulose synthase was dashed by the discovery that >95% of this labelled 18-kd polypeptide was found in the supernatant following centrifugation for 1 hr at 100,000 g (34). However, when labelling is performed prior to membrane isolation, we do find some of this protein (5–10%) associated with the membrane fraction (see Figure 3). Other recent results (34) indicate that the polypeptide is also about 18 kd in its native state; like lactalbumin, the 18-kd polypeptide is acidic and appears to be developmentally regulated, since ability to label this polypeptide increases substantially at the onset of secondary wall cellulose synthesis in the cotton fiber. The polypeptide is similar in size and charge to cotton calmodulin but is clearly distinct from this protein as judged by its mobility in nondenaturing acrylamide gels; it also does not show the shift in mobility in SDS gels in the presence of Ca^{2+} that is characteristic of calmodulin (17).

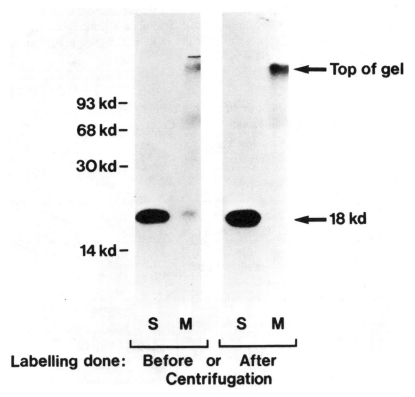

Figure 3 Identification of a DCB receptor from cotton fibers (34). Crude extracts of cotton fibers harvested at the time of active secondary wall cellulose synthesis were incubated in the dark for 10 min at 4°C with 0.5 μM (^{3}H)-2,6-dichlorophenylazine, a photoreactive analog of DCB. This was followed by 5-min illumination with UV light to allow covalent coupling of the analog with any adjacent protein molecules. Labelled proteins were separated on 15% SDS gels and detected by fluorography. The two tracks on the left represent proteins that were labelled as a crude extract *prior* to separation into soluble (S) and membrane (M) fractions by centrifugation for 45 min at 100,000 g. The two tracks on the right were labelled *after* separation into soluble and membrane fractions. It is evident that the 18-kd polypeptide, which is labelled, is largely soluble; however 5–10% of this polypeptide is found associated with the membrane fraction if labelling is done prior to, but not after, separation of soluble from membrane components.

MODEL FOR THE REGULATION OF CELLULOSE AND CALLOSE SYNTHESIS

The specificity of labelling, coupled with its developmental regulation, strongly favors the concept that the 18-kd polypeptide plays some specific role in the process of cellulose biosynthesis. At most only loosely associated with membranes under conventional techniques of isolation, it is unlikely to be the catalytic polypeptide of a cellulose synthase complex, which is expected to be an integral membrane component. In speculating on the possible function for the 18-kd polypeptide, we recalled the case of regulation of lactose synthase by lactalbumin and have developed a working model (see below) that assigns a regulatory role for the 18 kd-polypeptide in the biosynthesis of cellulose and callose.

Intact cell:

$$\text{UDP-Glc} \xrightarrow[\substack{\text{low } Ca^{2+} \\ \text{negative } \Delta\Psi}]{\substack{\text{Plasma-membrane-localized} \\ \text{glucosyltransferase} \sim 18 \text{ kd Complex}}} \text{cellulose}$$

Intact cell plus DCB:

18 Kd ~ DCB
dissociated or nonfunctional

$$\text{UDP-Glc} \xrightarrow[\substack{\text{low } Ca^{2+} \\ \text{negative } \Delta\Psi}]{\substack{\text{Plasma-membrane-localized} \\ \text{glucosyltransferase}}} \text{no product}$$

Damaged cell or isolated membranes:

18 Kd dissociated,
found in soluble fraction

$$\text{UDP-Glc} \xrightarrow[\substack{\text{elevated } Ca^{2+} \\ \text{dissipated } \Delta\Psi}]{\substack{\text{Plasma-membrane-localized} \\ \text{glucosyltransferase}}} \text{callose}$$

This model presumes that there is one plasma-membrane-associated UDP-glucose:glucosyltransferase that is capable of catalyzing the synthesis either of cellulose or callose. Conditions characteristic of an intact cell, e.g. sub-micromolar levels of Ca^{2+} and/or a negative electrical potential across the plasma membrane, favor the association of 18-kd polypeptide(s) with the glucosyltransferase. Under these conditions, the enzyme preferentially transfers glucose to the 4-OH of the terminal glucose in the growing glucan chain.

To assure protection against wounding, plant cells may need to keep high levels of callose synthase present in the plasma membrane at all times; the synthase is latent in intact cells having low free Ca^{2+} levels, but it is available to provide for rapid deposition of callose upon wounding. The rate of cellulose synthesis in such cells would then be modulated by the level of the 18-kd polypeptide; at limiting concentrations, only a few complexes would make cellulose and the rest, lacking 18 kd and excess Ca^{2+}, would make no product.

Similarly, the model proposes that DCB, by binding to the 18-kd polypeptide, either renders it nonfunctional or dissociated, and this also leads to inactive glucosyltransferase. Conditions associated with damage or membrane isolation, e.g. elevation of intracellular Ca^{2+} and/or dissipation of membrane potential, lead to dissociation of the 18-kd protein from the complex; provided sufficient Ca^{2+} is available, the glucosyltransferase now transfers glucose to the 3-OH of nascent glucan chains, resulting in synthesis of callose.

I am intrigued by the idea that there may be two different types of situations in which plants make callose. In one type, cell damage may occur during infection by pathogens or by environmental or mechanical stress. In these cases, localized elevation of Ca^{2+} levels results in localized deposition of callose. I think of this as "useful callose," since it may well serve the function of blocking wound sites or pathogen invasion (1). The other type of situation occurs when a localized concentration of Ca^{2+} is needed for other cellular functions: e.g. to establish the location of a cell plate (66) or to direct the movement and fusion of vesicles to the tips of growing pollen tubes (109, 121). In these cases, the callose may be deposited only as a consequence of other events; such "accidental callose" may actually have no function. The Ca^{2+} currents at the tip of pollen tubes may thus explain in part the unique wall structure of the pollen tube; according to the model presented here, it may be impossible to deposit cellulose at the tip, and cellulose may only be found as a secondary wall deposited well beyond the growing tip (116).

I emphasize again that this model for the regulation of cellulose and callose synthesis is largely unproven. It has the attractive features of explaining the lack of substantial cellulose synthesis in vitro, since the necessary regulatory protein would have been lost during membrane isolation, and it also provides an explanation for the developmental regulation of the 18-kd polypeptide and the mode of action of DCB. The effect of Ca^{2+} on stimulation of callose synthesis is now well-documented (82, 83); it seems logical that elevation of intracellular Ca^{2+} should also inhibit cellulose synthesis, although there is only limited experimental evidence available to support this concept (114). Similarly, although there is direct evidence that membrane potentials can affect glucan synthesis in *A. xylinum* (32) and less direct evidence also for higher plants (7, 21), the mechanism by which such regulation may occur remains obscure.

The model should be testable, provided that one can isolate sufficient quantities of purified 18-kd polypeptide and can find the appropriate conditions to favor its association with plasma membranes. Whether this model is even partially correct or not, it should be a high priority for workers in this field to purify not only this polypeptide but the callose synthase as well. Several recent studies (37, 58, 82) indicate that purification of the callose synthase should be feasible in the near future. Using UDP-pyridoxal (120), we have recently developed an affinity-labelling technique for this enzyme that offers hope that the catalytic polypeptide will soon be identified. If it is, only partial purification may be necessary to isolate a pure polypeptide from SDS gels for use in sequencing or antibody production. Such antibodies could be tested for reaction to terminal globules or rosettes. A logical extension of the above model might include the proposal that the terminal globules represent the glucosyltransferase, and the rosettes represent arrays of 18-kd polypeptides, all of which associate in the membrane to form the cellulose synthase complex.

Finally, we should perhaps be paying more attention to the possible need for a primer in glucan synthesis; it may be that the binding and orientation of such a molecule may also be critical for determination of the linkage formed in the final glucan. Blaschek et al (10) recently presented evidence that the $(1\rightarrow4)$-β-glucan synthesized in vitro, using membranes derived from tobacco, may be attached to an endogenous 4-linked primer glucan, although a direct chemical linkage between the glucan and primer was not proven. The significance of this potentially interesting finding with respect to cellulose biosynthesis is still not clear. Current evidence does not favor (but does not really disprove) a primer function for the β-glucoside activator required for callose synthesis in vitro (58, 94); isolation of the natural heat-stable activator/primer found in soluble fractions of plant extracts may be another important step in clarifying the relationship between the synthesis of cellulose and callose. It is certainly possible that the necessary conditions for in vitro synthesis of cellulose will require a combination of conditions that may include a precise concentration ratio of specific divalent cations such as Ca^{2+} and Mg^{2+}, 18-kd polypeptides, membrane potential, and appropriate primer.

I ended my last review with a plea to young scientists to join in the frustrating but fascinating study of cellulose biosynthesis. One of these days one of us will clone some gene involved in the process, and then we'll be joined by any number of new colleagues. Until that day, however, we can still benefit from new talent and ideas. The few of us who still persist are certainly having fun along the way, and we welcome new travelers to share in the adventure.

ACKNOWLEDGMENT

I thank Karen Long for her prompt and skillful preparation of this manuscript.

Literature Cited

1. Aist, J. R. 1983. Structural responses as resistance mechanisms. In *The Dynamics of Host Defense*, ed. J. A. Bailey, B. J. Deverall, pp. 30–70. Australia: Academic

2. Aloni, Y., Delmer, D. P., Benziman, M. 1982. Achievement of high rates of *in vitro* synthesis of 1,4-β-D-glucan: Activation by cooperative interaction of the *Acetobacter xylinum* enzyme system with GTP, polyethylene glycol, and a protein factor. *Proc. Natl. Acad. Sci. USA* 79:6448–52

3. Aloni, Y., Cohen, R., Benziman, M., Delmer, D. 1983. Solubilization of the UDP-glucose: 1,4-β-D-glucosyltransferase (cellulose synthase) from *Acetobacter xylinum*. *J. Biol. Chem.* 258:4419–23

4. Amino, S., Yoshihisa, T., Komamine, A. 1985. Changes in glucan synthase activities during the cell cycle in a synchronous culture of *Catharanthus roseus*. *Physiol. Plant.* 65:67–71

5. Atalla, R. H., Van der Hart, D. L. 1984. Native cellulose: a composite of two distinct crystalline forms. *Science* 223:283–85

6. Babczinski, P., Haselbeck, A., Tanner, W. 1980. Yeast mannosyl transferases requiring dolichyl phosphate and dolichyl phosphate mannose as substrate. Partial purification and characterization of the solubilized enzyme. *Eur. J. Biochem.* 105:509–15

7. Bacic, A., Delmer, D. P. 1981. Stimulation of membrane-associated polysaccharide synthetases by a membrane potential in developing cotton fibers. *Planta* 152:346–51

8. Benziman, M., Aloni, Y., Delmer, D. P. 1983. Unique regulatory properties of the UDP-glucose: β-1,4-glucan synthetase of *Acetobacter xylinum*. *J. Appl. Polym. Sci. Appl. Polym. Symp.* 37: 131–43

9. Benziman, M., Haigler, C. H., Brown, R. M. Jr., White, A. R., Cooper, K. M. 1980. Cellulose biogenesis: polymerization and crystallization are coupled processes in *Acetobacter xylinum*. *Proc. Natl. Acad. Sci. USA* 77:6678–82

10. Blaschek, W., Haass, D., Koehler, H., Semler, U., Franz, G. 1983. Demonstration of a β-1,4-primer glucan in cellulose-like glucan synthesized *in vitro*. *Z. Pflanzenphysiol.* 111:357–64

11. Blaschek, W., Semler, U., Franz, G. 1985. The influence of potential inhibitors on the *in vivo* and *in vitro* cell-wall β-glucan biosynthesis in tobacco cells. *J. Plant Physiol.* 120:457–70

11a. Bouligand, Y. 1972. Twisted fibrous arrangements in biological materials and cholesteric mesophases. *Tissue Cell* 4:189–217

12. Bourret, A., Chanzy, H., Lazaro, R. 1972. Crystallite features of *Valonia* cellulose by electron diffraction and dark-field electron microscopy. *Biopolymers* 11:893–98

13. Brown, R. M. Jr. 1985. Cellulose microfibril assembly and orientation: recent developments. *J. Cell Sci. Suppl.* 2:13–32

14. Brown, R. M. Jr., Montezinos, D. 1976. Cellulose microfibrils: visualization of biosynthetic and orienting complexes in association with the plasma membrane. *Proc. Natl. Acad. Sci. USA* 73:143–47

15. Brown, R. M. Jr., Willison, J. H. M., Richardson, C. L. 1976. Cellulose biosynthesis in *Acetobacter xylinum*: Visualization of the site of synthesis and direct measurement of the *in vivo* process. *Proc. Natl. Acad. Sci. USA* 73: 4565–69

16. Buchala, A. J., Meier, H. 1985. Biosynthesis of β-glucans in growing cotton (*Gossypium arboreum* L. and *Gossypium hirsutum* L.) fibres. In *Biochemistry of Plant Cell Walls*, ed. C. T. Brett, J. R. Hillman, pp. 221–41. Cambridge: Cambridge Univ. Press

17. Burgess, W. H., Jemiolo, D. K., Kretsinger, R. H. 1980. Interaction of calcium and calmodulin in the presence of sodium dodecyl sulfate. *Biochim. Biophys. Acta* 623:257–70

18. Callaghan, T., Benziman, M. 1984. High rates of *in vitro* synthesis of 1,4-β-D-glucan in cell-free preparations from *Phaseolus aureus*. *Nature* 311:165–67

19. Callaghan, T., Benziman, M. 1985. Corrigendum. *Nature* 314:383

20. Camirand, A., Torossian, K., Hayashi, T., Maclachlan, G. 1985. Are charged

lipid-linked intermediates involved in the biosynthesis of β-glucans? *Can. J. Bot.* 63:867–71

21. Carpita, N. C., Delmer, D. P. 1980. Protection of cellulose synthesis in detached cotton fibers by polyethylene glycol. *Plant Physiol.* 66:911–16

22. Chanzy, H., Henrissat, B. 1985. Unidirectional degradation of *Valonia* cellose microcrystals subjected to cellulase action. *FEBS Lett.* 184:285–88

23. Chao, H-Y., Maclachlan, G. A. 1978. Soluble factors in *Pisum* extracts which moderate *Pisum* β-glucan synthase activity. *Plant Physiol.* 61:943–48

24. Colvin, J. R. 1980. The biosynthesis of cellulose. In *Plant Biochemistry*, ed. J. Priess, 3:543–70. New York: Academic

25. Cook, J. A., Fincher, G. B., Keller, F., Stone, B. A. 1980. The use of specific β-glucan hydrolases in the characterization of β-glucan synthetase products. In *Mechanisms of Saccharide Polymerization and Depolymerization*, ed., J. J. Marshall, pp. 301–15. New York: Academic

26. Cooper, G., Delmer, D. P., Nitsche, C. 1986. Photoaffinity analog of herbicide inhibiting cellulose biosynthesis: synthesis of [³H]-2,6-dichlorophenylazide. *J. Labelled Compounds.* In press

27. Couso, R. O., Lelpi, L., Garcia, R. C., Dankert, M. A. 1982. Biosynthesis of polysaccharides in *Acetobacter xylinum.* Sequential synthesis of a heptasaccharide diphosphate prenol. *Eur. J. Biochem.* 123:617–28

28. Datema, R., Schwarz, R. T., Rivas, L. A., Pont Lezica, R. 1983. Inhibition of β-1,4-glucan biosynthesis by deoxyglucose. The effect on the glucosylation of lipid intermediates. *Plant Physiol.* 71:76–81

29. Delmer, D. P. 1977. Biosynthesis of cellulose and other plant cell wall polysaccharides. *Recent Adv. Phytochem.* 11:45–77

30. Delmer, D. P. 1983. Biosynthesis of cellulose. *Adv. Carbohydr. Chem. Biochem.* 41:105–53

31. Delmer, D. P., Benziman, M., Klein, A. S., Bacic, A., Mitchell, B., et al. 1983. A comparison of the mechanism of cellulose biosynthesis in plants and bacteria. *J. Appl. Polym. Sci., Appl. Polym. Symp.* 37:1–16

32. Delmer, D. P., Benziman, M., Padan, E. 1982. Requirement for a membrane potential for cellulose synthesis in intact cells of *Acetobacter xylinum. Proc. Natl. Acad. Sci. USA* 79:5282–86

33. Delmer, D. P., Cooper, G., Alexander, D., Cooper, J., Hayashi, T., et al. 1985.

New approaches to the study of cellulose biosynthesis. *J. Cell Sci. Suppl.* 2:33–50

34. Delmer, D. P., Cooper, G. 1987. Identification of a receptor protein for the herbicide 2,6-dichlorobenzonitrile. *Plant Physiol.* In press

35. Delmer, D. P., Thelen, M., Marsden, M. P. F. 1984. Regulatory mechanisms for the synthesis of β-glucans in plants. In *Structure, Function, and Biosynthesis of Plant Cell Walls*, ed., W. M. Dugger, S. Bartnicki-Garcia, pp. 133–49. Rockville, MD: Am. Soc. Plant Physiol.

36. Delmer, D. P., Stone, B. A. 1987. Biosynthesis of plant cell walls. In *Plant Biochemistry*, ed. J. Priess. New York: Academic. In press

37. Eiberger, L. L., Wasserman, B. P. 1987. Partial purification of β-glucan synthase from red beet root. *Plant Physiol.* In press

38. Emons, A. M. C. 1985. Plasma membrane rosettes in root hairs of *Equisetum hyemale. Planta* 163:350–59

39. Falconer, M. M., Seagull, R. W. 1985. Immunofluorescent and Calcofluor White staining of developing tracheary elements in *Zinnia elegans* L. cultures. *Protoplasma* 125:190–98

40. Fevre, M. 1983. Nucleotide effects on glucan-synthesis activities of particulate enzymes from *Saprolegnia. Planta* 159:130–35

41. Fevre, M., Rougier, M. 1981. β-1,3- and β-1,4-glucan synthesis by membrane fractions from the fungus *Saprolegnia. Planta* 151:232–41

42. French, A. D. 1985. Physical and theoretical methods for determining the supramolecular structure of cellulose. See Ref. 105, pp. 84–111

43. Fulcher, R. G., McCully, M. E., Setterfield, G., Sutherland, J. 1975. β-1,3-Glucans may be associated with cell plate formation during cytokinesis. *Can. J. Bot.* 54:539–42

44. Galbraith, D. W., Shields, B. A. 1982. The effects of inhibitors of cell wall synthesis on tobacco protoplast development. *Physiol. Plant.* 55:25–30

45. Garcia, R. C., Recondo, E., Dankert, M. A. 1974. Polysaccharide biosynthesis in *Acetobacter xylinum. Eur. J. Biochem.* 43:93–195

46. Giddings, T. H., Brower, D. L., Staehelin, L. A. 1980. Visualization of particle complexes in the plasma membrane of *Micrasterias denticulata* associated with the formation of cellulose fibrils in primary and secondary walls. *J. Cell Biol.* 84:327–39

47. Girard, V., Fevre, M. 1984. β-1,4- and β-1,3-glucan synthases are associated

with the plasma membrane of the fungus *Saprolegnia. Planta* 160:400–6

48. Glaser, L. 1958. The synthesis of cellulose in cell-free extracts of *Acetobacter xylinum. J. Biol. Chem.* 232:627–36

49. Gunning, B. E. S., Hardham, A. R. 1982. Microtubules. *Ann. Rev. Plant Physiol.* 33:651–98

50. Haass, D., Hackspacher, G., Franz, G. 1985. Orientation of cell wall β-glucan synthases in plasma membrane vesicles. *Plant Sci. Lett.* 41:1–9

51. Haigler, C. H. 1985. The functions and biogenesis of native cellulose. See Ref. 105, pp. 30–83

52. Haigler, C. H., Brown, R. M. Jr., Benziman, M. 1980. Calcofluor White ST alters *in vivo* assembly of cellulose microfibrils. *Science* 210:903–6

53. Harris, P. H., Henry, R. J., Blakeney, A. B., Stone, B. A. 1984. An improved procedure for the methylation analysis of oligosaccharides and polysaccharides. *Carbohydr. Res.* 127:59–73

54. Hayashi, T., Marsden, M. P. F., Delmer, D. P. 1986. Pea xyloglucan and cellulose. VI. Xyloglucan-cellulose interactions *in vitro* and *in vivo. Plant Physiol.* In press

55. Hayashi, T., Maclachlan, G. 1984. Pea xyloglucan and cellulose. I. Macromolecular organization. *Plant Physiol.* 75:596–604

56. Hayashi, T., Matsuda, K. 1981. Biosynthesis of xyloglucan in suspension-cultured soybean cells. Evidence that the enzyme system of xyloglucan synthesis does not contain β-1,4-glucan 4-β-D-glucosyltransferase activity (EC 2.4.1.12). *Plant Cell Physiol.* 22:1571–84

57. Hayashi, T., Nakajima, T., Matsuda, K. 1984. Biosynthesis of xyloglucan in suspension-cultured soybean cells. Processing of the oligosaccharide building blocks. *Agric. Biol. Chem.* 48:1023–27

58. Hayashi, T., Read, S. M., Bussell, J., Thelen, M. T., Lin, F-C., et al. 1986. Mung bean and cotton fiber UDP-glucose: (1→3)-β-glucan synthases: Differential effects of Ca^{2+} and Mg^{2+} on enzyme activity and conformation and on macromolecular structure of the reaction product. *Plant Physiol.* In press

59. Heath, I. B., Seagull, R. W. 1982. Oriented cellulose fibrils and the cytoskeleton. In *The Cytoskeleton in Plant Growth and Development*, ed. C. W. Lloyd, pp. 163–182. London: Academic

60. Hebert, J. J. 1984. Crystalline form of native cellulose. *Science* 224:79

61. Heiniger, U., Delmer, D. P. 1977. UDP-glucose: glucan synthetase in de-

veloping cotton fibers. *Plant Physiol.* 59:719–23

62. Henry, R. J., Stone, B. A. 1982. Solubilization of β-glucan synthases from the membranes of cultured ryegrass endosperm cells. *Biochem. J.* 203:629–36

63. Henry, R. J., Stone, B. A. 1982. Factors influencing β-glucan synthesis by particulate enzymes from suspension-cultured *Lolium multiflorum* endosperm cells. *Plant Physiol.* 69:632–36

64. Henry, R. J., Stone, B. A. 1985. Extent of β-glucan chain elongation by ryegrass (*Lolium multiflorum*) enzymes. *Carbohydr. Polymers* 5:1–12

65. Henry, Y., Pouphile, M., Gulik-Krzywicki, T., Wiessner, W., Lefort-Tran, M. 1985. Freeze-fracture study of ordered arrays of particles in the plasma membrane *Chlamydobotrys stillata* Korsch. (Volvocales). *Protoplasma* 126:100–13

66. Hepler, P. K., Wayne, R. O. 1985. Calcium and plant development. *Ann. Rev. Plant Physiol.* 36:397–439

67. Herth, W. 1983. Arrays of plasma membrane "rosettes" involved in cellulose microfibril formation of *Spirogyra. Planta* 159:347–56

68. Herth, W. 1984. Oriented "rosette" alignment during cellulose formation in mung bean hypocotyl. *Naturwissenschaften* 71:216–17

69. Herth, W. 1985. Plasma-membrane rosettes involved in localized wall thickening during xylem vessel formation of *Lepidium sativum* L. *Planta* 164:12–21

70. Herth, W., Weber, G. 1984. Occurrence of the putative cellulose-synthesizing "rosettes" in the plasma membrane of *Glycine max* suspension culture cells. *Naturwissenschaften* 71:153–54

71. Hieta, K., Kuga, S., Usuda, M. 1984. Electron staining of reducing ends evidences a parallel-chain structure in *Valonia* cellulose. *Biopolymers* 23:1807–10

72. Hill, R. L., Brew, K. 1975. Lactose synthetase. *Adv. Enzymol. Related Areas Mol. Biol.* 43:411–90

73. Hogetsu, T. 1983. Distribution and local activity of particle complexes synthesizing cellulose microfibrils in the plasma membrane of *Closterium acerosum* (Schrank) Ehrenberg. *Plant Cell Physiol.* 24:777–81

74. Hogetsu, T., Shibaoka, H., Shimodoriyama, M. 1974. Involvement of cellulose synthesis in actions of gibberellin and kinetin on cell expansion. 2,6-Dichlorobenzonitrile as a new cellulose-

synthesis inhibitor. *Plant Cell Physiol.* 15:389–93
75. Hopp, H. E., Romero, P. A., Daleo, G. R., Pont Lezica, R. 1978. Synthesis of cellulose precursors. The involvement of lipid-linked sugars. *Eur. J. Biochem.* 84:561–71
76. Hopp, H. E., Romero, P. A., Pont Lezica, R. 1978. On the inhibition of cellulose biosynthesis by coumarin. *FEBS Lett.* 86:259–62
77. Inouhe, M., Yamamoto, R., Masuda, Y. 1986. Inhibition of IAA-induced cell elongation in *Avena* coleoptile segments by galactose: its effect on UDP-glucose formation. *Physiol. Plant.* 66:370–76
78. Itoh, T., Brown, R. M. Jr. 1984. The assembly of cellulose microfibrils in *Valonia macrophysa* Kutz. *Planta* 160:372–81
79. Itoh, T., O'Neil, R. M., Brown, R. M. Jr. 1984. Interference of cell wall regeneration of *Boergesenia forbesii* protoplasts by Tinopal LPW, a fluorescent brightening agent. *Protoplasma* 123:174–83
80. Jacob, S. R., Northcote, D. 1985. *In vitro* glucan synthesis by membranes of celery petioles: the role of the membrane in determining the linkage formed. *J. Cell Sci. Suppl.* 2:1–11
81. Jaffe, M., Leopold, A. C. 1984. Callose deposition during gravitropism of *Zea mays* and *Pisum sativum* and its inhibition by 2-deoxy-D-glucose. *Planta* 161:20–26
82. Kauss, H. 1987. Some aspects of calcium-dependent regulation in plant metabolism. *Ann. Rev. Plant Physiol.* 38:47–72
83. Kauss, H. 1985. Callose biosynthesis as a Ca^{2+}-regulated process and possible relations to the induction of other metabolic changes. *J. Cell Sci. Suppl.* 2:89–103
84. Kauss, H., Jeblick, W. 1985. Activation by polyamines, polycations, and ruthenium red of the Ca^{2+}-dependent glucan synthase from soybean cells. *FEBS Lett.* 185:226–30
85. Kauss, H., Kohle, H., Jeblick, W. 1983. Proteolytic activation and stimulation by Ca^{2+} of glucan synthase from soybean cells. *FEBS Lett.* 158:84–88
86. Khan, A. W., Colvin, J. R. 1961. Synthesis of bacterial cellulose from labeled precursor. *Science* 133:2014–15
87. Kohle, H., Jeblick, W., Poten, F., Blaschek, W., Kauss, H. 1985. Chitosan-elicited callose synthesis in soybean cells as a Ca^{2+}-dependent process. *Plant Physiol.* 77:544–51
88. Kolpak, F. J., Weih, M., Blackwell, J.

1978. "Mercerization of cellulose. 1. Determination of the structure of mercerized cotton." *Polymer* 19:123–31
89. Lambright, D. G., Lee, T. K., Wong, S. S. 1985. Association-dissociation modulation of enzyme activity: Case of lactose synthase. *Biochemistry* 24:910–14
90. Lin, F. C., Brown, R. M. Jr., Cooper, J. B., Delmer, D. P. 1985. Synthesis of fibrils *in vitro* by a solubilized cellulose synthase from *Acetobacter xylinum*. *Science* 230:822–25
91. Lin, F. C., Brown, R. M. Jr., Delmer, D. P. 1987. In vitro assembly of cellulose microfibrils prepared from *Acetobacter xylinum*. *Biopolymers.* Manuscript in preparation
92. Lloyd, C. W. 1984. Toward a dynamic helical model for the influence of microtubules on wall patterns in plants. *Int. Rev. Cytol.* 86:1–51
92a. Lloyd, C. W. 1987. The plant cytoskeleton: the impact of fluorescence microscopy. *Ann. Rev. Plant Physiol.* 38:119–39
93. Lloyd, C. W., Seagull, R. W. 1985. A new spring for plant cell biology: microtubules as dynamic helices. *Trends Biochem. Sci.* 10:476–78
94. Maclachlan, G. 1982. Does β-glucan synthesis need a primer? In *Cellulose and Other Natural Polymer Systems*, ed. R. M. Brown Jr., pp. 327–39. New York: Plenum
95. Maclachlan, G. 1983. Studies on cellulose metabolism (1970–1990). In *The New Frontiers in Plant Biochemistry*, ed. T. Asahi, H. Imasaki, pp. 83–91. Tokyo: Jpn. Sci. Soc. Press
96. Maltby, D., Carpita, N. C., Montezinos, D., Kulow, C., Delmer, D. P. 1979. β-1,3-Glucan in developing cotton fibers. *Plant Physiol.* 63:1158–64
97. McCarthy, K. J., Wasserman, B. P. 1986. Regulation of plasma membrane β-glucan synthase from red beet root by phospholipids. *Plant Physiol.* 82:396–400
98. Meyer, Y., Herth, W. 1978. Chemical inhibition of cell wall formation and cytokinesis, but not of nuclear division, in protoplasts of *Nicotiana tabacum* L. cultivated *in vitro*. *Planta* 142:253–62
99. Mizuta, S. 1985. Assembly of cellulose synthesizing complexes on the plasma membrane of *Boodlea coacta*. *Plant Cell Physiol.* 26:1443–53
100. Montezinos, D., Brown, R. M. Jr. 1979. Cell wall biogenesis in *Oocystis*: Experimental alteration of microfibril assembly and orientation. *Cytobios* 23:119–39

101. Montezinos, D., Delmer, D. P. 1980. Characterization of inhibitors of cellulose synthesis in cotton fibers. *Planta* 148:305–11

102. Morrow, D. L., Lucas, W. J. 1986. (1→3)-β-Glucan synthase from sugar beet. I. Isolation and solubilization. *Plant Physiol.* 81:171–76

103. Mueller, S. C., Brown, R. M. Jr. 1980. Evidence for an intramembrane component associated with a cellulose microfibril synthesizing complex in higher plants. *J. Cell Biol.* 84:315–26

104. Mueller, S. C., Brown, R. M. Jr. 1982. The control of cellulose microfibril deposition in the cell wall of higher plants. *Planta* 154:489–515

105. Nevell, R. P., Zeronian, S. H. 1985. *Cellulose Chemistry and its Applications.* Chichester, UK: Horwood. 552 pp.

106. Neville, A. C., Levy, S. 1985. The helicoidal concept in plant cell wall ultrastructure and morphogenesis. See Ref. 16, pp. 99–124

107. Notario, V., Kawai, H., Cabib, E. 1982. Interaction between β-(1→3)-glucan synthetase and activating phosphorylated compounds. A kinetic study. *J. Biol. Chem.* 257:1902–5

108. Pattee, H. H. 1961. On the origin of macromolecular sequences. *Biophys. J.* 1:683–710

109. Picton, J. M., Steer, M. W. 1985. The effects of ruthenium red, lanthanum, fluorescein isothiocyanate and trifluoperazine on vesicle transport, vesicle fusion, and tip extension in pollen tubes. *Planta* 163:20–26

110. Pillonel, C. H., Meier, H. 1985. Influence of external factors on callose and cellulose synthesis during incubation *in vitro* of intact cotton fibres with [^{14}C]sucrose. *Planta* 165:76–84

111. Pizzi, A., Eaton, N. 1985. The structure of cellulose by conformational analysis. Crystalline and amorphous structure of cellulose I. *J. Macromol. Sci. Chem.* A22:139–60

112. Prieels, J-P., Monnom, D., Dolmans, M., Beyer, T. A., Hill, R. L. 1981. Copurification of the Lewis blood group *N*-acetylglucosaminide α-1→4 fucosyltransferase and an *N*-acetylglucosaminide α-1→3 fucosyltransferase from human milk. *J. Biol. Chem.* 256: 10456–63

113. Quader, H. 1985. Tunicamycin prevents cellulose microfibril formation in *Oocystis solitaria. Plant Physiol.* 75:534–38

114. Quader, H., Robinson, D. G. 1979. Structure, synthesis, and orientation of microfibrils. VI. The role of ions in microfibril deposition in *Oocystis solitaria. Eur. J. Cell Biol.* 20:51–56

115. Quader, H., Robinson, D. G. 1981. *Oocystis solitaria:* a model organism for understanding the organization of cellulose synthesis. *Ber. Dtsch. Bot. Ges.* 94:75–84

116. Rae, A., Harris, P. J., Bacic, A., Clarke, A. E. 1985. Composition of the cell walls of *Nicotiana alata* Link et Otto pollen tubes. *Planta* 166:128–33

117. Ray, P. M. 1979. Separation of maize coleoptile cellular membranes that bear different types of glucan synthetase activity. In *Plant Organelles,* ed. E. Reid, pp. 135–46. Chichester, UK: Horwood

118. Ray, P. M. 1980. Cooperative action of β-glucan synthetase and UDP-xylose xylosly transferase of Golgi membranes in the synthesis of xyloglucan-like polysaccharide. *Biochim. Biophys. Acta* 629: 431–44

119. Ray, P. M. 1985. Auxin and fusicoccin enhancement of β-glucan synthase in peas. *Plant Physiol.* 78:466–72

120. Read, S. M., Delmer, D. P. 1986. Identification of UDP-glucose-binding proteins in mung bean membranes. In *Cell Walls '86, Paris,* ed. B. Vian, D. Reis, R. Goldberg, pp. 308–12. Paris: Groupe Parois France

121. Reiss, H. D., Herth, W. 1985. Nifedipine-sensitive calcium channels are involved in polar growth of lily pollen tubes. *J. Cell Sci.* 76:247–54

122. Reiss, H. D., Schnepf, E., Herth, W. 1984. The plasma membrane of the *Funaria* caulonema tip cell: morphology and distribution of particle rosettes, and the kinetics of cellulose synthesis. *Planta* 160:428–35

123. Revol, J-F. 1982. On the cross-sectional shape of cellulose crystallites in *Valonia ventricosa. Carbohydr. Polymers* 2: 123–32

124. Revol, J-F., Goring, D. A. I. 1983. Directionality of the fibre c-axis of cellulose crystallites in microfibrils of *Valonia ventricosa. Polymer* 24:1547–50

125. Rivas, L. A., Pont Lezica, R. 1985. β-Glucan biosynthesis in synchronous cells of *Prototheca zopfii. Planta* 165: 348–53

126. Roberts, I. N., Lloyd, C. W., Roberts, K. 1985. Ethylene-induced microtubule reorientations: mediation by helical arrays. *Planta* 164:439–47

127. Robinson, D. G., Quader, H. 1981. Structure, synthesis, and orientation of microfibrils. IX. A freeze-fracture investigation of the *Oocystis* plasma mem-

brane after inhibitor treatment. *Eur. J. Cell Biol.* 25:278–88

128. Robinson, D. G., Quader, H. 1982. The microtubule-microfibril syndrome. See Ref. 59, pp. 109–26

129. Ross, P., Aloni, Y., Weinhouse, C., Michaeli, D., Weinberger-Ohana, P., et al. 1985. An unusual guanyl oligonucleotide regulates cellulose synthesis in *Acetobacter xylinum*. *FEBS Lett.* 186:191–96

130. Ross, P., Aloni, Y., Weinhouse, H., Michaeli, D., Weinberger-Ohana, P., et al. 1986. Control of cellulose synthesis in *Acetobacter xylinum*. A unique guanyl oligonucelotide is the immediate activator of the cellulose synthase. *Carbohydr. Res.* 149:101–17

131. Ross, P., Weinhouse, H., Aloni, Y., Michaeli, D., Weinberger-Ohana, P., et al. 1987. Regulation of cellulose synthesis in *Acetobacter xylinum* by cyclic diguanylic acid. *Nature* 325:279–81

132. Ryser, U. 1985. Cell wall biosynthesis in differentiating cotton fibres. *Eur. J. Cell Biol.* 39:236–56

133. Sarko, A., Okano, T. 1981. Structural aspects of cellulose mercerization. In *Proc. Ekman-Days. Int. Symp. Wood Pulping Chem.* 4:91–95

134. Sloan, M. E., Eiberger, L. L., Wasserman, B. P. 1986. Molecular size estimation of plasma membrane β-glucan synthase from red beet root. *Plant Physiol.* 80(Suppl.):30

135. Staehelin, L. A., Giddings, T. H. 1982. Membrane-mediated control of cell wall microfibrillar order. In *Developmental Order: Its Origin and Regulation*, pp. 133–47. New York: Liss

136. Stone, B. A. 1984. Noncellulosic β-glucans in cell walls. See Ref. 35, pp. 52–74

137. Stone, B. A., Evans, N. A., Bonig, I., Clarke, A. E. 1984. The application of Sirofluor, a chemically defined fluorochrome from aniline blue, for the histochemical detection of callose. *Protoplasma* 122:191–95

138. Sugiyama, J., Harada, H., Fujiyoshi, Y., Uyeda, N. 1985. Lattice images from ultrathin sections of cellulose microfibrils in the cell wall of *Valonia macrophysa* Kutz. *Planta* 166:161–68

139. Swissa, M., Aloni, Y., Weinhouse, H., Benziman, M. 1980. Intermediary steps in *Acetobacter xylinum* cellulose synthesis: Studies with whole cells and cell-free preparations of the wild type and a celluloseless mutant. *J. Bacteriol.* 143:1142–50

140. Thelen, M., Delmer, D. P. 1986. Gel-electrophoretic separation, detection, and characterization of plant and bacterial UDP-glucose glucosyltransferases. *Plant Physiol.* 81:913–18

140a. Vale, R. D., Schnapp, B. J., Mitchison, T., Steuer, E., Reese, T. S., Sheetz, M. P. 1985. Different axoplasmic proteins generate movement in opposite directions along microtubules *in vitro*. *Cell* 43:623–32

141. Vian, B., Reis, D., Mosiniak, M., Roland, J. C. 1986. The glucuronoxylans and the helicoidal shift in cellulose microfibrils in linden wood: Cytochemistry *in muro* and on isolated molecules. *Protoplasma* 131:185–99

142. Wada, M., Staehelin, L. A. 1981. Freeze-fracture observation on the plasma membrane, the cell wall and the cuticle of growing protonemata of *Adiantum capillus-veneris* L. *Planta* 151:462–68

143. Watkins, W. M. 1986. Glycosyltransferases. Early history, development, and future prospects. *Carbohydr. Res.* 149:1–12

144. Willison, J. H. M., Brown, R. M. Jr. 1978. Cell wall structure and deposition in *Glaucocystis nostochinearum*. *J. Cell Biol.* 77:103–19

145. Zaar, K. 1979. Visualization of pores (export sites) correlated with cellulose production in the envelope of the gram-negative bacterium *Acetobacter xylinum*. *J. Cell Biol.* 80:773–77

Ann. Rev. Plant Physiol. 1987. 38:291–315

PLANT VIRUS-HOST INTERACTIONS

Milton Zaitlin

Department of Plant Pathology, Cornell University, Ithaca, New York 14853

Roger Hull

Department of Virus Research, The John Innes Institute, Colney Lane, Norwich NR4 7UH, England

CONTENTS

INTRODUCTION 291
HOST RANGE 292
PLANT-TO-PLANT SPREAD 293
EARLY EVENTS IN THE PLANT 294
   Uncoating 294
   Gene Expression and Replication 295
SPREAD OF VIRUSES WITHIN THE PLANT 299
EFFECTS OF VIRUSES ON THEIR HOSTS 303
   Photosynthesis and the Chloroplast 304
   The Hypersensitive Reaction 305
   Host RNA-Dependent RNA Polymerases 306
TRANSGENIC PLANTS AND VIRUS RESISTANCE 308
CONCLUSIONS 309

INTRODUCTION

The tools of molecular biology have made possible major advances in understanding the biology of plant virus genomes and of virus replication. Unfortunately, this remarkable progress has not been accompanied by a similar expansion in knowledge of the effects of virus infection on the physiology of host plants. In fact, there has been little significant change in this area since reviews published in the 1950s and 1960s (9, 33, 50).

There are some seven hundred plant viruses and, as can be seen from Table

291

0066-4294/87/0601-0291$02.00

1, the majority have single-stranded messenger sense (or plus-strand) RNA. By contrast, a much higher proportion of viruses of vertebrates have DNA genomes, and fungi have mainly double-stranded (ds) RNA genomes. The reason for this difference has been the subject of speculation, but to date direct evidence is lacking. Because of the high proportion of RNA viruses in plants, we shall largely confine our discussion to this group.

This chapter is not intended to be an exhaustive review of plant virology. For detailed treatises on the subject the reader is referred to textbooks by Matthews (82) and by Gibbs & Harrison (46). Virologists are now beginning to ask more questions about the interactions between viruses and plants, and these clearly relate to physiology as well as to molecular biology. Our intent is to highlight the present state of knowledge about some of these interactions. We shall present a broad picture, and we cite other reviews as reference sources for readers who seek in-depth discussion of particular aspects.

HOST RANGE

It is obvious that the genotype of the plant has a major influence on the events following a virus entering a cell in that plant. Terms such as immunity, resistance, tolerance, and susceptibility are used to describe the expression of the plant genotype. For further details concerning the factors that affect the plant response the reader is referred to Matthews (82). If a plant produces overt symptoms to virus infection, or if it can be shown to support virus replication even without showing symptoms, it can be said to be a host of that virus. However, the concept of host range becomes more difficult to interpret in other cases, e.g. subliminal infections (110). Matthews (82) also pointed out that host range can be rather a meaningless term because (a) only a small proportion of plant species are tested, (b) nonhosts are not always reported, and (c) testing for symptomless infections is not as frequent as it should be.

Infection responses in the whole plant can provide a misleading impression about the ability of any one plant genome to support virus replication. As was pointed out by Hull (62), susceptibility and/or immunity can be considered as operating at four levels. These are not mutually exclusive, and a given host

Table 1 Types of nucleic acids comprising plant viral genomes

	Number of viruses	Percentage
Plus-stranded RNA	484	76.6
Minus-stranded RNA	82	13.0
Double-stranded RNA	27	4.3
Single-stranded DNA	26	4.1
Double-stranded DNA	13	2.0

can give different responses, depending upon the physiological conditions to which it has been subjected. (*a*) In total immunity the virus does not replicate, even in the initially entered cell. The reasons for this are not known but are likely to be complex. (*b*) Some viruses replicate in the initially infected cells but do not spread to adjacent cells; this has been termed subliminal infection (110). This phenomenon is discussed further in the section on the spread of viruses within the plant and is likely to be much more widespread than presently appreciated. (*c*) At the third level, the limited susceptibility of hosts restricts the virus to a few cells around the point of entry; this is discussed below in the section on hypersensitive reaction. (*d*) In total susceptibility, most of the cells of the plant become infected. This is the usual condition for virus-host combinations in which virus replication and virus-host interactions are studied. It should be emphasized however, that total susceptibility is rare; most plants are functionally resistant to most viruses.

PLANT-TO-PLANT SPREAD

Because of the physical structure of plants, the cuticle and the cellulose walls must be penetrated before virus particles can gain entry into the cell. Thus many plant viruses have formed associations with insects, nematodes, or other organisms that can induce the necessary wounds (see 57). In most cases these associations are highly specific, with well-defined relationships between the virus and the vector. There are two basic forms of interaction between the virus and the insect vector: circulative, in which the virus passes through the insect's gut wall and accumulates in the salivary glands, and noncirculative, in which the virus interacts with what is possibly a specific region of the insect's exoskeleton in the mouthparts or the anterior region of the alimentary canal. Capsid and noncapsid virus-encoded products are involved in these interactions. For further detailed discussion of virus-vector relationships the reader is referred to (56, 64). For successful transmission, in addition to the virus-vector interaction there are also interactions between the vector and the host and between the virus and the host. The latter point, host-range determination, was discussed in the previous section.

Other mechanisms for mechanical damage to plants include the breakage of leaf hairs implicated in the spread of viruses that occur in high concentration, e.g. TMV[1] and PVX, and the act of fertilization, which is used to spread certain viruses by pollen.

[1]Abbreviations of virus names: AlMV, alfalfa mosaic virus; BGMV, bean golden mosaic virus; BMV, brome mosaic virus; CaMV, cauliflower mosaic virus; CCMV, cowpea chlorotic mottle virus; CERV, carnation etched ring virus; CPMV, cowpea mosaic virus; CMV, cucumber mosaic virus; PVX, potato virus X; SMV, squash mosaic virus; TEV, tobacco etch virus; TMV, tobacco mosaic virus; TRV, tobacco rattle virus; TSV, tobacco streak virus; TVMV, tobacco vein mottle virus; and TYMV, turnip yellow mosaic virus.

All the available evidence points to a lack of specific receptors for virus particles at the plant plasmamembrane. The work of Burgess et al (17), Honda et al (61), Bancroft et al (6), and Watts et al (122) showed that plant protoplasts do not take up virus particles by pinocytosis. For successful entry, there has to be damage to the plasmamembrane, and the virus particles have to be of the correct charge. Both these requirements can be effected by adding a polycation such as poly-L-ornithine to the infection medium. It could be argued that protoplasts represent an artificial situation and that results obtained using them do not pertain to the whole plant. However, as noted later in the section on the spread of viruses within the plant, viruses do not have to traverse plasmamembranes to move from cell to cell. Thus, unlike animal viruses, there is no requirement for plant virus recognition at the plasmamembrane. Furthermore, the majority of plant viruses are simple in that they are made up of multiple copies of a single- or occasionally two-coat protein species. Relatively few plant viruses (the minus-stranded RNA viruses and some of the ds RNA viruses, Table 1) have lipid envelopes and/or glycoprotein spikes. These features provide one of the main mechanisms by which animal viruses interact with cell surfaces; plant viruses that have these features are transmitted by arthropods in the circulative manner (though there are other circulative viruses lacking these features), and it is thought that such properties are involved in entry into insect cells.

EARLY EVENTS IN THE PLANT

Uncoating

Once the virus particles enter the cell, they must be uncoated to release their nucleic acid. Until recently, the mechanisms involved in the uncoating of plant viruses were a mystery. One suggestion was that viruses were uncoated extracellularly (32). However, the pioneering work of Wilson and collaborators has shown more plausibly that at least some of the plus-stranded RNA viruses are uncoated by cytoplasmic ribosomes, which at the same time translate the 5' viral cistron (for recent reviews see 93, 124, 125). This phenomenon was first demonstrated for TMV, the rod-shaped particles of which are disassembled in in vitro translation systems (123). Structures comprising partially stripped virus particles associated with ribosomes (termed "striposomes") have been found in vivo (105).

The RNA in some viruses with isometric particles can also be translated concurrently with disassembly. Brisco et al (15) were able to correlate the ability of the virus particle to swell under physiological conditions (i.e. to relax protein-protein interactions) with the translatability in in vitro systems. However, in the case of other viruses the conditions for in vitro cotranslational disassembly have not yet been found. Some of these viruses have rod-shaped particles whose assembly is thought to commence at the 5' end, e.g. PVX; some viruses have isometric particles. With regard to the former, structures resembling striposomes have been found in vivo (125), which suggests that cotranslational disassembly is the mechanism by which these viruses are naturally uncoated. Among the latter are viruses whose particle populations contain a proportion of capsids without nucleic acid (empty particles), which suggests that the capsids are stabilized primarily by protein-protein interactions and that protein-RNA interactions are not involved in particle assembly. Particles of these viruses do not swell under physiological conditions. The rapid release of RNA into cells upon inoculation with one of these viruses, TYMV, has been demonstrated by Kurtz-Fritsch & Hirth (75) and by Matthews & Witz (83). Various in vitro treatments of TYMV, e.g. alkaline pH and 1 M KCl (71) or freezing and thawing (70), lead to release of about 5–8 protein subunits and the RNA; however, these treatments can scarcely be considered as physiological. The suggestion that stabilization of the particles of such viruses may be relaxed by interactions with structures such as membranes (40) does not appear to have been considered to any extent.

Gene Expression and Replication

The expression of viral genes and the replication of viral nucleic acid are closely interlinked. Much has been written about the translation of viral nucleic acids (see 30, 38, 47a) and the replication of viral genomes (see 15a, 64, 93). In this chapter we consider several general points that are not always highlighted in such reviews.

Virus genomes code for several proteins that, for most if not all viruses, include coat protein, one or more proteins involved in replication, and the cell-to-cell spread factor (see section on the spread of viruses within the plant). The constraints of translation by eukaryotic ribosomes dictate that, in a polycistronic mRNA, only the 5' cistron is translated; any cistron downstream of the 5' cistron is effectively closed. Plant viruses, in common with viruses of other eukaryotes, employ two basic strategies to overcome this problem; examples are shown in Figure 1. CPMV and TVMV (a potyvirus) translate all their information as one, or in the case of CPMV, two polyproteins, which are cleaved to give the eventual functional proteins (48). TMV and BMV divide their genomic RNA into monocistronic segments, which can then be translated individually. With TMV, the division probably occurs during transcription from the viral RNA and produces subgenomic RNAs (93); with BMV, only the coat protein cistron (RNA 4) is opened by transcription (from RNA 3); the other three genes either are on monocistronic RNAs (RNAs 1 and 2) or are 5' on a bicistronic RNA (RNA 3) (2). TYMV is an example of a virus that employs both strategies (see 38). In addition to these strategies, viruses utilize

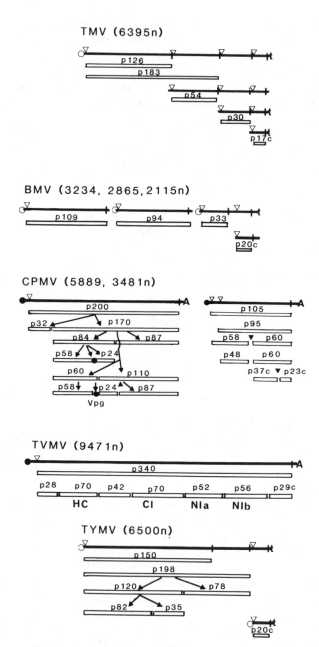

Figure 1   Plant virus translation strategies. The pathways of expression of the proteins of five plant viruses mediated by their genomic RNAs. The abbreviations for plant virus names are given in a footnote in the section on Plant-to-Plant Spread, and in each case is followed by the size of

various subtleties by which further expression is obtained; one example is the readthrough of an amber termination codon at the end of the 5' gene of TMV, which yields two amino co-terminal proteins (94).

The various viral gene products are required in different amounts. For instance, each TMV particle comprises about 2150 subunits of coat protein, yet it is likely that the replication complex producing the RNA encapsidated in those subunits utilizes, at the most, only a few copies of other viral gene products. The second translational strategy, by which there is independent translation of the various genes, is open to translational (or transcriptional) control. For instance, the amounts of the various subgenomic RNAs can vary; furthermore, the coat protein mRNA is able to outcompete the larger subgenomic RNAs for ribosomes and thereby provides the high level of coat protein that is required (16, 31). In contrast, regulation of the amounts of various proteins produced via the polyprotein strategy is much more difficult to envisage because the products are synthesized in equimolar amounts. In CPMV the polyprotein yielding the coat protein subunits is translated from a different RNA species (RNA 2) from that giving the polymerase (P87) and the 5' genome-linked protein, Vpg (Figure 1). Thus, some form of control can be effected by differential translation of the two viral RNAs. However, though other gene products from RNA 2 (P58 and P48) are produced in equimolar amounts to coat protein, the functions of these other gene products are unknown, though it has been suggested (97) that one at least may be involved in cell-to-cell spread. This form of translational control is not possible in potyviruses (e.g. TVMV, Figure 1, and TEV) in which the whole genome is translated as one polyprotein (3, 35). It is of interest to note that some of the TVMV gene products, P70, P52, and P56 (the functions of which are not known) accumulate in cells as cytoplasmic (CI) or nuclear (NI) inclusion bodies; the P54 of TEV is the nuclear inclusion protein. The aggregation of these gene products into large inclusion bodies may be a method of removing proteins that, in large quantities, may be detrimental to the functioning of the cell. The inclusion bodies may also be associated with viral replication

←———

the genome in nucleotides (TYMV has not yet been fully sequenced and the size given is estimated by gel electrophoresis). The top solid line(s) for each virus represent(s) the genomic RNA; other solid lines represent subgenomic RNAs derived from the genomic RNA. Symbols on RNA: o = 5' $m^7G^5ppp^5'$ cap sequence; ● = 5' Vpg (a covalently linked protein); t = 3' amino acylating sequence; A = poly(A); ▽ = translation initiation; | = translation termination. Open boxes represent proteins with the approximate molecular weight n ($\times$ 10^{-3}) given as pn; coat protein is designated pnc. Cleavage pathways are shown by arrows. The TVMV products mentioned in the text are: HC = aphid transmission helper component; CI = cytoplasmic inclusion protein; NI and NIb = nuclear inclusion proteins. Data from references 30, 35, 47a, and 48. The TMV p54 has not been found in vivo (100a).

complexes and/or may be the sites of virus assembly. TEV P54 shows amino acid homologies with viral-coded RNA-dependent RNA polymerases (3). However, in another virus, PVX, inclusion bodies appeared after the virus particles had been synthesized, and thus it was considered unlikely that they were the sites of virus assembly (102).

The points discussed above concerning the differential levels of production of viral gene products also apply to the temporal expression of viral genomes. The gene products involved in the replication of the viral nucleic acid are obviously required before those needed for virus assembly. Potentially, such temporal control can be achieved by the translation strategy, which involves subgenomic mRNAs, but not by the polyprotein strategy. Studies on temporal expression have to be performed in protoplasts, which theoretically give synchronous or near synchronous replication. In all the protoplast studies on subgenomic-strategy viruses, there are few clear-cut cases of temporal control of expression. Hull & Maule (64) summarize the published data for tripartite viruses such as BMV; for monopartite viruses, e.g. TMV, the data are reviewed (93). One hint of temporal expression comes from the observation by Brisco et al (15), in an in vitro cotranslational disassembly system in that RNAs 1 and 2 from the tripartite BMV and CCMV are translated to a greater extent than the other RNAs; it is the products from RNAs 1 and 2, at least from BMV, that are involved in viral nucleic acid replication (54, 73).

The major nucleic acid replication and transcription systems in the nuclei of healthy plants make DNA from DNA or RNA from DNA. Furthermore, the DNA-DNA replication occurs only once per cell cycle and, although transcription of DNA to give RNA is continuous, it is unidirectional. Thus viruses, which have many cycles of replication per cell cycle, cannot depend upon the cell's nucleic acid polymerase machinery for their increase. All the RNA viruses studied to date encode one or more proteins that are most probably involved in the RNA-dependent RNA polymerase enzyme complex responsible for viral genome replication (for review see 54). The full composition of the RNA virus replication complexes has not yet been elucidated, but there is evidence from studies on TYMV that the replicating enzyme contains both viral and host components (20, 87). Even in the well-studied BMV (85), CPMV (36), TYMV (87), and TMV (127) systems no one has yet synthesized new full-length virus-sense RNA from input virus-sense RNA. Thus, there may be several host or viral factors that are lost during extraction of the replication complexes.

The only group of plant viruses known to contain double-stranded DNA, the caulimoviruses, overcome the cell cycle problem by using a biphasic replication procedure. The input viral DNA is transcribed to RNA by the host DNA-dependent RNA polymerase II; this RNA is then the template for transcription to DNA using a virus-encoded reverse transcriptase (for recent reviews see 29, 60, 84).

In whole plants most viruses do not replicate continuously but go through one or more phases of replication. In the classic mosaic diseases, e.g. TMV, AlMV, BMV, and CMV, the concentration of infective virus rises to a maximum over a period of 10–14 days after inoculation. In TMV it remains approximately at this level, but in the other three viruses the concentration then falls. In CMV infections, after falling for a period (7–14 days) it rises again and then falls, continuing this cyclical behavior over a period of time. The factors affecting first the limitation of virus replication and then the reduction in virus concentration and the cyclical behavior of CMV are not completely understood. Among the ideas put forward are the depletion of pools of amino acids and/or nucleotides and the degradation of virus particles by host nucleases and proteases. In CMV the alternating presence of the ds form of a symptom-suppressing satellite RNA correlates with the alternation of layers of symptomatic and symptomless leaves seen in some infections (53). However, the situation for all these viruses must be extremely complex. There must be compartmentalization of disassembly and reassembly and there must be mechanisms to ensure that the replication of the virus and the consequent sequestering of amino acids and nucleotides do not result in the premature death of the host. Thus, as well as the possibility of host involvement in the actual replication of the virus, there must be close interaction between the virus and the host at all stages.

SPREAD OF VIRUSES WITHIN THE PLANT

In order to generate a productive infection and to be perpetuated in nature, a virus must move within its host plant and also must be transmitted to other uninfected plants by means of a vector, usually an arthropod. There are two forms of spread of viruses within a plant: short distance and long distance. The short-distance spread is from cell to cell, and recent evidence indicates that this form of spread is virus controlled. To move to an adjacent cell, a virus (or its nucleic acid) passes through the protoplasmic bridges, the plasmodesmata. This is not a passive process and plasmodesmata are selective in what will move through them. Gunning & Overall (52) have pointed out that ribosomes do not move through plasmodesmata, but isometric viruses of about the same size do. Movement of viruses within a plant has been reviewed in 1984 by Atabekov & Dorokhov (5).

The first evidence for a virus-coded "movement protein" came from the studies of Nishiguchi and colleagues (89) using a tomato strain of TMV (LS-1), which is temperature sensitive in its capacity to spread throughout the plant. The temperature-sensitive defect is not in the replication process per se;

the virus replicates normally in protoplasts at the restrictive temperature. There is good correlative evidence that the TMV-encoded 30,000 M_r (30K) protein is involved in the movement process in this case. Peptide maps show a change in this protein when LS-1 30K protein is compared with its "parental" strain (77), and sequence analysis of clones of this viral gene from LS-1 shows that a single base substitution results in a serine replacement for a proline residue of the parental strain (90). Direct evidence for the role of this protein has been provided by transforming plants with the parental 30K gene and showing that the defective strain LS-1 will now move in the transgenic plant at a restrictive temperature (C. M. Deom, M. J. Oliver, and R. N. Beachy, unpublished).

Proteins to which a movement function has been tentatively assigned have been identified in a number of plant viruses such as CPMV (97), AlMV (7), TRV (14), TSV (28), and CaMV (63, 65), to name a few. In the expectation that sequence homologies might indicate common function(s), the amino acid sequences of these proteins were compared. The protein of AlMV shows a "limited number of matches" to that of BMV but no homology to that of TMV (58). Moreover, antibodies raised to the AlMV putative movement protein do not react with putative movement proteins of three other viruses (115). On the other hand, dot matrix comparisons of sequences revealed a strong homology between the TMV, TSV, and TRV putative movement proteins (14, 28) and, surprisingly, between those of TMV and the ORF1 protein of the DNA viruses, CaMV and CERV (63, 65).

The TMV 30K protein was detected initially by in vitro translation but was only found in tissues recently (68, 119). Time course studies with radiolabeled infected protoplasts and leaves suggested that the protein is ephemeral, i.e. it is synthesized early in the infection and then its rate of synthesis and accumulation declines to below detectable levels, whereas the rate of viral synthesis continues to rise during this period. This pattern of synthesis is unexpected for a protein with a late function in the viral replication-pathogenesis process; it is probably explainable, however, by some very recent unpublished studies of C. Stussi-Garaud and her co-workers in Strasbourg, who have made a detailed study of the 32K putative movement protein of AlMV. They prepared antisera to the protein (using a synthetic polypeptide corresponding to the carboxy-proximal end of the molecule) and have used it to explore in detail the intracellular distribution of the protein. Interestingly, they have found that synthesis of that protein also appears to be ephemeral in leaves, as is the TMV movement protein, only if they examined the membrane-containing fraction of diseased-leaf homogenates. On the other hand, they determined that the protein ended up in the wall fraction of the homogenate as the infection progressed. Further, by immunogold labeling of infected tissues, they observed that the protein was localized in the middle lamella of the cell wall. Thus, if the TMV protein behaves in a similar manner, its ephemeral behavior could be explained. When leaf tissues are homogenized, the extract is normally clarified and wall fragments are discarded; thus, the 30K protein would not be detected in advanced infections. It is also likely that the 30K protein would be excreted from the wall-less isolated mesophyll protoplasts, which would account for the observed loss of the protein with time (119). Surprisingly, in tobacco protoplasts a nuclear site of accumulation has been proposed for this protein (120), a suggestion inconsistent with the experiments described above.

Having documented the case for the involvement of these "movement" proteins in intercellular virus movement, it would now be appropriate to discuss their mode of action. Unfortunately, we know nothing in this regard. The proteins have not been isolated from plants. Our knowledge of them comes from in vitro translation and from their amino acid sequences. Some are slightly basic (28) and the TMV 30K protein is phosphorylated (D. Zimmern, personal communication). The TMV and CaMV proteins show amino acid homology to yeast cytochrome b apoprotein and to plastocyanin (65, 130).

Even though it is generally accepted that viruses and/or their nucleic acids move through plasmodesmata, their possible effect on those structures is conjectural. Some workers, for instance, have shown that in virus-containing plasmodesmata, their central structure (the desmotubule) is perturbed (42, 74). Others (104) have seen no morphological manifestation of the viral presence in virion-containing plasmodesmata. Furthermore, the numbers of plasmodesmata per unit area were reduced by the replication of the defective strain LS-1 when replicating under movement-restrictive temperatures (104).

The action of the movement proteins is apparently not always virus-specific. Taliansky and colleagues (112) have shown that a movement-capable virus can potentiate the movement of an unrelated virus that cannot move in a given host, and furthermore, the host range of the virus can be expanded in this way by allowing it to spread within the tissues of an apparent "nonhost." Moreover, a virus that is normally phloem-limited in its host (BGMV) can be induced to move to adjacent mesophyll tissue when the plant is also infected with a helper virus (TMV) that normally moves in mesophyll tissue (22). It is not unreasonable to presume that a movement protein is functioning here, but a corollary to this presumption is that phloem-limited viruses spread from cell to cell by a different mechanism to mesophyll-inhabiting viruses, and that their movement proteins (if they have them) cannot normally function on mesophyll cells.

As noted earlier in the section on host range, the inability of a virus to move within a plant explains the resistance some apparent "nonhost" plants exhibit to a given virus. There are many examples of "subliminal" infections, i.e.

those in which very little virus can be recovered from the inoculated leaves and no symptoms may be seen (24). Sulzinski & Zaitlin (110) demonstrated that TMV replicated to a high level in two apparent nonhosts (cotton and cowpea), but only in those cells directly infected at the time the leaf was rubbed with inoculum. Many other plant host-virus combinations show similar responses (5). It is conceivable in plants showing subliminal infections that the host inactivates the movement protein, or that the movement protein will not function on that plant's plasmodesmata, or that it is not transported to the site of action in or near the cell wall, or that it needs to interact with a host protein that is not expressed in a suitable form. On the other hand, Atabekov & Dorokhov (5) caution that there are other interpretations to these observations and that the movement protein is not necessarily involved. They suggest that cells in tissues could have a different capacity for virus replication, or that in the preparation of protoplasts the loss of the wall and " . . . partial exchange of the cell contents with the medium and probable loss of some cytoplasmic contents . . ." enables viruses to replicate in protoplasts but not in tissues. We believe the movement protein hypothesis is more tenable, and it will soon become testable as antibodies to these proteins become available.

The region of the TMV genome encoding the 30K movement protein has two additional functions: In most TMV strains it contains the nucleotide sequence at which encapsidation of viral RNA by its coat protein is originated (51). Also, the 30K protein, or the RNA sequence in the 30K region of the genome, governs whether the infection of certain species or cultivars of plants will respond to that infection by producing a hypersensitive response (normally a necrotic local lesion) or a general systemic infection. Recently, K. W. Mundry and colleagues (unpublished results) have mapped the HR-inducing region of the TMV genome to the center of the 30K protein gene in experiments that were a fine-tuned extension of the earlier studies by Kado & Knight (69). These results suggest a 30K protein-host protein interaction. It is of interest here, if the 30K gene product per se is required for necrosis in N gene hosts, that protoplasts isolated from tobacco varieties carrying that gene do not express necrosis (91). It is reasonable to ask whether loss of the 30K protein from the protoplasts could explain this phenomenon; as discussed above, the migration of the protein to the cell wall in tissues could have caused the protein to be lost in wall-less cells.

Once a virus has left the site of the original infection, it will move eventually to distant parts of the plant—often very quickly (long-distance spread). This process was reviewed very thoroughly by Bennett in 1956 (10), and with the exception of some consideration of the nature of the entity that is transported, it has since received little attention. The nature of the infectious entity that is transported within a plant has been reviewed by Atabekov & Dorokhov (5). Basically, it is accepted that viruses move in the phloem and that generally they travel with the photosynthate; moreover, they can move through tissues to distant parts of the plants without necessarily replicating in those tissues. Interference with the photosynthate flow (by shading a leaf, for example) can interfere with virus movement to or from that leaf. Movement to the vascular tissue from the parenchyma must require unique circumstances beyond those needed for movement within the parenchyma, because most defective viruses (which do not have coat protein-encapsidated nucleic acids) can move between mesophyll cells but do not show long-distance movement (4). On the other hand, viroids, which have no coat protein, move long distances with the photosynthetic flow, apparently as well as do encapsidated viruses (P. Palukaitis, unpublished).

Some viruses also move to the reproductive organs of the plant and can effect their transmission between plants by infection via pollen or infected seed. The presence of viruses in seed or pollen is the exception; most viruses are neither seed nor pollen borne. The reason for the relatively few seed-borne viruses is a mystery. One of the determining factors is the inability of most viruses to invade male or female gametophytes, possibly because of the absence of plasmodesmatal connections between the embryo and adjacent cells (11). However, why seed transmission varies between viruses and hosts is unexplained by this postulate. Transmission of viruses through seed and pollen has been reviewed by Bennett (11), Shepherd (106), and Carroll (23).

EFFECTS OF VIRUSES ON THEIR HOSTS

The replication of a virus can have a dramatic effect on its plant host. The symptoms are most probably induced by secondary events rather than by virus replication per se; some viruses replicate to a substantial degree but elicit mild or even no symptoms. Moreover, environmental manipulations or growth-regulating sprays (113) can sometimes ameliorate symptoms without reducing virus replication. The question of virus-induced perturbations has been reviewed by Zaitlin (128). Over the years, plant physiologists and plant virologists have examined diseased plants in the hope of uncovering the cause(s) of their viral-induced distress, and there are multitudes of studies describing the dramatic effects of viruses upon respiration, photosynthesis, secondary metabolism, etc. Much of this work has been summarized in a recent edition of Goodman et al (49). It is not unexpected that diseased plants would show metabolic and catabolic disturbances. What is lacking, however, is an understanding of *how* the viruses initiate the often deleterious symptoms we observe. Virus infection releases a cascade of events resulting in the disease syndrome. Here we discuss only a few areas in which there has been recent research.

Photosynthesis and the Chloroplast

There is a renewed interest in the influence of the virus on the chloroplast, the organelle that is deleteriously affected in many host-virus interactions. The symptoms we describe as "mosaics," "green islands," "yellows," or "chlorosis" are all attributable largely to chloroplast aberrations (81). Many of these abnormalities are only manifested when the tissue is infected early in its development. Quite frequently, leaves infected when mature can replicate virus without showing symptoms. Goodman et al (49) describe viral-initiated chloroplast aberrations in detail. There have been many reports over the years of virus particles within chloroplasts, although there is little evidence to support the idea that the chloroplast is the principal site of viral replication; this is not to say that some replication does not occur there. The replication of one virus, TYMV, does have a close association with the chloroplast. Viral RNA is replicated in peripheral vesicles, and the chloroplasts aggregate to form structures termed polyplasts. However, the association is with the cytoplasmic side of the chloroplast membrane; thus replication is cytoplasmic rather than chloroplastic.

In tobacco infected with some strains of TMV, chloroplasts have been found to contain pseudovirions—short, virus-like particles (103) that contain host rather than viral RNA. Siegel and his associates (98) have shown that they contain chloroplast RNA. The presence of these short rods in chloroplasts (103) is certainly puzzling and their significance is unknown. Rochon & Siegel (98) suggest they could have an influence on either host range or symptomatology.

In recent years the technology for chloroplast isolation has improved dramatically with the introduction of silica sol gradients (95). With these procedures, the finding of virus components within chloroplasts has been substantiated. Both TMV coat protein (96) and TMV RNA (M. Zaitlin, unpublished) have been isolated from Percoll-purified chloroplasts. In addition, an unusual protein, "H protein" (25), now known to be a hybrid protein comprising TMV coat protein and a protein of the host, has been found to be concentrated in chloroplasts (R. Dietzgen, D. Dunigan, and M. Zaitlin, submitted). Furthermore, TMV coat protein has an immunological cross-reactivity and some amino acid sequence homology with the large subunit of ribulose-1,5-bisphosphate carboxylase (34). These observations suggest a close association of TMV with the chloroplast, although the significance of this is unknown. If it does reflect an effect on viral symptomatology, the effect will once again probably be indirect, because infections with mild strains can have more chloroplast-containing coat protein than those with severe strains (96, 98).

An intriguing but unanswered question is how do these viral components enter the chloroplasts? We know of no precedent for foreign RNA migration into the chloroplast; furthermore, the coat protein of TMV does not contain obvious signal sequences that would direct it there from its principal site of synthesis in the cytoplasm. Attempts to see if coat protein will migrate to the chloroplast in vitro, utilizing a protocol that will direct the rubisco small subunit (nuclear-encoded) to the chloroplast were unsuccessful (M. Zaitlin and D. Dunigan, unpublished). The finding of viral RNA in the chloroplast could suggest that viral coat protein might be synthesized on 7OS chloroplast ribosomes. Techniques are now available to test this hypothesis (43). There have been two reports of TMV coat protein synthesis with chloroplast ribosomes in vitro (19, 101) and 70S *Escherichia coli* ribosomes can translate internal cistrons on TMV RNA to yield coat protein (47). If 7OS chloroplast ribosomes were able to synthesize viral coat protein from the full-length genomic RNA, the need for the generation in chloroplasts of subgenomic RNAs for several of the TMV-coded proteins would be obviated.

In the chloroplast, one of the most dramatic and perhaps the most fundamental effects of virus infection is on the synthesis of chloroplast ribosomal RNA and of chloroplast proteins (59, 86). In contrast to the equivalent cytoplasmic functions, which are not significantly affected (108), ribosomal RNA and protein synthesis are shut down in chloroplasts of infected plants. The shutdown of the synthesis of these products could account for the reduction in chloroplast number often found in virus-infected leaves. Interestingly, in a study (80) of many in vitro parameters of photosynthesis (sensitivity of chloroplasts to inhibitors, activity of rubisco, rates of cyclic and noncyclic photophosphorylation, products of CO_2 fixation, etc), no difference was found between chloroplasts isolated from healthy leaves and those isolated from SMV-infected leaves. Chloroplast numbers were reduced, and in those from infected leaves there was a shift in biosynthesis from sugars to amino and organic acids. One can occasionally demonstrate dramatic effects on chloroplast photosynthetic processes in vitro, but these can be overcome with adequate nitrogen nutrition to the plant—circumstances that also enhance viral replication (129). These observations again show that the magnitude of viral replication itself is not the principal cause of viral symptomatology; the interaction between virus and plant must be more subtle!

The Hypersensitive Reaction

Many plants show a hypersensitive response to virus infection, which often results in necrotic local lesions several millimeters in diameter at the primary infection sites (see 94a). Virus is restricted to the necrotic lesion and to a few layers of cells at the outer edge. (The local lesion response is the basis for the assay of many plant viruses; the number of lesions produced can be directly proportional to the virus concentration (82)). It is important to note, however, that the necrosis is not directly responsible for virus localization. Often the viral-induced lesion is chlorotic rather than necrotic, and in those instances virus is also restricted to the primary infection site.

The presence of the lesions on the leaf induces a form of resistance; upon second inoculation with either the same or a different virus, new lesions cannot form close to old lesions (99); furthermore, upper, noninoculated leaves on that plant are less likely to form lesions, and those that form are smaller (100). Systemic acquired resistance, as the latter phenomenon is called, is not restricted to virus-plant interactions; plants with bacterial-, fungal-, or viral-induced necroses are resistant in part to subsequent infection by a systemic virus (12). Thus, a signal that moves in the plant must trigger the resistance induction in the upper noninoculated plant parts. The nature of this signal, which is graft transmissible (67), is unknown, although cytokinins have been implicated (49, p. 413; 111). In the necrotic reaction on the inoculated leaf, lignification is considered to be a necessary component of the induction (55). A virus-inhibiting substance (termed IVR) has been isolated from protoplasts of a hypersensitive-responding tobacco cultivar (78). How this relates to the signal described above is uncertain.

In a number of plant species, virus infection is accompanied by the synthesis of a number of acid-soluble proteins (termed PR for pathogenesis-related) that are coded for by the host and not by the virus (see van Loon (114) for a review). Their stimulated synthesis is not virus-specific; extracts of fungi or bacteria, plant hormones, and a variety of chemicals, notably polyacrylic acid and acetyl salicylic acid (aspirin), can induce them in a wide variety of plant species. Some authors note their correlation with the acquired resistance phenomenon just described (21, 109), whereas other evidence suggests they may not play an active role in the phenomenon (39, 44).

At least 10 PR proteins have been described, and these can be grouped serologically into several classes. The best studied is the PR-1 group, which comprises three members that share a 90% amino acid sequence homology (27). Analysis of genomic clones reveal that these proteins contain a transit peptide, which is probably the signal to transport them through the plasmalemma to the intercellular space where they are localized (114). Surprisingly, one of these proteins was found to have extensive homology to the very sweet-tasting protein thaumatin found in the tropical monocot, *Thaumatococcus daniellii* (26).

The hypersensitive response of various plants to assorted viruses is, as stated earlier, a form of resistance. Very often such resistance is controlled by a single dominant gene (45), which controls a series of reactions involving phenylpropanoid metabolism (76).

Host RNA-Dependent RNA Polymerases

As noted in the section on gene expression and replication, the genomes of RNA plant viruses are transcribed enzymatically by RNA-dependent RNA polymerases (RdRp), termed replicases. In many species of "healthy" plants however, there is an endogenous RdRp, the activity of which is markedly stimulated by many perturbations, including virus replication. The isolated, partially purified enzyme can be shown to transcribe RNA; its product is hydrogen-bonded to its RNA template; thus it synthesizes ds RNA. This virus-stimulated RNA synthesis persuaded many early replicase workers to conclude that they had extracted the replicase or a component of it. Although limited advocacy for this point of view (72) still exists, there is strong evidence to indicate that the host enzyme is not a viral replicase (37).

Plant viruses do indeed generate specific replicases as noted earlier. What then, is the in vivo function of this stress-induced host RdRp? There are suggestions that enzymes with parallel activities occur in animal cells and that they might amplify specific mRNAs. A recent report (116), for instance, suggests a cytoplasmic amplification of globin mRNA in mouse erythroleuke-mia cells. The implication is that minus-strand globin mRNA is synthesized outside of the nucleus, which in turn will act as template for the synthesis of additional mature globin mRNA. There is no evidence for such a scenerio in plants; in vitro, the plant enzyme catalyzes the synthesis of only low molecu-lar weight, polydisperse ds RNA. Furthermore, the only ds RNA that could be detected in healthy tobacco plants is also of low molecular weight and does not approach the size expected for ds forms of mRNA (66), although high molecular weight ds RNA was found in one cultivar of *Phaseolus vulgaris* but not in several others (117).

One further complication is the suggestion that an RdRp is responsible for some RNA synthesis in mitochondria of, at least, certain plant cultivars (for review see 79); whether this has any relevance to the situation in virus-infected plants is uncertain. Thus there is no evidence suggesting a realistic function for this enzyme. It is conceivable that what is being measured is an in vitro function of an enzyme bearing no relation to its in vivo role. On the other hand, the enzyme could be part of a stress-induced host defense against "foreign" RNA. The enzyme might transcribe such RNA; the transcript, if bound to its template, would then preclude the ability of the invader to serve as a mRNA. If this defense mechanism does apply, it would have to be very selective; only foreign RNA should be recognized so as not to interfere with the function of the cells' normal mRNA population. This inhibition might have a parallel in the inhibition of gene expression that was observed when specific "anti-sense" (or minus strand) RNAs were introduced into cells of a number of organisms, including mammals, and that was recently demon-strated with plant protoplasts (41).

One other intriguing aspect of this enzyme is worthy of note. The level of enzyme activity is enhanced by wounding a leaf by rubbing gently with buffer, as one would do during virus inoculation. It is of interest that the enzyme activity is enhanced not only in the rubbed leaf but also in other leaves of the plant (M. Zaitlin, unpublished observations). Once again, as in the case of the systemic acquired resistance discussed in the section on the hypersensitive reaction, a signal moves from a point of stress to other parts of the plant. It is also reminiscent of the stress-induced proteinase inhibitors that accumulate in distant parts of a plant in response to insect attack or mechanical wounding (118).

TRANSGENIC PLANTS AND VIRUS RESISTANCE

Infection of a plant by one virus can often interfere with the replication or pathogenesis of a second virus. The compelling characteristic of the challenging virus is that it has to be "related" to the resident one. This phenomenon has been termed "cross protection"; the strategy has been used in a few instances to protect or "immunize" crop plants against the effects of viral disease (45a). A recent review documents some examples and considers the theories that have been advanced to account for the phenomenon (92). Recently there has been a renewed interest in cross-protection, based in large measure on the newly acquired potential of plant scientists to transform plants and to introduce portions of the viral genome into the genome of the host plant. The transformation of plants with viral sequences is being explored as a means of mimicking the cross-protection phenomenon. Because noninfectious inserts are employed, any potential hazard to the environment is eliminated. The coat protein gene of TMV was first shown to be expressed (at very low levels) in transgenic tobacco by Bevan et al (13). Abel and colleagues (1) also transformed tobacco with the coat protein gene of TMV and found that those plants, when challenged with TMV, had a delayed expression of viral symptoms, and a few plants never developed the disease. They found, further, that the magnitude of the delay was related to the number of primary infection sites on the challenged leaves; moreover, there was no protection when viral RNA rather than the virus was used in the challenge inoculations (R. N. Beachy, personal communication). These results are most easily interpreted to suggest that the protection resulted from the inhibition of the uncoating of the viral RNA by the resident coat protein in the cell of the transformed plants. They are also consistent with the inhibition of co-translational disassembly by coat protein (126). There are probably several types of cross-protection; the coat protein protection mode is one form (107) but is probably not the mechanism underlying most cross-protection observations. We say this because viruses without functional coat proteins or viroids, which generate no coat protein, can cross-protect (18, 88).

The postulate of "negative strand capture" as a mechanism explaining generalized cross-protection has been put forth by Palukaitis & Zaitlin (92). They suggest that the RNA of the challenge virus is inhibited because the resident viral RNA binds to its complementary minus-strand RNA of the challenge virus as it starts its replication. Inserted sequences of viral RNA, in either the "sense" or "anti-sense" orientation, could be visualized as inhibiting viral replication by interfering with replicase binding or by blocking translation by hindering ribosome binding or translocation. Transgenic plants can be used to test these hypotheses, although if the expression of the inserted viral sequences is restricted to the nucleus, thereby preventing their encounter with the incoming viral RNA outside of the nucleus, no protection would be expected. Research in this area, now in progress in a number of laboratories (e.g. 78a), will soon provide answers to these questions.

Another potential means whereby plants may be protected from the effects of specific viral disease is derived from recent studies by Baulcombe et al (8), who have inserted DNA transcripts of the satellite of CMV into the genome of tobacco plants. The several different satellites of this virus are dependent on CMV for their replication; they are not required for CMV replication, but when present in the infection they affect the symptomatology, often ameliorating, but sometimes exacerbating, the CMV symptoms (121). Baulcombe and colleagues (8) found that dimeric DNA copies of the satellite in the transformed tobacco could be processed from the genome to generate infectious satellites only when the plant was subsequently infected with CMV. The satellite then affected symptoms in the expected way. This strategy might also be exploited as a means of disease control, although the problems inherent in introducing a potentially replicating agent into the environment may make it undesirable.

On the other hand, when we learn in molecular terms how the satellites influence symptom development, and why they affect some hosts differently from others, we should be able to modify them to ensure that their only function is symptom amelioration. Further, because the host genome of transgenic plants can replicate the satellite in the absence of CMV, it would be worthwhile to modify the satellite to preclude its CMV-dependent replication but still allow for symptom amelioration. In the field, there could be further problems because CMV-infected plants would be symptomless or nearly so. Such plants would make the "protected" crop a major source of infection for other crops; furthermore, there are possible deleterious synergistic effects if a second "unrelated" virus invades the "protected" crop.

CONCLUSIONS

Plant virologists have learned a great deal about plant viruses with respect to nucleotide sequence, genome organization, structure, protein function, etc; these studies are in logarithmic expansion. With the powerful techniques of molecular biology, we should be able to continue to extend such knowledge to encompass virtually every plant virus. We will also be able to determine which viral nucleotide sequences or protein products interact with the host to cause the plant to produce disease symptoms. Completely lacking, however, is a knowledge of where in the cell these interactions take place, just how they first perturb the host, why some plants are affected and others not, how environmental manipulations elicit host responses, and why some plants will support virus replication and others will not. Cooperative studies between viral molecular biologists, cell biologists, biochemists, and physiologists will provide answers to these problems. These, in turn, will lead to a further understanding of plants, both at the single-cell and whole plant level.

ACKNOWLEDGMENTS

We thank numerous colleagues for allowing us to use unpublished data. The work reported herein from the M. Z. laboratory was supported, in part, by grants from the National Science Foundation Biochemistry Program and from the Competitive Grants Program of the United States Department of Agriculture. The John Innes Institute is supported by a grant-in-aid from the Agriculture and Food Research Council.

Literature Cited

1. Abel, P. P., Nelson, R. S., De, B., Hoffmann, N., Rogers, S. G., Fraley, R. T., Beachy, R. N. 1986. Delay of disease development in transgenic plants that express the tobacco mosaic virus coat protein gene. *Science* 322:738–43

2. Ahlquist, P., Luckow, V., Kaesberg, P. 1981. Complete nucleotide sequence of brome mosaic virus RNA3. *J. Mol. Biol.* 153:23–38

3. Allison, R., Johnston, R. E., Dougherty, W. G. 1986. The nucleotide sequence of the coding region of tobacco etch virus genomic RNA: Evidence for the synthesis of a single polyprotein. *Virology* 154:9–20

4. Atabekov, J. G. 1977. Defective and satellite plant viruses. *Compr. Virol.* 11:143–200

5. Atabekov, J. G., Dorokhov, Y. L. 1984. Plant virus-specific transport function and resistance of plants to viruses. *Adv. Virus Res.* 29:313–64

6. Bancroft, J. B., Motoyoshi, F., Watts, J. W., Dawson, J. R. O. 1975. Cowpea chlorotic mottle and brome mosaic viruses in tobacco protoplasts. In *2nd John Innes Symposium. Modification of the Information Content of Plant Cells,* pp. 133–60. Amsterdam: North Holland

7. Barker, R. F., Jarvis, N. P., Thompson, D. V., Loesch-Fries, L. S., Hall, T. C. 1983. Complete nucleotide sequence of alfalfa mosaic virus RNA 3. *Nucleic Acids Res.* 11:2881–91

8. Baulcombe, D. C., Saunders, G. R., Bevan, M. W., Mayo, M. A., Harrison, B. D. 1986. Expression of biologically active viral satellite RNA from the nuclear genome of transformed plants. *Nature* 321:446–49

9. Bawden, F. C., Pirie, N. W. 1952. Physiology of virus diseases. *Ann. Rev. Plant Physiol.* 3:171–88

10. Bennett, C. W. 1956. Biological relation of plant viruses. *Ann. Rev. Plant Physiol.* 7:143–70

11. Bennett, C. W. 1969. Seed transmission of plant viruses. *Adv. Virus Res.* 14:221–61

12. Bergstrom, G. C., Johnson, M. C., Kuc, J. 1982. Effects of local infection of cucumber by *Colletotrichum lagenarium, Pseudomonas lachrymans* or tobacco necrosis virus on systemic resistance to cucumber mosaic virus. *Phytopathology* 72:922–26

13. Bevan, M. W., Mason, S. E., Goelet, P. 1985. Expression of tobacco mosaic virus coat protein by a cauliflower mosaic virus promoter in plants transformed by *Agrobacterium. EMBO J.* 4:1921–26

14. Boccara, M., Hamilton, W. D. O., Baulcombe, D. C. 1986. The organisation and interviral homologies of genes

at the 3' end of tobacco rattle virus RNA 1. *EMBO J.* 5:223–29

15. Brisco, M., Hull, R., Wilson, T. M. A. 1986. Swelling of isometric and of bacilliform plant virus nucleocapsids is required for virus-specific protein synthesis *in vitro. Virology* 148:210–17

15a. Bruening, G. 1981. Biochemistry of plant viruses. In *The Biochemistry of Plants: Proteins and Nucleic Acids*, ed. A. Marcus, 6:571–631. New York: Academic

16. Bruening, G., Beachy, R. N., Scalla, R., Zaitlin, M. 1976. *In vitro* and *in vivo* translation of the ribonucleic acids of a cowpea strain of tobacco mosaic virus. *Virology* 71:498–517

17. Burgess, J., Motoyoshi, F., Fleming, E. N. 1974. Structural changes accompanying infection of tobacco protoplasts with two spherical viruses. *Planta* 117:133–44

18. Cadman, C. H., Harrison, B. D. 1959. Studies on the properties of soil-borne viruses of the tobacco-rattle type occurring in Scotland. *Ann. Appl. Biol.* 47:542–56

19. Camerino, G., Savi, A., Ciferri, A. O. 1982. A chloroplast system capable of translating heterologous mRNAs. *FEBS Lett.* 150:94–98

20. Candresse, T., Mouches, C., Bové, J. M. 1986. Characterization of the virus encoded subunit of turnip yellow mosaic virus RNA replicase. *Virology* 152:322–30

21. Carr, J. P., Dixon, D. C., Klessig, D. F. 1985. Synthesis of pathogenesis-related proteins in tobacco is regulated at the level of mRNA accumulation and occurs on membrane-bound polysomes. *Proc. Natl. Acad. Sci. USA* 82:7999–8003

22. Carr, R. J., Kim, K. S. 1983. Evidence that bean golden mosaic virus invades non-phloem tissue in double infections with tobacco mosaic virus. *J. Gen. Virology* 64:2489–92

23. Carroll, T. W. 1981. Seedborne viruses: Virus-host interactions. In *Plant Diseases and Vectors: Ecology and Epidemiology*, ed. K. Maramorosch, K. F. Harris, pp. 293–317. New York/London: Academic

24. Cheo, P. C., Gerard, J. S. 1971. Differences in virus-replicating capacity among plant species inoculated with tobacco mosaic virus. *Phytopathology* 61:1010–12

25. Collmer, C. W., Vogt, V. M., Zaitlin, M. 1983. H protein, a minor protein of TMV virions, contains sequences of the viral coat protein. *Virology* 126:429–48

26. Cornelissen, B. J. C., Hooft van

Huijsduignen, R. A. M., Bol, J. F. 1986. A tobacco mosaic virus induced tobacco protein is homologous to the sweet-tasting protein thaumatin. *Nature* 321:531–32

27. Cornelissen, B. J. C., Hooft van Huijsduijnen, R. A. M., van Loon, L. C., Bol, J. F. 1986. Molecular characterization of messenger RNAs for 'pathogenesis-related' proteins 1a, 1b and 1c, induced by TMV infection of tobacco. *EMBO J.* 5:37–40

28. Cornelissen, B. J. C., Janssen, H., Zuidema, D., Bol, J. F. 1984. Complete nucleotide sequence of tobacco streak virus RNA 3. *Nucleic Acids Res.* 12:2427–37

29. Covey, S. N., Hull, R. 1985. Advances in cauliflower mosaic virus research. *Oxford Surveys Plant Mol. Cell. Biol.* 2:339–46

30. Davies, J. W., Hull, R. 1982. Genome expression of plant positive-strand RNA viruses. *J. Gen. Virol.* 61:1–14

31. Davies, J. W., Kaesberg, P. 1974. Translation of virus mRNA: Protein synthesis directed by several virus RNAs in a cell-free extract from wheat germ. *J. Gen. Virol.* 25:11–20

32. deZoeten, G. A., Gaard, G. 1984. The presence of viral antigen in the apoplast of systematically virus-infected plants. *Virus Res.* 1:713–25

33. Diener, T. O. 1963. Physiology of virus-infected plants. *Ann. Rev. Phytopathol.* 1:197–218

34. Dietzgen, R. G., Zaitlin, M. 1986. Tobacco mosaic virus coat protein and the large subunit of the host protein ribulose-1,5-bisphosphate carboxylase share a common antigenic determinant. *Virology* 155:262–66

35. Domier, L. L., Franklin, K. M., Shahabuddin, M., Hellmann, G. M., Overmeyer, J. H., Hiremath, S. T., Siaw, M. F. E., Lomonossoff, G. P., Shaw, J. G., Rhoads, R. E. 1986. The nucleotide sequence of tobacco vein mottling virus RNA. *Nucleic Acids Res.* 14:5417–30

36. Dorssers, L., van der Krol, S., van Kammen, A., Zabel, P. 1984. Purification of cowpea mosaic virus RNA replication complex: identification of a virus encoded 110,000 dalton polypeptide responsible for RNA chain elongation. *Proc. Natl. Acad. Sci. USA* 81:1951–55

37. Dorssers, L., van der Meer, J., van Kammen, A., Zabel, P. 1983. The cowpea mosaic virus RNA replication complex and the host-encoded RNA-dependent RNA polymerase-template

complex are functionally different. *Virology* 125:155–74

38. Dougherty, W. G., Hiebert, E. 1985. Genome structure and gene expression in plant RNA viruses. In *Molecular Plant Virology*, ed. J. W. Davies, 2:23–81. Boca Raton, Fla: CRC Press

39. Dumas, E., Gianinazzi, S. 1986. Pathogenesis-related (b) proteins do not play a central role in TMV localization in *Nicotiana rustica*. *Physiol. Mol. Plant Pathol.* 28:243–50

40. Durham, A. C. H., Hendry, D. A., von Wechmar, M. B. 1977. Does calcium ion binding control plant virus disassembly? *Virology* 77:524–33

41. Ecker, J. R., Davis, R. W. 1986. Inhibition of gene expression in plant cells by expression of antisense RNA. *Proc. Natl. Acad. Sci. USA* 83:5372–76

42. Esau, K. 1968. *Viruses in Plant Hosts*. Madison: Univ. Wisconsin Press

43. Fish, L. E., Jagendorf, A. T. 1982. High rates of protein synthesis in isolated chloroplasts. *Plant Physiol.* 70: 1107–14

44. Fraser, R. S. S. 1982. Are 'pathogenesis-related' proteins involved in acquired systemic resistance of tobacco plants to tobacco mosaic virus? *J. Gen. Virol.* 58:305–13

45. Fraser, R. S. S., Gerwitz, A. 1986. The genetics of resistance and virulence in plant virus disease. In *Genetics and Plant Pathogenesis*, ed. P. R. Day, G. J. Jellis, pp. 33–44. Oxford: Blackwell

45a. Fulton, R. W. 1986. Practices and precautions in the use of cross protection for plant virus disease control. *Ann. Rev. Phytopathol.* 24:67–81

46. Gibbs, A., Harrison, B. 1976. *Plant Virology: The Principles*. London: Arnold

47. Glover, J. F., Wilson, T. M. A. 1982. Efficient translation of the coat protein cistron of tobacco mosaic virus in a cell-free system from *Escherichia coli*. *Eur. J. Biochem.* 122:485–92

47a. Goldbach, R. W. 1986. Molecular evolution of plant RNA viruses. *Ann. Rev. Phytopathol.* 24:289–310

48. Goldbach, R., van Kammen, A. 1985. Structure, replication and expression of the bipartite genome of cowpea mosaic virus. See Ref. 38, pp. 83–120

49. Goodman, R. N., Kiraly, Z., Wood, K. R. 1986. *The Biochemistry and Physiology of Plant Disease*. Columbia, MO: Univ. Missouri Press

50. Goodman, R. N., Kiraly, Z., Zaitlin, M. 1964. *The Biochemistry and Physiology of Infectious Plant Disease*. Princeton, NJ: van Nostrand

51. Guilley, H., Jonard, G., Richards, K. S., Hirth, L. 1975. Sequence of a specifically encapsidated RNA fragment originating from the tobacco-mosaic-virus coat-protein cistron. *Eur. J. Biochem.* 54:135–44

52. Gunning, B. E. S., Overall, R. L. 1983. Plasmodesmata and cell-to-cell transport in plants. *Bioscience* 33:260–63

53. Habili, N., Kaper, J. M. 1981. Cucumber mosaic virus associated RNA 5. VII. Double-stranded form accumulation and disease attenuation in tobacco. *Virology* 112:250–61

54. Hall, T. C., Miller, W. A., Bujarski, J. J. 1982. Enzymes involved in the replication of plant viral RNAs. *Adv. Plant Pathol.* 1:179–211

55. Hammerschmidt, R., Kuc, J. 1982. Lignification as a mechanism for induced systemic resistance in cucumber. *Physiol. Plant Pathol.* 20:61–71

56. Harris, K. F. 1981. Arthropod and nematode vectors of plant viruses. *Ann. Rev. Phytopathol.* 19:391–426

57. Harrison, B. D., Murant, A. F. 1984. Involvement of virus-coded proteins in transmission of plant viruses by vectors. In *Vectors in Virus Biology*, ed. M. A. Mayo, K. A. Harrap, pp. 1–36. London: Academic

58. Haseloff, J., Goelet, P., Zimmern, D., Ahlquist, P., Dasgupta, R., Kaesberg, P. 1984. Striking similarities in amino acid sequence among non-structural proteins encoded by RNA viruses that have dissimilar genomic organization. *Proc. Natl. Acad. Sci. USA* 81:4358–62

59. Hirai, A., Wildman, S. G. 1969. Effect of TMV multiplication on RNA and protein synthesis in tobacco chloroplasts. *Virology* 38:73–82

60. Hohn, T., Hohn, B., Pfeiffer, P. 1985. Reverse transcription in CaMV. *Trends Biochem. Sci.* 4:205–9

61. Honda, Y., Matsui, C., Otsuki, Y., Takebe, I. 1974. Ultrastructure of tobacco mesophyll protoplasts inoculated with cucumber mosaic virus. *Phytopathology* 64:30–34

62. Hull, R. 1986. The pathogenesis of cauliflower mosaic virus. In *Genetics and Plant Pathogenesis*, ed. P. R. Day, G. J. Jellis, pp. 25–32. Oxford: Blackwell

63. Hull, R., Covey, S. N. 1985. Cauliflower mosaic virus: pathways of infection. *BioEssays* 3:160–63

64. Hull, R., Maule, A. J. 1985. Virus multiplication. In *The Plant Viruses*. Vol. 1. *Polyhedral Virions with Tripartite Genomes*, ed. R. I. B. Francki, pp. 83–115. New York: Plenum

65. Hull, R., Sadler, J., Longstaff, M. 1986. The sequence of carnation etched ring virus DNA: comparison with cauliflower mosaic virus and retroviruses. *EMBO J.* 5:3083–90

66. Ikegami, M., Fraenkel-Conrat, H. 1979. Characterization of double-stranded ribonucleic acid in tobacco leaves. *Proc. Natl. Acad. Sci. USA* 76:3637–40

67. Jenns, A. E., Kuc, J. 1979. Graft transmission of systemic resistance of cucumber to anthracnose induced by *Colletotricum lagenarium* and tobacco necrosis virus. *Phytopathology* 69:753–56

68. Joshi, S., Pleij, C. W. A., Haenni, A. L., Chapeville, F., Bosch, L. 1983. Properties of the tobacco mosaic virus intermediate length RNA-2 and its translation. *Virology* 127:100–11

69. Kado, C. I., Knight, C. A. 1966. Location of a local lesion gene in tobacco mosaic virus RNA. *Proc. Natl. Acad. Sci. USA* 55:1276–83

70. Katzouzian-Safadi, M., Berthet-Colominas, C. 1983. Evidence for the presence of a hole in the capsid of turnip yellow mosaic virus after RNA release by freezing and thawing. Decapsidation of turnip yellow mosaic virus *in vitro*. *Eur. J. Biochem.* 137:47–55

71. Keeling, J., Matthews, R. E. F. 1982. Mechanism of release of RNA from turnip yellow mosaic virus at high pH. *Virology* 119:214–18

72. Khan, Z. A., Hiriyanna, K. T., Chavez, F., Fraenkel-Conrat, H. 1986. RNA-directed RNA polymerases from healthy and from virus-infected cucumber. *Proc. Natl. Acad. Sci. USA* 83:2383–86

73. Kiberstis, P. A., Loesch-Fries, L. S., Hall, T. C. 1981. Viral protein synthesis in barley protoplasts inoculated with native and fractionated brome mosaic virus RNA. *Virology* 112:804–8

74. Kitajima, E. W., Lauritis, J. A. 1967. Plant virions in plasmodesmata. *Virology* 37:681–85

75. Kurtz-Fritsch, C., Hirth, L. 1972. Uncoating of two spherical plant viruses. *Virology* 47:385–96

76. Legrand, M., Fritig, B., Hirth, L. 1976. Enzymes of the phenyl-propanoid pathway and the necrotic reaction of hypersensitive tobacco to tobacco mosaic virus. *Phytochemistry* 15:1353–59

77. Leonard, D. A., Zaitlin, M. 1982. A temperature-sensitive strain of tobacco mosaic virus defective in cell-to-cell movement generates an altered viral-coded protein. *Virology* 117:416–24

78. Loebenstein, G., Gera, A. 1981. Inhibitor of virus replication released from tobacco mosaic virus-infected protoplasts of a local lesion-responding tobacco cultivar. *Virology* 114:132–39

78a. Loesch-Fries, L. S., Halk, E., Merlo, D., Jarvis, N., Nelson, S., Krahn, K., Burhop, L. 1986. Expression of alfalfa mosaic virus coat protein gene and antisense cDNA in transformed tobacco tissue. In *UCLA Symp. Proc. Mol. Cell. Biol.* In press

79. Lonsdale, D. M. 1986. Viral RNA in mitochondria. *Nature* 323:299

80. Magyarosy, A. C., Buchanan, B. B., Schurmann, P. 1973. Effect of a systemic virus infection on chloroplast function and structure. *Virology* 55:426–38

81. Martelli, G. P., Russo, M. 1985. Virus-host relationships: Symptomatological and ultrastructural aspects. See Ref. 64, pp. 163–205

82. Matthews, R. E. F. 1981. *Plant Virology*, pp. 12–17. New York: Academic. 2nd ed.

83. Matthews, R. E. F., Witz, J. 1985. Uncoating of turnip yellow mosaic virus RNA in vivo. *Virology* 144:318–27

84. Maule, A. J. 1985. Replication of cauliflower mosaic virus in plants and protoplasts. See Ref. 38, pp. 161–90

85. Miller, W. A., Bujarski, J. J., Dreher, T. W., Hall, T. C. 1986. Minus-strand initiation by brome mosaic virus replicase within the 3' tRNA-like structure of native and modified RNA templates. *J. Mol. Biol.* 187:537–46

86. Mohamed, N. A., Randles, J. W. 1972. Effect of tomato spotted wilt virus on ribosomes, ribonucleic acid and Fraction 1 protein in Nicotiana tabacum leaves. *Physiol. Plant Pathol.* 2:235–45

87. Mouches, C., Candresse, T., Bové, J. M. 1984. Turnip yellow mosaic virus RNA replicase contains host and virus encoded subunits. *Virology* 134:78–90

88. Niblett, C. L., Dickson, E., Fernow, K. H., Horst, R. K., Zaitlin, M. 1978. Cross protection among four viroids. *Virology* 91:198–203

89. Nishiguchi, M., Motoyoshi, F., Oshima, N. 1978. Behaviour of a temperature sensitive strain of tobacco mosaic virus in tomato leaves and protoplasts. *J. Gen. Virol.* 39:53–61

90. Ohno, T., Takamatsu, N., Meshi, T., Okada, Y., Nishiguchi, M., Kiho, Y. 1983. Single amino acid substitution in 30K protein of TMV defective in virus transport function. *Virology* 131:255–58

91. Otsuki, Y., Shimomura, T., Takebe, I. 1972. Tobacco mosaic virus multiplication and expression of the N gene in necrotic responding tobacco varieties. *Virology* 50:45–50

92. Palukaitis, P., Zaitlin, M. 1984. A model to explain the "cross protection" phenomenon shown by plant viruses and viroids. In *Plant Microbe Interactions: Molecular and Genetic Aspects*, ed. T. Kosuge, E. W. Nester, 1:420–29. New York/London: Macmillan

93. Palukaitis, P., Zaitlin, M. 1986. Tobacco mosaic virus: Infectivity and replication. In *Plant Viruses*, ed. M. H. V. van Regenmortel, H. Fraenkel-Conrat, 2:105–31. New York: Plenum

94. Pelham, H. R. B. 1978. Leaky UAG termination codon in tobacco mosaic virus RNA. *Nature* 272:469–71

94a. Ponz, F., Bruening, G. 1986. Mechanisms of resistance to plant viruses. *Ann. Rev. Phytopathol.* 24:355–83

95. Price, C. A., Reardon, E. M. 1982. Isolation of chloroplasts for protein synthesis from spinach and *Euglena gracilis* by centrifugation in silica sols. In *Methods in Chloroplast Molecular Biology*, ed. M. Edelman, R. B. Hallick, N.-H. Chua, pp. 189–209. Amsterdam/New York/Oxford: Elsevier

96. Reinero, A., Beachy, R. N. 1986. Association of TMV coat protein with chloroplast membranes in virus-infected leaves. *Plant Mol. Biol.* 6:291–301

97. Rezelman, G., Franssen, H. J., Goldbach, R. W., Ie, T. S., van Kammen, A. 1982. Limits to the independence of bottom component RNA of cowpea mosaic virus. *J. Gen. Virol.* 60:335–42

98. Rochon, D., Siegel, A. 1984. Chloroplast DNA transcripts are encapsidated by tobacco mosaic virus coat protein. *Proc. Natl. Acad. Sci. USA* 81:1719–23

99. Ross, A. F. 1961. Localized acquired resistance to plant virus infection in hypersensitive hosts. *Virology* 14:329–39

100. Ross, A. F. 1961. Systemic acquired resistance induced by localized virus infection in plants. *Virology* 14:340–58

100a. Saito, T., Watanabe, Y., Meishi, T., Okada, Y. 1986. Preparation of antibodies that react with the large nonstructural proteins of tobacco mosaic virus by using *Escherichia coli* expressed fragments. *Mol. Gen. Genet.* 205:82–89

101. Sela, I., Kaesberg, P. 1969. Cell-free synthesis of tobacco mosaic virus coat protein and its combination with ribonucleic acid to yield tobacco mosaic virus. *J. Virol.* 3:89–91

102. Shalla, T. A., Petersen, L. J. 1973. Infection of isolated plant protoplasts with potato virus X. *Phytopathology* 63:1125–30

103. Shalla, T. A., Petersen, L. J., Guichedi,

L. 1975. Partial characterization of virus-like particles in chloroplasts of plants infected with the U5 strain of TMV. *Virology* 66:94–105

104. Shalla, T. A., Petersen, L. J., Zaitlin, M. 1982. Restrictive movement of a temperature-sensitive virus in tobacco leaves is associated with a reduction in numbers of plasmodesmata. *J. Gen. Virol.* 60:355–58

105. Shaw, J. G., Plaskitt, K. A., Wilson, T. M. A. 1986. Evidence that tobacco mosaic virus particles disassemble cotranslationally *in vivo*. *Virology* 148:326–36

106. Shepherd, R. J. 1972. Transmission of viruses through seed and pollen. In *Principles and Techniques in Plant Virology*, ed. C. I. Kado, H. O. Agrawal, pp. 267–92. New York: van Nostrand

107. Sherwood, J. L., Fulton, R. W. 1982. The specific involvement of coat protein in tobacco mosaic virus cross protection. *Virology* 119:150–58

108. Siegel, A., Hari, V., Kolacz, K. 1978. The effect of tobacco mosaic virus infection on host and virus-specific protein synthesis in protoplasts. *Virology* 85:494–503

109. Somssich, I. E., Schmelzer, E., Bollmann, J., Hahlbrock, K. 1986. Rapid activation by fungal elicitor of genes encoding "pathogenesis-related" proteins in cultured parsley cells. *Proc. Natl. Acad. Sci. USA* 83:2427–30

110. Sulzinski, M. A., Zaitlin, M. 1982. Tobacco mosaic virus replication in resistant and susceptible plants: In some resistant species virus is confined to a small number of initially infected cells. *Virology* 121:12–19

110a. Sulzinski, M. A., Gabard, K. A., Palukaitis, P., Zaitlin, M. 1985. Replication of tobacco mosaic virus VIII. Characterization of a third subgenomic TMV RNA. *Virology* 145:132–40

111. Sziráki, I., Balázs, E., Király, Z. 1980. Role of different stresses in inducing systemic acquired resistance to TMV and increasing cytokinin level in tobacco. *Physiol. Plant Pathol.* 16:277–84

112. Taliansky, M. E., Malyshenko, S. I., Pshennikova, E. S., Atabekov, J. G. 1982. Plant virus-specific transport function. II. A factor controlling virus host range. *Virology* 122:327–31

113. Tomlinson, J. A., Faithfull, E. M., Ward, C. M. 1976. Chemical suppression of the symptoms of two virus diseases. *Ann. Appl. Biol.* 84:31–41

114. van Loon, L. C. 1985. Pathogenesis-related proteins. *Plant Mol. Biol.* 4:111–16

115. van Tol, R. G. L., van Vloten-Doting, L. 1981. Lack of serological relationship between the 35K non-structural protein of alfalfa mosaic virus and the corresponding proteins of three other plant viruses with a tripartite genome. *Virology* 109:444–47

116. Volloch, V. 1986. Cytoplasmic synthesis of globin RNA in differentiated murine erythroleukemia cells: Possible involvement of RNA-dependent RNA polymerase. *Proc. Natl. Acad. Sci. USA* 83:1208–12

117. Wakarchuk, D. A., Hamilton, R. I. 1985. Cellular double-stranded RNA in *Phaseolus vulgaris. Plant Mol. Biol.* 5:55–63

118. Walker-Simmons, M., Holländer-Czytko, H., Andersen, J. K., Ryan, C. A. 1984. Wound signals in plants: A systemic plant wound signal alters plasma membrane integrity. *Proc. Natl. Acad. Sci. USA* 81:3737–41

119. Watanabe, Y., Emori, Y., Ooshika, I., Meshi, T., Ohno, T., Okada, Y. 1984. Synthesis of TMV-specific RNAs and proteins at the early stage of infection in tobacco protoplasts: Transient expression of the 30K protein and its mRNA. *Virology* 133:18–24

120. Watanabe, Y., Ooshika, I., Meshi, T., Okada, Y. 1986. Subcellular localization of the 30k protein in TMV-inoculated tobacco protoplasts. *Virology* 152:414–20

121. Waterworth, H. E., Kaper, J. M., Tousignant, M. E. 1979. CARNA 5, the small cucumber mosaic virus-dependent replicating RNA, regulates disease expression. *Science* 204:845–47

122. Watts, J. W., Dawson, J. R. O., King, J. M. 1981. The mechanism of entry of viruses into plant protoplasts. *Ciba Found. Symp.* 80:56–71

123. Wilson, T. M. A. 1984. Cotranslational disassembly of tobacco mosaic virus in vitro. *Virology* 137:255–65

124. Wilson, T. M. A. 1985. Nucleocapsid disassembly and early gene expression by positive-strand RNA viruses. *J. Gen. Virol.* 66:1201–7

125. Wilson, T. M. A., Shaw, J. G. 1986. Cotranslational disassembly of filamentous plant virus nucleocapsids in vitro and in vivo. In *Positive Strand RNA Viruses. UCLA Symp. Mol. Cell. Biol.*, (NS), Vol. 54, ed. M. A. Brinton, R. Rueckert. New York: Liss. In press

126. Wilson, T. M. A., Watkins, P. A. C. 1986. Influence of exogenous viral coat protein on the cotranslational disassembly of tobacco mosaic virus (TMV) particles in vitro. *Virology* 149:132–35

127. Young, N. D., Zaitlin, M. 1986. An analysis of tobacco mosaic virus replicative structures synthesized in vitro. *Plant Mol. Biol.* 6:455–65

128. Zaitlin, M. 1979. How viruses and viroids induce disease. In *Plant Disease,* ed. J. G. Horsfall, E. B. Cowling, 4:257–71. New York: Academic

129. Zaitlin, M., Jagendorf, A. T. 1960. Photosynthetic phosphorylation and Hill reaction activities of chloroplasts isolated from plants infected with tobacco mosaic virus. *Virology* 12:477–86

130. Zimmern, D., Hunter, T. 1983. Point mutation in the 30K open reading frame of TMV implicated in temperature-sensitive assembly and local lesion spreading of mutant Ni2519. *EMBO J.* 2:1893–1900

PLANTS IN SPACE

Thora W. Halstead

Life Sciences Division, National Aeronautics and Space Administration, Washington, DC 20546

F. Ronald Dutcher

Science Communication Studies, The George Washington University, Washington, DC 20037

CONTENTS

INTRODUCTION 318
DEVELOPMENT, GROWTH, AND REPRODUCTION 319
   Germination and Growth 319
   Cell Division 320
   Chromosomal Effects 322
   Cell Size and Shape 326
   Differentiation 326
   Reproduction 328
CELLULAR AND SUBCELLULAR CHANGES 329
   Plastids 329
   Nuclei 331
   Endoplasmic Reticulum and Ribosomes 331
   Mitochondria 332
   Dictyosomes 332
   Cell Walls 333
   Essential Elements 333
PLANT ORIENTATION AND MOVEMENT 334
   Orientation 334
   Tropisms 335
   G-Threshold Values 335
   Nutation 336
   Epinasty 337
CONCLUSIONS 337

0066-4294/87/0601-0317$02.00

INTRODUCTION[1]

Space may be, as some have called it, our last frontier. As such, it provides novel, even unique research opportunities. Plants are sure to figure significantly in these activities. The ability to manipulate the force of gravity from near zero to 1 g affords fresh opportunities to investigate gravity's physiological effects as well as a means of probing gravi- and phototropism, thigmomorphogenesis, and other environmental effects in a state uncompromised by gravity.

The resource of space as an experimental tool has barely been tapped. Its promise, however, and recent progress in increased basic ground-based plant research (especially that involving calcium) have brought a new wave of interest in gravity's role in regulating plant growth and development. Current research is providing new insight into light modulation of gravitropism. Interrelationships between the cellular mechanisms controlling photomorphogenesis and gravitropism are under intensive investigation. A new hypothesis, yet to be explored, proposes that plants respond to all mechanical forces, including gravity, by means of similar cellular mechanisms (46). These developments have been summarized in excellent recent reviews on the role of calcium in mediating cellular functions (59, 118), and of calcium, hormones, protons (110, 111), and light (117) in the regulation of gravitropism. Gravity-sensing systems (93, 151) and the biophysical control of plant growth (27) have also been authoritatively reviewed. In an earlier paper we described space-flight vehicles, the range of apparatus used to grow plants in space, and the broad profile of space-flight environmental conditions, and we summarized the results then available of space experiments on plants (58).

A great deal can be learned about gravitropism in ground-based experiments, but some key questions can only be answered in a weightless environment. Even the few relatively unsophisticated experiments conducted to date have produced surprises, and more will undoubtedly arise as longer space experiments are coupled with more advanced techniques. It is becoming evident that the physiological effects of gravity, though subtle, are substantial and embrace many other cellular events in addition to those solely associated with gravitropism. Such physiological effects can only be identified in a weightless environment where the force of gravity can be removed or manipulated between zero and 1 g through centrifugation.

We have tried to make this review comprehensive, but it has been sharply limited by a lack of detail in a substantial portion of the literature. For details

[1]*Abbreviations used:* STS-2, Space Transportation System (Space Shuttle), 2nd mission; STS-3, Space Shuttle, 3rd mission; STS 61-C, Space Shuttle, 24th mission; SL-2, Space Shuttle Spacelab laboratory-2, 19th mission; D-1, Space Shuttle, 1st German mission; Cosmos, Soviet unmanned satellite; Salyut, Soviet manned space station.

we have therefore relied heavily on investigations flown on US spacecraft, even as we have attempted to fill in gaps encountered in the Soviet literature with such data as could be gleaned from Soviet news releases.

In this review we aim primarily to consider phenomenology, a goal that befits the state of our knowledge from space experiments. We intend to provide grist for future ground-based and space experiments and to reveal the potential for scientific discovery in this area.

DEVELOPMENT, GROWTH, AND REPRODUCTION

Germination and Growth

Recent experiments (17, 30, 83, 88, 131, 132, 149) have confirmed previous observations that space flight does not affect germination (55, 80, 84, 90, 116); but effects on growth are less clear, for while growth measured by gross observation frequently seemed to be unaffected by space flight, cytological studies would argue to the contrary. Cowles et al (28, 30) generally found decreased stem and root growth of pine (*Pinus elliotti* Engelm.), oat (*Avena sativa* L. cv 'Garry'), and mung bean [*Vigna radiata* (L.) Wilczek] seedlings grown on Shuttle missions STS-3 and SL-2, compared with ground controls. Fresh and dry weights of the SL-2 seedlings were lower also, with mung bean showing the greatest decreases in length and weight. However, several experiments showed that hypocotyls of lettuce (*Lactuca sativa* L.), *Arabidopsis thaliana* (L.) Heynh., and garden cress (*Lepidium sativum* L.) grown in microgravity were longer than those grown on a flight centrifuge (86–88). At the same time, flight-grown hypocotyls and roots, whether centrifuged or not, were considerably shorter than the corresponding ground controls. From these results, the investigators inferred that not weightlessness but some other factor(s) associated with space flight had inhibited growth. By contrast, cucumber seedlings (*Cucumis sativus* L.) grown in a stationary mode in space had shorter roots and hypocotyls than flight-centrifuged seedlings, and lettuce seedlings grown on Cosmos 1667 had longer hypocotyls than ground controls (83, 130). The space-flown lettuce hypocotyls were thicker and longer, owing to an increase in the area occupied by cortical parenchyma cells (83).

Amazingly, the root cap, the gravity-sensing region in roots, apparently does not regenerate in microgravity. Decapped roots of maize (*Zea mays* L.) seedlings taken into space on STS 61-C did not regenerate their root caps during the flight, even though their roots continued to elongate. While on Earth, roots similarly treated completely regenerated their caps within 4.8 days (96). The same result was obtained from a student-initiated experiment, also with maize seedlings, flown on the Shuttle (6). Such experiments suggest that while a continuous gravitational stimulus is not required for root growth, gravity does play an essential and still undefined role in root cap formation.

Moreover, gravity may indirectly affect the gross morphology of the root system (see Differentiation section, below).

Development beyond the seedling stage and reproduction in microgravity are more problematical. Such experiments have been carried out largely by Soviet investigators, who had access to long space-flight missions. Hundreds of growth tests have been conducted over the years (58, 105, 128).

Many species have grown well in weightlessness for limited (and generally unspecified) periods, including peas, wheat, onions, and various other vegetables (136a,b; 137b,e). An orchid, *Epidendrum radicans,* grew for nearly 6 months in space and produced new leaves and roots, although the new leaf and shoot growth was less than that of ground controls (25).

Many plants grew normally for short periods but then exhibited a slowing or cessation of growth and frequently died. Peas grown on Salyut 4 followed this scenario, though their death was attributed to problems in maintaining suitable water supply and root aeration (43, 122). Death at the flowering stage was most common; peas, wheat, and a number of other plants died at this stage. Orchids brought onto the Salyut 6 space station immediately lost their flowers and would not flower in space, though they could subsequently be brought to flower on Earth. Tulips were taken into space ready to flower, but their buds would not open (54, 136c, 137c,d,f).

While methods for sustaining plant growth have gradually improved, news reports indicate that difficulties are still being encountered in growing plants in space. On Salyut 7, in 1985, cotton seedlings died soon after they had begun to grow (137g).

Cell Division

Reduced or inhibited cell division has been observed in a number of plants grown in space—e.g. in roots of wheat [*Triticum vulgare* (Vill.) Host. var. 'Georgia 1123'] seedlings grown on Biosatellite II and pea (*Pisum sativum* L.) and *Crepis capillaris* (L.) Wallr. seedlings grown on Soviet flights (43, 55, 84, 133). Roots of sunflower (*Helianthus annuus* L. cv. 'Teddy Bear') seedlings grown on Shuttle missions STS-2 and STS-3 and oat seedlings grown on STS-3 and SL-2 possessed approximately one tenth as many dividing cells as ground controls, while mung bean seedlings grown on STS-3 exhibited a smaller decrease (77; A. D. Krikorian, personal communication). Merkys found a lowered mitotic index in root meristems of lettuce seedlings grown in weightlessness, compared with those grown on a 1-g flight centrifuge. The mitotic index for both flight samples was lower than that in corresponding ground controls (85). In other instances no evidence of any inhibition of mitotic activity has been reported (83, 129).

Little is known about why mitotic activity might decline. Krikorian, who reported reduced cell-division activity in the roots of space-grown sunflower, oat, and mung bean seedlings, found few cells in anaphase and telophase in roots not exposed to colchicine treatment. He suggested that the usual range of cells at varying stages of cell division was not present prior to recovery back on the ground. This was taken as evidence that the system was running downhill and that cell division would have been still further reduced had the duration of the space flight been extended. The bulk of observed dividing cells were arrested in metaphase even with low levels of colchicine, a result which suggests that few cells were in other stages of the mitotic cycle when the first division cycle resumed upon return to Earth. The presence of cells at different stages would have reflected the sort of asynchronous cell-division activity typical of meristems on Earth (77; A. D. Krikorian, personal communication).

While these results do not confirm a gravitational effect on any particular stage of mitosis, the results presented by Merkys et al (85) for lettuce suggest that in the in-flight weightlessness condition where cell-division activity was most reduced a considerably higher proportion of dividing cells were in telophase, compared with controls. This suggests a block during the last stages of mitosis.

Other cytological abnormalities indicating disturbances in the mitotic process during space flight have been observed in some but by no means all experiments (83, 129). Multiple nuclei have been encountered in *Crepis* and pea seedlings grown on Soviet flights (84, 139), and aneuploidy (2n-1=33) was seen in root cells of two sunflower seedlings grown on STS-2 (77). Events symptomatic of mitotic disturbances were consistently observed in microspores of *Tradescantia paludosa* inflorescences carried into space (34, 35, 39, 41). These events included abnormal migration of the nucleus, chromosomes bunching instead of lying in the equatorial plate at metaphase, an alteration in the direction of the spindle axis, chromosome nondisjunction and lagging of individual unseparated chromatids, and the presence of 3- and 4-pole mitoses. Incorrectly separated chromatids can produce aneuploids like those observed in sunflower. Nondisjunction has also been reported in space-flown *Drosophila* (44).

Disturbances in the mitotic spindle mechanism were also observed in *Tradescantia* clone (designated 02) inflorescences flown on Biosatellite II (126). Embryo sacs from flight samples exhibited a higher frequency of fused, clumped, or scattered nuclei, presumably because of spindle malfunctioning. Flight-grown root tip cells also exhibited more abnormalities reflective of spindle disturbances, including multinucleate cells and abnormally shaped nuclei. Space-flown microspores exhibited similar disturbances. Ground simulation tests produced no significant increase in such effects as a consequence of vibrational stresses or of the space vehicle atmosphere, nor was any interaction with γ-radiation detected. Weightlessness was judged (by both US and Soviet investigators) to be the most likely cause of the mitotic disturbances (34, 126).

Chromosomal Effects

Chromosomes have been damaged in numerous plants carried on space flights. In many cases, increases in the frequency of chromosomal aberrations were observed in seeds carried into space and later germinated on Earth (13, 32, 37, 38, 40, 45, 47–49, 52, 65, 66, 74, 98, 101–103, 112, 115, 139–143, 145, 146). These aberrations were generally scored in root meristem cells and consisted of chromosome fragments and recombinations. Increased levels of chromosomal aberration were found in *Tradescantia* microspores as well (34, 35, 39, 41).

In an extensive series of flight experiments, small increases in the frequency of chromosomal aberration in *Crepis capillaris* seeds and decreases in germination and subsequent fertility and viability in *Arabidopsis thaliana* seeds were observed (7–12, 74, 75, 138–142, 144–146, 148). Similar effects were observed in the reproductive organs and seeds of *Arabidopsis* plants grown through most or all of their life cycles in space (see the Reproduction section, below, for details). The decreases in germination, survival, and fertility found in *Arabidopsis* were attributed to the occurrence of chromosomal aberrations in meristematic embryonic cells. Because these aberrations were similar to those observed directly in *Crepis* seeds, a common mechanism of cell damage is suspected.

Not unexpectedly, growing or actively metabolizing systems appear more sensitive to space flight than dormant systems like seeds. *Crepis* seeds stored on spacecraft for 3–234 days and then germinated and fixed in space consistently exhibited more aberrations than seeds similarly exposed to flight but germinated on the ground (74). Significant chromosomal abnormalities, including fragmented chromosomes and chromosomal bridges, were found in roots of sunflower seedlings grown on STS-2 and roots of oat seedlings grown on STS-3 and SL-2 (77; A. D. Krikorian, personal communication). Figure 1 shows several types of damage observed in oats.

Not all in-flight growth experiments have disclosed an increase in chromosomal aberrations, however. Seedlings of pine (*Pinus sylvestris* L.), *Crepis*, and lettuce exhibited no such increase (83, 116, 133). Similarly, it was not possible to show chromosomal aberrations in roots of mung bean germinated in space, but this may have been a technical problem caused by their tiny size (77).

The reason for the increases in chromosomal disturbance is unknown. Cosmic radiation is one possibility, since ionizing radiation is known to cause such disturbances; but this factor is unlikely to be responsible in short flights, during which radiation exposures would have been quite low (105). Further-

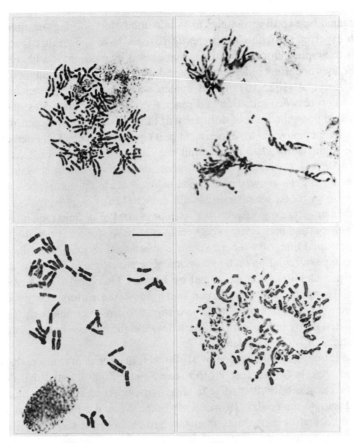

Figure 1   Chromosomes from root tip cells of oats (2n=42) grown on the ground and in space on SL-2. The top left panel shows a metaphase preparation typical of ground-grown control materials; other panels show a range of abnormal division figures encountered in "space root" cells undergoing their first division on Earth after recovery. The top right panel shows a metaphase with longitudinal breaks and fractures in chromatids; the bottom left, a metaphase with extreme fragmentation; the bottom right, a telophase with fragmentation and a bridge. Scale bar = 10 μm. Photographs courtesy of A. D. Krikorian and S. A. O'Connor, SUNY at Stony Brook.

more, all but one of several Soviet experiments comparing the known γ-radiation sensitivities of the seeds of various plants to the observed aberration frequencies failed to show any meaningful correlation (37, 40, 66, 123). The use of radioprotectors like cysteine, aminoethylisothiourea, and 5-methoxytryptamine failed to reduce the level of flight-induced aberrations in barley seeds, a finding that also speaks against radiation as the cause of chromosomal damage (100, 103). These data seem to eliminate radiation as the direct and primary cause of the aberrations encountered.

Possible interactions between radiation and space flight have been investigated by exposing samples to γ-radiation before, during, or after flight. Such experiments with *Crepis capillaris* seeds have shown that space flight tends to amplify chromosomal damage caused by preflight irradiation; flight either does not affect or lessens damage due to postflight irradiation (10, 140). Nuzhdin et al (102, 103) found evidence in barley seeds of an increased interaction between radiation and space flight in certain instances; the effect depended upon the state of dormancy of the seeds and the radiosensitivity of the variety employed. An experiment best equipped to detect an interaction of space flight with radiation but limited by its short duration of 2 days—the Biosatellite II *Tradescantia* experiment, in which γ-radiation was applied during the flight—showed no increase in levels of chromosomal aberrations in space-flown roots, whether irradiated or not (126).

Other investigators suspect that one component of cosmic radiation—HZE (high charge and high energy) particles—may have an effect on biological and chromosomal integrity. Indeed, there is substantial concern about the effects of this type of radiation on human nonregenerative tissue (135). The effect of these particles has been examined by layering biological materials such as seeds, bacterial spores, protozoan cysts, and brine shrimp and insect eggs between particle-track detectors. Investigators can determine which objects have been hit by individual HZE particles, and the hits can be correlated with the extent of biological damage.

The European group headed by H. Bücker, in a series of tests conducted on Apollo, Apollo-Soyuz Test Project, and Shuttle flights, found that dormant seeds were generally insensitive to HZE particles (compared with some of the tested animal materials). However, in some cases the frequency of certain developmental abnormalities in plants grown from hit seeds did increase. Examples include the increased frequencies of *A. thaliana* plantlets with multiple stems, and cotyledonary abnormalities in tobacco (*Nicotiana tabacum* L.). One maize plant grown from a seed with two hits contained large yellow stripes in its leaves. Also, fewer hit tobacco seeds germinated (14, 18–22, 109). Seedlings grown from seeds hit by HZE particles on Soviet Cosmos flights 610 and 690 showed considerably decreased viability (3, 4).

A joint series of Soviet-French tests attempted to determine whether HZE particle hits increased chromosomal aberration frequencies. These tests, which employed lettuce seeds, produced inconsistent results: Either (*a*) similar aberration increases in seeds hit and not hit in flight (50, 57, 97) or (*b*) a larger aberration increase in seeds hit in flight (81, 99) was observed, compared with ground controls. More consistent was the finding of more multiple chromosomal aberrations (more than one aberration per cell) in seeds hit by HZE particles in flight (50, 81, 97, 99). Seeds kept longest in space (308 days) exhibited very high levels of aberration, both single and multiple; but for this flight exposure, seeds hit during space flight exhibited only slightly more aberrations than those not hit (81). The frequency of aberration was highest when the root meristem or hypocotyl was hit (81, 97, 99), and it has been suggested that some experiments showed no difference between hit and nonhit seeds because they involved seeds in which the root meristem or embryo was not directly struck (56, 57). When tobacco seeds were scored for germination and frequency of morphological abnormalities in these Soviet-French experiments the results were also inconsistent (15, 51, 56).

Attempts have been made to correlate the extent of chromosomal aberration with the duration of flight. Such a correlation would support a role for cosmic radiation and/or weightlessness as a cause of this damage, since their effects presumably accumulate over time, in contrast to the one- or two-time vibrational stresses associated with spacecraft launch and landing. Extensive series of tests with *Crepis, Arabidopsis,* and various other seeds have failed to show a correlation. An exception occurred during the longest exposure to flight, 827 days, where large increases in chromosomal aberrations in *Crepis* seeds (13.83% versus 3.63% for controls) and germination in *Arabidopsis* seeds (nearly zero in flown seeds compared with 73% in controls) were encountered. I. D. Anikeeva and coworkers (9, 141) felt that the extensive damage observed in the 827-day seeds could not have been due solely or primarily to radiation; rather, they believed that weightlessness plays a role, perhaps by accelerating the aging process of seeds through an effect on cellular DNA-repair mechanisms. A correlation was observed between flight duration and the degree to which flight increased the damage due to preflight irradiation (9, 74, 139, 142). It has also been suggested that weightlessness or space flight may affect chromosomal structure not directly but by increasing the likelihood of damage by altering sensitivity to other mutagenic factors (130).

Use of an onboard centrifuge during Cosmos 782 failed to shed any light on a role for weightlessness, since results with *Crepis* and lettuce seeds were inconsistent (98, 144). Comparison of types of chromosomal aberrations observed (chromosomal, chromatid) with known effects of radiation and vibration failed to support a role for either factor as the cause of the flight-induced aberrations (147). The factors of vibration and acceleration associated with spacecraft launch and landing cannot be ignored, however, since ground-based tests have shown that these factors can cause increased chromosomal abnormalities (143); and an analysis of the timing of occurrence pattern of chromosomal aberrations observed in *Tradescantia* microspores led Delone et al (34) to conclude that some aberrations were caused by vibrational stresses. The abnormality increases observed in seeds carried on long flights, however, argue against these vibrational factors as the primary cause.

In interpreting these and all of the other data, one would prefer more thorough descriptions of the physiological state and storage conditions of both test and control seeds, since varying environmental conditions such as temperature, humidity, and aeration would certainly affect the results. Inadequacies in the methods employed have been mentioned (105, 128, 130).

The observed chromosomal effects show that either exposure to weightlessness, factors in the space environment like radiation, or other space-flight factors can cause chromosomal damage. No airtight case can yet be made that weightlessness is primary among these causal factors. What should be emphasized is that the response(s) have been observed repeatedly, at least in some cells of some plants grown in space. The reason(s) for these disturbances must be determined, and as wide a range of species as possible must be tested.

Cell Size and Shape

Cells larger, smaller, and the same size as ground controls have been found in space. Larger cells were found in root or other cells of wheat, pea, maize, and *Crepis* (55, 84, 90, 122, 133). Cotyledonary cells of pine seedlings and columella cells of maize seedlings showed no difference in size. Root, shoot, and leaf cells of flight-grown orchids were smaller (25, 94, 116). In the wheat and pea seedlings, a depression in the rate of mitosis and an increase in cell elongation apparently combined to produce roots of normal length (55, 84). Smaller cells were found in flight-grown chloronemata and rhizoids of the moss *Funaria hygrometrica*. Insufficient carbohydrate formation may have disturbed cell elongation; partial destruction of chloroplasts and reduced starch were cited as support for this hypothesis (69; also see the sections below on Plastids, Cell Walls, and Dictyosomes).

Cells rounder than controls were reported in cotyledonary cells of pine and root meristem cells of maize and *Crepis* seedlings grown in space. This rounding was attributed to the greater role for surface tension under microgravity and decreased mutual mechanical pressures of cells on each other under gravity-free conditions (116, 133). A. D. Krikorian (personal communication) observed rounded cells in shoot apices of flight-grown maize seedlings. *Funaria hygrometrica* samples grown in space contained increased numbers of pear- or dumbbell-shaped cells, in contrast to the normal cylindrical shape of cells seen on Earth (69).

Differentiation

There is some evidence that space flight may alter patterns of differentiation or the rate at which differentiation proceeds. This was observed most recently in an experiment conducted on Shuttle mission D-1, in which aseptically cultured, undifferentiated cell clusters of anise (*Pimpinella anisum* L.), induced to undergo somatic embryogenesis during flight, differentiated more rapidly in space. In space some 90% of the cell clusters showed polarity and leaf or root primordia after 4 days, compared with 6 days for ground controls. Leaves of plantlets generated in flight formed chlorophyll earlier, even though the cultures in flight were exposed to only half as much light as ground controls (134) (a comparison at equal light levels would have been more meaningful). More rapid development in space was also suggested by Butenko et al (23), since she and her coworkers reported longer roots in carrot somatic embryos *(Daucus carota* L.) and plantlets formed from totipotent cells and cell clusters grown aseptically in agar in petri dishes during Cosmos 782 (78). The increase in root length was attributed to a larger zone of differentiated cells, since meristematic and elongation zones showed no change. They concluded that weightlessness caused earlier and more rapid development of embryonic organs. Accelerated development of the male gametophyte in grape hyacinth *(Muscari racemosum)* bulbs carried into space has also been reported (70).

Altered patterns of root growth and differentiation were encountered in several experiments. The length of the elongation zone of lettuce roots grown in space was found to be highly reduced, compared with roots grown on a flight centrifuge and ground controls. Those cells that did not elongate normally differentiated prematurely and sprouted root hairs early and closer to the root tip. It was concluded that weightlessness was capable of speeding both organ formation and development (85). Analogous results included a significant increase in the number of seedlings that formed lateral roots in weightlessness, compared with flight-centrifuged and ground control seedlings (88). Similar results were reported with *Crepis capillaris* seedlings grown on Cosmos 936, in which the distribution of root hairs was shifted towards the root tip (107), and oat seedlings grown on SL-2, in which increased numbers of lateral roots were found. This last result reflected damage to, or inhibition of growth of the primary root tip, due perhaps to the severely reduced cell division observed in root tips of oats grown in flight (see the Cell Division section, above). A consequential hormonal imbalance and the increased lateral root branching that would naturally follow this damage are well-known phenomena (A. D. Krikorian, personal communication). Pea seedling internodes grown in space were noticeably elongated, compared with controls (90).

Moore (95) found evidence that in root caps of maize, microgravity or other space factors alter the allocations of cellular volume to various organelles (see the section on Cellular and Subcellular Changes, below). Tairbekov et al (132) found virtually no differences, however, in cytological characteristics of root cap and meristem cells of flight-grown maize seedlings.

The question of whether the space environment affects the timing of developmental stages and aging has been raised before. For example, the aging process appeared to be accelerated in *Drosophila* flown in space (91). Some of the experiments just mentioned (e.g. totipotent cell cultures of anise and carrot and bulbs of *Muscari*) suggest that rapid aging may occur in plants as well. Analogous effects were observed in samples grown on the ground after an exposure to space flight, including earlier root growth of onion bulbs, earlier flowering of gladiolus bulbs, and a "rejuvenation" of lost germination capacity in old seeds (33, 36, 38, 137a). Retarded development has been observed as well. Orchid plants brought into space in flower lost their flowers immediately and ceased growing. In space, normal vegetative growth of orchid stems, roots, and leaves eventually resumed, but flowering never occurred, even after 6 months (54, 137f). *Arabidopsis* plants were slower to develop, requiring 28 days to form a four-leaf rosette and 44 days to initiate flowering, compared with corresponding 13- and 36-day periods for ground controls (82).

Reproduction

The most extensive tests examining growth and development in space were a series of experiments with *Arabidopsis thaliana*. Until 1982, no plant in space, including *Arabidopsis*, had ever passed successfully from the vegetative to the reproductive stage, flowered, and set seed, despite a number of attempts (54, 86). In an earlier experiment, *Arabidopsis* seedlings carried into space at the cotyledonary stage succeeded in growing and flowering, although seeds were not produced in flight and "elements" of the androecium and gynoecium of these plants were degenerated and sterile. The gross morphology of the reproductive organs was not affected, and sterility was attributed to processes occurring at the cellular level, such as vacuolization and lysis of inner integument cells (68, 72, 73). However, a later report of this same experiment attributed the infertility to insufficient illumination (74). When these plants were returned to Earth they completed flowering, but analysis of fruits developed from axillary floral buds revealed a reduction in plant fertility (more sterile siliques, unfertilized ovules, etc), a marked increase in the frequency of recessive mutants, and reduced germination of seeds collected from these plants (74, 75).

Finally, in 1982, *Arabidopsis* plants grown from seeds sown in space were brought through a complete life cycle and produced fertile seeds (82, 87). Increased illumination (7000 lux) and improved ventilation conditions were said to contribute to the success. The plants grew slower in flight, were shorter, and produced fewer leaves when in the rosette stage and smaller siliques and seeds after flowering. Of seven flight-grown plants, two produced a total of 11 siliques with aborted ovules and five plants produced 22 siliques containing about 200 seeds total, while eight ground control plants produced 34 siliques. The F_1 seeds obtained in flight, when germinated on Earth, showed reduced germination (72.4% of the experimentals vs 87.3% of the controls), an increased frequency of embryonic lethals (42.0% among the experimentals vs 22.5% among the controls), and seedlings with shorter hypocotyls and cotyledonary leaves, compared with controls. Fewer of the flight-derived seeds were biologically "complete"—i.e. capable of developing into fertile plants (42.0% of the flight-derived seeds vs 67.6% of the controls). [In an earlier experiment, similar effects—increased levels of sterile ovules and embryonic lethals and reduced levels of fertile seeds—were produced by an *Arabidopsis* plant that had already begun to flower when it was carried into space on Cosmos 1129 (106).] F_2 seeds when collected, germinated, and grown to maturity on Earth showed no differences in developmental morphology or in any of the other observed parameters.

Merkys et al (82, 87) felt that the effects they observed were not unusual for plants grown in altered environmental conditions, especially since conditions of cultivation in space are still being improved. The important point, they felt, was that the experiment demonstrated that plants were capable of performing a full cycle of development under space-flight conditions, undergoing fertilization, embryogenesis, and seed maturation, and of producing fully viable seeds. This experiment should be repeated and also tried with species other than *Arabidopsis*. A great deal of long-duration plant physiological investigation in the space environment rests on our ability to rear plants from "seed to seed" over multiple generations.

CELLULAR AND SUBCELLULAR CHANGES

Space-flight experiments have demonstrated that weightlessness significantly affects the morphology and cytology of a number of plants, ranging from the unicellular alga *Chlorella* to the angiosperms. But there have been few studies of the physiological changes that occur, and fewer still in which biochemical and cytological examinations have been coordinated, much less conducted on the same specimen. Correlation of cellular, structural, and biochemical changes is consequently still highly speculative.

Plastids

One consistent observation has been the random distribution of amyloplasts in root-cap central statenchyma cells of all plants grown under microgravity— e.g. cucumber, pea, *A. thaliana* (130), lettuce (85, 86), maize (96, 132), cress (149), and lentil (*Lens culinaris* L.) (108). Clinostating of plants also results in randomly distributed amyloplasts, while in both Earth-grown and space-flown control plants centrifuged at 1 *g*, the putative statoliths are located in the distal end of the cells (85, 86, 108, 130, 149).

Merkys (89) attempted to correlate root curvature with the intracellular spatial distribution of amyloplasts in lettuce grown in space at 0, 0.01, 0.1, and 1 *g* to test for a direct relationship between amyloplast location and the perception of gravity. He found no direct relationship, for at 1 and 0.1 *g* amyloplasts were distributed within 50.6 and 66.7% of the cell volume, respectively, although root curvature was identical. Conversely, at both 0.01 and 0 g, amyloplasts were distributed within three fourths of each cell, but root curvature dropped from 50° to 10°, respectively.

Gravity affects more than just the distribution of amyloplasts within the columella cells. Sytnik et al (130) reported that in 7-day-old space-grown pea seedlings, the stroma of the amyloplasts became clearer, the peripheral plastid reticulum was reduced to varying degrees, and the volume of starch grains was reduced. By 18 days, some plastids contained centrally located single large starch granules, while others contained a few small grains of starch or none at all.

Within the stroma of flight-grown *Arabidopsis* plastids, circular membrane formations were seen and the starch grains were smaller than normal (130). The columella cells of 4.8-day-old maize seedlings grown on the Space Shuttle contained a significantly smaller starch volume than Earth-grown controls. While space- and Earth-grown seedlings contained comparable numbers of starch grains per plastid and amyloplasts per cell, both the plastids and starch grains were smaller in the flight-grown plants. In fact, there was 61% more amyloplast starch in the ground control plants (94). Volkmann et al observed a reduction in the starch content of space-grown cress amyloplasts also (149). Consistent evidence thus compels one to conclude that starch in root cell amyloplasts is depleted under weightless conditions. The volume—possibly even the structure and integrity—of the plastid membrane may be altered in microgravity.

Information about leaf and stem cell plastids is minimal. Leaves of pea plants grown for 29 days under weightless conditions aboard the Salyut 7 spacecraft were analyzed cytologically and biochemically. The ultrastructure of the palisade parenchyma of the space-grown pea leaves revealed significant morphological changes in the structural organization of the chloroplasts not seen in the ground control plants. Space-grown plants showed disintegration or destruction of the grana, disorientation or separation of the intergrana, shrinkage of the membrane comprising the grana stacks, and formation of electron-transparent vesicles. Specimens from the experimental group also either lacked starch reserves or contained very few small grains. Higher chlorophyll a and b levels indicated to the investigators that chlorophyll synthesis was stimulated in flight, and a shift occurring in the low-temperature fluorescence peak was apparently due to an increase in the percentage of long-wave aggregated forms (2, 5). On the other hand, data from 24-day-old space-grown leaves and stems of pea indicated that the space-grown plants contained significantly less chlorophyll a and b than the ground control plants (79). A later report on pea and the orchid *Epidendrum radicans* grown for 29, 42, and 110 days, respectively, on Salyut 7 indicated maximum cellular changes in the chloroplasts where grana morphology was modified and reserve carbohydrate and Feulgen-positive zones were decreased (1). Chloroplasts of *Chlorella vulgaris* and *C. pyrenoidosa* and the protonemal cells of *Funaria hygrometrica* cultivated on long flights (e.g. 96 days) had starch grains of reduced size and volume, and partial swelling of the thylakoids in the grana and stroma (130).

Nuclei

As already mentioned, plants grown in space under near-weightlessness have shown decreases in mitotic level as well as chromosomal aberrations. Researchers have also found that the nuclei of root cells of the central statenchyma of pea and *Arabidopsis* grown in space contain an abnormal distribution and increased volume of condensed chromatin (71, 130). Platonova et al (116) noted a reduction in the number of nucleoli in space-grown pine cells. Nuclear volume was reported to be increased in wheat seedlings (55) but unaffected in maize root cap cells (95). However, the location of the nucleus within the central statenchyma of cress (149), lentil (108), and lettuce (85) cells was unaffected by weightlessness, implying that the location of the nucleus in such cells may be genetically programmed. The increased volume of condensed chromatin might indicate reduced functional activity, which in turn would affect protein synthesis.

Studies of the effect of weightlessness on proteins in plants have been limited in number and scope, so it would be presumptuous to do more than suggest that gravity may affect protein metabolism. Cherevchenko et al (26) examined the effects of long-term weightlessness on the electrophoretic mobility of soluble and structural proteins and the activity of the enzyme D-ribulose-1,5-bisphosphate carboxylase in leaves of five orchid species grown in space for 6 months. Measurements made after the flight showed no consistency in increases or decreases of specific proteins or carboxylase activity, and all changes were reversible upon return to Earth. One species, *Doritis pulcherrima*, after return to Earth, exhibited intensive growth and flowering and a doubled carboxylase activity. The investigators considered these latter results indicative of an effect of weightlessness on orchid metabolism.

Endoplasmic Reticulum and Ribosomes

The endoplasmic reticulum (ER) is of special interest not only because of its role in protein synthesis but also because of its close proximity to putative statolith amyloplasts in root cap cells. Sytnik and coworkers found that the electron-dense content of ER cisternae was reduced in the columella cells of pea, cucumber, and *Arabidopsis* seedlings. Other researchers found both a reduction and rounding of the cisternae and the disappearance of rough ER and ribosomes in *Funaria* protonema (130). The amount of ER in cress grown under microgravity conditions appeared to be increased and had a less parallel and more convoluted loose arrangement, similar to that seen in clinostat experiments or after inversion of the plant (149). Moore et al (96) noted that maize seedlings germinated on Earth but grown in space retained the normal peripherally located ER distribution pattern seen in ground control columella cells, but seedlings germinated in space contained spherical and ellipsoidal masses of ER clustered near the cell periphery. These results suggest that not only the integrity of the ER but also its arrangement and distribution within the cell are affected by gravity.

Mitochondria

Swollen mitochondria in root cells of space-grown maize seedlings were reported by Tairbekov and coworkers, who also noted that the mitochondria were characterized by fewer and less-ordered internal membranes (133). Weightlessness effects on pea and *Arabidopsis* root cap cells were characterized by swollen mitochondria with an electron-dense matrix and well-developed, regularly arranged cristae (71). Similar swollen mitochondria were seen in root cells of oat and mung bean seedlings flown on the Space Shuttle, but the researchers interpreted the changes as fixation artifacts (124). Mitochondrial volume within the calyptrogen cells of maize seedling roots was found by Moore (95) to be 31% less than that in ground control cells.

Observations of mitochondrial changes in cells of plants exposed to weightlessness led to the development of a space flight "biocalorimeter" to measure energy expenditure in plants at an early stage of development (131). Maize seedlings were flown in this device aboard a Cosmos biosatellite, but the 7-day experiment was compromised since germination and possibly the first 2 days of the experiment were spent on the ground before launch. Even so, results were interpreted to indicate that seedling metabolism was unaffected by weightlessness.

Dictyosomes

Dictyosomes of pea root meristem cells under weightlessness displayed changes in the morphology of the cisternae, especially in the peripheral cisterna of the secretory pole (129). Moore examined the effect of microgravity on the relative volume of dictyosomes in relation to the types of cells representing cellular differentiation in the maize root cap. The calyptrogen, columella, and peripheral cells of the cap all contained 50–90% less dictyosomal volume than the cells of Earth-grown seedlings. This correlated well with the observed reduction of mucilage secreted by the flight-grown plants (95). A similar observation of decreased mucilage secretion around root caps of space-grown maize seedlings was noted by Tairbekov et al (132). Since polysaccharides are supplied to the cell wall by dictyosomes, it is tempting to draw a correlation between the reduction in dictyosome activity and the thinning of cell walls.

Cell Walls

A question that naturally arises when considering the effects of gravity on plants is the role it plays in the synthesis of the structural polymers cellulose and lignin. It was reported that under weightless conditions cell walls of higher plants were thinner, and that in *Funaria hygrometrica* walls were one third the thickness of walls of control cells. The organization of the cellulose microfibrils was also said to be affected (130). Pea plants grown for 24 days on the Salyut 4 space station yielded 54% less cellulose than comparable ground control plants (79). Cell wall synthesis was also examined in pine, oat, and mung bean seedlings grown aboard the Space Shuttle during two separate flights. Seedlings analyzed biochemically revealed significant reduction (18%) in lignin content in mung beans in the first mission and significant decreases (7–25%) in all three species in the second. Measurements of activities of the enzymes involved in the lignin biosynthetic pathway generally showed decreases, although not in every instance. Cellulose and protein contents were somewhat lower also (3–12% and 6–21%, respectively) in the flight seedlings (28–30). Collectively, these findings support the hypothesis that cell walls, and in particular cellulose and lignin content, are reduced in microgravity.

Essential Elements

The growing recognition of the importance of calcium in the gravitropic perception-transduction-response mechanism has led to at least two space-flight experiments. Pea plants grown for 24 days aboard a Salyut space station and analyzed on the ground shortly after flight indicated a disruption of mineral balance in the shoots. This was manifested by a significant increase in phosphorus and potassium content, accompanied by a sharp reduction in calcium, magnesium, manganese, zinc, and iron. The results were interpreted as a disruption in the control by root cells of the intake of ions (79).

Cytochemical studies were conducted to determine the activity and localization of membrane-associated adenosine triphosphatases (ATPases) in cells of 5-day-old pea seedling roots grown on a clinostat to simulate hypogravity. The investigators concluded that hypogravity inhibits the activation of ATPase by Ca^{2+}, Mg^{2+}, and K^+ ions, caused by a significant decrease in the proteins of the plasmalemma. They hypothesized that since these changes would not affect the action of the H^+ pump, which is activated by Mg^{2+} and K^+ ions, hypogravity must affect the Ca^{2+} ion transport system. They further postulated that the reduced ATPase activity in the presence of Ca^{2+} ions was due to a reduction in the number of Ca^{2+}-dependent ATPase protein molecules. They also observed that while Ca^{2+} ions were located in the plasmalemma and plasmodesmata of the 1-g control cells, in clinostated plants Ca^{2+} disappeared from plasmalemma and appeared in the ER, dictyosomes, plastids, and nuclear membrane. Therefore, the reduced Ca^{2+}-ATPase activity and protein content of the plasmalemma disrupts both passive and active transport of Ca^{2+} ions. This in turn may effect a change in the concentration of Ca^{2+} ions within the cell and the activation of membrane phospholipases (104).

A follow-up experiment was conducted on pea seedlings aboard Salyut 6 by some of the same scientists. Details of the experiment are too few to ascertain the procedures used. The interesting results and conclusions concern an exciting area, but they must be viewed with caution. The investigators found an increase in the concentration of membrane-bound calcium and a change in the localization of Ca^{2+}-ATPase. They postulated a rise in the concentration of cytoplasmic calcium under hypogravic conditions and a change in the state of calcium from free to bound (67).

Unfortunately, in the preceding papers (67, 79, 104) the cytochemical methods and assays were nonspecific, the purity of membrane preparations untested, and untoward clinostat effects uncontrolled. These ground and flight experiments should be repeated, however, because of the importance of the questions they address.

PLANT ORIENTATION AND MOVEMENT

Plants respond to directions from both endogenous and exogenous cues. Their movements are generated by either irreversible differential growth or temporary and reversible changes in turgor pressure. The movements of nutation, epinasty, and gravitropism have all been examined under the near-weightless conditions of space flight to ascertain the role of gravity in their control.

Orientation

The orientation of roots and shoots appears to be directed totally by external stimuli. When external stimuli are absent, the shape and position of the embryo within the seed and the orientation of the seed with respect to the substrate determine the direction of root and shoot growth. Platonova and coworkers demonstrated this with a variety of seeds germinated in space (113, 114, 116). Pine (*Pinus sylvestris* L.) and wheat (*Triticum compactum*) seedlings maintained a linear shape with the root and shoot growing as straight extensions of the radicle and coleoptile; tomato (*Lycopersicon esculentum* L.) seedlings maintained their characteristic embryonic horseshoe shape; and pea seedlings assumed the arc or comma-like shape characteristic of pea embryos. Volkmann et al (149) made a similar observation of cress roots that grew in a straight line from the tip of the embryonic radicle. Tairbekov et al (133) reported similar findings with *Crepis capillaris* and *Zea mays*. This same root and shoot orientation behavior was observed in seedlings of wheat, pea, and cress when seedlings were clinostated on Earth (113, 149).

Tropisms

Space-flight experiments suggest that gravity directs the orientation of root growth but shoot orientation is guided by both gravity and light. Platonova et al (113) found that in the absence of a gravistimulus during space flight, light but not nutrients in the substrate influenced the orientation of wheat and pea shoots, while roots remained disoriented. Growth of the wheat shoots toward light was more pronounced than that of the pea shoots. Cabbage, flax, *Crepis*, maize (60, 122), pea (92), oat, and mung bean (30) shoots of seedlings germinated and grown in space were oriented by light to varying degrees, depending upon the species. All these flight results suggest that in space, light alone may provide an adequate directional stimulus to monocotyledonous (wheat, maize, oat) shoots but only limited directional guidance to the shoots of dicotyledons. A student experiment with rice seedlings germinated and grown on Skylab 4, however, showed no phototropic effect under a variety of light levels (127). Yet in a ground test of the response of clinostated oat coleoptiles to unilateral gravity or light stimulation, Shen-Miller & Gordon (120) noted a greater degree of tropic curvature in shoots exposed to light than in those exposed to 1 *g*. Space experiments can provide the means to probe the still-unknown mechanisms of phototropism and the causes of the differences in phototropic responses of shoots of different species.

Ground-based research by many investigators indicates that the root cap is necessary for root gravitropism and that the gravity-detection mechanism in plants is located within the root cap (62, 64, 150). Recently a startling observation was made that could modify this view. With the absence of a gravistimulus in space flight, root caps were affected dramatically. In two separate experiments, maize seedling roots (*Zea mays* 'Pioneer' and cv. 'Bear Hybrid') (6, 96), whose caps had been removed prior to 7 and 4.8 days of space flight, respectively, failed to regenerate new caps in space, while ground controls fully regenerated caps well within the time frame of the experiments. Moore suggests that the capless root tip must also sense gravity since regeneration of the root cap is an indicator of graviresponsiveness.

G-Threshold Values

In order to measure tropic growth, plants have been displaced with regard to the gravity vector, by either simple horizontal placement, the use of clinostats, or a unique combination of clinostat and centrifuge. Threshold values for *presentation time* (the minimum stimulation time required for a gravitational force to elicit a detectable curvature response) and for *graviperception* (the magnitude of acceleration required to induce a growth response) have been calculated for a variety of species on the ground (see 53, 121, 125).

Merkys et al (83, 87) have found a striking correlation between the threshold values for graviperception in lettuce seedling hypocotyls and roots measured in space and those of oat (*Avena sativa* cv. 'Victory') coleoptiles and roots measured on the ground (121). A series of space experiments aboard Salyut 7 were conducted in a "biogravistat" centrifuge. In one set of experiments, lettuce seeds were germinated and grown for 84 hr under constant exposure to 0, 0.01, 0.1 or 1 g. Calculations of root and shoot curvature in response to the levels of acceleration indicated a threshold value for gravitropic response of hypocotyls of 2.9×10^{-3} g and for roots 1.5×10^{-4} g (83, 87). Earlier clinostat experiments conducted by Shen-Miller et al estimated the minimally effective acceleration threshold for oat coleoptiles and roots to be respectively 1.4×10^{-3} g and 1.3×10^{-4} g. Measurements were made after clinostating for 65–70 hr, and the product of acceleration and time was therefore about 340 $g \cdot \sec$ for shoots and 30 $g \cdot \sec$ for roots, values which the authors felt were only rough approximations (121). Using continuous centrifugal stimulation without gravity compensation, Czapek (31) observed the beginning of a graviresponse between 5×10^{-4} and 10^{-3} g in *Lupinus, Vicia, Pisum,* and *Zea* roots and *Helianthus* hypocotyls.

Volkmann et al (149) attempted to determine the presentation-time threshold of cress roots in an experiment carried out on the Space Shuttle. A dose of 15–30 $g \cdot \sec$ was applied to the seedlings by accelerating them for 30 sec with a hand centrifuge. Threshold values of 12 $g \cdot \sec$ had been estimated for cress roots in ground clinostat experiments, but the acceleration applied in space was insufficient to initiate the perception-transduction-response process. Experimental difficulties encountered in space beyond the investigators' control cast doubts upon the accuracy of these results, however, and the test-stimuli range was small, allowing little room for error.

Nutation

On Earth, the elongation of plant roots and shoots occurs even as rotational differential growth along the long axis of the growing plant organ takes place. This cyclic differential growth leads to oscillations at a frequency of about one cycle every 1–2 hr. These nutational movements were present in microgravity in both sunflower shoot (16) and cress root tips (149), as shown by analysis of timed sequential photographs taken during flight. The Darwinian view that one or more internal oscillators might regulate the cyclic growth pattern of circumnutation was strengthened, although no endogenous oscillator has been identified.

Epinasty

Only one experiment has been flown in space to test nastic responses in reduced gravity. Pepper plants (*Capsicum annuum* L.) were flown and photographed on Biosatellite II to test the gravity dependence of the classic epinastic response of specimens exposed to horizontal clinostat rotation. Both flight-grown and Earth-grown clinostated plants exhibited marked epinasty, but the amplitude and kinetics of the response and the recovery from epinasty upon return to 1 *g* were detectably different. Subsequent ground tests suggested that some of these differences may have been provoked by vibration (63).

CONCLUSIONS

Ground-based research on the physiological effects of gravity is quite properly focused on the mechanism of gravitropism because little is known about the other effects of gravity. By and large, gravity's effects on growth and development can only be revealed by eliminating its influence, as in the weightless environment of space. In this review we have documented and assessed the more significant observations made on plants grown in space and tried to identify the effects of weightlessness.

Data collected from space-grown plants is sparse; it is less than ideal from a statistical viewpoint, and has frequently had to be interpreted without reference to adequate controls. At this point in the development of experimental designs suited to studying plants in space, nearly all experiments have suffered technical difficulties to some degree. Few would draw unequivocal conclusions from similar data collected on Earth. Therefore, the interpretations and conclusions reported here are provisional, meant primarily to stimulate new research by raising exciting questions.

Dormant seeds and germination do not appear to be gravity sensitive; but as seedlings develop, both mitosis and cytokinesis appear to be affected by the lack of a directional *g* force. Reports of reduced or inhibited cell division, lowered mitotic index, a disproportionate ratio of cells in the various stages of mitosis, abnormal division figures such as aneuploidy, chromatid breaks and fragmentation, and bridges all indicate the mitotic process is disturbed. Even in developing microspores, disturbances have been encountered, including multiple pole mitoses, bunched chromosomes not aligned along the cell plate at metaphase, and chromosome nondisjunction. These abnormalities and others are generally attributed to spindle disturbances.

There is growing evidence that Ca^{2+} plays a major role in the regulation of karyokinesis and cytokinesis, and low Ca^{2+} concentrations result in reduced division rates. It has been proposed that the targets may be the cell plate and/or mitotic apparatus (59). Since Ca^{2+} transport and redistribution can be induced by gravity (117), it is tempting to speculate that the observed mitotic abnormalities may have been brought about by a reduction or redistribution of endogenous Ca^{2+} as a result of weightlessness. The effects of gravity on mitosis and cytokinesis, independent of other untoward effects, could also account for noted decreases in survival, flowering, and fertility.

Chromosomal damage reported in nonmetabolizing systems like dormant seeds, especially with extended space exposure, appears to be due not to gravity but to radiation or other environmental stresses; but even this is not certain.

Accumulated evidence suggests space flight also alters differentiation or the rate at which it proceeds. In most relevant instances, differentiation was accelerated. The most plausible explanation for increased lateral root formation in flight-grown oat seedlings attributes precocious root branching and development to hormonal imbalances precipitated by reduced cell division in the primary root tip (A. D. Krikorian, personal communication).

We still do not fully understand the role of hormones in gravitropism, although hormones, especially auxin, have been recognized for many years as an essential effector in the mechanism. We have no space-flight data to support either the suspected effect of gravity on hormones or the role of hormones in mediating plant physiological changes due to gravity; but the observations reviewed here, together with results from ground-based research, strongly suggest a relationship between gravity and hormones that extends beyond gravitropism.

Histological evidence supports the idea that gravity affects polysaccharide metabolism. Starch has been consistently depleted in amyloplasts of roots grown under weightless conditions and reserve carbohydrates have been depleted in chloroplasts. Dictyosomes, which provide polysaccharides to cell walls, have decreased in volume in root cap cells to 10–50% of that in Earth-grown cells, accompanied by a significant decrease in mucilage secretion. Cell walls have been thinner in space, while levels of cellulose, lignin, and the enzymes of the lignin biosynthetic pathway have all been reduced.

Organelle membranes, too, appear to be affected. Reports indicate reduced peripheral plastid reticulum, smaller plastids, and circular membrane formations in root cells. Disintegration or destruction of grana, disorientation or separation of the thylakoids of the stroma, and shrinkage of the membranes of the grana stacks of leaf chloroplasts are seen in space-grown seedlings. ER distribution pattern, arrangement, form, and electron-dense content changes have also been observed, as well as mitochondrial swelling, and chromatin aggregation and multiple nucleoli in nuclei.

All these changes, if they are gravity related, suggest an extensive influence of gravity on all aspects of plant growth and development, above and beyond the expected tropisms, including regulation of protein and mineral metabolism, Ca^{2+}-binding proteins, Ca^{2+}-mediated processes, phosphorylation, and even gene expression. An additional attractive feature of experimentation in space is that one can study the physiological effects of light and mechanical stimuli on plants without the complications of gravity. How these stimuli interact with gravity to effect physiological changes, and if and how cellular mechanisms controlling the responses are interrelated, similar, or the same, are exciting questions. With the prospects of manipulating gravity in space with variable-g centrifuges and conducting longer space-flight experiments, we can anticipate asking more penetrating questions on these and other basic aspects of plant growth and development. The next several years should prove to be particularly enlightening.

Literature Cited

1. Abilov, Z. K. 1986. Adaptative, physiological and morphological changes in chloroplasts of plants, different periods of time cultivated at "Salyut-7" station. In *Plenary Meeting COSPAR, Abstr., 26th, Toulouse, 1986*, p. 301. Paris: COSPAR (Abstr.)
2. Abilov, Z. K., Alekperov, U. K., Mashinskiy, A. L., Fadeyeva, S. I., Aliyev, A. A. 1985. The morphological and functional state of the photosynthetic system of plant cells grown for varying periods under space flight conditions. See Ref. 76, pp. 29–32
3. Abramov, B., Abramov, S. 1977. *Znaniye-Sila*, No. 10, pp. 19–22 (In Russian)
4. Abramova, V. M., Asaturyan, V. I., Benevolenskii, V. N., Vasil'eva, N. G. 1977. The "Biostack-2" experiment. Report 3. Investigation of biological effects of heavy galactic cosmic radiation on *Arabidopsis thaliana* seeds. *Radiobiology* 17(6):80–84
5. Aliyev, A. A., Abilov, Z. K., Mashinskiy, A. L., Ganiyeva, R. A., Ragimova, G. K., et al. 1985. Ultrastructural and some physiological features of photosynthetic apparatus of garden pea cultivated for 29 days in "Salyut-7" space station. *Izv. Akad. Nauk Az. SSR Ser. Biol. Nauk*, No. 6, pp. 18–23 (In Russian)
6. Amberg, S. 1986. Statoliths in corn seed root caps: Final report of Shuttle Student Involvement Project Experiment SE82-3-001. 19 pp.
7. Anikeeva, I. D. 1984. Effect of space flight factors on the air-dry seeds of *Arabidopsis thaliana* (L.) Heynh.

See Ref. 42, pp. 60–63
8. Anikeeva, I. D., Kostina, L. N., Vaulina, E. N. 1978. The modifying influence of physical factors of space flight on radiation effect of additional gamma-irradiation of air-dry seeds of *Arabidopsis thaliana* (L.) Heynh. See Ref. 152, pp. 44–52
9. Anikeeva, I. D., Kostina, L. N., Vaulina, E. N. 1983. Experiments with air-dried seeds of *Arabidopsis thaliana* (L.) Heynh. and *Crepis capillaris* (L.) Wallr., aboard Salyut 6. *Adv. Space Res.* 3(8):129–33
10. Anikeeva, I. D., Kostina, L. N., Vaulina, E. N. 1984. Effect of space flight factors on the radiation effects from prior and subsequent gamma-irradiation of air-dry seeds. See Ref. 42, pp. 102–8
11. Anikeeva, I. D., Vaulina, E. N. 1979. Modification of the radiation effect in *Arabidopsis thaliana* by factors of space flight. See Ref. 61, pp. 184–90
12. Anikeeva, I. D., Vaulina, E. N., Kostina, L. N. 1979. The action of space flight factors on the radiation effects of additional γ-irradiation of seeds. *Life Sci. Space Res.* 17:133–37
13. Antipov, V. V., Delone, N. L., Nikitin, M. D., Parfenov, G. P., Saksenov, P. P. 1968. *Radiobiological experiments on Cosmos-110 biosatellite*. Presented at 19th Int. Astronaut. Fed. Congr. 38 pp. (In Russian)
14. Barbier, M., Dulieu, H. C. 1978. Biological study of tobacco seeds flown in the joint Apollo-Soyuz Test-Project. *Life Sci. Space Res.* 16:143–46
15. Bayonove, J., Burg, M., Delpoux, M., Mir, A. 1984. Biological changes ob-

served on rice and biological and genetic changes observed on tobacco after space flight in the orbital station Salyut-7 (Biobloc III experiment). *Adv. Space Res.* 4(10):97–101

16. Brown, A. H., Chapman, D. K. 1984. Circumnutation observed without a significant gravitational force in spaceflight. *Science* 225:230–32

17. Brown, A. H., Chapman, D. K. 1984. Experiments on plants grown in space: A test to verify the biocompatibility of a method for plant culture in a microgravity environment. *Ann. Bot.* 54 (Suppl. 3):19–31

18. Bücker, H. 1974. The Biostack experiments I and II aboard Apollo 16 and 17. *Life Sci. Space Res.* 12:43–50

19. Bücker, H. 1975. Biostack—a study of the biological effects of HZE galactic cosmic radiation. In *Biomedical Results of Apollo*, ed. R. S. Johnston, L. F. Dietlein, C. A. Berry, pp. 343–54. NASA SP-368. Washington, DC: NASA

20. Bücker, H., Delpoux, M., Fogel, S., Freeling, M., Graul, E. H., et al. 1977. Biostack III: Experiment MA-107. In *Apollo-Soyuz Test Project: Summary Science Report*, 1:211–26. NASA SP-412. Washington, DC: NASA

21. Bücker, H., Horneck, G. 1975. The biological effectiveness of HZE-particles of cosmic radiation studied in the Apollo 16 and 17 Biostack experiments. *Acta Astron.* 2:247–64

22. Bücker, H., Horneck, G., Facius, R., Reitz, G., Schafer, M., et al. 1984. Radiobiological advanced Biostack experiment. *Science* 225:222–24

23. Butenko, R. G., Dmitrieva, N. N., Ongko, V., Basyrova, L. V. 1979. The effect of weightlessness on somatic embryogenesis. See Ref. 61, pp. 118–25

24. Calvin, M., Gazenko, O. G., eds. 1975. *Foundations of Space Biology and Medicine*, Vol. II, Book 2. Washington, DC: NASA

25. Cherevchenko, T. M., Mayko, T. K. 1983. Some results from studies on the effects of weightlessness on the growth of epiphytic orchids. *Visn. Akad. Nauk Ukr. RSR*, No. 1, pp. 31–35 (In Ukrainian)

26. Cherevchenko, T. M., Shmigovs'kaya, V. V., Kosakovs'kaya, I. V., Chernyad'yev, I. I. 1984. Effects of prolonged weightlessness on Orchidaceae proteins. *Dopov. Akad. Nauk Ukr. RSR, Ser. B*, No. 5, pp. 75–77 (In Ukrainian)

27. Cosgrove, D. 1986. Biophysical control of plant cell growth. *Ann. Rev. Plant Physiol.* 37:377–405

28. Cowles, J., Jahns, G., LeMay, R., Omran, R. 1986. Growth characteristics of plants grown in microgravity on Spacelab 2. *Plant Physiol.* 80(Suppl. 4):9 (Abstr.)

29. Cowles, J., LeMay, R., Omran, R., Jahns, G. 1986. Cell wall related synthesis in plant seedlings grown in the microgravity environment of the space shuttle. *Plant Physiol.* 80(Suppl. 4):9 (Abstr.)

30. Cowles, J. R., Scheld, H. W., LeMay, R., Peterson, C. 1984. Experiments on plants grown in space: Growth and lignification in seedlings exposed to eight days of microgravity. *Ann. Bot.* 54(Suppl. 30):33–48

31. Czapek, F. 1895. Untersuchungen über Geotropismus. *Jahrb. Wiss. Bot.* 27:243–339 (In German) [cited in (121)]

32. Delone, N. L., Antipov, V. V. 1967. Further investigation of the effect of space flight conditions on the chromosomes of radicles in the seeds of some higher plants. *Cosmic Res.* 5(2):274

33. Delone, N. L., Antipov, V. V., Davydov, B. I. 1982. The effect of space flight factors on quiescent nuclei of certain plant and animal test objects. *Kosm. Issled.* 20(3):489–92 (In Russian)

34. Delone, N. L., Bykovskii, V. F., Antipov, V. V., Parfenov, G. P., Vysotskii, V. G., et al. 1964. Effect of space-flight factors on *Tradescantia paludosa* microspores on board the satellite spaceships Vostok 5 and 6. *Cosmic Res.* 2(2):268–77

35. Delone, N. L., Egorov, B. B., Antipov, V. V. 1966. Effect of the factors of cosmic flight of the satellite ship Voskhod on the microspore *Tradescantia paludosa*. *Cosmic Res.* 4(1):139–43

36. Delone, N. L., Morozova, E. M., Antipov, V. V. 1971. A cytological study of some higher plants whose seeds and bulbs were on board 5, 6, and 7 probes. *Life Sci. Space Res.* 9:111

37. Delone, N. L., Morozova, E. M., Antipov, V. V. 1971. Effect of conditions of space flight on station "Zond-5" on seeds, onions, and *Tradescantia* plants. *Cosmic Res.* 9(1):146–48

38. Delone, N. L., Morozova, E. M., Antipov, V. V., Parfenov, G. P., Trusova, A. S. 1967. Stimulation of growth of the onion *Allium cera* after space flight of the bulbs on satellite-ship "Kosmos-110". *Cosmic Res.* 5(6):794–97

39. Delone, N. L., Popovich, P. R., Antipov, V. V., Vysotskii, V. G. 1963. Effect of space-flight factors on *Tradescantia paludosa* microspores on board

the satellite space-ships "Vostok-3" and "Vostok-4". *Cosmic Res.* 1(2):257–68

40. Delone, N. L., Rudneva, N. A., Antipov, V. V. 1965. Effect of space-flight conditions in satellite spaceships "Vostok-5" and "Vostok-6" on chromosomes of radicles in seeds of some higher plants. *Cosmic Res.* 3(3):371–78

41. Delone, N. L., Trusova, A. S., Morozova, E. M., Antipov, V. V., Parfenov, G. P. 1968. The effect of space flight on Cosmos-110 on the microspores of *Tradescantia paludosa*. *Cosmic Res.* 6(2):250–53

42. Dubinin, N. P., ed. 1984. *Biologicheskiye Issledovaniya na Orbital'nykh Stantsiyakh "Salyut"*. Moscow: Nauka. 248 pp. (In Russian)

43. Dubinin, N. P., Glembotsky, Ya. L., Vaulina, E. N., Merkys, A. J., Laurinavichius, R. S., et al. 1977. Biological experiments on the orbital station Salyut 4. *Life Sci. Space Res.* 15:267–72

44. Dubinin, N. P., Kanavets, O. L. 1962. Factors of cosmic flight and primary non-disjunction of chromosomes. In *Problemy Kosmicheskoy Biologii*, Vol. 1, ed. N. M. Sisakyan. Moscow: USSR Acad. Sci. (In Russian)

45. Dubinina, L. G., Chernikova, O. P. 1970. Effect of space-flight factors on *Crepis capillaris* seeds. *Cosmic Res.* 8(1):146–48

46. Edwards, K. L., Pickard, B. G. 1986. Detection and transduction of physical stimuli in plants. In *Cell Surface and Signal Transduction,* ed. H. Greppin, B. Millet, E. Wagner. Berlin/Heidelberg/NY: Plenum. In press

47. Farber, Yu. V., Nevzgodina, L. V., Pap'yan, N. M., Soboleva, T. N. 1971. Effect of flight factors on dormant lettuce seeds. *Space Biol. Med.* 5(6):36–45

48. Garina, K. P., Romanova, N. I. 1970. Effect of space flight factors on barley seeds. *Cosmic Res.* 8(1):149–51

49. Garina, K. P., Romanova, N. I. 1971. Influence of space-flight factors and ethylenimine on barley seeds. *Cosmic Res.* 9(6):873–76

50. Gaubin, Y., Kovalev, E. E., Planel, H., Nevzgodina, L. V., Gasset, G., et al. 1979. Development capacity of *Artemia* cysts and lettuce seeds flown in Cosmos 936 and directly exposed to cosmic rays. *Aviat. Space Environ. Med.* 50(2):134–38

51. Gaubin, Y., Planel, H., Gasset, G., Pianezzi, B., Clegg, J., et al. 1983. Results on *Artemia* cysts, lettuce and tobacco seeds in the Biobloc 4 experiment flown aboard the Soviet biosatellite Cosmos 1129. *Adv. Space Res.* 3(8):135–40

52. Gordon, L. K., Delone, N. L., Antipov, V. V., Vysotskii, V. G. 1963. Effect of space-flight conditions in satellite spaceship Vostok-3 on the seeds of higher plants. *Cosmic Res.* 1(1):149–52

53. Gordon, S. A., Shen-Miller, J. 1971. Simulated weightlessness studies by compensation. In *Gravity and the Organism,* ed. S. A. Gordon, M. J. Cohen, pp. 415–26. Chicago: Univ. Chicago

54. Gorkin, Yu., Mashinskiy, A., Yazdovskiy, V. 1980. Birthday flowers: "Salyut-6", our commentary. *Pravda,* Oct. 27, p. 3 (In Russian)

55. Gray, S. W., Edwards, B. F. 1971. Experiment P-1020: The effect of weightlessness on the growth and orientation of roots and shoots of monocotyledonous seedlings. See Ref. 119, pp. 123–65

56. Grigoriev, Yu. G., Planel, H., Delpoux, M., Gaubin-Blanquet, Y., Nevzgodina, L. V., et al. 1978. Radiobiological investigations in Cosmos 782 space flight (Biobloc SF1 experiment). *Life Sci. Space Res.* 16:137–42

57. Grigoriev, Yu. G., Nevzgodina, L. V. 1978. The effect of space radiation heavy ions on lettuce seeds exposed on satellite "Cosmos-782" (Experiment "Bioblock"). See Ref. 152, pp. 52–58

58. Halstead, T. W., Dutcher, F. R. 1984. Experiments on plants grown in space: Status and prospects. *Ann. Bot.* 54: (Suppl. 3) 3–18

59. Hepler, P. K., Wayne, R. O. 1985. Calcium and plant development. *Ann. Rev. Plant Physiol.* 36:397–439

60. Il'in, Ye. A. 1973. *Osnovnyye Itogi i Perspektivy Sovetskikh Biologicheskikh Eksperimentov v Kosmose.* Moscow: USSR Acad. Sci. 90 pp. (In Russian)

61. Il'in, Ye. A., Parfenov, G. P., eds. 1979. *Biologicheskiye Issledovaniya na Biosputnikakh "Kosmos".* Moscow: Nauka. 239 pp. (In Russian)

62. Jackson, M. B., Barlow, P. W. 1981. Root geotropism and the role of growth regulators from the cap: A re-examination. *Plant Cell Environ.* 4:107–23

63. Johnson, S. P., Tibbitts, T. W. 1971. Experiment P-1017: The liminal angle of a plagiogeotropic organ under weightlessness. See Ref. 119, pp. 223–48

64. Juniper, B. E., Groves, S., Landau-Schachar, B., Audus, L. J. 1966. Root cap and the perception of gravity. *Nature* 209:93–94

65. Khvostova, V. V., Gostimskii, S. A., Mozhaeva, V. S., Nevzgodina, L. V. 1963. A further study of the effect of space-flight conditions on chromosomes in the radicles of pea and wheat seeds. *Cosmic Res.* 1(1):153–57

66. Khvostova, V. V., Prokof'yeva-Bel'govskaya, A. A., Sidorov, B. N., Sokolov, N. N. 1962. The effect of space flight conditions on the seeds of higher plants and actinomycetes. In *Problemy Kosmicheskoy Biologii*, ed. N. M. Sisakyan, V. I. Yazdovskiy, 2:153–62. Moscow: USSR Acad. Sci. (In Russian)

67. Kordyum, E. L., Belyavskaya, N. A., Nedukha, E. M., Palladina, T. A., Tarasenko, V. A. 1984. The role of calcium ions in cytological effects of hypogravity. *Adv. Space Res.* 4(12):23–26

68. Kordyum, E. L., Chernyaeva, I. I. 1982. Peculiarities in formation of *Arabidopsis thaliana* (L.) Heynh. generative organs under space flight conditions. *Dopov. Akad. Nauk Ukr. RSR, Ser. B*, No. 8, pp. 67–70 (In Ukrainian)

69. Kordyum, E. L., Nedukha, E. M., Sytnik, K. M., Mashinskiy, A. L. 1981. Optical and electron-microscopic studies of the *Funaria hygrometrica* protonema after cultivation for 96 days in space. *Adv. Space Res.* 1(14):159–62

70. Kordyum, E. L., Popova, A. F., Mashinskiy, A. L. 1979. Influence of orbital flight conditions on formation of genitals in *Muscari racemosum* and *Anethum graveolens*. *Life Sci. Space Res.* 17:301–4

71. Kordyum, E. L., Sytnik, K. M. 1983. Biological effects of weightlessness at cellular and subcellular levels. *Physiologist* 26(6):S141–42(Suppl.)

72. Kordyum, E. L., Sytnik, K. M., Chernyaeva, I. I. 1983. Peculiarities of genital organ formation in *Arabidopsis thaliana* (L.) Heynh. under spaceflight conditions. *Adv. Space Res.* 3(9):247–50

73. Kordyum, E. L., Sytnik, K. M., Chernyaeva, I. I., Anikeeva, I. D., Vaulina, E. N. 1984. Characteristics of the formation of androecium and gynoecium in *Arabidopsis thaliana* under space flight conditions. See Ref. 42, pp. 81–96

74. Kostina, L., Anikeeva, I., Vaulina, E. 1984. The influence of space flight factors on viability and mutability of plants. *Adv. Space Res.* 4(10):65–70

75. Kostina, L. N., Anikeeva, I. D., Vaulina, E. N. 1986. Experiments with developing plants aboard Salyut-5, Salyut-6 and Salyut-7 orbital stations. *Space Biol. Aerosp. Med.* 20:73–78

76. Kovrov, B. G., Kordyum, V. A., eds. 1985. *Mikroorganizmy v Iskusstvennykh Ekosistemakh*. Novosibirsk: Nauka. 192 pp. (In Russian)

77. Krikorian, A. D., O'Connor, S. A. 1984. Experiments on plants grown in space: Karyological observations. *Ann. Bot.* 54(Suppl. 3):49–63

78. Krikorian, A. D., Steward, F. C. 1978. Morphogenetic responses of cultured totipotent cells of carrot (*Daucus carota* var. *carota*) at zero gravity. *Science* 200:67–68

79. Laurinavichius, R. S., Yaroshius, A. V., Marchyukaytis, A., Shvegzhdene, D. V., Mashinskiy, A. L. 1984. Metabolism of pea plants grown under space flight conditions. See Ref. 42, pp. 96–102

80. Lyon, C. J. 1971. Experiment P-1096: Growth physiology of the wheat seedling in space. See Ref. 119, pp. 167–88

81. Maksimova, Ye. N. 1985. Effect on seeds of heavy charged particles of galactic cosmic radiation. *Space Biol. Aerosp. Med.* 19(3):103–7

82. Merkys, A. J., Laurinavichius, R. S. 1983. Complete cycle of individual development of *Arabidopsis thaliana* (L.) Heynh. plants on board the "Salyut-7" orbital station. *Dokl. Akad. Nauk SSSR* 271(2):509–12 (In Russian)

83. Merkys, A. J., Laurinavichius, R. S., Bendoraityte, D. P., Shvegzhdene, D. V., Rupainene, O. J. 1986. Interaction of growth-determining systems with gravity. Presented at *COSPAR Meeting, 26th, Toulouse, 1986*

84. Merkys, A. J., Laurinavichius, R. S., Mashinskiy, A. L., Yaroshius, A. V., Savichene, E. K., et al. 1976. Effect of weightlessness and its simulation on the growth and morphology of cells and tissues of pea and lettuce seedlings. In *Organizmy i Sila Tyazhesti. Materialy I Vsesoyuznoy Konferentsii "Gravitatsiya i Organism"*, ed. A. J. Merkys, pp. 238–46. Vilnius: Inst. Botany, Lithuanian SSR Acad. Sci. and Inst. Gen. Genet., USSR Acad. Sci. (In Russian)

85. Merkys, A. J., Laurinavichius, R. S., Rupainene, O. J., Savichene, E. K., Yaroshius, A. V., et al. 1983. The state of gravity sensors and peculiarities of plant growth during different gravitational loads. *Adv. Space Res.* 3(9):211–19

86. Merkys, A. J., Laurinavichius, R. S., Rupainene, O. J., Shvegzhdene, D. V., Yaroshius, A. V. 1981. Gravity as an obligatory factor in normal higher plant growth and development. *Adv. Space Res.* 1(14):109–16

87. Merkys, A. J., Laurinavichius, R. S., Shvegzhdene, D. V. 1984. Plant growth, development and embryogenesis during Salyut-7 flight. *Adv. Space Res.* 4(10):55–63

88. Merkys, A. J., Laurinavichius, R. S.,

PLANTS IN SPACE 343

Shvegzhdene, D. V. 1984. Spatial orientation and the growth of plants in weightlessness and in an artificial gravitational field. See Ref. 42, pp. 72–81

89. Merkys, A. J., Laurinavichius, R. S., Shvegzhdene, D. V., Yaroshius, A. V. 1985. Investigations of higher plants under weightlessness. *Physiologist* 28(6): S43–S46 (Suppl.)

90. Merkys, A. J., Mashinskiy, A. L., Laurinavichius, R. S., Nechitailo, G. S., Yaroshius, A. V., et al. 1975. The development of seedling shoots under space flight conditions. *Life Sci. Space Res.* 13:53–57

91. Miquel, J., Philpott, D. E. 1978. Effects of weightlessness on the genetics and aging process of *Drosophila melanogaster*. In *Final Reports of U.S. Experiments Flown on the Soviet Satellite Cosmos 936*, ed. S. N. Rosenzweig, K. A. Souza, pp. 32–59. NASA TM-78526. Moffett Field, CA: NASA

92. Milov, M., Rusakova, G. 1980. Greenhouses in space. *Aviatsiia i Kosmonavtika*, No. 3, pp. 36–37 (In Russian)

93. Moore, R., Evans, M. L. 1986. How roots perceive and respond to gravity. *Am. J. Bot.* 73(4):574–87

94. Moore, R., Fondren, W. M., Koon, E. C., Wang, C.-L. 1986. The influence of gravity on the formation of amyloplasts in columella cells of *Zea mays* L. *Plant Physiol.* 82:867–68

95. Moore, R., Fondren, W. M., McClelen, C. E., Wang, C.-L. 1987. The influence of microgravity on cellular differentiation in root caps of *Zea mays* L. *Am. J. Bot.* In press

96. Moore, R., McClelen, C. E., Fondren, W. M., Wang, C.-L. 1987. The influence of microgravity on root-cap regeneration and the structure of columella cells in *Zea mays*. *Am. J. Bot.* 74:216–21

97. Nevzgodina, L. V., Gaubin, Y., Kovalev, E. E., Planel, H., Clegg, J., et al. 1984. Changes in developmental capacity of *Artemia* cyst and chromosomal aberrations in lettuce seeds flown aboard Salyut-7 (Biobloc III Experiment). *Adv. Space Res.* 4(10):71–76

98. Nevzgodina, L. V., Grigoriev, Yu. G. 1979. Spontaneous and induced gamma radiation mutagenesis of *Lactuca sativa*. See Ref. 61, pp. 190–94

99. Nevzgodina, L. V., Maksimova, Ye. N. 1982. Cytogenetic effects of heavy charged particles of galactic cosmic radiation in experiments aboard Cosmos-1129 biosatellite. *Space Biol. Aerosp. Med.* 16(4):103–8

100. Nuzhdin, N. I., Dozortseva, R. L. 1972.

Genetic changes induced by space flight factors in barley seeds on "Soyuz-5" and "Soyuz-9" craft. *Zh. Obshch. Biol.* 33(3):336–46 (In Russian)

101. Nuzhdin, N. I., Dozortseva, R. L., Pastushenko-Strelets, N. A., Samokhvalova, N. S. 1965. The effect of space flight factors on seeds of the spindle tree (*Euonymus europaea* L.). *Izv. Akad. Nauk SSSR, Ser. Biol.*, No. 4, pp. 576–80 (In Russian)

102. Nuzhdin, N. I., Dozortseva, R. L., Pastushenko-Strelets, N. A., Samokhvalova, N. S., Chudinovskaya, G. A. 1970. Chromosome mutations induced by space flight factors in barley seeds during the circumlunar flights of the automatic stations "Zond 5" and "Zond 6". *Zh. Obshch. Biol.* 31(1):72–83 (In Russian)

103. Nuzhdin, N. I., Dozortseva, R. L., Samokhvalova, N. S., Nechaev, I. A., Petrova, L. E. 1975. Genetic damages of seeds caused by their two-stage stay in space. *Zh. Obshch. Biol.* 36(3):432–40 (In Russian)

104. Palladina, T. O., Kordyum, E. L., Bilyavs'ka, N. O. 1984. Activity and localization of transport adenosine triphosphatases in cells of pea seedling root cells in hypogravity conditions. *Ukr. Bot. Zh.* 41(5):54–57 (In Ukrainian)

105. Parfenov, G. P. 1975. Biologic guidelines for future space research. See Ref. 24, pp. 707–39

106. Parfenov, G. P., Abramova, V. M. 1981. Flowering and maturing of *Arabidopsis* seeds in weightlessness: Experiment on the biosatellite "Kosmos-1129". *Dokl. Akad. Nauk SSSR* 256(1):254–56 (In Russian)

107. Parfenov, G. P., Platonova, R. N., Tairbekov, M. G., Zhvalikovskaya, V. P., Mozgovaya, I. E., et al. 1979. Biological experiments carried out aboard biological satellite Cosmos-936. *Life Sci. Space Res.* 17:297–99

108. Perbal, G., Driss-Ecole, D., Sallé, G. 1986. Perception of gravity in the lentil root. *Naturwissenschaften* 73:444–46

109. Peterson, D. D., Benton, E. V., Tran, M., Yang, T., Freeling, M., et al. 1977. Biological effects of high-LET particles on corn-seed embryos in the Apollo-Soyuz Test Project—Biostack III experiment. *Life Sci. Space Res.* 15:151–55

110. Pickard, B. G. 1985. Early events in geotropism of seedling shoots. *Ann. Rev. Plant Physiol.* 36:55–75

111. Pickard, B. G. 1985. Roles of hormones, protons and calcium in geotro-

pism. In *Encyclopedia of Plant Physiology, (N S), Hormonal Regulation of Development. III. Role of Environmental Factors,* ed. R. P. Pharis, D. M. Reid, 11:193–281. Berlin/Heidelberg/NY: Springer-Verlag

112. Platonova, R. N. 1973. Cytogenetic analysis of seeds of diploid and autotetraploid forms of *Crepis capillaris* after flight on the artificial earth satellite "Kosmos-368". *Cosmic Res.* 11(3):420–23

113. Platonova, R. N., Lyubchenko, V. Yu., Devyatko, A. V., Malysheva, G. I., Panova, S. A., et al. 1983. The influence of photo- and chemotropism on the orientation of higher plants in the absence of geotropism. *Izv. Akad. Nauk SSSR, Ser. Biol.,* No. 1, pp. 51–59 (In Russian)

114. Platonova, R. N., Lyubchenko, V. Yu., Devyatko, A. V., Malysheva, G. I., Tairbekov, M. G. 1980. Orientation of tomato seedlings grown in weightlessness (investigation on satellite "Cosmos-1129"). *Izv. Akad. Nauk SSSR, Ser. Biol.,* No. 6, pp. 891–96 (In Russian)

115. Platonova, R. N., Ol'khovenko, V. P., Parfenov, G. P., Lukin, A. A., Chuchkin, V. G. 1977. Effects of space flight factors and increased temperature on seeds of diploid and tetraploid buckwheat. *Izv. Akad. Nauk SSSR, Ser. Biol.,* No. 1, pp. 65–72 (In Russian)

116. Platonova, R. N., Parfenov, G. P., Zhvalikovskaya, V. P. 1979. Orientation of plants in weightlessness. See Ref. 61, pp. 149–61

117. Roux, S. J., Serlin, B. S. 1987. Cellular mechanisms controlling light-stimulated gravitropism: Role of calcium. *CRC Crit. Rev. Plant Sci.* In press

118. Roux, S. J., Slocum, R. D. 1982. Role of calcium in mediating cellular functions important for growth and development in higher plants. In *Calcium and Cell Function,* ed. W. Y. Cheung, 3:409–53. New York: Academic

119. Saunders, J. F., ed. 1971. *The Experiments of Biosatellite II.* NASA SP-204. Washington, DC: NASA. 352 pp.

120. Shen-Miller, J., Gordon, S. A. 1967. Gravitational compensation and the phototropic response of oat coleoptiles. *Plant Physiol.* 42:352–60

121. Shen-Miller, J., Hinchman, R., Gordon, S. A. 1968. Thresholds for georesponse to acceleration in gravity-compensated *Avena* seedlings. *Plant Physiol.* 43:338–44

122. Shepelev, Ye. Ya. 1981. Results of experiments with plants in space. Presented at *12th US/USSR Joint Working Group Meet. Space Biol. Med., Washington, DC*

123. Sidorov, B. N., Sokolov, N. N. 1962. Influence of space-flight conditions on seeds of *Allium fistulosum* (spring onion) and *Nigella damascena* (nutmeg flower). In *Problemy Kosmicheskoy Biologii,* ed. N. M. Sisakyan, 1:248–51. Moscow: USSR Acad. Sci. (In Russian)

124. Slocum, R. D., Gaynor, J. J., Galston, A. W. 1984. Experiments on plants grown in space: Cytological and ultrastructural studies on root tissues. *Ann. Bot.* 54(Suppl. 3):65–76

125. Sobick, V., Sievers, A. 1979. Responses of roots to simulated weightlessness on the fast-rotating clinostat. *Life Sci. Space Res.* 17:285–90

126. Sparrow, A. H., Schairer, L. A., Marimuthu, K. M. 1971. Experiment P-1123: Radiobiologic studies of *Tradescantia* plants orbited in Biosatellite II. See Ref. 119, pp. 99–122

127. Summerlin, L. B., ed. 1977. *Skylab, Classroom in Space,* pp. 67–73. NASA SP-401. Washington, DC: NASA. 182 pp.

128. Sytnik, K. M., Kordyum, E. L. 1980. Botanical research in space. *Ukr. Bot. Zh.* 37(1):1–10 (In Ukrainian)

129. Sytnik, K. M., Kordyum, E. L., Belyavskaya, N. A., Nedukha, E. M., Tarasenko, V. A. 1983. Biological effects of weightlessness and clinostatic conditions registered in cells of root meristem and cap of higher plants. *Adv. Space Res.* 3(9):251–55

130. Sytnik, K. M., Kordyum, E. L., Nedukha, E. M., Sidorenko, P. G., Fomicheva, V. M. 1984. *Rastitel'naya Kletka pri Izmenenii Geofizicheskikh Faktorov.* Kiev: Naukova Dumka. 135 pp. (In Russian)

131. Tairbekov, M. G., Devyatko, A. V. 1985. Energy exchange of plants in weightlessness. *Dokl. Akad. Nauk SSSR* 280(2):509–12 (In Russian)

132. Tairbekov, M. G., Grif, V. G., Barmicheva, E. M., Valovich, E. M. 1986. Cytomorphology and ultrastructure of the root of the maize meristem in weightlessness. *Izv. Akad. Nauk SSSR, Ser. Biol.,* No. 5, pp. 680–87 (In Russian)

133. Tairbekov, M. G., Parfenov, G. P., Platonova, R. N., Zhvalikovskaya, V. P. 1979. Study of plant cells using the "Biofiksator-1" instrument. See Ref. 61, pp. 161–69

134. Theimer, R. R., Kudielka, R. A., Rösch, I. 1986. Induction of somatic embryogenesis in anise in microgravity. *Naturwissenschaften* 73:442–43

135. Tobias, C. A., Grigoriev, Yu. G. 1975. Ionizing radiation. See Ref. 24, pp. 473–531

136. USSR press releases: Moscow Home Service (a) April 8, 1979; (b) July 13, 1979; (c) August 18, 1980

137. USSR press releases: TASS (a) February 21, 1979; (b) March 23, 1979; (c) May 14, 1979; (d) July 25, 1979; (e) September 7, 1979; (f) December 8, 1980; (g) October 14, 1985

138. Vaulina, E. N. 1976. The effect of weightlessness on hereditary structures. In Problemy Kosmicheskoy Biologii: Gravitatsiya i Organizm, ed. N. P. Dubinin, 33:174–99. Moscow: Nauka (In Russian)

139. Vaulina, E. N. 1978. Genetic studies. In Vliyaniye Kosmicheskogo Poleta na Razvivayushchiyesya Organizimy, ed. V. A. Kordyum, pp. 40–53. Kiev: Naukova Dumka (In Russian)

140. Vaulina, E., Anikeeva, I., Kostina, L. 1984. Radiosensibility after space flight. Adv. Space Res. 4(10):103–7

141. Vaulina, E. N., Anikeeva, I. D., Kostina, L. N. 1985. The viability and mutability of plants after space flight. See Ref. 76, pp. 5–10

142. Vaulina, E. N., Anikeeva, I. D., Kostina, L. N., Kogan, I. G., Palmbakh, L. R., et al. 1981. The role of weightlessness in the genetic damage from preflight gamma-irradiation of organisms in experiments aboard the Salyut 6 orbital station. Adv. Space Res. 1(14):163–69

143. Vaulina, E. N., Kostina, L. N. 1975. Modifying effect of dynamic space flight factors on radiation damage of air-dry seeds of Crepis capillaris (L.) Wallr. Life Sci. Space Res. 13:167–72

144. Vaulina, E. N., Kostina, L. N. 1979. Genetic studies of the effects of an artificial force of gravity on Crepis capillaris. See Ref. 61, pp. 181–83

145. Vaulina, E. N., Kostina, L. N. 1984. Research with air-dry seeds and seedlings of Crepis capillaris (L.) Wallr. See Ref. 42, pp. 68–72

146. Vaulina, E. N., Kostina, L. N., Anikeeva, I. D., Balayeva, A. V. 1986. The results of the experiments with plant seeds on the artificial Earth satellites Cosmos-1514 and Cosmos-1667. See Ref. 1, p. 300

147. Vaulina, E. N., Kostina, L. N., Mashinskiy, A. L. 1976. Cytogenetic analysis of seeds of Crepis capillaris (L.) Wallr. exposed on board the Earth artificial satellite Cosmos 613. Life Sci. Space Res. 14:201–4

148. Vaulina, E. N., Palmbakh, L. R., Antipov, V. V., Anikeeva, I. D., Kostina, L. N., et al. 1979. Biological investigations on the orbital station "Salyut-5". Life Sci. Space Res. 17:241–46

149. Volkmann, D., Behrens, H. M., Sievers, A. 1986. Development and gravity sensing of cress roots under microgravity. Naturwissenschaften 73:438–41

150. Wilkins, M. B. 1979. Growth-control mechanisms in gravitropism. In Encyclopedia of Plant Physiology, (NS), Physiology of Movements, ed. W. Haupt, M. E. Feinleib, 7:601–26. Berlin/Heidelberg/NY: Springer-Verlag

151. Wilkins, M. B. 1984. Gravitropism. In Advanced Plant Physiology, ed. M. B. Wilkins, pp. 163–85. London: Pitman

152. Yurov, S. S., Akhmadieva, A. Kh., Vaulina, E. N., eds. 1978. Uspekhi Kosmicheskoy Biofiziki. Pushchino: USSR Acad. Sci. 88 pp. (In Russian)

MEMBRANE-PROTON INTERACTIONS IN CHLOROPLAST BIOENERGETICS:
Localized Proton Domains

Richard A. Dilley, Steven M. Theg and William A. Beard

Department of Biological Sciences, Purdue University, West Lafayette, Indiana 47907

CONTENTS

INTRODUCTION.. 348
WHAT IS A LOCALIZED PROTON INTERACTION DOMAIN?...................... 349
THYLAKOID MEMBRANE METASTABLE PROTON BUFFERING
    DOMAINS.. 350
    General Properties 350
    Acetic Anhydride as a Probe for the Sequestered Domains........................... 351
    Is the Metastable Proton Domain Sequestered or Lumen-Exposed?................... 352
THYLAKOID MEMBRANE PROTEINS ASSOCIATED WITH BURIED
    DOMAINS.. 354
LOCALIZED DOMAINS, THYLAKOID H^+ PUMPS AND
    SITE SPECIFICITY.. 358
    Early Experiments Suggesting Site Specificity.. 358
    Experiments with Acetic Anhydride 360
    Other Photosystem-Specific Effects; Electric Field and Fluorescence Data.......... 361
EFFECT OF DOMAIN-SEQUESTERED PROTONS ON CHLOROPLAST
    ELECTRON TRANSPORT FUNCTIONS... 363
    Release of Cl^- from the Oxygen-Evolving Complex.................................... 363
    The Active Site of the Oxygen-Evolving Complex Is Located in the Domains........ 364
    Studies of Proton Deposition into the Lumen Using Neutral Red..................... 365
LOCALIZED DOMAINS AND MEMBRANE ENERGIZATION FOR
    ATP FORMATION 367
    The Coupling Mode Depends on the Quality of the Sample Preparation 367
    Permeable Buffer Effects on ATP Formation Lags: New Evidence for Dual
        Proton Gradient Coupling... 368

347

0066-4294/87/0601-0347$02.00

Protons in the Domains are Utilized to Energize ATP Formation...................... 375
Kinetics of Bulk Phase and Localized Proton Gradient–Driven ATP
    Formation.. 377
Uncoupler Effects Possibly Related to Localized Coupling Via
    Membrane Domains.. 378
CONCLUDING REMARKS AND SPECULATIONS...................................... 379
Speculations on a Physiological Role for Localized and Delocalized
    Proton Gradients... 381

INTRODUCTION

Membrane phenomena in biological systems are not yet well understood. This is particularly true of the bioenergetics of ATP formation processes in chloroplasts, mitochondria, and bacteria. Most workers agree that proton electrochemical potential gradients are an essential feature of membrane bioenergetics. Oxidation-reduction reactions in thylakoids release protons in the oxidation of water by PSII and the oxidation of PQH_2 by the PSI-linked cytochrome f. These reactions are called proton pumps because H^+ ions are a primary, obligatory reaction product. Moving protons into and around the membrane with its associated aqueous phases must involve compensating movements of such ions as Mg^{2+}, K^+, and Cl^- (19, 23, 40), but the latter movements can be regarded as secondary because both the type and proportion of these compensating ion movements vary (23).

The question of how ATP formation is driven thus focuses on the proton gradient and its dissipation through the CF_0-CF_1 energy-coupling complex. Research in this area began in 1959 with Mitchell's suggestion that energy coupling involves a reversible, H^+-pumping ATP synthetase–ATPase plugged through the thylakoid, mitochondrial, or bacterial membrane (64), and that the ATP synthetase is coupled to the redox H^+ pumps via bulk phase–to–bulk phase delocalized proton electrochemical potential gradients. About the same time, Williams suggested that redox-linked proton fluxes localized in or constrained to the membrane phase might link proton gradients to ATP formation (107). For 26 years the question of whether proton coupling is localized or delocalized has remained an unsettled issue.

Many reviews have covered the subject (e.g. 5, 12, 25, 37, 47, 57, 78, 105). A balanced, in-depth treatment of the experimental background for the localized-vs-delocalized proton gradient issue is available in Ferguson's review (25). Here we cover proton-thylakoid interactions from the viewpoint of how the proton gradients interact with the membrane macromolecules and how such proton-membrane interactions influence energy-coupling reactions such as ATP formation. Some of those interactions imply the existence of sequestered, proton-buffering domains associated with membrane proteins. The possible role of such domains in energy coupling is being studied, but the picture is not yet clear. Enough is known to formulate a working hypothesis that such putative domains may be importantly involved in linking energy-coupling reactions (ATP formation) to proton diffusion in the membrane or at the membrane interface.

WHAT IS A LOCALIZED
PROTON INTERACTION DOMAIN?

We define a localized domain as a space containing potentially mobile protons. These are distinguishable experimentally from protons in the two aqueous bulk phases (the outer suspending phase and the inner or lumen phase) because they are not in rapid equilibrium with protons in the bulk phases. The thylakoid membrane is about 50% protein and 50% lipid (69), and exchangeable protons can be expected to be associated mainly with protein acid-base groups. Thylakoids have virtually no amino lipids. The lipid composition for spinach, for example, is 48% monogalactosyl diacylglycer-ide, 25% digalactosyl diacylglyceride (neither of which has fast-exchanging protons), 13% phosphatidylglycerol (having an exchangeable phosphate POH proton), 8% sulfoquinovosyldiacylglyceride (with an SO_3 proton), 2% phos-phatidyl choline (with a POH proton), and 4% other phospholipids (76). Whether the phosphate or sulfonate groups may be in sufficiently hydrophob-ic environments to have their pKs shifted up toward the pH 5–7 region is an open question. In any event, protein functional groups such as $-NH_2$, $-COOH$, $C-OH$, and $-SH$ provide far more acid-base, exchangeable pro-ton sites than do lipids.

Ryrie & Jagendorf (80) showed that the CF_1 complex contains as many as 90 H atoms exchangeable with ^3H_2O in the energized conformation but not exchangeable in the de-energized state—a cogent example of a sequestered, hydrophilic domain. These investigators concluded that conformational changes of the multisubunit CF_1 occur during the energy-linked proton fluxes through the CF_1. "Buried" or slowly exchanging protons are known to occur in proteins. H-tritium exchange techniques have revealed a slowly exchanging component (24) in some of the peptide-bond amide protons and non-amide side-chain protons of rhodopsin. Englander (24a) has detected about 27 exchangeable protons in deoxyhemoglobin that have half-times from 100 min to 50 hr. The association of ribonuclease with its subtilisin-cleaved 20-amino-acid residue segment can bury 5 or 6 exchangeable protons and slow the half-time of exchange from tens of minutes to tens of hours (84). Combining H-deuterium exchange with neutron diffraction analysis of trypsin gave evi-dence that hydrogen bonding interactions—particularly in β sheet regions—slow down the exchange rate of certain protons (56). Kossiakoff estimated that 64% of the hydrophilic groups involved in the trypsin β sheet structure were unexchanged (56). Woodward and colleagues have also studied the accessibility of slowly exchangeable protons in proteins (reviewed in 108). In the chloroplast system, Siefermann-Harms & Ninnemann (89) have shown that the isolated light-harvesting pigment protein complex (LHC) of lettuce thylakoids has buried chlorophyll-containing regions that are protected from interaction with aqueous-phase protons. Heat, urea, or Triton X-100 denaturing treatments made the chlorophyll accessible to external acid.

The above examples point out that even in soluble proteins, but certainly in membrane proteins, structural factors can combine to shield or bury certain exchangeable proton association-dissociation sites. Therefore, it should not be surprising that the complex thylakoid structure shows sequestered exchangeable protons. The theme of this review is that buried, exchangeable proton-buffering and/or hydrogen-bonding groups may have a role in the thylakoid energy-transduction functions.

THYLAKOID MEMBRANE METASTABLE PROTON-BUFFERING DOMAINS

Thylakoid preparations suspended at pH 8.3–8.5 in the dark for periods long enough to equilibrate the lumen bulk phase pH with the external medium show a loss of 20–50 nmol H^+ (mg chl)$^{-1}$ to the suspending medium after addition of protonophoric uncouplers (7, 8, 49, 74, 94). The proton efflux can be directly detected with a sensitive recording pH meter (8, 19a, 74, 94).

Before discussing the measurements of protons associated with domains in thylakoids and their specific effects, we set forth the general properties of the sequestered domains.

General Properties

The domains appear to be separated from the bulk phases by barriers that are relatively impermeable to protons. Protons tend to remain in the domains in the dark unless the proton permeability of the thylakoid membrane is increased. This can be brought about by the addition of uncouplers (7), a mild heat treatment (75), or removal of CF_1 from the coupling-factor complex (51; D. Bhatnagar, R. A. Dilley, unpublished results). We have yet to discover an uncoupler that did not deplete the domains of protons under appropriate conditions; the list of uncouplers tested includes gramicidin, nigericin, $(NH_4)_2SO_4$, CCCP, desaspidin, and A23187 (8, 49, 93b, 94). Valinomycin and the neutral divalent cationophore ETH1001 were rather ineffective (93b).

Where tested, the pH dependence of proton release from the domains (or more commonly, the pH dependence of a phenomenon linked to the protonation state of the domains) titrated with a steep curve showing a pK between 7.2 and 7.8. The pH profile is such that any effects of domain protons on a particular process are manifested at and above pH 8.0, but not below pH 7.0.

The domains can be emptied by any of the above-mentioned treatments and refilled by proton pumping by either the redox (7, 74a, 75, 93, 93a) or ATPase H^+ pump (8). This property, combined with the sequestration of the protons in buried domains, makes the proton domains interesting to investigators of bioenergetics.

The size of the buffer pool in the domains appears to be sensitive to the osmotic conditions in the respective assay buffers (49, 59, 74). The number of H^+ ions detected as effluxing from the domains decreases as the osmotic strength increases. This phenomenon has received little attention and is inadequately understood. The effect may be due to the high osmotic pressure (low water potential) that lowers the water content of the membrane, causing a "tightening" of the structures, perhaps keeping protons more tightly associated with salt bridges.

The properties of domains and domain-associated protons outlined above provide criteria by which to judge whether a phenomenon is influenced by domain protons. First, the phenomenon must be affected by the addition of uncouplers to dark-adapted chloroplasts. Second, any such effects should be apparent at alkaline pH and disappear at or below pH 7.0. Third, the effects should be reversible by sufficient proton pumping activity. Finally (though this criterion is largely untested), the effects should decrease when the osmotic strength of the assay medium is increased.

Acetic Anhydride as a Probe for the Sequestered Domains

Acetic anhydride has been an important experimental probe for studying the thylakoid sequestered domains. Baker et al (7) correlated the dark uncoupler-induced proton efflux with an equivalent increase in groups acetylatable by acetic anhydride, probably amines located behind the membrane permeability barrier. One or more of the acetylatable groups is required for PSII water oxidation activity, as shown by the correlation between the extent of acetylation and inhibition of oxygen evolution (7). To measure the protonation state of the special-domain amine groups two parameters have been used, singly or together: the uncoupler-dependent inhibition of oxygen evolution and the uncoupler-dependent increase in labeling of the membrane proteins with [^{3}H]acetic anhydride (7, 8, 58, 59, 75). Acetic anhydride is highly lipid soluble and thus passes readily across the membrane. The anhydride carbonyl carbon reacts avidly with the lone pair electrons of amine nitrogen but not with protonated amines (63). The overall rate of derivatization is a function of both the rate of the nucleophilic carbonyl addition reaction and the rate of the dissociation of the protonated base to the free base (55). In short reaction times (20–30 s in our studies) and alkaline conditions, the extent of acetylation correlates positively with the proportion of the unprotonated amine species (63). Hence, acetylation extents can report the relative pH around reactive amine groups.

The proton efflux and the accompanying acetylation increase due to uncoupler addition are observable after tens of minutes of dark incubation at the high pH (59). Heating thylakoids to 30°C for 1 min (7) or removing the CF_1 complex (D. Bhatnagar and R. A. Dilley, unpublished data) also results in dumping the metastable proton pool, as indicated by an increase in [^{3}H]acetic anhydride incorporation. The pK of the metastable proton pool detected by the acetic anhydride probe is near 7.6–7.8 (59). This is below the normal pK of lysine ϵ amino groups (about 10.2) but is the expected pK resulting when a positive charge is close to a lysine. Lys-Lys pairs commonly have one of the pKs in the pH 6–8 range owing to the electrostatic effect (81). While hydrophobic effects can raise the pK of carboxyl groups 3–4 pH units, there is little or no pK effect (down shift) on amine groups (15). The reason for the difference between the two classes of groups is that R-COOH (neutral) ionizes to R^- + H^+, and the dipole nature of H_2O solvates and stabilizes the separated charges. Low dielectric strength (hydrophobic) solvents cannot provide solvation effects to shield the cation and anion; hence the acid form is the more stable in such solvents. The amine ionization reaction has a positive charge on either side of the reaction $RNH_3^+ \rightleftarrows RNH_2 + H^+$ so the system does not change charge state during ionization. Thus, lysine residues in a positively charged environment are reasonable candidates for providing the pK $\simeq$ 7.8 buffering groups that make up the metastable proton pool and the acetylation sites.

One such lysine has been identified as the Lys 48 of the 8-kd CF_0 proton-channel portion of the coupling complex (92). The three Tris-releasable proteins associated with the proton source in PSII water oxidation (M_rs of 33, 22, and 18) all contain nearly 10 mol percent lysine and show a large contribution to the acetylation increase upon uncoupler dissipation of the metastable proton pool (58). Cytochrome f, involved with the proton-releasing plastohydroquinone oxidation, shows a large component of buried, acetylatable groups (T. Allnutt and R. A. Dilley, in preparation). Thus, the intriguing situation is revealed, that buried proton-interacting groups occur at the sources of protons and at the CF_0 sink. Such correlations have stimulated our interest in these groups as candidates for proton hopping sites in sequestered domains (64c, 64d), possibly involved with proton movement during energy coupling in ATP formation (21).

Is the Metastable Proton Domain Sequestered or Lumen-Exposed?

It is clear that the uncoupler-releaseable proton pool, measured by the Dilley and Homann groups, is a metastable, normally protonated buffer domain. A critical issue for understanding membrane structure-function relationships is whether the domain is in the inner aqueous space or in a sequestered or membrane-localized domain. We favor the concept that it is a sequestered domain for the following reasons:

1. Long term (> 30 min) dark equilibration of thylakoids, at pH values nearly 1 pH unit above the pK of the acetylatable groups, does not lead to the $RNH_3^+ \rightleftharpoons RNH_2 + H^+$ dissociation unless protonophoric uncouplers are added (59, see also 93b, 94) or mild heating or CF_1 removal treatments are given (7). The permeability of thylakoids to H^+ at pH 8.5 is high enough to equilibrate the lumen bulk phase with the suspending phase within 3 min or less. The $t_{\frac{1}{2}}$ of 30 s for dark H^+ efflux at 4°C is a conservatively slow estimate, and > 99% equilibration will occur after 7 half lives (3.5 min). The metastable pool is retained for tens of minutes with only 19% or so loss of the bound protons in 10 min (59). Equilibration of the lumen pH with the pH 8.6 suspending medium was tested directly by measuring the pH_{in} by the distribution of radiolabeled amines, and the results show that the pH_{in} was within 0.1 pH unit of the external pH (59). Such "pH clamp" experiments are the strongest evidence supporting a sequestered location for the acetic anhydride-sensitive buffering groups.

2. A surface-charge effect keeping protons in a Gouy-Chapman layer at the exposed lumen-membrane interface is unlikely because the metastable pool was present, though attenuated somewhat, when 425 mM KCl was present (with or without valinomycin) in the buffer rather than the usual 50 mM KCl (59). Similarly, Theg et al (94) found that adding 100 mM KCl or 50 mM $MgCl_2$ had no effect on the uncoupler release of H^+ ions. Retention of the sequestered protons in the Gouy-Chapman layer is also excluded by noting that the sequestered protons would be in equilibrium with, albeit at a higher concentration than, lumen protons. The fact that the sequestered protons move in response to uncoupler addition indicates that they are not in equilibrium with the bulk phases.

3. A membrane-phase location of the bound proton domain is suggested by the finding that the metastable protons were more completely lost (94) and the acetylation was of greater extent (59) after uncoupler addition to thylakoids swollen in a low-osmotic-potential medium.

The swollen thylakoids' larger lumen volume, containing more of the added buffers used in the suspension, can be an artifactual source of protons if the pK of the storage buffer is within 1 pH unit of the proton-release assay pH. Having carefully considered the issue of added buffer present in the lumen volume, Homann's group concluded that if Mes (pK 6.1) was used in the resuspension-storage buffer and the proton release assay medium was at pH 8.5–8.6, there was no measurable proton release due to the internally located Mes when a 2-min incubation at pH 8.5 was given prior to adding uncoupler

(74). Hepes (pK 7.5) at 5 mM could contribute a significant number of protons in the proton release assay, unless a sufficient pH 8.5 equilibration time was given.

The above evidence, interpreted as support for a membrane-buried location of the proton-buffering domain, should eventually be tested by other experiments capable of giving structural data. The ultimate understanding of the location of hydrogen bonding and acid-base dissociation groups in membranes will come from a combination of neutron or X-ray diffraction structural data (17) and the use of such probes as chemical modification (50a), controlled proteolysis, and antibody interactions (80b).

THYLAKOID MEMBRANE PROTEINS ASSOCIATED WITH THE BURIED DOMAINS

The metastable proton-binding domain is largely accounted for by protein side-chain groups of thylakoid polypeptides (58), probably lysine residues (58, 92). Acetic anhydride also acetylates acetone-soluble thylakoid materials, but the acetylated products have not been identified beyond the point of noting that they behave as neutral lipids in thin layer or HPLC silicic acid chromatographic systems (R. Dilley and J. Laszlo, unpublished observations). Insight into the functional role of the proton domains may come from knowledge of the particular thylakoid proteins that constitute the domains. Before considering the proteins that contribute to the localized domains, some comments on the structure of integral membrane proteins is in order, as well as a summary of the protein complexes that appear to constitute the thylakoid electron-transport, energy-conversion apparatus.

The electron-transport "z" scheme is generally accepted as a valid model of chloroplast redox reactions. Recent concepts put most of the redox components in complexes, with PSII as one multiprotein aggregate, the cytochrome b/f nonheme iron protein complex as another (91), and the PSI reaction center as a third. The CF_0-CF_1 complex is also a mobile unit (63a). Figure 1 shows a diagram of a typical model for thylakoid structure. Some of the light-harvesting pigment protein complexes (LHC) are aggregates that associate closely with the photosystems, and at least one (LHCII) is mobile between the PSI or PSII reaction centers, depending on conditions (96). As recently compiled in the *Encyclopedia of Plant Physiology*, New Series, Volume 19 (91), the five complexes account for about 30 polypeptides visible on SDS-PAGE gels, with another 7 polypeptides evidently not tightly associated with a complex [plastocyanin, Fd-NADP oxidoreductase (2 polypeptides), perhaps three protein kinases, and the 34-kd phosphorylated protein]. As we note below, proteins from all the complexes and some not-

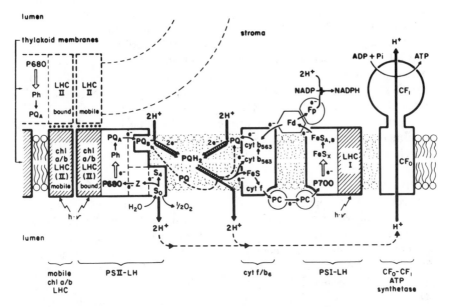

Figure 1   Organization of chloroplast membrane components participating in the electron coupling reactions of photosynthetic electron transport. Solid, thin arrows indicate electron transfer reactions; open arrows, chemical transitions; thick arrows, proton movements; dashed arrows, recycling routes for PQ. Three structurally distinct protein complexes participate in the linear electron transport pathway from water to NADP: a PSII complex, which is linked by a lipid-soluble pool of plastoquinone molecules to a cytochrome f/b_6 complex, and a PSI complex, which receives electrons from the cytochrome complex via the water-soluble protein, plastocyanin. Also indicated is a quinone electron transport cycle, or Q cycle, which causes the translocation of one additional proton for every two electrons passing through the linear electron transport chain. Protons deposited within the thylakoid lumen exit through the proton-translocating CF_0-CF_1 ATP synthetase complex. Light-harvesting pigment-protein complexes (LHC) serve both photosystems; these preferentially associate with either PSI or PSII and are designated LHCI or LHCII, respectively. One population of the LHCII is mobile and can serve either PSI or PSII by traveling laterally between the stroma lamellae (enriched in PSI centers) and grana stacks (enriched in PSII). Abbreviations: *chl a/b LHC,* chlorophyll *a/b* light-harvesting complex; *S* and *Z,* water-splitting and O_2-evolving enzymes that donate electrons to P680; P680, reaction center of PSII; *Ph,* bound pheophytin *a* (primary electron acceptor of PSII); PQ_A and PQ_B, special bound plastoquinone molecules (PQ_B is associated with the herbicide binding protein and can exchange with pool PQ); PQ and PQH_2, plastoquinone and reduced plastoquinone; FeS, Rieske iron sulfur protein; *cyt f* and *cyt b_{563},* cytochromes *f* and b_{563} (b_6); PC, plastocyanin; P700, reaction center of photosystem I; FeS_x, special FeS center that serves as the primary acceptor of photosystem I; $FeS_{A,B}$, two special FeS centers; *Fd,* ferredoxin; *Fp,* flavoprotein (ferredoxin-NADP reductase); NADP, nicotinamide adenine dinucleotide phosphate; CF_0 and CF_1, basepiece and headpiece of ATP synthetase; *ADP* and *ATP,* adenosine di- and triphosphate; *h·ν,* light energy. (Reprinted by permission of the authors, A. Staehelin and C. Arntzen, 1983, Regulation of chloroplast membrane function. *J. Cell. Biol.* 97:1327–37.)

yet-identified thylakoid proteins contribute proton buffering groups to the special domain(s). Hence, a complicated pattern is emerging, and we do not yet have a clear idea about how the domains are structured.

The arrangement of membrane proteins is an active, ongoing research area, and little exact structural information is available. For the purposes of this review it is important to consider what is known of the secondary and tertiary membrane-protein structures vis-à-vis the lipid bilayer. Three classes of integral membrane proteins are known. (a) Many membrane proteins have large extramembrane domains on one or both sides of the bilayer linked by a transmembrane segment of ~ 25–30 amino acids, often thought to be in the α-helical configuration (17, 63b). Examples are cyt f (2, 106) and the LHCII (54, 96) of chloroplasts and glycophorin A of red blood cells (87). The ADP-ATP carrier protein of mitochondria (102) and the DCCD-sensitive proteolipid of chloroplast F_0 (86) and mitochondrial F_0 (61) are of this class but with somewhat less extramembrane mass. (b) Another class of membrane proteins is mostly buried within the membrane, having multiple α-helixes spanning the membrane and connected by variable lengths of the polypeptide chain sticking out into the aqueous phases at either surface [bacteriorhodopsin (38), cyt oxidase subunit III (4)]. (c) In the third class of membrane protein, β-sheet structures could form oligomers that span a membrane, as recently suggested by Tobkes et al (98) for the α-toxin of S. aureus, a channel-forming protein.

SDS-PAGE gel analysis of thylakoids labeled with [H^3]acetic anhydride before dumping the proton domain, compared to those labeled after dumping,

Table 1 Thylakoid proteins contributing acetic anhydride labeling sites (probably low-pK lysines).

Polypeptide M_r	Identification	Comments
33, 22, 17	Tris-releasable proteins involved in water oxidation	extrinsic, tightly bound at inner (lumen) membrane surface
24, 28	light harvesting pigment proteins	membrane-spanning, with large mass exposed on lumen side
33	cyt f[a]	membrane-spanning with large mass exposed on lumen side
11	plastocyanin[a]	extrinsic, loosely bound at lumen surface
15	unknown	
8	DCCD sensitive CF$_0$ component	probably membrane-spanning, very hydrophobic

[a] F. C. T. Allnutt and R. A. Dilley, work in progress

showed about nine polypeptides with increased acetylation following un-coupler-induced conversion of $-NH_3^+$ to $-NH_2$ groups (58). Through the differential labeling of amine groups, this experiment identified the proteins involved with the buried domains.

Table 1 lists the major polypeptides that contribute to the metastable proton domain, their identity (if known), and the type of membrane–protein interaction. Cytochrome f, the 8-kd CF_0, an unknown 15-kd protein, the LHC, plastocyanin, and the three PSII-associated extrinsic proteins are major contributors to the proton-buffering domains. Cyt f (2, 106), the 8-kd CF_0 (92), and LHC (54) are membrane-spanning proteins, while the three PSII-associated proteins (1) and PC (34) are more or less bound at the membrane lumen surface. So far, the common denominator of the "domain" proteins is that part or all of their mass is at the lumen surface. Cytochrome f and the three PSII-associated proteins have most of their mass exposed to the lumen. That they adhere closely to the surface, rather than form a diffuse fibrous network as spectrin does in red blood cells, is suggested by the absence of fibers in electron micrographs in stained, thin-sectioned or freeze-fractured chloroplasts. It is not known whether PC is mostly bound or freely mobile in the lumen. The large interfacial mass of protein occurring in PSII proteins and with cytochrome f, both close to sites of proton release, suggests that these interfacial protein masses may provide a sequestered domain linked to the proton release sites. Our current working hypothesis is that the sequestered-domain proton-buffering groups, identified by the techniques discussed above, are associated with those interfacial protein masses, as diagrammed in Figure 4a.

This model avoids the problem of postulating an array of acid-base, hydrogen bonding groups completely buried in the lipid bilayer. However, such an alternative sequestered space cannot be ruled out. Membrane-buried charged groups are known to occur, for example, in bacteriorhodopsin (38) and subunit III of cytochrome oxidase (4); however, interfacial structures occluded by the mass of the lumen-exposed parts of the proteins may provide a better means for long-range H^+ diffusion via localized domains. This hypothesis remains speculative for the present, and more work is needed to define the location of the proton-binding groups vis-à-vis the membrane structure before a definitive model can be devised. An attractive feature of this model is the ease with which the localized proton domain could be opened to the lumen phase, allowing a delocalized proton gradient to form. As we explain below, new strong evidence indicates that both localized and delocalized proton gradients, involved in driving ATP formation, can occur; and the conditions that convert one to the other are consistent with the model.

LOCALIZED DOMAINS, THYLAKLOID H$^+$ PUMPS, AND SITE SPECIFICITY

Early Experiments Suggesting Site Specificity

The decade of the 1960s saw increasing acceptance of Mitchell's chemiosmotic coupling hypothesis as the paradigm best explaining energy transduction from the redox energy sources into ATP formation. The Mitchell concept proposed bulk phase–to–bulk phase $\Delta\bar{\mu}_{H^+}$ as the direct driving force acting on the CF$_1$ to link the dissipative flow of protons through the CF$_1$ to ATP formation (25, 64). Hind & Jagendorf's seminal work on postillumination phosphorylation (39, 48) gave enormous support to the chemisomotic hypothesis. Those kinetic experiments (39, 48, 65) and the work of Izawa (44) for the first time directly implicated redox-linked proton accumulation and passive proton efflux as being closely related to the high energy state responsible for ATP formation. The acid-base driven ATP formation experiments from Jagendorf's group (47, 99) showed that ATP formation utilized protons from added succinic acid contained in the lumen bulk aqueous phase, ruling out the direct participation of redox turnovers as obligatory steps in ATP formation. Thus by the mid 1960s it was reasonable to put forward models for energy transduction, based on phosphorylation data, consistent with bulk phase–to–bulk phase $\Delta\bar{\mu}_{H^+}$ as the direct driving force for ATP synthesis. Consistent evidence from several laboratories appeared, supporting that concept of the chloroplast system (18b, 22, 52, 62, 79, 85, 100). Other work with mitochondria and bacteria, also generally supportive of Mitchell's hypothesis, is not reviewed here; see references 25 and 78 for details.

A report by Izawa et al (46) in 1974 questioned that a bulk-phase delocalized $\Delta\bar{\mu}_{H^+}$ was the sole driving force for ATP formation. That work showed a different pH dependence for the phosphorylation efficiency (P/2e$^-$ ratio) when either a PSI or a PSII partial electron transport reaction energized ATP formation. Such a difference is not consistent with the simple chemiosmotic view, which pictured each proton pump site dumping protons into a common aqueous phase, from which the $\Delta\bar{\mu}H^+$ alone determines the rate of ATP formation. Photosystem or site specificity implies a more complex relationship between the redox-driven proton pumps and energization of the membrane for ATP formation. Localized proton gradient formation could be one such complexity.

Following that report, several laboratories published work testing for other photosystem-specific proton gradient or phosphorylation effects, with the results falling on both sides of the issue. Bradeen et al (14), following early work of Izawa & Good (45), showed that HgCl$_2$ inhibited PSI-linked

(DCIPH$_2$ → MV, plus DCMU) but not PSII-linked (H$_2$O → PD$_{ox}$, plus DBMIB) ATP formation. Neither did HgCl$_2$ inhibit postillumination ATP formation nor ATPase activity (13). The mechanism of the Hg^{2+} effect is not clear (13, 32), but Gould (32) attributed the Hg^{2+} resistance of PSII-linked ATP formation to a relative inaccessibility of the water-soluble Hg^{2+} cations to the more hydrophobic environment of PSII. Antisera to CF$_1$ equally inhibited PSI and PSII-linked ATP formation, so the Hg^{2+} effect seems subtler than just a difference in exposure of part of the CF$_1$. The Hg^{2+} effects could be the consequence of differences in proton-membrane interactions associated with the two photosystems. Grebanier & Jagendorf (33a) found no site specificity with the phosphorylation inhibitors SO$_4^{2-}$ or N-ethylmaleimide, both agents acting at the CF$_1$ complex; this finding is consistent with the CF$_1$-antibody results mentioned above. Thus, site specificity, when it has been observed, seems to reflect bioenergetic steps prior to the terminal ATP formation reactions at the CF$_1$.

Work by Dilley and colleagues provided evidence for different patterns of membrane-protein functional group reactivity with chemical modification probes, depending on whether PSI or PSII partial reactions were used to energize the thylakoids (26–28). Proton release in the water oxidation reactions (H$_2$O → methyl viologen) caused greater reactivity of thylakoid components toward diazobenzene sulfonate (DABS) than did a slightly greater rate of electron flow from I → MV, a PSII oxidation reaction not releasing protons (28). DABS-labeling studies are plagued with problems (60); but with careful attention to controls, the experiments yielded reliable, though perplexing data (28). Cyclic electron transport mediated by PMS, which may not include PQ turnover (27), or by the menadione cyclic system, which does include PQ, did not elicit the greater DABS binding (26). PMS without DCMU, a system with PSII functioning, did give the 4–5-fold DABS-binding increase (27).

Those data clearly implicated PSII water oxidation activity as necessary to give some type of conformational change reflected in greater DABS labeling. That proton release was the critical step rather than an electron transfer step was shown by comparing the DABS binding to H$^+$ uptake (measured by NH$_4^+$ accumulation) and phosphorylation with the partial reaction H$_2$O → silicomolybdate (SiMo) (28). Only under conditions where PSII water oxidation resulted in H$^+$ accumulation did the greater DABS binding occur. Under the conditions used, SiMo reduction in the presence of DCMU did not allow either H$^+$ accumulation or ATP formation, and the extra DABS binding did not occur. SiMo did not inhibit PMS cyclic phosphorylation, so the effects were subtler than a simple uncoupling action. The observed requirement for PSII proton accumulation to give DABS binding changes, combined with the lack of a PSI-dependent proton accumulation effect, suggested PSII proton-specific effects on thylakoid conformational changes. Thus, some type of localized proton-membrane interactions were hypothesized.

Other chemical modification probes were used to test the hypothesis posed by the DABS experiments. Iodoacetate covalently labels SH groups and to a lesser extent, amine groups (63). The amount of label incorporated in thylakoid membranes was less than 2 nmol (mg chl)$^{-1}$—i.e. about stoichiometric with the electron transport chains (75). The light-dependent changes in iodoacetate binding ranged from 25 to 50% increases in the light, and the pattern again suggested that PSII water oxidation was required to produce the effect (75). The reaction was not characterized further. As discussed above, acetic anhydride proved to be an interesting reagent, providing data consistent with the existence of membrane-localized, chemically reactive amine groups (7, 8, 58, 59, 75, 92).

Experiments with Acetic Anhydride

When the metastable sequestered proton-buffering array is protonated (the usual state), the amines do not react so readily with acetic anhydride as after dumping the protons into pH $\simeq$ 8.5 media by uncoupler addition or a brief heat treatment (30–35°C for $\leq$ 1 min). The greater anhydride reactivity was accompanied by more inhibition of PSII oxygen evolution [Figure 2 in (7)]. Either the redox (7) or the ATPase (8) H^+ pump readily restored the amine pool to the protonated, anhydride-unreactive state. The rapid protonation of the emptied buffering domains via the H^+ pumps is a property that makes the domains interesting and potentially relevant to the bioenergetics of proton gradient formation and utilization. Some early experiments (7) indicated that the partial redox reactions of PSII were more effective than partial reactions of PSI in delivering protons to the emptied domains. Later experiments showed that either partial redox system would restore the protonated domain (6). The conditions are not yet understood whereby more or less PSII site specificity may occur for refilling the domains. Work is still in progress to clarify the issue. Our present notion is that the localized domains make connection between the PSII and PSI (more exactly the plastohydroquinone—cyt b/f oxidoreductase) H^+ pump site, as diagrammed in Figure 4a. This model, though a rough working hypothesis, fits the data presently under consideration and provides a frame of reference for the phosphorylation data discussed below. Moreover, some recent experiments give considerable support to the notion of proton exchange between PSII and PSI (the cyt b/f site) via proton-conducting channels.

Experiments with dicyclohexylcarbodiimide (DCCD) and triphenyltin chloride (TPT) suggest that proton exchange between PSI-produced protons and PSII reaction centers occurs via pathways that can be blocked by the H^+

channel-blocking agents. Theg et al (93) obtained evidence of this using a delayed-light assay. Under certain conditions delayed light from PSII can be modulated (increased) by protons produced in PSI-specific reactions (see 93a for reference to the original work by Bowes & Crofts). We showed that DCCD and TPT blocked the PSI-proton-dependent effects on PSII, suggesting that the protons from PSI can normally reach PSII centers via specific channels.

A similar conclusion was recently reached (R. A. Dilley, in preparation) by assaying with acetic anhydride inhibition of PSII, an approach similar to that used by Baker et al (6–8). Thylakoids were exposed to a dark H^+-dumping heat treatment, returned to 18°C, and exposed to either PSII + PSI electron (proton) transport or PSI only ($\pm$ DCCD), prior to addition of acetic anhydride. The final step was an assay for PSII oxygen evolution activity. The results indicated that PSI-only proton accumulation could partially restore PSII anhydride-sensitive $-NH_2$ groups to the NH_3^+, unreactive state. DCCD present during the PSI-only proton-accumulation step greatly reduced the capacity of the PSI system to restore the NH_2 groups to the NH_3^+ state. PSII + PSI electron (proton) transport protection against acetic anhydride inhibition (at a proton-accumulation level somewhat lower than that given by the PSI-only system) was only slightly affected by the DCCD. Those results support the results of Theg et al (93) and the hypothesis that PSI-produced protons reach PSII via specific proton-conductive channels. Understanding the phenomena better will require more work, but the possibility is evident that specific sites for localized proton movement may be identified and studied in much greater detail. Sane et al (80a) have reported DCCD inhibition of electron transport near the plastoquinone locus, and our results are consistent with that observation.

Other Photosystem-Specific Effects; Electric Field and Fluorescence Data

If there is a relationship between protons in the buried domains and localized proton gradients driving ATP formation (see below), then protons from either photosystem-linked proton-pump mechanism should occupy a common membrane-localized domain. Indeed, Hangarter & Ort (35) found that during energization of ATP formation, many electron-transport complexes cooperate to form the ΔpH required to initiate ATP formation; but this threshold pool did not include the lumen aqueous phase. Moreover, it is surely possible, even likely, that the proton release sites of the water oxidation apparatus are in a buried domain surrounded by membrane proteins quite different from those at the site where PQH_2 is oxidized. Justification for this statement is found, for instance, in the recent work from several laboratories on PSII reconstitution with the three PSII-associated interfacial proteins (the 33-, 22-, and 17-kd proteins discussed in Section 4). The prevailing interpretations indicate that the three proteins occlude, or sequester, the manganese ions thought to be involved in the S-state generator of the oxygen evolving complex (cf 15a, for a review of the relevant work up to 1985 and diagrams showing the buried water oxidation site). Hence, it is possible that different proton–protein interactions could occur in the two proton-release mechanisms. These could involve different physical-chemical interactions close to the proton-release sites and over the diffusion path that may take the protons first to a common membrane-localized domain and thence to the $CF_0\text{-}CF_1$ complex. Of course, the evidence is clear that ultimately either proton source can lead to proton accumulation in the lumen bulk phase. The question of whether the lumen is a secondary proton repository is discussed below.

Using a different experimental approach, Junge and his coworkers (51, 74a, 95) have described the selective buffering of PSII-derived, but not PSI-derived, protons after uncoupler-induced deprotonation of the domains in alkaline media. These experiments are discussed below but are mentioned here because they provide strong evidence for a distinction between protons originating in the two photosystems.

Other studies have suggested differences between PSII and PSI mediation of membrane-related phenomena. For example, Remish et al (77) measured electrical potentials elicited by PSI or PSII partial electron-transfer reactions using microelectrodes embedded in grana stacks; they found different kinetics of the potential change, with PSII reactions giving slower kinetics than PSI. To the extent that the electrical potentials measured with chloroplasts impaled on electrodes have the same origin as the P515 electrochromic absorption change, those results are relevant to site-specific charge effects and to possible localized proton gradients. Vredenberg and colleagues (101, 104) have shown correspondence between the P515 bandshift and the impaled electrode signal. Peters et al (72) have suggested that the P515 signal 2 (the slow phase of the rise in absorbance after a single-turnover flash) is caused by localized proton interactions near the FeS, cytochrome b/f complex.

In a related study, the occurrence of the P515 slow-phase signal was influenced by the light intensity during plant growth. Growth in low light resulted in a decrease of both the slow P515 signal and the ratio of mono- to digalactosyldiacylglycerol (73). Membrane structural organization, reflected in the lipid composition, may be related to the stability of proton-containing microcompartments associated with proteins. Monogalactosyldiacylglycerol may be involved in special packing associations with membrane proteins (43). The slow phase of the P515 bandshift is known to be a sensitive parameter (104), easily lost if thylakoids are subjected to an aging treatment or to a brief temperature shock. Interestingly, the stability of the metastable proton-binding domain shows a similar sensitivity to aging and temperature shock

(7). The slow phase of the P515 bandshift is not yet fully understood; but despite significant differences in details of interpretation (31, 88), there is general agreement that it is associated with proton transport (50, 70). In a theoretical treatment of an electric field produced by a reaction center dipole, Zimanyi & Garab (109) proposed that the slow phase is associated with movement of protons from localized, low-dialectric membrane domains into the more conductive, adjacent delocalized phase.

The protons involved in the fast and slow phases of P515 change can be contributed by either the redox proton pumps or the ATPase H^+ pump (71, 83). Similarly, the metastable proton-buffering pool can be filled from either the redox or the ATPase H^+ pump (8). This matter is discussed further below.

Another parameter that may reflect localized proton interactions is the reverse electron flow driven by ATP hydrolysis. Prompt fluorescence and delayed fluorescence signals are associated with ATPase-dependent reverse electron flow, the latter being particularly sensitive to low levels of uncouplers (82a). Schreiber (82) has described two phases in ATP-induced fluorescence, one of which he interpreted as indicating a type of direct, localized coupling of proton flow between CF_0-CF_1 units and the proton-producing electron transport reactions of "non-B" PSII units.

More definitive experiments are needed to clarify site-specific proton-membrane interactions in thylakoids. But sufficient data of diverse sorts indicate that photosystem-specific interactions occur, related to membrane bioenergetic functions.

EFFECT OF DOMAIN-SEQUESTERED PROTONS ON CHLOROPLAST ELECTRON TRANSPORT FUNCTIONS

As discussed above, acetic anhydride inhibits electron transport in chloroplasts from which the sequestered protons had been released by uncouplers. If the uncoupler concentration is kept low enough, acetic anhydride addition results in much less inhibition in the light than in the dark, owing to proton accumulation via redox H^+ pumps (7). The inhibition by acetic anhydride was found to be located on the oxidizing side of PSII by studying its effects on the partial reactions (7). Below we describe a number of other electron transport–related phenomena believed to be dependent on the protonation state of the sequestered domains.

Release of Cl^- from the Oxygen-Evolving Complex

Chloride ions are required on the oxidizing side of PSII for maximal rates of electron transport (15b, 45a). Symptoms of Cl^- deficiency could not be evoked simply by washing chloroplasts in the absence of Cl^-; harsher, somewhat damaging conditions, such as mild heating, aging, or a CF_1-extracting EDTA wash were required (45a). Cl^- depletion became severer in the presence of certain uncouplers and could be observed only at alkaline pH (45a).

The conditions required for Cl^- release from the oxygen-evolving complex were similar to those required to release protons from their metastable domain pools (7). Theg & Homann hypothesized that Cl^- would remain associated with the oxygen-evolving system as long as protons remained in the sequestered domains (93b), and a number of predictions based on that idea have been tested and realized. Specifically, (a) Cl^- loss occurred only when thylakoids were uncoupled by addition of a wide variety of uncouplers to dark-adapted thylakoids, or after extraction of CF_1, and only at alkaline pH (93b); (b) even in an uncoupled state, thylakoids retained Cl^- at the oxygen-evolving complex unless they were suspended in alkaline media; and (c) provided that the uncoupler concentration was kept low enough, Cl^- release could be prevented by light—i.e. proton accumulation—via PSII. These observations meet the criteria set forth above, by which a process can be linked to protons in the domains.

One interesting difference between conditions required for inhibition of electron transport by acetic anhydride and those needed for Cl^- release from PSII is the amount of uncoupler that can be tolerated before light is no longer able to make the thylakoids behave as though their domains were protonated. In the first case, uncoupler concentrations sufficient to release electron transport from photosynthetic control completely were sufficient to cause significant sensitivity toward acetic anhydride inhibition in the light. The same uncoupler concentration still allowed Cl^- to be retained, in the light, at its binding site on the oxygen-evolving complex. This phenomenon can be explained by postulating that the Cl^- binding site is closer to the site of H^+ release during water oxidation than is the critical lysine group acetylated by acetic anhydride.

The Active Site of the Oxygen-Evolving Complex Is Located in the Domains

If protons liberated by the redox enzymes of the thylakoid electron transport chain are deposited initially into the domains, and if the Cl^- binding site on the oxygen-evolving complex is in the domains, then at least part of the complex (the active site) must also face the domains. Consistent with this view is the finding of Theg et al (93b) that the stability of the oxygen-evolving complex towards two treatments that cause it to lose Mn was affected by the addition of uncouplers to dark-adapted samples. Thylakoids were exposed in the dark to either damaging heat or high concentrations of Mg^{2+} in the presence or absence of uncouplers. These workers noted that at alkaline pH uncouplers caused a marked increase in the rate of loss of potential Hill activity. The uncoupler-mediated increase in the inhibition rate was not observed at or below pH 7.

The effects of uncouplers on the stability of the oxygen-evolving complex can be explained when one considers that the complex is more susceptible to denaturation at alkaline pH. If the target groups for heat and Mg^{2+} deactivation face the normally protonated sequestered domains, then uncouplers will render the complex less stable in alkaline media by raising the pH of the local environment. The uncoupler effect should have disappeared (and did) as the medium pH approached that of the protonated domains—in these experiments, around pH 7.

Evidence that the S-states of the oxygen-evolving complex are located in the domains was obtained by examining the dependence of the S-state deactivation rate on the protonation state of the domains (93a). Bowes & Crofts (11a, b) demonstrated that the deprotonated S-states [S_3 and S_4 (cf 11b)] would deactivate more quickly when their environment was acidified by PSI-dependent proton pumping. That is, the reaction $S_i \rightleftarrows S_{i+1} + H^+ + e^-$ is driven to the left by increasing acidity. Monitoring deactivation as an increase in millisecond delayed fluorescence, Theg et al (93a) demonstrated that this reaction was slowed by deprotonating the domain pools in alkaline media prior to the measurement. As with the other effects of uncouplers added in the dark, this effect was absent at pH 6.5. Since protons in the lumen were always allowed to equilibrate with those in the external medium before the measurement, those results suggested that the pH in the domains, not the lumen pH, governed the rate of the deactivation reaction. Thus the S-states must be exposed to the domains, not to the lumen.

These experimental approaches also allowed for investigation of the ability of protons released during PSI-mediated electron transport to equilibrate with the site of the PSII protolytic reactions. The S-state deactivation rate was not affected by increasing the buffering capacity of the lumen (93a), which should have slowed the acidification of the oxygen-evolving complex if the PSI-derived protons traveled through the lumen to reach the PSII location. Moreover, as discussed above, the proton-channel inhibitors dicyclohexylcarbodiimide (DCCD) and triphenyltin seemed to block the PSI proton-dependent acceleration of the deactivations (93a). This is consistent with proton movement between the PSI proton release site and the S-state system, via proton-conducting channels.

Studies of Proton Deposition into the Lumen Using Neutral Red

Junge and coworkers have used spectroscopic techniques to study the deposition of protons into the thylakoid lumen using the lipid-soluble, pH-indicating dye neutral red. They examined the effect on proton deposition of unloading the proton-sequestering domains by two separate methods—by the addition of nanomolar concentrations of uncouplers, which remained present during the subsequent measurements (95, 74a), or by the mild extraction of a small percentage of CF_1 by incubation with EDTA (51). In both cases, they observed that a significant fraction (i.e. during the first seven or eight flashes) of PSII-derived protons were rendered undetectable by neutral red after the treatment. Neutral red is thought to report lumen pH changes, even though it is bound primarily to the membrane surface (40a). When the domains were initially filled with protons, the first seven flashes gave the rapid neutral red response. The undetected protons could be found neither in the lumen at high kinetic resolution ($20\mu s$ per address) nor in the external medium using the water-soluble indicator cresol red (95). Thus it appeared that the undetected protons had been buffered within the membrane domains.

Remarkably, the effect of unloading the proton domains was restricted to the fastest of the two kinetic phases of the neutral red signal. This phase had previously been shown to arise solely from PSII-derived protons (2a). Proton deposition during PSI-mediated electron transport, detected as the slow phase ($\sim$ 20 ms) of the neutral red signal, was not altered by deprotonating the membrane domains. This is another example of photosystem specificity related to proton effects. The effect of mild uncoupling treatments on the fast phase of the neutral red signal was pH-dependent. Using both uncoupling treatments, the effect titrated between pH 7.2 and 7.8, the pH range consistent with domain effects. A difference was noted, however, in the slope of the titration curves depending on the method of uncoupling. When gramicidin was used, a Hill coefficient of 2 was required to fit the data points to a calculated titration curve (95). When uncoupling was caused by mild CF_1 extraction, a Hill coefficient of 6 was required for a good fit to the data (51).

The loss of the rapid phase of proton deposition by incubation with gramicidin was also shown to be reversible with illumination (74a, 95). When a train of flashes was delivered at 8.3 Hz to an uncoupler-containing sample, the fraction of the rapid phase relative to the total neutral red signal at a given flash increased from < 10% to > 90% in seven flashes. This provides a rough estimate of the size of the PSII-associated buffering pool, which after saturation no longer traps protons "out of view" of neutral red. The number of flashes required to reach saturation is expected to be a function of pH; unexpectedly, it has recently been shown to be dependent on the redox poise of the assay medium, increasing under oxidizing conditions (74a). A similar redox dependence has been observed for the inhibition of electron transport by acetic anhydride (S. M. Theg and R. A. Dilley, unpublished).

Polle & Junge (74a) recently showed that more than one electron transport chain could feed into a particular PSII domain. They demonstrated the 6–7 flash recovery of the fast phase of the neutral red signal in the presence of gramicidin and then repeated the experiment in the presence of an approximatly half-saturating concentration of DCMU. They reasoned that if a single electron transport chain fed protons to a single domain, the number of flashes required to recover the neutral red signal in the presence of gramicidin would be independent of the presence of DCMU; if more than one electron transport chain had access to a particular domain, then more flashes would be required to restore the signal in the presence of DCMU. Their finding that DCMU delayed the recovery of the signal beyond 15 flashes suggested the latter alternative. The exact number of chains providing protons to a given domain could not be determined by their technique, but the proton delivery must have occurred via the membrane core since the protons were not detected with neutral red.

The results of neutral red experiments performed with samples uncoupled with low concentrations of uncouplers differed slightly from those using mild CF_1 extraction (51). As mentioned above, the pH dependencies titrated with different Hill coefficients. In addition, gramicidin addition resulted in the loss of the entire fast phase of the neutral red signal (53, 95). CF_1 extraction caused the loss of only 50% of the fast phase and a slight ($\sim 10\%$) decrease in the amplitude of the flash-induced P515 electrochromic shift relative to controls (51). Based on these differences, Junge et al (51) and Theg & Junge (95) interpreted their experiments differently. Theg & Junge concluded that the protons that escaped detection by neutral red had been buffered in sequestering domains like those reviewed here, while Junge et al (51) postulated that the protons missing after mild CF_1 extraction had been trapped by buffering groups specifically in CF_0. In our view, both data sets suit a single explanation. An acetic anhydride–reactive lysine of the 8-kd CF_0 polypeptide has been identified as one of the localized domain buffering groups (92) but cannot alone account for the size of the sequestered pool. Perhaps 6 of a total of about 30 nmol(mg chl)$^{-1}$ can be accounted for by the 8-kd CF_0 (92). On the other hand, CF_1 extraction can cause unloading of all the sequestered domains (D. Bhatnagar and R. A. Dilley, unpublished). We suggest that this occurred in the Junge et al (51) experiments and that the protons that escaped detection by neutral red were buffered by other amines associated with the non-CF_0 portions of the sequestered domains.

Neutral red and other pH-indicating dyes will continue to be important tools for analyzing proton fluxes. An important advance would be finding a spectrophotometric probe that reported proton concentration in the putative thylakoid localized domains.

LOCALIZED DOMAINS AND MEMBRANE ENERGIZATION FOR ATP FORMATION

Protons contained in the thylakoid lumen bulk phase can be dissipated through the CF_0–CF_1, energizing ATP formation (3, 47, 64f, 99). However, it appears that localized proton gradients can also drive ATP formation (10, 33, 36, 41,

67, 68, 90). If so, we need a dual coupling hypothesis to account for both types of proton gradients. Such a hypothesis has been suggested by several groups for the chloroplast system (10, 20, 36, 41) and by Rottenberg (who used the term "parallel coupling") from work with mitochondria (78).

Before discussing the domains (as we experimentally define them) in regard to phosphorylation, we briefly touch on recent developments supporting localized coupling. Ferguson's excellent review (25) covers the experimental basis for the controversy over a delocalized vs a localized protonmotive force.

The Coupling Mode Depends on the Quality of the Sample Preparation

It now appears that the quality of the thylakoid preparation and the conditions of the experimental assay can alter the mode of energy coupling. As discussed above, Schreiber (82) has interpreted an ATP-induced fluorescence fast phase as indicating a direct coupling of proton flow between CF_0-CF_1 complexes and "non-B" PSII units. Schreiber typically uses intact chloroplasts that are hypotonically shocked prior to assay. The rapid fluorescence phase he refers to is extremely sensitive to factors that affect the state of the thylakoid membrane. In class C chloroplasts (i.e. naked lamellae), this rapid phase is greatly diminished (82).

Junge and coworkers have routinely used the pH-indicating dye neutral red as a probe to follow flash-induced proton release in the internal aqueous phase when the external phase was buffered by albumin. Hong & Junge (40a) found that thylakoid preparation affected both the response of neutral red to hydrophilic buffers and shifts in apparent pK upon addition of salts. With freeze-thawed thylakoids, neutral red indicated changes in surface pH that rapidly equilibrated with the pH of an internal aqueous bulk phase. Interestingly, with freshly prepared chloroplasts neutral red was not accessible to added hydrophilic buffers nor was the surface potential modified by salts. This led them to postulate that an extended internal bulk aqueous phase may not exist in freshly prepared material. This is difficult to reconcile with the fact that thylakoids are well-behaved osmometers (20a) and that the water potential in the lumen must be at equilibrium with the water potential in the suspending phase. It may be more likely that neutral red binds to the interfacial membrane structures differently in the two types of thylakoid preparations. Other indications that membrane integrity may influence the mode of energy coupling are noted in the following sections.

Permeable Buffer Effects on ATP Formation Lags: New Evidence for Dual Proton Gradient Coupling

Recent developments in this area have significantly improved the experimental base on which the localized-coupling viewpoint depends.

The $\Delta\psi$ component of $\Delta\bar{\mu}_{H^+}$ in thylakoids can readily be suppressed (e.g. with valinomycin plus KCl) leaving the ΔpH as the sole driving force for ATP formation (67). The application of permeable buffers in combination with short illumination times provides a system with which to test the chemiosmotic-hypothesis tenet that buffering the lumen bulk phase should delay the onset of ATP formation in proportion to the amount of added buffer. Experiments using this approach have given divided results. Ort et al (68) found little lag extension due to added buffers and suggested a localized-coupling model. Using similar techniques, Vinkler et al (103) and Davenport & McCarty (16) came to the opposite conclusion, that bulk phase proton gradients, sensitive to permeable buffers, energized ATP formation. Horner & Moudrianakis (41, 42) have applied rapid mixing and quenching techniques in the investigation of the effect of the permeable buffer imidazole on the onset of ATP formation and postillumination ATP formation. They observed, in agreement with Ort et al (68), that while imidazole had no effect on the lag length for light-driven phosphorylation, postillumination ATP formation was severely inhibited. This led them to postulate that the increase in lag induced by imidazole in the earlier work (16, 103) may have been the result of the procedure used to plot the data to derive curves indicating lags, a procedure that did not take into account the effect of imidazole on postillumination ATP formation.

A DISCREPANCY IS RESOLVED   We think this issue of localized or delocalized proton gradient coupling is now resolved, and either point of view can be correct for a given situation. As was recently shown by Beard and Dilley (10, 11) thylakoids that normally show the localized coupling mode, typical of the Ort et al (68) results (this laboratory has prepared thylakoids in the same way since the days of Ort's work), can be induced to show permeable buffer extensions of the ATP onset lag (i.e. delocalized coupling) simply by washing the membranes in 100 mM KCl prior to the phosphorylation experiment. In that work (10, 11) we compared the length of the ATP-formation onset lags without and with pyridine (a permeable buffer with a suitable pK, 5.4) in thylakoids prepared and washed in the normal low-salts medium, or thylakoids resuspended in 100 mM KCl for about 30 min prior to dilution into the phosphorylation medium. ATP formed by 5Hz single-turnover flashes was detected in the cuvette by the luciferin-luciferase technique (see 21 for details on the technique). The technique is a particularly convenient one for detecting differences in the energization lag, which is given by the number of flashes required to initiate ATP formation. The results clearly showed that after KCl washing, pyridine caused an extension of the ATP onset lag, consistent with bulk phase proton gradient coupling. In Figure 2 the reader can see the pronounced effect of pyridine on the lag in the +KCl case. In the control case, with sucrose rather than KCl as the osmoticum, there was no lag extension due to added pyridine, in keeping with our earlier

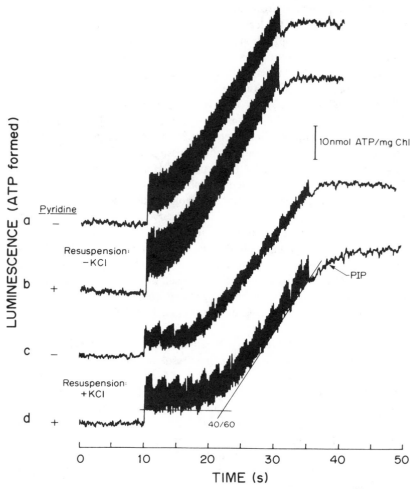

Figure 2 The effect of pyridine on single-turnover flash-initiated phosphorylation with thylakoids resuspended in the absence or presence of 100 mM KCl. Thylakoids were washed and resuspended in 5 mM Hepes pH 7.5, 2 mM $MgCl_2$, 0.5 mg/ml BSA, and 200 mM sucrose (*a* and *b*: $-KCL$) or 100 mM KCl in place of the sucrose (*c* and *d*: $+KCl$). Flashes (100 in *a* and *b*; 125 in *c* and *d*) were delivered at a rate of 5 Hz to thylakoids containing 14 μM chlorophyll suspended at 10°C in 1 ml of reaction mixture containing 50 mM Tricine-KOH pH 8.0, 10 mM sorbitol, 3 mM $MgCl_2$, 1 mM KH_2PO_4, 5 mM DTT, 0.1 mM ADP, 0.1 mM methylviologen, 400 nM valinomycin, and 5 μM diadenosine pentaphosphate. The reaction mixture was pH adjusted at 10°C with (*b* and *d*) or without (*a* and *c*) 5 mM pyridine. The vertical spike is the result of a light leak and served as a useful event marker. The different flash-spike heights are an artifactual result of the sampling time of the signal averager. The flash lag for the onset of ATP formation was determined with the aid of two criteria: 1. the first detectable rise in luminescence (40 in *d*) and 2. the back-extrapolation of the steady rise in the flash-induced luminescence increase to the x-axis (60 in *d*). The increase in luminescence after the last flash was due to postillumination phosphorylation. See (10) for details. [Reproduced from (10) with permission of the publisher.]

results (68). It is significant that the experiments of Vinkler et al (103) and Davenport & McCarty (16) were performed with 50 mM or more KCl. This may explain the difference vis-à-vis the Ort et al (68) results. The mechanism of the KCl effect is under active investigation.

The luciferin-luciferase technique allowed direct observation of postillumination phosphorylation (PIP), even in the case where ATP was made during the flash train. The bottom trace of Figure 2 identifies the PIP component. That is, one component of ATP formation occurred after the last flash and can only have occurred as a result of the dissipation of stored energization (efflux of energetically competent protons). An extremely important observation was made, that pyridine did not affect the PIP ATP yield in low salt–treated thylakoids at external pH of either 8.0 (10) or 7.0 (11). However, the KCl-treated thylakoids showed the predicted pyridine effects under both pH regimes. These results and the steady-state phosphorylation results discussed below strongly support the concept that the lumen phase does not participate as part of the proton diffusion pathway in the low-KCl case but does so after the KCl treatment. The thylakoids revert back to the pyridine-insensitive condition after the KCl-treated membranes are washed in the sucrose buffer (W. A. Beard and R. A. Dilley, in preparation), so the KCl effect is rather like a reversible switch than a permanent change in the system.

This point is important because the traditional way to measure PIP is to illuminate in the absence of ADP and P_i, building up energization, and then to add ADP and P_i in the dark. Such a protocol can always acidify the lumen enough to allow lumen protons to drive ATP formation because in the light, protons do not leave through the CF_1 in the linked H^+-efflux–ATP-formation step. That lumen acidification occurs in a basal electron transport mode can be shown with the luciferin technique we used. In that case withholding ADP and P_i until after the flash train gives a PIP ATP yield larger than that following a flash train with ADP and P_i present. Moreover, using membranes treated in high sucrose and low salt in the former case produced the predicted effects of a bulk phase buffer (pyridine greatly increased the PIP yield); but in the latter case, PIP was unaffected by pyridine (W. A. Beard and R. A. Dilley, in preparation).

The KCl washing step was suggested by recent work of Sigalat et al (90). They measured the rate of steady-state ATP formation coupled either to a PSI cyclic or a PSII noncyclic redox system as a function of the apparent ΔpH (estimated by the 9-aminoacridine method), using that as a protocol for testing for behavior consistent with chemisomotic behavior. They found that raising the KCl concentration to 50 mM caused the system to shift from a nonchemiosmotic pattern to behavior predicted by a simple bulk phase protonic coupling model. That group also observed a delocalized behavior, not normally observed, when their thylakoid preparation was hypotonically shocked prior to assay, or was treated with low concentrations of amines or nigericin (17a).

The prediction that the KCl treatment caused the proton gradient to equilibrate more rapidly with the lumen bulk phase was tested by another experiment, the intention of which was to determine whether the two different responses to permeable buffer addition could be observed in steady-state phosphorylation conditions as well as during the initial events of ATP formation. Steady-state phosphorylation was established in a cuvette with luciferin-luciferase present to report the ATP level continuously. At 40 s of illumination, a permeable buffer, hydroxyethane morpholine (HEM, pK $\simeq$ 6.3), was added to the stirred cuvette. If the proton gradient responsible for the steady-state ATP formation equilibrated rapidly with the lumen bulk phase in both types of thylakoids there should have been a transient slowing of the phosphorylation rate after HEM addition. On the other hand, if localized proton gradients were maintained in the steady state, HEM addition should have produced no effect. High KCl–treated thylakoids showed the transient slowing of ATP formation rate, while HEM addition did not affect the control thylakoids (Figure 3). Thus, the localized proton gradient effects are maintained in the steady state as well as in the initial energization stages. The details of the experiment will be reported elsewhere (S. M. Theg and R. A. Dilley, in preparation). The data in Figure 3 also demonstrate the effect of permeable buffer on extending the onset lag in the high KCl–treated thylakoids, as discussed above for pyridine used with single-turnover flashes.

The pattern seems clear. (a) Carefully prepared thylakoids, kept in low-salt media and provided with all the phosphorylation components, will drive ATP formation via localized proton coupling. Results from three different experimental protocols supported this conclusion: lack of permeable-buffer effects on the length of the onset lag (10, 11, 68), effects on the postillumination ATP formation (10, 11) and uncoupler and light intensity variation effects on phosphorylation driven by the partial electron transport systems of PSI or PSII (17a, 90). (b) Similar chloroplasts energized without ADP and P_i present will, given sufficient turnovers of the redox chain, acidify the lumen and show delocalized proton gradient energy coupling in postillumination phosphorylation experiments (10, 11). (c) Chloroplasts prepared with 100 mM KCl in place of 200 mM sucrose in the resuspension buffer show delocalized gradient coupling by all three criteria mentioned above (10, 11, 17a, 90).

While the origin of the high-KCl effect is not yet understood, it is important to be aware that thylakoids prepared from the same leaves can exhibit either localized or delocalized proton gradient effects. We can expect a variety of studies to be carried out with thylakoids treated to show one or the other effect, with the goal of trying to discover what is involved in the mechanism

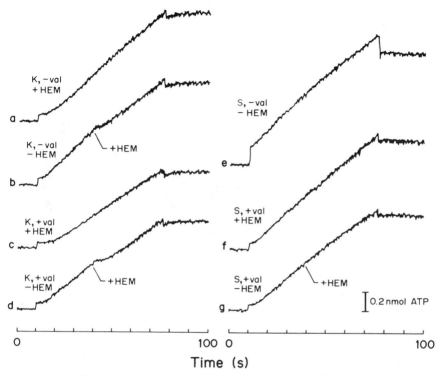

Time (s)

Figure 3   Effects of the permeable buffer hydroxyethane morpholine (HEM) addition on steady-state phosphorylation rates. Thylakoids were prepared and resuspended as in (10) with either 200 mM sucrose, 5 mM Hepes pH 7.5, 3 mM MgCl$_2$ (S in figure) or 100 mM KCl, 5 mM Hepes pH 7.5, 3 mM MgCl$_2$ (K in figure). The luciferin-luciferase phosphorylation assay was as in (10), except that 1 μg chlorophyll was added in the 1 ml assay volume. The light intensity was saturating. Traces *a–d* show the effect of HEM addition to KCl-treated thylakoids, and *e–g* are with sucrose-treated thylakoids. With 0.3 μM valinomycin to collapse the Δψ component, HEM added after 40 s in the light caused a distinct decrease in the rate of ATP formation in KCl-treated thylakoids (trace *d*), not observed with the sucrose-treated thylakoids (trace *g*). The inhibitory effects of HEM addition on steady-state ATP formation rate in the KCl-treated case are consistent with operation of a delocalized proton gradient coupling mode over the entire time course from the initial energization out to 40 s of illumination time. Also evident in the KCl-treated, but not with the sucrose-treated sample, was a greater lag in the onset of ATP formation (compare *c* to *f*), consistent with effects reported in (10, 11).

that maintains sequestered proton diffusion pathways on the one hand, and what constitutes the switch from localized to delocalized proton gradient coupling on the other. Figure 4a, b depicts a possible model expressing localized or delocalized coupling. We suggest that KCl treatment in some way switches the H$^+$ gradient so as to permit rapid equilibration with the lumen bulk phase. In Figure 4b this is shown diagrammatically; the proteins that

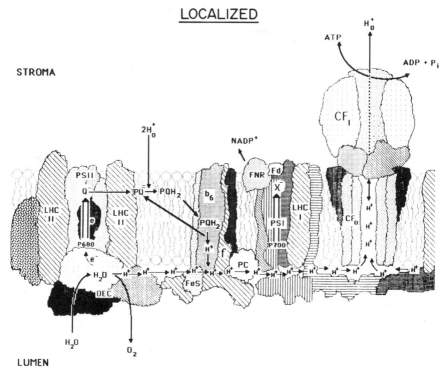

LUMEN

Figure 4 a *(above)*: A model for thylakoid membrane protein arrangement depicting hypothesized localized proton buffering domains for proton diffusion from the PSII and plastohydro-
quinone-cyt *b/f* oxidoreductase proton pumps to the CF_0-CF_1 complex. The localized proton-
binding domains are shown as formed between the membrane lipid bilayer lumen surface and
overlaying portions of diverse membrane proteins that protrude out into the lumen. The oxygen-
evolving complex (OEC) is shown comprised of the three Tris-releasable polypeptides of 33, 22,
and 18 M_r (see text for details and references). Cytochrome *f* and plastocyanin (PC) are shown as
participating in the domain, as are three polypeptides to the right of PC. Membrane proteins that
are believed to contribute buffering groups to the sequestered domain have been identified
through acetic anhydride chemical modification studies, as discussed in the text.

    b *(opposite)*: This model depicts a possible way of switching the localized proton-
processing domains into a lumen bulk-phase, delocalized proton gradient. As discussed in the
text, treating isolated thylakoids with 100 M KCl leads to phosphorylation data consistent with
delocalized proton gradient coupling. The models were drawn using computer graphics by Dr. F.
C. T. Allnutt.

normally adhere closely to the lumen surface are detached by the high-salt
treatment and expose more of their mass to the lumen bulk phase.

    According to de Kouchkovsky et al (17a), low concentrations of amines
such as NH_4Cl, hexylamine, or hypotonic shock can also shift localized
proton gradients to the delocalized mode. This could explain why McCarty's
group (16a, 74b) found such consistent evidence for delocalized $\Delta\bar{\mu}_{H^+}$ gradients energizing ATP formation: They routinely introduced 25 μM hexylamine into all the samples to permit ΔpH measurement. In this light, and

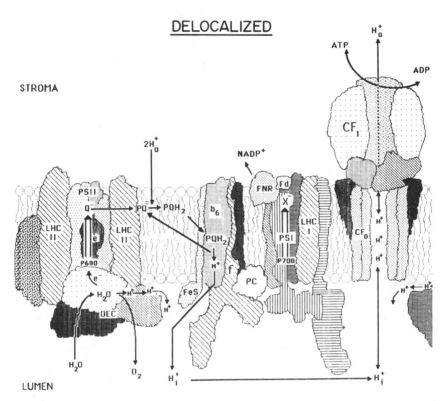

considering the demonstrated shift in coupling mode described above for our spinach chloroplast preparations, we suggest work demonstrating one coupling mode does not contradict that showing the other. The results obtained depend on the thylakoid treatment conditions. We discuss the biological implications of these two modes of proton gradient coupling at the end of our review.

Protons in the Domains Are Utilized to Energize ATP Formation

Some interesting properties of the sequestered proton-buffering domains relate to proton gradient coupling. The question here is whether the special domain proton-binding groups are part of a localized diffusion pathway required in energy coupling, or if the buffering groups are only an incidental structural feature. An experiment consistent with the former showed what happens to the length of the onset lag in ATP formation when the metastable H^+ pool is dumped (21). Beginning the flash sequence with the domains reversibly dumped of protons (by CCCP followed by addition of bovine serum albumin to bind the CCCP) caused an extension in the lag from about

18 flashes out to 28. A second cycle of flashes given after three minutes of dark produced a lag near 18 flashes, typical of the control thylakoids that had the metastable H^+ domain full at the beginning of the first flash sequence. The experiment was done with valinomycin and K^+ present so that only the ΔpH component of protonmotive force contributed to energization. Ten flashes are equivalent to about 30–35 nmol H^+ $(mg\ chl)^{-1}$, just the size of the metastable H^+ domain pool. Thus, before the membranes became energized for ATP formation, the special pool-buffering groups absorbed protons. This experiment did not demonstrate that the domain protons were those "in transit" to the CF_0-CF_1. The buffering groups may be on a dead-end side path that can absorb protons but not be mechanistically involved as a required pathway for proton diffusion to the CF_0-CF_1.

An experiment to test this point was done utilizing both the ΔpH and the $\Delta\psi$ components to energize ATP formation. The idea was that the $\Delta\psi$ field, rapidly established in the flash, and quickly expressed as a transmembrane field, would exert its force on H^+ ions nearest the CF_0 channel. We wondered, in other words, whether the protons "next in line" to pass through the CF_0-CF_1 came from the domains or from the lumen. The bulk lumen phase was equilibrated with the external phase (pH 8.4) by 3 min (or more in some cases) of dark incubation, time enough at that pH to allow pH equilibration. By using CCCP/BSA and sufficiently long ($\geq$ 3 min) incubation times, we could present the domain-exposed portion of the CF_0 with either protonated or empty domains, while we presented the lumen-exposed portion, at the beginning of the flash sequence, with a single nonvarying proton concentration. Under these conditions, a flash lag extension caused by dumping the domains would have been evidence for the participation of domain protons in ATP synthesis. If the empty domains were part of a dead-end sink, not connected to the CF_0, then there should have been the same energization onset lags for both cases. The data showed that with $\Delta\psi$ contributing to the energization, there was an additional 10 flash lag in the onset for the dumped H^+ pool case, compared to a control that began with the domains full of protons (93a). We concluded that the evidence favors a connection between the domains and the CF_0. The domains are therefore indistinguishable from an obligatory pathway utilized by protons as they proceed from the redox H^+ sources to the CF_0.

Recently Junge et al (53) came to just the opposite conclusion on the basis of neutral red experiments, namely that protons pass through the lumen en route to the coupling factor during ATP formation. Study led us to conclude that their data pattern could have been produced by either localized or delocalized gradient coupling, so we do not feel that their experiments robustly distinguish the two cases. Moreover, one might infer that the chloroplasts used by Junge et al more closely resembled the "high-KCl" samples discussed above, while the samples used by Theg & Dilley were "low-KCl" thylakoids. The neutral red measurements were made in a hypotonic assay medium that induces a delocalized coupling behavior (17a, 90). Although PSII domains were demonstrated (53), the possibility cannot be excluded that the "localized" pathway from the domains to the coupling factor complex was perturbed closer to the CF_0-CF_1. Presumably the conflict between these experiments will be resolved when they are performed under identical conditions.

Kinetics of Bulk Phase and Localized Proton Gradient–Driven ATP Formation

One of the clearest indicators that two proton diffusion pathways can function during ATP formation would be kinetic differences. Kinetic analysis of Beard & Dilley's experiment [Figure 2 of (11)] indicated that when thylakoids were isolated in the presence of KCl rather than sucrose the postillumination ATP (PIP) formation at pH 7.0 decayed three times slower [apparent $t_{1/2} \simeq 1.2$ s ($-$KCl); apparent $t_{1/2} \simeq 3.9$ s ($+$KCl)]; the postillumination ATP yield was nearly twice as large in the KCl-treated case. Thus an additional (or different) buffering pool contributed to the proton flux through the CF_0-CF_1 complex in the KCl-treated samples. Our interpretation is that the protons enter the lumen in that case, thereby gaining access to a different array of endogenous buffering groups and a larger volume in which to diffuse.

Recent work indicates that changing the internal buffer can affect proton gradient decay kinetics and in that way influence the kinetics of postillumination ATP yield. Schoenfeld & Kopeliovitch (81a) have shown that the dark proton efflux kinetics are strongly dependent on the internal buffering capacity. Whitmarsh (105a) has recently made a theoretical analysis of this phenomenon, and his results support that concept. His analysis suggests that an internally located buffer with a pK considerably lower than the external pH will decrease the proton efflux rate constant. This occurs because the buffer prevents the pH_{in} from changing as fast as it would in the absence of the buffer. Junge & McLaughlin (51a) have analyzed buffer effects on H^+ diffusion based on Fick's Second Law of Diffusion and reached a generally similar conclusion.

The inhibition of the postillumination ATP yield by pyridine in the KCl-treated thylakoids at pH 7.0 (11) indicates that the protons contributing to the gradient driving PIP were in the lumen. That is, as the proton accumulation spilled over into the lumen, the pyridine delayed the drop of internal pH down to near 4.7, thus reducing the amount of energetically competent protons available to drive PIP ATP formation. (With an energetic threshold ΔpH of 2.3 required, with valinomycin present to suppress the $\Delta \psi$ component, an internal pH of 4.7 is the threshold pH when the external pH was 7.0.) Pyridine had no effect on the apparent kinetic rate constant of the PIP decay in either control or KCl-treated thylakoids. That is understandable, because in the control case ($-$KCl), lumenal pyridine would not appear to be accessible to the energetically competent protons if they were in localized domains. With the high KCl–isolated thylakoids, pyridine was not expected to alter the decay kinetics, even though the pyridine was in the same phase (lumen) receiving the protons, since the ATP formation is the result of the efflux of energetically competent protons (those more acidic than pH 4.7); in the case where the external pH was 7.0, lumenally protonated pyridine could not serve as a large reservoir of energetically competent protons.

When the external pH was 8.0, however, kinetic analysis of the PIP [Figure 2 of (10)] from KCl-treated thylakoids indicated that when pyridine could serve as a source of energetically competent protons, it would increase not only the PIP yield but also the apparent $t_{1/2}$ (without pyridine, $t_{1/2} \simeq 1$ s; with pyridine, $t_{1/2} \simeq 2$ s). Following Whitmarsh's (105a) analysis, we think the increased $t_{1/2}$ (slower first-order reaction) was due to the pyridine in the lumen, which slowed down the relaxation of the system as it decayed to equilibrium. The pyridine-released protons below pH 5.7 (the energetic threshold) contributed to the PIP yield, as well as slowing the kinetics. The PIP (pH 8.0) kinetics and extent were unaffected by pyridine with thylakoids isolated in the absence of KCl. Thus protons for ATP formation came not primarily from the internal aqueous phase but from membrane buffering groups that were neither in equilibrium with the lumen nor accessible to pyridine.

Horner and Moudrianakis (41) observed dramatic effects of imidazole on PIP formed under conditions that allowed for proton accumulation under coupled conditions. Beard & Dilley (10) did not observe pyridine effects on PIP under similar coupling conditions when thylakoids were isolated in the absence of KCl. This apparent discrepancy may be due to the mode of energization used and/or the effect of imidazole, but not pyridine, in causing a delocalization of the proton gradient (17a). Horner & Moudrianakis gave their thylakoids 10 ms to 1 s illumination with PMS/ascorbate as cyclic cofactors, while Beard & Dilley controlled the amount of energization by using single-turnover flashes (10 μs) delivered at 5 Hz with methylviologen as the terminal electron acceptor. The different effects of permeable buffers on PIP yields are therefore probably a result of a much larger proton accumulation that occurred with cyclic electron flow, combined with the proton gradient-delocalizing effects of imidazole, which "swamped" the localized proton domains as well as shunting protons to the lumen, leading to the development of a large transmembrane ΔpH not observed with single-turnover flash-initiated phosphorylation.

Uncoupler Effects Possibly Related to Localized Coupling Via Membrane Domains

Other experiments that appear to implicate sequestered membrane domains in localized coupling mechanisms include those by Opanasenko et al (66), who found a greater sensitivity toward gramicidin of phosphorylation driven by noncyclic electron flow from water to methylviologen than by the PMS cyclic partial reaction. They suggested that low gramicidin concentrations specifically inhibited ATP formation associated with PSII and the localized H^+ binding domains. DCMU inhibited only that portion of PMS cyclic phosphorylation sensitive to low gramicidin. Higher gramicidin concentrations inhibited the part of PMS cyclic phosphorylation that remained in the presence of DCMU. These data are consistent with localized, site-specific proton processing linked to energy coupling.

Another anomolous uncoupler effect stemmed from work in Heber's group (97) with CO_2-fixing intact chloroplasts. Low concentrations of uncouplers stimulated CO_2 fixation, which is limited by ATP availability, and did not decrease the measured $\Delta\bar{\mu}_{H^+}$. The investigators suggested that a localized $\Delta\bar{\mu}_{H^+}$ may explain the data. Giersch (29) later found that phosphorylation could be stimulated, at constant $P/2e^-$, by low concentrations of certain uncouplers, particularly amines. It is known that electron transport is maximum at an "internal" pH of near 5–5.3 (9), and electron transfer rates can be limited because the "internal" pH is more acidic than the optimum. (The internal pH is that near the PQH_2 oxidation site, in our view a site buried in the special proton processing domain.) One explanation we propose for Giersch's effects is that low concentrations of uncouplers could act on the localized domain near the rate-limiting plastohydroquinone oxidation site to carry protons from the domains into the lumen bulk phase. That would have the dual effect of slightly accelerating the rate-limiting step by raising the local pH near the PQH_2 oxidation site while delivering the protons to the lumen where they can exert their ΔpH effect and contribute to the driving force for ATP formation via the bulk phase. Alternatively, the amine base could enter the localized domain and become protonated to the cation form, thereby raising the local pH but maintaining the total protonmotive force via an increase in $\Delta\psi$. De Kouchkovsky et al (17a) might explain the phenomenon somewhat differently, by attributing the uncoupler-increased ATP-formation rate to the uncoupler decreasing the proton diffusion resistance of localized proton flow pathways. Higher concentrations of uncoupler may dissipate the proton gradient in the usual way, inhibiting ATP formation. Further work with these interesting effects should be helpful in gaining a better understanding of proton gradient interactions with thylakoid membranes during energy coupling.

CONCLUDING REMARKS AND SPECULATIONS

Chloroplast thylakoid membranes have sequestered proton-buffering domains. The functional significance of the domains remains to be elucidated, although the evidence is compelling that the proton content of the domains strongly influences the build-up of the energization required to drive ATP formation. "Empty domains" correlate with a lower energization state, and protons produced by the electron transport chain turnovers must fill the domains before ATP formation begins (21, 93). Because most of the approximately 30–40 nmol H^+ (mg chl) $^{-1}$ contained in the domains associate with non-CF_0-CF_1 proteins (58, 92), we expect the domain structures to function in the delivery of protons to the CF_0-CF_1 complex, rather than to act solely at the CF_0-CF_1 level of the ATP-formation reactions. Figure 4 diagrams a working hypothesis for localized coupling; it also indicates how a delocalized coupling mode can occur after conformational rearrangements of one or more proteins switch the domain to allow direct deposition of protons into the lumen. The model is speculative, of course, as little exact membrane structural information is available; but some of the data underlying the model were discussed above.

Figure 4a shows lumen-exposed parts of membrane proteins closely adhering to the membrane surface so as to form sequestered domains along which protons might diffuse to the CF_0-CF_1 complex while remaining occluded from direct interaction with the lumen phase by a barrier provided by the protein material. Proton movement along the interface could be rapid and vectorially directed, as in a hydrogen-bonded chain mechanism of the sort proposed by Nagle and colleagues (64c, 64d). Protein hydrogen bonding groups and/or chains of oriented water molecules associated with the interfacial region could provide the H^+-bonding sites for such a mechanism. We believe that some sort of proton diffusion barrier must be present to maintain localized proton gradients, as discussed by Nagle & Dilley (64c). Teissie et al (92a) and Prats et al (74c) have demonstrated significant increases (20-fold) in the propagation of a chemically induced pH change along a simple phospholipid monolayer on an aqueous surface. Haines (34a) has speculated on the possible role in proton diffusion of polar head groups of phospholipids. Thylakoids contain only about 14% charged diacylglycerolipids, so these suggestions may not be applicable to proton diffusion in the thylakoid system. Our experiments have implicated low-pK amines (58) associated with the domain buffering action. These could provide proton hopping sites for a putative proton diffusion pathway of the type discussed by Nagle and colleagues (64c, d, e). The DCCD effects on proton interaction between PSI and PSII complexes discussed above (93a; R. A. Dilley, in preparation) suggest that carboxyl groups may be involved in the hypothesized proton-diffusion pathways. DCCD is known to inhibit proton conduction through the F_0 channel in chloroplasts, mitochondria, and bacteria (86). Recent studies on the F_0 proton channel in $E.$ $coli$ led Cox et al (15a) to conclude that a subunit a transmembrane helix together with the 8-kd DCCD-sensitive proteolipid unit (subunit c) could form a hydrogen-bonded chain conduction pathway. Zundel and colleagues (cf 110 for a review) have used infrared spectroscopy to study proton polarizability of hydrogen bonds in model systems, seeking to understand the possible relevance of the studies to proton conduction in membranes. Some of their speculations are in accord with hydrogen-bonded chains functioning as proton conducting pathways in membranes. Cain & Simoni (14a) have analyzed *E. coli* F_0 mutants and they speculate that the larger F_0 component, subunit *a*, could form a proton-conducting channel of the Nagle-Morowitz type by interaction of two of the proposed five membrane-spanning helixes. More research is required to test whether these notions form the correct paradigm for understanding proton diffusion-conduction along or across membranes. At present, the hypothesis seems to be attracting serious attention.

Figure 4b shows the domains open to the lumen, expressing the notion that thylakoids can, in certain circumstances (such as after high-KCl treatment), show delocalized protonic coupling during energization of phosphorylation (10, 11). We believe the available data supports the concept that both localized and delocalized protonic coupling can occur in isolated thylakoids, depending on factors such as the salt environment (10, 90), freeze-thaw cycles (40a), and osmotic conditions (90).

Speculations on a Physiological Role for Localized and Delocalized Proton Gradients

Does such a dual coupling possibility occur in intact chloroplasts and *in vivo?* And what biological advantage might accrue to a plant if thylakoids can switch between these two options? Future work must decide the former question. An indication that localized coupling may be the primary coupling mode is the fact that localized phenomena and permeable buffer–insensitive ATP-formation onset lags are best observed in the most carefully prepared thylakoids. Heating (7), freeze-thaw cycles (40a), low concentrations of amines (17a), and osmotic stress (90) attenuate or abolish domain-related effects, leading to delocalized coupling.

A possible advantage to plants in having membrane-localized rather than lumen-delocalized proton gradients as the primary coupling mode would be to avoid overacidifying the lumen in the light, an event that can have deleterious effects on chloroplasts. The initial ionic and osmotic events occurring upon illumination of isolated thylakoids are H^+ uptake (48) linked to Mg^{2+} and K^+ efflux (18, 23) with a loss of water (23). Light-induced shrinkage was first documented by comparing light-scattering, Coulter counter, and packed cell volume techniques (20a, 23), and freeze-fracture electron microscopy (19c). The model, suggested by Dilley & Vernon (23) and further elaborated by Murakami & Packer (64b), visualizes the dark, resting thylakoid having carboxyl groups neutralized by mobile cations Mg^{2+} and K^+ [this ion can be replaced by Na^+, depending on growth conditions (23)]. Proton accumulation by the redox or ATPase H^+ pumps acidifies the membrane, particularly under basal, nonphosphorylating conditions, causing -COOH formation and Mg^{2+}, K^+, and H_2O efflux. It was already observed by light-scattering studies that phosphorylating conditions greatly diminished the signal (18a), to be understood now as there being a lesser H^+ accumulation under the ATP-forming conditions, leading to less -COOH formation and therefore less mobile cation and H_2O efflux. As an aside, it would be of interest to compare exchanges of H^+ and Mg^{2+} (or K^+) in thylakoids phosphorylating under the two kinds of conditions described above—those in which localized and those in which delocalized proton gradient coupling is observed. The latter conditions result in significantly more H^+ accumulation during the pre–ATP formation onset flashes (W. A. Beard and R. A. Dilley, in preparation), probably causing more Mg^{2+} efflux prior to the threshold energization.

Rather than the greater or lesser small-amplitude thylakoid shrinkage mentioned above, a large-amplitude thylakoid swelling has been carefully documented (64a, 65a, 68a), particularly when thylakoids are suspended in high-NaCl ($\sim$ 300 mM) or high-Tris-Cl concentrations. The swelling can be disruptive to thylakoid structure owing to greatly increased lumen volume caused by the net uptake of Na^+ or $Tris^+$, Cl^-, and osmotically driven H_2O (64a). The light- and NaCl-dependent swelling was not always observable in thylakoids (19b). Apparently, subtle factors in the preparation and/or handling of the spinach thylakoids determine whether swelling will occur. A thylakoid H^+:Na^+ (or K^+) antiporter with Cl^- passively following was proposed in 1966 (18a, cf also 64a) to explain data on thylakoid ion exchange and osmotic responses. If such an exchanger were regulated and/or if the Cl^- (or other anions such as HPO_4^{2-}) permeability were controlled, then one could account for the fact that thylakoids do not necessarily always swell in high salt concentrations and illumination. Recent work has suggested that anion (Cl^-) permeability in thylakoids may be influenced by a membrane protein (100a).

In fact, Siegenthaler (89a) has identified C_{18} unsaturated fatty acids as potent inducers of the light-induced thylakoid swelling response (cf 64a). Such fatty acids are the most abundant in thylakoid lipids. Thus, localized proton currents might normally keep the lumen from overacidifying, and factors such as excessive lipid hydrolysis may promote delocalization of the proton fluxes into the lumen. This, combined with stimulation of an H^+_{in}:(Na^+ or K^+)$_{out}$ exchange with Cl^- ion and water fluxes following (18a, 64a), could account for a regulated swelling response. Interestingly, Murakami & Nobel (64a) and Siegenthaler (89a) found that shorter-chain fatty acids either inhibited (perhaps owing to a general protonophoric uncoupling action of the fatty acids) or had little effect on light-induced swelling—a result consistent with a possible specific effect of the C_{18} unsaturated fatty acids. This remains to be clarified. It is relevant that Rottenberg & Hashimoto (78a) have suggested that fatty acids uncouple mitochondria by blocking intramembrane proton transfer to the F_0-F_1 complex.

Little is known about thylakoid cation and anion exchangers and their regulation, while much is known or is being learned about these topics in mitochondria (25a), bacteria (57a), and red blood cells (14a). Cation: H^+ exchangers are ubiquitous in cells and are extremely important in cell physiology (cf 14a). Obviously, the chloroplast thylakoid and the outer envelope have no cell wall to resist their swelling; they must osmoregulate their volume just as do mitochondria and red cells. Hence, we can expect increased attention to be given this topic in chloroplast research in the years ahead.

Literature Cited

1. Akerlund, H. E., Jansson, C., Andersson, B. 1982. Reconstitution of photosynthetic water splitting in inside-out thylakoid vesicles and identification of a participating polypeptide. *Biochim. Biophys. Acta* 681:1–10

2. Alt, J., Herrmann, R. G. 1984. Nucleotide sequence of the gene for pre-apocytochrome f in the spinach plastid chromosome. *Curr. Genet.* 8:551–57

2a. Auslánder, W., Junge, W. 1975. Neutral red, a rapid indicator for pH changes in the inner phase of thylakoids. *FEBS Lett.* 59:310–15

3. Avron, M. 1971. The relation of light-induced reactions of chloroplasts to proton concentrations. In *Proc. 2nd. Int. Congr. Photosynth.*, ed. G. Forti, M. Avron, B. A. Melandri, 2:861–71. The Hague: Junk

4. Azzi, A. 1980. Cytochrome c oxidase. Towards a clarification of its structure, interactions and mechanism. *Biochim. Biophys. Acta* 594:231–52

5. Baccarini-Melandri, A., Casadio, R., Melandri, B. A. 1981. Electron transfer, proton translocation and ATP synthesis in bacterial chromatophores. *Curr. Top. Bioenerg.* 12:197–258

6. Baker, G. M. 1982. *Use of acetic anhydride as a chemical modification probe for photosystem-specific proton-membrane interaction in chloroplasts.* PhD thesis. Purdue Univ.

7. Baker, G. M., Bhatnagar, D., Dilley, R. A. 1981. Protein release in photosynthetic water oxidation. Evidence for proton movement in a restricted domain. *Biochemistry* 20:2307–15

8. Baker, G. M., Bhatnagar, D., Dilley, R. A. 1982. Site-specific interaction of ATPase-pumped protons with photosystem II in chloroplast thylakoid membranes. *J. Bioenerg. Biomembr.* 14:249–64

9. Bamberger, E. S., Rottenberg, H., Avron, M. 1973. Internal pH, ΔpH and the kinetics of electron transport in chloroplasts. *Eur. J. Biochem.* 34:557–63

10. Beard, W. A., Dilley, R. A. 1986. A shift in chloroplast energy coupling by KCl from localized to bulk phase delocalized proton gradients. *FEBS Lett.* 201:57–62

11. Beard, W. A., Dilley, R. A. 1986. Further evidence that KCl-thylakoid isolation induces a localized proton gradient energy coupling to become bulk phase delocalized. *7th Int. Congr. Photosynth.* In press

11a. Bowes, J. M., Crofts, A. R. 1978. Interactions of protons with transitions of the water splitting enzyme of photosystem II as measured by delayed fluorescence. *Z. Naturforsch. Teil C* 33:271–75

11b. Bowes, J. M., Crofts, A. R. 1981. The role of pH and membrane potential in the reactions of photosystem II as measured by effects on delayed fluorescence. *Biochim. Biophys. Acta* 637:464–72

12. Boyer, P. D., Chance, B., Ernster, L., Mitchell, P., Racker, E., Slater, E. C. 1977. Oxidative phosphorylation and photophosphorylation. *Ann. Rev. Biochem.* 46:955–1026

13. Bradeen, D. A., Winget, G. D. 1974. Site-specific inhibition in isolated spinach chloroplasts by Hg. *Biochim. Biophys. Acta* 33:331–42

14. Bradeen, D. A., Winget, G. D., Gould, J. M., Ort, D. R. 1973. Site-specific inhibition of photophosphorylation in isolated spinach chloroplasts by mercuric chloride. *Plant Physiol.* 52:680–82

14a. Cain, B. D., Simoni, R. D. 1986. Impaired proton conductivity resulting from mutations in the a subunit of F_1F_0 ATPase in *E. coli*. *J. Biol. Chem.* 261:10043–50

14b. Cala, P. M. 1980. Volume-sensitive ion fluxes in Amphiuma red blood cells: general principles governing Na^+-H^+ and K^+-H^+ exchange transport and Cl^--

HCO$_3$ exchange coupling. *Curr. Top. Membr. Transp.* 27:193–218

15. Cleland, W. W. 1977. Determining chemical mechanisms of enzyme-catalyzed reactions by kinetic studies. *Adv. Enzymol.* 45:273–387

15a. Cox, G. B., Fimmel, A. L., Gibson, F., Hatch, L. 1986. The mechanism of ATP synthase: a reassessment of the functions of the b and a subunits. *Biochim. Biophys. Acta* 849:62–69

15b. Critchley, C. 1985. The role of chloride in photosystem II. *Biochim. Biophys. Acta* 811:33–46

16. Davenport, J. W., McCarty, R. E. 1980. The onset of photophosphorylation correlates with the rise in transmembrane electrochemical proton gradients. *Biochim. Biophys. Acta* 589:353–57

16a. Davenport, J. W., McCarty, R. E. 1986. Relationship between rates of steady-state ATP synthesis and the magnitude of the proton-activity gradient across the thylakoid membranes. *Biochim. Biophys. Acta* 851:136–45

17. Deisenhofer, J., Epp, O., Miki, K., Huber, R., Michel, H. 1985. Structure of the protein subunits in the photosynthetic reaction centre of *R. viridis* at 3A resolution. *Nature* 318:618–24

17a. De Kouchkovsky, Y., Sigalat, C., Haraux, F. 1986. Delocalization of energy coupling in thyalkoids by amines. See Ref. 11

18. Dilley, R. A. 1964. Light-induced potassium efflux from spinach chloroplasts. *Biochem. Biophys. Res. Commun.* 17:716–22

18a. Dilley, R. A. 1966. Ion and water transport processes in spinach chloroplasts. *Brookhaven Symp. Biol.* 19:258–80

18b. Dilley, R. A. 1969. Evidence for the requirement of H$^+$ transport in photophosphorylation by spinach chloroplast. In *Progress in Photosynthesis Research, Proc. 3rd Int. Congr. Photosynth.*, ed. H. Metzner, 3:1354–60. Tübingen: H. Laupp Jr.

19. Dilley, R. A. 1971. Coupling of ion and electron transport in chloroplasts. *Curr. Top. Bioenerg.* 4:237–71

19a. Dilley, R. A., Baker, G. M., Bhatnagar, D., Millner, P., Laszlo, J. 1981. Chloroplast membrane protein-proton interactions. In *Energy Coupling in Photosynthesis*, ed. B. Selman, S. Selman-Reimer, pp. 47–58. New York: Elsevier–North Holland

19b. Dilley, R. A., Deamer, D. 1971. Light-dependent chloroplast volume changes in chloride media. *Bioenergetics* 2:33–38

19c. Dilley, R. A., Park, R. B., Branton, D. 1967. Ultrastructural studies of the light-induced chloroplast shrinkage. *Photochem. Photobiol.* 6:407–12

20. Dilley, R. A., Prochaska, L. J., Baker, G. M., Tandy, N. E., Millner, P. A. 1982. Proton-membrane interactions in chloroplast bioenergetics. *Curr. Top. Membr. Transp.* 16:345–46

20a. Dilley, R. A., Rothstein, A. 1967. Chloroplast membrane characteristics. *Biochim. Biophys. Acta* 135:427–43

21. Dilley, R. A., Schreiber, U. 1984. Correlation between membrane-localized protons and flash-driven ATP formation in chloroplast thylakoids. *J. Bioenerg. Biomembr.* 16:173–93

22. Dilley, R. A., Shavit, N. 1968. On the relationship of H$^+$ transport to photophosphorylation in spinach chloroplasts. *Biochim. Biophys. Acta* 162:86–96

23. Dilley, R. A., Vernon, L. P. 1965. Ion and water transport processes related to the light-dependent shrinkage of chloroplasts. *Arch. Biochem. Biophys.* 11:365–75

24. Englander, J. J., Downer, N. W., Englander, S. W. 1982. Reexamination of rhodopsin structure by hydrogen exchange. *J. Biol. Chem.* 257:7982–86

24a. Englander, S. W. 1975. Measurement of structural and free energy changes in hemoglobin by hydrogen exchange methods. *Ann. NY Acad. Sci.* 244:10–27

25. Ferguson, S. J. 1985. Fully delocalized chemiosmotic or localized proton flow pathways in energy coupling? A scrutiny of experimental evidence. *Biochim. Biophys. Acta* 811:47–95

25a. Garlid, K. D. 1980. On the mechanism and regulation of the mitochondrial K$^+$/H$^+$ exchanger. *J. Biol. Chem.* 255:11273–79

26. Giaquinta, R. T., Dilley, R. A., Anderson, B. J., Horton, P. 1974. A chloroplast membrane conformational change activated by electron transport between the region of PSII and plastoquinone. *Bioenergetics* 6:167–77

27. Giaquinta, R. T., Dilley, R. A., Selman, B. R., Anderson, B. J. 1974. Chemical modification studies of chloroplast membranes. *Arch. Biochem. Biophys.* 162:200–9

28. Giaquinta, R. T., Ort, D. R., Dilley, R. A. 1975. The possible relationship between a membrane conformational change and photosystem II dependent on H$^+$ ion accumulation and ATP synthesis. *Biochemistry* 14:4392–96

29. Giersch, C. 1981. Photophosphorylation by chloroplasts: effects of low con-

centrations of ammonia and methylamine. *Z. Naturforsch. Teil C* 37:242–50

30. Giersch, C., Heber, U., Kobayashi, Y., Inoue, Y., Shibata, K., Heldt, H. W. 1980. Energy charge, phosphorylation potential and proton motive force in chloroplasts. *Biochim. Biophys. Acta* 590:59–73

31. Girvin, M. E., Cramer, W. A. 1984. A redox study of the electron transport pathway responsible for generation of the slow electrochromic phase in chloroplasts. *Biochim. Biophys. Acta* 767:29–38

32. Gould, J. M. 1975. Inhibition of photosystem II–dependent phosphorylation in chloroplasts by mercurials. *Biochem. Biophys. Res. Commun.* 64:673–80

33. Graan, T., Flores, S., Ort, D. R. 1981. The nature of ATP formation associated with single turnovers of the electron transport carriers in chloroplasts. See Ref. 19a, pp. 25–34

33a. Grebanier, A. E., Jagendorf, A. T. 1977. Lack of site-specificity of spinach chloroplast coupling factor 1. *Biochim. Biophys. Acta* 459:1–9

34. Haehnel, W., Berzborn, R. J., Andersson, B. 1981. Location of the reaction side of plastocyanin from immunological and kinetic studies with inside-out thylakoid vesicles. *Biochim. Biophys. Acta* 637:389–99

34a. Haines, T. H. 1983. Anionic lipid headgroups as a proton-conducting pathway along the surface of membranes: a hypothesis. *Proc. Natl. Acad. Sci. USA* 80:160–64

35. Hangarter, R., Ort, D. R. 1985. Cooperation among electron-transfer complexes in ATP synthesis. *Eur. J. Biochem.* 149:503–10

36. Haraux, F., Sigalat, C., Moreau, A., de Kouchkovsky, Y. 1983. The efficiency of energized protons for ATP synthesis depends on the membrane topography in thylakoids. *FEBS Lett.* 155:248–52

37. Harold, F. 1977. Membranes and energy transduction in bacteria. *Curr. Top. Bioenerg.* 6:85–149

38. Henderson, R., Unwin, P. N. T. 1975. Three-dimensional model of purple membrane obtained by electron microscopy. *Nature* 257:28–32

39. Hind, G., Jagendorf, A. T. 1963. Separation of light and dark stages in photophosphorylation. *Proc. Natl. Acad. Sci. USA* 49:715–22

40. Hind, G., Nakatani, H. Y., Izawa, S. 1974. Light-dependent redistribution of ions in suspensions of chloroplast thylakoid membranes. *Proc. Natl. Acad. Sci. USA* 71:1484–88

40a. Hong, Y. Q., Junge, W. 1983. Localized or delocalized protons in photophosphorylation? On the accessibility of the thylakoid lumen for ions and buffers. *Biochim. Biophys. Acta* 722:197–208

41. Horner, R. D., Moudrianakis, E. N. 1983. The effect of permeant buffers on initial ATP synthesis by chloroplasts using rapid mix-quench techniques. *J. Biol. Chem.* 258:11643–47

42. Horner, R. D., Moudrianakis, E. N. 1985. Millisecond kinetics of ATP synthesis driven by externally imposed electrochemical potential in chloroplasts. *J. Biol. Chem.* 260:6153–59

43. Israelachvila, J. N., Marcelja, S., Horn, R. G. 1980. Physical principles of membrane organization. *Q. Rev. Biophys.* 13:121–200

44. Izawa, S. 1970. The relation of postillumination ATP formation capacity (Xe) to H⁺ accumulation in chloroplasts. *Biochim. Biophys. Acta* 223:165–73

45. Izawa, S., Good, N. E. 1969. Effect of p-chloromercuribenzoate and Hg⁺⁺ on chloroplast photophosphorylation. See Ref. 18b, pp. 1288–98

45a. Izawa, S., Heath, R. L., Hind, G. 1969. The role of chloride in photosynthesis. *Biochim. Biophys. Acta* 180:388–98

46. Izawa, S., Ort, D. R., Gould, J. M., Good, N. E. 1974. Electron transport reactions, energy conservation reactions and phosphorylation in chloroplasts. *Proc. 3rd Int. Congr. Photosynthesis,* ed. M. Avron, pp. 449–61. Amsterdam: Elsevier

47. Jagendorf, A. T. 1975. Mechanisms of photophosphorylation. In *Bioenergetics of Photosynthesis,* ed. Govindjee, pp. 414–92. New York: Academic

48. Jagendorf, A. T., Hind, G. 1963. Studies on the mechanism of photophosphorylation. In *Photosynthetic Mechanisms of Green Plants,* ed. B. Kok, A. T. Jagendorf, pp. 599–610. Washington, DC: Natl. Acad. Sci. Natl. Res. Counc. Publ. 1145

49. Johnson, J. D., Pfister, V. R., Homann, P. H. 1983. Metastable proton pools in thylakoids and their importance for the stability of photosystem II. *Biochim. Biophys. Acta* 723:256–65

50. Joliot, P., Joliot, A. 1986. Proton pumping and electron transfer in the cytochrome b/f complex of algae. *Biochim. Biophys. Acta* 849:211–22

50a. Juliano, R. L. 1978. Techniques for the analysis of membrane glycoproteins. *Curr. Top. Membr. Transp.* 11:107–44

51. Junge, W., Hong, Y. Q., Qian, L. P., Viale, A. 1984. Cooperative transient trapping of photosystem II protons by CF_0 after mild extraction of four-subunit CF_1. Proc. Natl. Acad. Sci. USA 81: 3078–82

51a. Junge, W., McLaughlin, S. 1987. The role of fixed and mobile buffers in the kinetics of proton movement. Biochim. Biophys. Acta. In press

52. Junge, W., Schliephake, W. D., Witt, H. I. 1969. Experimental evidence for the chemisosmotic hypothesis. See Ref. 45, pp. 1384–91

53. Junge, W., Schoenknecht, G., Lill, H. 1986. Complete tracking of proton flows mediated by CF_0-CF_1 and by CF_0. See Ref. 11

54. Karlin-Neumann, G. A., Kohorn, B. D., Thornber, J. P., Tobin, E. M. 1985. A chlorophyll a/b protein encoded by a gene containing an intron with characteristics of a transposable element. J. Mol. Appl. Genet. 3:45–61

55. Knowles, J. R. 1976. The intrinsic pKa values of functional groups in enzyme. CRC Crit. Rev. Biochem. 4:165–73

56. Kossiakoff, A. A. 1982. Protein dynamics investigated by the neutron diffraction–hydrogen exchange technique. Nature 296:713–21

57. Kozlov, A., Skulachev, V. P. 1982. An H^+-ATP synthetase: A substrate translocation concept. Curr. Top. Membr. Transp. 16:285–301

57a. Krulwich, T. A. 1983. Na^+/H^+ antiporters. Biochim. Biophys. Acta 726: 245–64

58. Laszlo, J. A., Baker, G. M., Dilley, R. A. 1984. Chloroplast thylakoid proteins having buried amine buffering groups. Biochim. Biophys. Acta 764:160–69

59. Laszlo, J. A., Baker, G. M., Dilley, R. A. 1984. Nonequilibrium of membrane-associated protons with the internal aqueous space in dark-maintained chloroplast thylakoids. J. Bioenerg. Biomembr. 16:37–51

60. Lockau, W., Selman, B. 1975. Correlation of the photosynthetic reduction of diazobenzene sulfonate with the increased binding of the probe to the thylakoid membrane. Z. Naturforsch. Teil C 31:48–54

61. Mao, D., Wachter, E., Wallace, B. A. 1982. Folding of the mitochondrial proton adenosinetriphosphatase proteolipid channel in phospholipid vesicles. Biochemistry 21:4960–68

62. McCarty, R. E. 1970. The stimulation of post-illumination ATP synthesis by valinomycin. FEBS Lett. 9:313–16

63. Means, G. E., Feeney, R. E. 1971.

Chemical Modification of Proteins, p. 70. San Francisco: Holden-Day

63a. Michel, H., Weyer, K. A., Gruenberg, H., Dunger, I., Oesterhelt, D., Lottspeich, F. 1986. The light and medium subunits of the photosynthetic reaction center from R. viridis: isolation of the genes, nucleotide and amino acid sequence. EMBO J. 5:1149–58

63b. Miller, K. R., Staehelin, L. A. 1976. Analysis of the thylakoid outer surface. Coupling factor is limited to unstacked membrane regions. J. Cell Biol. 68:30–47

64. Mitchell, P. 1966. Chemiosmotic coupling in oxidative and photosynthetic phosphorylation. Biol. Rev. Cambridge Philos. Soc. 41:445–540

64a. Murakami, S., Nobel, P. S. 1967. Lipids and light-dependent swelling of isolated spinach chloroplasts. Plant Cell Physiol. 8:657–71

64b. Murakami, S., Packer, L. 1970. Protonation and chloroplast membrane structure. J. Cell Biol. 47:332–51

64c. Nagle, J. F., Dilley, R. A. 1986. Models of localized energy coupling. J. Bioenerg. Biomembr. 18:55–64

64d. Nagle, J. F., Morowitz, H. J. 1978. Molecular mechanisms for proton transport in membranes. Proc. Natl. Acad. Sci. USA 75:298–302

64e. Nagle, J. F., Tristram-Nagle, S. 1983. Hydrogen–bonded chain mechanisms for proton conduction and proton pumping. J. Membr. Biol. 74:1–14

64f. Nelson, N., Nelson, H., Naim, Y., Neumann, J. 1971. Effect of pyridine on the light-induced pH rise and post-illumination ATP synthesis in chloroplasts. Arch. Biochem. Biophys. 145: 263–67

65. Neumann, J., Jagendorf, A. T. 1964. Light-induced pH changes related to phosphorylation by chloroplasts. Arch. Biochem. Biophys. 107:109–19

65a. Nishida, K., Tamai, N., Ryoyama, K. 1966. Light-induced high-amplitude swelling and shrinking in isolated spinach chloroplasts. Plant Cell Physiol. 7:415–28

66. Opanasenko, V. K., Red'ko, T. P., Kuz'mina, V. P., Yaguzhinsky, L. E. 1985. The effect of gramicidin on ATP synthesis in pea chloroplasts: two modes of phosphorylation. FEBS Lett. 187: 257–69

67. Ort, D. R., Dilley, R. A. 1976. Photophosphorylation as a function of illumination time. I. Effects of permeant cations and permeant anions. Biochim. Biophys. Acta 449:95–107

68. Ort, D. R., Dilley, R. A., Good, N. E.

1976. Photophosphorylation as a function of illumination time. II. Effects of permeant buffers. *Biochim. Biophys. Acta* 449:108–24

68a. Packer, L., Siegenthaler, P. A., Nobel, P. S. 1965. Light-induced high-amplitude swelling of spinach chloroplasts. *Biochem. Biophys. Res. Commun.* 18:474–77

69. Park, R. B., Pon, N. G. 1963. Chemical composition and the substructure of lamellae isolated from *Spinach oleracea* chloroplasts. *J. Mol. Biol.* 6:105–14

70. Peters, F. A. L. J., Van der Pal, R. H. M., Peters, R. L. A., Vredenberg, W. J., Kraayenhof, R. 1984. Studies on well-coupled PSI enriched vesicles. Discrimination of flash-induced fast and slow electric potential components. *Biochim. Biophys. Acta* 766:169–78

71. Peters, R. L. A., Rossen, M., van Kooten, O., Vredenberg, W. J. 1983. On the correlation between the activity of ATP hydrolase and the kinetics of the flash-induced P515 electrochromic bandshift in spinach chloroplasts. *J. Bioenerg. Biomembr.* 15:335–46

72. Peters, R. L. A., van Kooten, O., Vredenberg, W. J. 1984. The effect of uncouplers (F)CCCP and NH$_4$Cl on the kinetics of the flash-induced P515 electrochromic bandshift in spinach chloroplasts. *FEBS Lett.* 177:11–16

73. Peters, R. L. A., van Kooten, O., Vredenberg, W. J. 1984. The kinetics of P515 in relation to the lipid composition of the thylakoid membrane. *J. Bioenerg. Biomembr.* 16:283–94

74. Pfister, V. R., Homann, P. H. 1986. Intrinsic and artifactual pH buffering in chloroplast thylakoids. *Arch. Biochem. Biophys.* 246:525–30

74a. Polle, A., Junge, W. 1986. Transient and intramembrane trapping of pumped protons in thylakoids. *FEBS Lett.* 198:263–67

74b. Portis, A. R., McCarty, R. E. 1976. Quantitative relationships between phosphorylation, electron flow and internal hydrogen ion concentrations in spinach chloroplasts. *J. Biol. Chem.* 251:1610–17

74c. Prats, M., Tocanne, J. F., Teissie, J. 1985. Lateral proton conduction at a lipid/water interface. *Eur. J. Biochem.* 149:663–68

75. Prochaska, L. J., Dilley, R. A. 1978. Chloroplast membrane conformational changes measured by chemical modification. *Arch. Biochem. Biophys.* 187:61–71

76. Quinn, P. J., Williams, W. P. 1983. The structural role of lipids in photosynthetic membranes. *Biochim. Biophys. Acta* 737:223–66

77. Remish, D., Bulychew, A. A., Kurella, G. A. 1981. Light-induced changes of the electrical potential in chloroplasts associated with the activity of PSI and PSII. *J. Exp. Bot.* 32:979–87

78. Rottenberg, H. 1985. Proton-coupled energy conversion: chemiosmotic and intramembrane coupling. In *Modern Cell Biology*, 4:47–83. New York: Liss

78a. Rottenberg, H., Hashimoto, K. 1986. Fatty acid uncoupling of oxidative phosphorylation in rat liver mitochondria. *Biochemistry* 25:1747–55

79. Rumberg, B., Reinwald, E., Schröder, H., Siggel, U. 1969. Correlation between electron transfer, proton translocation and phosphorylation in chloroplasts. See Ref. 18b, pp. 1374–82

80. Ryrie, I. J., Jagendorf, A. T. 1972. Correlation between a conformational change in the CF$_1$ protein and high energy state in chloroplasts. *J. Biol. Chem.* 247:4453–59

80a. Sane, P. V., Johanningmeier, U., Trebst, A. 1979. The inhibition of photosynthetic electron flow by DCCD. *FEBS Lett.* 108:136–40

80b. Sayre, R. T., Andersson, B., Bogorad, L. 1986. Experimental determination of the secondary/tertiary protein structure of the 32 KD herbicide binding thylakoid protein of photosystem II. See Ref. 11

81. Schmidt, D. F., Westheimer, F. H. 1971. pK of the lysine amino group at the active site of acetoacetate decarboxylase. *Biochemistry* 10:1249–53

81a. Schoenfeld, M., Kopeliovitch, B. S. 1985. Kinetics of dark proton efflux in chloroplast. *FEBS Lett.* 193:79–92

82. Schreiber, U. 1984. Comparison of ATP-induced and DCMU-induced increases of chlorophyll fluorescence. *Biochim. Biophys. Acta* 767:80–86

82a. Schreiber, U., Avron, M. 1979. Properties of ATP-driven reverse electron flow in chloroplasts. *Biochim. Biophys. Acta* 546:436–47

83. Schreiber, U., Rienits, K. G. 1982. Complimentarity of ATP-induced and light-induced absorbance changes around 515 nm. *Biochim. Biophys. Acta* 682:115–23

84. Schreier, A. A., Baldwin, R. L. 1976. Concentration dependent hydrogen exchange kinetics of ^{3}H-labeled S-peptide in ribonuclease-S. *J. Mol. Biol.* 105:409–26

85. Schuldiner, S., Rottenberg, H., Avron, M. 1973. Stimulation of ATP synthesis

by a membrane potential in chloroplasts. *Eur. J. Biochem.* 39:455–62

86. Sebald, W., Hoppe, J. 1981. On the structure and genetics of the proteolipid subunit of the ATP synthase complex. *Curr. Top. Bioenerg.* 12:2–65

87. Segrest, J. P., Kahane, I., Jackson, R. L., Marchesi, V. T. 1973. Major glycoprotein of the human erythrocyte membrane: Evidence for an amphipathic molecular structure. *Arch. Biochem. Biophys.* 155:167–83

88. Selak, M. A., Whitmarsh, J. 1982. Kinetics of the electrogenic step and cytochrome b_6 and f redox changes in chloroplasts. *FEBS Lett.* 150:286–92

89. Siefermann-Harms, D., Ninnemann, H. 1982. Pigment organization in the LHC of lettuce chloroplasts. Evidence obtained from protection of the chlorophyll against proton attack and from excitation energy transfer. *Photochem. Photobiol.* 35:719–31

89a. Siegenthaler, P. A. 1972. Aging of the photosynthetic apparatus. IV. Similarity between the effects of aging and unsaturated fatty acids on isolated spinach chloroplasts as expressed by volume changes. *Biochim. Biophys. Acta* 275:182–91

90. Sigalat, C., Haraux, F., de Kouchkovsky, F., Hung, S. P. N., de Kouchkovsky, Y. 1985. Adjustable microchemiosmotic character of the proton gradient generated by systems I and II for photosynthetic phosphorylation in thylakoids. *Biochim. Biophys. Acta* 809:403–13

91. Staehelin, L. A. 1986. Chloroplast structure and supramolecular organization of photosynthetic membranes. In *Encyclopedia Plant Physiology* (NS), ed. L. A. Staehelin, C. J. Arntzen, 19:1–84. Berlin: Springer-Verlag

92. Tandy, N. E., Dilley, R. A., Hermodson, M. A., Bhatnagar, D. 1982. Evidence for an interaction between protons released in chloroplast photosystem II water oxidation and the 8000 M_r hydrophobic subunit of the energy coupling complex. *J. Biol. Chem.* 257:4301–7

92a. Teissie, J., Prats, M., Soucaille, P., Tocanne, J. F. 1985. Evidence for conduction of protons along the interface between water and a polar lipid monolayer. *Proc. Natl. Acad. Sci. USA* 82:3217–21

93. Theg, S. M., Belanger, K. M., Dilley, R. A. 1986. Interaction of photosystem I–derived protons with the water-splitting enzyme complex. Evidence for localized domains. *J. Bioenerg. Biomembr.* In press

93a. Theg, S. M., Dilley, R. A. 1986. Protons contained in the thylakoid sequestered domains are utilized for energizing ATP synthesis. See Ref. 11

93b. Theg, S. M., Homann, P. 1982. Light, pH and uncoupler-dependent association of chloride with chloroplast thylakoids. *Biochim. Biophys. Acta* 679:221–34

94. Theg, S. M., Johnson, J. D., Homann, P. H. 1982. Proton efflux from thylakoids induced in darkness and its effect on photosystem II. *FEBS Lett.* 145:25–29

95. Theg, S. M., Junge, W. 1983. The effect of low concentrations of uncouplers on the detectability of proton deposition in thylakoids. Evidence for subcompartmentation and preexisting pH differences in the dark. *Biochim. Biophys. Acta* 723:294–307

96. Thornber, J. P. 1986. Biochemical characterization and structure of pigment proteins of photosynthetic organisms. See Ref. 91, pp. 98–142

97. Tillberg, J. E., Giersch, C., Heber, U. 1977. CO_2 reduction by intact chloroplasts under a diminished proton gradient. *Biochim. Biophys. Acta* 461:31–47

98. Tobkes, N., Wallace, B. A., Bayley, H. 1985. Secondary structure and assembly mechanism of an oligomeric channel protein. *Biochemistry* 24:1915–20

99. Uribe, E. G., Jagendorf, A. T. 1967. On the location of organic acids in acid-induced ATP synthesis. *Plant Physiol.* 42:697–705

100. Uribe, E. G., Li, B. C. Y. 1973. Stimulation and inhibition of membrane-dependent ATP synthesis in chloroplasts by artificially induced K^+ gradients. *Bioenergetics* 4:435–44

100a. Vambutas, V., Beattie, D. S., Bittman, R. 1984. Isolation of protein(s) containing chloride ion transport activity from thylakoid membranes. *Arch. Biochem. Biophys.* 232:538–48

101. van Kooten, O., Leermakers, A. M., Peters, R. L. A., Vredenberg, W. J. 1984. Indications for the chloroplast as a tri-compartment system: microelectrode and P515 measurements imply semilocalized chemiosmosis. In *Advances in Photosynthesis Research*, 2:265–68. The Hague: Junk

102. Vignais, P. V., Block, M. R., Boulay, F., Brandolin, G., Lauquin, G. J. M. 1985. Molecular aspects of structure-function relationships in mitochondrial adenine nucleotide carrier. In *Structure and Properties of Cell Membranes*, ed. C. Bengha, 2:139–79. Boca Raton, Fla: CRC Press

103. Vinkler, C., Avron, M., Boyer, P. D.

1980. Effects of permeant buffers on the initial time course of photophosphorylation and post-illumination phosphorylation. *J. Biol. Chem.* 255:2263–66

104. Vredenberg, W. J. 1981. P515: A monitor of photosynthetic energization in chloroplast membrane. *Physiol. Plant* 55:598–602

105. Westerhoff, H. V., Melandri, B. A., Venturoli, G., Azzone, G. D., Kell, D. B. 1984. A minimal hypothesis for membrane-linked free-energy transduction. *Biochim. Biophys. Acta* 768:257–92

105a. Whitmarsh, C. J. 1986. Proton decay kinetics for vesicles and thylakoids containing buffers—analytical solution. *Photosynthetica.* In press

106. Willey, D. L., Auffret, A. D., Gray, J. C. 1984. Structure and topology of cyt f in pea chloroplast membranes. *Cell* 36:555–62

107. Williams, R. J. P. 1961. Possible functions of chains of catalysts. *J. Theor. Biol.* 1:1–17

108. Woodward, C. K., Hilton, B. D. 1979. Hydrogen exchange kinetics and internal motions of proteins and nucleic acids. *Ann. Rev. Biophys. Bioeng.* 8:99–127

109. Zimanyi, L., Garab, G. 1982. Configuration of the light-induced electric field in thylakoid and its possible role in the kinetics of the 515 nm absorbance change. *J. Theor. Biol.* 95:811–21

110. Zundel, G. 1986. Proton polarizability of hydrogen bonds: infrared methodology, and relevance to electrochemical and biological systems. *Meth. Enzymol.* 127:439–55

Ann. Rev. Plant Physiol. 1987. 38:391–418

EVOLUTION OF HIGHER-PLANT CHLOROPLAST DNA-ENCODED GENES: IMPLICATIONS FOR STRUCTURE-FUNCTION AND PHYLOGENETIC STUDIES

Gerard Zurawski

Department of Molecular Biology, DNAX Research Institute, 901 California Avenue, Palo Alto, California 94304

Michael T. Clegg

Department of Botany and Plant Science, University of California, Riverside, California 92521

CONTENTS

INTRODUCTION 392
THE NATURE OF NUCLEOTIDE SUBSTITUTIONS IN CHLOROPLAST
    DNA EVOLUTION 393
    Estimating the Number of Nucleotide Substitutions for rbcL, atpBE, and rp12 394
IDENTIFICATION OF FUNCTIONALLY IMPORTANT SEQUENCES
    USING THE EVOLUTIONARY FILTERING PROCEDURE 399
    Promoters 399
    Leaders 402
    Ribosome-Binding Sites 402
    Protein-Coding Regions 404
    Intervening Sequences 407
    Transcription Terminators 409
ESTIMATING PHYLOGENETIC TOPOLOGIES FROM CHLOROPLAST-
    DNA SEQUENCE DATA 411
    Protein-Coding Regions 412
    Noncoding Regions 413
CONCLUSIONS 414

391

INTRODUCTION

Our present knowledge of the organization of genetic information on chloro-plast DNA rests almost exclusively on the application of molecular cloning and nucleotide sequencing technologies. Restriction mapping and cloning confirmed the circular nature of the DNA molecule and established that the size of the DNA varies between 120 kb and 190 kb in flowering plant species (27, 42). This range in size is largely accounted for by differences in the extent of reiteration of a significant portion of the genome, which is referred to as the inverted repeat region. The repeated region is unusual only in that it includes the genes for the 4.5S, 5S, 16S, and 23S rRNAs, and its absence in at least one genome indicates that it is not an essential feature of the genome (27). Nucleotide sequencing of cloned chloroplast DNAs is complete for two genomes (26, 34) and reveals the presence of about 50 protein-coding genes and genes for 30 different tRNAs. Sequences for defined regions of many other genomes have also been completed (see below).

The above studies have defined the nature and organization of genes encoded by chloroplast DNA to the extent of fixing their primary structure and, by identifying coding regions, have revealed the organization of genes relative to their neighboring genes. Analysis of chloroplast RNA, together with structural data, further defined the relative organization by revealing those sequences that are cotranscribed and, where relevant, processed.

Given a completeness in structural information, a full understanding of the intricacies of the chloroplast genome requires experimental identification of (*a*) sequences that direct the initiation and regulation of transcription of each gene or set of genes, (*b*) sequences that direct processing of transcripts, (*c*) sequences that direct the initiation and regulation of translation of each protein-coding region, and (*d*) sequences that specify termination of transcrip-tion of each gene or set of genes.

In this review we start with the assumption that the nature of sequence divergence of chloroplast DNA between species is such that regions under functional constraint are more highly conserved than regions under no con-straint. Using this assumption we select sets of functionally equivalent chloro-plast DNA coding regions for which sequence data are available from many higher plant species. We then analyze these sets of data by a procedure we term evolutionary filtering: We align functionally equivalent (i.e. coding for the same gene or intergenic spacer) sequences from various species in such a way as to reveal which nucleotides have altered least through evolution and are therefore presumably under functional constraint. Our use of the word "filtering" is meant to convey a sieving process through which sequences that are evolutionarily relatively variable are discarded and relatively conserved sequences are retained. We argue that this sequence-comparison method can be a powerful tool in the identification of regulatory signals that are encoded by each gene, and that such information can serve as a firm basis for directing further experimentation. We cite a number of cases where such further experimentation has substantiated the method.

We first address the nature of chloroplast DNA evolution as defined by comparative sequencing studies. We examine the rates and biases of nucleotide substitutions in protein-coding genes, the differences in mutational events in protein-coding versus noncoding regions, and heterogeneity in the process of nucleotide substitution in different protein-coding genes. Second, we consider the use of evolutionary filtering to identify and define promoters, ribosome-binding sites, protein-coding regions, terminators, and other sequences that may be involved in the expression of chloroplast-encoded genes. In our third section we examine the application of chloroplast sequence data to phylogenetic studies. When suitable genes are selected, sequence data promise to significantly aid phylogenetic reconstruction.

We focus on the comparison of higher-plant chloroplast DNAs. We have selected data that best exemplify the functional, structural, and evolutionary analysis of chloroplast genes.

THE NATURE OF NUCLEOTIDE SUBSTITUTIONS IN CHLOROPLAST DNA EVOLUTION

The nature of chloroplast DNA evolution has been increasingly defined by comparative DNA-sequence analyses. These have addressed a wide variety of problems, including the estimation of rates of molecular evolution, the molecular nature of mutational changes, the characteristic of mutational biases in nucleotide substitutions, and the detection of statistical heterogeneity in rates of evolution for different genes or between different regions within genes.

The methods of comparative DNA-sequence data analysis involve several operations. Each operation involves certain assumptions (some not explicit) about the phylogenetic history of the sequences and the kinds of genetic changes that have occurred. For example, when aligning two or more sequences, different weights or penalty functions are usually assigned to gaps (additions/deletions) versus nucleotide differences (41). The choice of weights often reflects an investigator's assumption about the likelihood of addition/deletion events relative to that of nucleotide substitutions. Here we adopt an alignment algorithm that minimizes the total number of evolutionary events (n) between compared sequences, where each addition/deletion is regarded as a single event. Because the probability of a particular realization of evolutionary differences is the product of n independent and unlikely events, the minimization of n is regarded as the most probable match. This is tantamount to minimizing addition/deletion changes while maximizing percentage of nucleotide sequence homology. Typically the number of differences between two chloroplast DNA sequences is relatively small and presents little or no difficulty in the application of the alignment rule.

The assumption that the set of sequences under comparison may be traced back to a common ancestral sequence underlies all sequence comparisons. Once sequences have been aligned, it is possible to count the number of pairwise differences between sequences (either as nucleotide differences or as addition/deletion differences). The resulting data are typically arranged in a matrix. The number of differences between any pair of species is assumed to reflect the accumulation of evolutionary events since the separation of the pair of sequences from a common ancestor.

Three major problems can be addressed from the matrix of nucleotide differences. The first problem is to estimate rates of nucleotide substitution from nucleotide differences. Nucleotide substitutions must be estimated because two or more substitutions may have occurred at a site and thus the count of differences underestimates the true number of substitution events. The second problem is to measure the rate of addition/deletion changes relative to substitution events. The third problem is to infer the phylogenetic topology of a set of sampled sequences. It is then assumed that the resulting phylogenetic topology represents the phylogeny of the taxa that provided the sequence data.

The number of nucleotide substitutions between homologous sequences must be estimated using a mathematical model of the substitution process. The simple estimator of Jukes & Cantor (15) assumed that the substitution process was the same for all categories of mutations and that all nucleotides were equally frequent at equilibrium. Many refinements of this estimation model have subsequently been published [reviewed by Li et al (21)].

One should select an estimation model that allows for differences in substitution rates for transitions (purine to purine or pyrimidine to pyrimidine) and for transversion events (purine to pyrimidine or pyrimidine to purine) because these rates are almost always found to differ when nucleic acid sequence data are analyzed. Here we employ an estimator of Kimura (17) that includes three separate parameters accounting for the three categories of mutational events.

Estimating the Number of Nucleotide Substitutions for rbcL, atpBE, and rp12

An intensively studied region of chloroplast DNA from several species is that which contains the adjacent and divergently transcribed genes *rbcL* and *atpBE* (Figure 1). These genes encode, respectively, the large subunit of ribulose 1,5-bisphosphate carboxylase and the β and ε subunits of ATPase. Complete DNA sequence data for *rbcL* and *atpB* are available for five species of

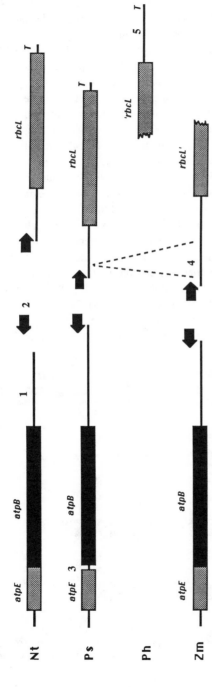

Figure 1 Comparative structural organization of the chloroplast DNA *rbcL*, *atpBE* region. The regions encoding the large subunit of ribulose 1,5-bisphosphate carboxylase (*rbcL*) and the β and ε subunits of ATPase (*atpBE*) are shown for *Nicotiana tabacum* (Nt), *Pisum sativum* (Ps), *Phaseolus albus* (Ph), and *Zea mays* (Zm) chloroplast DNAs. Features that have varied through evolution and that are discussed in the text are (numbered in the figure) *1* the nontranslated, transcribed leader regions; *2* the promoter regions (shown as arrows); *3* the *atpB-atpE* intergenic regions; *4* a 133-bp insertion-deletion in the *rbcL* leader region; and *5* the *rbcL* transcription terminator (T) region. The non-protein-coding regions are approximately to scale, while the protein-coding regions are at half scale.

flowering plants (Table 1): barley [Hv (46, 47)], maize [Zm (19, 24)], spinach [So (43, 44)], tobacco [Nt (32, 33)], and pea [Ps (48, 50)]. A second intensively studied region resides in the inverted repeats and contains the genes for tRNA$_{ile}$ *(trnI)* and for the ribosomal proteins L23 *(rpl23)*, L2 *(rpl2)*, and S19 *(rps19)* (Figure 2). Complete DNA sequence data for *rpl2* are available from the nine flowering plants listed in Table 1.

Before considering rates of evolution, we must investigate biases in the kinds of mutational events that are observed when protein-coding regions are compared. A common bias is an excess of transition substitutions over transversion substitutions. A convenient measure of the transition bias is the ratio of transition to transversion events (T_0), normalized so that no bias yields a ratio of one (30). For barley versus maize, the value of T_0 is 1.55 and 1.63 for *rbcL* and *atpB,* respectively, indicating a moderate excess of transition events. When the comparison is extended to all monocot versus all dicot sequence data, the ratios are $T_0 = 1.58$ and 1.67 for *rbcL* and *atpB,* respectively. Measured in terms of average numbers of third-codon position substitutions, the evolutionary distance between the monocot and dicot sequences is about four times that between the barley and maize sequences. It therefore appears that the transition bias is relatively stable over long periods of evolutionary time (from 50 to 110 million years). In contrast, T_0 for *rpl2* is 1.33 and 1.27 for the tobacco-spinach and dock-spinach comparisons, respectively. These values are significantly different from the *rbcL* and *atpB* values

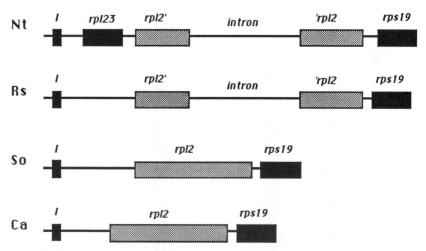

Figure 2   Comparative structural organization of the region encoding tRNA[Ile] (I) and ribosomal proteins L23 *(rpl23)*, L2 *(rpl2)*, and S19 *(rps19)* on chloroplast DNA. The species shown are *Nicotiana tabacum* [Nt(40)], *Rumex sp.* (Rs), *Spinacea oleracea* (So), and *Cerastium arvensae* (Ca). The coding regions are bars and are approximately to scale. The direction of transcription is from left to right.

Table 1   Systematic relationships among the flowering plant genera that provided *rpl2*, *rbcL*, and *atpB* sequence data.

Class	Family	Genus and species	Name[a]
		rpl2	
Dicots	Chenopodiaceae	*Spinacia oleracea*	So, spinach
		Kochia americana	Ka
		Chenopodium murale	Cm
		Beta vulgaris	Bv
	Amaranthaceae	*Amaranthus salicifolius*	As
	Caryophyllaceae	*Cerastium arvense*	Ca
	Polygonaceae	*Rumex* sp.	Rs, dock
	Solanaceae	*Nicotiana tabacum*	Nt, tobacco
		Nicotiana debneyi	Nd
		rbcL and *atpB*	
	Chenopodiaceae	*Spinacia oleracea*	So, spinach
	Fabaceae	*Pisum sativum*	Ps, pea
		Phaseolus albus	Pa
	Solanaceae	*Nicotiana tabacum*	Nt, tobacco
Monocots	Gramineae	*Hordeum vulgare*	Hv, barley
		Zea mays	Zm, maize
		Pennisetum americanum	Pe
		Cenchrus setigerus	Cs
		Triticum aestivum	Ta, wheat

[a] Abbreviation and/or common name used in text

and suggest that the substitutional bias may differ among different protein-coding regions.

To investigate further the potential causes for differences in estimates of T_0 between genes it is necessary to consider the two classes of transversion: (*a*) type-I events which do not change the number of hydrogen bonds ($A \rightleftharpoons T$ or $G \rightleftharpoons C$), and (*b*) type-II events, which alter the hydrogen bonding ($A \rightleftharpoons C$ or $G \rightleftharpoons T$). A ratio of the two classes of transversion events (T_1), calculated as twice the type-I events divided by the sum of both kinds of events, is expected to be 1 when there is no bias. For the monocot–dicot comparison this ratio is 1.01 and 1.28 for *rbcL* and *atpB*, respectively; for the tobacco-spinach and dock-spinach comparisons for *rpl2* it is 0.0 and 0.4, respectively. Again, there is a significant difference between *rpl2* and *rbcL-atpB* in the pattern of transversion events. In the case of *rpl2*, type-I transversions occur at a rate much below that expected assuming no bias.

An analysis of the transition-transversion bias based upon partitioning the *rbcL-atpB* data by codon position reveals that estimates of T_0 are heterogeneous over codon position, with the third codon positions showing the greatest excess of transition events. With respect to transversion events, the estimates of T_1 are heterogeneous over codon position for *rbcL*, but not for *atpB;* and there is no consistent pattern over codon positions between the two genes (30). In the case of *rpl2*, type-I transversion events are almost completely absent for all three codon positions. These results indicate that the process of nucleotide substitution is extremely biased for *rpl2*.

Estimates of the average number of nucleotide substitutions per nucleotide site, accounting for the bias in transition and transversion rates, and partitioned by codon position, are given in Table 2. The estimates of substitution numbers are denoted K_1, K_2, and K_3 for first, second, and third codon positions, respectively. The values of K are heterogeneous over codon positions within a gene, with K_3 averaging about 9.5 times K_2 for *rbcL* and *atpB*. The large excess of third-position events almost certainly reflects the fact that most such events are synonymous, whereas all second-position events cause missense changes and are likely to be subject to strong selective constraints. Considered within codon positions, the values for K are homogeneous for *rbcL* and *atpB*. Thus *rbcL* and *atpB* appear to be evolving at approximately the same rate in the few flowering plants for which data are available.

The situation with *rpl2*, again, stands in marked contrast to that of *rbcL* and *atpB*. The tobacco-spinach comparison reveals that $K_2 > K_3$; moreover, the estimate for K_3 is more than tenfold below the estimates from *rbcL* and *atpB* in the same plant taxa. The above analysis of substitution events reveals striking anomalies between the rates and patterns of nucleotide change for *rbcL/atpBE* versus *rpl2*. There are at least two potential explanations for this anomaly. The first stems from the location of *rpl2* in the inverted repeats, a region known to be subject to a copy correction mechanism. A problem with

Table 2 Estimates of the per-nucleotide number of substitution events for comparisons between plant taxa for *rbcL, atpB,* and *rpl2*[a]

Taxa compared	Gene	Codon position		
		1st	2nd	3rd
Monocots-dicots		0.083	0.038	0.496
Tobacco-spinach	*rbcL*	0.061	0.043	0.213
Barley-maize		0.036	0.019	0.135
Monocots-dicots		0.095	0.039	0.501
Tobacco-spinach	*atpB*	0.050	0.020	0.224
Barley-maize		0.023	0.014	0.112
Tobacco-spinach	*rpl2*	0.015	0.031	0.015

[a] Estimates of the number of nucleotide substitutions were obtained using the formula of Kimura (17).

this explanation is that it does not account for the differential reduction in third-position substitutions relative to second-position substitutions in *rp12*. It is not obvious how a copy correction mechanism would retard third-position substitutions while allowing the rate of second-position substitutions to remain at that seen for *rbcL/atpBE*. A comparative analysis of protein-coding genes other than *rp12* from the inverted repeat region would help determine whether these anomalies are an inverted repeat-linked phenomenon. The second potential explanation stems from the presence of a large open reading frame on the opposite strand of much of the *rp12* coding region. This open reading frame, if functional, could impose a substantial selective constraint on third-position substitutions in *rp12*. For this reading frame to be functional, however, two introns must be invoked—one to provide a translation start codon and another to splice out the *rp12* intron in those species with this intron. Another problem is that the retarded rate of third-position change in *rp12* is maintained outside of this open reading frame.

IDENTIFICATION OF FUNCTIONALLY IMPORTANT SEQUENCES USING THE EVOLUTIONARY FILTERING PROCEDURE

For the purpose of discussion a protein-coding gene is considered to be composed of the following parts: the promoter, or site of transcription initiation; the leader, or transcribed and nontranslated 5' upstream region; the ribosome-binding site, or region defining initiation of translation; the coding region; and the terminator, or region that specifies transcription termination. The discussions below highlight the utility of comparative analysis for identifying these important sequences and for revealing otherwise unrecognized aspects of these features.

Promoters

The definition of chloroplast DNA sequences upstream from regions encoding different proteins, together with studies that defined the 5' end of the relevant mRNAs, enabled comparison of sequences adjacent to the proposed sites of transcription initiation of chloroplast genes (42). These comparative sequence studies were unconvincing in several respects. First, chloroplast mRNA's often undergo posttranscriptional processing, thus making possible the misassignment of sites of transcription initiation (6, 25). Second, the nature of prokaryotic promoter regions is such that, although consensus sequences can be accurately derived, most positions have substantial tolerance to change from the consensus sequence (31, 36). This can sometimes result in functional promoters that have little apparent homology to the consensus sequence (31). Third, both chloroplast 5' upstream regions and the prokaryotic consensus promoter are particularly AT rich (42). Since most of the sometimes quite large 5' upstream region is presumably not required for promoter function, the significance of observed homologies with prokaryotic promoters was more difficult to establish.

An alternate and perhaps more powerful method of comparative analysis is to compare the same region between species. Such a study is illustrated in Figure 3, which compares the 5' upstream sequences for *rbcL* from six higher plants. These sequences have been aligned for maximal homology consistent with minimal addition/deletion events. As is typical of such regions, a dominant feature of the divergence pattern is addition/deletion ranging from single nucleotides to 100 or more nucleotides. Such change, when considered together with nucleotide substitutions, can be used to identify regions that are evolving less rapidly than the surrounding DNA. The analysis in Figure 3 identifies such a region (between positions -345 and -287). Most of this approximately 60-bp region is proximal to the proposed sites of transcription initiation for five of the seven species with two sites unassigned (Ta and Cs). The highly conserved nature of this $\sim$60-bp sequence and its location relative to the proposed site of transcription initiation strongly suggest a promoter function for this region. This suggestion has been tested directly in a chloroplast in vitro transcription system by demonstrating that the region from position -336 to -297 (Figure 3) is capable of driving efficient transcription of a heterologous tester gene (12). The evolutionary filtering method has also been applied and confirmed using similar techniques for the promoter regions of *atpBE, psbA,* and *trnM1* (11, 12).

The identification and in vitro expression of such promoter sequences have facilitated the further characterization of a typical chloroplast promoter as containing two critical sequence elements called cpt1 and cpt2 (Figure 3) (11). The cpt1 element is analogous to the prokaryotic consensus "-35" sequence and is usually TTGACA or a near homologue. The cpt2 element is analogous to the prokaryotic consensus "-10" sequence and is usually a homologue of TATAAT. Mutational analyses, including the construction of hybrid promoters (12, 23), are providing an increased understanding of the roles these sequence elements play in determining the level of gene expression provided by individual chloroplast promoters.

These successes point to the value of evolutionary filtering in locating potential promoter regions, especially those for which RNA mapping data are unavailable and for those with poor homology to consensus sequences. The choice of compared sequences may, however, sometimes influence the conclusions from the filtering process. In the above case of the *rbcL* promoter, the promoter is sufficiently conserved across a wide range of higher plants to enable its tentative identification. In contrast, the *atpBE* promoter is not sufficiently conserved between the monocots and dicots to enable its

```
      -350      -340          -330          -320          -310         -300          -290            -280            -270          -260
Zm  aagattagggtTTGGGTTGCGCtATAtcTATcAAAGAGTATAcAATAATgATGgATTTggtgAATCAAAT  CAcGGTttaat  AAcgaaccgtgttaacttaccataa
Hv  gggatta  atTTGGGTTGCGCtATAccTATcAAAGAGTATAcAATAATgATGgATTTggtaAATCAAAT  CAtGGTttaat  AAcgaaccgtgttaacttaccataa
Ta  ggaatta  atTTGGGTTGCGCtATAtcTATcAAAGAGTATAcAATAATtATGgATTTggtaAATCAAAT cCAtGGTttaat  AAcgaaccgtgttaacttaccataa
Cs  gggatt  ggtTTGGGTTGCGCtATAtcTATcAAAGAGTATAcAATAATgATGgATTTggtgAATCAAAT cCAtGGTttaat  AAcgaaccgtgttaacttaccataa
Ps  aaaaaaaacggTTGGGTTGCGCcATAcaTATgAAAGAGTATAgAATAATgATGtATTTcccaAATCAAATatCAtGGTaaaatttAAcaatga
Nt  aaaaagaaaaaTTGGGTTGCGCtATAtaTATgAAAGAGTATAcAATAATgATGtATTTggcaAATCAAATacCAtGGTctaa  tAAtcaaa
So  aacggttacggTTGGGTTGCGCcATAtaTATgAAAGAGTATAcAATAATgATGtATTTggcgAATCAAATa CAtGGTcta  ttAAcgaaa
                   ctp1                    ctp2

          -240      -230                                                                            -220      -210      -200      -190
Zm  caacaactcaattcctatcgaattc                                                                 ctatagtaaaattcctataggatagaacgtacacagggtgt
Hv  caacaactcaattcctatcgaattc                                                                 ctatagtggaattcctataggatagaacatacacagggtgt
Ta  caacaactc                                                                                         gaattcctataggatagaacatacacagggtgt
Cs  caacaactcaattcctatcgaattcttttcctatcgaattcctattcctatcgaattcctattcctatctatagtataattcctatagaatagaacgtacacatggtgt
Ps              [  α  ]          [  α  ]     [  α  ][  α  ]     [  α  ][  α  ]           [  α  ]
Nt             [        β       ]  [        β        ]  [        β        ]
So

     -180         -170      -160      -150      -140      -130      -120                     -110      -100
Zm  atac   ataaatgaatcaaacatattacttaacttaagcatactcctttttttatttaATgagTTGATA          ttaattaaatatcatttttttttt
Hv  acgcattatatatgaatgaaacatattcattaacctaagcatgccctcaattttctttaATgagTTGATA     ttatattaattgaatatccttttttgttttacg
Ta  acgcattatatatgaatgaaacatattcattaacctaagcatgccctcaattttctttaATgagTTGATA     ttatattaattgaatatccttttttgttttacg
Cs  acgcattatatatgactgaaacatattcattaacttaagcatactcctttttttatttaATgaaTTGATA          ttaattgaatatcttttttttttttttt
Ps                                                   cccattcggATtcaTTGATA     atattagttgatggatcatttgta
Nt                                                     cattctgATtagTTGATAatattagtattagttggaaattttgtgaaagattccta
So                                                    ccattttgATtagTTGATA     atattaattgagaatttgatgaaagattgcta

          -90        -80           -70        -60          -50          -40        -30        -20        -10
Zm        agatttttttgcaAAgGTTTc    tttcgcctAAtcc taTCGAGTtgtCCcTGTgc   ttgtgtgAAttcTTAAttcatga GTTGTAGGGagggacTTATG
Hv        agattttttgctAAaGTTTc   atttacgcctAAttaacaTCGAGTagaCCcTGTta   tggtgagAAttcTTAAttcaaga GTTGTAGGGagggacTTATG
Ta        agatt tttgctAAaGTTTc   atttacgcctAAttaacaTCGAGTagaCCcTCTtattg   tgagAAttcnnnnnnnnnnnnnnnnnnnnnnnnnnnnnnnnnnn
Cs  tttttttagatt tttgcaAAgGTTTc   atttacgcctAAtccataTCGAGTagaCCcTGTcgttg   tgagAAttcTTAAttcatga GTTGTAGGGagggacTTATG
Ps                 aaAAgGTTT atattaactcctAAgttatgTCGAGTagaCCtTGTtgttgtggctatAAttcTTAAttcatgc GTTGTAGGGggagatTTATG
Nt                tgaAAaGTTTcatattaacacggAAttcgtgTCGAGTagaCCtTGTtg   tggtgagAAttcTTAAttcatga GTTGTAGGGagggatTTATG
So                taaAAgGTTTcatattaaggcctAAtttatgTCGAGTagaCCtTGTtgct tggttgtAAaaaTTAAaatttgaaGTTGTAGGGagggacTTATG
                                                                                                         rbs
```

Figure 3 Comparative sequences of *rbcL* promoter and leader regions. The sequences represented are *Zea mays* [Zm(24)], *Hordeum vulgare* [Hv (46)], *Triticum aestivum* [Ta(13)], *Cenchrus setigerus* (Cs), *Pisum sativum* [Ps(48)], *Nicotiana tabacum* [Nt(32)], and *Spinacia oleracea* [So(43)]. The sequences commence just prior to the promoter region (overlined) and end at the ATG codon that specifies the start of the *rbcL* protein-coding region. The conserved promoter sequence elements, cpt1 and cpt2 (11) are underlined. The sequences are aligned to display maximal homology consistent with the fewest insertion-deletion changes. Regions where two or more continuous nucleotides are in common with all species are in uppercase; α and β indicate particular repeat sequences. The numbering is for the *Zea mays* sequence and is relative to the first *rbcL* ATG codon. Sites identified as encoding the 5' ends of *rbcL* mRNA are underlined and are located at positions −306 to −308 and −63 to −64; n are undetermined nucleotides.

identification by evolutionary filtering. Figure 4 shows that, by comparing only the dicot species, a highly conserved region can be located. This identified region has been found to sustain efficient transcription of a heterologous tester gene in the in vitro system (12). The degree of divergence in the equivalent region of the five available (graminaceous) monocot sequences is insufficient to identify the promoter region. The *atpBE* promoter data highlight the advantage of having available for the analysis sequences from species that have various degrees of relatedness. The analysis can then proceed by sequentially including data from more distantly related species until conserved sequences are identified.

Leaders

The transcribed, but untranslated, 5' upstream regions of chloroplast DNA transcriptional units vary between several to several hundred nucleotides (42). The part of this leader region that is immediately proximal to a protein-coding region probably contributes to ribosome binding (see the next section). However, the remainder has no clear role. In those cases that have been critically examined [*rbcL* and *atpBE* (43; this work)] these regions do not seem to encode the ability to form secondary structures that in prokaryotic systems signify regulatory functions (29). Examination of such chloroplast DNA regions by evolutionary filtering (e.g. Figure 3) reveals few conserved sequences. Two equally plausible possibilities can account for these observations. First, these regions may have no function. Second, the nature of their function may be such that only relatively small stretches of particular sequences are critical to that function. In this latter case, for example a regulatory protein binding sequence, the identification of such sequences by evolutionary filtering would only be plausible if data were available from very many more species and is more likely to be accomplished by direct genetic or biochemical means.

Ribosome-Binding Sites

Since the translational machinery of the chloroplast is typically prokaryotic, it was anticipated that sequences adjacent to translation start codons of chloroplast genes would encode ribosome-binding functions similar to those identified in prokaryotes. The prominent feature of prokaryotic ribosome-binding sites is the "Shine-Dalgarno" (SD) sequence, which is a short nucleotide stretch complementary to the 3' end of 16S rRNA occurring 4–10 nucleotides proximal to the initiator ATG (10, 39). Examination of putative chloroplast DNA ribosome-binding sites reveals that many have identifiable SD sequences that seem to follow the same rules regarding extent of complementarity and location relative to the initiator ATG as those established for gram-negative prokaryotes (2, 48). Examples include *rbcL,* where the SD sequence

```
        -400      -390      -380      -370      -360      -350      -340      -330      -320      -310      -300
                                                       ctp1
Ps  AAaAAagatattcttgacCTTGACAGTGatcTATGTTGTATATGTAaATCCTAGATGTaAAAATcggcaGAATTttcccaatAAaagaaaAAttggtaaaatgatatgc
So  AAtAAttcgaaattagtCTTGACAGTGgtaTATGTTGTATATGTAtATCCTAGATGTgAAAATatgcaGAATTctctcatgAAaggataAAagaataggctactcataa
Nt  AAtAAtaagaacttccccCTTGACAGTGgtaTATGTTGTATATGTAaATCCTAGATGTgAAAATatacgGAATTcctctatgAAtctatgAAaggtataaaaagaaaga
```

```
        -340    ctp1  -320      -310  ctp2  -300      -290      -280      -270      -260      -250
Hv  ATACTAAtAAAATTCTtTGTTGACAGCAATCATGCTTCACAGTAGTATATATTTTGTATATCGAAGTCcTAGATAGGAAaGTAGAGTAGGCACAGATCCTcCACAAAG
Cs  ATACTAAgAAAATTCTtTGTTGACAGCAATCATGCTTCACAGTAGTATATATTTTGTATATCGAAGTCtTAGATAGGAAgGTAGAGTAGGCACAGATCCTtCACAAAG
Zm  ATACTAAgAAAATTCTcTGTTGACAGCAATCATGCTTCACAGTAGTATATATTTTGTATATCGAAGTCcTAGATAGGAAaGTAGAGTAGGCACAGATCCTtCACAAAG
Pe  ATACTAAgAAAATTCTtTGTTGACAGCAATCATGCTTCACAGTAGTATATATTTTGTATATCGAAGTCtTAGATAGGAAgGTAGAGTAGGCACAGATCCTtCACAAAG
Ta  ATACTAAtAAAATTCTtTGTTGACAGCAATCATGCTTCACAGTAGTATATATTTTGTATATCGAAGTCcTAGATAGGAAgGTAGAGTAGGCACAGATCCTcCACAAAG
```

Figure 4   Comparative analysis of the apBE promoter region. (*top*) The sequences from *Pisum sativum* [Ps(50)], *Spinacia oleracea* [So(44)], and *Nicotiana tabacum* [Nt(32)]. The numbers refer to nucleotides proximal to the translation initiation codon for Ps apB. The overlined region represents the functional synthetic So sequence. The mature 5' end coding residues for Ps and So apBE mRNAs are underlined, as is the cpt1 chloroplast promoter sequence element. Uppercase nucleotides show two or more conserved nucleotides in a row. (*bottom*) The sequences from *Hordeum vulgare* [Hv(47)], *Cenchrus setigerus* (Cs), *Zea mays* [Zm(19)], *Pennisetum americanum* (Pe), and *Triticum aestivum* [Ta(13)]. The numbers refer to nucleotides proximal to the translation initiation codon for Hv apB. The mature 5' end coding residues for Hv apBE mRNAs are underlined. The cpt1 and cpt2 chloroplast promoter sequence elements are overlined.

GGAGG is conserved in all species but pea (Figure 3), which has an AGG or an GGAG SD sequence, and *rps19,* where the AGGAG SD sequence is conserved in all species and is strikingly surrounded by highly variable sequences (Figure 5).

There are, however, examples in chloroplast DNA of translation initiation regions that have no recognizable prokaryotic-like ribosome-binding sequences. For example, there is no obvious SD sequence proximal to the proposed translation initiation codon for *atpB,* and furthermore, this sequence is poorly conserved between species and has undergone extensive addition/ deletion change (Figure 6). The nature of ribosomal binding to such sites is more likely to be governed by the highly conserved sequences immediately after the initiation codon. This suggestion is consistent with data from prokaryotes that also implicate such distal sequences in ribosome-binding events at some sites (5). The spinach *atpB* site is known to function efficiently in *E. coli* (50).

Protein-Coding Regions

Since the N-termini of only a few chloroplast-encoded proteins have been determined, other criteria have been applied to define the start of protein-coding regions. These criteria include (*a*) the first ATG of the open reading frame, (*b*) the degree of conservation between the proposed protein sequence and protein(s) of homologous function, and (*c*) proximity of the proposed start site to a likely SD site. These criteria suffer if the protein initiates with GTG (e.g. Figure 5), if the true initiation site does not have an obvious SD sequence, or if a homologous protein sequence is unavailable. Fortunately, since protein-coding and non-protein-coding regions evolve differently, the evolutionary filtering method can be used as a further criterion in defining the coding region. An example in which evolutionary filtering has been used effectively is the assignment of the first of two likely ATG codons as the translation initiation codon for the chloroplast gene *psbA* (22, 45).

An instructive case of uncertain assignment of a protein-coding region is that of *rpl23,* which was assigned on the basis of 23% amino acid homology between the predicted product of the coding region and the *E. coli* L23 ribosomal protein (40). Confidence in this assignment increases when it is noted that genes encoding ribosomal proteins L23, L2, S19, L22, S3, and L16 are adjacent and colinear in both chloroplast and *E. coli* DNAs (40). An examination of the chloroplast *rpl23* region by evolutionary filtering of the nine available higher-plant sequences, however, yields a surprising result. Figure 8 shows that this region, in all but two genomes examined, has undergone the extensive addition/deletion change that is characteristic of non-protein-coding regions. That these changes have not occurred in multiples of three nucleotides indicates that this region does not encode the L23

rpl2
```
            700        710        720        730        740        750        760        770        780        790        800
So   GGGTGGTGCAAGGGAGGGCCCCAATTGGTAGAAAAAGCCCTACAACCCCTTGGGGTTATCCTGCACTTGGAAGAAGAAGTAGAAAAAGGAATAAATATAGTGATAATTTTATTATTCGTCG
Cm   GGGTGGTGCAAGGGAGGGCCCCAATTGGgAGAAAAAacCCcCACAACCCCTTGGGGTTTATCCTGCACTTGGAAGAAGAAGTAGAAAAAGGAATAAATATAGTGATAATTTTATTATTCGTCG
As   GGGTGGTGaGGGGAGGGCCCCAATTGGTAGAAAAAacCCCACAACCCCTTGGGGTTTATCCTGCACTTGGAAGAAGAAGTAGAAAAAGGAATAAATATAGTGATAATTTTATTATTCGTCG
Bv   GGGgGGTGCAAGGGAGGGCCCCAATTGGTAGAAAAAaaCCCCACAACCCCTTGGGGTTTATCCTGCACTTGGAAGAAGAAGTAGAAAAAGGAATAAATATAGTGATAATTTTATTATTCGTCG
Ca   GGGTGGTGCAAGGGAGGGCCCCAATTGGTAGAAAAAacCCCACAACCCCTTGGGGTTATCCTGCACTTGGAAGAAGAAGTAGAAAAAGGAATAAATATAGTGATAATTTTATTATTCGTCG
Nt   GGGTGGTGCAAGGGAGGaGCCCCAATTGGTAGAAAAAaaaCCCACAACCCCTTGGGGTTATCCTGCACTTGGAGaAGAAGAAGTAGAAAAAGGAATAAATATAGTGATAATTTgATTcTTCGTCG
Nd   GGGTGGTGCAAGGGAGGGCCCCAATTGGTAGAAAAAaaaCCCACAACCCCTTGGGGTTATCCTGCACTTGGAGaAGAAGAAGTAGAAAAAGGAATAAATATAGTGATAATTTgATTcTTCGTCG
Rs   GGGTGGTGCAAGGGAGGGCCCCAATTGGTAGAAAAAAGCCCcaCAAtaaTGGGGcTtCTCCTGCACTTGGAAGAAGAAGTAGAAAAAGGAATAAATATAGTGATAATTTgATTATTCGTCG
```

```
            810        820        830        840        850        860
So   ACGTAGTAAATAGGAAAGAAAATGAAA    ATAGAATTAGTTTCTTCGTCTTTTACAT
Cm   ACGTAGTAAATAGGAAAGAAAATGAAA    ATAGAATTAGTTTCTTCGTtTTTACAT
As   ACGTAGTAAATAGGAAAGAAAATGAAA    ATAGAATTAGTTTCTTCGTCTTTTACAT̃
Bv   ACGTAGTAAATAGGACAGAAAATGAAA    ATAGAATTAGTTTCTTCGTCTTTTACAT̃
Ca   ACGTAGTAAATAGGAAA             AAGTctAATTAGTTTCTTCGTCTTTTACAT
Nt   cCGTAGTAAATAGGAgAGAAAATcgAATTAaAttcTTcGTTTtTaC
Nd   cCGTAGTAAATAGGAgAGAAAATcgAATTAaATcgATTTTaC
Rs   ACGTAGTAAATAGGAgAGAAAAtTgAATTAgttccTcGTcTtT
```

rbs   rps19
```
                                                        10         20
So   AAAAAAAATAGGAGTAATTAACTGTGACACGTTCACTAAAAAAATCCTTT
Cm   AAAAAAAtAtAGGAGTAATTAgCTGTGACACGTTCACTAAAAAAAcCCTTT
As   AAAAAAAAAcAAAtAtAGGAGTAATTAACTGTGACACGTTCACTAAAAAAATCCTT
Bv   AAAAAAAtAtAGGAGTAATTAgcTGTGACACGTTCACTAAAAAAAAcCCTTT
Ca   AAAAAtttttAGGAGTAATTAACTGTGACACGTTCACTAAAAAAATCCTTT
Nt   AAAAAAAAAAAAAAAATAGGAGTAA    gctGTGACACGTTCACTAAAAAAATCCTT
Nd   AAAAAAAAAAAAAAAATAGGAGTAA    gctGTGACACGTTCACTAAAAAAATCCCTT
Rs   AAAtAAAAAAATAGGAGTAATTAAccGTGACACGTTCACTAAAAAAATCCTTT
```

Figure 5 Comparative analysis of the *rp12-rps19* intergenic region. The numbers refer to nucleotides distal to the translation initiation codons for *rp12* and *rps19* in the *Spinacea oleracea* (So) sequence (49). The sequences are aligned to show maximal homology consistent with minimal addition/deletion change. Nucleotides not conserved with respect to So are shown in lowercase. The translation stop codons for *rp12* and the SD sequences (called rbs) and start codons for *rps19* are underlined. The key to the species represented is contained in Table 1.

```
Ps  ttTTtcaaaaaaaaaccgatatTTTgcaa                                          TATGAcAAtaacTCCTccccCTTCtgatactgaggtTTCtgtactt
So  atTTtggattcgataatttcattTTTgcaaaaa attccgacata        ctttactatat atTATGAgAAtcaaTCCTactaCTTCtgatctggggtTTCcacactt
Nt  atTTtgaattcgata        attTTTgcaaaaacattccgacata tttatttatttattatTATGAgAAtcaaTCCTactaCTTCtggttctggggtTTCcacgctt
Hv  taTTcaacaataaaaaagaaaaaTTTcgacaaattccttttttt        aattatgtgataatTATGAgAAccaaTCCTactaCTTCtcgtcccggggtTTCcacaagt
Ta  taTTcaacaataaaaaagaaaaaTTTcgacaaattccttttttt        aattatgtgataatTATGAgAAccaaTCCTactaCTTCtcctcccggggcTTCcacaatt
Zm  taTTcaacaagaaaaaa      aTTTcgacaaattcttttttttgaaaattatgtgtaatTATGAgAAccaaTCCTactaCTTCgcgtcccgggatTTCcacaatt
        -50        -40        -30        -20        -10        10        20        30        40
```

Figure 6 Comparative analysis of the *atpB* translation initiation region. The numbers refer to nucleotides relative to the start codon for Zm *atpB* and nucleotides are in uppercase where two or more in a row are conserved in all species. The key to the species represented is contained in Table 1.

protein in all species examined except for the two *Nicotiana* species (Figure 2 and Figure 7) and for the liverwort *Marchantia polymorpha* (26). This result may indicate either a loss of requirement for L23 protein in chloroplast ribosomes or a translocation of this gene to the nuclear genome of at least some plants.

The amino acid sequence itself is also of interest in that it specifies the structure and, therefore, the function of the gene product. Comparative sequence analysis of proteins is of value as a first step in defining the most important structural features of the particular protein. However, given the slow pace of evolution of protein-coding regions in general, it is preferable to compare widely diverged sequences such as a chloroplast protein and its prokaryotic homologue. Examples of the usefulness of such studies include the *rbcL* gene product (4) where sequences implicated by biochemical criteria as being critical to the active site are more highly conserved than other regions. The degree of confidence in such studies is especially sensitive to the size and nature of the database. For example, a comparison between tobacco chloroplast ribosomal protein L2 and its *E. coli* counterpart shows that the C-terminal region of this protein is relatively conserved (40). However, examination of further (in this case chloroplast) L2 proteins shows that the *rpl2* coding region can be up to 22 residues longer at the C-terminus in some species (Figure 5). This result suggests that the additional C-terminal residues are not essential to L2 function in those proteins that contain them.

Intervening Sequences

A final complication in correctly assigning coding regions for the chloroplast genome stems from the presence of intervening sequences in some genes. In the tobacco chloroplast genome, 15 genes are split, 7 of which encode tRNAs and 8 of which encode proteins (34). These introns seem to be representatives of at least two groups of introns (35). Group I have conserved internal sequences consistent with secondary structure (e.g. *trnL*) and are self-excised in other systems as linear molecules (20). Group II also have conserved secondary structures and conserved boundaries and may be self-spliced as circular molecules (28). An example of the latter class is the intron in *trnVl* for which Clegg et al (3) have shown that the rate of evolution is heterogeneous when compared across spinach, tobacco, maize, and barley. In particular, a central region of 30 nucleotides in the ~600 nucleotide intron are rigidly conserved, with just one nucleotide difference between monocot and dicot species. The identification of such regions by the evolutionary filtering method provides a basis for further experiments that will address the nature of intron processing in chloroplasts as compared to introns in other genomes.

One higher-plant chloroplast gene, *rpl2*, is of particular interest in that it is not split by an intron in all species. Utilizing *rpl2* intron-specific probes, we

```
                  rpl23     10        20        30              40        50        60        70         80        90
Nt GGGAAGAAAATcGATTTATGGATGGAATCAAATATGCAGTATTTACAGACAAAAGT      ATTCgGTTATTGGGGAAAAAtCAATATACTTCTAATGTCGAATCAGGATCAACTAG
Nd GGGAAGAAAATcGATTTATGGATGGAATCAAATATGCAGTATTTACAGACAAAAGT      ATTCgGTTATTGGGGAAAAAtCAATATACTTCTAATGTCGAATCAGGATCAACTAG
So GGGAAGAAAATAGATTTATGGATGGAATCAAATATGCAGTATTTACAGACAAAAGT      ATTCAGTTATTGGGGAAAAAGCAATATACTTCTAATGTCGAATCAAGATCAACTAG
Ka GGGAAGAAAATAGATTTATGGATGGAATCAAATATGCAGTATTTACAGACAAAAGTAAAAGTATTCgGTTATTGGGGAAAAAtCAATATACTTCTAATGTCGAATCAAGATCAACTAG
Cm GGGAAGAAAATAGATTTATGGATGGAATCAAATATGtAGTATTTACAGACAAAAGT      ATTCAGTTATTGGGGAAAAAGCAATATcCTTCTAATGTCGAATCAAGATCAACTAG
As GGGAAGAAAATAGATTTATGGATGGAATCAAATATGtAGTATTTACAGACAAAAGT      ATTCgGTTATTGGGGAAAAAtCAATATACTTCTAATGTCGAATCAAGATCAACTAG
Bv GGGAAGAAAATcGATTTATGGATGGAATCAAATATGtAGTATTTACAGACAAAAGT      ATTCgGTTATTGGaaAAAAAtCAATATACTTCTAATGTCGAATCAAGATCAACTAG
Ca                                                        TAAAAGTATcagGTTATTtaGtAAAAAtCAcTATACT     TGTCGAATtCAAccCAgCTAG
Rs GGGAAGAAA        ATaGATaGAATCAcATATGCAaTATaTgCAGACAAAA          TCgGTTATTGGGGAAAAAtCAATATACTTCTAATGTCGAActAGGATCAACTAG

         100       110       120       130                                  140       150       160       170       180
Nt GACaGAAATAAAGCATTGGGTCGAACTcTTCTTTGGTGTCAAG                         GtAATAGCTATGAATAGTCATCGACTtCCGGGAAAGaGTAGAAGaATGGGA
Nd GACaGAAATAAAGCATTGGGTCGAACTcTTCTTTGGTGTCAAG                         GtAATAGCTATGAATAGTCATCGACTtCCGGGAAAGaGTAGAAGaATGGGA
So GACGGAAATAAAGCATTGGGTCGAACTTT                                       GGAATAGCTATGAATAGTCATCGACTCCCGGGAAAGGGTAGAAGGATGGGA
Ka GACGGAAATAAAGCATTGGGTCGAACTTT                                       GGAATAGCTATGAATAGTCATCGACTCCCGGGAAAGGGTAGAAGaATGGGA
Cm GACGGAAATAAAGCATTGGGTCGAACTTT                                       GtAATAGCTATGAATAGTCATCGACTCCCGGGAAAGGGTAGAAGGATGGGA
As GACGGAAATAAGCATTGaGTCGAACTTTT                                       GtAATAGCTATGAATAGTCATCGACTCCCGaGAAAGGGTAGAAGaATGGGA
Bv GACGGAAATAAGCATTGGGTCGAACTTTT                                       GtAATAGCTATGAATAGTCATCGACTCCCGGGAAAGGGTAGAAGaATGGGA
Ca GACGGAAAaAt   ATTGGGctGAACTcTTCTTTGGTGTCAAGGCAGCGGTATAGTATATAAAAATAC                GTCgTCGACTCCtctGAAAGGGTAGAAGaATGGGA
Rs GACaGAAATAAAGCATTGGGTCGAACTcTTCTTTGGTGTCAAG                         GtAATAGCTATGAATAGTCATCGACTCCCGGGAAAGaGTAGAAGaATGGGA

       190           200        210       220       230       240       250       260       270                      rpl2
Nt CCTATTATG     GGACATACAATGCATTACAGACGTATGATCATTACGCTTCAAcCGgGTTATTCTATTCCACCTCTTAGAAAGAAAAGAACTTAAAaaAAAATACTTAATAGCATG
Nd CCTATTATG     GGACATACAATGCATTACAGACGTATGATCATTACGCTTCAAcCGgGTTATTCTATTCCACCTCTTAGAAAGAAAAGAACTTAAAaaAAAATACTTAATAGCATG
So CCTATTATG     GGACATACAATGCATTACAGACGTATGATCATTACGCTTCAATCGAGTTATTCTATTCCACCTCTTAGAAAGAAAAGAACTTAAATCAAAATACTTAATAGCATG
Ka CCTATTATGTTATGTGACATACAATGCATTACAGACGTATGATCATTACGCTTCAATCGAGTTATTCTATTCCACCTCTTAGAAAGAAAAGAACTTAAATCAAAATACTTAATAGCATG
Cm CCTATTATG     GGACATACAATGCATTACAGACGTATGATCATTACGCTTCAATCGAGTTATTCTATTCCACCTCTTAGAAAGAAAAGAACTTAAATCAAAATACTTAATAGCATG
As CCTATTATG     GGACATACAATGCATTACAGACGTATGATCATTACaCTTCAATCGAGTTATTCTATTCCACCTCTTAGAAAGAAAAGAACTTAAATCAAAATACTTAATAGCATG
Bv CCTATTATG     GGACATACAATGCATTACAGACGTATGATCATTACGCTTCAATCGAGTTATTCTATTCCACCTCTTAGAAAGAAAAGAACTTAAATCAAAATACTTAATAGCATG
Ca CaTATTATG     GaACATACAATGCATTACAGACGTATGATCATTACGCTTCAATCGAGTTATTCTATTCCACCTCTTAGAAAGAAAAGAACTTAAAgCAAAATACTTAATAGCATG
Rs CCTATTATG     GGACATACAATGCATTACAGACGTATGATCATTACGCTTCAAcCGgGTTATTCTATTCCACCTCTTAGAAAGAAAAGAACTgAAATaAAAATACTTAATAGCATG
```

Figure 7 Comparative sequences proximal to the *rp12* coding region. The sequences represented are *Nicotiana tabacum* [Nt(40)], *Nicotiana debneyi* [Nd(49)], *Spinacia oleracea* [So(49)], *Kochia americana* (Ka), *Chenopodium murale* (Cm), *Amaranthus salicifolius* (As), *Beta vulgaris* (Bv), *Cerastium arvense* (Ca), and *Rumex sp*. (Rs). The sequences commence proximal to the ATG initiaton codon for *rp123* and end at the ATG initiaton codon for *rp12*. The numbers are relative to the ATG initiaton codon for Nt *rp123*. The sequences are aligned to reveal maximal homology consistent with the fewest addition/deletion changes. Nucleotides that are conserved between all species are in uppercase. The initiation (1) and termination (278) codons for *rp123* and the initiation codon (297) for *rp12* are underlined. Also underlined are the first in-frame stop codons of the *rp123* reading frame in those species in which the *rpl23* gene is nonfunctional.

(G. Zurawski, A. Bang, J. Palmer, J. Watson, unpublished) have shown that this intron is present in five higher-plant orders surveyed but is absent in all nine species surveyed (from four families) of order Caryophyllales (Figure 2). This result suggests the disappearance of the intron at the time this group of plants diverged and may reflect similar losses that have occurred as the chloroplast has evolved. It is of interest that algal chloroplast genomes seem to have many more introns. For instance *rbcL* and *psbA* in *Euglena* (16, 18) and the 23S rRNA gene in *Chlamydomonas* (1) have introns while these genes have no introns in all higher-plant genomes examined. This indicates either that new introns are arising in the evolution of such algae or that higher plants have selected against previously existing introns. Alternately, the algae may represent a separate lineage. A comparative examination of the nature of sequence divergence of both algal and higher-plant introns should help resolve these questions.

Transcription Terminators

Transcription at the end of prokaryotic genes or operons ceases after sequences that are characterized by their ability to encode the formation of secondary "stem-loop" structure in the mRNA (29). In some cases two classes of structural signals are present, those that act on the RNA polymerase independently of the ρ termination protein and those that act only in the presence of ρ (29). The latter class is less well characterized but can occur at the end of transcriptional units to serve as a second terminator to complete residual transcription.

An examination by evolutionary filtering of the sequence adjacent to the region that encodes the 3' end of *rbcL* mRNA is shown in Figure 8. This analysis identifies a highly conserved region immediately proximal to the 3' end coding sequence. This conserved region encodes in the mRNA the ability to form a stem-loop structure. The sequences adjacent to this region and part of the loop-coding region are much less conserved. Similar conserved structural features also occur adjacent to the 3' end coding regions of other chloroplast transcriptional units (22, 45, 47). It is possible that such sequences serve as direct terminators in chloroplast transcription, although the sequences lack the characteristic polyU coding region that immediately follows typical ρ-independent prokaryotic terminators (29). It is equally likely, however, that the conserved sequences encode RNA processing signals or RNA degradation endpoints as is sometimes seen in prokaryotes (29). These possibilities can be directly tested in either the homologous in vitro transcription system or in suitable *E. coli* test systems.

The *Phaseolus rbcL* C-terminal sequence (Pa, Figure 8) is of particular interest in that in this plant the *rbcL* mRNA is known to terminate some several hundred nucleotides further downstream (J. Watson, W. Thompson,

```
Ph  nnn nnn nnn nnn nnn gAa  TTc gaA GCa aTG GAt act tTg gat          TAa
Ps  TGG AAg GAa ATC aaA TTt gAa  TTc ccA GCa aTG GAt act tTg          TAa
So  TGG AAg GAa ATC aaA TTt gAa  TTc ccA GCa aTG GAt aca gTc          TAg
Nt  TGG AAa GAg ATC gtA TTt aAt  TTt gcA GCa gTG GAc gtt tTg gat aag  TAa aaa
Zm  TGG AAg GAg ATC aaA TTc gAt ggt  TTc aaA GCg aTG GAt acc aTa      TAa aat
```

```
Ph
Ps                                                              tccaGTAATTatcaTtcgtT
So                                                              gctaaGTAATTaatgTccggT
Nt  cagtagacattagcagataaattagcaggaaataaagaaggataaggagaaagaactcaaGTAATTatccTtcgtT
Zm  aaaaaaaagcaaaatatgaagtgaaaaaataagttatgaaatgaaatgaacGTAATT         cTttatT
```

```
Ph         TTgcaATTaAAtTCGGCaCaATCTTTTcCTAAAA       gaagGATGAGCCGAAtacaaagatac
Ps  CtatTAA  TTtccATTaAAcTCGGCcCcAATCTTTTaCTAAAA    gGAtTGAGCCGAAtactgtacaca
So  CtctTAAtataaTTgtaATTaAAcTCGGCcCcAATCTTTTaCTAAAA gGAtTGAGCCGAAtacaattattg
Nt  CtctTAAttgaaTTgcaATTaAAcTCGGCcCcAATCTTTTaCTAAAA gGAtTGAGCCGAAtacaacaaaga
Zm  CctcTAAttga TTgcaATTcAAtTCGGCtC ATCTTTT CTAAAAaaaaaaaaaGAcTGAGCCGAAaagaaaaagat
```

Figure 8 Comparative sequences in the *rbcL* transcription termination region. The sequences represented are *Phaseolus albus* (Ph), *Pisum sativum* [Ps(48)], *Spinacia oleracea* [So(43)], *Nicotiana tabacum* [Nt(33)], and *Zea mays* [Zm(24)]. The sequences commence in the C-terminal protein-coding region (arranged as codons) and end after the transcription termination regions (underlined) that have been identified for Ps, So, and Nt. The sequences are aligned to reveal maximal homology consistent with the fewest addition/deletion changes. Nucleotides that are conserved between Ps, So, Nt, Zm, and, where present, Ph are in uppercase; n are undetermined nucleotides. Conserved sequences involved in the formation of the putative stem-loop structure at the 3' end of the mRNA are overlined.

and G. Zurawski, unpublished). Thus, this plant can be considered as a natural *rbcL* terminator mutant. Figure 8 shows that *Phaseolus rbcL* mRNA retains the capacity to form the characteristic conserved stem-loop structure. This indicates that this structure of itself is probably not sufficient to ensure termination and that as yet unidentified ancilliary sequence(s) are required. Although the evolutionary filtering analysis was able to identify the stem-loop region, in this case, it lacked the power to identify potential ancilliary termination sequences. As with all such analyses, an increased number of sequences to be compared increases the power of the method, especially in the case of functional sequences that are only several nucleotides in length.

ESTIMATING PHYLOGENETIC TOPOLOGIES FROM CHLOROPLAST-DNA SEQUENCE DATA

Table 3 presents crude estimates for rates of molecular evolution at the third codon position. These results are based on the average *rbcL* and *atpB* estimates and use times of divergence obtained from Jones & Luchsinger (14) and Stebbins (38). The rates given in Table 3 should be regarded as approximate owing to the limited data base and to the uncertainties in determining times of divergence among major plant categories. Despite these caveats, the estimated rates resemble those obtained for a number of mammalian protein-coding genes which average 4.7×10^{-9} synonymous substitutions per site per year (21). The limited chloroplast data suggest a somewhat reduced rate of chloroplast gene evolution, but this impression will need to be confirmed with much larger samples. However, the rate and nature of chloroplast DNA evolution are such that it offers a potentially powerful tool for estimating phylogenetic topologies.

Many techniques exist for constructing phylogenetic topologies from DNA sequence data. The simplest algorithm is the unweighted pair group method (UPGM), which clusters sequences based upon similarity (37). A drawback of the UPGM algorithm is that the statistical error associated with the inferred topology is not evaluated, because this method is not based on a probabilistic model. It is important to realize that the accumulation of mutational differences is a stochastic process and that pairs of sequences that have been separated for the same length of time may have different numbers of nucleotide substitutions simply because the occurrence of a mutation is a random event. It is therefore desirable to estimate the statistical error associated with the process of nucleotide substitution.

Felsenstein (7, 8) has developed a maximum likelihood (MLE) algorithm for the estimation of phylogenetic topologies that is based on a probabilistic model of the substitution process and that allows the calculation of confidence intervals for branch lengths. A major drawback of this method is that it is

Table 3 Times of divergence and synonymous substitution estimates for various plant taxa[a]

Taxa	Divergence time	3rd Position	Synonymous substitution rate
Monocot-dicot	110×10^6 yr	0.499	1.9×10^{-9}/yr
Spinach-tobacco	70×10^6	0.217	1.3×10^{-9}/yr
Barley-maize	50×10^6	0.135	1.2×10^{-9}/yr

[a] All comparisons are for the *rbcL* and *atpB* sequences. The mean rate of divergence is 1.47×10^{-9} synonymous substitutions per site per year.

computationally difficult and the computer time required for convergence to the MLE topology is quite large for a moderate number of taxa (i.e. greater than 20) or a moderate number of site differences. Resampling methods, such as the bootstrap method, are beginning to be used for sequence analysis and promise to yield a quick approach to investigating the statistical error of phylogenetic topologies (9, 30).

Protein-Coding Regions

Ritland & Clegg (30) have used MLE techniques to reconstruct phylogenetic topologies from the *rbcL* sequence data. Their analysis partitions protein-coding sequences by codon position to investigate the consistency of phylogenetic reconstructions when nucleotide substitution rates differ over codon positions. They also investigate the consistency of phylogenetic reconstructions over different protein-coding genes and over noncoding regions of the chloroplast genome. It is reassuring that phylogenetic topologies are consistent over codon positions and over different chloroplast genes. Inferences drawn from chloroplast sequence comparisons can usually be expected to yield a topologically consistent picture of plant phylogeny, independent of the genes selected for sequence comparison, although *rpl2* is an exception (see below).

The fact that chloroplast genes evolve at a slow rate makes them ideal candidates for the molecular investigation of the major features of plant evolution. Figure 9 presents an unrooted tree that depicts the phylogenetic relationships between flowering plants, green algae, and cyanobacteria, based on a MLE analysis of the *rbcL* DNA sequence data pooled over codon positions. Ritland & Clegg (30) present similar trees partitioned by codon position that yield the same topology, although evolutionary rates along specific branches vary among taxa for particular codon positions.

The available *rpl2* coding region data has also been examined for its use in estimating phylogenic topologies. Such topologies constructed from the *rpl2* matrix of nucleotide differences are inconsistent with a unique loss of the *rpl2*

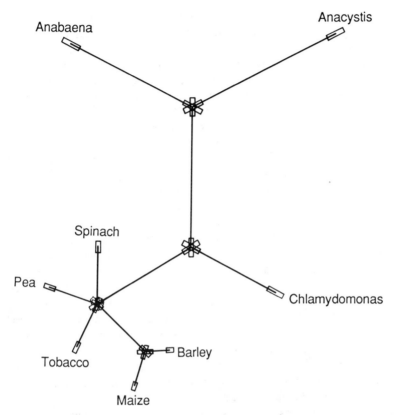

Figure 9 MLE topology based on *rbcL* sequence data from five angiosperm taxa, a green alga *(Chlamydomonas)*, and two cyanobacteria [after Ritland & Clegg (30)]. The branch lengths are proportional to evolutionary distance, and the boxes indicate 95% confidence intervals for the branch lengths and node positions.

intron and appear to conflict with generally accepted classification schemes. We believe this result can be traced to the very low number of nucleotide differences and to the anomolous pattern of nucleotide substitution for the *rp12* data. Ritland & Clegg (30) have, however, shown that phylogenetic analyses are consistent for various other coding and noncoding chloroplast DNA sequences. It is clearly important to collect more comparative data for chloroplast genes to help understand why the *rp12* data are exceptional.

Noncoding Regions

As discussed above, the evolutionary comparison of noncoding regions is useful for preliminary searches to detect differentially conserved regions that may be subject to unknown functional constraints. The functional constraints in these regions are different from those that apply to protein-coding regions, and this difference is reflected in a fundamentally different pattern of sequence divergence.

Additions and deletions of short sequences that range from two to about eight nucleotides occur frequently in noncoding regions of the chloroplast genome. In a comparison between barley and maize of the noncoding region separating *rbcL* and *atpB*, Zurawski et al (46) noted that about 40% of these events were associated with direct repeats. They also showed that the distribution of these events was nonrandom along the sequence and that additions and deletions were excluded from the region surrounding the initiation of transcription for both genes.

The fact that additions and deletions tend to occur at direct repeats and also tend to occur nonrandomly along the noncoding region raises the possibility that these events reoccur relatively rapidly in evolution at a restricted number of potential sites. To investigate this possibility M. Durbin and M. T. Clegg (unpublished) have sequenced the noncoding region separating *rbcL* and *atpB* from the closely related grass genera *Pennisetum* and *Cenchrus*. The ratio of addition/deletion events to nucleotide differences is 7/5 = 1.4 for these taxa. Moreover, one addition of 42 nucleotides in *Cenchrus* is composed of a complex series of internal repeats that may represent a compound of several separate events (see Figure 3). A similar comparison between the more distantly related taxa *Hordeum* and *Pennisetum* yields a ratio of 9 addition/ deletion events to 50 nucleotide differences = 0.18. If addition/deletion events were occurring at a constant rate relative to nucleotide differences, a constant ratio would be expected. The fact that the relative number of addition/deletion events decreases with taxonomic distance suggests that these events reoccur at nearly the same sites on a rapid time scale relative to the occurrence of nucleotide differences. This quite different dynamic can confound the use of restriction-fragment comparisons for evolutionary studies, to the extent that restriction sites are involved in these more rapid changes. In this context, it it interesting to note that the 42-bp addition in *Cenchrus* contains two *Eco*RI sites.

The rapid rate of addition/deletion change in noncoding regions offers the potential for addressing phylogenetic questions at a level different from that offered by substitution events. The number of addition/deletion changes between *Pennisetum* and *Cenchrus* suggests that this category of change provides the appropriate level of resolution for such closely related taxa.

CONCLUSIONS

The evolutionary analysis of DNA sequence data provides a powerful means of addressing a variety of questions regarding gene function. This is particularly the case for chloroplast encoded genes where genetic analyses are generally unavailable. Sequence comparisons also reveal much about evolutionary rates, mutational processes, and phylogenetic relationships. It is now well established that the chloroplast genome evolves at a conservative rate. Present data indicate an average synonymous rate of approximately 1.5 × 10^{-9} substitutions per site per year. This value is somewhat below synonymous rates for animal genes. A fundamental advantage of a slow rate of evolution is that it yields a degree of phylogenetic resolution that is ideal for the investigation of the deepest branches of plant evolution. The application of chloroplast DNA sequence analysis to the study of plant phylogeny is in its infancy; however, these methods promise to have a profound influence in our understanding of plant evolution.

Two features of the mutational process are revealed by sequence analysis. First, the substitution process is widely biased in favor of transition events, and the degree of bias appears to remain relatively constant over long periods of evolutionary time. Second, addition/deletion events are a common feature of noncoding regions. These events tend to occur on a more rapid time scale than substitution events, and addition/deletion events may tend to recur at approximately the same sites. Another feature of addition/deletion change is a tendency to be associated with short direct repeats. These features are probably all related to a common underlying mechanism of mutation associated with slippage during the replication process.

The analysis of chloroplast DNA sequence data has a further dimension. It can be used to address basic issues of organization, expression, and regulation of the genome. The technique of comparing the analogous sequence from various species (termed evolutionary filtering) can be used to identify important gene structural features. Promoter regions for at least some chloroplast genes can be correctly assigned solely on this basis, and further experimentation has revealed that such regions are structural homologues of typical prokaryotic promoters. Evolutionary filtering analysis of the sometimes long transcribed but untranslated 5' leader regions of chloroplast genes suggests no conserved structural or sequence features except for prokaryotic-like ribosome-binding sites immediately prior to the protein-coding region. Evolutionary filtering is a powerful tool for the assignment of protein-coding regions, especially in cases where sequences of analogous proteins from other systems are unavailable. Sequence comparisons are a useful way of identifying those conserved structural features within both intron and transcription terminator regions, which are the most likely targets for further study.

Evolutionary comparisons are open-ended in the sense that the process is unbiased. It may serve not only to verify preconceived hypotheses about gene function but also occasionally to reveal unexpected features that may lead to new hypotheses. Examples of unexpected findings concerning the evolution of the chloroplast genome that are discussed in this review include the absence of functional *rp123* coding sequences in many, but not all, higher-plant chloroplast genomes. This result suggests that this gene is presently being fixed into the nuclear genome. The absence of the *rp12* intron from only a limited group of higher plants suggests reductive evolution of this intron in the chloroplast genome; together with the *rp123* data, it dispels the present image of conservation of chloroplast genome intron and gene content across the higher-plant kingdom. At a nucleotide level, sequence comparisons have also revealed anomolies in the evolution of certain regions of the chloroplast genome that must be addressed before a full understanding of the evolution of this genome is achieved.

ACKNOWLEDGMENT

The authors thank Dr. A. H. D. Brown for comments on an earlier version of the manuscript. MTC acknowledges the support of NSF grant BSR-8500206.

Literature Cited

1. Allet, B. Rochaix, J. D. 1979. Structural analysis at the ends of the intervening DNA sequences in the chloroplast 23S ribosomal genes of *C. reinhardii*. *Cell* 18:55–60

2. Bogorad, L., Crossland, L. D., Fish, L. E., Krebbers, E. T., Kuck, U. et al. 1984. The organization of the maize plastid chromosome: properties and expression of its genes. In *Biosynthesis of the Photosynthetic Apparatus: Molecular Biology, Development, and Regulation*, ed. J. P. Thornber, L. A. Staehelin, R. B. Hallick, pp. 257–72. NY: Alan R. Liss

3. Clegg, M. T., Ritland, K., Zurawski, G. 1986. Processes of chloroplast DNA evolution. In *Evolutionary Processes and Theory*, ed. S. Karlin, E. Nevo, pp. 275–94. NY: Academic

4. Curtis, S. E., Haselkorn, R. 1983. Isolation and sequence of the gene for the large subunit of ribulose-1,5-bisphosphate carboxylase from the cyanobacterium *Anabaena* 7120. *Proc. Natl. Acad. Sci. USA* 80:1835–39

5. Dunn, J. J., Buzash-Pollert, E., Studier, F. W. 1978. Mutations of bacteriophage T7 that affect initiation of synthesis of the gene 0.3 protein. *Proc. Natl. Acad. Sci. USA* 75:2741–45

6. Erion, J. L. 1985. Characterization of the messenger RNA transcripts of the maize ribulose-1,5-bisphosphate carboxylase large subunit gene. *Plant Mol. Biol.* 4:169–80

7. Felsenstein, J. 1981. Evolutionary trees from DNA sequences: a maximum likelihood approach. *J. Mol. Evol.* 17:368–76

8. Felsenstein, J. 1983. Inferring evolutionary trees from DNA sequences. In *Statistical Analysis of DNA Sequence Data*, ed. B. S. Weir. NY: Marcel Dekker

9. Felsenstein, J. 1985. Confidence limits on phylogenies: an approach using the bootstrap. *Evolution* 39:783–91

10. Gold, L., Pribnow, D., Schneider, T., Shinedling, S., Singer, B. S., Stormo, G. 1981. Translation initiation in prokaryotes. *Ann. Rev. Microbiol.* 35:365–403

11. Gruissem, W., Zurawski, G. 1985. Identification and mutational analysis of the promoter for a spinach transfer RNA gene. *EMBO J.* 4:1637–44

12. Gruissem, W., Zurawski, G. 1985. Analysis of promoter regions for the spinach chloroplast *rbcL, atpB* and *psbA* genes. *EMBO J.* 4:3375–83

13. Howe, C. J., Fearnley, I. M., Walker, J. E., Dyer, T. A., Gray, J. C. 1985. Nucleotide sequences of the genes for the alpha, beta and epsilon subunits of wheat chloroplast ATP synthase. *Plant Mol. Biol.* 4:333–45

14. Jones, S. B., Luchsinger, A. E. 1986. *Plant Systematics*. NY: McGraw-Hill

15. Jukes, T. H., Cantor, C. R. 1969. Evolution of protein molecules. In *Mammalian Protein Metabolism*, ed. H.

N. Munro, pp. 21–132. NY: Academic

16. Karabin. G. D., Farley, M., Hallick, R. B. 1984. Chloroplast gene for M_r 32000 polypeptide of photosystem II in *Euglena gracilis* is interrupted by four introns with conserved boundary sequences. *Nucleic Acids Res.* 12:5801–12

17. Kimura, M. 1981. Estimation of evolutionary distances between homologous nucleotide sequences. *Proc. Natl. Acad. USA* 78:454–58

18. Koller, B., Gingrich, J. C., Steigler, G. L., Farley, M. A., Delius, H., Hallick, R. B. 1984. Nine introns with conserved boundary sequences in the *Euglena gracilis* chloroplast ribulose-1,5-bisphosphate carboxylase gene. *Cell* 36:545–53

19. Krebbers, E. T., Larrinua, I. M., McIntosh, L., Bogorad, L. 1982. The maize chloroplast genes for the beta and epsilon subunits of the photosynthetic coupling factor CF_1 are fused. *Nucleic Acids Res.* 10:4985–5002

20. Kruger, K., Grabowski, P. J., Zaug, A. J., Sands, J., Gottschling, D. E., Cech, T. R. 1982. Self-splicing RNA: autoexcision and autocyclization of ribosomal RNA intervening sequence of *Tetrahymena. Cell* 31:147–57

21. Li, W.-H., Luo, C.-C., Wu, C.-I. 1985. Evolution of DNA sequences. In *Molecular Evolutionary Genetics*, ed. R. J. MacIntyre, pp. 1–94. NY: Plenum

22. Link, G., Langridge, U. 1984. Structure of the chloroplast gene for the precursor of the M_r 32,000 photosystem II protein from mustard (*Sinapis alba* L.). *Nucleic Acids Res.* 12:945–58

23. Link, G. 1984. DNA sequence requirements for the accurate transcription of a protein-coding plastid gene in a plastid in vitro system from mustard (*Sinapis alba* L.). *EMBO J.* 3:1679–1704

24. MacIntosh, L., Poulsen, C., Bogorad, L. 1980. Chloroplast gene sequence for the large subunit of ribulose bisphosphate carboxylase of maize. *Nature* 288:556–60

25. Mullet, J. E., Orozco, E. M. Jr., Chua, N.-H. 1985. Multiple transcripts for higher plant *rbcL* and *atpB* genes and localization of the transcription initiation site of the *rbcL* gene. *Plant Mol. Biol.* 4:39–54

26. Ohyama, K., Fukuzawa, H., Kohchi, T., Shirai, H., Sano, T. et al. 1986. Chloroplast gene organization deduced from complete sequence of liverwort *Marchantia polymorpha* chloroplast DNA. *Nature* 322:572–74

27. Palmer, J. D. 1985. Comparative organ-

ization of chloroplast genomes. *Ann. Rev. Genet.* 19:325–54

28. Peebles, C. L., Perlman, P. S., Mecklenberg, K. L., Petrillo, M. L., Tabor, J. H. 1986. A self-splicing RNA excises an intron lariat. *Cell* 44:213–23

29. Platt, T. 1986. Transcription termination and the regulation of gene expression. *Ann. Rev. Biochem.* 55:339–72

30. Ritland, K., Clegg, M. T. 1987. Evolutionary analysis of plant DNA sequences. *Am. Nat.* In press

31. Rosenberg, M., Court, D. 1979. Regulatory sequences involved in the promotion and termination of RNA transcription. *Ann. Rev. Genet.* 13:319–53

32. Shinozaki, K., Sugiura, M. 1982. Sequence of the intercistronic region between the ribulose 1,5-bisphosphate carboxylase/oxygenase large subunit and the coupling factor β-subunit gene. *Nucleic Acids Res.* 10:4923–33

33. Shinozaki, K., Sugiura, M. 1982. The nucleotide sequence of the tobacco chloroplast gene for the large subunit of ribulose-1,5-bisphosphate carboxylase/oxygenase. *Gene* 20:91–102

34. Shinozaki, K., Ohme, M., Tanaka, M., Wakasugi, T., Hayashida, H. et al. 1986. The complete nucleotide sequence of the tobacco chloroplast genome: its gene organization and expression. *EMBO J.* 5:2043–49

35. Shinozaki, K., Deno, H., Sugita, M., Kuramitsu, S., Sugiura, M. 1986. Intron in the gene for the ribosomal protein S16 of tobacco chloroplast and its conserved boundary sequences. *Mol. Gen. Genet.* 202:1–5

36. Siebenlist, U., Simpson, R. B., Gilbert, W. 1980. *E. coli* RNA polymerase interacts homologously with two different promoters. *Cell* 20:269–81

37. Sneath, P. H. A., Sokal, R. R. 1973. *Numerical Taxonomy.* San Francisco: W. H. Freeman

38. Stebbins, G. L. 1981. Coevolution of grasses and herbivores. *Ann. Mo. Bot. Gard.* 68:75–86

39. Stormo, G. D., Schneider, T. D., Gold, L. M. 1982. Characterization of translation initiation sites in *E. coli. Nucleic Acids Res.* 10:2971–96

40. Tanaka, M., Wakasugi, T., Sugita, M., Shinozaki, K., Sugiura, M. 1986. Genes for the eight ribosomal proteins are clustered on the chloroplast genome of tobacco (*Nicotiana tabacum*) similarly to the S10 and *spc* operons of *Escherichia coli.* 1986. *Proc. Natl. Acad. Sci. USA* 83:6030–34

41. Weir, B. S. 1985. Statistical analysis of

molecular genetic data. *J. Math. Appl. Med. Biol.* 2:1–39

42. Whitfeld, P. R., Bottomley, W. 1983. Organization and structure of chloroplast genes. *Ann. Rev. Plant Physiol.* 34:279–310

43. Zurawski, G., Perrot, B., Bottomley, W., Whitfeld, P. R. 1981. The structure of the gene for the large subunit of ribulose 1,5-bisphosphate carboxylase from spinach chloroplast DNA. *Nucleic Acids Res.* 9:3251–69

44. Zurawski, G., Bottomley, W., Whitfeld, P. R. 1982. Structures of the genes for the β and ε subunits of spinach chloroplast ATPase indicates a dicistronic mRNA and an overlapping translation stop/start signal. *Proc. Natl. Acad. Sci. USA* 79:6260–64

45. Zurawski, G., Bohnert, H. J., Whitfeld, P. R., Bottomley, W. 1982. Nucleotide sequence of the gene for the M_r 32,000 thylakoid membrane protein from *Spinacea oleracea* and *Nicotiana debneyi* predicts a totally conserved primary translation product of M_r 38,950. *Proc. Natl. Acad. Sci. USA* 79:7699–7703

46. Zurawski, G., Clegg, M. T., Brown, A. H. D. 1984. The nature of nucleotide sequence divergence between barley and maize chloroplast DNA. *Genetics* 106:735–49

47. Zurawski, G., Clegg, M. T. 1984. The barley chloroplast DNA *atpBE*, *trnM2* and *trnV1* loci. *Nucleic Acids Res.* 12:2549–59

48. Zurawski, G., Whitfeld, P. R., Bottomley, W. 1986. Sequence of the gene for the large subunit of ribulose 1,5-bisphosphate carboxylase from pea chloroplasts. *Nucleic Acids Res.* 14:3975

49. Zurawski, G., Bottomley, W., Whitfeld, P. R. 1984. Junctions of the large single copy region and the inverted repeats in *Spinacea oleracea* and *Nicotiana debneyi* chloroplast DNA: sequence of the genes for tRNA[His] and the ribosomal proteins S19 and L2. *Nucleic Acids Res.* 12:6547–58

50. Zurawski, G., Bottomley, W., Whitfeld, P. R. 1986. Sequence of the genes for the β and ε subunits of ATP synthase from pea chloroplasts. *Nucleic Acids Res.* 14:3974

Ann. Rev. Plant Physiol. 1987. 38:419–65
Copyright © 1987 by Annual Reviews Inc. All rights reserved

GIBBERELLIN BIOSYNTHESIS AND CONTROL

Jan E. Graebe

Pflanzenphysiologisches Institut der Universität Göttingen, Untere Karspüle 2,
D-3400 Göttingen, Federal Republic of Germany

CONTENTS

INTRODUCTION 419
ent-KAURENE BIOSYNTHESIS 420
   Pathway and Enzymes 420
   New Cell-Free Systems and Other Aspects 422
FROM *ent*-KAURENE TO GIBBERELLIN A_{12}-ALDEHYDE 424
   Pathway and Enzymes 424
   Other Cell-Free Systems 427
THE PATHWAYS AFTER GIBBERELLIN A_{12}-ALDEHYDE 428
   The General Pathway 428
   Individual Pathways 431
   Summary of Pathways After GA_{12}-Aldehyde 438
   Enzymes 438
PHYSIOLOGY OF GIBBERELLIN CONTROL 441
   Genetic Control of Internode Elongation 441
   Other Physiological Aspects 446
   Plant Growth Retardants 452
CONCLUDING REMARKS 456

INTRODUCTION

The study of GA[1] biosynthesis offers many facets of interest to the organic chemist as well as to the biochemist and the plant physiologist. The metabolic

[1]*Abbreviations used:* ancymidol, α-cyclopropyl-α(4-methoxyphenyl)-5-pyrimidine methanol; CPP, copalyl pyrophosphate; GA(s), gibberellin(s); GA_n, gibberellin A_n; GC-MS, combined gas-liquid chromatography–mass spectrometry; GC-RC, gas chromatography-radio counting; GC-SIM, gas chromatography–selected ion monitoring; GGPP, geranylgeranyl pyrophosphate; HPLC, high performance liquid chromatography; MVA, mevalonate; paclobutrazol (=PP333), (2*RS*,3*RS*)-1-(4-chlorophenyl)-4,4-dimethyl-2-(1,2,4-triazol-1-yl)pentan-3-ol; tetcyclacis, 5-(4-chlorophenyl)-3,4,5,9,10-pentaazatetracyclo-5,4,1,0^{2,6},0^{8,11}-dodeca-3,9-diene; TLC, thin layer chromatography.

419

pathway is long and complicated, containing unusual reaction mechanisms, branchings of unknown function, and a wealth of enzyme reactions. GA biosynthesis is activated and inactivated during the life cycle of a plant; the pathways are different in different species and even in different organs of the same species. The proper functioning of GA biosynthesis is essential to the normal development of higher plants, except for those cases where mutation has rendered the plants independent of GAs.

The last review on GA metabolism in this series was in 1978 (82), and another comprehensive review on the same topic appeared in the same year (67). Of the many reviews on GA metabolism published since then, several chapters in a two-volume treatise edited by Crozier (see 12) deserve special attention because they present a particularly complete picture of GA metabolism research up to 1983, both in higher plants and the fungus *Gibberella fujikuroi*. Many of the other reviews on GA metabolism (35, 46, 148, 165), on special aspects of GA metabolism (10, 63–66, 75, 117–119, 134, 136, 137, 176), and on closely related topics (11, 76, 133, 198, 199) provide supplementary information. A CRC handbook contains much useful information on all aspects of GA research (188).

This review highlights the main developments in the field after the 1978 reviews. The emphasis is on GA biosynthesis in higher plants because this is where the most progress has been made in this period. GA biosynthesis can be seen in three stages: (*a*) the biosynthesis of *ent*-kaurene, (*b*) the biosynthesis of GA_{12}-aldehyde, and (*c*) the biosynthesis after GA_{12}-aldehyde; and this is the order in which I have discussed pathways and enzymes. I have also included information on steps lying beyond the actual biosynthesis of GAs (catabolite formation, conjugation) but not steps lying before the formation of GGPP, belonging to the general terpenoid pathway. What is known about the biochemical control of GA biosynthesis is strewn in with the pathway work; the rest falls under the headings of physiological and plant growth regulator control of GAs. Much of the information on GA relations in reproductive structures discussed by Pharis & King only two years ago in this series (133) is relevant to this article also. I have tried to avoid repetitions, but some of the work cited by Pharis & King is mentioned here again in a different context or in the light of new evidence. Listings of GA structures in numerical order are available (11, 12, 67, 82, 188).

ent-KAURENE BIOSYNTHESIS

Pathway and Enzymes

The pathway of *ent*-kaurene biosynthesis was well known at the time of the previous reviews (67, 82). MVA is first converted via the terpenoid pathway to GGPP, which is converted further to *ent*-kaurene in a two-step sequence catalyzed by *ent*-kaurene synthetase (Scheme 1). In the first step (A-activity) of this sequence, GGPP is partially cyclized to form the bicyclic diterpene CPP, which is cyclized again in the second step (B-activity) to give *ent*-kaurene. As new results in this part of the pathway, the stereochemistry of the ring C formation has been experimentally defined by the analysis of *ent*-kaurene biosynthesized from ^{3}H-labelled substrates in a known enzyme system from *Marah macrocarpus* (28). The previously suggested stereochemistry of the ring D formation has been confirmed by ^{13}C NMR spectroscopy analyis of *ent*-[^{13}C]kaurene produced by incubation of [3,6-^{13}C$_2$]MVA with a high-speed supernatant of *Gibberella fujikuroi* (89).

As for the enzymes involved, *ent*-kaurene synthetase had been partially purified and characterized prior to 1978. Preparations had been obtained from *M. macrocarpus* immature seeds, *Ricinus communis* germinating seeds, and *G. fujikuroi* cultures. It was known that both the A- and the B-activity appear in the supernatant after high-speed centrifugation and that divalent metal ions, preferably Mg^{2+}, are required for activity. There were indications that the two steps are catalyzed by separate enzymes, and preparations of the B-activity had been obtained from *R. communis* seedlings; but it had not been possible to get preparations of the A-activity alone. As a major advance, Duncan & West (45) recently succeeded in separating the two activities in *M. macrocarpus* endosperm preparations by anion exchange gel chromatography or polyacrylamide gel electrophoresis of partially purified protein fractions. The molecular weight for the A- and B-enzymes is ~82,000 each. The overall activity, i.e. the conversion of GGPP to *ent*-kaurene, is reconstituted by combining the two enzymes, and this reconstitution has interesting properties. Thus, although the B-activity is directly proportional to the protein concentration, the overall activity shows a second-order dependence on the protein concentration. Furthermore, the reconstituted system shows a channelling effect: When [^{14}C]GGPP and [^{3}H]CPP are used simultaneously as substrates for the AB-enzyme, the ^{14}C:^{3}H ratio in the *ent*-kaurene produced is 10–13 times greater than theoretically expected if the (intermediately formed) [^{14}C]CPP had equilibrated completely with the exogenous pool. The authors interpreted these results to mean that *ent*-kaurene synthetase consists of two

geranylgeranyl   copalyl   ent-kaurene
pyrophosphate   pyrophosphate

Scheme 1   *ent*-Kaurene synthetase A- and B-activities.

distinct enzymes that associate during *ent*-kaurene synthesis from GGPP, and that this association might play a role in the enzyme activity regulation (45; see also 29, 200).

A curious story, which may have a bearing on the regulation of *ent*-kaurene synthetase, involves the presence of the B-activity, but not the A-activity, in lysed chloroplast preparations from several plant species (45, 149, 200). When such chloroplast preparations from pea shoots are fractionated into a thylakoid and a stromal fraction, both fractions are needed for full *ent*-kaurene synthetase B-activity (149). Railton et al suggested that *ent*-kaurene synthetase may be a soluble stromal enzyme that normally requires weak association with plastid membranes, perhaps thylakoids, for full activity in catalyzing the conversion of CPP to *ent*-kaurene. They also suggested that such an association between *ent*-kaurene synthetase and plastid membrane proteins might facilitate the production of the lipophilic *ent*-kaurene directly at a plastid membrane site for further oxidation by membrane-bound enzymes. Whereas the possible significance of the half *ent*-kaurene synthetase activity in chloroplasts remains obscure, the hypothesis of *ent*-kaurene synthesis at a membrane site would be very attractive, if applied to the endoplasmic reticulum instead. For, at least in pea and pumpkin seeds, the enzymes that catalyze the further conversion of *ent*-kaurene along the GA pathway are associated with the endoplasmic reticulum and not with plastid membranes (see later discussion). Since the chloroplast fractions were purified by differential centrifugation only (149), and no marker enzyme activities were measured, it seems likely that the thylakoid fraction might have contained fragments of the endoplasmic reticulum as well. The production of *ent*-kaurene at the proper site for further conversion would, of course, obviate a separate carrier protein.

As Railton et al (149) pointed out, the evidence for GA biosynthesis in chloroplasts is scant. This point deserves to be emphasized, because it is sometimes stated that evidence for GA biosynthesis in chloroplasts is accumulating. There is no proof by modern means for GA biosynthesis in properly purified and identified plastids, although a few single steps of *ent*-kaurene oxidation with low activity have been observed, primarily in etioplasts (see 29, 67, 82, 174, 183). The presence of these steps and, possibly, *ent*-kaurene synthesis in plastids may encourage further investigations by conclusive methods, but it should be remembered that *ent*-kaurene biosynthesis is not GA biosynthesis, by far. There is also no conclusive identification of GAs in properly identified and purified chloroplast fractions (see 174, for a critical appraisal of the usually cited evidence). This, however, is no reason to dismiss the entire phenomenon (183).

New Cell-Free Systems and Other Aspects

In addition to the objects mentioned above, cell-free systems from *Cucurbita maxima* endosperm, pea seeds, and pea seedling shoots were known in 1978 to biosynthesize *ent*-kaurene from MVA. Several new cell-free systems with different properties can now be added. In accord with the idea that *ent*-kaurene synthesis would be a logical point of regulation for the entire GA biosynthesis (e.g. 200), the capacity for *ent*-kaurene synthesis in cell-free systems from developing *Pharbitis nil* seeds is maximal during rapid seed development, which coincides with the maximum in GA contents (4). Similar results have been obtained with pea seed cell-free systems, in which both *ent*-kaurene synthesis and the conversion of other GA biosynthesis intermediates have characteristic maxima during seed development (see 63). Other recent results are the conversion of MVA to *ent*-kaurene both in intact suspensors of *Phaseolus coccineus* seeds and in cell-free systems from these suspensors (22, 25). MVA is also converted to *ent*-kaurene in cell-free systems from *P. coccineus* cotyledons from a later stage of seed development, when the suspensors have disintegrated (191), and in cell-free systems from endosperm and cotyledons of developing *Sechium edule* seeds (20).

Most cell-free systems for the study of *ent*-kaurene biosynthesis in higher plants have been prepared from developing seeds, because these organs contain much more GA than vegetative plant tissues. However, the function of GAs in seeds, if any, is unknown. It is therefore desirable to extend the studies to vegetative stages of plant development, where the GAs have a definite role. However, enzyme extracts from vegetative stages are much less active and often contain inhibitory substances. Work with small quantities of *ent*-kaurene is particularly problematic because this hydrocarbon is volatile and (which is less known) soluble in liquid nitrogen. The concentration of a solution at room temperature or the storage of *ent*-kaurene samples in liquid nitrogen may therefore lead to large, uncontrolled losses. In spite of these difficulties, pioneering work is being done. In pea seedlings, the conversion of MVA to *ent*-kaurene (identified by TLC) is most active in cell-free extracts prepared from the shoot tips, petioles, and stipules near the young elongating internodes, and the activity decreases as these organs get older (27). A cell-free system from maize seedlings converts GGPP to *ent*-kaurene and other diterpene hydrocarbons, identified by argentation TLC; and the incorporation is enhanced 50- to 100-fold by fungal infection of the seedlings prior to preparation of the cell-free extract (122).

As mentioned above, the *ent*-kaurene synthetase B-activity is more often found in cell-free preparations than the A-activity (200). Various explanations for this seemingly unlogical distribution have been suggested, including higher lability or higher sensitivity of the A-activity to inhibitors. Indeed, a cell-free system from etiolated *Helianthus annuus* seedlings contains dialyzable material that inhibits the AB-activity but not the A-activity of *ent*-kaurene synthetase in cell-free systems from *M. macrocarpus* endosperm (166). Dialysis followed by storage of the sunflower enzyme preparations in liquid nitrogen increases the *ent*-kaurene synthetase activity and also decreases the inhibitory activity (167). With preparations thus activated, evidence was obtained that *ent*-kaurene synthetase is more active in etiolated than in green seedlings (168). Seemingly opposite results have been reported for pea shoot enzyme preparations, in which the capacity for *ent*-kaurene biosynthesis from $[2\text{-}^{14}C]MVA$ was absent if the shoots had been dark grown, and increased with increasing irradiation time (26). An action spectrum seemed to show that phytochrome was involved in the process; but in view of the extremely low incorporation (25 dpm mg^{-1} protein at the peak of the curve), and the difficulties in measuring small amounts of this volatile hydrocarbon (see above), the time and the methods do not seem ripe for definite conclusions regarding the effect of light on *ent*-kaurene biosynthesis. In fact, the comparison of endogenous amounts of GAs in dark- and light-grown pea seedlings do not favor the view that overall GA biosynthesis is photoinduced (see the section below on genetic control).

Cell-free systems from *Zea mays* seedlings (83) and from *Arabidopsis thaliana* siliques (5), which also synthesize *ent*-kaurene from MVA, will be discussed together with the genetic control of internode elongation.

FROM *ent*-KAURENE TO GIBBERELLIN A_{12}-ALDEHYDE

Pathway and Enzymes

The pathway from *ent*-kaurene to GA_{12}-aldehyde is identical in all plants investigated to date and may therefore be universal. The main course of this pathway (Scheme 2) had been worked out before 1978 by experimenting with cell-free systems from *Marah macrocarpus, Cucurbita maxima,* and *Gibberella fujikuroi*. It was known that these reactions are catalyzed by microsomal monooxygenases, requiring NADPH as the only cofactor and probably containing cytochrome P-450 (see 77, 198). The first one or two steps after GA_{12}-aldehyde, which may be different in different species, are also often catalyzed by microsomal enzymes, or by both microsomal and soluble enzymes. The latter is the case in the formation of GA_{12} in *C. maxima* (see Scheme 2). As a new result, the hydrogen atom that is removed when *ent*-kaurenol is oxidized to *ent*-kaurenal has recently been defined (169).

The pathway shown in Scheme 2 has two branches, the kaurenolide branch and the *ent*-6α,7α-dihydroxykaurenoic acid branch. The position of the first branch, leading from *ent*-kaurenoic acid to the kaurenolides, and the intermediacy of the *ent*-kaura-6,16-dienoic acid, have been shown in recent work with the *C. maxima* cell-free system (78). It was previously believed that the kaurenolide branch emerged from *ent*-7α-hydroxykaurenoic acid (see 67, 82). The pathway apparently branches at the same point in *G. fujikuroi* also, except that 7β,18-dihydroxykaurenolide is the end product instead of 7β,12α-dihydroxykaurenolide. However, in this fungus, the intermediate

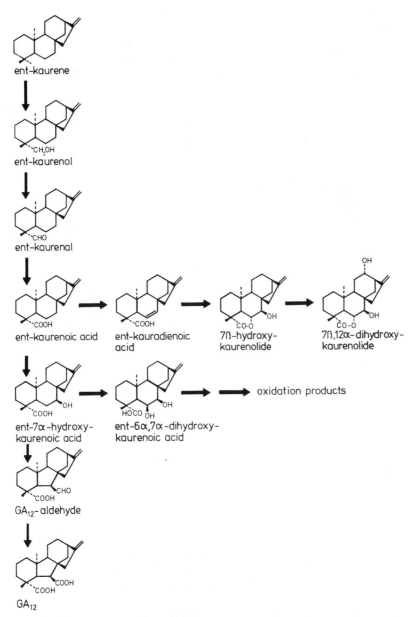

Scheme 2 The microsomal part of GA biosynthesis in the *Cucurbita maxima* endosperm cell-free system at pH 7.6. Below pH 6.5, 12α-OH GA₁₂-aldehyde is the last product instead of GA₁₂ (see the text section on Individual Pathways, Other Species).

ent-kauradienoic acid could not be found, presumably because it is too rapidly converted to 7β-hydroxykaurenolide (7). This conversion of the *ent*-kauradienoic acid to 7β-hydroxykaurenolide probably occurs via *ent*-6α,7α-epoxykaurenoic acid, although this intermediate has not been found in either system. Most likely, the presence of the C-19 carboxyl group favors rapid epoxide opening and lactone formation (7). In support of this view, the formation of an analogous *ent*-6α,7α-epoxide was obtained by feeding to *G. fujikuroi* a modified substrate (*ent*-3β,18-dihydroxykaur-6,16-diene), in which the oxidation of C-19 to a carboxylic acid and therefore also lactone formation were structurally prevented (50). Bearder (12) provides a detailed discussion of the kaurenolide branch in the GA pathway.

The final oxidation product (not shown) of the branch proceeding via *ent*-6α,7α-dihydroxykaurenoic acid has been tentatively identified in the *C. maxima* system in cooperation with Prof. J. MacMillan (see 66), and the intermediates in its formation have been identified in unpublished experiments (see 77). In the *C. maxima* cell-free system, ~2–3% of the *ent*-kaurenoic acid formed is diverted into the kaurenolide branch, and almost 50% of the *ent*-7α-hydroxykaurenoic acid formed is diverted into the *ent*-6α,7α-dihydroxykaurenoic acid branch. The products of these two branches are not converted to GAs, and the intermediates are lost to GA biosynthesis.

The GA biosynthesis enzymes of *G. fujikuroi* apparently have low substrate specificity, because cultures of this fungus metabolize analogues of GA intermediates to various degrees. Such metabolism of *ent*-kaurene analogues is interesting for the elucidation of the corresponding reaction mechanisms in GA biosynthesis. The formation of a 6,7-epoxide has already been mentioned. Another interesting finding is that the presence of an *ent*-7α-hydroxy group in the substrate before oxidation at C-19, as in *ent*-7α-hydroxykaurene, prevents the formation of kaurenolides while allowing the formation of GAs (51, 52). Conversely, the presence of a 6,7 double bond in the substrate, as in *ent*-kaura-6,16-diene, directs the pathway entirely into kaurenolide formation, so that no GAs are formed (50). Of the several other conversions of kaurenoid analogues that have been shown with *G. fujikuroi,* the conversion of steviol (*ent*-13-hydroxykaurenoic acid) to 13-hydroxylated C_{20}- and C_{19}-GAs (see 82) is interesting, because 13-OH GAs are typical for higher plants but do not occur naturally in the fungus. The conversion of steviol has been used to prepare labelled 13-hydroxylated substrates for GA metabolism studies with higher plants (60, 127). Further results obtained by feeding kaurenoid analogues to *G. fujikuroi* have been comprehensively reviewed (12; see also 75 and the recent publications 9, 39, 57, 171, 195).

Studying the subcellular localization of GA biosynthesis, Hafemann (73; see also 64) has identified the microsomal fraction that catalyzes the second stage of GA biosynthesis in immature peas and pumpkin endosperm as endoplasmic reticulum. Thus, the conversions of *ent*-kaurenol, *ent*-7α-hydroxykaurenoic acid and GA_{12}, all to GA_{53}, were located in the same fractions as the activity of NADH:cytochrome *c*–reductase, the marker enzyme for endoplasmic reticulum, after centrifugation of pea microsomal membranes on continuous sucrose gradients. Both the GA biosynthesis and the marker enzyme activities displayed Mg^{2+}-shifting when high- and low-Mg^{2+} concentrations were used for the gradients (cf 156). Plasma membranes, plastid envelopes, thylakoid membranes, Golgi fragments, and mitochondrial membranes could be excluded as carriers of GA biosynthesis in peas. Essentially the same results were obtained with the microsomal fraction from *C. maxima* endosperm, using *ent*-7α-hydroxykaurenoic acid as the substrate. Although the Mg^{2+}-shifting of the endoplasmic reticulum is much less pronounced in this material, the GA biosynthesis enzyme activities were again found at the location of endoplasmic reticulum on the density gradients, thus confirming the results obtained with pea microsomes.

Other Cell-Free Systems

The main part of the pathway shown in Scheme 2 has now also been shown in cell-free systems from immature pea seeds (160), *Phaseolus coccineus* suspensors (except for GA_{12}-aldehyde; 24), *P. coccineus* cotyledons (191), and *Sechium edule* immature seeds (20). Cell-free systems from apple seeds also convert *ent*-kaurene to several products including *ent*-7α-hydroxykaurenoic acid, GA_{12}-aldehyde, and GA_9. Only the embryos of these seeds were active; the endosperm was inactive (80). As with *ent*-kaurene biosynthesis, it is difficult to obtain the second part of the GA pathway (Scheme 2) from seedlings. Wurtele et al (202) showed the conversion of *ent*-kaurenol to *ent*-kaurenal and *ent*-19-acetoxykaurene (not illustrated), and the further conversion of *ent*-kaurenal to *ent*-kaurenoic acid in extracts of etiolated maize seedling shoots. The identification of the products was by TLC, gas-liquid chromatography-radio counting and co-crystallization to constant specific radioactivity. The stepwise conversion of *ent*-kaurene to *ent*-kaurenol, *ent*-kaurenal, *ent*-kaurenoic acid, *ent*-7α-hydroxykaurenoic acid, and a hydroxy-kaurenolide was shown in a cell-free systems from *C. maxima* seedling roots (85; see also 64). GA_{12}-aldehyde was converted to 12α-OH GA_{12}-aldehyde, 12β-OH GA_{12}-aldehyde (neither illustrated), and GA_{12}. Since no conversion of MVA to *ent*-kaurene and of *ent*-7α-hydroxykaurenoic acid to GA_{12}-aldehyde was obtained in spite of several attempts, the experiments neither prove nor disprove that there is GA biosynthesis in *C. maxima* roots. All products, except *ent*-kaurenol and *ent*-kaurenoic acid, were identified by GC-MS.

The *C. maxima* endosperm cell-free system is the most active system to date both for the conversion of MVA to *ent*-kaurene and for the parts of the pathway discussed in this and the next section. The intermediates were identified by GC-MS in the early work (see 67, 77, 82), and the preparation is simple and rapid. It has therefore been used extensively since its discovery in 1969 to produce *ent*-kaurene, GA_{12}-aldehyde, GA_{12}, and other intermediates of GA biosynthesis for further use in metabolic studies (e.g. 15, 37, 62, 79, 81, 100–102, 120, 160, 190, 191). A claim for improved enzymatic synthesis of GA_{12}-aldehyde and GA_{12} in this system (16) is based on a comparison with arbitrary yields reported by Graebe and associates for an identification experiment in 1972, rather than on a controlled comparison of the methods in question. The near original version of the method routinely used by us (e.g. 81, 190) is simpler, costs less, and yields amounts of GA_{12}-aldehyde and GA_{12} equal to or in excess of those reported by Birnberg et al (16). The separation of some of the intermediates between *ent*-kaurene and GA_{12} by HPLC is described by Birnberg et al (16) and, with more components identified, by Turnbull et al (191).

THE PATHWAYS AFTER GIBBERELLIN A_{12}-ALDEHYDE

The General Pathway

OVERVIEW   On the basis of work done with cell-free systems and feeding experiments with intact plants, we can now formulate a general reaction sequence for the formation of GAs after GA_{12}-aldehyde as shown in Scheme 3 (I–VII). In the first step, the C-7-aldehyde (I) is oxidized to a dicarboxylic acid (II). This dicarboxylic acid is oxidized at C-20 to a hydroxy-acid (III) and then to a C-20 aldehyde (IV). The aldehyde (IV) is again oxidized at C-20, which results in the formation of either a tricarboxylic acid (V) or a C_{19}-GA (VI). At the end of the pathway, the C_{19}-GA is generally 2β-hydroxylated (VII). Superimposed on this sequence, which represents the general course of GA biosynthesis, are species- and organ-specific hydroxylation reactions leading to individual pathways and interesting variations. Since the non-, 3β-, 13-hydroxylated, and 3β, 13-dihydroxylated GAs are the ones most frequently found in the objects studied to date, and also the ones discussed in this section, they are listed in the order of the sequence (I–VII) in Scheme 3 for easy reference. All the GAs listed occur naturally in higher plants (see 11, 188). As for the aldehydes (top row of Scheme 3), GA_{12}-aldehyde has been identified as endogenous to mature pea seeds (see 65), and GA_{53}-aldehyde and GA_{14}-aldehyde have been obtained in vitro; but GA_{18}-aldehyde has not yet been identified.

According to current views, the biological activity resides mainly with the C_{19}-GAs represented by stage VI, whereby the hydroxylation pattern has a profound influence on the activity, as discussed below. The C_{20}-GAs repre-

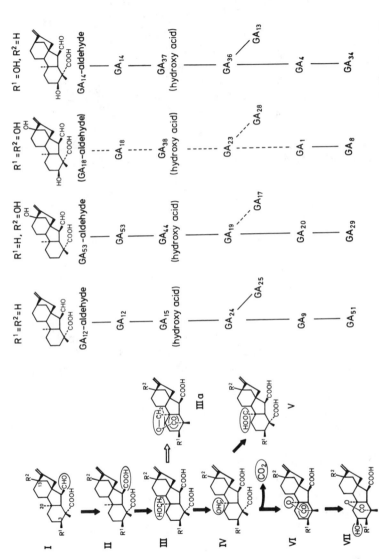

Scheme 3 General GA biosynthesis after GA$_{12}$-aldehyde. *Left*: the general scheme. *Body*: non-hydroxylated, 13-hydroxylated, 3β,13-dihydroxylated, and 3β-hydroxylated GAs listed in accord with the general scheme. The vertical lines show GA metabolic relationships. Broken lines indicate possible reactions, not yet directly demonstrated. Arrows have been avoided, because the actual pathways are connected by hydroxylations at different stages in different species (see e.g. Schemes 4 and 5).

sented by structures I–IV are also often bioactive, but this may be due to their conversion to C_{19}-GAs (VI) in the test objects. The last step of the sequence shown in Scheme 3 is an inactivation reaction, since 2β-OH GAs (VII) are bioinactive. Many other hydroxylation patterns yield GAs that are inactive or have low activity in the standard bioassays, and 2β-hydroxylation may occur at other stages in the pathway; but only 2β-hydroxylation at the end of the pathway is general enough to justify its inclusion in this scheme. The 2β-OH GAs can be further oxidized to catabolites; and both 2β-hydroxylated and other GAs may become covalently bound to smaller molecules, mostly glucose, to form GA conjugates, as discussed below.

NEW FEATURES   The C_{20}-GAs with C-20 at the alcohol oxidation level, as in structure III, are obtained exclusively as 19–20 lactones (IIIa) when they are extracted and purified. It is now clear that the hydroxyl group, at least in several cases, must be free for further oxidation to occur at C-20. Thus, Bearder et al (see 12), by the use of $[19\text{-}^{18}O]GA_{12}$, showed that GA_{15} in its lactone form cannot be an intermediate in GA biosynthesis in *Gibberella fujikuroi*. Furthermore, in the *Cucurbita maxima* cell-free system, GA_{15} in its normal, lactone form is converted only to a by-product, whereas base hydrolyzed GA_{15} is converted further along the main pathway (79). Likewise, GA_{15} and GA_{44} are converted by the soluble enzymes of a pea seed cell-free system only if they are hydrolyzed prior to incubation (100). In these systems, therefore, the hydroxy-acids are the true but unstable intermediates, lactonizing to δ-lactones on extraction. In contrast, cell-free systems from spinach leaves and pea epicotyls accept the lactone forms of GA_{15} and GA_{44} as substrates (62, 204). Obviously, these cell-free extracts, originating from vegetative plant parts, either contain enzymes that can hydrolyze the δ-lactones or they are capable of oxidizing these lactones directly.

Two advances have been made since 1978 with respect to the conversion of C_{20}-GAs to C_{19}-GAs (stage III to stage VI, in Scheme 3). First, it has been demonstrated that this occurs when C-20 is at the aldehyde oxidation stage (IV), whereas the tricarboxylic acids (V) are not converted to C_{19}-GAs. This was shown both in the *C. maxima* cell free system, where GA_{36}, but not GA_{13}, was converted to GA_4 (66), and in the pea seed cell-free system (see next section), where GA_{24}, but neither GA_{25} nor its anhydride, was converted to GA_9 and GA_{51} (100). The conversion of C_{20}-GAs to C_{19}-GAs at the aldehyde stage had been suggested by Durley et al (47). An observed incorporation of GA_{13}-7-aldehyde into C_{19}-GAs in *G. fujikuroi* (40) was apparently based on substrate misidentification, for neither synthetic GA_{13}-7-aldehyde, its anhydride, nor GA_{13}-7-alcohol was converted to C_{19}-GAs in *G. fujikuroi* in critically conducted experiments (12, 13).

Second, it has been shown that C-20 is lost as carbon dioxide (CO_2) and not as formaldehyde (CHO) or formic acid (HCOOH) when C_{20}-GAs are converted to C_{19}-GAs (101). Thus, when GA_{12}, labeled with ^{14}C at C-20 and three other positions, was incubated with the pea seed cell-free system, $^{14}CO_2$, originating from C-20, was produced at the same rate as ^{14}C-labeled C_{19}-GAs (101). Control measurements of formic acid dehydrogenase activity excluded the possibility that the CO_2 released had been formed via HCOOH. Previous work (41) had shown that C-20 appeared as CO_2 when *ent*-kaurene labeled at C-20 was converted to C_{19}-GAs in *G. fujikuroi* cultures. No labeled HCHO or HCOOH was produced in these fermentations either, but the possibility could not be excluded that the CO_2 released was derived from HCHO or HCOOH by oxidation within the cells. The same is true for a fermentation of $[3'-^{13}C]MVA$ with *G. fujikuroi*, in which $^{13}CO_2$ was trapped as $[^{13}C]$bicarbonate, which was identified by ^{13}C NMR (115). The latter study (115) is interesting because it is the first in which the formation of $[^{13}C]GAs$ could be observed in vivo by ^{13}C NMR.

Although the first step in GA biosynthesis after GA_{12}-aldehyde and some of the hydroxylations may be catalyzed by microsomal enzymes, the steps lying after GA_{12}-aldehyde are generally catalyzed by soluble enzymes. It was a major advance when Hedden discovered that these soluble enzymes are 2-oxoglutarate-dependent dioxygenases (79). The properties of these dioxygenases are discussed further below in the section on enzymes. I first turn to the individual pathways on which the general scheme is based.

Individual Pathways

PISUM SATIVUM   Scheme 4 shows GA metabolism after GA_{12}-aldehyde in *Pisum sativum*. In this species there are two parallel pathways, connected by 13-hydroxylation at the C_{20} stage, and a partial third pathway, which arises through 3β-hydroxylation at the C_{19} stage. The two first pathways correspond to the two first series of GAs listed in Scheme 3, complete except for GA_{25}. Each pathway ends with a catabolite. The pathways from GA_{12} to GA_{51} and from GA_{53} to GA_{29} were worked out in cell-free systems from immature pea seeds. Each step was defined by incubating with ^{14}C-labeled precursors and identifying the products by GC-MS. In the earliest experiments, homogenates of whole peas were used as the enzyme preparation, and GA_{12}-aldehyde was converted to GA_{12}, GA_{53}, and GA_{44}, but not further (160). The rest of the pathway was obtained by omitting the seed coats from the preparations and concentrating the enzyme extracts (100). The oxidation of GA_{12}-aldehyde to GA_{12} and the 13-hydroxylations (Scheme 4) are catalyzed by NADPH-dependent microsomal enzymes, while all other reactions are catalyzed by 2-oxoglutarate-dependent soluble enzymes (100). The 13-non-hydroxylation and 13-hydroxylation pathways can therefore be studied separately by using a high-speed supernatant and GA_{12} or GA_{53}, respectively, as the substrate. The pea system is useful for the preparation of labeled 13-OH GAs (see e.g. 62, 102, 187). As Scheme 4 shows, 13-hydroxylation can occur on at least three different levels in the cell-free system. Theoretically, it may also occur at the level of GA$_{12}$-aldehyde, in which case the first 13-hydroxlated product would be GA$_{53}$-aldehyde as in beans (see Scheme 5), but the formation of GA$_{53}$-aldehyde is not regularly observed in the pea system. In the absence of the microsomal fraction, GA$_{12}$-aldehyde is only converted to its glucosyl ester (see 100).

The products obtained in the cell-free system are identical with the GAs endogenous to immature pea seeds, except for a few minor discrepancies:

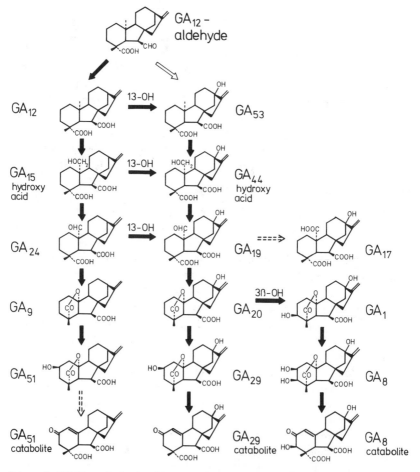

Scheme 4   GA biosynthesis after GA$_{12}$-aldehyde in *Pisum sativum* immature seeds and shoots. 3β-Hydroxylation only occurs in young, growing regions of the shoots and possibly in very young fruits. Broken line arrows show reactions not yet directly demonstrated.

GA_{17} has not been shown in the cell-free system, whereas GA_{12}, GA_{15}, and GA_{24} do not accumulate in vivo (cf 56, 120, 176). The pathways obtained in vitro are therefore considered representative of the biosynthesis in vivo. The 13-non-hydroxylation pathway probably plays a minor role in vivo, since the 13-OH GAs predominate as endogenous components. Indeed, the 13-non-hydroxylation pathway may be merely accidental in the sense that it may arise by the conversion of non-hydroxylated GAs by the same enzymes as catalyze the 13-hydroxylation pathway. As will be seen below, the soluble 2-oxoglutarate-dependent enzymes are often rather unspecific.

The 2β-hydroxylation steps from GA_9 to GA_{51} and from GA_{20} to GA_{29} were first found by feeding GA_9 and GA_{20} to *P. sativum* seeds and seedlings in vivo (see 82, 174). The further oxidation to catabolites was also first indicated by feeding studies (48, 179, 180), which soon led to their identification as endogenous constituents in both seeds and seedlings (180). The structures have been confirmed by synthesis (58). In developing pea seeds, 2β-hydroxylation can occur both in the embryo and the testa, whereas the further oxidation to the GA_{29}-catabolite occurs in the testa only (175). This is also where the bulk of endogenous catabolite is found. Catabolite formation does not strictly belong to GA biosynthesis, but it is included here because these substances are major endogenous constituents of both peas and other plants (1). Interestingly, the catabolites of seedlings are particularly found in the roots (90, 91), and the roots may therefore be disposal sinks for used-up GAs.

In seedlings, GA_{20} is not only converted to GA_{29} and GA_{29}-catabolite, but it is also 3β-hydroxylated to GA_1, which is subsequently converted further to GA_8 and GA_8-catabolite (Scheme 4; 90–93, 177). This formation of GA_1 is most significant for seedling growth, for GA_1 is the main, perhaps the only, GA controlling internode elongation in pea and maize and perhaps in many other species as well (see the section below on Genetic Control of Internode Elongation). In peas, the 3β-hydroxylation has been conclusively demonstrated in the growing regions of seedling shoots only (see below), but GA_1, GA_8, and GA_8-catabolite have recently been identified by GC-MS in seeds and pods at the very young stage of 5 days after pollination (54). Indirect evidence suggested that these 3β-hydroxylated GAs may have originated in the young fruits, but they may also have been imported from vegetative parts of the plant. No 3β-OH GAs were found in 21-day-old developing seeds (56).

One might ask why 3β-hydroxylation occurs at the stage of C_{19}-GAs only (see Scheme 4). In fact, this has never been shown. The impression that 3β-hydroxylation only occurs at the level of GA_{20} is based on the circumstances that no other GAs have been tried as precursors in the 3β-hydroxylation experiments and that no 3β,13-dihydroxy C_{20}-GAs have been identified in pea shoots. But the 13-hydroxylated C_{20}-GAs are usually also found in small amounts and on special search only (38, 56, 120, 141). Durley et al (47) fed [³H]GA₁₄ to etiolated pea seedlings and identified the members of a putative 3β,13-dihydroxylation pathway, GA₁, ₈, ₁₄, ₂₃, ₂₈, ₃₈ (Scheme 3), as products by GC-RC and also by GC-MS. These results have been considered artefacts (see e.g. 67) because pea plants at that time were believed not to contain 3β-OH GAs. It now seems possible that the conversions observed (47) could have been natural. If 3β-hydroxylation does indeed occur only late in the pathway, as shown in Scheme 4, this may be due to unequal distribution of the enzymes in the plant. Thus there is some indication that GA₂₀ is biosynthesized in young leaves and then translocated to the growing regions, where it becomes 3β-hydroxylated to GA₁ (see Genetic Control of Internode Elongation, below).

PHASEOLUS SPECIES   Earlier studies of C₁₉-GA metabolism and the identification of endogenous GAs in beans have recently been reviewed (133, 174). Scheme 5 shows new results on GA biosynthesis after GA₁₂-aldehyde in cell-free extracts from immature *Phaseolus vulgaris* seeds. Each step was defined by the incubation of stable isotope intermediates and GC-MS identification of the products (102, 187). The pathway complex is superim-

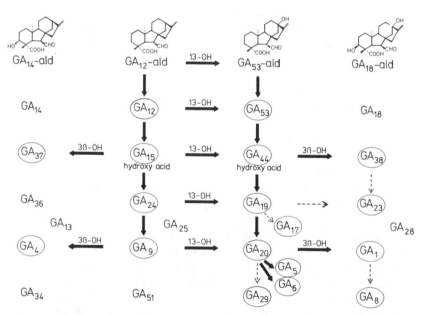

Scheme 5   GA biosynthesis after GA₁₂-aldehyde in a cell-free system from *Phaseolus vulgaris* immature seeds. The circles mark endogenous GAs. Broken line arrows show reactions, not obtained in the cell-free system, although some of them have been demonstrated in other systems (see text).

posed on the GAs listed in Scheme 3 (scrolled one step to the right) for comparison. Like the peas, the beans have a 13-non-hydroxylation and a 13-hydroxylation pathway, and 13-hydroxylation can occur at several points. The differences to the peas are: (*a*) the definite formation of GA_{53} via GA_{53}-aldehyde, (*b*) the 3β-hydroxylation in the seeds (and not only in seedlings), (*c*) the 3β-hydroxylation at the stage of the C_{20}-GAs, and (*d*) the formation of GA_5 and GA_6 (see structures in Scheme 6). A second route to GA_1 via GA_4, which has been shown by feeding GA_4 to intact bean seeds (see 174), was not found in the cell-free system. Independent 3β-hydroxylation or 3β,13-dihydroxylation pathways were neither found nor excluded in the beans.

As in peas, there is close agreement between the GAs formed in the cell-free system and the endogenous GAs (encircled in Scheme 5). These were determined in the same lot of seeds. No 2β-hydroxylation occurred in the cell-free system, although 2β-OH GAs were present and although 2β-hydroxylations of GA_{20} to GA_{29} and of GA_1 to GA_8 have been demonstrated both with intact bean seeds (see 174) and with purified enzymes from germinating beans (172). This was probably because young seeds in the middle of rapid development were used for the cell-free system, and 2β-hydroxylation is very active only later in seed development (see 174). Also for the pea cell-free system, seeds from older stages had to be used as starting material to demonstrate 2β-hydroxylation (100). Of the two endogenous C_{20}-GAs that have not been obtained metabolically, GA_{17} probably is formed via GA_{19} while GA_{23} could come either from GA_{38} or GA_{19}. The molar ratio of endogenous 13-hydroxylated to 13-non-hydroxylated GAs was 8:1, which shows that the 13-hydroxylation pathway, as in the peas, dominates in vivo.

Metabolic work done with *Phaseolus coccineus* indicates a pathway in this species similar or identical to that in *P. vulgaris*. Thus *ent*-7α-hydroxy-kaurenoic acid was converted to $GA_{1, 5, 8}$ in a cell-free preparation from suspensors of *P. coccineus* seeds (23), and [^{14}C]GA_{12}-aldehyde was con-

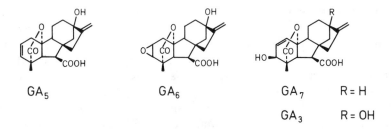

GA_5 GA_6 GA_7 R = H

                                    GA_3 R = OH

Scheme 6   Structures of important GAs not shown in Scheme 3.

verted to 13 labeled products in a cell-free system from *P. coccineus* immature embryos (190). These 13 products, identified by GC-MS, were the same as in the *P. vulgaris* preparation (Scheme 5), except that GA_{17} was obtained in the *P. coccineus* cell-free system, whereas $GA_{9, 12, 53}$ were not. Once more, the products formed in vitro corresponded to the endogenous GAs of the same species (cf 1, 174). Of the endogenous GAs in immature *P. coccineus* seeds, GA_8 is by far the most abundant, accumulating in the testa, where it probably would be conjugated or converted to the GA_8-catabolite at a later stage of seed development (1).

OTHER SPECIES   *Spinacia oleracea* seedlings convert [²H]GA_{53} to $GA_{44, 19, 20}$ (59), and a cell-free system from the leaves converts [¹⁴C]GA_{12} sequentially to [¹⁴C]$GA_{53, 44, 19, 20}$, as identified by GC-MS (62). This sequence is typical for the 13-hydroxylation pathway (cf Scheme 3); and the GAs obtained are identical with the endogenous GAs of the spinach plant (123), except that GA_{29} was not obtained with the enzyme preparations. All the enzymes catalyzing this sequence are soluble and have the typical cofactor requirements of 2-oxoglutarate-dependent dioxygenases. The lack of 2β-hydroxylation is unexplained. The endogenous amounts of GA_{29} are substantial and 2β-OH GA_{12} is formed as a minor metabolite from GA_{12} in the cell-free system, which shows that 2β-hydroxylation can take place. The most interesting point with the spinach system is that some of the enzyme activities are under photoperiodic control (see Physiology of Gibberellin Control, below).

The very active cell-free system from *Cucurbita maxima* endosperm, already discussed under *ent*-kaurene and GA_{12}-aldehyde biosynthesis, is also active for the part of the pathway lying after GA_{12}-aldehyde, as was well known before 1978 (67, 77, 82). The pathway of GA metabolism proceeds partly via 3β-non-hydroxylated, partly via 3β-hydroxylated intermediates (the fourth series in Scheme 3). Main products are the tricarboxylic C_{20}-GAs GA_{13} and GA_{43} (=2β-OH GA_{13}, not shown in Scheme 3), whereas the C_{19}-GA GA_4 is formed in small quantities only. In this case, the pathway is diverted into the formation of the tricarboxylic acid (stage V, Scheme 3), and the C_{19}-GA (stage VI, Scheme 3) is a very minor product. In recent experiments, a pattern of 12α-hydroxylation has been obtained with the *C. maxima* cell-free extracts (81). This 12α-hydroxylation is microsomal and pH-dependent: Above pH 6.5, GA_{12}-aldehyde conversion is active and the product is almost exclusively GA_{12}, whereas below pH 6.5, the conversion is less active and the main product is 12α-OH GA_{12}-aldehyde. Once formed, 12α-OH GA_{12}-aldehyde is converted further by soluble enzymes along the general pathway (Scheme 3) to form 12α-OH GA_{14}, -GA_{15}, and -GA_{37} as well as several unidentified products (not illustrated here; 81). It was also investigated whether 12α-hydroxylation can occur before the formation of

GA_{12}-aldehyde in the *C. maxima* system (81). The result was negative; *ent*-12β-hydroxykaurenoic acid doubly labeled with ^{13}C and ^3H was converted in high yields to *ent*-7α,12β-dihydroxykaurenoic acid; but ring contraction, which would have resulted in formation of the corresponding 12α-OH GA_{12}-aldehyde, did not occur (cf Scheme 2). Extensive identification of the endogenous GAs in the endosperm of *C. maxima* (8, 17) showed that, as in the other objects, the GAs obtained in vitro correspond to the endogenous GAs. However, two of the main endogenous 12α-OH GAs in the endosperm, GA_{58} and GA_{49} (neither illustrated), have not yet been obtained in the cell-free extracts. As a point of interest, the identification of 12-oxygenated compounds in *C. maxima* by GC-MS was made possible by production and characterization of the appropriate reference compounds by fermentation of 12-oxygenated *ent*-kaurenoid compounds with *Gibberella fujikuroi* (57).

Some aspects of the GA relations in *Pharbitis nil* were reviewed by Pharis & King (133). A study of [^3H]GA_5 metabolism in developing *P. nil* seeds (110) is interesting with respect to the pathways, because GA_3 was obtained; the origin of GA_3 in higher plants is not well understood. [^3H]GA_5 injected into developing seed capsules was metabolized to [^3H]$GA_{1,\ 3,\ 6,\ 8,\ 22,\ 29}$, as identified by a series of chromatographic steps followed by capillary GC-SIM. The main metabolites were GA_1, GA_3, and their purported glucosyl conjugates. The authors postulated a sequence in which GA_5 is converted to GA_1, GA_3, and GA_{29} via GA_6. Similar results were obtained by feeding [^3H]GA_5 to cell suspension cultures derived from the hypocotyl tissue of a germinated *P. nil* seed (108). Since the appearance of Pharis & King's review the endogenous GAs of the related *Pharbitis purpurea* have been determined (53); they are nearly the same as in *P. nil*.

Only two filamentous fungi, the ascomycetes *Gibberella fujikuroi* and *Sphaceloma manihoticola* (the causal agent of the superelongation disease of Cassava), are known to produce GAs. Most of the GA biosynthesis pathway in *G. fujikuroi* was known before 1978 (see 12). Although the details of the pathway after GA_{12}-aldehyde are still not known, it is clear that the 3-non-hydroxylation and 3β-hydroxylation pathways are operating much as they do in the higher plants, except that 2β-hydroxylation does not occur in the fungus. The pathway in *G. fujikuroi* is interesting because 3β-hydroxylation is microsomal and occurs at the beginning of the pathway, whereas 13-hydroxylation is catalyzed by soluble enzymes and occurs at the end of the pathway. This is opposite to the situation in the *Pisum sativum* and *Phaseolus vulgaris* systems. The main product of GA biosynthesis in *G. fujikuroi* is GA_3 (see Scheme 6), which is formed via GA_4 and GA_7; but in *S. manihoticola*, GA_4 is the main product and GA_7 and GA_3 are not formed (66, 145, 206). Obviously, 1,2-dehydrogenation and 13-hydroxylation, both necessary for the conversion of GA_4 to GA_3, do not belong to the repertoire of *S. manihoticola*.

Summary of Pathways after GA_{12}-Aldehyde

The cell-free work has shown that GA biosynthesis follows pathways of the general type shown in Scheme 3 connected by hydroxylations. If the hydroxylations occur at the stage of GA_{12}-aldehyde or GA_{12}, entire pathways arise, as in the cases of the 13-hydroxylation pathways in peas and beans (Schemes 4, 5) and the 3β-hydroxylation pathway in *G. fujikuroi* (see 12). In all these species, hydroxylation is incomplete, which results in parallel non-hydroxylation pathways. Hydroxylations occurring at intermediate position in the pathway, such as the 3β-hydroxylation in *Phaseolus vulgaris* (Scheme 5) and 3β-hydroxylation in *Cucurbita maxima* (79), result in hydroxylation of both C_{20}- and C_{19}-GAs. Late hydroxylations, on the other hand, such as 3β-hydroxylation in pea shoots (illustrated in Scheme 4) and 13-hydroxylation in *G. fujikuroi*, affect the C_{19}-GAs only. The same seems to be the case for the 12β- and 18-hydroxylations found by feeding studies with *Hordeum vulgare* in vivo (61). 2β-Hydroxylation, finally, is often the last reaction in the pathway of the true GAs (cf Schemes 3–5). The early hydroxylations are often catalyzed by microsomal monooxygenases, while the intermediate and late hydroxylations tend to be catalyzed by soluble dioxygenases. The end products of the pathways—especially the 2β-OH GAs and the catabolites, but sometimes also the tricarboxylic acids—often accumulate in vivo, whereas the intermediates remain at low or undetectable levels.

With this knowledge and the help of Scheme 3, the results of conversions in which the single steps have not been studied (e.g. 190), or the results of experiments in which all the endogenous GAs of an object have been identified, can be interpreted and arranged in hypothetical but sensible biosynthetic sequences (e.g. 17, 55, 103, 203). For a more complete coverage of the many feeds and identifications done since 1978 and earlier, the reader is referred to a discussion of hydroxylation patterns (119), recent extensive reviews (133, 174), and listings (11, 188).

Enzymes

In 1978, it was well known that the six-step conversion of GA_{12}-aldehyde to GA_{43} in the cell-free system of *Cucurbita maxima* was catalyzed by soluble enzymes that required Fe^{2+} for activity, were erratically stimulated by NADPH, and were inhibited by Mn^{2+}. It was also known that GA 2β-hydroxylation in *Phaseolus vulgaris* cotyledons was catalyzed by the soluble fraction and required O_2, Fe^{2+}, and NADPH or ascorbate for activity. The enzyme activity was already low after ammonium sulfate precipitation, and no further attempts at purification of the enzyme were made (see 67, 82).

Further purification became possible when it was found that the soluble enzymes catalyzing the steps from GA_{12}-aldehyde to GA_{43} in a gel-filtered *C.*

maxima endosperm preparation had a specific requirement for 2-oxoglutarate in addition to Fe^{2+} (79). The activity was further stimulated by ascorbate. Crude homogenates contained sufficient endogenous 2-oxoglutarate to support the reaction fully, and the requirement for Fe^{2+} could only be shown by removal of Fe^{2+} with chelating agents. The same cofactor requirements have since been established for the series of conversions in cell-free systems from pea cotyledons (Scheme 4) and spinach leaves (see previous section), and for purified enzymes from bean, pea, and pumpkin seeds (see below). The same cofactors support GA biosynthesis in cell-free systems from *P. vulgaris* seeds (Scheme 5), *Phaseolus coccineus* seeds, and, as far as the requirement for Fe^{2+} is concerned, the conversion of *ent*-kaurene to C_{19}-GAs in a cell-free system from *Sechium edule* seeds, all of which have been discussed already. It is therefore a general feature that the soluble enzymes of the GA pathway directly or closely after GA_{12}-aldehyde are of the 2-oxoglutarate-dependent kind.

The cofactor requirements mentioned above are identical to those of several oxygenase reactions in which 2-oxoglutarate is oxidized to succinate and CO_2 at the same time as one oxygen atom is incorporated into the substrate. These reactions and their enzymes have been most extensively studied in animals, but also in bacteria, fungi, and higher plants (see 72, 118, 120a, 155). The most prominent enzyme of this kind is prolyl hydroxylase, which is involved in cell wall synthesis in higher plants and collagen synthesis in animals. A stereochemical concept for the reaction mechanism has been proposed that accounts for many of the known properties of prolyl hydroxylase (74).

The first 2-oxoglutarate-dependent enzymes of the GA pathway to be successfully purified were 2β-hydroxylases. One 2β-hydroxylase was purified from the cotyledons of mature imbibed *P. vulgaris* seeds by ion exchange and gel filtration chromatography (87, 172). This preparation catalyzed the 2β-hydroxylation of GA_1, GA_4, and, at a much lower rate, GA_9 and GA_{20} (cf Scheme 3). Since the relative rates of these activities did not change throughout purification, the authors concluded that the hydroxylase activities might all reside in one and the same protein, although they did not exclude the possibility that copurification of discrete enzymes could have occurred. Another 2β-hydroxylase was purified from the cotyledons of germinating pea seeds by methanol precipitation, ion exchange, and gel filtration chromatography (173). Like the bean preparation, the pea preparation catalyzed 2β-hydroxylation of $GA_{1, 4, 9, 20}$, but the specificity was the opposite. Thus GA_9 and GA_{20} were more efficiently 2β-hydroxylated than GA_1 and GA_4. Comparison of the relative purification factors and thermal denaturation rates indicated that at least two different proteins were at work, one with higher specificity for GA_9 and GA_{20} and another more specific for GA_1 and GA_4; but attempts to separate these activities further by protein HPLC were not successful.

In the meantime, enzymes from the earlier part of the GA pathway were also purified and studied from *C. maxima* endosperm and developing pea cotyledons. The proteins of *C. maxima* endosperm were fractionated by ammonium sulfate precipitation, ion exchange chromatography, hydroxylapatite chromatography, and preparative isoelectric focusing (192). One of the protein fractions thus obtained catalyzed the conversion of GA_{12}-aldehyde to GA_{12} and was therefore a C-7 aldehyde oxidase. This fraction also catalyzed the conversion of GA_{14}-aldehyde to GA_{14} (cf Scheme 3), but the rate of this reaction was lower than the oxidation of GA_{12}-aldehyde by an estimated factor of 10. Another fraction from the preparative isoelectric focusing turned out to be a 3β-hydroxylase. It did not show a strict substrate specificity and catalyzed the hydroxylation of GA_{12}-aldehyde to GA_{14}-aldehyde, GA_{12} to GA_{14}, GA_{15} to GA_{37}, GA_{20} to GA_1, and GA_{53} to GA_{18}, all as identified by TLC. The highest affinity of the enzyme preparation was for GA_{15}. Finally, a GA_{53} C-20 hydroxylase has been purified from extracts of the cotyledons of developing pea seeds by ammonium sulfate precipitation, gel filtration, and anion exchange chromatography (112). This preparation oxidized GA_{53} to GA_{44} and GA_{19}, the ratio of these two products remaining constant throughout purification. After the final purification step, all these enzymes showed only one or two major bands on sodium dodecyl sulfate polyacrylamide gel electrophoresis, but at least in the case of the GA_{53} C-20 hydroxylase these principal bands could not be identical with the enzyme. Thus semi-preparative ultra thin-layer isoelectric focusing separated the bulk of the protein in the fraction from the enzyme activity, which appeared at a pI of 5.6–5.9. In this range, analytical ultra thin-layer isoelectric focusing showed only two very weak bands lying closely together at a pI of about 5.8. These bands may therefore represent the C-20 hydroxylase activities (112).

The enzymes discussed above are all present in very low amounts in the tissues. Since they are also very hydrophobic and very unstable, they are difficult to isolate. Like the crude preparations, the purified enzymes require 2-oxoglutarate, Fe^{2+}, and ascorbate; and residual Fe^{2+} may have to be removed by chelating agents before the requirement can be shown. For the GA_{53} C-20 hydroxylase, Fe^{2+} can only be removed under incubation conditions, i.e. when the substrates and the cofactors are present (112). Catalase and bovine serum albumin stimulate some of the enzymes but not others. Dithioerythritol or dithiothreitol is needed throughout the purification of some of the enzymes, but their effects during incubation are complicated and may even be inhibitory (112).

The relative molecular weights, as estimated by gel filtration, lie in the range of 36000–45000 for all the enzymes. The quantitative kinetic parameters obtained are widely different and difficult to compare because the enzymes were all purified by different methods and the activities determined by different assays. Thus, the 2β-hydroxylase activities were measured by the use of [2β-^{3}H]-labeled substrates and determination of the rate of formation of tritiated water. Therefore, the rate measurements include a primary tritium isotope effect of unknown magnitude. The 3β- and GA$_{53}$ C-20 hydroxylases and the C-7 aldehyde oxygenase activities were measured by the use of ^{14}C-labeled substrates and radio counting of the products after separation by HPLC or TLC. The 2β-hydroxylase from beans and the GA$_{53}$ C-20 hydroxylase from peas were obtained in reasonable yields, and their specific activities incresed some 300–500-fold during purification (112, 172); but the 2β-hydroxylase from peas and the C-7 aldehyde oxygenase from *C. maxima* were extremely labile and lost most of the activity during purification (173, 192)

PHYSIOLOGY OF GIBBERELLIN CONTROL

Genetic Control of Internode Elongation

Dwarf cultivars, varieties, and mutants have been involved in GA research on higher plants since its very beginning (see historical aspects in 135), and the use of single gene mutants for GA research had already been recommended by 1960 (140). The greatest progress in our understanding of the connections between GA metabolism and dwarfism in plants, however, has been made during the last few years, after reliable methods for the analysis of endogenous GAs had been developed, the biosynthetic pathways had become known, and new mutants had been classified. The usefulness of mutants for physiological studies in general was recently reviewed (121).

Dwarf plants have shorter internodes than their tall counterparts (usually classified as "normals"). Depending on the genotypes, dwarf plants may either respond or not respond to exogenously added GA by increased growth. The dwarfs that do respond (GA-responsive dwarfs) may have an altered GA metabolism, so that they produce no GA, less GA, or GA of another kind than their tall counterparts. For several of these mutants, blocks in the GA biosynthesis pathway have been established in the last few years. The dwarf plants that do not respond to exogenously added GA (GA-insensitive dwarfs) obviously have an altered sensitivity to GA (e.g. 104, 154, 184). In a third type of mutant, the slender type, growth is altogether independent of GA and the plants grow tall whether they have a supply of GA or not (see 121, 144, 161). Of these three types, the GA-responsive mutants are of interest in connection with GA biosynthesis and are discussed below. The GA-independent mutants are of more interest in connection with GA action and are mentioned incidentally in this article.

GA-responsive mutants of pea, maize, and rice have been particularly intensively studied (see 136 for a stimulating review). All three species apparently synthesize their GAs via the 13-hydroxylation pathway supplemented with 3β-hydroxylation to form GA_1, as shown for peas in Scheme 4. Phinney (136, 138) has suggested that, at least in these species, GA_1 is the only GA causing internode elongation per se, and that the other GAs of the 13-hydroxylation pathway only appear active because they are converted to GA_1 in the assay plants. This concept was developed during work with dwarf mutants of maize, but I will first turn to the pea mutants, because they have been studied in more detail.

PISUM SATIVUM   Seven genetic loci are known to regulate internode elongation in pea seedlings; four of these, *Na, Le, Lh,* and *Ls,* do so by influencing GA metabolism (see 152, 153 for descriptions of the mutant types; 92, 136, 144 for photographs of the impressive phenotypic differences). The loci *Lh* and *Ls* have only just been discovered (154) and are not discussed further here, but the loci *Na* and *Le* have been extensively studied, providing a host of information.

The recessive allele *na* blocks GA biosynthesis at some early step(s) in the pathway (142). The phenotype, nana, of the double recessive genotype *(na)* has extremely short internodes, small darkly colored leaves, and no bioactive GAs. The yield of fruits and seeds in nana is puny, unless the plants are treated with a little exogenous GA a few times during the vegetative stage. The plants then grow tall and have a normal fruit set. In spite of the severe dwarfing, there seems to be a "leak" in the genetic block, for the end point metabolites of GA biosynthesis, GA_8- and GA_{29}-catabolites (see Scheme 4), have been found in ~4-week-old *na* genotype seedlings, particularly in the roots (90, 91). Obviously, these products can accumulate even at a very slow rate of GA biosynthesis. The *na* mutation does not affect later steps in the biosynthetic pathway. This fact has been utilized to study the conversion of GA_{20} in the near absence of endogenous GAs, which greatly facilitates the detection and quantitative estimation of products (90, 91, 142). Interestingly, the *na* mutation is not expressed in the seeds, which contain a normal supplement of GAs, at least by bioassay (141, 142).

The other locus, *Le,* controls 3β-hydroxylation in pea seedlings and with that the conversion of GA_{20} to GA_1 (cf Scheme 4). The phenotype of the double recessive genotype *(le)* is a semi-dwarf in which the upper internodes are reduced in length by 40–60% and the stem has a typical zig-zag appearance (see 153). The dwarfing is much less severe than in nana. The yields of seeds are normal or high. The true meaning of this locus has only recently been realized; for even at the beginning of this decade no 13-OH GAs had been conclusively identified in peas, and it seemed certain that GA_{20} was the main active GA in this species. Earlier reports of a GA_1-like component found in pea shoots were based on bioassay and therefore appeared less conclusive.

The problem gained new interest when Potts and associates (142, 143), again by bioassay, but now introducing genetically defined lines, demonstrated the presence of a GA_1-like component in extracts from Le genotype seedlings, which was absent in extracts from le genotype seedlings. Subsequently, this component was conclusively identified by GC-MS as GA_1, and the conversion of GA_{20} to GA_1 in Le genotype seedlings was shown by feeding $[^3H]GA_{20}$ and identifying the products by TLC (93). Final proof that 3β-hydroxylation occurs in pea shoots and is controlled by the Le locus was obtained by feeding $[^{13}C,^3H]GA_{20}$ and identifying both products and endogenous GAs by GC-MS (92). GA_1 is found in the elongating apical region of the shoot (143) and in very young fruits (54) but not in older (21 days) developing seeds (56), which certainly was the reason it was not discovered earlier.

The influence of the Le locus on 3β-hydroxylation in pea seedlings and the significance of this reaction for internode elongation have been studied by feeding $[^{13}C,^3H]GA_{20}$ to different genotypes and analyzing the products by GC-MS (90–92, 177). The use of a substrate, doubly labeled with a radioactive and a stable isotope, allows for a maximum of information: Recovery and percentage conversion can be calculated, the metabolites can be conclusively identified, and the amounts of endogenous, unlabeled metabolites and substrate can be determined (178; see also 117, 174). Using this technique, Ingram et al (91) found that the growth-promoting activity of GA_{20} could be explained by its conversion to GA_1 as proposed by Phinney (136), rather than by an intrinsic biological activity of GA_{20}. Thus the growth response of different genotype mutants to $[^{13}C,^3H]GA_{20}$ was proportional to the different amounts of $[^{13}C,^3H]GA_8$ accumulating in these mutants, and not to the total amount of $[^{13}C,^3H]GA_{20}$ reaching the growing tissue. $[^{13}C,^3H]GA_8$ was used as a measure of 3β-hydroxylation, and hence of GA_1-formation, since GA_1 was rapidly turned over and did not accumulate (cf Scheme 4). The authors (91) concluded that their results provided strong evidence for a relation between internode growth and GA_1 concentration in peas, but it may be good to keep in mind that it was the turnover of GA_1 that was measured and not its concentration.

Some confusion with respect to the significance of GA_1 for internode elongation has been caused by the fact that the le mutation does not block 3β-hydroxylation completely. Thus GA_{20} is not only converted to GA_{29} and GA_{29}-catabolite in the le genotype dwarf cultivar Progress No. 9, but it is also slowly converted to GA_1, GA_8, and GA_8-catabolite (177). Furthermore, this cultivar contains as much endogenous GA_1 as the tall cultivar Alaska (56, 177). The same has been found for other le mutants with different genotypic background, and it is believed that this measurable production of GA_1 in le mutants is the reason they are less severely dwarfed than na mutants (90, 91). The paradox that the dwarf cultivar contains as much GA_1 as the tall cultivar may be resolved by assuming that GA_1 is metabolized faster in rapidly elongating than in resting cells. This could be because of 2β-hydroxylation in the enlarging vacuoles of the elongating cells, since metabolism of GA_1 to GA_8 has been observed in isolated vacuoles of barley (150). A more rapid metabolism of GA_1 would preclude the accumulation of GA_1 in the overall cell population of the growing tissue, even if high concentrations might be reached in single cells before they elongate. In the dwarfs, on the other hand, slow accumulation below a threshold level in many cells might be measurable, without leading to elongation and accelerated metabolism.

There are other indications of a more rapid turnover of GAs in actively elongating tissues. Thus the content of GA_1 does not increase when the content of GA_{19} declines in vigorously growing shoots of the rice cultivar Nihonbare (186). Further, Potts et al (144) found less GA-like activity in extracts from rapidly growing segregates with slender phenotypes (genotype cry^s) than in dwarf phenotypes (genotype Cry), both on the background of *le Na la*. The locus Cry/cry^s does not affect GA metabolism, but only the dependence on GA for growth (144). The difference in GA-like activity, therefore, does not seem to be based on genetic regulation of GA biosynthesis, but could be the consequence of a more rapid turnover of GAs (in this case GA_{20}) in the rapidly growing phenotype. A further example is the faster metabolism of added $[^3H]GA_1$ in tall *(rht3)* than in gibberellin-insensitive dwarf *(Rht3)* wheat seedlings (184). It is true that the slower metabolism in the dwarf seedlings was shown to be partly caused by dilution with endogenous GA—the endogenous pool by radioimmunoassay was 11.8-fold greater in the dwarf than in the tall—but this accumulation of GA in the dwarf is in itself an indication of decreased GA metabolism; and even after correction for the pool size, the rate of $[^3H]GA_1$ metabolism was higher in the tall seedlings. The suggestion made here, that rapid growth may result in more active GA metabolism, is fully compatible with the conclusion reached by Stoddart (184), that the *Rht3* gene does not appear to operate through an enhanced metabolism of GAs.

An interesting point to keep in mind is that precursors of GA_1, including its immediate precursor GA_{20}, seem to be mobile in the tissue, whereas GA_1 itself is not. Thus, the effect of the *Na* locus is graft transmissible while the effect of the *Le* locus is not. In the former case, precursors of GA_1 may be translocated into the *na* genotype graft partner and there converted to GA_1 on a *Le* genotype background. In the latter case, GA_1 would have to be translocated into the *le* graft partner to effect growth, and this apparently does not happen (153, and references cited therein). Furthermore, when $[^{13}C,^3H]GA_{20}$ is applied to the leaves of intact plants, $[^{13}C,^3H]GA_1$ is found in the growing portions of the shoot only, suggesting that it is formed directly within the growing region and remains there (92). Another interesting point, albeit of unknown significance, is that more GA_{20} is transported to the growing region in dwarf than in tall pea seedlings (91, 92).

Using segregating progeny and near isogenic lines, Reid (151) showed that the genes at the loci *Le* and *Na* exert their effects on internode length in both the light and the dark. He also observed that light has a shortening effect that does not work via GA metabolism: It affected the growth of slender segregates as well, and their growth is independent of GA metabolism. Sponsel (177), on the other hand, found that 3β-hydroxylation of $[^{13}C,^{3}H]GA_{20}$ in the dwarf cultivar Progress No. 9 *(le)* is more active in the dark than in the light, and attributed to this the long-known fact that Progress No. 9 has a tall phenotype when grown in complete darkness. The differences in response to light of the *le* genotypes used in the two studies have been blamed on differences in the genetic background at loci other than the *Le* locus.

ZEA MAYS   The evidence for the 13-hydroxylation pathway in *Zea mays* comes from the identification of endogenous GAs and feeding studies (84, 137, 158, 181). Of several dwarfing genes in this species, the simple recessive non-allelic genes *d1, d2, d3,* and *d5* of the phenotypes dwarf-1 (d-1), d-2, d-3, and d-5, respectively, influence GA metabolism (see 136, 137). In d-5, genetic regulation of *ent*-kaurene synthesis is implied, since cell-free systems from this mutant convert MVA, GGPP, and CPP (see Scheme 1) primarily to *ent*-isokaurene. Cell-free systems from d-5 therefore have a greatly lowered *ent*-kaurene production compared to those from "normal" seedlings (83). Since *ent*-isokaurene is not a GA precursor, d-5 has no measurable endogenous GAs; and because the block is early in the pathway, d-5 responds to both C_{20}- and C_{19}-GAs (138). This makes it a favorite plant for GA bioassays.

By the expert use of bioassays, Phinney & Spray (138, 139) obtained evidence for the site of action of the other genes in the GA pathway as well. Thus, d-3 seedlings are probably blocked in the 13-hydroxylation reaction, because they respond to GA_{53}-aldehyde, GA_{53}, GA_{20}, and GA_1, but not to GA_{12}-aldehyde (cf Scheme 5). Dwarf-2 seedlings respond to GA_1, GA_{20}, and GA_{53}, but not to GA_{12}-aldehyde or GA_{53}-aldehyde, and thus may be blocked in the C-7 aldehyde oxidation step. Dwarf-1, finally, responds fully to GA_1 only, which suggests a block in the 3β-hydroxylation. This latter block has been conclusively proven by showing that d-1 seedlings do not convert $[^{13}C,^{3}H]GA_{20}$ to $[^{13}C,^{3}H]GA_1$, whereas "normal" and d-5 seedlings do (181). The ^{13}C-label from GA_{20} was diluted by endogenous GAs in the "normal" and d-1 but not in the d-5 seedlings, thus confirming that the *d5* block lies early in the GA pathway (see above).

OTHER SPECIES   *Oryza sativa* also contains the GAs of the 13-hydroxylation pathway and, on the basis of GC-SIM, GA_1 (111, 186). Immature seeds and flowering ears also contain members of the 13-non-hydroxylation pathway and the 3β-OH GAs, GA_4 and GA_{34} (cf Scheme 3; 103). In rice, too, bioassay studies with dwarf mutants have suggested genetic blocks, located in the early part of the pathway for the well-known cultivar Tan-ginbozu (dwarfing gene dx), and between GA_{19} and GA_1 for cultivar Waito-C and others (dwarfing gene dy; 126). The situation in rice is different from the one in maize in that GA_{19} accumulates at certain developmental stages in both dwarf and tall cultivars. It has therefore been suggested that GA_{19} may function as a "pool" GA in rice, and that the level of active GAs, such as GA_1, might be regulated by the metabolic conversion of GA_{19} (111).

Many GA-responsive dwarf mutants of *Arabidopsis thaliana*, some of which will not even germinate without exogenously added GA, have been isolated (105, see also 106). At least one of the mutant alleles, *ga-1*, appears to block GA biosynthesis before *ent*-kaurene, because cell-free systems from young *ga-1* mutant siliques do not convert $[2\text{-}^{14}C]$MVA to *ent*-$[^{14}C]$kaurene, whereas cell-free systems from the wild type siliques do (5). In accord with this, the *ga-1* siliques do not contain significant amounts of GA-like activity by bioassay, while the wild type siliques contain sizable amounts.

Other Physiological Aspects

PHOTOPERIODIC CONTROL   Stem growth in *Spinacia oleracea* and *Agrostemma githago* is photoperiod dependent. Under short-day conditions (SD), the plants have a rosette growth habit and remain vegetative; under long-day conditions (LD), the stems elongate, and flower primordia are produced. In both species the effect on stem elongation appears to be mediated through photoperiodic control of GA metabolism. In the spinach plant, identification of endogenous GAs and feeding studies in vivo have suggested that one point of photoperiodic control is at the conversion of GA_{19} to GA_{20} in the 13-hydroxylation pathway (cf Scheme 3; see 96, 133). New results obtained with the cell-free system from spinach (see the section on Individual Pathways, above) suggest that both the oxidation of GA_{53} to GA_{44} and the oxidation of GA_{19} to GA_{20} are under photoperiodic control (62). Thus, preparations from plants grown in LD catalyze the whole sequence $[^{14}C]GA_{53} \rightarrow GA_{44} \rightarrow GA_{19} \rightarrow GA_{20}$, whereas preparations from plants grown in SD and harvested at the end of the 16-hr dark period only oxidize GA_{44} to GA_{19}. The GA_{53}- and GA_{19}-oxidizing activities appear within 4–9 hr after the end of the dark period of plants grown in SD and continue to rise up to ~30 hr when the plants are transferred to continuous light. Conversely, if plants exposed to 8 days of continuous light are placed in darkness again, the activities of the GA_{53}- and GA_{19}-oxidizing enzymes decrease during the first hours and then remain very low or negligible for at least 32 hr in darkness. The activity of the conversion of GA_{44} to GA_{19} is not under photoperiodic control, remaining constant and high under all conditions. The discrepancy to the in vivo studies, where only one step seemed to be under photperiodic control, may be due to the longer incubation times involved in the in vivo experiments (6 weeks in the identification experiments, and 48 hr in the feeding experiments, vs 1 hr in the cell-free system), which might negate any differences in the conversion rate of GA_{53} in LD and SD. Otherwise, the changes in enzyme activities in the cell-free system agree with the results obtained in vivo and explain the higher GA_{20} content found in LD than in SD. The more active production of GA_{20} is presumed to be the cause of stem elongation in LD.

In view of the results obtained with dwarf peas and dwarf corn (see previous section) one might expect GA_1 to be responsible for stem elongation in spinach, but despite careful examination GA_1 has not been found in this species (96). Strangely, 3-epi $[^3H]GA_1$ has been obtained as a product of $[^3H]GA_{20}$ feeds (124), but the significance of this compound, which has not been found endogenously in spinach and which does not induce stem elongation in spinach in SD, is unknown. GA metabolism is photoperiod dependent in A. *githago* also, but the situation is less clear than in spinach. A. *githago* contains both endogenous GA_1 and 3-epi GA_1 (see 96, 133).

An interesting case of photoperiodic control of GA metabolism is associated with the prevention of senescence in the G2 line of peas. In this line, apical senescence occurs after several fruits have developed in LD, whereas growth is indeterminate in spite of fruit development in SD. In the original bioassay work, these physiological differences were attributed to both quantitative and qualitative differences in the endogenous GA complement in LD and SD (144a), but no qualitative differences were found by later GC-MS analysis (38). Nevertheless, a few uncertainties remained (38), and new work shows differences in the $[^{14}C]GA_{12}$-aldehyde metabolism in SD and LD. These differences may result in the accumulation of different intermediates under the two conditions (37). Most of the products identified belonged to the 13-hydroxylation pathway, but others were unknown; their ultimate identification may throw light on both the senescence problem and GA biosynthesis in pea shoots.

SEED DEVELOPMENT AND GERMINATION   Pharis & King (133) have thoroughly reviewed the available information on GAs in relation to fruit and seed development. Immature seeds generally contain higher GA levels than any other tissue. These levels seem to be regulated, for they increase over several orders of magnitude at specific stages and then decrease again, often to reach zero at seed maturity. However, no special function can be found for these GAs in the seeds, and they may in fact be redundant. In peas, a function for the seed GAs in early fruit development, immediately after fertilization, has been suggested, but the main increase in GA contents of the seeds occurs when the fruits are fully developed. Wang & Sponsel (cited in 133) have correlated this rise of GAs in maturing pea seeds with the mobilization of assimilate to the rapidly growing seeds, thus suggesting a function of the seed GAs in this process. However, in older experiments the GA contents of pea seeds cultivated in vitro could be drastically reduced with plant growth retardants without greatly affecting seed development, which seems to show that the bulk of the GA is not necessary for the seed development either (see 133).

New experiments done with pea explants, consisting of 10-cm pea stem segments with ovary and adjacent leaf, support the view that endogenous GAs play a role in early pod development, although they do not reveal whether these GAs are made in the pods, the seeds, or the leaf (54). Fertilized ovaries on such explants, standing in water, develop into pods with peas and GAs; but if pollination is prevented by emasculation before anthesis, both pod development and the GA levels are reduced. A certain correlation between endogenous GA_1 and GA_{20} in the fruits and growth of the pods was found for treatment variations. Conversely, exogenously added GA_1 and GA_{20} promoted parthenocarpic growth of the emasculated ovaries. This suggested a function for these GAs in pod growth; but as in earlier experiments, no function was found for the main rise in seed GAs that occurred later in development. Thus, when the GA content of the peas was reduced to almost zero by the use of growth retardants, the seeds still developed and attained 80% of their normal fresh weight.

Further experiments with the explants even seemed to suggest that there is no independent GA biosynthesis within the seeds. Thus, the plant growth retardant paclobutrazol, labeled with ^{14}C and applied through the stem segments of fertilized ovary explants, effectively inhibited GA accumulation in the developing pea seeds, although it did not seem to enter into the seeds itself. The authors therefore suggested that the high rate of GA biosynthesis normally found in the seeds might be dependent on the import of GA precursors from the pod. However, in longer-term experiments, enough paclobutrazol did enter the seeds to dwarf the seedlings when the seeds were subsequently germinated, and it seems likely that this amount would have sufficed to inhibit GA biosynthesis within the seeds themselves. Since all the enzymes needed for GA biosynthesis are present in the seeds, and these enzyme activities show maxima that coincide with the most rapid accumulation of GA in the seeds (29, 63, 173), it is likely that there is an autonomous GA biosynthesis in the seeds.

Independent experiments with excised 4-day-old pods growing on a piece of peduncle (without leaf) in a complete medium suggest that both GA and cytokinins are needed for normal pod development, but they contradict the idea that the factors needed for pod development would come from the seeds (6). Thus the best pod development, including full inflation, was obtained when 10^{-6} M GA$_3$ and 10^{-5} M benzylaminopurine were added together to the basal medium. Interestingly, this treatment reduced both the seed number and the seed weight by about 50%. Full seed development was obtained with the basal medium without added plant hormones, but under these conditions the pods did not develop. The author (6) suggested that the GA and cytokinin supplied may either inhibit seed development, thereby increasing the sink activity of the pod, or stimulate and maintain the pod as the major sink to the detriment of the seeds. However, the plant hormones must have affected something more than simple sink competition, since the plants were supplied with a complete nutrient medium. In any case, since the seeds developed fully on the basal medium, whereas the pods did not, any GAs and cytokinins that may be produced in the seeds do not maintain pod growth (6). Furthermore, other experiments indicate that GAs are not extensively translocated from the seeds to the pod. Thus, the pods of *na* genotype peas (see previous section) contain almost no GA-like activity and the pods of *Na* genotype peas contain only moderate amounts of GA-like activity, although the seeds of both genotypes are rich in GAs (141). All evidence taken together, it looks like the GA and cytokinin needed for pod growth come from neither the seeds nor the pods but rather from the vegetative parts of the plant.

If GA is not needed for seed development in explants, this could be because the seeds are removed from the competition of other sinks in the plant, but experiments with intact plants point in the same direction. Thus, GA-deficient seeds develop and attain almost normal weight on GA-deficient *Arabidopsis* dwarfs, where the shortened siliques become crowded with seeds (5). This is further evidence that GA is needed for the development of seed-bearing structures, rather than for the seeds themselves. The significance, if any, of the strong GA biosynthesis in developing seeds remains unknown.

The role of GA in seed germination may be different in different species. Pea seeds germinate normally even if GA biosynthesis inhibitors are added at the time of imbibition (65, 175). Germination is therefore independent of de novo GA biosynthesis. However, seedlings emerging from such seeds are shortened to a measurable degree 4 days after germination. Thus by this time at the latest, elongation is dependent on GA biosynthesis. In contrast, some induced, GA-free mutants of *Arabidopsis,* tomato, and other plants do not germinate unless GA is added (105, 106).

CONJUGATION   GA conjugates are covalent associations of GAs with other low molecular weight compounds, usually glucose in the pyranose form. The sugar may either be linked to a hydroxyl group of the GA moiety, as in GA glucosides (GA glucosyl ethers), or to the GA C-7 carboxyl function, as in

GA glucosyl esters. GA conjugates occur generally and they are formed in large amounts in almost any GA feeding experiment, except in peas, which are low in conjugates. An extensive review of this vast field has been published recently (163), and only one aspect, the seed storage hypothesis, is discussed here. GA conjugates are generally inactive in bioassays (the most recent work is 164), but enzymatic hydrolysis releases the GA moiety, which may or may not be bioactive according to its structure. Since conjugates were first found in seeds, where they preferentially accumulate at the end of maturation, the hypothesis arose that they might be storage forms for GAs, capable of rapidly releasing GA by hydrolysis when the seeds germinate (see 163). Experiments in which radiolabeled GAs were fed to maturing seed and the fate of the label was followed through maturation and germination of the treated seeds seemed to support this hypothesis; but these experiments have been criticized because they were based on relative proportions of incompletely identified metabolites only (see 67, 82, 174).

Similar experiments have been reported recently, in which [^{3}H]GA$_4$ or [^{3}H]GA$_{20}$ was injected into the shanks of maize cobs during rapid grain filling (157, 159). The labeled precursors were partly metabolized, yielding a mixture of ^{3}H-labeled free and conjugated GAs. These were analyzed by silicic acid column chromatography and HPLC at the time of maturity, after imbibition, and at different times after germination. The analyses showed that the proportion of conjugated [^{3}H]GAs was high in the mature kernels and decreased on imbibition. This seemed to support the hypothesis that GAs become conjugated during seed maturation and that free, bioactive GAs are released again during imbibition. Unfortunately, the conclusions are based on the ratio of the labeled products obtained, without any relation to the endogenous amounts. This is symptomatic for many feeding experiments with ^{3}H-labeled precursors. The label may be used to show that a reaction can take place, and it may be used to trace the endogenous components. But the ratio of the labeled products is not a measure of the corresponding endogenous compounds, unless the precursor equilibrates with an endogenous pool and maintains a constant specific radioactivity throughout the experiment. In particular, the ratio of free and conjugated GAs forming after the administration of a single pulse of [^{3}H]GA$_{20}$ at an arbitrary time would not be expected to represent the ratios arising in the kernels over the whole period of their development. The widely different amounts of ^{3}H-label reaching the individual kernels (see the data in 157) show that the GAs and GA conjugates must have had widely varying specific radioactivities at the end, making any average incorporation meaningless. Furthermore, because the total amounts were too variable for statistical treatment, the total recovery of label in each sample was arbitrarily adjusted to 100%, whereupon the ratios of free and conjugated GAs were compared. The danger of using this approach when no

balance can be drawn has been pointed out before (67). If ratios are compared, any loss of label from one component automatically turns up as a
complementary gain in the others, giving the false impression that the label is
transferred from one fraction to another. The possibilities for uncontrolled and
specific losses of label from either the free or conjugated GAs are many—e.g.
leaking of conjugates into the medium during imbibition, loss of label due to
hydroxylation (the label was partially present at the sites getting hydroxy-
lated), or simply less recovery of the more polar conjugates.

To demonstrate the conjugation-deconjugation hypothesis it would be
necessary to show that endogenous conjugates of bioactive GAs are present in
the mature seeds, that these conjugates can get hydrolyzed, and that there is a
concomitant increase in endogenous free GAs. This can hardly be done
without label, but it must be ascertained that the label represents the pool of
endogenous material.

TISSUE CULTURES   Tissue cultures can metabolize GAs, and they also
contain endogenous GAs, which in all probability are biosynthesized within
the cultured cells. The [^{3}H]GA$_5$ metabolism in *Pharbitis nil* cell suspension
cultures was mentioned in the section on Individual Pathways, above. Somatic embryos derived from grape anther callus cultures contain GA$_4$, GA$_9$,
ent-kaurene, and *ent*-kaurenoic acid, as identified by GC-MS (189). In the
same cultures, the endogenous GA-like substances, as measured by bioassay,
undergo characteristic changes during somatic embryogenesis and as a response to chilling. Somatic anise and carrot suspension cultures convert
[^{3}H]GA$_4$ to [^{3}H]GA$_1$, [^{3}H]GA$_8$, and many conjugates, all as identified by
column chromatography and HPLC (107, 109, 129). Tomato cell suspension
cultures 2β-hydroxylate [^{3}H]GA$_1$ to [^{3}H]GA$_8$ and 13-hydroxylate [^{3}H]GA$_9$ to
[^{3}H]GA$_{20}$, as identified by TLC in combination with chemical and enzymatic
methods (116). A study by Liebisch (116) shows that the cell cultures retain a
certain species specificity, similar to that of the intact plant, with regard to the
hydroxylation pattern. Several GAs were identified by GC-SIM in cell cultures derived from both normal and crown gall tobacco tissues (132).

STRUCTURE-ACTIVITY RELATIONSHIPS   The structure-activity relations of
GA derivatives with substituents other than hydroxyl groups at C-2 were first
investigated with the rationale that prevention of 2β-hydroxylation might
produce GAs that remain active longer in the plants (87, 88). The results were
interesting and the studies were extended to include GA$_1$ and GA$_4$ derivatives
with substituents at C-1 and C-17 (10). Among other results, it was found that
2α-alkyl-, especially 2α-ethyl, GA$_1$ and GA$_4$ derivatives are much more
active in the d-5 maize and Tan-ginbozu rice bioassays than either GA itself or
the corresponding 2β-alkyl derivatives. For the 1-methyl derivatives, the

relation of activity to stereochemistry is reversed: The 1β-methyl GA_1 and GA_4 derivatives are much more active than either GA itself or the corresponding 1α-methyl derivatives. Although the C-1 and C-2 substituted GAs may indeed persist longer in the plants than the unsubstituted GAs, their high activities in the bioassays seem to be independent of this. The structure-activity relationships of the derivatives in inducing α-amylase synthesis in *Avena fatua* aleurone cells and in protoplasts prepared from these cells are different from the ones in the d-5 maize and Tan-ginbozu rice assays (10).

Plant Growth Retardants

Plant growth retardants are synthetic compounds that, when applied to higher plants, generally result in shorter more stable internodes. No other functions of the plants are adversely affected by modern plant growth retardants, but the leaves may be smaller and of a deeper green color, the main roots may be longer and thicker, and the yield of cereals and fruits may be higher. Because of these properties, plant growth retardants are used extensively in agriculture to control lodging, and the search for new and better structures with plant growth retarding activity has become an important branch of the chemical industry (see e.g. 36, 98, 147).

EFFECT ON GA BIOSYNTHESIS   Many plant growth retardants act at least in part by inhibiting endogenous GA biosynthesis in the treated plants. The evidence for this is that (*a*) treated plants have lowered endogenous GA contents, (*b*) the morphological effects of the treatment are relieved by simultaneous application of GA, and (*c*) GA biosynthesis in cell-free systems from higher plants is inhibited by plant growth retardants. These effects have been described in various combinations (e.g. 18, 19, 36, 95, 97, 99, 125, 147, 182, 196, 197, 205).

Many plant growth retardants, mostly quaternary ammonium compounds, such as AMO 1618 [(2-isopropyl-5-methyl-4-trimethylammonium chloride)-phenyl-1-piperidiniumcarboxylate], have been known for a long time to inhibit the activity, particularly the A-activity, of *ent*-kaurene synthetase (cf Scheme 1). This has been shown both with higher plant cell-free systems and with *Gibberella fujikuroi* cultures (see 67), and the relative activities of many analogues have been determined (see 29). Other well-known retardants, e.g. CCC [(2-chlorethyl)trimethylammonium chloride], do not seem to inhibit GA biosynthesis or lower the GA contents of higher plants (see 29, 67, 133).

Three new classes of compounds with plant growth retardant activities have been found in recent years. The first class may be described as substituted pyrimidines, of which ancymidol is the best-known representative (31 and references cited therein; 33). The second class are the norbornenodiazetine derivatives, of which tecyclacis (LAB 102 883, BAS 106.. W) has become the best known representative (98, 147). The third class consists of triazole derivatives, with compounds, such as paclobutrazol (PP 333; see 36), triapenthenol [RSW 0411, (E)-(R,S)-1-cyclo-hexyl-4,4-dimethyl-2-(1H-1,2,4-triazol-1-yl)pent-1-en-3-ol; 113], uniconazole [S-3307, (E)-1-(4-chlorophenyl-4,4-dimethyl-2-(1,2,4-triazol-1-yl)-1-penten-3-ol; 95], and several newly developed triazoles still without names (97, 99, 146) as the most promising members. Another new compound with similar properties, but not belonging to any of the classes mentioned, is 1-n-decylimidazole (193, 194). The members of the new classes of plant growth retardants are much more active and have broader spectra for action than the older types of structures (98, 146). At least the norbornenodiazetine and triazole types of plant growth retardants are exclusively translocated in the xylem and therefore are more active when they are applied via the roots or stems than when they are sprayed onto the leaves (see 36, 147).

All the new types of plant growth retardants inhibit the first three steps of *ent*-kaurene oxidation in cell-free systems of higher plants. Thus, the formation of *ent*-kaurenol, *ent*-kaurenal, and *ent*-kaurenoic acid is inhibited, whereas steps lying after *ent*-kaurenoic acid in the pathway are not affected (cf Scheme 2; 31, 33, 64, 80, 86, 94, 147; see also preliminary results in 36).

INHIBITION OF STEROL BIOSYNTHESIS   Douglas and Paleg have shown that several of the well-known old-type retardants, including Amo 1618 and CCC, not only inhibit *ent*-kaurene biosynthesis but also sterol biosynthesis, and that both sterols and GA can relieve the plant growth inhibition caused by these compounds (42–44 and references cited therein). Conversely, hypocholesterolemic agents also retard plant growth (44). As a logical consequence, Douglas & Paleg have postulated that the inhibition of sterol biosynthesis in plants might account for at least part of the growth retardation induced by these compounds.

The new types of growth retardants also affect sterol metabolism. Paclobutrazol alters the sterol metabolism in seedlings by decreasing the proportions of C-4 desmethyl sterols while accumulating sterols with C-4 and C-14 methyl groups (18, 19). This is not surprising, since triazole fungicides are known to inhibit 14α-methyl sterol demethylase in various fungi (see 14). Tetcyclacis has a similar effect on sterol metabolism in maize cell cultures (69). Ancymidol at first seemed very specific in its action. It was shown to inhibit the three steps of GA biosynthesis (see above) in microsomal preparations from *Marah macrocarpus* endosperm, but not other cytochrome P-450 dependent reactions, such as *trans*-cinnamate oxidation by *Sorghum* microsomes and *ent*-kaurene oxidation by rat liver and *G. fujikuroi* microsomes (33). Later, it was found that ancymidol inhibits both growth in several fungal species (2) and *ent*-kaurene oxidation in microsomal preparations from *G. fujikuroi*, although higher concentrations are needed than with microsomal preparations from higher plants (32). Further lack of specificity is indicated by the fact that ancymidol and several other plant growth retardants, including Amo 1618, 1-*n*-decylimidazole, and paclobutrazol, all inhibit abscisic acid biosynthesis in the fungus *Cercospora rosicola* without reducing mycelial mass (130, 131). This lack of specificity must be kept in mind when the plant growth retardants are used to probe the effects of GA biosynthesis inhibition in an object.

Interesting insights into the relative importance of GA and sterol biosynthesis for plant growth have been gained by comparison of the effects of growth retardants in cell suspension cultures and seedlings. Tetcyclacis and other plant growth retardants inhibit cell division in cell suspension cultures with approximately the same relative efficiency as they inhibit growth in seedlings (36, 68–71, 128). However, the inhibition of cell division in the cell cultures seems to be due entirely to the influence on sterol biosynthesis, since these cultures are not dependent on GA for growth. The addition of sterols to cell cultures treated with tetcyclacis also restores cell division completely, whereas the addition of GA$_3$ and *ent*-kaurenoic acid does not (69, 71). In seedlings, tetcyclacis has a dual effect: At concentrations below 10^{-6} M it mainly inhibits cell elongation, while at concentrations above 10^{-6} M it also inhibits cell division (128). Combining this result with the knowledge gained with cell cultures, the authors suggested that tetcyclacis at lower concentrations primarily inhibits GA biosynthesis, while at higher concentrations it also inhibits sterol biosynthesis. This would account for the fact that higher tetcyclacis concentrations are needed to inhibit growth in cell suspension cultures than to inhibit seedling growth (128). It might also account for the observation that the application of GA can overcome the effect of lower, but not higher, concentrations of ancymidol (34). Direct proof for this hypothesis is pending.

STRUCTURE-ACTIVITY RELATIONSHIPS, FUNGITOXICITY   Rademacher & Jung (146) and Jung & Rademacher (98) have compared the action of old and new types of growth retardants on rice seedlings, finding that the quaternary ammonium compounds were the least effective, whereas ancymidol and the norbornenodiazetine derivative tetcyclacis were the most effective. The triazoles used in these studies had intermediate activity, but later structure-activity studies (97, 99, 147) involving several plant species and newly developed triazole compounds show that these compounds can be as effective as ancymidol and tetcyclacis. The effectiveness of five triazoles in reducing growth and GA contents of cereal seedlings and tomato plants has also been compared (18).

Most of the plant growth retardants are also fungitoxic. In fact, the triazoles were first developed as fungicides and the plant growth retarding properties were discovered for two fungicides (19). Coolbaugh et al (32, 34) compared the effectiveness of ancymidol and several ancymidol analogues as inhibitors of (*a*) pea internode elongation, (*b*) *ent*-kaurene oxidation in higher plant microsomal preparations, (*c*) fungal growth, and (*d*) GA production in *G. fujikuroi*. In general, the compounds that had been developed as growth retardants were most effective for (*a*) and (*b*), whereas the fungicides were most effective for (*c*) and (*d*). Fletcher et al (49) obtained slightly different results when they compared fungitoxic and plant growth regulating properties of five triazole derivatives, three of which were categorized as fungicides while the other two had been developed as plant growth regulators. The plant growth regulators inhibited growth of higher plants most efficiently, but they were also more fungitoxic than some of the fungicides. Thus the correlation of the classes is not perfect.

Sauter has thoroughly reviewed the structural requirements for plant growth retarding activity and species specificity in several types of compounds from a chemical viewpoint (162). In other studies an interesting point has emerged from the resolution of paclobutrazol into its enantiomers (185). The 2*S*,3*S* enantiomer reduces growth of apple seedlings more efficiently than the 2*R*,3*R* enantiomer does, whereas the 2*R*,3*R* isomer is the better fungicide. Computer modelling showed that the paclobutrazol 2*S*,3*S* and 2*R*,3*R* enantiomers can be superimposed on *ent*-kaurene and lanosterol, respectively—i.e. on the substrates for the inhibited enzymes in the GA and sterol biosynthesis pathways, respectively. In accord with this, the paclobutrazol 2*S*,3*S* enantiomer inhibits the oxidation of *ent*-kaurene in cell-free extracts from apple embryos more efficiently than does the 2*R*,3*R* enantiomer (80).

Other compounds also have plant growth retardant activity. Thus, several cytokinins inhibit *ent*-kaurene oxidation in microsomal preparations from *Marah oreganus* (30). N^6-cyclohexanemethyladenosine was the most active of these, inhibiting GA biosynthesis to 50% at $2 \cdot 10^{-6}$ M. Isoamyladenine, benzyladenine, and dihydrozeatin were also active, whereas zeatin, kinetin, adenine, benzyladenosine and isopentenyladenosine were not. Mevinolin and compactin are two structurally related fungal metabolites and hypocholesterolemic agents. They are competitive inhibitors of hydroxymethylglutaryl coenzyme-A reductase and thus prevent MVA biosynthesis. They inhibit higher plant growth, especially root growth; and the addition of MVA, but not of GA_3, can overcome the inhibition (3, 21, 170). The relation of these compounds to GA biosynthesis in higher plants has not been investigated, but both mevinolin and compactin inhibit the GA_3 production in *G. fujikuroi* cultures (114, 170). Compactin inhibits both acetate and MVA incorporation into GA_3, which shows that the HMGCoA reductase inhibition is not fully specific (114). A great variety of compounds affect the *G. fujikuroi* GA production (see 12, 77).

OTHER COMPOUNDS   A series of papers on the effect of several chloroaceta-
mide and thiocarbamate herbicides on GA precursor biosynthesis in a cell-free
system from sorghum seedlings may, at first glance, seem to show that these
herbicides inhibit GA biosynthesis at different defined sites (201 and references cited therein). However, the methodology and the concepts displayed in
these papers are such that it would be desirable to repeat the experiments by
acceptable methods before the results are quoted as facts. In these investigations the cell-free system from sorghum seedlings is used without any
characterization of the activity parameters and without other identification
than by TLC of the complicated product pattern obtained on incubation with
[2-^{14}C]MVA. The reliability of the identifications is questionable, if only
because putative products of the microsomal part of the GA pathway are
formed and quantitated in supernatants, centrifuged at 40,000 g for 1 hr (in all
but the two most recent publications). The incorporation into products is
variable, so the conclusions are drawn on the basis of relative changes only
(see discussion above). Finally, as the author points out, the herbicides
studied have many effects that are not explained by GA biosynthesis inhibition. Indeed, the more obvious symptoms of treatment with these herbicides,
such as total loss of the pigmentation and death of the seedlings, are neither
typical for GA deficiency nor reversed by GA treatment.

CONCLUDING REMARKS

The topic of GA metabolism has been reviewed regularly since 1957 in this
series. If the 1950s was the decade of discoveries, basic structural elucidation,
and paper chromatography, the 1960s may be called the decade of the
bioassays, plant growth retardants, beginning biosynthesis, and TLC. The
number of known GAs rose from 9 to 29. The 1970s saw the fruits of isotope
work and the use of GC-MS: The biosynthesis up to GA$_{12}$-aldehyde was
clarified in cell-free systems, the terminal steps of GA metabolism were
studied in higher plants, GA metabolism in *G. fujikuroi* was exhaustively
investigated, and immature seeds were analyzed. At the time of the most
recent previous review (82), 52 GAs were known. The period covered by the
present review has seen the advent of HPLC, the development of capillary
GC-MS, and the use of heavy isotopes. These techniques have contributed to
the ever more detailed analysis of endogenous GAs and GA metabolism, at
first in seeds, but then also in seedlings. New defined mutants, particularly of
pea, have entered the picture; the pathways after GA$_{12}$-aldehyde have been
defined; and enzymes of these pathways have been purified. The number of
defined GA structures has reached 72 (119).

   The work on GA biosynthesis is now moving into the range of im-
munochemistry and molecular biology. Antibodies and immunoaffinity pro-

cedures for the quantitation and isolation of GAs are being developed. Antibodies are being raised against purified enzymes for cellular location of GA biosynthesis in the plants. In another approach, work is in progress to identify, isolate, and clone the dwarfing genes from mutator dwarfs in maize, produced by the use of transposable elements (137). Sequencing of these genes is expected to define the differences between mutant and wild type genes. Subsequent transcription and translation of the genes should yield enough enzyme to permit chemical and biochemical analyses. The results of these studies will probably be the main topics for the next review in this series.

ACKNOWLEDGMENTS

I thank Mrs. Anke Schopf for drawing the schemes. Work in our laboratory is supported by the Deutsche Forschungsgemeinschaft.

Literature Cited

1. Albone, K. S., Gaskin, P., MacMillan, J., Sponsel, V. M. 1984. Identification and localization of gibberellins in maturing seeds of the cucurbit *Sechium edule*, and a comparison between this cucurbit and the legume *Phaseolus coccineus*. *Planta* 162:560–65
2. Ali, A., Hall, R., Fletcher, R. A. 1979. Inhibition of fungal growth by plant growth retardants. *Can. J. Bot.* 57:458–60
3. Bach, T. J., Lichtenthaler, H. K. 1983. Inhibition by mevinolin of plant growth, sterol formation and pigment accumulation. *Physiol. Plant.* 59:50–60
4. Barendse, G. W. M., Dijkstra, A., Moore, T. C. 1983. The biosynthesis of the gibberellin precursor *ent*-kaurene in cell-free extracts and the endogenous gibberellins of Japanese morning glory in relation to seed development. *J. Plant Growth Regul.* 2:165–75
5. Barendse, G. W. M., Kepczynski, J., Karssen, C. M., Koornneef, M. 1986. The role of endogenous gibberellins during fruit and seed development: Studies on gibberellin-deficient genotypes of *Arabidopsis thaliana*. *Physiol. Plant.* 67:316–19
6. Barratt, D. H. P. 1986. In vitro pod culture of *Pisum sativum*. *Plant Sci.* 43:223–28
7. Beale, M. H., Bearder, J. R., Down, G. H., Hutchison, M., MacMillan, J., et al. 1982. The biosynthesis of kaurenolide diterpenoids by *Gibberella fujikuroi*. *Phytochemistry* 21:1279–87
8. Beale, M. H., Bearder, J. R., Hedden,

P., Graebe, J. E., MacMillan, J. 1984. Gibberellin A_{58} and *ent*-6α,7α,12α-trihydroxykaur-16-en-19-oic acid from seeds of *Cucurbita maxima*. *Phytochemistry* 23:565–67
9. Beale, M. H., Bearder, J. R., MacMillan, J., Matsuo, A., Phinney, B. O. 1983. Diterpene acids from *Helianthus* species and their microbiological conversion by *Gibberella fujikuroi*, mutant B1-41a. *Phytochemistry* 22:875–81
10. Beale, M. H., Hooley, R., MacMillan, J. 1986. Gibberellins: Structure-activity relationships and the design of molecular probes. In *Plant Growth Substances 1985*, ed. M. Bopp, pp. 65–73. Berlin/Heidelberg/New York/Tokyo: Springer-Verlag. 420 pp.
11. Bearder, J. R. 1980. Plant hormones and other growth substances—their background, structures and occurrence. In *Encyclopedia of Plant Physiology*, New Series, Vol. 9: *Hormonal Regulation of Development. I. Molecular Aspects of Plant Hormones*, ed. J. MacMillan, pp. 9–112. Berlin/Heidelberg/New York: Springer-Verlag. 681 pp.
12. Bearder, J. R. 1983. *In vivo* diterpenoid biosynthesis in *Gibberella fujikuroi:* the pathway after *ent*-kaurene. In *The Biochemistry and Physiology of Gibberellins*, ed. A. Crozier, 1:251–387. New York: Praeger. 568 pp.
13. Bearder, J. R., MacMillan, J., Phinney, B. O., Hanson, J. R., Rivett, D. E. A., et al. 1982. Gibberellin A_{13} 7-aldehyde: a proposed intermediate in the fungal

biosynthesis of gibberellin A_3. *Phytochemistry* 21:2225–30

14. Benveniste, P. 1986. Sterol biosynthesis. *Ann. Rev. Plant Physiol.* 37:275–308

15. Birnberg, P. R., Brenner, M. L., Mardaus, M. C., Abe, H., Pharis, R. P. 1986. Metabolism of gibberellin A_{12}-7-aldehyde by soybean cotyledons and its use in identifying gibberellin A_7 as an endogenous gibberellin. *Plant Physiol.* 82:241–46

16. Birnberg, P. R., Maki, S. L., Brenner, M. L., Davis, G. C., Carnes, M. G. 1986. An improved enzymatic synthesis of labeled gibberellin A_{12}-aldehyde and gibberellin A_{12}. *Anal. Biochem.* 153:1–8

17. Blechschmidt, S., Castel, U., Gaskin, P., Hedden, P., Graebe, J. E. et al. 1984. GC/MS analysis of the plant hormones in seeds of *Cucurbita maxima*. *Phytochemistry* 23:553–58

18. Buchenauer, H., Kutzner, B., Koths, T. 1984. Effect of various triazole fungicides on growth of cereal seedlings and tomato plants as well as on gibberellin contents and lipid metabolism in barley seedlings. *J. Plant Dis. Protect.* 91:506–24 (In German, with English abstract)

19. Buchenauer, H., Röhner, E. 1981. Effect of triadimefon and triadimenol on growth of various plant species as well as on gibberellin content and sterol metabolism in shoots of barley seedlings. *Pestic. Biochem. Physiol.* 15:58–70

20. Ceccarelli, N., Lorenzi, R. 1983. Gibberellin biosynthesis in endosperm and cotyledons of *Sechium edule* seeds. *Phytochemistry* 22:2203–5

21. Ceccarelli, N., Lorenzi, R. 1984. Growth inhibition by competitive inhibitors of 3-hydroxymethylglutarylcoenzyme A reductase in *Helianthus tuberosus* tissue explants. *Plant Sci. Lett.* 34:269–76

22. Ceccarelli, N., Lorenzi, R., Alpi, A. 1979. Kaurene and kaurenol biosynthesis in cell-free system of *Phaseolus coccineus* suspensor. *Phytochemistry* 18:1657–58

23. Ceccarelli, N., Lorenzi, R., Alpi, A. 1981. Gibberellin biosynthesis in *Phaseolus coccineus* suspensor. *Z. Pflanzenphysiol.* 102:37–44

24. Ceccarelli, N., Lorenzi, R., Alpi, A. 1981. Kaurene metabolism in cell-free extracts of *Phaseolus coccineus* suspensors. *Plant Sci. Lett.* 21:325–32

25. Ceccarelli, N., Lorenzi, R., Alpi, A. 1981. Kaurene biosynthesis in intact *Phaseolus coccineus* suspensors. *Experientia* 37:478

26. Choinski, J. S. Jr., Moore, T. C. 1980. Relationship between chloroplast development and *ent*-kaurene biosynthesis in peas. *Plant Physiol.* 65:1031–35

27. Chung, C. H., Coolbaugh, R. C. 1986. *ent*-Kaurene biosynthesis in cell-free extracts of excised parts of tall and dwarf pea seedlings. *Plant Physiol.* 80:544–48

28. Coates, R. M., Cavender, P. L. 1980. Stereochemistry of the enzymatic cyclization of copalyl pyrophosphate to kaurene in enzyme preparations from *Marah macrocarpus*. *J. Am. Chem. Soc.* 102:6358–59

29. Coolbaugh, R. C. 1983. Early stages of gibberellin biosynthesis. See Ref. 12, pp. 53–98

30. Coolbaugh, R. C. 1984. Inhibition of *ent*-kaurene oxidation by cytokinins. *J. Plant Growth Regul.* 3:97–109

31. Coolbaugh, R. C., Hamilton, R. 1976. Inhibition of *ent*-kaurene oxidation and growth by α-cyclopropyl-α-(p-methoxyphenyl)-5-pyrimidine methyl alcohol. *Plant Physiol.* 57:245–48

32. Coolbaugh, R. C., Heil, D. R., West, C. A. 1982. Comparative effects of substituted pyrimidines on growth and gibberellin biosynthesis in *Gibberella fujikuroi*. *Plant Physiol.* 69:712–16

33. Coolbaugh, R. C., Hirano, S. S., West, C. A. 1978. Studies on the specificity and site of action of α-cyclopropyl-α-[p-methoxyphenyl]-5-pyrimidine methyl alcohol (ancymidol), a plant growth regulator. *Plant Physiol.* 62:571–76

34. Coolbaugh, R. C., Swanson, D. I., West, C. A. 1982. Comparative effects of ancymidol and its analogs on growth of peas and *ent*-kaurene oxidation in cell-free extracts of immature *Marah macrocarpus* endosperm. *Plant Physiol.* 69:707–11

35. Crozier, A. 1981. Aspects of the metabolism and physiology of gibberellins. *Adv. Bot. Res.* 9:33–149

36. Dalziel, J., Lawrence, D. K. 1984. Biochemical and biological effects of kaurene oxidase inhibitors, such as paclobutrazol. In *Biochemical Aspects of Synthetic and Naturally Occurring Plant Growth Regulators*, ed. R. Menhenett, D. K. Lawrence, 11:43–57. Wantage: Br. Plant Growth Regulator Group. 121 pp. Monogr.

37. Davies, P. J., Birnberg, P. R., Maki, S. L., Brenner, M. L. 1986. Photoperiod modification of $[^{14}C]$gibberellin A_{12} aldehyde metabolism in shoots of pea, line G2. *Plant Physiol.* 81:991–96

38. Davies, P. J., Emshwiller, E., Gianfag-

na, T. J., Proebsting, W. M., Noma, M. et al. 1982. The endogenous gibberellins of vegetative and reproductive tissue of G2 peas. *Planta* 154:266–72

39. Diaz, C. E., Fraga, B. M., Gonzalez, A. G., Gonzalez, P., Hanson, J. R. et al. 1984. The microbiological transformation of some trachylobane diterpenoids by *Gibberella fujikuroi*. *Phytochemistry* 12:2813–16

40. Dockerill, B., Evans, R., Hanson, J. R. 1977. Removal of C-20 in gibberellin biosynthesis. *J. Chem. Soc. Chem. Commun.* 1977:919–21

41. Dockerill, B., Hanson, J. R. 1978. The fate of C-20 in C_{19} gibberellin biosynthesis. *Phytochemistry* 17:701–4

42. Douglas, T. J., Paleg, L. G. 1978. AMO 1618 and sterol biosynthesis in tissues and sub-cellular fractions of tobacco seedlings. *Phytochemistry* 17: 705–12

43. Douglas, T. J., Paleg, L. G. 1978. Amo 1618 effects on incorporation of ^{14}C-MVA and ^{14}C-acetate into sterols in *Nicotiana* and *Digitalis* seedlings and cell-free preparations from *Nicotiana*. *Phytochemistry* 17:713–18

44. Douglas, T. J., Paleg, L. G. 1981. Inhibition of sterol biosynthesis and stem elongation of tobacco seedlings induced by some hypocholesterolemic agents. *J. Exp. Bot.* 32:59–68

45. Duncan, J. D., West, C. A. 1981. Properties of kaurene synthetase from *Marah macrocarpus* endosperm: evidence for the participation of separate but interacting enzymes. *Plant Physiol.* 68:1128–34

46. Durley, R. C. 1983. Biosynthesis of gibberellins in higher plants. In *Aspects of Physiology and Biochemistry of Plant Hormones*, ed. S. S. Purohit, pp. 93–123. New Delhi: Kalyani

47. Durley, R. C., Railton, I. D., Pharis, R. P. 1974. Conversion of gibberellin A_{14} to other gibberellins in seedlings of dwarf *Pisum sativum*. *Phytochemistry* 13:547–51

48. Durley, R. C., Sassa, T., Pharis, R. P. 1979. Metabolism of tritiated gibberellin A_{20} in immature seeds of dwarf pea, cv. Meteor. *Plant Physiol.* 64:214–19

49. Fletcher, R. A., Hofstra, G., Gao, J. 1986. Comparative fungitoxic and plant growth regulating properties of triazole derivatives. *Plant Cell Physiol.* 27:367–71

50. Fraga, B. M., Gonzalez, A. G., Gonzalez, P., Hanson, J. R., Hernandez, M. G. 1983. The microbiological transformation of some *ent*-kaur-6,16-dienes by *Gibberella fujikuroi*. *Phytochemistry* 22:691–94

51. Fraga, B. M., Hanson, J. R., Hernandez, M. G. 1978. The microbiological transformation of epicandicandiol, *ent*-7α,18-dihydroxykaur-16-ene, by *Gibberella fujikuroi*. *Phytochemistry* 17: 812–14

52. Fraga, B. M., Hanson, J. R., Hernandez, M. G., Sarah, F. Y. 1980. The microbiological transformation of some *ent*-kaur-16-ene 7-, 15- and 18-alcohols by *Gibberella fujikuroi*. *Phytochemistry* 19:1087–91

53. Fujisawa, S., Yamaguchi, I., Park, K. H., Kobayashi, M., Takahashi, N. 1985. Qualitative and semi-quantitative analyses of gibberellins in immature seeds of *Pharbitis purpurea*. *Agric. Biol. Chem.* 49:27–33

54. Garcia-Martinez, J. L., Sponsel, V. M., Gaskin, P. 1986. Gibberellins in developing fruits of *Pisum sativum* cv. Alaska: studies on their role in pod growth and seed development. *Planta.* 170:130–37

55. Gaskin, P., Gilmour, S. J., Lenton, J. R., MacMillan, J., Sponsel, V. M. 1984. Endogenous gibberellins and kauranoids identified from developing and germinating barley grain. *J. Plant Growth Regul.* 2:229–42

56. Gaskin, P., Gilmour, S. J., MacMillan, J., Sponsel, V. M. 1985. Gibberellins in immature seeds and dark-grown shoots of *Pisum sativum*. *Planta* 163:283–89

57. Gaskin, P., Hutchison, M., Lewis, N., MacMillan, J., Phinney, B. O. 1984. Microbiological conversion of 12-oxygenated and other derivatives of *ent*-kaur-16-en-19-oic acid by *Gibberella fujikuroi*, mutant B1-41a. *Phytochemistry* 23:559–64

58. Gaskin, P., Kirkwood, P. S., MacMillan, J. 1981. Partial synthesis of *ent*-13-hydroxy-2-oxo-20-norgibberella-1(10), 16-diene-7,19-dioic acid, a catabolite of gibberellin A_{29}, and of related compounds. *J. Chem. Soc. Perkin Trans. I* 1981:1083–91

59. Gianfagna, T., Zeevaart, J. A. D., Lusk, W. J. 1983. Effect of photoperiod on the metabolism of deuterium-labeled gibberellin A_{53} in spinach. *Plant Physiol.* 72:86–89

60. Gianfagna, T., Zeevaart, J. A. D., Lusk, W. J. 1983. Synthesis of [^{2}H]-gibberellins from steviol using the fungus *Gibberella fujikuroi*. *Phytochemistry* 22:427–30

61. Gilmour, S. J., Gaskin, P., Sponsel, V. M., MacMillan, J. 1984. Metabolism of gibberellins by immature barley grain. *Planta* 161:186–92

62. Gilmour, S. J., Zeevaart, J. A. D.,

Schwenen, L., Graebe, J. E. 1986. Gibberellin metabolism in cell-free extracts from spinach leaves in relation to photoperiod. *Plant Physiol.* 82:190–95

63. Graebe, J. E. 1980. GA-biosynthesis: the development and application of cell-free systems for biosynthetic studies. In *Plant Growth Substances 1979,* ed. F. Skoog, pp. 180–87. Berlin/Heidelberg/New York: Springer-Verlag. 527 pp.

64. Graebe, J. E. 1982. Gibberellin biosynthesis in cell-free systems from higher plants. In *Plant Growth Substances 1982,* ed. P. F. Wareing, pp. 71–80. London: Academic. 683 pp.

65. Graebe, J. E. 1986. Gibberellin biosynthesis from gibberellin A_{12}-aldehyde. See Ref. 10, pp. 74–82

66. Graebe, J. E., Hedden, P., Rademacher, W. 1980. Gibberellin biosynthesis. In *Gibberellins—Chemistry, Physiology and Use,* ed. J. R. Lenton, 5:31–47. Wantage: Brit. Plant Growth Regul. Group. 143 pp. Monogr.

67. Graebe, J. E., Ropers, H. J. 1978. Gibberellins. In *Phytohormones and Related Compounds—A Comprehensive Treatise,* ed. D. S. Letham, P. B. Goodwin, T. J. V. Higgins, 1:107–204. Amsterdam/Oxford/New York: Elsevier/North Holland Biomed. 641 pp.

68. Grossmann, K., Rademacher, W., Jung, J. 1982. Plant cell suspension cultures as model systems for investigating growth regulating compounds. *Plant Cell Rep.* 1:281–84

69. Grossmann, K., Rademacher, W., Jung, J. 1983. Effects of NDA, a new plant growth retardant, on cell culture growth of *Zea mays* L. *J. Plant Growth Regul.* 2:19–29

70. Grossmann, K., Rademacher, W., Sauter, H., Jung, J. 1984. Comparative potency of different plant growth retardants in cell cultures and intact plants. *J. Plant Growth Regul.* 3:197–205

71. Grossmann, K., Weiler, E. W., Jung, J. 1985. Effects of different sterols on the inhibition of cell culture growth caused by the growth retardant tetcyclacis. *Planta* 164:370–75

72. Gunsalus, I. C., Pederson, T. C., Sligar, S. G. 1975. Oxygenase-catalyzed biological hydroxylations. *Ann. Rev. Biochem.* 44:377–407

73. Hafemann, C. 1985. *Kompartimentierung der membrangebundenen Gibberellin-Biosyntheseschritte in Erbsensamen und Kürbisendosperm.* PhD thesis. Univ. Göttingen, F.R.G. 109 pp.

74. Hanauske-Abel, H. M., Günzler, V. 1982. A stereochemical concept for the

catalytic mechanism of prolylhydroxylase. *J. Theor. Biol.* 94:421–55

75. Hanson, J. R. 1983. Aspects of diterpenoid and gibberellin biosynthesis in *Gibberella fujikuroi. Biochem. Soc. Trans.* 11:522–28

76. Hedden, P. 1979. Aspects of gibberellin chemistry. In *Plant Growth Substances,* ed. N. B. Mandava, ACS Symp. Ser. 111:19–56. Washington, DC: Am. Chem. Soc. 310 pp.

77. Hedden, P. 1983. *In vitro* metabolism of gibberellins. See Ref. 12, pp. 99–149

78. Hedden, P., Graebe, J. E. 1981. Kaurenolide biosynthesis in a cell-free system from *Cucurbita maxima* seeds. *Phytochemistry* 20:1011–15

79. Hedden, P., Graebe, J. E. 1982. Cofactor requirements for the soluble oxidases in the metabolism of the C_{20}-gibberellins. *J. Plant Growth Regul.* 1:105–16

80. Hedden, P., Graebe, J. E. 1985. Inhibition of gibberellin biosynthesis by paclobutrazol in cell-free homogenates of *Cucurbita maxima* endosperm and *Malus pumila* embryos. *J. Plant Growth Regul.* 4:111–22

81. Hedden, P., Graebe, J. E., Beale, M. H., Gaskin, P., MacMillan, J. 1984. The biosynthesis of 12α-hydroxylated gibberellins in a cell-free system from *Cucurbita maxima* endosperm. *Phytochemistry* 23:569–74

82. Hedden, P., MacMillan, J., Phinney, B. O. 1978. The metabolism of the gibberellins. *Ann. Rev. Plant Physiol.* 29:149–92

83. Hedden, P., Phinney, B. O. 1979. Comparison of *ent*-kaurene and *ent*-isokaurene synthesis in cell-free systems from etiolated shoots of normal and *dwarf-5* maize seedlings. *Phytochemistry* 18:1475–79

84. Heupel, R. C., Phinney, B. O., Spray, C. R., Gaskin, P., MacMillan, J. et al. 1985. Native gibberellins and the metabolism of [^{14}C]gibberellin A_{53} and of [17-^{13}C,17-^{3}H$_2$]gibberellin A_{20} in tassels of *Zea mays. Phytochemistry* 24:47–53

85. Heyser, M. 1984. *Untersuchungen zur Gibberellin-Biosynthese in Keimlingswurzeln.* PhD thesis. Univ. Göttingen, F.R.G. 134 pp.

86. Hildebrandt, E. 1982. *Der Einfluss von ausgewählten Wachstumshemmern auf die Gibberellinbiosynthese in zellfreien Systemen.* Diplom thesis. Univ. Göttingen, F.R.G. 87 pp.

87. Hoad, G. V., MacMillan, J., Smith, V. A., Sponsel, V. M., Taylor, D. A. 1982. Gibberellin 2β-hydroxylases and

biological activity of 2β-alkyl gibberellins. See Ref. 64, pp. 91–100

88. Hoad, G. V., Phinney, B. O., Sponsel, V. M., MacMillan, J. 1981. The biological activity of sixteen gibberellin A$_4$ and gibberellin A$_9$ derivatives using seven bioassays. *Phytochemistry* 20: 703–13

89. Honda, K., Shishibori, T., Suga, T. 1980. Biosynthesis of (−)-kaurene. [13]C n.m.r. spectroscopic evidence for the mechanism of formation of ring D. *J. Chem. Res.–S* 1980:218–19

90. Ingram, T. J., Reid, J. B., MacMillan, J. 1985. Internode length in *Pisum sativum* L. The kinetics of growth and [3H]gibberellin A$_{20}$ metabolism in genotype *na Le*. *Planta* 164:429–38

91. Ingram, T. J., Reid, J. B., MacMillan, J. 1986. The quantitative relationship between gibberellin A$_1$ and internode growth in *Pisum sativum* L. *Planta* 168:414–20

92. Ingram, T. J., Reid, J. B., Murfet, I. C., Gaskin, P., Willis, C. L., et al. 1984. Internode length in *Pisum*. The *Le* gene controls the 3β-hydroxylation of gibberellin A$_{20}$ to gibberellin A$_1$. *Planta* 160:455–63

93. Ingram, T. J., Reid, J. B., Potts, W. C., Murfet, I. C. 1983. Internode length in *Pisum*. IV. The effect of the *Le* gene on gibberellin metabolism. *Physiol. Plant.* 59:607–16

94. Izumi, K., Kamiya, Y., Sakurai, A., Oshio, H., Takahashi, N. 1985. Studies of sites of action of a new plant growth retardant (E)-1-(4-chlorophenyl)-4,4-dimethyl - 2 - (1,2,4 - triazol - 1 - yl) - 1 - penten -3-ol (S-3307) and comparative effects of its stereoisomers in a cell-free system from *Cucurbita maxima*. *Plant Cell Physiol.* 26:821–27

95. Izumi, K., Yamaguchi, I., Wada, A., Oshio, H., Takahashi, N. 1984. Effect of a new plant growth retardant (E)-1-(4-chlorophenyl)-4,4-dimethyl-2-(1,2,4-triazol-1-yl)-1-penten-3-ol (S-3307) on the growth and gibberellin content of rice plants. *Plant Cell Physiol.* 25:611–17

96. Jones, M. G. 1984. Endogenous gibberellins and stem extension. See Ref. 36, pp. 33–42

97. Jung, J., Luib, M., Sauter, H., Zeeh, B., Rademacher, W. 1987. Growth regulation in crop plants with new types of triazole compounds. *J. Agron. Crop Sci.* 158: In press

98. Jung, J., Rademacher, W. 1983. Cereal grains. In *Plant Growth Regulating Chemicals*, ed. L. G. Nickell, 1:253–71. Boca Raton: CRC Press. 280 pp.

99. Jung, J., Rentzea, C., Rademacher, W. 1986. Plant growth regulation with triazoles of the dioxanyl type. *J. Plant Growth Regul.* 4:181–88

100. Kamiya, Y., Graebe, J. E. 1983. The biosynthesis of all major pea gibberellins in a cell-free system from *Pisum sativum*. *Phytochemistry* 22:681–89

101. Kamiya, Y., Takahashi, N., Graebe, J. E. 1986. The loss of carbon-20 in C$_{19}$-gibberellin biosynthesis in a cell-free system from *Pisum sativum* L. *Planta* 169:524–28

102. Kamiya, Y., Takahashi, M., Takahashi, N., Graebe, J. E. 1984. Conversion of gibberellin A$_{20}$ to gibberellins A$_1$ and A$_5$ in a cell-free system from *Phaseolus vulgaris*. *Planta* 162:154–58

103. Kobayashi, M., Yamaguchi, I., Murofushi, N., Ota, Y., Takahashi, N. 1984. Endogenous gibberellins in immature seeds and flowering ears of rice. *Agric. Biol. Chem.* 48:2725–29

104. Koornneef, M., Elgersma, A., Hanhart, C. J., van Loenen-Martinet, E. P., van Rijn, L. et al. 1985. A gibberellin insensitive mutant of *Arabidopsis thaliana*. *Physiol. Plant* 65:33–39

105. Koornneef, M., van der Veen, J. H. 1980. Induction and analysis of gibberellin sensitive mutants in *Arabidopsis thaliana* (L.) Heynh. *Theor. Appl. Genet.* 58:257–63

106. Koornneef, M., van der Veen, J. H., Spruit, C. J. P., Karssen, C. M. 1981. Isolation and use of mutants with an altered germination behaviour in *Arabidopsis thaliana* and tomato. In *Induced Mutations as a Tool for Crop Plant Improvement*, 251:227–32. Vienna: IAEA-SM

107. Koshioka, M., Douglas, T. J., Ernst, D., Huber, J., Pharis, R. P. 1983. Metabolism of [3H]gibberellin A$_4$ in somatic suspension cultures of anise. *Phytochemistry* 22:1577–84

108. Koshioka, M., Hisajima, S., Pharis, R. P., Murofushi, N. 1985. Metabolism of [3H]gibberellin A$_5$ in cell suspension cultures of *Pharbitis nil*. *Agric. Biol. Chem.* 49:2627–31

109. Koshioka, M., Jones, A., Koshioka, M. N., Pharis, R. P. 1983. Metabolism of [3H]gibberellin A$_4$ in somatic suspension cell cultures of carrot. *Phytochemistry* 22:1585–90

110. Koshioka, M., Pharis, R. P., King, R. W., Murofushi, N., Durley, R. C. 1985. Metabolism of [3H]gibberellin A$_5$ in developing *Pharbitis nil* seeds. *Phytochemistry* 24:663–71

111. Kurogochi, S., Murofushi, N., Ota, Y., Takahashi, N. 1979. Identification of

gibberellins in the rice plant and quantitative changes of gibberellin A_{19} throughout its life cycle. *Planta* 146: 185–91

112. Lange, T. 1986. *Reinigung und Charakterisierung von Enzymen des 3. Abschnittes der Gibberellinbiosynthese in reifenden Erbsen.* Diplom thesis. Univ. Göttingen, F.R.G., 138 pp.

113. Lembrich, H., Dengel, H. J., Lürsen, K., Reiser, R. 1984. RSW 0411, ein Wachstumsregulator zur Verbesserung der Standfestigkeit im Winterraps, sowie in Grassamenvermehrungsbau. *Mitt. Biol. Bundesanst. Land- Forstwirtsch. Berlin-Dahlem* 233:315–16

114. Lewer, P., MacMillan, J. 1983. Effect of compactin on the incorporation of mevalonolactone into gibberellic acid by *Gibberella fujikuroi. Phytochemistry* 22:602–3

115. Lewer, P., MacMillan, J. 1984. An NMR study of the loss of carbon-20 in the biosynthesis of gibberellin A_3 by *Gibberella fujikuroi. Phytochemistry* 23:2803–11

116. Liebisch, H. W. 1980. Comparative investigations on the metabolism of GA_1, GA_3, and GA_9 in cell suspension cultures of *Lycopersicon esculentum* and in different intact plants. *Biochem. Physiol. Pflanz.* 175:797–805 (In German)

117. MacMillan, J. 1983. Gibberellins in higher plants. *Biochem. Soc. Trans.* 11:528–34

118. MacMillan, J. 1984. Analysis of plant hormones and metabolism of gibberellins. In *The Biosynthesis and Metabolism of Plant Hormones,* ed. A. Crozier, J. R. Hillman, Soc. Exp. Biol. Semin. Ser. 23:1–16. Oxford: Oxford Univ. Press. 288 pp.

119. MacMillan, J. 1985. Gibberellins: Metabolism and function. In *Current Topics in Plant Biochemistry and Physiology,* ed. D. D. Randall, D. G. Blevins, R. L. Larson, 4:53–66. Columbia: Univ. Missouri. 258 pp.

120. Maki, S. L., Brenner, M. L., Birnberg, P. R., Davies, P. J., Krick, T. P. 1986. Identification of pea gibberellins by studying $[^{14}C]GA_{12}$-aldehyde metabolism. *Plant Physiol.* 81:984–90

120a. Malmström, B. G. 1982. Enzymology of oxygen. *Ann. Rev. Biochem.* 51:21–59

121. Marx, G. A. 1983. Developmental mutants in some annual seed plants. *Ann. Rev. Plant Physiol.* 34:389–417

122. Mellon, J. E., West, C. A. 1979. Diterpene biosynthesis in maize seedlings in response to fungal infection. *Plant Physiol.* 64:406–10

123. Metzger, J. D., Zeevaart, J. A. D. 1980. Identification of six endogenous gibberellins in spinach shoots. *Plant Physiol.* 65:623–26

124. Metzger, J. D., Zeevaart, J. A. D. 1982. Photoperiodic control of gibberellin metabolism in spinach. *Plant Physiol.* 69:287–91

125. Mita, T., Shibaoka, H. 1984. Effects of S-3307, an inhibitor of gibberellin biosynthesis, on swelling of leaf sheath cells and the arrangement of cortical microtubules in onion seedlings. *Plant Cell Physiol.* 25:1531–39

126. Murakami, Y. 1972. Dwarfing genes in rice and their relation to gibberellin biosynthesis. In *Plant Growth Substances 1970,* ed. D. J. Carr, pp. 166–74. Berlin/Heidelberg/New York: Springer-Verlag. 837 pp.

127. Murofushi, N., Shigematsu, Y., Nagura, S., Takahashi, N. 1982. Metabolism of steviol and its derivatives by *Gibberella fujikuroi. Agric. Biol. Chem.* 46:2305–11

128. Nitsche, K., Grossmann, K., Sauerbrey, E., Jung, J. 1985. Influence of the growth retardant tetcyclacis on cell division and cell elongation in plants and cell cultures of sunflower, soybean, and maize. *J. Plant Physiol.* 118:209–18

129. Noma, M., Huber, J., Ernst, D., Pharis, R. P. 1982. Quantitation of gibberellins and the metabolism of $[^3H]$gibberellin A_1 during somatic embryogenesis in carrot and anise cell cultures. *Planta* 155:369–76

130. Norman, S. M., Bennett, R. D., Poling, S. M., Maier, V. P., Nelson, M. D. 1986. Paclobutrazol inhibits abscisic acid biosynthesis in *Cercospora rosicola. Plant Physiol.* 80:122–25

131. Norman, S. M., Poling, S. M., Maier, V. P., Orme, E. D. 1983. Inhibition of abscisic acid biosynthesis in *Cercospora rosicola* by inhibitors of gibberellin biosynthesis and plant growth retardants. *Plant Physiol.* 71:15–18

132. Park, K. H., Fujisawa, S., Sakurai, A., Yamaguchi, I., Takahashi, N. 1983. Gibberellin production in cultured cells of *Nicotiana tabacum. Plant Cell Physiol.* 24:1241–49

133. Pharis, R. P., King, R. W. 1985. Gibberellins and reproductive development in seed plants. *Ann. Rev. Plant Physiol.* 36:517–68

134. Phinney, B. O. 1979. Gibberellin biosynthesis in the fungus *Gibberella fujikuroi* and in higher plants. See Ref. 76, pp. 57–78

135. Phinney, B. O. 1983. The history of gibberellins. See Ref. 12, pp. 19–52

136. Phinney, B. O. 1984. Gibberellin A_1, dwarfism and the control of shoot elongation in higher plants. See Ref. 118, pp. 17–41

137. Phinney, B. O., Freeling, M., Robertson, D. S., Spray, C. R., Silverthorne, J. 1986. Dwarf mutants in maize—the gibberellin biosynthetic pathway and its molecular future. See Ref. 10, pp. 55–64

138. Phinney, B. O., Spray, C. 1982. Chemical genetics and the gibberellin pathway in *Zea mays* L. See Ref. 64, pp. 101–10

139. Phinney, B. O., Spray, C. 1983. Gibberellin biosynthesis in *Zea mays:* the 3-hydroxylation step GA_{20} to GA_1. In *Pesticide Chemistry, Human Welfare and the Environment (IUPAC),* ed. J. Miamoto, P. C. Kearney, pp. 81–86. Oxford: Pergamon

140. Phinney, B. O., West, C. A. 1960. Gibberellins as native plant growth regulators. *Ann. Rev. Plant Physiol.* 11:411–36

141. Potts, W. C. 1986. Gibberellins in light-grown shoots of *Pisum sativum* L. and the influence of reproductive development. *Plant Cell Physiol.* 27:997–1003

142. Potts, W. C., Reid, J. B. 1983. Internode length in *Pisum.* III. The effect and interaction of the *Na/na* and *Le/le* gene differences on endogenous gibberellin-like substances. *Physiol. Plant.* 57:448–54

143. Potts, W. C., Reid, J. B., Murfet, I. C. 1982. Internode length in *Pisum.* I. The effect of the *Le/le* gene difference on endogenous gibberellin-like substances. *Physiol. Plant.* 55:323–28

144. Potts, W. C., Reid, J. B., Murfet, I. C. 1985. Internode length in *Pisum.* Gibberellins and the slender phenotype. *Physiol. Plant.* 63:357–64

144a. Proebsting, W. M., Davies, P. J., Marx, G. A. 1978. Photoperiod-induced changes in gibberellin metabolism in relation to apical growth and senescence in genetic lines of peas (*Pisum sativum* L.). *Planta* 141:231–38

145. Rademacher, W., Graebe, J. E. 1979. Gibberellin A_4 produced by *Sphaceloma manihoticola,* the cause of the super-elongation disease of cassava *(Manihot esculenta). Biochem. Biophys. Res. Commun.* 91:35–40

146. Rademacher, W., Jung, J. 1981. Comparative potency of various synthetic plant growth retardants on the elongation of rice seedlings. *Z. Acker- Pflanzenbau* 150:363–71

147. Rademacher, W., Jung, J., Graebe, J. E., Schwenen, L. 1984. On the mode of action of tetcyclacis and triazole growth retardants. See Ref. 36, pp. 1–11

148. Railton, I. D. 1982. Gibberellin metabolism in plants. *Cell Biol. Int. Rep.* 6:319–37

149. Railton, I. D., Fellows, B., West, C. A. 1984. *ent*-Kaurene synthesis in chloroplasts from higher plants. *Phytochemistry* 23:1261–67

150. Rappaport, L., Adams, D. 1978. Gibberellins: Synthesis, compartmentation and physiological process. *Philos. Trans. R. Soc. London Ser. B.* 284:521–39

151. Reid, J. B. 1983. Internode length in *Pisum.* Do the internode length genes effect growth in dark-grown plants?. *Plant Physiol.* 72:759–63

152. Reid, J. B. 1986. Internode length in *Pisum.* Three further loci, *lh, ls* and *lk. Ann. Bot.* 57:577–92

153. Reid, J. B., Murfet, I. C., Potts, W. C. 1983. Internode length in *Pisum.* II. Additional information on the relationship and action of loci *Le, La, Cry, Na* and *Lm. J. Exp. Bot.* 34:349–64

154. Reid, J. B., Potts, W. C. 1986. Internode length in *Pisum.* Two further mutants, *lh* and *ls,* with reduced gibberellin synthesis, and a gibberellin insensitive mutant, *lk. Physiol. Plant.* 66:417–26

155. Robinson, D. G., Andreae, M., Sauer, A. 1985. Hydroxyproline-rich glycoprotein biosynthesis: A comparison with that of collagen. In *Biochemistry of Plant Cell Walls,* ed. C. T. Brett, J. R. Hillman, Soc. Exp. Biol. Semin. Ser. 28:155–76. Cambridge: Cambridge Univ. Press. 312 pp.

156. Robinson, D. G., Eberle, M., Hafemann, C., Wienecke, K., Graebe, J. E. 1982. Mg^{2+}-Shifting of plasma membrane associated glucan synthetase activity. *Z. Pflanzenphysiol.* 105:323–30

157. Rood, S. B., Beall, F. D., Pharis, R. P. 1986. Photocontrol of gibberellin metabolism *in situ* in maize. *Plant Physiol.* 80:448–53

158. Rood, S. B., Koshioka, M., Douglas, T. J., Pharis, R. P. 1982. Metabolism of tritiated gibberellin A_{20} in maize. *Plant Physiol.* 70:1614–18

159. Rood, S. B., Pharis, R. P., Koshioka, M. 1983. Reversible conjugation of gibberellins *in situ* in maize. *Plant Physiol.* 73:340–46

160. Ropers, H. J., Graebe, J. E., Gaskin, P., MacMillan, J. 1978. Gibberellin biosynthesis in a cell-free system from immature seeds of *Pisum sativum. Biochem. Biophys. Res. Commun.* 80:690–97

161. Ross, J. J., Reid, J. B. 1986. Internode length in *Pisum*. The involvement of ethylene with the gibberellin-insensitive erectoides phenotype. *Physiol. Plant.* 67:673–79
162. Sauter, H. 1984. Chemical aspects of some bioregulators. In *Bioregulators Chemistry and Uses*, ed. R. L. Ory, F. R. Rittig, ACS Symp. Ser. 257:9–21. Washington, DC: Am. Chem. Soc.
163. Schneider, G. 1983. Gibberellin conjugates. See Ref. 12, pp. 389–456
164. Schneider, G., Sembdner, G., Phinney, B. O. 1984. Synthesis of GA_{20} glucosyl derivatives and the biological activity of some gibberellin conjugates. *J. Plant Growth Regul* 3:207–15
165. Sembdner, G., Gross, D., Liebisch, H. W., Schneider, G. 1980. Biosynthesis and metabolism of plant hormones. See Ref. 11, pp. 281–444
166. Shen-Miller, J., West, C. A. 1982. *Ent*-kaurene biosynthesis in extracts of *Helianthus annuus* L. seedlings. *Plant Physiol.* 69:637–41
167. Shen-Miller, J., West, C. A. 1984. Kaurene synthetase activity in *Helianthus annuus* L. Increases in enzyme actiivity after storage of seedlings in liquid nitrogen. *Plant Physiol.* 74:439–41
168. Shen-Miller, J., West, C. A. 1985. Distribution of *ent*-kaurene synthetase in *Helianthus annuus* and *Marah macrocarpus*. *Phytochemistry* 24:461–64
169. Sherwin, P. F., Coates, R. M. 1982. Stereospecificity of the oxidation of *ent*-kauren-19-ol to *ent*-kaurenal by a microsomal enzyme preparation from *Marah macrocarpus*. *J. Chem. Soc. Chem. Commun.* 1982:1013–14
170. Shiao, M. S. 1983. Inhibition of gibberellin biosynthesis in *Gibberella fujikuroi* and germination of *Oryza sativa* by mevinolin. *Bot. Bull. Acad. Sinica* 24:135–43
171. Shigematsu, Y., Murofushi, N., Takahashi, N. 1982. Structures of the metabolites from steviol methyl ester by *Gibberella fujikuroi*. *Agric. Biol. Chem.* 46:2313–18
172. Smith, V. A., MacMillan, J. 1984. Purification and partial characterization of a gibberellin 2β-hydroxylase from *Phaseolus vulgaris*. *J. Plant Growth Regul.* 2:251–64
173. Smith, V. A., MacMillan, J. 1986. The partial purification and characterisation of gibberellin 2β-hydroxylases from seeds of *Pisum sativum*. *Planta* 167:9–18
174. Sponsel, V. M. 1983. *In vivo* gibberellin metabolism in higher plants. See Ref. 12, pp. 151–250

175. Sponsel, V. M. 1983. The localization, metabolism and biological activity of gibberellins in maturing and germinating seeds of *Pisum sativum* cv. Progress No. 9. *Planta* 159:454–68
176. Sponsel, V. M. 1985. Gibberellins in *Pisum sativum*—their nature, distribution and involvement in growth and development of the plant. *Physiol. Plant.* 65:533–38
177. Sponsel, V. M. 1986. Gibberellins in dark- and red-light-grown shoots of dwarf and tall cultivars of *Pisum sativum:* The quantification, metabolism and biological activity of gibberellins in Progress No. 9 and Alaska. *Planta* 168:119–29
178. Sponsel, V. M., MacMillan, J. 1977. Further studies on the metabolism of gibberellins (GAs) A_9, A_{20} and A_{29} in immature seeds of *Pisum sativum* cv. Progress No. 9. *Planta* 135:129–36
179. Sponsel, V. M., MacMillan, J. 1978. Metabolism of gibberellin A_{29} in seeds of *Pisum sativum* cv. Progress No. 9; use of [^{2}H] and [^{3}H]GAs, and the identification of a new GA catabolite. *Planta* 144:69–78
180. Sponsel, V. M., MacMillan, J. 1980. Metabolism of [$^{13}C_1$]gibberellin A_{29} to [$^{13}C_1$]gibberellin catabolite in maturing seeds of *Pisum sativum* cv. Progress No. 9. *Planta* 150:46–52
181. Spray, C., Phinney, B. O., Gaskin, P., Gilmour, S. J., MacMillan, J. 1984. Internode length in *Zea mays* L. The dwarf-1 mutation controls the 3β-hydroxylation of gibberellin A_{20} to gibberellin A_1. *Planta* 160:464–68
182. Steffens, G. L., Byun, J. K., Wang, S. Y. 1985. Controlling plant growth via the gibberellin biosynthesis system—I. Growth parameter alterations in apple seedlings. *Physiol. Plant.* 63:163–68
183. Stoddart, J. L. 1983. Sites of gibberellin biosynthesis and action. In *The Biochemistry and Physiology of Gibberellins*, ed. A. Crozier, 2:1–55. New York: Praeger. 452 pp.
184. Stoddart, J. L. 1984. Growth and gibberellin-A_1 metabolism in normal and gibberellin-insensitive *(Rht3)* wheat (*Triticum aestivum* L.) seedlings. *Planta* 161:432–38
185. Sugavanam, B. 1984. Diastereoisomers and enantiomers of paclobutrazol: their preparation and biological activity. *Pestic. Sci.* 15:296–302
186. Suzuki, Y., Kurogochi, S., Murofushi, N., Ota, Y., Takahashi, N. 1981. Seasonal changes of GA_1, GA_{19} and abscisic acid in three rice cultivars. *Plant Cell Physiol.* 22:1085–93

187. Takahashi, M., Kamiya, Y., Takahashi, N., Graebe, J. E. 1986. Metabolism of gibberellins in a cell-free system from immature seeds of *Phaseolus vulgaris* L. *Planta* 168:190–99

188. Takahashi, N., Yamaguchi, I., Yamane, H. 1986. Gibberellins. In *Chemistry of Plant Hormones*, ed. N. Takahashi, pp. 57–151. Boca Raton: CRC Press. 277 pp.

189. Takeno, K., Koshioka, M., Pharis, R. P., Rajasekaran, K., Mullins, M. G. 1983. Endogenous gibberellin-like substances in somatic embryos of grape (*Vitis vinifera* × *Vitis rupestris*) in relation to embryogenesis and the chilling requirement for subsequent development of mature embryos. *Plant Physiol.* 73:803–8

190. Turnbull, C. G. N., Crozier, A., Schwenen, L., Graebe, J. E. 1985. Conversion of [^{14}C]gibberellin A$_{12}$-aldehyde to C$_{19}$- and C$_{20}$-gibberellins in a cell-free system from immature seed of *Phaseolus coccineus* L.. *Planta* 165:108–13

191. Turnbull, C. G. N., Crozier, A., Schwenen, L., Graebe, J. E. 1986. Biosynthesis of gibberellin A$_{12}$-aldehyde, gibberellin A$_{12}$ and their kaurenoid precursors from [^{14}C]mevalonic acid in a cell-free system from immature seed of *Phaseolus coccineus*. *Phytochemistry* 25:97–101

192. Volger, H., Graebe, J. E. 1985. Isolation and characterization of soluble enzymes involved in gibberellin metabolism. *Book of Abstr., Int. Conf. Plant Growth Substances, 12th, Heidelberg,* Abstr. 12, p. 10

193. Wada, K. 1978. New gibberellin biosynthesis inhibitors, 1-*n*-decyl- and 1-geranylimidazole: Inhibitors of (−)-kaurene 19-oxidation. *Agric. Biol. Chem.* 42:2411–13

194. Wada, K., Imai, T. 1980. Effect of 1-*n*-decylimidazole on gibberellin biosynthesis in Tan-ginbozu, a dwarf variety of rice. *Agric. Biol. Chem.* 44:2511–12

195. Wada, K., Imai, T., Yamashita, H. 1981. Microbial production of plant gibberellins and related compounds from *ent*-kaurene derivatives in *Gibberella fujikuroi*. *Agric. Biol. Chem.* 45:1833–42

196. Wample, R. L., Culver, E. B. 1983. The influence of paclobutrozol, a new growth regulator, on sunflowers. *J. Am. Soc. Hortic. Sci.* 108:122–25

197. Wang, S. Y., Byun, J. K., Steffens, G. L. 1985. Controlling plant growth via the gibberellin biosynthesis system—II. Biochemical and physiological alterations in apple seedlings. *Physiol. Plant.* 63:169–75

198. West, C. A. 1980. Hydroxylases, monooxygenases, and cytochrome P-450. In *The Biochemistry of Plants—A Comprehensive Treatise*, ed. D. D. Davies, 2:317–64. New York/London/Toronto/Sydney/San Francisco: Academic

199. West, C. A. 1981. Biosynthesis of diterpenes. In *Biosynthesis of Isoprenoid Compounds*, ed. J. W. Porter, S. L. Spurgeon, 1:375–411. New York/Chichester/Brisbane/Toronto: Wiley-Interscience

200. West, C. A., Shen-Miller, J., Railton, I. D. 1982. Regulation of kaurene synthetase. See Ref. 64, pp. 81–90

201. Wilkinson, R. E. 1986. Diallate inhibition of gibberellin biosynthesis in sorghum coleoptiles. *Pest. Biochem. Physiol.* 25:93–97

202. Wurtele, E. S., Hedden, P., Phinney, B. O. 1982. Metabolism of the gibberellin precursors *ent*-kaurene, *ent*-kaurenol, and *ent*-kaurenal in a cell-free system from seedling shoots of normal maize. *J. Plant Growth Regul.* 1:15–24

203. Yamane, H., Yamaguchi, I., Kobayashi, M., Takahashi, M., Sato, Y., et al. 1985. Identification of ten gibberellins from sporophytes of the tree fern, *Cyathea australis*. *Plant Physiol.* 78:899–903

204. Zander, M. 1986. *Gibberellinbiosynthese in keimenden Samen und jungen Keimlingen von Pisum sativum*. Diplom thesis. Univ. Göttingen, F.R.G. 114 pp.

205. Zeevaart, J. A. D. 1985. Inhibition of stem growth and gibberellin production in *Agrostemma githago* L. by the growth retardant tetcyclacis. *Planta* 166:276–79

206. Zeigler, R. S., Powell, L. E., Thurston, H. D. 1980. Gibberellin A$_4$ production by *Sphaceloma manihoticola*, causal agent of cassava superelongation disease. *Phytopathology* 70:589–93

Ann. Rev. Plant Physiol. 1987. 38:467–86

AGROBACTERIUM-MEDIATED PLANT TRANSFORMATION AND ITS FURTHER APPLICATIONS TO PLANT BIOLOGY

Harry Klee, Robert Horsch, and Stephen Rogers

Plant Molecular Biology, Corporate Research Laboratories, Monsanto Company, St. Louis, Missouri 63198

CONTENTS

AGROBACTERIUM BIOLOGY 467
VECTORS FOR *AGROBACTERIUM*-MEDIATED PLANT TRANSFORMATION..... 470
SELECTABLE MARKERS FOR PLANT TRANSFORMATION 471
AGROBACTERIUM HOSTS FOR TRANSFORMATION 473
TRANSFORMATION OF PLANTS 475
BEYOND GENE TRANSFER 476
CONCLUSIONS... 482

This review is intended as a perspective on *Agrobacterium*-mediated plant transformation and special applications of the *Agrobacterium* system that go beyond simple gene transfer. Since several recent reviews have dealt with both the vectors used in transformation and the mechanisms of T-DNA transfer, we do not cover these topics comprehensively. Rather, we explain the basic features of *Agrobacterium* transformation as they pertain to rational design of plant transformation systems. We discuss how knowledge of *Agrobacterium*-mediated transformation can extend our capacities beyond the simple generation of transgenic plants. *Agrobacterium* is a tool that can help us to understand some of the most fundamental processes of plant biology.

AGROBACTERIUM BIOLOGY

The ability to transform plant cells is correlated with the presence in *Agrobacterium* of either a tumor-inducing (Ti) or root-inducing (Ri) plasmid. Two important regions in the Ti and Ri plasmids are essential for transformation by transferred DNA (T-DNA). These are the T-DNA itself and the virulence region (vir). The T-DNA is defined by flanking 25-base-pair (bp) directly

467

repeated sequences (57, 59, 61). These sequences are required in *cis* for T-DNA transfer and are the recognition sequences for a site-specific endonuclease encoded by the *virD* operon (47, 60). Cleavage results in a linear, single-stranded molecule that is presumed to be an intermediate in T-DNA transfer to plant cells. What is critical from a perspective of vector design is that the border sequences must flank the DNA to be transferred and that they are the only *cis*-acting elements required for T-DNA transfer. No other T-DNA genes play a necessary role in transfer, and all of the essential transfer functions act in *trans* on the border sequences. Any DNA placed between the borders will be transferred to a plant.

The T-DNA is capable of inducing tumor formation in a transformed plant cell. This is accomplished by the products of three genes, *iaaM (tms1)*, *iaaH (tms2)*, and *ipt (tmr)*. It has now been clearly established by biochemical means that these genes are involved in the synthesis of phytohormones (Figure 1). The *iaaM* and *iaaH* genes encode respectively a tryptophan

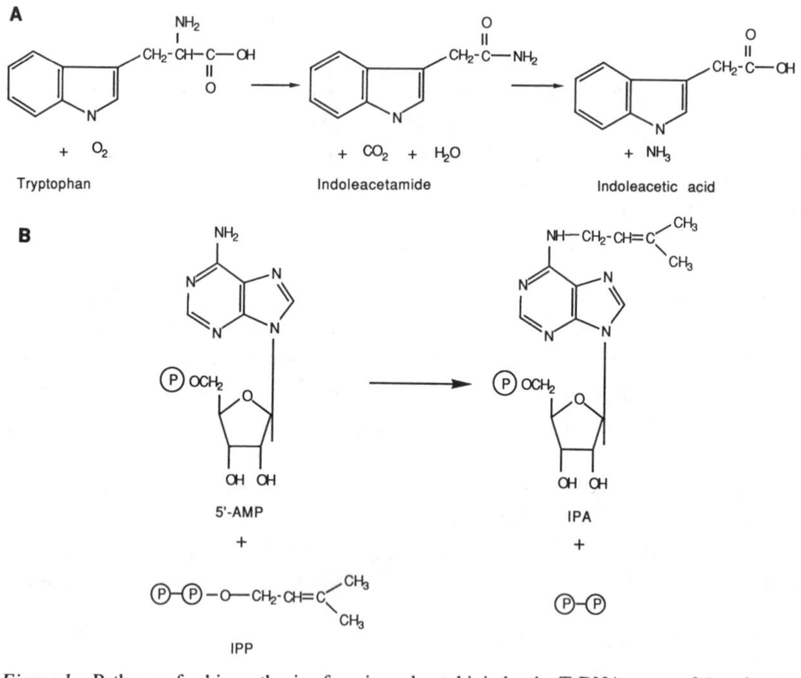

Figure 1 Pathways for biosynthesis of auxin and cytokinin by the T-DNA genes of *Agrobacterium tumefaciens*. (*a*) The auxin pathway involves conversion of tryptophan to indoleacetamide and then to indoleacetic acid by tryptophan monooxygenase and indoleacetamide hydrolase, respectively. (*b*) The cytokinin reaction involves the attachment of an isopentenyl group to 5'-adenosine monophosphate by the enzyme isopentenyl transferase.

PtiA6 Virulence Loci

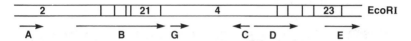

Figure 2 An EcoRI restriction endonuclease map of the virulence genes from PtiA6, an octopine-type Ti plasmid. The locations of the operons and directions of transcription are indicated by the arrows. The virA, virB, virD, and virG loci are absolutely essential for virulence. Mutations in virC or virE result in a weakly virulent phenotype.

monooxygenase (51) and an indoleacetamide hydrolase (41, 50) which, together, synthesize the auxin indoleacetic acid. The *ipt* gene encodes an isopentenyl transferase (2, 4), which uses isopentenyl pyrophosphate and adenosine monophosphate to synthesize the cytokinin isopentenyl adenosine. The combination of auxin and cytokinin biosynthesis directed by the T-DNA leads to the tumorous morphology of transformed plant tissue. These phyto-hormone biosynthetic genes are almost certainly bacterial in origin. The auxin biosynthetic pathway utilized by *Agrobacterium* is not normally used by plants. These T-DNA genes, while bacterial in origin, are powerful tools to the plant biologist trying to understand the roles of phytohormones in plant development (see below).

The other major portion of the Ti plasmid involved in T-DNA transfer is the vir region. This region of about 35 kb encompasses at least six operons (26, 46). The map of the vir region of PtiA6 is shown in Figure 2. While most of these operons are essential for T-DNA transfer (24), they need not be physi-cally linked to the T-DNA (11, 20). In practice this has led to the development of binary transformation vectors (see below). The virulence functions, since they encompass such a large segment of DNA, are left intact on the Ti plasmid; and the T-DNA is placed on a much smaller, easily manipulable plasmid that replicates autonomously.

The regulation of the vir genes has turned out to be an interesting and potentially useful story. These genes are expressed at very low levels in *Agrobacterium* under normal conditions. When they are exposed to plant cells or exudates they are turned on to varying degrees (45). This induction is a slow process, taking 8–16 hr to reach maximal levels of expression. Several phenolic compounds capable of induction have been identified and can be used in pure form to stimulate expression of the various vir operons. Two of the vir operons, *virA* and *virG,* have been implicated in the induction process (48). This work clearly illustrates that the *Agrobacterium* and the plant cell interact. The bacteria sense the presence of plant cells, and only then do they express the genes necessary for T-DNA excision and transfer. To date there have been no reports of enhanced transformation of any plant species by using bacteria exposed to the enhancer compounds. However, it is possible that induction of the vir genes could be a limiting step in efficient plant transformation in some plant species.

VECTORS FOR *AGROBACTERIUM*-MEDIATED PLANT TRANSFORMATION.

Plant transformation vectors based on *Agrobacterium* can generally be divided into two categories: those that cointegrate into a resident Ti plasmid and those that replicate autonomously (the binary vectors). All of the vectors have several common features imposed upon them by the requirements of *Agrobacterium*.

As mentioned above, the T-DNA is delimited by direct 25-bp imperfect repeats. No other element is required in *cis* to cause T-DNA excision and transfer. A transformation vector must, then, include at least one border (see below) or be capable of cointegrating in such a manner that it becomes flanked by borders. There is good evidence for the existence of a second sequence, termed "overdrive," which is found in the vicinity of the border sequence (36). While this sequence is not essential for excision and transfer, it does appear to stimulate the process. Most vectors utilize a segment of a Ti plasmid that includes both the border and the overdrive sequences. Some vectors, however, use only the 25-bp border sequence (for an example, see below), and these seem to work efficiently for plant transformations.

Besides a border, all vectors include a selectable marker for identification of transformed plant cells and a bacterial selectable marker essential for introduction of the vector into *Agrobacterium*. Some vectors also include a scorable marker such as nopaline synthase. This allows for early verification of transformation by a simple assay for nopaline. Nopaline is produced from alpha-ketoglutarate and arginine upon catalysis by nopaline synthase.

Cointegrating Vectors

Cointegrating transformation vectors must include a region of homology between the vector plasmid and the Ti plasmid. This requirement for homology means that the vector is capable of integrating into a limited number of Ti plasmids. The vector is usually designed to cointegrate into one or a few specific Ti plasmids. Two cointegrating systems are in use today. The first utilizes the disarmed *Agrobacterium* Ti plasmid pGV3850 (62). In this plasmid the phytohormone genes of the C58 plasmid have been excised and replaced by pBR322 sequence. Any plasmid containing the pBR322 sequence homology can be cointegrated into the disarmed Ti plasmid. The border sequences as well as a nopaline synthase gene are part of the Ti plasmid, and the cointegration places the new sequence between the T-DNA borders.

A different approach to a cointegrating vector was used by Fraley et al (14). In this system, the right border and all of the phytohormone genes are removed from the Ti plasmid. A left border and a small part of the original T-DNA, referred to as the Limited Internal Homology (LIH), remain intact. The vector to be introduced into *Agrobacterium* contains the LIH region for homologous recombination as well as a right border. The cointegrated DNA reconstructs a functional T-DNA with a right border and a left border. This system has been used extensively for introduction of many genes into plants. The cointegrate systems, while more difficult to use, do offer advantages. Once the cointegrate has been formed, the plasmid is stable in *Agrobacterium* and is virtually impossible to lose. Binary vectors, on the other hand, are not completely stable in *Agrobacterium* in the absence of drug selection. There is also evidence that a cointegrating vector can transform tomato at a higher frequency than a binary vector (33).

Binary Vectors

Binary transformation vectors are somewhat different from cointegrating vectors. Instead of a region of homology with the Ti plasmid, they contain origins of replication from a broad host-range plasmid. These replication origins permit autonomous replication of the vector in *Agrobacterium*. Since the plasmid does not need to form a cointegrate, these plasmids are considerably easier to introduce into *Agrobacterium*. The frequency of introduction into *Agrobacterium* is about 10^{-1} whereas cointegrate formation is usually about 10^{-5}. Since these vectors do not cointegrate, they must contain the T-DNA border sequence(s). Many vectors contain two borders that delimit the portion of the plasmid transferred to the plant. However, a single border is sufficient for transfer (23). In this case transfer initiates at the single border, and the entire plasmid becomes the T-DNA.

A major advantage to binary vectors is their lack of dependence on a specific Ti plasmid. The vector may be introduced into virtually any *Agrobacterium* host containing any Ti or Ri plasmid, as long as the vir helper functions are provided. This may be important in the transformation of some plant species since different *Agrobacterium* strains exhibit major differences in their abilities to infect different plant species (see below).

SELECTABLE MARKERS FOR PLANT TRANSFORMATION

Several selectable marker genes are widely available today for plant transformation. These are discussed below. Several requirements must be considered in the development of a truly useful selectable marker system. It is most critical that the selective agent be inhibitory to plant cells. However, not all compounds toxic to plant cells are necessarily useful as selective agents. Cells that are not transformed can be killed in such a manner that they become toxic to adjacent transformed cells. This presumably happens because of leakage of toxic compounds, such as phenols, from the dying cells. If this occurs, even high-level expression of a resistance gene in the transformed cells is insufficient to rescue these cells. The best selective agents are compounds that arrest growth of nontransformed cells or slowly kill them.

By far the most widely used selectable marker has been the neomycin phosphotransferase, type II (NPTII) enzyme, which was originally isolated from the prokaryotic transposon Tn5 (6). This enzyme detoxifies aminoglycoside compounds such as kanamycin and G418 by phosphorylation. This gene, fused to constitutive plant transcriptional promoters, has been used successfully to transform a large number of plant species (see Table 1) and has been incorporated into numerous plant transformation vectors (3, 7, 14, 15, 23, 27, 28, 32, 52, 54).

For several reasons a single antibiotic resistance gene, even one as versatile as NPTII, has not fulfilled all of the needs of plant molecular biologists. Probably the single most important reason is that this marker does not work in all plant species. This can be the consequence either of the lack of toxicity of kanamycin (G418) or of the failure of the enzyme to confer selectability in transformed cells. A good example of this lack of selectability is *Arabidopsis thaliana* var. Columbia (30). While the NPTII gene is clearly expressed and can, in fact, be used to select transformed, germinating seeds, it does not work as an efficient selectable marker for primary transformation. Another

Table 1 Transgenic plants produced with *A. tumefaciens*-mediated transformation

Species	Reference
Nicotiana plumbaginifolia	21
petunia	22
tobacco	9a
tomato	33
lettuce	Michelmore R., pers. comm.
celery	Michelmore R., pers. comm.
poplar	13
Arabidopsis thaliana	30
Brassica napus	16
sunflower	Everett, N., pers. comm.
cotton	Firoozabady E., pers. comm.

good reason for development of alternative selectable markers is related to the need for introducing more than one gene into a plant. If one desires to introduce multiple genes into a single plant, one simple approach is retransformation. This requires additional, independent selectable markers.

The need for alternative markers has led to the development of two useful systems. The first of these is also a bacterial phosphotransferase, one encoding resistance to hygromycin (40, 53, 55). This selectable marker has been demonstrated in a number of plant species. It has been particularly useful in the development of a transformation system for *Arabidopsis*, where the NPTII gene has not worked well (30). The second generally useful marker system is based on the enzyme dihydrofolate reductase (DHFR). In this system the activity of DHFR is blocked by methotrexate. A DHFR enzyme with a 500-fold lower affinity for methotrexate has been isolated from mouse (40). When the resistant enzyme is fused to the CaMV 35S promoter, transformed plant cells are highly resistant to methotrexate. This marker has worked well in a number of plant species (H. Klee, R. Horsch, unpublished results). Not only do these selectable markers extend the transformation systems beyond those used with NPTII but they have also been useful for retransformation of plants already resistant to kanamycin (N. Hoffmann, R. Horsch, unpublished). It is likely that other selectable marker systems will be developed in the near future. One promising system should be the acetohydroxy acid synthase enzyme described by Haughn & Somerville (19). This mutant enzyme is tolerant to a 300-fold higher level of chlorsulfuron than is required to inhibit wild-type enzyme.

AGROBACTERIUM HOSTS FOR TRANSFORMATION

Disarmed Strains

Wild-type *Agrobacterium* Ti and Ri plasmids are capable of transferring to a plant cell genes that encode hormone biosynthetic activities. Obviously cells synthesizing high levels of auxin and cytokinin will be difficult, if not impossible, to regenerate into whole plants. In general there have been two approaches to separating these hormone genes from a foreign gene of interest. The simplest has been to disarm the Ti plasmid. Using homologous recombination, the hormone genes can easily be removed from the Ti plasmid, leaving behind a "disarmed" *Agrobacterium*. This recombination can be accomplished to remove the entire T-DNA, as is the case for LBA4404 (34). A Ti plasmid with its entire T-DNA removed can then be used in conjunction with virtually any binary transformation vector. Alternatively, a portion of the T-DNA (not including the hormone genes) can be left behind as homologous DNA for cointegrating vectors. This is the approach used in the SEV (split end vectors) system of Fraley et al (14). In either case, the essential vir functions remain intact in the *Agrobacterium,* and T-DNA transfer can occur. Transformed cells can then be regenerated using standard protocols.

Nondisarmed Strains

An alternative approach to separating the hormone biosynthetic genes from the gene of interest can be used with binary vectors. When two independent T-DNAs are present within a single *Agrobacterium,* as in the case of a cell containing a wild-type Ti plasmid and a binary vector, there is a high probability that the two T-DNAs will become integrated into a single plant cell (10, 37, 44). However, some transformed cells will contain only one of the T-DNAs. Since some selective agent such as kanamycin is always used during the transformation/regeneration protocol, only the cells transformed with the binary T-DNA will survive. Some of these cells will contain only the binary T-DNA, while others will contain both T-DNAs. If fertile, transformed plants can be recovered, any wild-type T-DNA can be eliminated by outcrossing, assuming that the two T-DNAs are not linked.

The problem with this approach is that the cells transformed with the wild-type T-DNA will synthesize phytohormones that can still interfere with the regeneration protocol. The main advantage is related to host-range effects observed with different *Agrobacterium* strains (see below). For the most part, disarmed Ti plasmids have been constructed from only a few common laboratory strains such as pTiC58 (62) and pTiT37 (14). Construction of a disarmed Ti plasmid, while straightforward, requires a detailed knowledge of the plasmid. If one wishes to utilize an *Agrobacterium* strain that happens to transform a particular plant species very well one need not do all of the work to construct a disarmed Ti plasmid.

So far we have stressed the disadvantages to regeneration of plant tissues when transformed, hormone-autonomous cells are present. There is one case where there may actually be an advantage in such an arrangement. When tissue from many plants is transformed with *Agrobacterium rhizogenes,* it differentiates into transformed roots. Frequently these roots give rise spontaneously to plantlets that can become fully differentiated plants. These plants still contain at least a portion of the Ri T-DNA (37, 44). If they have been cotransformed with a binary vector containing some gene, plants can be obtained with that gene. The problem with this approach is that the plants still contain at least a portion of the Ri T-DNA. The plants obtained in this way frequently show abnormalities suggestive of phytohormone imbalances. Nonetheless, this cotransformation/regeneration protocol could be of use in some plant species that are difficult to regenerate.

Host Range

A growing body of evidence indicates that the tumorigenicity exhibited by different strains of *Agrobacterium tumefaciens* and *A. rhizogenes* is highly dependent on the plant host (8, 35, 42). This means that different strains may vary significantly in their abilities to infect a cultivar of a given plant species. It should be noted that tumorigenicity is not the same as T-DNA transformation. For example, Facciotti et al (12) have observed transformed cells in wounded soybean seedlings despite the lack of tumor production. Most of the assays done to date have, by necessity, used tumor formation as an assay for infectibility. Many factors are involved in tumor induction. Most of these are probably related to T-DNA phytohormone production and the sensitivity of the plant to those hormones. Nonetheless, the observations related to *Agrobacterium* host range are important to consider, especially in plants not traditionally used as *Agrobacterium* hosts.

The problem of host range is probably best dealt with by the use of binary vectors. These vectors can be readily mobilized into any *Agrobacterium* capable of providing vir functions in *trans*. This allows the experimenter to assay for T-DNA transfer by selecting the marker on the binary vector. With this approach, one should be able to screen large numbers of *Agrobacterium* strains on a plant host and determine which of these is capable of the highest level of transformation.

TRANSFORMATION OF PLANTS

Cocultivation

Two basic approaches have been used to obtain transgenic plants: cocultivation of regenerating protoplasts (21, 31, 58) and the leaf disc procedure (22). Cocultivation was the first procedure used successfully to generate a transgenic plant. The procedure involves *Agrobacterium* transformation of regenerating protoplasts, followed by selection and regeneration to plants. For solanaceous plants such as petunia and tobacco this can be an efficient way to generate large numbers of independent transformants. Because it requires a good regeneration protocol for protoplasts of the plant species, the procedure is not useful for many important plant species.

Leaf Disc Transformation

In the leaf disc procedure, surface-sterilized leaf pieces, or other axenic explants, are cocultured on regeneration medium for 2–3 days with *Agrobacterium*. The best choice of explant is usually one that regenerates well in tissue culture for the species of interest. For example, in tomato, cotyledons are used as the explant source (33), while in *Brassica napus,* stem segments seem to work well (16). During cocultivation, the vir genes are induced in *Agrobacterium,* bacteria bind to plant cells around the wounded edge of the explant, and T-DNA transfer occurs. A nurse culture of tobacco cells is usually used during cocultivation to increase transformation frequency. This presumably results in better induction of the vir genes. Following coculture, the explants are transferred to regeneration/selection medium. This medium contains carbenicillin to kill the *Agrobacterium* and the appropriate antibiotic to select for transformed plant cells. During the next several weeks, transformed callus grows and differentiates into shoots. The shoots are then excised, rooted on an appropriate medium in the presence of the selective agent, and transferred to soil.

The leaf disc procedure and variants thereof are a great improvement over cocultivation for several reasons. The leaf disc procedure requires far less tissue culture expertise than does protoplast preparation. It is far more generally applicable because a protoplasting procedure is not required. And transgenic plants are obtained far more quickly—often within 4–6 weeks of coculture.

One technical issue concerning leaf disc transformation should be considered, especially with new plant species. Some of the shoots that regenerate on selective medium do not contain T-DNA. The reasons for these escapes are not clear but may include loss of T-DNA or incomplete selection due to cross-protection of wild-type cells by nearby transformed cells. The problem of escapes is most efficiently dealt with by a second selection for the ability to form roots in the presence of the selective agent. A scorable marker such as nopaline synthase, if present, can also be useful for identification of transformed shoots. Once a transformed shoot has also rooted on selective medium, the T-DNA insertion appears to be stable. T-DNAs have been characterized genetically for a number of generations and behave as normal Mendelian traits (56).

BEYOND GENE TRANSFER

Our understanding of how *Agrobacterium* causes crown gall disease has led to major advances in the ability to transform plants. The elucidation of T-DNA's role and of the various *cis*- and *trans*-acting factors necessary for this transformation event to occur has permitted intelligent design of plant transformation systems and has provided clues toward improving the systems further. At this point the technology for routine introduction of genes into many plant species is firmly established. The technology is now rapidly evolving past the stage of simple gene introduction. We can now use what we know about *Agrobacterium* biology to address more fundamental questions of plant biology.

Use of the T-DNA as a Mutagen

The concept of using the T-DNA as a mutagen in an insertional inactivation scheme is an obvious one. If T-DNA insertion is essentially random, then a certain percentage of insertional events will alter genes. This technique in itself would be greatly inferior to chemical or ultraviolet mutagenesis (for reasons discussed below), except that the occurrence of a T-DNA-induced mutation provides a built-in tag that identifies the affected gene.

Use of T-DNA in a gene tagging scheme involves several problems. The first is the infrequency of T-DNA transfer. While transformation is efficient in some species, only one or a few such transfers occur per transformed cell. Generation of many transgenic plants thus becomes labor intensive. The second, related problem concerns the nature of the plant genome. Most plants have large genomes with a great deal of repetitive DNA. Therefore, many T-DNA insertion events will not disrupt genes. The random insertion of T-DNA into most plant genomes is therefore an inefficient way to produce mutations. Such mutations do, however, occur at a low frequency. In the course of introducing genes into plants we have found at least one T-DNA insertion in *Nicotiana plumbaginifolia* that cannot be maintained in a homozygous state. This homozygous lethal mutation appears to be linked to a T-DNA (R. Horsch, unpublished). In petunia and tobacco, with their large genomes, this is a rare class of mutation.

One way around the problem of the large number of T-DNA insertions needed to screen for mutations is to use a plant with a simple genome such as *Arabidopsis thaliana*. *Arabidopsis* has a small genome relative to most plants (29) and has considerably less repetitive DNA as well (38). Rough calculations of genome size and the number of genes in a plant indicate that approximately one in four T-DNA insertions should be into a gene. Thus, if the T-DNA inserts randomly, it should be straightforward to generate a number of mutant lines; but this transformation and regeneration system, although reliable, is still labor intensive (30). Such an approach will certainly yield many interesting T-DNA–tagged mutants. Because of the frequency and labor involved, however, it is probably unrealistic to think about using T-DNA to generate a specific mutation, unless it can be identified or selected for in tissue culture.

Mutagenesis with Promoterless Markers

Is there a way to use our knowledge of T-DNA transfer to make the concept of insertional mutagenesis more generally applicable to plants with large, complex genomes? The answer is probably yes. All that defines a T-DNA in *Agrobacterium* is an imperfect direct 25-bp repeat. This sequence is recognized as the border of the T-DNA, and in almost all cases the plant-bacterial junction occurs within this sequence (59, 61). Furthermore, transfer is always unidirectional from the right border. Several vectors have been designed to take advantage of this knowledge (49; H. Klee, unpublished). One vector that we have constructed, pMON557, is illustrated in Figure 3. A binary

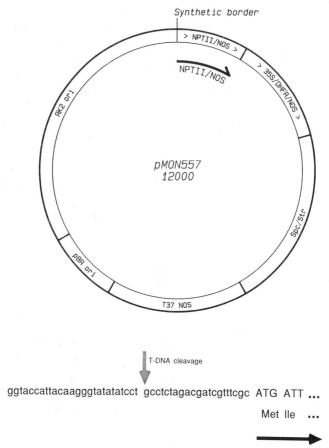

Synthetic border

pMON557
12000

T-DNA cleavage

ggtaccattacaagggtatatatcct gcctctagacgatcgtttcgc ATG ATT ...

Met Ile ...

Figure 3 Structure of a plasmid, pMON557, designed for insertional mutagenesis. The vector contains a gene encoding kanamycin resistance (NPTII/NOS) but lacking a transcriptional promoter. This vector also contains a chimeric gene encoding methotrexate resistance in transformed plant cells (35S/DHFR/NOS), a bacterial gene encoding spectinomycin and streptomycin resistance in *Agrobacterium* and *E. coli* (Spc/Str), a nopaline synthase gene (NOS), and plasmid replication origins which function in *E. coli* (pBR ori) and *Agrobacterium* (RK2 ori). The DNA sequence of the synthetic T-DNA border and the adjacent start codon of the NPTII gene are shown at figure bottom.

transformation vector requires only a single 25-bp border sequence for efficient transfer to plants (23). The single border acts as both the start and stop signal for transfer. We placed a gene encoding kanamycin resistance, the NPTII gene, just inside of the border. This NPTII gene lacks a transcriptional promoter. When pMON557 is used to transform plant cells for kanamycin resistance, the only cells that should grow are those in which a fusion between a plant gene and the NPTII gene has occurred. Therefore, any tissue that grows on kanamycin, barring T-DNA rearrangements, must have an insertion into a gene.

One limitation of this approach is that the gene into which the T-DNA has inserted must be expressed to confer kanamycin resistance. Fortunately, this is not a major limitation since many genes that are stringently regulated in a plant appear to be expressed constitutively in callus culture. A good example of this is the gene encoding the 7S soybean storage protein, which we have found to express at a low level in callus (25). A T-DNA insertion into such a gene would confer kanamycin resistance in transformed callus but not in regenerated shoots or roots. For this reason, we constructed pMON557 with a second selectable marker gene, one encoding resistance to methotrexate. Following an initial selection on kanamycin, tissue is transferred to methotrexate for regeneration.

This approach should be more general than simple T-DNA insertional inactivation since it eliminates the need to work with plants with simple genomes such as *Arabidopsis*. Insertion events into nontranscribed regions will not lead to kanamycin resistance and will never be observed. The only limitation, then, is the necessity for a high-frequency transformation system to generate large numbers of insertion events that confer kanamycin resistance.

Gene Rescue by Shotgun Cloning

Gene cloning by complementation of a mutant phenotype or transfer of a dominant phenotype to a new host has been used successfully by bacteriologists for a number of years. This approach has allowed for the cloning of many regulatory genes that could never have been cloned by more standard approaches. Such an approach has not heretofore been possible in plants because of technical difficulties. The major limitations have been the large genome sizes of most plants and the lack of a high-frequency transformation system to generate the numbers of transgenic plants that must be screened to cover a genome equivalent of a typical plant.

The development of small, highly efficient binary plant transformation vectors has allowed for construction of plant genomic banks that can be transformed directly back into plants. A bank constructed from *Arabidopsis* would require only 7,000–10,000 clones, depending on insert size, to have a 99% probability of containing any sequence. Improvements in plant transformation systems such as the leaf disc procedure now provide the means to generate the thousands of independent transformants needed to screen a genomic bank, at least in the case of *Arabidopsis*.

We set up a model system for determining the efficacy of gene rescue in plants (H. Klee and M. Hayford, submitted). The starting material consisted of *Arabidopsis* tissue that had been transformed with pMON200. The single T-DNA in this tissue conferred kanamycin resistance as well as the ability to synthesize nopaline on the tissue. A genomic bank was constructed from this tissue in a vector that could be mobilized into *Agrobacterium* and used to retransform plant cells directly. The entire bank was then used to transform petunia leaf discs for kanamycin resistance. Kanamycin-resistant nopaline-positive tissue was obtained at a low frequency (approximately one in twenty leaf discs), indicating that a single-copy gene can be rescued from a genomic pool.

There are obvious limitations to such an approach. The gene to be rescued must have an easily identified or selected phenotype. The gene must be isolated from a plant with a relatively small genome, such as *Arabidopsis*. There must be retransformants enough to screen to insure a reasonable probability of introducing the gene of interest. While these are major limitations, the gene rescue approach should be valuable for cloning a number of genes that have been unavailable.

Agrobacterium *and Plant Viruses*

In the past few years a great deal of progress has been made using *Agrobacterium*-mediated plant transformation in conjunction with the study of plant viral genetics. A major difficulty with viral genetics has been the inefficiency with which viruses, especially those containing mutations in some genes, can be introduced back into a plant and develop into a systemic infection. This problem can, in many cases, be alleviated by introducing viruses back into plants as part of a T-DNA. This approach has been successful for cauliflower mosaic virus (18), tomato golden mosaic virus (39), and maize streak virus (17). With this approach it should be possible to carry out detailed genetic analyses that will allow unambiguous determination of gene functions.

The ability to introduce a virus into a plant cell via a T-DNA has additional benefits. These viruses, once in a plant cell, appear to be capable of excision and autonomous replication (18, 39). In theory, a single transformed plant cell can give rise to a systemic viral infection. The appearance of the systemic infection can then be used as a sensitive assay for T-DNA transfer. This has been shown for T-DNA transfer in maize with maize streak virus (17). Whether the introduced virus will ultimately be useful as a plant transformation vector has yet to be determined.

Phytohormones

The observation that the *Agrobacterium* T-DNA contains genes whose products are capable of synthesizing phytohormones was particularly exciting.

The molecular and biochemical characterization of these genes has provided unique tools for dissecting the roles of auxins and cytokinins in plant growth and differentiation. Our current knowledge of the roles of auxins and cytokinins has been formulated as a result of many experiments involving exogenous application. It has been difficult to study the mechanisms of hormone action because of problems with uptake and transport of exogenously applied hormones. Many studies have been performed with excised segments of plants to minimize these problems, but this is a somewhat artificial approach since it disrupts the interactions of organs with tissues. The T-DNA auxin and cytokinin genes provide tools for a unique approach to the study of phytohormones. We can now take these genes, fuse them to any transcriptional promoter we choose, reintroduce the chimeric genes, and examine their effects on the transgenic plants. In this manner, auxin and cytokinin can be synthesized endogenously in a controlled, predictable manner.

At this point, the hormone genes have been placed in transgenic plants under the control of strong, constitutively expressing promoters (5, 25). The effects of auxin or cytokinin overproduction on the plants have been dramatic. Transgenic petunias overproducing cytokinins are extremely difficult to regenerate. Overproduction of cytokinin by transformed shoots almost completely suppresses root formation (Figure 4) (5). The plants that do survive in soil have almost no root systems, show a lack of apical dominance, and are very green. These effects have all been correlated with an excess of cytokinin. Transgenic petunias overproducing auxins show a number of interesting effects that are correlated with auxin excess as well as auxin-induced ethylene production (25). Figure 5 shows the morphology of a transgenic petunia plant with an auxin level ten times that of a wild-type plant. These plants exhibit an almost complete apical dominance, greater internode distance, leaf curling, a doubling in the amount of xylem and phloem production, and parthenocarpic fruit development.

While the constitutive overproduction of auxin and cytokinin has been interesting, the real promise of endogenous phytohormone overproduction will be in conjunction with promoter elements that allow tissue-specific and inducible expression of these genes. At present, the approach is limited only by the availability of appropriate transcriptional promoters for construction of chimeric genes. Overproduction of auxin and cytokinin in specific cell types or in response to some induction will allow for precise experiments impossible by exogenous application. The availability of other biosynthetic genes involved in synthesis of ethylene, gibberellins, and abscisic acid would also greatly expand the possible applications of this approach. The potential of altering a plant's growth habits to suit a specific need are likely to have a major impact on agriculture in the immediate future.

Figure 4 The effect of constitutive overproduction of cytokinin in transformed, regenerating petunia shoots. The shoot on the right was transformed with a control vector lacking the *ipt* gene while the shoot on the left was transformed with the wild-type *ipt* gene from pTiT37 (5). The shoots are approximately the same age. The control shoot was transferred to, and maintained in, soil for approximately six weeks. The *ipt*-expressing shoot never produced roots on sterile rooting medium.

CONCLUSIONS

Our knowledge of *Agrobacterium* biology has led to remarkable break-throughs in our ability to manipulate plant genomes. The ability of *Agrobac-*

Figure 5 The effect of constitutive overproduction of auxin in transgenic petunia. The two plants are both F1 progeny from a plant that was transformed with a chimeric gene consisting of the cauliflower mosaic virus 19S promoter and the coding sequence for tryptophan monooxygenase. The plant on the left does not contain the T-DNA with the chimeric gene while the plant on the right contains an active T-DNA. The plants are exactly the same age and grown under identical conditions.

terium to be a natural genetic engineer has given plant molecular biologists an important genetic tool. Plants have been engineered to be resistant to herbicides (9, 43) and viruses (1). Transformation techniques have been worked out for many plant species. Vectors, based on the Ti plasmid, are now available for routine introduction of genes into plants. Our rapidly increasing understanding of how *Agrobacterium* transforms plants has been essential for this rapid development. But this understanding has provided us with the tools to do much more than introduce genes back into plants. It has given us a means to understand basic plant biology and viral genetics. We hope that this review has illustrated that *Agrobacterium* can be far more than a tool to generate transformed plants.

Literature Cited

1. Abel, P. P., Nelson, R. S., De, B., Hoffmann, N., Rogers, S. G., et al. 1986. Delay of disease development in transgenic plants that express the tobacco mosaic virus coat protein. *Science* 232:738–43
2. Akiyoshi, D. E., Klee, H., Amasino, R. M., Nester, E. W., Gordon, M. P. 1984. T-DNA of *Agrobacterium tumefaciens* encodes an enzyme of cytokinin biosynthesis. *Proc. Natl. Acad. Sci. USA* 81:5994–98
3. An, G., Watson, B. D., Stachel, S., Gordon, M. P., Nester, E. W. 1985. New cloning vehicles for transformation of higher plants. *EMBO J.* 4:277–84
4. Barry, G. F., Rogers, S. G., Fraley, R. T., Brand, L. 1984. Identification of a cloned cytokinin biosynthetic gene. *Proc. Natl. Acad. Sci. USA* 81:4776–80
5. Barry, G. F., Rogers, S. G., Hein, M., Niedermeyer, J., Hoffmann, N., et al. 1985. Identification of cytokinin genes and transfer into plants. *Curr. Top. Plant Biochem. Physiol.* 4:101–9
6. Beck, E., Ludwig, G., Auerswald, E. A., Reiss, B., Schaller, H. 1982. Nucleotide sequence and exact localization of the neomycin phosphotransferase gene from transposon Tn5. *Gene* 19:327–36
7. Bevan, M. 1984. Binary *Agrobacterium* vectors for plant transformation. *Nucleic Acids Res.* 12:8711–21
8. Byrne, M., McDonnell, R., Wright, M., Carnes, M. 1987. Strain and genotype specificity in *Agrobacterium*-soybean interaction. *Plant Cell Tissue Organ Culture*. In press
9. Comai, L., Facciotti, D., Hiatt, W. R., Thompson, G., Rose, R. E., Stalker, D. M. 1985. Expression in plants of a mutant *aroA* gene from *Salmonella typhimurium* confers tolerance to glyphosate. *Nature* 317:741–43
9a. De Block, M., Herrera-Estrella, L., Van Montagu, M., Schell, J., Zambryski, P. 1984. Expression of foreign genes in regenerated plants and in their progeny. *EMBO J.* 3:1681–89
10. de Framond, A., Back, E., Chilton, W., Kayes, L., Chilton, M.-D. 1986. Two unlinked T-DNAs can transform the same tobacco plant cell and segregate in the F1 generation. *Mol. Gen. Genet.* 202:125–31
11. de Framond, A., Barton, K. A., Chilton, M.-D. 1983. Mini-Ti; a new vector strategy for plant genetic engineering. *Bio-Technology* 1:262–69
12. Facciotti, D., O'Neal, J. K., Lee, S., Shewmaker, C. K. 1985. Light-inducible expression of a chimeric gene in soybean tissue transformed with *Agrobacterium*. *Bio-Technology* 3:241–46
13. Fillatti, J., Sellmer, J., McCown, B. 1986. Regeneration and transformation of *Populus*. *Proc. 6th Int. Congr. Plant Tissue Cell Culture*
14. Fraley, R. T., Rogers, S. G., Horsch, R. B., Eicholtz, D. A., Flick, J. S., et al. 1985. The SEV system: a new disarmed Ti plasmid vector for plant transformation. *Bio-Technology* 3:629–35
15. Fraley, R. T., Rogers, S. G., Horsch, R. B., Sanders, P. R., Flick, J. S., et al. 1983. Expression of bacterial genes in plant cells. *Proc. Natl. Acad. Sci. USA* 80:4803–7
16. Fry, J., Horsch, R. 1986. Transformation of *Brassica napus* by *Agrobacterium tumefaciens* based vectors. *Proc. 6th Int. Congr. Plant Tissue Cell Culture*
17. Grimsley, N., Hohn, T., Davis, J. W., Hohn, B. 1987. *Agrobacterium*-mediated delivery of infectious maize streak virus into maize plants. *Nature* 325:177–79
18. Grimsley, N., Hohn, B., Hohn, T. and Walden, R. 1986. "Agroinfection," an alternative route for viral infection of plants by using the Ti plasmid. *Proc. Natl. Acad. Sci. USA* 83:3282–86
19. Haughn, G., Somerville, C. 1986. Sulfonylurea-resistant mutants of *Arabidopsis thaliana*. *Mol. Gen. Genet.* 204:430–34
20. Hoekema, A., Hirsch, P. R., Hooykaas, P. J., Schilperoort, R. A. 1983. A binary plant vector strategy based on separation of vir and T-region of the *Agrobacterium*. *Nature* 303:179–81
21. Horsch, R. B., Fraley, R. T., Rogers, S. G., Sanders, P. R., Lloyd, A., Hoffmann, N. 1984. Inheritance of functional foreign genes in plants. *Science* 223:496–98
22. Horsch, R. B., Fry, J. E., Hoffmann, N., Eichholtz, D., Rogers, S. G., Fraley, R. T. 1985. A simple and general method for transferring genes into plants. *Science* 227:1229–31
23. Horsch, R. B., Klee, H. J. 1986. Rapid assay of foreign gene expression in leaf discs transformed by *Agrobacterium tumefaciens:* Role of T-DNA borders in the transfer process. *Proc. Natl. Acad. Sci. USA* 83:4428–32

24. Horsch, R. B., Klee, H. J., Stachel, S., Winans, S. C., Nester, E. W., et al. 1986. Analysis of *Agrobacterium tumefaciens* virulence mutants in leaf discs. *Proc. Natl. Acad. Sci. USA* 83:2571–75

25. Klee, H., Horsch, R., Hinchee, M., Hein, M., Hoffmann, N. 1987. The effects of overproduction of two *Agrobacterium tumefaciens* T-DNA auxin biosynthetic gene products in transgenic petunia plants. *Genes Dev.* In press

26. Klee, H. J., White, F. F., Iyer, V. N., Gordon, M. P., Nester, E. W. 1983. Mutational analysis of the virulence region of an *Agrobacterium tumefaciens* Ti plasmid. *J. Bacteriol.* 153:878–83

27. Klee, H. J., Yanofsky, M., Nester, E. W. 1985. Vectors for transformation of higher plants. *Bio-Technology* 3:637–42

28. Koncz, C., Schell, J. 1986. The promoter of T-L-DNA gene 5 controls the tissue-specific expression of chimeric genes carried by a novel type of *Agrobacterium* binary vector. *Mol. Gen. Genet.* 204:383–96

29. Leutwiler, L. S., Hough-Evans, B. R., Meyerowitz, E. M. 1984. The DNA of *Arabidopsis thaliana*. *Mol. Gen. Genet.* 194:15–23

30. Lloyd, A., Barnason, A., Rogers, S. G., Byrne, M., Fraley, R. T., Horsch, R. B. 1986. Transformation of *Arabidopsis thaliana* with *Agrobacterium tumefaciens*. *Science* 234:464–66

31. Marton, L., Wullems, G. J., Molendijk, L., Schilperoort, R. A. 1979. In vitro transformation of cultured cells from *Nicotiana tabacum* by *Agrobacterium tumefaciens*. *Nature* 277:129–31

32. Matzke, A., Matzke, M. 1986. A set of novel Ti plasmid-derived vectors for production of transgenic plants. *Plant. Mol. Biol.* 7:357–65

33. McCormick, S., Niedermeyer, J., Fry, J., Barnason, A., Horsch, R., Fraley, R. 1986. Leaf disc transformation of cultivated tomato (*L. esculentum*) using *Agrobacterium tumefaciens*. *Plant Cell Rep.* 5:81–84

34. Ooms, G., Hooykaas, P. J., Moolenaar, G., Schilperoort, R. A. 1981. Crown gall plant tumors of abnormal morphology, induced by *Agrobacterium tumefaciens* carrying mutated octopine Ti plasmids; analysis of T-DNA functions. *Gene* 14:33–50

35. Owens, L. D., Cress, D. E. 1985. Genotypic variability of soybean response to *Agrobacterium* strains harboring Ti or Ri plasmids. *Plant Physiol.* 77:87–94

36. Peralta, E. G., Hellmiss, R., Ream, W.

1986. Overdrive, a T-DNA transmission enhancer on the *A. tumefaciens* tumor-inducing plasmid. *EMBO J.* 5:1137–42

37. Petit, A., Berkaloff, A., Tempe, J. 1986. Multiple transformation of plant cells by *Agrobacterium* may be responsible for the complex organization of T-DNA in crown gall and hairy root. *Mol. Gen. Genet.* 202:388–93

38. Pruitt, R. E., Meyerowitz, E. M. 1986. Characterization of the genome of *Arabidopsis thaliana*. *J. Mol. Biol.* 187:169–83

39. Rogers, S. G., Bisaro, D. M., Horsch, R. B., Fraley, R. T., Hoffmann, N. L., et al. 1986. Tomato golden mosaic virus A component replicates autonomously in transgenic plants. *Cell* 45:593–600

40. Rogers, S., Klee, H., Byrne, M., Horsch, R., Fraley, R. 1987. Improved vectors for plant transformation: Expression cassette vectors and new selectable markers. *Methods Enzymol.* In press

41. Schroder, G., Waffenschmidt, S., Weiler, E. W., Schroder, J. 1983. The T-region of Ti plasmids codes for an enzyme synthesizing indole-3-acetic acid. *Eur. J. Biochem.* 138:387–91

42. Sederoff, R., Stomp, A.-M., Chilton, W. S., Moore, L. W. 1986. Gene transfer into loblolly pine by *Agrobacterium tumefaciens*. *Bio-Technology* 4:647–49

43. Shah, D. M., Horsch, R. B., Klee, H. J., Kishore, G. M., Winter, J. A., et al. 1986. Engineering herbicide tolerance in transgenic plants. *Science* 233:478–81

44. Simpson, R. B., Spielmann, A., Margossian, L., McKnight, T. D. 1986. A disarmed binary vector from *Agrobacterium tumefaciens* functions in *Agrobacterium rhizogenes*. *Plant Mol. Biol.* 6:403–15

45. Stachel, S. E., Messens, E., Van Montagu, M., Zambryski, P. 1985. Identification of the signal molecules produced by wounded plant cells that activate T-DNA transfer in *Agrobacterium tumefaciens*. *Nature* 318:624–29

46. Stachel, S. E., and Nester, E. W. 1986. The genetic and transcriptional organization of the vir region of the A6 Ti plasmid of *Agrobacterium tumefaciens*. *EMBO J.* 5:1445–54

47. Stachel, S. E., Timmerman, B., Zambryski, P. 1986. Generation of single-stranded T-DNA molecules during the initial stages of T-DNA transfer from *Agrobacterium tumefaciens* to plant cells. *Nature* 322:706–12

48. Stachel, S. E., Zambryski, P. 1986. *VirA* and *virG* control the plant-induced

activation of the T-DNA transfer process of *Agrobacterium tumefaciens*. *Cell* 46:325–33

49. Teeri, T., Herrera-Estrella, L., Depicker, A., Van Montagu, M., Palva, E. 1986. Identification of plant promoters in situ by T-DNA-mediated transcriptional fusions to the *npt-II* gene. *EMBO J.* 5:1755–60

50. Thomashow, L. S., Reeves, S., Thomashow, M. F. 1984. Crown gall oncogenesis: evidence that a T-DNA gene from the *Agrobacterium* Ti plasmid pTiA6 encodes an enzyme that catalyzes synthesis of indoleactic acid. *Proc. Natl. Acad. Sci. USA* 81:5071–75

51. Thomashow, M. F., Hugly, S., Buchholz, W., Thomashow, L. S. 1986. Molecular basis for the auxin independent phenotype of crown gall tumor tissues. *Science* 231:616–18

52. van den Elzen, P., Lee, K. Y., Townsend, J., Bedbrook, J. 1985. Simple binary vectors for DNA transfer to plant cells. *Plant Mol. Biol.* 5:149–54

53. van den Elzen, P., Townsend, J., Lee, K. Y., Bedbrook, J. 1985. A chimeric hygromycin resistance gene as a selectable marker in plant cells. *Plant Mol. Biol.* 5:299–302

54. Velten, J., Schell, J. 1985. Selection-expression plasmid vectors for use in genetic transformation of higher plants. *Nucleic Acids Res.* 13:6981–98

55. Waldron, C., Murphy, E. B., Roberts, J. L., Gustafson, G. D., Armour, S. L., Malcolm, S. K. 1985. Resistance to hygromycin-B. *Plant Mol. Biol.* 5:103–8

56. Wallroth, M., Gerats, A. G. M., Rogers, S. G., Fraley, R. T., Horsch, R. B.

1986. Chromosomal localization of foreign genes in *Petunia hybrida*. *Mol. Gen. Genet.* 202:6–15

57. Wang, K., Herrera-Estrella, L., Van Montagu, M., Zambryski, P. 1984. Right 25-bp terminus of the nopaline T-DNA is essential for and determines direction of DNA transfer from *Agrobacterium* to the plant genome. *Cell* 38:455–62

58. Wullems, G., Molendijk, L., Ooms, G., Schilperoort, R. A. 1981. Differential expression of crown gall tumor markers in transformants obtained after in vitro *Agrobacterium tumefaciens*–induced transformation of cell wall regenerating protoplasts derived from *Nicotiana tabacum*. *Proc. Natl. Acad. Sci. USA* 78:4344–48

59. Yadav, N. S., Vanderleyden, J., Bennett, D. R., Barnes, W. M., and Chilton, M.-D. 1982. Short direct repeats flank the T-DNA on a nopaline Ti plasmid. *Proc. Natl. Acad. Sci. USA* 79:6322–26

60. Yanofsky, M., Porter, S., Young, C., Albright, L., Gordon, M., Nester, E. 1986. The *virD* operon of *Agrobacterium tumefaciens* encodes a site-specific endonuclease. *Cell* 47:471–77

61. Zambryski, P., Depicker, A., Kruger, D., Goodman, H. 1982. Tumor induction by *Agrobacterium tumefaciens*: analysis of the boundaries of T-DNA. *J. Mol. Appl. Genet.* 1:361–70

62. Zambryski, P., Joos, H., Genetello, C., Leemans, J., Van Montagu, M., Schell, J. 1983. Ti plasmid vector for the introduction of DNA into plant cells without alteration of their normal regeneration capacity. *EMBO J.* 2:2143–50

AUTHOR INDEX

A

Abe, H., 428
Abeles, F. B., 162
Abilov, Z. K., 330
Abramov, B., 324
Abramov, S., 324
Abramova, V. M., 324, 329
Adams, D., 444
Adams, T. L., 245
Adato, I., 155, 159
Adeli, K., 237
Affolter, H. U., 75
Ahlquist, P., 295, 300
Akazawa, T., 77, 213
Akerlund, H. E., 357
Akiyoshi, D. E., 469
Albernaz, J., 127, 134
Albersheim, P., 157, 171
Alberte, R. S., 17
Alberti, M., 27, 30
Albone, K. S., 433, 436
Albright, L., 468
Alden, R. A., 207
Alekperov, U. K., 330
Alexander, D., 59, 62, 274, 275, 277, 280
Alexander, D. G., 162, 164
Alhadeff, M., 18, 19, 25
Ali, A., 453
Ali, Z. M., 162, 163, 170
Alibert, G., 80
Aliyev, A. A., 330
Allen, J. F., 78, 86
Allison, R., 297, 298
ALONI, R., 179-204; 180, 182, 183, 185-97
Aloni, Y., 265, 266
Alpi, A., 423, 427, 435
Alt, J., 356, 357
Altosaar, I., 237
Amasino, R. M., 223, 469
Ambard-Bretteville, F., 79
Amberg, S., 319, 335
Amderton, B. H., 130, 132
Amesz, J., 23, 26, 32, 33, 35
Amino, S., 275
An, G., 246, 472
Ananthan, J., 239, 240
Ancora, G., 206
Andersen, J. K., 308
Anderson, B. J., 359
Anderson, J. D., 165, 168
Anderson, J. M., 15, 16, 18, 20, 24
Anderson, M. A., 63, 214
Anderson, O. D., 145, 148, 150

Andersson, B., 15, 16, 18, 20, 23, 354, 357
Andreae, M., 439
Angeles, G., 192
Anikeeva, I. D., 322, 324, 325, 328
Anstine, W., 207
Antipov, V. V., 321-23, 325, 328
Ao, L. D., 170
Apel, K., 208, 224-26, 231
Apelbaum, A., 169
Apostolakos, P., 121, 129, 132
Arad, S., 163, 164
Arfmann, H. A., 75
Argos, P., 235
Armour, S. L., 473
Arnott, S. M., 186
Arntzen, C. J., 13-15, 17, 18, 22, 23, 78, 79
Asaturyan, V. I., 324
Ashburner, M., 239
Ashley, C. C., 50, 99, 103, 104, 108, 109, 111
Aspart, L., 235
Astin, R., 145
Atabekov, J. G., 299, 301-3
Atal, C. K., 196
Atalla, R. H., 264
Atkey, P. T., 160
Atmar, V. J., 75
Audus, L. J., 335
Auerswald, E. A., 472
Auffret, A. D., 356, 357
Auslánder, W., 366
Avanzi, S., 185
Avron, M., 358, 363, 367, 369, 371, 379
Awad, M., 164, 170
Azzi, A., 356, 357
Azzone, G. D., 348

B

Baas, P., 186, 188
Babczinski, P., 266
Baccarini-Melandri, A., 348
Bach, T. J., 455
Bachofen, R., 27, 31
Bacic, A., 63, 265, 275, 277, 283, 284
Back, E., 474
Badley, R. A., 120
Bailey, I. W., 186, 190
Bailey, J. A., 243
Baines, C. R., 158
Bajer, A. S., 121, 125
Baker, B., 223

Baker, G. M., 350, 357, 360, 361, 363, 364, 368, 380, 381
Baker, J., 169
Baker, J. E., 165, 168
Bakhuizen, R., 121, 122, 133
Bakker, J. G. C., 32
Balayeva, A. V., 322
Balaźs, E., 244, 306
Baldwin, R. L., 349
Baltimore, B., 15, 16
Bamberger, E. S., 379
Banasik, O. J., 150
Bancroft, J. B., 294
Bangerth, F., 49, 158
Barak, L. S., 127
Barber, J., 22, 23, 78
Barber, R. F., 165
Barbier, M., 324
Barendse, G. W. M., 423, 424, 446, 449
Barker, R. F., 235, 244, 300
Barlow, P. W., 335
Barmicheva, E. M., 319, 327, 329, 332
Barnason, A., 471-73, 475, 477
Barnes, W. M., 468, 477
Barr, C. E., 101, 102, 106
Barr, J. K., 101, 102, 106
Barr, R., 52, 54
Barratt, D. H. P., 449
Barry, G. F., 469, 481, 482
Bart, K. M., 51
Barta, A., 228
Bartels, D., 144-48, 150
Bartholomew, D., 97
Bartley, I. M., 165, 169, 170
Barton, K. A., 469
Bassett, M. J., 163
Basyrova, L. V., 327
Bathgate, B., 162
Batschauer, A., 224-26, 231, 234
Batt, T., 159
Baulcombe, D. C., 213, 238, 300, 309
Baumann, G., 240, 241
Baumeister, W., 31
Baumlein, H., 235
Bawden, F. C., 291
Bayer, A., 51
Bayley, H., 356
Bayonove, J., 325
Beach, L. R., 238
Beachy, R. N., 236, 237, 297, 304, 308
Beale, M. H., 420, 426, 428, 436, 437, 451, 452

Beale, S. Y., 210
Beall, F. D., 450
BEARD, W. A., 347-89; 367-
    73, 377, 378, 381
Bearder, J. R., 420, 426, 428,
    430, 437, 438, 455
Beattie, D. S., 382
Beavo, J. A., 73
Beck, E., 472
Becker, D. W., 54
Becker, M., 28, 29
Bedbrook, J., 226, 232, 234,
    472, 473
Begusch, H., 27, 30
Behrens, H. M., 319, 329-32,
    334-36
Beilby, M. J., 101, 102, 110
Belanger, K. M., 351, 361, 380
Bell, A. A., 243
Bell, J. N., 243
Belton, P., 102
Belyavskaya, N. A., 320, 321,
    332, 334
Ben-Aire, R., 165, 170
Benayoun, J., 191
Bendall, R., 12
Bendoraityte, D. P., 319-22,
    336
Benevolenskii, V. N., 324
Bengis, C., 17
Bennett, A. B., 161, 162, 164
Bennett, C. W., 302, 303
Bennett, D. C., 224
Bennett, D. R., 468, 477
Bennett, J., 22, 78, 86, 225,
    226
Bennett, M. D., 146
Bennett, R. D., 454
Ben-Sasson, R., 183
Benton, E. V., 324
Benveniste, P., 453
Benvenuto, E., 206
Benziman, M., 264, 265, 283
Berestovsky, G. N., 100, 102
Berg, P., 247
Bergmann, H., 242
Bergstrom, G. C., 306
Berkaloff, A., 474
Berlin, J., 209
Bernadin, J. E., 143
Berridge, M. J., 85, 88
Berry, J. O., 225
Berry-Lowe, S. L., 224, 232
Berthet-Colominas, C., 295
Bertl, A., 105, 107
Berzborn, R. J., 357
Beslow, D. T., 191, 193
Betti, J. A., 35
Bevan, M., 223, 240, 241, 472
Bevan, M. W., 223, 245, 308,
    309
Bewley, J. D., 240
Beyer, T. A., 279

Bhatnagar, D., 350, 357, 360,
    361, 363, 364, 367, 380,
    381
Biale, J. B., 158-60, 166, 168
Bienz, M., 240
Biggs, M. S., 162
Bilyavs'ka, N. O., 334
Binns, A. N., 236
Birnberg, P. R., 428, 433, 447
Bisaro, D. M., 247, 480
Bishop, D. G., 165
Bishop, P., 157, 171
Bisseling, T., 242
Bisson, M. A., 97-99, 101
Bittman, R., 382
Black, C. M., 121, 129-31
Black, M. T., 22, 23, 78, 86
Blackman, F. F., 156, 166
Blackwell, J., 264
Blakeney, A. B., 274
Blankenship, R. E., 33-36
Blankenship, S. M., 158
Blaschek, W., 49, 50, 55-59,
    62, 64, 274, 275, 277,
    279, 280, 284
Blechschmidt, S., 437, 438
Bliedung, H., 225
Blinks, L. R., 96
Blithe, D. L., 86
Block, M. R., 356
Blowers, D. P., 87, 88
Blumenthal, D. K., 73
Blumwald, E., 52
Boccara, M., 300
Böcher, M., 75, 77
Boggs, S. S., 247
Bogorad, L., 354, 396, 401-3,
    410
Bohlool, B. B., 242
Bohnert, H. J., 404, 409
Bol, J. F., 243, 300, 301, 306
Boller, T., 96
Bollini, R., 235
Bollmann, J., 243, 306
Bolognesi, M. C., 35
Bolwell, G. P., 209, 243
Bonaventura, C., 24
Bonig, I., 274
Bonnemain, J. L., 197
Bonnerjea, J., 17
Bonzon, M., 54
Bopp, M., 208
Borisy, G. G., 128, 130-32
Borochov, A., 165
Bosch, L., 300
Bose, S., 23, 24
Boss, W. F., 211, 212
Bottin, H., 23
Bottomley, W., 392, 396, 399-
    405, 408-10
Bouchet, M., 207
Bouck, G. B., 7
Boudet, A., 82

BOUDET, A. M., 73-93; 74,
    81-83
Boulay, F., 356
Bouligand, Y., 272
Boulter, D., 235
Bouman, H., 239
Bourret, A., 262, 270
Boutry, M., 227, 230, 232,
    234, 235, 244
Bové, J. M., 298
Bowes, J. M., 365
Boyer, N., 213
Boyer, P. D., 348, 369, 371
Braat, P., 122, 126, 133
Bradeen, D. A., 358, 359
Bradford, K. J., 160
BRADY, C. J., 155-78; 157,
    158, 160-64, 166-68, 170,
    171
Braithwaite, A., 207, 208
Brand, J. J., 54
Brand, L., 469
Brandolin, G., 356
Branlard, G., 150
Branton, D., 381
Braun, R., 75
Bray, E., 237
Brecht, J. K., 158, 164
Brennan, T., 165
Brenneman, T., 191
Brenner, M. L., 428, 433, 447
Breton, J., 31
Brew, K., 35, 279
Brewin, N. J., 214
Briantais, J. M., 78, 79
Briggs, W. R., 86, 225, 226,
    233, 239
Brisco, M., 294, 298
Brisson, N., 242, 247
Broch-Due, M., 35
Brockmann, H., 36
Broglie, K. E., 243
Broglie, R., 224, 226-28, 231,
    232, 234
Broglie, R. M., 243
Brooks, J. L., 165, 169
Brower, D. L., 268, 270, 271
Brown, A. H. D., 149, 319,
    336, 396, 401, 414
Brown, C. L., 187, 192
Brown, D. L., 125, 129, 136
Brown, E. G., 48
Brown, J. S., 13, 18
Brown, R. M. Jr., 264, 266,
    268, 270, 272, 279
Brown, S. O., 104, 105
Bruce, W. B., 246
Bruck, D. K., 183, 191, 192
Bruemmer, J. H., 170
Bruening, G., 295, 297, 305
Bruinsma, J., 165, 166, 184
Brulfert, J., 84, 85
Brummer, B., 80

Brunisholz, R. A., 27, 29-31, 35
Brunt, J. V., 100
Bryant, D. A., 24, 25, 36
Bryant, J. A., 186
Buchala, A. J., 277
Buchanan, B. B., 78, 305
Buchenauer, H., 452, 453, 455
Buchholz, W. G., 239, 469
Bücker, H., 324
Buckhout, T. J., 52
Buescher, R. W., 163, 164, 171
Buffard, D., 213
Bufler, G., 158
Bujarski, J. J., 298
Bullivant, S., 80
Bulychew, A. A., 362
Bunt, A. H., 212, 213
Burdon-Sanderson, J., 101
Bureau, T. E., 207
Burg, E. A., 157
Burg, M., 325
Burg, S. P., 157
Burgess, J., 123, 124, 128, 294
Burgess, W. H., 281
Burgoon, A. C., 169
Burhop, L., 309
Burke, J. J., 15, 17
Burnell, J. N., 83, 84, 86
Bush, D. R., 52
Bush, D. S., 51
Bussell, J., 276-79, 284
Butenko, R. G., 327
Butterfield, B. G., 183
Buzash-Pollert, E., 404
Byerly, L., 99, 103
Bygrave, F., 88
Bykovskii, V. F., 321, 322, 325
Bylina, E. J., 27, 30, 32
Byon, K. Y., 99
Byrne, M., 472, 473, 475, 477
Byun, J. K., 452

C

Cabib, E., 265, 267
Caboche, M., 247
Cadman, C. H., 308
Cain, B. D., 381, 383
Cairns, E., 206
Cairns, W. L., 206, 209, 210, 215
Caldwell, J. H., 100
Callaghan, T., 275, 278
Callaham, D. A., 125, 127
Callahan, A. M., 162
Camerino, G., 305
Camirand, A., 275
Camm, E. L., 14
Campbell, A. K., 50, 58
Campillo, A. J., 32

Candresse, T., 298
Cantor, C. R., 394
Cantwell, M., 164
Carafoli, E., 53, 54
Carbonell, J., 49, 213
Carley, W. W., 127
Carlquist, S., 186, 188, 190
Carmi, A., 183
Carnes, M. G., 428, 475
Caron, M. G., 85
Carpita, N. C., 277, 279, 284
Carr, J. P., 225, 306
Carr, R. J., 301
Carratu, G., 80
Carroll, B. J., 242
Carroll, T. W., 303
Carter, G. H., 245
Casadio, R., 348
Case, J. F., 8
Cashmore, A. R., 230, 232
Castel, U., 437, 438
Castenholz, R. W., 35, 36
Castillo, F. J., 64
Castresana, C., 227, 229
Cataldo, A. D., 191
Catesson, A.-M., 208
Cavalié, G., 74
Cavender, P. L., 421
Ceccarelli, N., 423, 427, 435, 455
Cecchini, J.-P., 208, 210
Cech, T. R., 407
Cederstrand, C., 5
Cella, R., 207, 209
Chalmers, D. J., 183
Chalmers, R., 5
Chaloner, W. G., 188
Chalutz, E., 155
Chan, K. Y., 212, 213
Chance, B., 348
Chanzy, H., 262, 263, 270
Chao, H-Y., 278
Chapeville, F., 300
Chapman, D. K., 319, 336
Chapman, K. S. R., 75
Chappell, J., 64
Charles, C., 207
Chavez, F., 307
Cheesbrough, V., 147, 148
Chen, K.-C., 144, 148
Chen, N. J., 170
Chen, Y. M., 239, 240
Chen, Y.-R., 76, 86, 233
Chen, Z.-L., 236
Cheng, H. C., 86
Chenglee, L., 188
Cheo, P. C., 302
Chereskin, B. M., 25
Cherevchenko, T. M., 320, 326, 331
Chernikova, O. P., 322
Chernyad'yev, I. I., 331
Chernyaeva, I. I., 328

Cherry, J. H., 159, 163, 165
Chibbar, R. N., 207-10, 212-14
Child, R. D., 158
Chilton, M.-D., 223, 245, 468, 469, 474, 477
Chilton, W., 474
Chilton, W. S., 475
Chock, P. B., 73, 74
Choi, C.-H., 246
Choinski, J. S. Jr., 424
Chollar, S., 15
Chrispeels, M. J., 235, 237
Christeller, J. T., 77
Christoffersen, R. E., 160, 161, 164
CHUA, N.-H., 221-57; 224-32, 234, 235, 244, 245, 399
Chuchkin, V. G., 322
Chudinovskaya, G. A., 322, 324
Chui, C.-F., 232
Chung, C. H., 423
Ciferri, A. O., 305
Clark, A. E., 214
Clark, W. G., 226, 232
Clarke, A. E., 63, 274, 277, 283
Clarke, J., 211
Clayton, L., 121, 129-32
Clegg, J., 324, 325
CLEGG, M. T., 391-418; 396, 398, 401, 403, 407, 409, 412-14
Cleland, R. E., 238
Cleland, W. W., 352
Clement-Metral, J. D., 25
Coates, R. M., 421, 424
Cogdell, R. J., 25, 27, 29-32
Cohen, D., 191
Cohen, M. H., 197
Cohen, P., 73-75, 85, 103
Cohen, R., 265, 266
Cohen-Bazire, G., 25, 36
Colbert, J. T., 186, 224, 226
Cole, K. S., 101
Collmer, C. W., 304
Colvin, J. R., 260, 264, 266
Comai, L., 483
Comer, A. E., 193
Cook, J. A., 273
Coolbaugh, R. C., 422, 423, 448, 452-55
Coombe, B. G., 155
Coombs, J., 12
Cooper, G., 59, 62, 274, 275, 277, 280, 281
Cooper, J., 59, 62, 274, 275, 277, 280
Cooper, J. A., 73, 88
Cooper, J. B., 266
Cooper, K. M., 264, 265
Cooper, P., 240
Corfield, K. G., 142

490 AUTHOR INDEX

Corke, F. M. K., 129
Cornelissen, B. J. C., 243, 300, 301, 306
Coruzzi, G., 224, 226-28, 231, 232, 234
Cosgrove, D., 318
Costantino, P., 206
Coster, H. G. L., 99, 101-3
Coughlan, S. J., 79
Court, D., 399
Couso, R. O., 266
Cove, D. J., 129, 133, 134
Covey, S. N., 299, 300
Cowles, J., 319, 333
Cowles, J. R., 319, 333, 335
Cox, G. B., 362, 380
Cramer, C. L., 243
Cramer, G. R., 49, 50
Cramer, W. A., 363
Crane, F. L., 52, 54
Crawford, D. L., 192
Creber, G. T., 188
Cress, D. E., 475
Crettaz, M., 86
Critchley, C., 363
Crofts, A. R., 365
Cronshaw, J., 181
Crookes, P. R., 161, 164, 170
Crossland, L. D., 402
Crouch, M. L., 237
Croy, R. R. D., 235
Crozier, A., 420, 423, 427, 428, 436, 438
Cruden, D. L., 35
Culver, E. B., 452
Currier, H. B., 62
Curtis, H. J., 101
Curtis, S. E., 407
Czaninski, Y., 208
Czapek, F., 336
Czarnecka, E., 240, 241
Czernilofsky, A. P., 223

D

Dainty, J., 107
Daleo, G. R., 274
Dalessandro, G., 191
Dalziel, J., 452-54
Damm, D., 245
Dandekar, A. M., 238
Dankert, M. A., 266
Danko, J. S., 77
Dann, I. R., 183
Darbyshire, B., 211
Darvill, A. G., 157, 171
Das, R., 86, 88
Dasgupta, R., 300
Datema, R., 274
Datta, N., 76, 86, 87, 233
Datta, P. C., 185, 197
Daurat-Larroque, S. T., 35

Davenport, J. W., 369, 371, 374
Davies, E., 160, 161
Davies, J. N., 160
Davies, J. R., 86
Davies, J. W., 295, 297
Davies, P. J., 428, 433, 447
Davis, G. C., 428
Davis, J. W., 480
Davis, M. S., 36
Davis, R. W., 223, 239, 307
Davydov, B. I., 328
Dawkins, D. J., 32
Dawson, J. R. O., 294
Dawson, P. J., 119, 121, 127, 128, 132
De, B., 308
Deamer, D., 382
Dean, C., 226, 232, 234
De Block, M., 472
de Brabander, M., 121, 130
Debus, R. J., 27
de Framond, A., 469, 474
DeGroote, D. K., 191
Deichgraber, G., 124, 133
Deikman, J., 49
Deisenhofer, J., 14, 28, 29, 31, 354, 356
de Kouchkovsky, F., 368, 371, 372, 377, 381
de Kouchkovsky, Y., 367, 368, 371, 372, 374, 377, 381
Delauney, A. J., 235
Deleteuhr, R., 230
Delius, H., 409
DellaPenna, D., 162, 164
DELMER, D. P., 259-90; 59, 62, 260-62, 265-67, 272-81, 283
Delone, N. L., 321-23, 325, 328
De Lorenzo, R. J., 50
Delpoux, M., 324, 325
Delvare, S., 82
DeMaggio, A. E., 193, 194
De Mairan, M., 4
Demarty, M., 49
de Mey, J., 121, 130, 131
De Michelis, M. I., 51, 52, 167
Deng, X., 15, 23
Dengel, H. J., 453
den Hollander, W. T. F., 32
Denne, M. P., 186, 188, 191, 195
Deno, H., 407
Denton, R. M., 54
Depicker, A., 468, 477
Depka, B., 14
Deputy, J., 170
Derksen, J., 122, 133
Desbiez, M. O., 213
de Visser, R., 167

de Vos, G. J., 35
Devyatko, A. V., 319, 332, 334, 335
dezoeten, G. A., 294
Dhindsa, P. P., 165, 169
Dhindsa, R. J., 165, 169
Diaz, C. E., 426
Dickinson, H. G., 121
Dickinson, L. C, 35
Dickson, E., 308
Diener, T. O., 291
Dieter, P., 48, 51, 52, 54
Dietzgen, R. G., 304
Digby, J., 184, 187, 194, 196
Dijkstra, A., 423
Dillenschneider, M., 83
Dilley, D. R., 160
DILLEY, R. A., 347-89; 348, 350-54, 357-61, 363-65, 367-73, 375-78, 380-82
Diner, B. A., 14
Ditto, C. L., 15
Dixon, D. C., 306
Dixon, M. A., 182
Dixon, R. A., 209, 243
Dmitrieva, N. N., 327
Dockerill, B., 430, 431
Dodd, R. S., 195
Dodds, J. H., 185
Dolan, E., 13, 14, 16
Doll, H., 149
Dolmans, M., 279
Domier, L. L., 297
Doonan, J. H., 121, 127-134
Dopere, F., 103
Döring, G., 13
Döring, H.-J., 99
Dorne, A. M., 78
Dörnemann, D., 16
Dorokhov, Y. L., 299, 302
Dorr, I., 231
Dorssers, L., 298, 307
Dostal, H. C., 158, 159, 163, 165
Dougherty, W. G., 295, 297, 298
Douglas, T. J., 445, 451, 453
Down, G. H., 426
Downer, N. W., 349
Dozortseva, R. L., 322-24
Dreher, T. W., 298
Drews, G., 25, 30, 31, 34, 35
Driss-Ecole, D., 329, 331
Drouet, A., 167, 168
Dubertret, C., 79
Dubinin, N. P., 320, 321
Dubinina, L. G., 322
Dudler, R., 240
Dudley, R. K., 244, 247
Dulieu, H. C., 324
Dumas, E., 306
Dunahay, T. G., 15

Duncan, J. D., 421, 422
Dunford, H. B., 206, 213
Dunger, I., 27, 28, 354
Duniec, J., 121, 126, 128, 130-32
Dunn, J. J., 404
Dunn, L. J., 162
Dunsmuir, P., 16, 226, 232, 234
Duranton, J., 17
Dure, L. III, 235, 237
Durham, A. C. H., 295
Durley, R. C., 420, 430, 433, 434, 437
Durr, A., 222
DUTCHER, F. R., 317-45; 318, 320
Duysens, L. M. N., 32
Dvořák, J., 144, 148
Dyck, R., 159, 169
Dyer, T. A., 401, 403
Dynan, W. S., 225, 231

E

Eaks, I. L., 157, 160
Eaton, N., 264
Ebashi, S., 110
Eberle, M., 427
Ebert, P. R., 246
Ecker, J. R., 223, 307
Edelman, A. M., 73
Edelman, L., 239
Edelman, M., 14
Edwards, B. F., 319, 320, 326, 331
Edwards, C., 224, 226, 228, 231, 232, 234
Edwards, K. L., 63, 161, 162, 243, 318
Edwards, S. L., 207
Egorov, B. B., 321, 322
Eiberger, L. L., 278, 284
Eichholtz, D. A., 223, 469, 471-75
Elgersma, A., 441
Elliott, D. C., 88
Ellis, J. W. S., 150
Ellis, R. J., 224, 244
Ellmore, G. S., 182, 188
Emerson, R., 5
Emons, A. M. C., 122, 126, 133, 268
Emori, Y., 300, 301
Emshwiller, E., 433, 447
Engelhardt, H., 31
Englander, J. J., 349
Englander, S. W., 349
Epp, O., 28, 29, 31, 354, 356
Erdmann, H., 75, 77
Erion, J. L., 399
Ernst, D., 231, 451

Ernster, L., 348
Esau, K., 180, 181, 184, 192-94, 301
Eskin, M., 165, 169
Espelie, K. E., 211, 213, 214
Euteneuer, U., 131
Evans, D. E., 48
Evans, M. C. W., 14, 16, 17
Evans, M. L., 53, 318
Evans, N. A., 55, 274
Evans, R., 430
Everett, M., 224, 226
Evert, R. F., 181, 186, 193
Ewers, F. W., 182, 188, 193, 194

F

Facciotti, D., 227, 475, 483
Facius, R., 324
Fadeyeva, S. I., 330
Fahn, A., 181, 183, 186, 188, 194
Faithfull, E. M., 303
Fajer, J., 36
Falconer, M. M., 123, 124, 131, 132, 272
Falk, H., 119
Falke, L., 63
Farber, Yu. V., 322
Farkas, D. L., 31
Farley, M., 409
Farley, M. A., 409
Fassler, J., 227, 229
Faulks, A. J., 145
Favinger, J. L., 36
Fearnley, I. M., 401, 403
Feeney, R. E., 351, 352, 360
Fegel, A. C., 186, 190
Feher, G., 27
Feick, R. G., 35
Feigenbaum, P., 191, 196, 197
Feldman, M., 143-46, 149
Felix, G., 239
Felle, H., 105, 107
Fellows, B., 422
Felsenstein, J., 411, 412
Fenna, R. E., 35
Ferguson, I. B., 160, 170
Ferguson, S. J., 348, 358, 368
Fernandez, D. E., 212
Fernandez, J. M., 50
Fernow, K. H., 308
Fevre, M., 267
Fewson, C. A., 84, 85
Field, J. M., 145
Fieldes, M. A., 206, 211
Fillatti, J., 472
Fimmel, A. L., 362, 380
Fincher, G. B., 273
Findlay, G. P., 101, 103, 104
Finkelstein, R., 237

Finkelstein, R. R., 237
Fish, L. E., 305, 402
Fitch, W. M., 207
Flavell, R. B., 145, 146, 148, 244
Fleck, J., 222
Fleckenstein, A., 99
Fleming, E. N., 294
Fletcher, R. A., 453, 455
Flick, J. S., 469, 471-74
Flores, S., 367
Flowers, T. J., 96
Fluhr, R., 225-29, 231, 232, 234, 235, 245
Fogel, S., 324
Fomicheva, V. M., 319, 325, 326, 329-31, 333
Fondren, W. M., 319, 326, 327, 329-32, 335
Forde, B. G., 236, 242
Forde, J., 145, 150
Fork, D. C., 13, 23, 54
Forman, A., 36
Fortin, M. G., 242
Fosket, D. E., 135, 136, 185, 191, 207
Foster, V., 242
Foyer, Ch., 77, 78, 83
Fraenkel-Conrat, H., 307
Fraga, B. M., 426
Fraley, R. T., 223, 226-29, 232, 236, 245, 308, 469, 471-77, 480
Franceschi, V. R., 105, 214
Francis, D., 186
Franke, W. W., 119
Franklin, K. M., 297
Franssen, H. J., 297, 300
Franz, G., 59, 275, 280, 284
Fraser, R. S. S., 306
Freeling, M., 224, 324, 420, 445, 457
Freer, S. T., 207
French, A. D., 263
Frenkel, C., 159, 160, 163, 165, 169, 170
Friend, J., 155
Fritig, B., 306
Fritsch, C., 222
Fromm, M., 223
Fry, J. E., 223, 471, 472, 475
Fry, S. C., 214
Fuchs, Y., 155
Fügistaller, P., 30, 31
Fujii, S., 104, 105
Fujisawa, S., 437, 451
Fujita, M., 101
Fujiyoshi, Y., 262, 270, 271
Fukuda, H., 185, 214
Fukushima, T., 159
Fukuzawa, H., 392, 407
Fulcher, R. G., 277

Fuller, R. C., 33-36
Fulton, R. W., 308
Furuya, M., 224
Fushimi, T., 223

G

Gaard, G., 294
Gaba, V., 14
Gad, A. E., 182, 195
Gadal, P., 84, 85
Gaffey, C. T., 101
Gahan, P. B., 180
Galatis, B., 121, 129, 132
Galau, G. A., 235, 237
Galbraith, D. W., 280
Galili, G., 143-46, 149
Gallagher, T. F., 224, 244
Galliard, T., 165
Galston, A. W., 207, 332
Ganiyeva, R. A., 330
Gantt, E., 25
Gao, J., 455
Garab, G., 363
Garcia, R. C., 266
Garcia-Martinez, J. L., 433, 448
Garlid, K. D., 383
Garrett, R., 243
Garrison, S. A., 163
Gaskin, P., 426-28, 431, 433, 436-38, 442-45, 448
Gaspar, Th., 64, 206, 207, 213
Gasset, G., 324, 325
Gatehouse, J. A., 235
Gaubin, Y., 324, 325
Gaubin-Blanquet, Y., 325
Gautier, M.-F., 148
Gayler, K. R., 208
Gaynor, J. J., 243, 332
Gebhardt, C., 242
Gelvin, S. B., 236
Genetello, C., 223, 245, 470, 474
Gepstein, S., 234
Gera, A., 306
Gerard, J. S., 302
Gerats, A. G. M., 476
Gerola, P. D., 34, 35
Gersani, M., 191
Gerwitz, A., 306
Gest, H., 36
Geuze, H. J., 212
Geytenbeek, M., 88
Ghirardi, M. L., 14, 18, 21, 22, 24
Gianfagna, T. J., 426, 433, 436, 447
Gianinazzi, S., 306
Giannattasio, M., 80
Giaquinta, R. T., 359
Gibbs, A., 292
Gibson, F., 362, 380

Giddings, T. H., 268, 270-72
Giersch, C., 379
Giesbrecht, P., 31
Gilbert, S. F., 190, 239
Gilbert, W., 399
Gill, A. M., 184
Gillham, D. J., 171
Gilmour, S. J., 428, 430, 432, 433, 436, 438, 443, 445, 446
Gilroy, J., 235
Gilroy, S., 51
Gingrich, J. C., 409
Giovannoni, S. J., 36
Girard, V., 267
Girvin, M. E., 363
Glaser, L., 265
Glasziou, K. T., 208
GLAZER, A. N., 11-45; 16-19, 24, 25
Glembotsky, Ya. L., 320
Glick, R. E., 18, 21, 22, 24
Gloudemans, T., 242
Glover, J. F., 305
Gluzman, Y., 229
Goelet, P., 300, 308
Gold, L., 402
Gold, L. M., 402
Gold, M. H., 213
Goldbach, R. W., 295, 297, 300
Goldberg, A. L., 239, 240
Goldberg, R., 208
Goldberg, R. B., 236
Goldsbrough, P. B., 236
Goldsmith, M. H. M., 191
Golecki, J. R., 31, 34-36
Gollmer, I., 224
Gonzalez, A. G., 426
Gonzalez, P., 426
Good, N. E., 358, 368, 369, 371, 372
Goodenough, P. W., 162, 166
Goodman, H., 468, 477
Goodman, R. N., 291, 303, 304, 306
Goodwin, B. C., 197
Goosen-de Roo, L., 121, 122, 133
Gordon, L. K., 322
Gordon, M. P., 223, 468, 469, 472
Gordon, S. A., 335, 336
Goring, D. A. I., 262, 264, 270
Goris, J., 103
Gorkin, Yu., 320, 328
Gostimskii, S. A., 322
Gottschling, D. E., 407
Gould, J. M., 358, 359
Govers, F., 242
Graan, T., 367
Grabowski, P. J., 407

GRAEBE, J. E., 419-65; 420, 422-24, 426-28, 430-32, 434-41, 446, 448-55
Graham, J. R., 18, 19, 22
Granick, S., 210
Graul, E. H., 324
Gray, J. C., 356, 357, 401, 403
Gray, S. W., 319, 320, 326, 331
Graziana, A., 74, 82, 83
Grebanier, A. E., 359
Green, B. R., 14
GREEN, P. J., 221-57
Greene, F. C., 148
Greenidge, K. N. H., 182
Greenway, H., 96
Greenway, S. C., 235
Greenwood, A. D., 12
Greenwood, J. S., 235, 237
Greer, R. L., 181
Gregg, R. G., 247
Greppin, H., 54, 64, 206, 207, 212, 213
Gresshoff, P. M., 242
Grierson, D., 155, 158, 160-62, 164, 168, 170
Grif, V. G., 319, 327, 329, 332
Griffith, M., 25
Grigoriev, Yu. G., 322, 324, 325
Grimbly, P. E., 159
Grimes, H., 212
Grimsley, N., 480
Gripshover, B., 87, 211, 212
Gronenborn, B., 247
Gross, D., 420
Gross, K. C., 157
Grossmann, K., 453, 454
Grotha, R., 62, 63
Groves, S., 335
Gruenberg, H., 27, 28, 354
Gruissem, W., 22, 161, 400-2
Gruss, P., 231
Gubler, U., 75
Guglielmi, G., 36
Guichedi, L., 304
Guilfoyle, T. J., 75, 239
Guilley, H., 244, 302
Guitton, C., 78
Gulik-Krzywicki, T., 270
Gunning, B. E. S., 119-21, 127-30, 132, 271, 299
Gunsalus, I. C., 439
Günzler, V., 439
Gurley, W. B., 240, 241, 246
Gustafson, G. D., 473
Gutknecht, J., 99

H

Haass, D., 275, 284
Habili, N., 299

Hackspacher, G., 275
Haehnel, W., 13, 357
Haenni, A. L., 300
Hafemann, C., 426, 427
Hagen, G., 239
Hagiwara, S., 99, 103
Hahlbrock, K., 64, 243, 306
Hahne, G., 136
Haigler, C. H., 260, 262, 264, 265
Hain, R., 223, 240
Haines, T. H., 380
Halevy, A. H., 165
Halford, N. G., 145, 150
Halk, E., 309
Hall, M. A., 155
Hall, R., 453
Hall, T. C., 235, 298, 300
Hallgren, L., 237, 238
Hallick, R. B., 409
Halperin, W., 191, 193, 194
HALSTEAD, T. W., 317-45; 318, 320
Hamilton, B., 210
Hamilton, R. I., 307, 452, 453
Hamilton, W. D. O., 300
Hammerschmidt, R., 306
Hanagata, T., 239
Hanauske-Abel, H. M., 439
Hand, J. M., 232
Handa, A. K., 162, 163
Hangarter, R., 361
Hanhart, C. J., 441
Hannah, L. C., 163
Hanney, C. E. A., 184
Hansen, A., 237, 238
Hanson, J. B., 49, 50, 84
Hanson, J. R., 420, 426, 430, 431
Hanson, L. K., 36
Harada, H., 223, 262, 270, 271
Haraux, F., 367, 368, 371, 372, 374, 377, 381
Harberd, N. P., 144-48, 150
Hardham, A. R., 119, 121, 271
Hardison, L. K., 87
Hari, V., 305
Harman, J., 159
Harold, F. M., 100, 348
Harpster, M. H., 224
Harriman, R. W., 162
Harris, K. F., 293
Harris, K. M., 50
Harris, P. H., 274
Harris, P. J., 63, 277, 283
Harrison, B. D., 292, 293, 308, 309
Hartley, M. R., 226
Hartmann, C., 167, 168
Hartmuth, K., 228
Harvey, G. W., 18, 22
Haschke, R. H., 212, 213
Haselbeck, A., 266

Haselkorn, R., 407
Haseloff, J., 300
Hashimoto, H., 223
Hashimoto, K., 382
Hastings, D. S., 99
Hastings, J. W., 6
Hatch, L., 362, 380
Hatch, M. D., 83, 84, 86
Hatfield, R., 161
Haug, A., 52
Haughn, G., 473
Haupt, W., 225
Hauptmann, R. M., 240
Hauska, G., 35
Hawes, C., 126, 127
Haworth, P., 15, 18, 22, 23
Haxo, F. T., 5, 7
Hayama, T., 99, 101, 102, 104, 108, 109
Hayashi, T., 59, 62, 265, 272, 274-80, 284
Hayashida, H., 392, 407
Hayward, G., 101, 102
Hazlet, M. A., 245
Heald, J. K., 192
Hearst, J. E., 27, 30
Heath, I. B., 271
Heath, R. L., 363, 364
Heathcote, P., 16
Heber, U., 379
Hebert, J. J., 264
Hedden, P., 420, 422, 424, 426-28, 430, 431, 433, 436-39, 445, 450, 453, 455, 456
Hedgcoth, C., 148
Heidecker, G., 236
Heikkila, J. J., 240
Heil, D. R., 454, 455
Heim, S., 77
Heimsch, C., 185
Hein, M., 479, 481, 482
Heiniger, U., 274
Hejgaard, J., 237, 238
Hejnowicz, A., 188
Hejnowicz, Z., 184
Hellebust, J. A., 96
Hellmann, G. M., 297
Hellmiss, R., 223, 470
Hemmings, B. A., 103
Hemrika-Wagner, A. M., 108
Henderson, R., 356, 357
Hendrickson, W., 230
Hendriks, T., 211
Hendrix, J. E., 181
Hendry, D. A., 295
Henrissat, B., 263
Henry, R. J., 273-75, 277
Henry, Y., 270
Hepler, P. K., 47-51, 53, 58, 63, 86, 125, 127, 131, 170, 283, 318, 338
Herbert, M., 213

Hering, G. E., 128, 130-32
Hermodson, M. A., 352, 354, 357, 360, 367, 380
Hernalsteens, J.-P., 223
Hernandez, M. G., 426
Herner, R. C., 156
Herrera-Estrella, L., 223-27, 229-32, 234, 235, 468, 472, 477
Herrmann, R. G., 356, 357
Hershberger, W. L., 162
Hershey, H. P., 224, 226
Herth, W., 51, 58, 119, 124, 125, 268, 270, 276, 280, 283
Hess, B., 158
Hess, T., 196
Hetherington, A., 74, 80, 87, 88
Heupel, R. C., 445
Heyser, M., 427
Heyworth, A., 236
Hiatt, W. R., 483
Hiebert, E., 295
Hieta, K., 263
Higgins, T. J. V., 235, 237
Hildebrandt, E., 453
Hill, R., 12
Hill, R. L., 279
Hill, S. E., 95
Hillis, W. E., 192
Hilton, B. D., 350
Hinchee, M., 479, 481
Hinchman, R., 336
Hind, G., 79, 348, 358, 363, 364, 381
Hirai, A., 305
Hirano, S. S., 452, 453
Hiremath, S. T., 297
Hiriyanna, K. T., 307
Hiron, R. W., 159
Hironaka, C. M., 226, 232
Hirono, C., 101
Hirota, A., 16
Hirsch, P. R., 469
Hirth, L., 222, 295, 302, 306
Hisajima, S., 437
Hiyama, T., 16
Ho, T.-H. D., 240
Hoad, G. V., 158, 439, 451
Hobson, G. E., 155, 159, 160, 163, 170, 171
Hodges, M., 78
Hodgkin, A. L., 95
Hoekema, A., 469
Hoekman, D. A., 188
Hofbauer, R., 210
Hoffman, F., 136
Hoffman, L. M., 237
Hoffman, N. E., 155, 157-59, 163, 169, 192
Hoffman-Falk, H., 14
Hoffmann, N. J., 236

Hoffmann, N. L., 223, 247, 308, 472, 475, 479-83
Hofinger, M., 213
Hofstra, G., 455
Hogetsu, T., 126, 129, 268, 280
Hohn, B., 299, 480
Hohn, T., 223, 247, 299, 480
Hokin, L. E., 85
Holländer-Czytko, H., 308
Holmes, M. G., 224
Holmes-Siedle, A. G., 205
Holo, H., 35
Holt, L. M., 142-51
Holten, D., 29
Holtum, J. A. M., 54, 55
Homann, P. H., 350, 351, 353, 354, 364
Honda, K., 421
Honda, Y., 294
Hong, Y. Q., 350, 362, 366-68, 381
Hongu, A., 16
Hooft van Huijsduijnen, R. A. M., 306
Hooley, R., 420, 451, 452
Hooykaas, P. J. J., 223, 469, 473
Hooykaas-Van Slogteren, G. M. S., 223
Hope, A. B., 101, 103, 104
Hopp, H. E., 274
Hoppe, J., 356, 380
Hörmann, G., 108
Horn, R. G., 362
Hornberg, C., 238
Horneck, G., 324
Horner, R. D., 367-69, 378
HORSCH, R., 468-86; 471-73, 479, 481
Horsch, R. B., 468-86; 223, 227, 236, 245, 469, 471-78, 480, 483
Horst, R. K., 308
Horton, P., 22, 23, 78, 86, 359
Houck, D. F., 191, 193
Hough-Evans, B. R., 477
Hovenkamp, P. H., 133
Howe, C. J., 401, 403
Howell, S. H., 247
Hoyne, P. A., 55
Hruschka, W. R., 165, 168
Hsu, D., 161
Hsu, M.-Y., 227, 230, 235
Hu, C., 212, 213
Hu, T. S., 242
Huang, J.-K., 238
Huber, D. J., 169, 170
Huber, J., 451
Huber, R., 14, 28, 29, 31, 354, 356
Hubermann, M., 64
Huebner, F. R., 142, 150

Hughes, D. W., 237
Hughes, J. E., 121
Hughes, W. A., 51
Hughes, W. G., 146
Hugly, S., 239, 469
HULL, R., 291-315; 292-95, 297-301
Hulme, J. S., 121, 127, 128, 132
Hung, S. P. N., 368, 371, 372, 377, 381
Hunter, C. N., 32
Hunter, T., 73, 88, 301
Hurt, E. C., 35
Hutchinson, J., 146
Hutchison, M., 426, 437
Huxley, A. F., 96
Huynh, T. V., 239
Hyer, R. C., 32
Hyldig-Nielsen, J. J., 243

I

Ida, S., 207
Ie, T. S., 297, 300
Iino, M., 225, 233
Ikegami, I., 14, 16
Ikegami, M., 307
Iki, K., 160
Il'in, Ye. A., 335
Imai, T., 426, 453
Imhoff, J. F., 26
Imlay, K. R. C., 161
Indig, E. F., 190
Ingram, T. J., 433, 442-45
Innocenti, A. M., 185
Inoue, I., 104
Inoue, M., 209
Inouhe, M., 276
Ismail, S., 30, 32
Israelachvila, J. N., 362
Itamura, H., 159
Itoh, T., 268, 270
Itoo, S., 170
Iyer, V. N., 469
Izawa, S., 348, 358, 363, 364
Izumi, K., 452, 453

J

Jabben, M., 224
Jackson, E. A., 142-47, 149-51
Jackson, M. B., 335
Jackson, R. L., 356
Jackson, W. T., 131
Jacob, J. S., 31
Jacob, S. R., 274-76, 279
Jacobs, M., 190, 239, 247
Jacobs, W. P., 180, 190-93, 196
Jacobsen, J. V., 207, 238
Jaffe, M. J., 64, 275

Jagendorf, A. T., 305, 348, 358, 359, 367, 381
Jahns, G., 319, 333
Janssen, H., 300, 301
Jansson, C., 357
Jarret, R. L., 163
Jarvis, N. P., 300, 309
Jay, F., 27, 29-31
Jay, F. A., 30, 31
Jeblick, W., 49, 50, 52, 55-58, 60-62, 64, 66, 274, 277-79
Jefferies, R. L., 99, 100
Jeffs, R. A., 193
Jeje, A. A., 182
Jemiolo, D. K., 281
Jenkins, G. I., 134, 224, 226
Jennings, A. C., 148
Jennings, R. C., 20
Jenns, A. E., 306
Jensen, E. O., 243
Jensen, J. S., 243
Jerebzoff, S., 89
Jerebzoff-Quintin, S., 89
Jerie, P. H., 155, 183
Jochimsen, B. U., 242
Jofuku, K. D., 236
Johanningmeier, U., 361
Johnson, J., 64
Johnson, J. D., 350, 351, 353
Johnson, M. C., 306
Johnson, S. P., 337
Johnston, R. E., 297, 298
Joliot, A., 363
Joliot, P., 363
Jonard, G., 244, 302
Jones, A., 451
Jones, K. A., 231
Jones, M. G., 446, 447
Jones, R. G. W., 49
Jones, R. L., 49, 51, 213, 238
Jones, S. B., 411
Joos, H., 223, 245, 470, 474
Jorgensen, K. G., 237, 238
Jorgensen, R. A., 224, 226
Joshi, S., 300
Jukes, T. H., 394
Juliano, R. L., 354
Jung, J., 452-54
Junge, W., 350, 351, 358, 362, 366-68, 376, 377, 381
Juniper, B. E., 335

K

Kader, A. A., 156, 158, 164
Kadir, G. O., 191, 197
Kado, C. I., 302
Kaesberg, P., 295, 297, 300, 305
Kaestner, K. H., 111
Kafatos, F. C., 235
Kahane, I., 356
Kahl, G., 76, 77, 86

Kahn, C. R., 86
Kaiser, W., 29
Kalra, S. K., 165
Kami-ike, U., 103
Kamikubo, T., 224, 226
Kamiya, N., 96, 97, 104, 108
Kamiya, Y., 428, 430-32, 434, 435, 453
Kanavets, O. L., 321
Kaper, J. M., 299, 309
Karabin. G. D., 409
Karlin-Neumann, G. A., 356, 357
Karn, J., 191
Karssen, C. M., 424, 446, 449
Kasarda, D. D., 143, 146, 148, 150
Kasuga, M., 86
Kataoka, H., 98
Katoh, S., 14, 16, 24, 25
Katoka, M., 31
Katsaros, C., 121, 129, 132
Katz, J. J., 28
Katzouzian-Safadi, M., 295
Kauffman, J. M., 232
Kaufman, L. S., 225, 226, 244
Kaulen, H., 230, 243
Kausch, A. P., 227, 229, 232
KAUSS, H., 47-72; 49, 50, 52, 55-58, 60-69, 74, 98, 260, 274, 276-79, 283, 284
Kawai, H., 265, 267
Kawamura, G., 107
Kay, E., 207
Kay, S., 230, 234, 235
Kay, S. A., 227, 230, 235
Kayes, L., 474
Ke, B., 13, 14, 16
Keeling, J., 295
Kehres, L. A., 33
Keifer, D. W., 105
Keith, B., 228
Keith, C. H., 51
Kell, D. B., 348
Keller, F., 273
Keming, C., 188
Kemp, J. D., 235
Kepczynski, J., 424, 446, 449
Key, J. L., 75, 76, 239-41
Keyser, H. H., 242
Khan, A. W., 266
Khan, K., 150
Khan, Z. A., 307
Khudairi, A. K., 159
Khvostova, V. V., 322, 323
Kiberstis, P. A., 298
Kiho, Y., 300
Kikuyama, M., 96, 97, 100-4, 109
Kim, K. S., 301
Kimura, M., 394, 398
King, J. M., 294

King, R. W., 420, 434, 437, 438, 446-48, 452
Kingston, E. E., 48
Király, Z., 291, 303, 304, 306
Kirino, Y., 105
Kirkwood, P. S., 433
Kirmaier, C., 29
Kirshner, M., 122
Kirst, G. O., 98, 99, 101
Kishimoto, U., 98, 101, 103, 108, 110
Kishore, G. M., 483
Kislev, N., 165, 170
Kitajima, E. W., 301
Kitamura, T., 159
Kitasato, H., 104, 105, 108
Kiyosawa, K., 98
Kjellbom, P., 49, 52
Klambt, D., 239
Kleczkowski, K., 76
KLEE, H., 468-86; 469, 473, 479, 481
Klee, H. J., 223, 245, 469, 471, 472, 478, 483
Klein, A. O., 224
Klein, A. S., 265
Klein, I., 160
Kleinschmidt, A., 239
Kleinsmith, L. J., 75
Klessig, D. F., 225, 306
Klevanik, A. V., 13
Klimov, V. V., 13
Kloppstech, K., 208, 224, 225
Knaff, D. B., 35
Knee, M., 159, 169, 170
Knegt, E., 184
Knight, C. A., 302
Knowles, J. R., 351
Köhle, D., 66, 67
Köhle, H., 49, 50, 55, 62-64
Körner, L. E., 49
Kobatake, Y., 104
Kobayashi, M., 16, 437, 438, 446
Koehler, H., 275, 284
Kogan, I. G., 322, 325
Kohchi, T., 392, 407
Kohle, H., 274, 277-79
Kohorn, B. D., 356, 357
Kokkinakis, D. M., 169
Kolacz, K., 305
Kolattukudy, P. E., 211, 213, 214
Koller, B., 409
Kolpak, F. J., 264
Komamine, A., 185, 214, 275
Koncz, C., 472
Kondorosi, A., 242
Kondorosi, E., 242
Kongsamut, S., 50
Konig, M., 231
Koon, E. C., 326, 330

Koornneef, M., 424, 441, 446, 449
Kopeliovitch, B. S., 377
Kopeliovitch, E., 164, 170, 171
Kopus, M. M., 150
Koralewski, M. A., 247
Kordyum, E. L., 319-21, 325-34
Kosakovs'kaya, I. V., 331
Koshiba, T., 206, 213
Koshioka, M., 437, 445, 450, 451
Koshioka, M. N., 451
Kossatz, V. C., 209-11
Kossiakoff, A. A., 349, 350
Kostina, L. N., 322, 324, 325, 328
Koths, T., 452-54
Kovalev, E. E., 324, 325
Koziel, M. G., 245
Kozlov, A., 348
Kozlowski, T. T., 187, 192
Kozukue, N., 159
Kraayenhof, R., 108, 363
Krahn, K., 309
Kramer, A., 63
Kramer, H. J. M., 23, 32
Kramer, P. J., 187
Krause, G. H., 78, 79
Krebbers, E. T., 396, 402, 403
Krebs, E. G., 73
Kreimer, G., 54, 55
Kreis, M., 236
Kretsinger, R. H., 281
Kreuz, K., 231
Kreuzaler, F., 230, 243
Krick, T. P., 428, 433
Krikorian, A. D., 320-22, 327
Kruger, D., 468, 477
Kruger, I., 213
Kruger, K., 407
Krulwich, T. A., 383
Kuang, T. Y., 15, 22
Kubiak, J., 130
Kuc, J., 306
Kucherlapati, R. S., 247
Kuck, U., 402
Kudielka, R. A., 327
Kuehn, G. D., 75
Kuga, S., 263
Kühlbrandt, W., 15, 31
KUHLEMEIER, C., 221-57; 225, 227-31, 234, 235
Kulow, C., 277
Kundu, B. C., 185
Kuo, J., 193
Kura-Hotta, M., 25
Kuramitsu, S., 407
Kurata, K., 160
Kurella, G. A., 362
Kuroda, K., 96, 97, 104
Kurogochi, S., 444-46
Kurtz-Fritsch, C., 295

Kusaba, S., 162
Kutzner, B., 452-54
Kuz'mina, V. P., 378
Kyle, D. J., 15, 22-24, 79

L

Labavitch, J., 158
Lachaud, S., 184, 197
Lafiandra, D., 148, 150
Lagarias, G. C., 86
Lagoutte, B., 17
Lai, V., 185
Laing, W. A., 77
Lam, E., 15, 18
Lamb, C. J., 198, 224, 243
Lambeir, A.-M., 213
Lambers, H., 167
Lambert, A.-M., 121, 130, 131
Lambillotte, M., 30, 31
Lambright, D. G., 279, 280
LaMotte, C. E., 191, 193, 196
Lamppa, G., 228, 234
Lancelle, S. A., 125, 127
Landau-Schachar, B., 335
Lange, T., 440, 441
Langridge, U., 404, 409
Larkins, A. P., 214
Larkins, B. A., 236
Laroche, M., 235
Larrinua, I. M., 396, 403
Larsen, K., 242
Larson, P. R., 183, 187, 191, 192
Larsson, C., 49, 52
Larsson, U. K., 15, 23
Laszlo, J. A., 350-54, 357, 360, 380
Laties, G. G., 160, 161, 164, 166
Latzko, E., 54, 55
Laub, O., 247
Läuchli, A., 49, 50
Lauquin, G. J. M., 356
Laurinavichius, R. S., 319-22, 326-31, 333, 334, 336
Lauritis, J. A., 301
Law, C. N., 142-48, 150, 151
Lawrence, D. K., 452-54
Lawrence, G. J., 143, 144, 146, 148, 150
Lawton, M. A., 224, 243
Lazaro, R., 262, 270
Lazarus, C. M., 213, 238
Lé, T., 208
Lee, E., 161, 162, 164, 168
Lee, K. Y., 472, 473
Lee, S., 227, 475
Lee, T. K., 279, 280
Leemans, J., 223, 245, 470, 474
Leermakers, A. M., 362
Lefkowitz, R. J., 85

LefortTran, M., 270
Legge, R. L., 169
Legocka, J., 239
Legrand, M., 306
Lehmann, H., 63
Leigh, R. A., 111
Lelpi, L., 266
Le Maréchal, P., 84, 85
LeMay, R., 319, 333, 335
Lembrich, H., 453
Lenton, J. R., 438
Leonard, D. A., 300
Leonardo, L., 193
Leopold, A. C., 165, 275
Lerner, H. R., 58
Lesham, Y. Y., 169
Lett, M. C., 222
Leutwiler, L. S., 477
Levin, W., 214
Levitan, I. B., 85, 86, 103, 111
Levy, S., 272
Lew, J. Y., 207, 211
Lewer, P., 431, 455
Lewis, E. D., 236
Lewis, N., 426, 437
Ley, A. C., 15
Li, B. C. Y., 358
Li, J., 16
Li, W.-H., 394, 411
Liang, Y. F., 170
Libbenga, A. K., 133
Libbenga, K. R., 122, 133, 185, 239
Lichtenthaler, H. K., 455
Lieberman, M., 169
Liebisch, H. W., 420, 451
Liedgens, W., 207
Lill, H., 367, 376, 377
Lim, S-S., 128, 130-32
Lin, C.-Y., 239, 240
Lin, F-C., 266, 276-79, 284
Lin, Z. F., 79
Lindau, M., 50
Lindsay, J. G., 30, 32
Link, G., 400, 404, 409
Linstead, P., 123, 124, 128
Lipinski, A., 36
Liskova, D., 191
List, A., 185
Litts, J. C., 148
Liu, Y., 157, 163
Lloyd, A., 223, 472, 473, 475, 477
LLOYD, C. W., 119-39; 119-34, 136, 271, 272
Lobarzewski, J., 209, 211-13
Lobler, M., 239
Lobo, M., 163
Lockau, W., 16, 52, 359
Loebenstein, G., 306
Loesch-Fries, L. S., 298, 300, 309
LoGullo, M. A., 181

Lomax, T. L., 239
Lomonossoff, G. P., 297
Longstaff, M., 300, 301
Lonsdale, D. M., 307
Lorenzi, R., 423, 427, 435, 455
Lorz, H., 223, 224
Lottspeich, F., 27, 28, 354
Loukari, A., 132
Lowe, S. B., 120, 136
Lu, Z.-S., 159
Lucas, W. J., 59, 62, 105-7, 277
Lucero, H. A., 79
Luchsinger, A. E., 411
Luckow, V., 295
Ludwig, F. R., 30, 31
Ludwig, G., 472
Luh, B. S., 162
Lühring, H., 53, 102
Luib, M., 452-54
Lukin, A. A., 322
Lundborg, T., 52
Lundell, D. J., 16, 17, 24
Lunevsky, V. Z., 100, 102
Lunt, O. R., 49
Luo, C.-C., 394, 411
Lurie, S., 165
Lusk, W. J., 426, 436
Lüssen, K., 453
Lycett, G. W., 235
Lyon, C. J., 319
Lyubchenko, V. Yu., 334, 335

M

MacAlpine, G., 160, 162, 164, 170
MacDonald, G., 120
Mache, R., 78
Macherel, D., 77
MacIntosh, L., 396, 401, 410
Macklon, A. E. S., 51
Maclachlan, G., 260, 265, 272, 275, 284
Maclachlan, G. A., 278
MacMillan, J., 238, 420, 422, 424, 426-28, 430, 431, 433, 435-39, 441-43, 445, 448, 450-52, 455, 456
MacRobbie, E. A. C., 110
Mäder, M., 208
Madhavan, S., 234
Maenhaut, R., 224, 226, 227, 232
Magnuson, J. A., 165
Magyarosy, A. C., 305
Maier, V. P., 454
Maines, S. L., 214
Maity, C., 197
Majmudar, G., 159
Maki, S. L., 428, 433, 447
Maksimova, Ye. N., 324, 325
Malawer, C. L., 185

Malcolm, S. K., 473
Maldonado, B. A., 208, 209, 213
Malkin, R., 15-18, 35
Malmström, B. G., 439
Malpica, J.-M., 145, 150
Maltby, D., 277
Malyshenko, S. I., 301
Malysheva, G. I., 334, 335
Mandak, V., 223
Manodori, A., 18, 19, 21, 22, 24, 25
Mansfield, R. W., 16
Mansson, P.-E., 161
Mao, D., 356
Marcelja, S., 362
Marchesi, V. T., 356
Marchyukaytis, A., 330, 333, 334
Marcker, K. A., 243
Marcy, B., 2, 3
Mardaus, M. C., 428
Marder, J. B., 14
Margoliash, E., 207
Margossian, L., 474
Marimuthu, K. M., 321, 324
Marishima, I., 212
Markert, C. L., 206, 213
Markwell, J. P., 77
Marmé, D., 48, 51, 52, 54, 80, 82, 85, 88
Marriott, J., 155
Marsden, M. P. F., 265, 272, 277
Martelli, G. P., 304
Martienssen, R. A., 213, 238
Marton, L., 475
Marx, G. A., 441, 447
Mashinskiy, A. L., 319-21, 325-28, 330, 333, 334
Mason, S. E., 308
Masuda, Y., 276
Mathis, P., 23
Matile, P., 169
Matlick, H. A., 8
Matoo, A. K., 14
Matsuda, K., 275
Matsui, C., 294
Matsumoto, H., 53
Matsuo, A., 426
Matthews, B. W., 35
Matthews, R. E. F., 292, 295, 305
Matzenauer, S., 88
Matzke, A. J. M., 228, 236, 472
Matzke, M. A., 228, 236, 472
Maule, A. J., 293, 295, 298, 299
Maunders, M., 161
Maunders, M. J., 161, 162
Mauro, V. P., 242
Mauzerall, D. C., 15

Maxfield, F. R., 51
Mayak, S., 169
Mayer, M. L., 100
Mayfield, S., 18
Mayfield, S. P., 224, 231
Mayko, T. K., 320, 326
Mayo, M. A., 309
Mazur, B. J., 232
Mazza, G., 207
McCarthy, K. J., 278
McCarty, R. E., 358, 369, 371, 374
McCauley, S. W., 14, 18, 21, 22, 24
McClelen, C. E., 319, 327, 329, 331, 332, 335
McCombs, P. J. A., 86
McCormack, J. G., 54
McCormick, S., 471, 472, 475
McCown, B., 472
McCully, M. E., 277
McDonald, C. E., 150
McDonnell, R., 475
McGlasson, W. B., 155, 157-60, 162-64, 170, 171
McIntosh, J. R., 131
McIntosh, L., 396, 403
McKee, J. M. T., 194
McKeon, T., 164
McKersie, B. D., 165
McKnight, T. D., 232, 474
McLaughlin, S., 377
McMurchie, E. J., 157
McNeil, D. L., 242
McNeil, M., 157, 171
Meagher, R. B., 224, 232
Means, G. E., 351, 352, 360
Mecham, D. K., 146
Mecklenberg, K. L., 407
Meekes, H., 122, 126, 133
Mehlhorn, R. J., 239
Meicenheimer, R. D., 191
Meier, H., 277, 280
Meins, F. Jr., 239
Meishi, T., 297
Melandri, B. A., 348
Melanson, D., 75
Meldrum, S. K., 162-64, 170, 171
MELIS, A., 11-45; 13, 14-22, 24
Melkonian, M., 54, 55
Mellon, J. E., 423
Melroy, D., 49
Mennes, A. M., 239
Menzel, D., 127, 134
Merkys, A. J., 319-22, 326-29, 331, 336
Merlevede, W., 103
Merlo, D. J., 235, 309
Meshi, T., 300, 301
Messens, E., 469
Messing, J., 236

Metakovsky, E. V., 150
Metzger, J. D., 436, 447
Metzler, M., 148
Meyer, G., 225
Meyer, R., 31
Meyer, Y., 208, 235, 280
Meyerowitz, E. M., 477
Miassod, R., 208, 210
Michaeli, D., 265, 266
Michel, H., 14, 27-29, 31, 354, 356
Micucci, V., 80, 86
Middendorf, D., 28
Miflin, B. J., 141, 142, 145, 242
Miki, K., 28, 29, 31, 213, 354, 356
Miller, A. R., 192
Miller, C., 130, 132
Miller, J., 245
Miller, K. R., 31, 356
Miller, R. J., 50
Miller, S. M., 149
Miller, T. E., 148, 149
Miller, W. A., 298
Millner, P., 350
Millner, P. A., 368
Milov, M., 335
MIMURA, T., 95-117; 102, 103, 105, 106
Minocha, S. C., 191, 193, 194
Miquel, J., 327
Mir, A., 325
Misler, S., 63
Mita, T., 125, 452
Mitchell, B., 265
Mitchell, P., 348, 358
Mitchison, T. J., 122, 273
Mitsui, T., 31, 101
Miyachi, S., 54, 55
Miyata, S., 213
Mizrahi, Y., 159, 163-65
Mizuguchi, K., 101
Mizuno, S., 158, 159
Mizuta, S., 268
Modi, V. V., 159
Moeremans, M., 121
Moerman, M., 242
Mohamed, N. A., 305
Mohanty, P., 54
Mohnen, D., 239
Mohr, H., 226
Mole-Bajer, J., 121, 125, 126, 131
Molendijk, L., 475
Moll, B. A., 79
Mollenhauer, H. H., 212
Möller, I. M., 49
Monger, T. G., 32
Monnom, D., 279
Monroe, A., 87
Montezinos, D., 264, 268, 277, 279, 280

Moody, W. J., 103
Moolenaar, G., 473
Moore, A. L., 51-55
Moore, L. W., 475
Moore, R., 318, 319, 326, 327, 329-32, 335
Moore, T. C., 423, 424
Moreau, A., 367, 368
Moreau, F., 166
Morejohn, L. C., 135, 136, 207
Morelli, G., 226-28, 232, 234, 245
Morgan, J. M., 96
Morgens, P. H., 162
Morita, Y., 207
Moriyasu, Y., 107, 111
Morowitz, H. J., 352, 380
Morozova, E. M., 321-23, 328
Morré, D. J., 52, 80, 86, 87, 211, 212
Morré, J. T., 80, 86, 87
Morris, D. A., 190, 191, 197
Morris, G. E. L., 194
Morris, R., 148, 150
Morrison, N. A., 242
Morrow, D. L., 59, 62, 277
Morvan, C., 49
Moses, P., 226, 228, 232, 234
Moshrefi, M., 162
Mösinger, E., 168, 224-26, 231
Mosiniak, M., 272
Mosquera, L. A., 240
Motoyoshi, F., 294, 299
Mouches, C., 298
Moudrianakis, E. N., 367-69, 378
Mousdale, D. A., 159
Mowlah, G., 170
Mozgovaya, I. E., 327
Mozhaeva, V. S., 322
Mueller, S. C., 268, 270, 272
Mühlethaler, K., 30, 31
Mukai, M. K., 169
Mullet, J. E., 14, 15, 17, 399
Mullins, L. J., 101
Mullins, M. G., 451
Munck, L., 237, 238
Mundy, J., 237, 238
Munns, R., 96
Munske, G. R., 165
Murai, N., 235
Murakami, S., 381, 382
Murakami, Y., 446
Murant, A. F., 293
Murashige, T., 238
Murata, N., 16
Murfet, I. C., 433, 441-45
Murofushi, N., 426, 437, 438, 444-46
Murphy, E. B., 473
Murray, M. G., 75, 76, 235
Muthukrishnan, S., 238

Muto, S., 54, 55, 78, 109-11, 126, 130-32
Mutschler, M. A., 163
Mutumba, M. C., 192
Myers, J., 18, 19, 22, 24

N

Nag, P., 197
Nagai, R., 97, 98, 104
Nagao, R. T., 240, 241
Nagatani, A., 224
Nagle, J. F., 352, 380
Nagura, S., 426
Nagy, F., 225, 227, 228, 230, 231, 234, 235, 245
Naim, Y., 367
Nakagawa, H., 160, 162
Nakagawa, S., 98
Nakajima, T., 275
Nakatani, H. Y., 14, 348
Nakazato, M., 16
Narayana, S. V. L., 235
Nastuk, W. L., 95
Natarajan, L. V., 35
Neale, P. J., 18, 21, 22, 24
Nechaev, I. A., 322-24
Nechitailo, G. S., 319, 326, 327
Nedukha, E. M., 319-21, 325, 326, 329-34
Negrutiu, I., 247
Nelson, H., 367
Nelson, M. D., 454
Nelson, N., 17, 27, 367
Nelson, N. D., 192
Nelson, R. S., 308
Nelson, S., 309
Nelson, T., 224
Ness, P. J., 158
Nessler, R. D., 235
Nester, E. W., 223, 245, 468, 469, 472
Neumann, J., 358, 367
Nevell, R. P., 260
Neville, A. C., 272
Nevins, D. J., 161
Nevzgodina, L. V., 322, 324, 325
Newton, R. P., 48
Nguyen, T., 242
Niblett, C. L., 308
Nichols, R., 160
Niedermeyer, J., 471, 472, 475, 481, 482
Nielsen, N. C., 235
Nikitin, M. D., 322
Nikolau, B. J., 225
Nimmo, G. A., 84, 85
Nimmo, H. G., 84, 85
Ninnemann, H., 350
Nishida, K., 382

Nishiguchi, M., 299, 300
Nishizaki, Y., 104
Nishizuka, Y., 85, 88
Nitsche, C., 280
Nitsche, K., 454
Nobel, P. S., 382
Noma, M., 433, 447, 451
Norman, S. M., 454
Norris, J. R., 28
Northcote, D. H., 193, 194, 274-76, 279
Notario, V., 265, 267
Nothnagel, E. A., 127
Notsani, B. E., 17
Novoselskaya, A. Yu., 150
Nugent, J. H. A., 14
Nuzhdin, N. I., 322-24

O

Oaks, A., 53
O'Connell, P. B. H., 158, 160, 167, 168
O'Connor, S. A., 320-22
Oda, K., 101
Odell, J. T., 245
O'Dell, M., 146, 244
Oelmüller, R., 226
Oelze, J., 25, 31, 36
Oesterhelt, D., 27, 28, 231, 354
Ogata, K., 106, 110
Ogawa, S., 212
Ogawa, T., 233
Ogden, R. C., 27
Ögren, E., 23
Ogura, N., 162
Ohad, I., 79
Ohme, M., 392, 407
Ohno, T., 300, 301
Ohyama, K., 392, 407
Oka, K., 101, 102
Okada, Y., 297, 300, 301
Okamura, M. Y., 27
Okamuro, J. K., 236
Okano, T., 263
Okazaki, Y., 98-100
Okita, T. W., 147, 148
Oliver, J. E., 242
Ol'khovenko, V. P., 322
Olson, J. M., 34, 35
Omran, R., 319, 333
O'Neal, J. K., 227, 475
O'Neil, R. M., 268, 270
Ongko, V., 327
Ooms, G., 473, 475
Ooshika, I., 300, 301
Opanasenko, V. K., 378
Öquist, G., 23
Orme, E. D., 454
Ormerod, J. G., 35
Orozco, E. M. Jr., 399

Ort, D. R., 19, 358, 359, 361, 367-69, 371, 372
Ortiz, W., 15, 17, 18
Osborn, M., 119, 120, 128, 130
Oshima, N., 299
Oshio, H., 452, 453
Osterhout, W. J. V., 95
Ota, Y., 438, 444-46
Otsuki, Y., 294, 302
Otto, V., 224
Overall, R. L., 299
Overmeyer, J. H., 297
Owens, L. D., 475

P

Paaren, H. E., 237
Packard, M. J., 132
Packer, L., 381, 382
Padan, E., 283
Pakrasi, H. B., 14
Paleg, L. G., 453
Palejwala, V. A., 159
Paliyath, G., 59, 80, 87, 165, 170
Palladina, T. A., 334
Palmbakh, L. R., 322, 325
Palmer, J. D., 224, 226, 392
Paludan, K., 243
Palukaitis, P., 294, 295, 298, 308
Palva, E., 477
Panova, S. A., 334, 335
Pant, N., 158
Paolillo, D. J. Jr., 183, 191, 192
Pap'yan, N. M., 322
Papp, J. E., 240
Parfenov, G. P., 320-22, 325-29
Parija, P., 156, 166
Parish, R. W., 80
Park, K. H., 437, 451
Park, R. B., 12, 349, 381
Parke, J., 130, 132
Parker, C. S., 231
Parmar, S., 145, 148, 149
Parson, W. W., 28, 29, 31, 32
Parthasarathy, M. V., 127, 134
Pastushenko-Strelets, N. A., 322, 324
Paszkowski, J., 223, 245, 247
Pate, J. S., 193
Patel, R. N., 186
Pattee, H. H., 279
Paul, K.-G., 207, 208, 211
Pauli, D., 240
Paull, R. E., 170
PAYNE, P. I., 141-53; 142-51
Peace, G. W., 127, 136
Peachell, P. T., 50

Pearce, F. L., 50
Pearce, G., 157, 171
Pearson, J. A., 164, 170, 171
Pederson, T. C., 439
Peebles, C. L., 407
Pegg, A. E., 61
Pelham, H., 239
Pelham, H. R. B., 240, 297
Penel, C., 64, 206, 207, 212, 213
Penel, C. E., 212
Penning de Vries, F. W. T., 167
Penon, P., 208, 210, 235
Penswick, J. R., 247
Peralta, E. G., 223, 470
Perbal, G., 329, 331
Perdue, T. D., 127, 134
Perlman, P. S., 407
Perrot, B., 396, 401, 402, 410
Pesacreta, T. C., 127
Peters, F. A. L. J., 363
Peters, J., 30, 31
Peters, R. L. A., 362, 363
Petersen, L. J., 298, 301, 304
Peterson, C., 319, 333, 335
Peterson, D. D., 324
Peterson, D. M., 145, 148, 150
Petit, A., 474
Petrillo, M. L., 407
Petrouleas, V., 14
Petrova, L. E., 322-24
Petruska, J., 223, 245
Pfeffer, S., 52
Pfeiffer, P., 299
Pfennig, N., 25, 26, 29
Pfister, V. R., 350, 351, 354
Pharis, R. P., 420, 428, 430, 433, 434, 437, 438, 445-48, 450-52
Philipson, W. R., 183
Phillips, R., 185, 186
Philpott, D. E., 327
Phinney, B. O., 420, 422, 424, 426-28, 430, 433, 436-38, 441-43, 445, 450, 451, 456, 457
Pianezzi, B., 325
Pickard, B. G., 53, 63, 64, 318
Picquot, P., 130, 131
Picton, J. M., 283
Piechulla, B., 161
Pierson, B. K., 35, 36
Pierson, E. S., 133
Pillonel, C. H., 280
Pinck, M., 222
Pinna, M. H., 208, 210
Pirie, N. W., 291
Pitman, M. G., 99, 100
Pizzi, A., 264
Pizzolato, T. D., 185
Planel, H., 324, 325

Plaskitt, K. A., 294
Platonova, R. N., 319, 320, 322, 326, 327, 331, 332, 334, 335
Platt, T., 402, 409
Platt-Aloia, K. A., 164, 165, 170
Pleij, C. W. A., 300
Plotkin, T., 182
Polans, N. O., 224, 226
Poling, S. M., 454
Polito, V., 49, 50
Poljakoff-Mayber, A., 58
Polle, A., 351, 362, 366
Polya, G. M., 80, 86
Pon, N. G., 349
Pont Lezica, R., 274
Ponz, F., 305
Ponzi, G., 80
Poole, R. J., 52, 111
Poovaiah, B. W., 59, 80, 87, 165, 170
Poperelya, F. A., 143, 146, 147, 151
Popova, A. F., 327
Popovich, P. R., 321, 322
Porter, S., 468
Portis, A. R., 374
Poten, F., 49, 50, 55-58, 62, 64, 274, 277, 279
Potrykus, I., 223, 245, 247
Potter, D., 186
Potts, W. C., 433, 441-44, 449
Poulos, T. L., 207
Poulsen, C., 232, 396, 401, 410
Pouphile, M., 270
Powell, A. J., 120, 127
Powell, L. E., 437
Prats, M., 380
Pratt, H. K., 158
Pratt, L. H., 86, 224
Pressey, R., 162
Prezelin, B. B., 8
Pribnow, D., 402
Price, C. A., 304
Prieels, J-P., 279
Prince, A. C., 36
Prochaska, L. J., 350, 351, 360, 368
Proebsting, W. M., 433, 447
Prokof'yeva-Bel'govskaya, A. A., 322, 323
Prosser, I. M., 166
Proudlove, M. O., 51-55
Pruitt, R. E., 477
Pshennikova, E. S., 301
Pugliarello, M. C., 51, 52
Purohit, S., 242
Purton, M., 162
Pustell, J., 235

Pyle, J. B., 162
Pywell, J., 236

Q

Qian, L. P., 350, 362, 366,
    367
Quader, H., 124, 268, 271,
    274, 275, 283
Quail, P. H., 86, 224, 226, 233
Qualset, C. O., 143, 146
Quatrano, R. S., 237
Queiroz, O., 84, 85
Quinn, P. J., 349

R

Racker, E., 79, 348
Rademacher, W., 420, 426,
    430, 437, 452-54
Rae, A., 277, 283
Rae, H., 158
Rafalski, A., 145, 150
Rafalski, J. A., 148
Raghothama, K. G., 80, 87,
    170
Ragimova, G. K., 330
Rai, R. M., 158
Railton, I. D., 420, 422, 423,
    430, 434
Rajasekaran, K., 451
Ralph, R. K., 80, 86
Ramming, D. W., 164
Ramshaw, J. A. M., 208
Randall, D. D., 81, 82
Randles, J. W., 305
RANJEVA, R., 73-93; 74, 80-
    83
Ranty, B., 74
Rao, P. K., 81, 82
Rapp, B., 81, 82
Rappaport, L., 444
Raschke, E., 240, 241
Rasi-Caldogno, F., 51, 52, 167
Ratan, R., 51
Rausch, U., 68, 69
Raven, J. A., 111
Ravi, K., 212, 213
Raviv, A., 192
Rawlins, D., 127, 128, 130-33
Ray, P. M., 239, 267, 275,
    276, 278
Rea, P. A., 111
Read, S. M., 276-79, 284
Reader, S. M., 149
Ream, W., 223, 470
Reardon, E. M., 304
Recondo, E., 266
Reddi, P. S., 212, 213
Red'ko, T. P., 378
Redlinger, T., 25
Reed, L. J., 81
Reeder, R. H., 244

Reese, T. S., 273
Reeves, C. D., 147, 148
Reeves, S., 469
Refeno, G., 81, 82
Reichert, N. A., 235
Reid, J. B., 433, 441-45
Reid, R. J., 99, 100
Reik, L. M., 214
Reinders-Gouwentak, C. A.,
    183, 190
Reinero, A., 304
Reinwald, E., 358
Reis, D., 272
Reiser, R., 453
Reiss, B., 472
Reiss, H. D., 51, 58, 133, 268,
    283
Reitz, G., 324
Remish, D., 362
Remy, R., 79
Renganathan, V., 213
Renger, G., 13
Rentzea, C., 452-54
Resink, T. J., 103
Reuveni, M., 58
Revol, J-F., 262, 264, 270
Reynaud, J., 207
Rezelman, G., 297, 300
Rhee, S. G., 73, 74
Rhoads, R. E., 297
Rhodes, M. J. C., 155, 166
Ricard, J., 207, 208, 210
Richards, D. M., 235
Richards, K. E., 244
Richards, K. S., 302
Richardson, C., 171
Richardson, C. L., 264
Richardson, D. G., 158
Richmond, A., 160, 166, 168
Richter, G., 226
Rienits, K. G., 363
Rier, J. P., 190, 191, 193, 194,
    196
Riley, G. A., 103
Rincon, M., 49, 50, 84
Rippka, R., 36
Ritland, J., 396, 398, 407,
    412, 413
Rivas, L. A., 274
Rivett, D. E. A., 430
Robbins, M. P., 209
Robert, L. S., 237
Roberts, I. N., 121-24, 126,
    128, 272
Roberts, J. A., 164
Roberts, J. K., 240
Roberts, J. L., 473
Roberts, K., 122-24, 126, 128,
    272
Roberts, L. L., 225, 226
Roberts, L. W., 180, 191,
    192
Roberts, M. S., 149

Robertson, D. S., 420, 445,
    457
Robertson, J. G., 214
Robertson, N. G., 155, 158,
    160
Robinson, D. G., 212, 268,
    271, 275, 283, 427, 439
Rochaix, J. D., 408, 409
Rochester, D. E., 240
Rochon, D., 304
Roe, B., 170
Rogers, J. C., 238
ROGERS, S., 468-86; 473
Rogers, S. A., 49, 50, 84
Rogers, S. G., 223, 227-29,
    236, 245, 247, 308, 469,
    471-77, 480-83
Röhner, E., 452, 453, 455
Roland, J. C., 272
Romani, R., 158, 165, 166,
    168
Romani, R. J., 156, 158
Romberger, J. A., 184, 197
Romero, P. A., 274
Rood, S. B., 445, 450
Ropers, H. J., 420, 422, 424,
    427, 428, 431, 434, 436,
    438, 450-52
Rösch, I., 327
Rose, R. E., 483
Rosenberg, M., 399
Ross, A. F., 306
Ross, J. J., 441
Ross, P., 265, 266
Rossen, M., 363
Rothstein, A., 368, 381
Rottenberg, H., 348, 358, 368,
    379, 382
Rougier, M., 267
Roux, J., 54
Roux, S. J., 76, 86, 87, 233,
    318, 338
Rubenstein, I., 236
Rubery, P. H., 239
Rubin, P. M., 81, 82
Ruden, D. M., 231
Rudneva, N. A., 322, 323
Rudolph, U., 133
Ruis, H., 210
Rumberg, B., 358
Rupainene, O. J., 319-21, 327-
    29, 331, 336
Rusakova, G., 335
Russo, M., 304
Rutherford, A. W., 14
Rutter, W. J., 247
Ryan, C. A., 157, 171, 308
Ryan, D. E., 214
Rybalka, A. I., 144, 146
Ryder, T. B., 243
Ryoyama, K., 382
Ryrie, I. J., 16, 349
Ryser, U., 277

S

Saarelainen, R., 242
Sacher, J. A., 155, 156, 160,
   165, 167, 168
Sachs, T., 180, 183, 190-92,
   195, 196
Sadler, J., 300, 301
Saedler, H., 240
Saftner, R. A., 165
Saito, K., 104
Saito, T., 297
Sakai, S., 239
Sakano, K., 104, 111
Sakihama, T., 224, 226
Saks, Y., 183, 186, 188, 191,
   196, 197
Saksonov, P. P., 322
Sakurai, A., 451, 453
Sakurai, H., 17
Salimath, B. P., 48, 51, 52,
   54, 80, 83, 85
Sallé, G., 329, 331
Salleo, S., 181
Saltveit, M. E., 160
Salvi, L., 54
Samokhvalova, N. S., 322-24
Samuelsson, G., 8
Sandelius, A. S., 52
Sanders, D., 97, 98, 101, 110
Sanders, P. R., 223, 472, 475
Sandrin, M. S., 214
Sands, J., 407
Sane, P. V., 12, 361
Sang, H. W. W. F., 108
Sanio, K., 186, 190
Sano, T., 392, 407
San Pietro, A., 17
Sarah, F. Y., 426
Sardana, R., 244
Sarko, A., 263
Sasaki, Y., 224, 226
Sassa, T., 433
Sato, T., 162
Sato, Y., 438
Satoh, K., 13, 23, 25
Sauer, A., 439
Sauerbrey, E., 454
Saul, M., 223
Saul, M. W., 223, 245
Saunders, B. C., 205
Saunders, G. R., 309
Saunders, M. J., 51
Sauter, H., 452-55
Savi, A., 305
Savichene, E. K., 319-21, 326,
   327, 329, 331
Savidge, R. A., 180, 183, 188,
   192
Sax, K., 192
Saxton, W. O., 31
Sayre, R. T., 354
Scalla, R., 297

Scandalios, J. G., 206, 207
Scarborough, G. A., 52
Schachter, J., 212
Schäfer, A., 88
Schafer, E., 224-26, 233
Schafer, M., 324
Schafer, W., 76, 77, 86
Schairer, L. A., 321, 324
Schaller, H., 472
Schapendonk, H. C. M., 108
Schatzmann, H. J., 51, 52
Scheets, K., 148
Scheld, H. W., 319, 333, 335
Schell, J., 223, 224, 226, 227,
   230, 232, 234, 235, 240,
   243, 245, 470, 472, 474
Schibeci, A., 80
Schilperoort, R. A., 223, 469,
   473, 475
Schindler, C. B., 164
Schlesinger, M. J., 239
Schliephake, W. D., 358
Schliwa, M., 127, 134
Schmelzer, E., 243, 306
Schmid, M. F., 35
Schmidt, D. F., 352
Schmidt, K., 35
Schmitt, A.-C., 121
Schnapp, B. J., 273
Schnarrenberger, C., 213
Schneider, B., 125
Schneider, G., 420, 450
Schneider, H. A. W., 207
Schneider, T., 402
Schneider, T. D., 402
Schnepf, E., 124, 133, 268
Schoeder, H. U., 16
Schoenfeld, M., 377
Schoenknecht, G., 367, 376,
   377
Schoffl, F., 240, 241
Schopfer, P., 168
Schram, A. W., 211
Schreiber, U., 352, 363, 368,
   369, 375, 380
Schreier, A. A., 349
Schroder, G., 469
Schröder, H., 358
Schroder, J., 469
Schuch, W., 161, 162,
   243
Schuldiner, S., 358
Schuler, M., 236
Schultz, G. A., 240
Schumaker, K. S., 52
Schürmann, P., 54, 305
Schuster, A. M., 160, 161
Schwarz, R. T., 274
Schwenen, L., 423, 427, 428,
   430, 432, 436, 438, 446,
   452-54
Scofield, S., 237
Scouten, W. H., 211

Seagull, R. W., 120, 122-24,
   128, 130-32, 271, 272
Sears, E. R., 143
Sebald, W., 356, 380
Sederoff, R., 475
Seebeck, T., 75
Seftor, R. E. B., 27, 35
Segrest, J. P., 356
Seib, E., 119
Sekiguchi, K., 160
Sela, I., 305
Selak, M. A., 363
Sellmer, J., 472
Selman, B. R., 359
Sembdner, G., 420, 450
Semler, U., 59, 275, 280, 284
Sen, S., 185
Senda, M., 104
Senger, H., 16, 225
Sengupta-Gopalan, C., 235, 237
Sequeira, L., 243
Serlin, B. S., 233, 318, 338
Setif, P., 17
Setterfield, G., 125, 129, 136,
   277
Sevier, E. D., 212
Seydewitz, H. H., 30
Shacter, E., 74
Shah, D. M., 226, 232, 240,
   483
Shahabuddin, M., 297
Shalla, T. A., 298, 301, 304
Shannon, L. M., 207, 211, 212
Shapiro, S. L., 32
Shaul, Y., 247
Shavit, N., 358
Shaw, C. H., 245
Shaw, E. R., 13
Shaw, J. G., 294, 295, 297
Shaw, P. J., 127, 128, 130-33
Sheetz, M. P., 110, 134, 273
Shelanski, M. L., 51
Sheldon, J. M., 121
Shen-Miller, J., 335, 336, 422-
   24
Shenk, T., 229
Shepelev, Ye. Ya., 320, 326,
   335
Shepherd, K. W., 142, 144,
   146, 148-50
Shepherd, R. J., 303
Sherwin, P. F., 424
Sherwood, J. L., 308
Shewmaker, C. K., 227, 475
Shewry, P. R., 141, 142, 145,
   148-50
Shiao, M. S., 455
Shibaoka, H., 125, 126, 280,
   452
Shields, B. A., 280
Shigematsu, Y., 426
Shigo, A. L., 182
Shiina, T., 100, 103, 106, 111

Shillito, R. D., 223, 245
SHIMMEN, T., 95-117; 96-111, 134
Shimodoriyama, M., 280
Shimogawara, K., 54, 78
Shimomura, T., 302
Shinedling, S., 402
Shininger, T. L., 179, 183, 185, 196
Shinitzky, M., 165
Shinozaki, K., 224, 226, 392, 396, 401, 403, 404, 407, 408, 410
Shinshi, H., 239
Shirai, H., 392, 407
Shiro, Y., 212
Shirsat, A. H., 235
Shishibori, T., 421
Shmigovs'kaya, V. V., 331
Shopes, R. J., 28
Short, T. W., 190
Showalter, A. M., 243
Shumway, J. E., 237
Shuvalov, V. A., 13, 16
Shvegzhdene, D. V., 319-22, 327, 328, 330, 334, 336
Siaw, M. F. E., 297
Sibaoka, T., 101
Sidler, W., 30, 31
Sidorenko, P. G., 319, 325, 326, 329-31, 333
Sidorov, B. N., 323
Siebenlist, U., 399
Sieber, M., 186
Siebers, A. M., 183
Siefermann-Harms, D., 350
Siegel, A., 304, 305
Siegel, B. Z., 207
Siegel, N., 52
Siegenthaler, P. A., 382
Sievers, A., 319, 329-32, 334-36
Sigalat, C., 367, 368, 371, 372, 374, 377, 381
Siggel, U., 358
Sihra, C. K., 16
Silverthorne, J., 224, 244, 420, 445, 457
Simmonds, D. H., 125, 129, 136
Simon, A., 237
Simon, E. W., 165
Simon, M. I., 27
Simon, P., 54
Simoni, R. D., 381, 383
Simpson, J., 225, 229-31, 234, 235
Simpson, R. B., 399, 474
Singer, B. S., 402
Singh, N. K., 144, 149
Siracusano, L., 181
Sistrom, W. R., 25
Skinner, J. D., 88

Skogland, U., 207
Skoog, F., 238
Skulachev, V. P., 348
Slabas, A. R., 120, 127
Slater, A., 161, 162
Slater, E. C., 348
Sligar, S. G., 439
Slighton, J. L., 235
Sloan, M. E., 278
Slocum, R. D., 318, 332
Slot, J. W., 212
Slovin, J. P., 225
Sluiman-denHertog, F. A., 121, 133
Smith, B. N., 234
Smith, F. A., 101, 105, 111
Smith, K. M., 33
Smith, S. M., 224
Smith, T. A., 61
Smith, V. A., 435, 439, 441, 448, 451
Smithies, O., 247
Smydzuk, J., 158
Snape, J. W., 146
Sneath, P. H. A., 411
Snozzi, M., 31
Sobick, V., 336
Sobko, T. A., 144, 149, 150
Soboleva, T. N., 322
Sokal, R. R., 411
Sokolov, N. N., 323
Solberg, L., 237
Soll, D. G., 145, 148, 150
Soll, J., 77, 78
Solomos, T., 155, 157, 166, 167, 169
Somerville, C., 473
Sommergruber, K., 228
Somssich, I. E., 243, 306
Sopory, S. K., 86, 88
Soucaille, P., 380
Sozinov, A. A., 143, 144, 146, 147, 150, 151
Spangfort, M., 18, 20
Spano, L., 206
Spanswick, R. M., 104, 105
Sparrow, A. H., 321, 324
Spear, D. G., 101, 102, 106
Speirs, J., 161, 162, 164, 166, 168
Spena, A., 240
Sperry, J. S., 181
Spielmann, A., 474
Spierer, A., 240
Sponsel, V. M., 420, 422, 433-36, 438, 439, 443, 445, 448-51
Sprague, S. G., 25, 36
Spray, C. R., 420, 442, 445, 457
Springer, B., 224
Spruit, C. J. P., 446, 449
Spudich, J. A., 110

Spyropoulos, C. S., 101, 102
Srivastava, L. M., 185
Srivastava, O. P., 206
Stabel, P., 223
Staby, G. L., 156
Stachel, S., 223, 469, 472
Stachel, S. E., 468, 469
Stack, S. M., 132
Stackebrandt, E., 33
Stadel, J. M., 85
Stadtman, E. R., 73, 74
Staehelin, L. A., 13, 15, 34, 35, 212, 268, 270-72, 354, 356
Stahl, D. A., 36
Stalker, D., 161
Stalker, D. M., 483
Stämpfli, R., 96
Stanier, R. Y., 25, 35
Stanley, J., 242
Stant, M. Y., 196
Stark, B. P., 205
Stark, W., 31
Starlinger, P., 224
Staswick, P., 235
Stebbins, G. L., 411
Steer, M. W., 283
Steffens, G. L., 452
Steigler, G. L., 409
Stein, G., 75
Stein, J., 75
Steinback, K. E., 22-24, 78
Steinbiss, H. H., 223
Steiner, L. A., 27
Steinmuller, K., 234
Stephan, D., 208, 209, 211, 213
Steudle, E., 99
Steuer, E., 273
Steward, F. C., 327
Sticher, L., 212
Stieber, J., 184
Stiehl, H. H., 13
Stiekema, W., 242
Stiekema, W. J., 224
Stigbrand, T., 207, 208, 211
Stillman, J. S., 206
Stoddart, J. L., 422, 441, 444
Stomp, A.-M., 475
Stone, B. A., 260, 267, 273-75, 277, 279, 280
Stormo, G., 402
Stratton, B. R., 75
Stroobant, P., 52
Strous, G. J., 212
Studier, F. W., 404
Su, L-Y., 164
Subramani, S., 247
Suga, T., 421
Sugavanam, B., 455
Sugawara, Y., 223
Sugita, M., 404, 407, 408

Sugiura, M., 396, 401, 403, 404, 407, 408, 410
Sugiyama, J., 262, 270, 271
Sugiyama, T., 145, 150
Sulzinski, M. A., 292, 293, 302
Summerlin, L. B., 335
Susani, M., 236
Sussex, I. M., 237
Sussman, M. R., 84
Suter, F., 27, 29-31, 33, 35
Sutherland, J., 277
Sutton, D. W., 235
Suzuki, Y., 444, 445
Swanson, D. I., 454, 455
SWEENEY, B. M., 1-9; 3-8
Swegle, M., 238
Swissa, M., 265
Syono, K., 223
Sytnik, K. M., 319-21, 325, 326, 328-33
Szaniawski, R. K., 167
Sze, H., 52, 111
Szent-Gyorgyi, A. G., 110, 134
Sziráki, I., 306

T

Tabor, G. J., 226, 232
Tabor, J. H., 407
Tada, T., 160
Tadros, M. H., 30
Tairbekov, M. G., 319, 320, 322, 326, 327, 329, 332, 334, 335
Taiz, L., 53
Takahashi, M., 428, 432, 434, 438
Takahashi, N., 420, 426, 428, 431, 432, 434, 437, 438, 444-46, 451-53
Takahashi, Y., 16
Takamatsu, N., 300
Takebe, I., 294, 302
Takeda, K., 126
Takeno, K., 451
Takeshige, K., 104, 106
Takeuchi, U., 103
Taliansky, M. E., 301
Tamai, N., 382
Tamaki, S., 226, 232, 234
Tanaka, M., 392, 404, 407, 408
Tanchak, M., 125, 129
Tandy, N. E., 352, 354, 357, 360, 367, 368, 380
Tanner, W., 266
Tarasenko, V. A., 320, 321, 332, 334
Tarkowska, J. A., 130
Tasaki, I., 95, 96, 101, 102
Taylor, D. A., 439, 451
Taylor, L. P., 223

Taylor, S. E., 15, 18
Taylor, W. C., 224, 231
TAZAWA, M., 95-117; 53, 96-98
Teeri, T., 477
Teissere, M., 208, 210
Teissie, J., 380
Telewski, F. W., 64
Telfer, A., 23, 78
Tellez-Ionen, M. T., 103
Tempe, J., 474
Tenbarge, K., 237
Tenbarge, K. M., 237
Tener, G., 86
Terai, H., 158
Terry, N., 15, 18
Tetour, M., 66
Teulières, C., 80
Tewari, J. D., 158
THEG, S. M., 347-89; 350, 351, 353, 361, 362, 364-67, 376, 380
Theiler, R., 30, 31
Theimer, R. R., 327
Thelen, M., 267, 277, 278
Thelen, M. T., 276-79, 284
Thellier, M., 49
Themmer, A. P. N., 170
Theologis, A., 239
Theuvenet, A. P., 108
Thia-Toong, L., 223
Thielen, A. P. G. M., 14, 15, 17, 18
Thimann, K. V., 3
Thomas, A. G., 190
Thomas, P. E., 214
Thomas, T. L., 235
Thomashow, L. S., 239, 469
Thomashow, M. F., 239, 469
Thompson, D., 228
Thompson, D. V., 300
Thompson, G., 483
Thompson, J. E., 165, 169
Thompson, N. P., 190, 191, 193
Thompson, R. D., 144-48, 150
Thompson, W. F., 224-26, 244
Thomson, K. S., 66, 67
Thomson, W. W., 164, 165, 170
Thornber, J. P., 27, 35, 36, 354, 356, 357
Thorpe, T. A., 165, 169, 206, 207, 213
Thorsch, J., 181
Thurston, H. D., 437
Tibbitts, T. W., 337
Tieman, D. M., 162
Tigchelaar, E. C., 158, 163, 164, 171
Tillberg, J. E., 379
Timko, M. P., 227, 229, 230, 232

Timmerman, B., 468
Tissieres, A., 239, 240
Tiwari, S. C., 127
Tjian, R., 225, 231
Tobias, C. A., 324
Tobin, E. M., 224, 225, 244, 356, 357
Tobkes, N., 356
Tocanne, J. F., 380
Tocchi, L. P., 207
Tomaszewski, M., 188
Tominaga, Y., 99, 109-11
Tomlinson, J. A., 303
Tomlinson, P. B., 186
Tomura, H., 206, 213
Topol, J., 231
Topper, Y. J., 185
Torossian, K., 275
Torres, H. N., 103
Torrey, J. G., 185, 191
Tournois, J., 4
Tousignant, M. E., 309
Townsend, J., 472, 473
Traas, J. A., 122, 126-28, 130-33
Tran, M., 324
Travers, A. A., 240
Trebst, A., 14, 361
Trewavas, A. J., 51, 74, 75, 80, 87, 88, 159, 238
Trip, P., 191
Triplett, B. A., 237
Triplett, E. W., 242
Tristram-Nagle, S., 380
Tritthart, H., 99
Troke, P. F., 96
Tromp, M., 243
Tronrud, D. E., 35
Trüper, H. G., 25, 26, 29
Trusova, A. S., 321, 322, 328
Tsay, L. M., 159
Tseng, C. K., 4
Tucci, G. F., 80
Tucker, G. A., 155, 158, 160, 164, 170
Tucker, M. L., 160, 161, 164, 167, 168
Tumer, N. E., 226, 232
Tung, H. Y. L., 103
Turcon, G., 208, 211
Turgeon, R., 185
Turnbull, C. G. N., 423, 427, 428, 436, 438
Tyree, M. T., 182
Tyson, H., 206, 211

U

Uchimiya, H., 223
Ueda, T., 104
Ueda, Y., 160, 162, 164, 170
Ueki, T., 31
Umrath, K., 95

Unwin, P. N. T., 356, 357
Uribe, E. G., 358, 367
Usuda, M., 263
Uyeda, N., 262, 270, 271

V

Vale, R. D., 110, 134, 273
Valentine, J., 30
Valovich, E. M., 319, 327, 329, 332
Vambutas, V., 382
Van Alfen, N. K., 181
van Boom, J. H., 243
vandenBerg, B. M., 206, 209-11, 213
van den Broeck, G., 224, 226, 227, 232
van den Elzen, P., 226, 232, 234, 472, 473
Vandenheede, J. R., 103
van de Poll, D., 32
Van der Hart, D. L., 264
Vanderhoef, L. N., 238
van der Kooij, F. W., 32
van der Krol, S., 298
Vanderleyden, J., 468, 477
van der Linde, P. C. G., 239
van der Meer, J., 307
Van der Pal, R. H. M., 363
van der Veen, J. H., 446, 449
van Dorssen, R. J., 35
van Eldik, L. J., 130, 131
van Gorkom, H. J., 13-15, 17, 18
van Grondelle, R., 32
van Huijsduidnen, R. A. M. H., 243
VAN HUYSTEE, R. B., 205-19; 206-15
van Kammen, A., 242, 295, 297, 298, 300, 307
Van Kerkhof, P., 212
van Kooten, O., 362, 363
van Loenen-Martinet, E. P., 441
Van Loon, L. C., 75, 243, 306
Van Montagu, M., 223-27, 229-35, 245, 468, 469, 470, 472, 474, 477
Van Netten, C., 102
van Rijn, L., 441
van Spronsen, P. C., 121, 122, 133
Vantard, M., 121, 130, 131
van Tol, R. G. L., 300
van Vloten-Doting, L., 300
Varga, A. R., 25
Varner, J. E., 207, 243
Varnold, R. L., 80, 86, 87
Vasil'eva, N. G., 324
Vasmel, H., 35
Vater, J., 13

Vaulina, E. N., 320-22, 324, 325, 328
Veen, H., 160
Velten, J., 472
Veluthambi, K., 80, 87, 170
Vendrell, M., 159, 169
Venis, M. A., 87
Venturoli, G., 348
Venverloo, C. J., 121, 133
Verma, D. P. S., 242, 243
Vernet, T., 222
Vernon, L. P., 348, 381
Vernotte, C., 78, 79
Viale, A., 77, 350, 362, 366, 367
Vian, B., 272
Vickery, R. S., 165, 166
Vidal, J., 84, 85
Viemken, V., 31
Vierling, E., 17
Vierstra, R. D., 233
Vignais, P. V., 356
Vincentz, M., 244
Vinkler, C., 369, 371
Vitale, A., 235
Voellmy, R., 239, 240
Vogt, V. M., 304
Volger, H., 440, 441
Volkmann, D., 319, 329-32, 334-36
Volloch, V., 307
Vonderhaar, B. K., 185
von Wechmar, M. B., 295
Vos, M., 32
Vostrikov, I. Y., 100, 102
Vredenberg, W. J., 32, 108, 362, 363
Vysotskii, V. G., 321, 322, 325

W

Wachter, E., 356
Wada, A., 452, 453
Wada, K., 426, 453
Wada, M., 268
Wade, N. L., 155, 158, 159, 165
Waffenschmidt, S., 469
Wagner, K. G., 75
Wakarchuk, D. A., 307
Wakasugi, T., 392, 404, 407, 408
Wakim, B., 36
Walbot, V., 222, 223
Walden, R., 480
Waldron, C., 473
Walker, J. C., 239
Walker, J. E., 401, 403
Walker, L. L., 247
Walker, M. D., 247
Walker, N. A., 101, 105
Walker-Simmons, M., 308
Wall, J. S., 142, 148, 150

Wallace, B. A., 356
Wallner, S. J., 157
Wallroth, M., 476
Walsh, D. A., 86
Wample, R. L., 452
Wang, C.-L., 319, 326, 327, 329-32, 335
Wang, K., 223, 468
Wang, M. X., 170
Wang, R. T. K., 18, 19, 22, 24
Wang, S., 169
Wang, S. Y., 452
Wang, Y., 111
Ward, C. M., 303
Ward, J. M., 183
Wareing, P. F., 180, 183, 184, 187, 192, 194, 196
Warm, E., 161
Warren Wilson, J., 183
Wasserman, B. P., 278, 284
Watada, A. E., 156
Watanabe, T., 16
Watanabe, Y., 297, 300, 301
Waterbury, J. B., 36
Waterworth, H. E., 309
Watkins, P. A. C., 308
Watkins, W. M., 279
Watson, B. D., 472
Watson, J. C., 244
Watson, J. L., 15, 18, 22
Watson, M. D., 245
Watts, J. W., 127, 128, 130-33, 294
Wayne, R. O., 47-51, 53, 58, 63, 86, 170, 283, 318, 338
Webb, W. W., 127
Weber, D. F., 242
Weber, G., 270, 276
Weber, K., 119, 120, 128, 130
Webster, D. H., 62
Webster, G. D., 27
Wechsler, T., 33, 35
Weeda, A. J., 133
Weerdenburg, C. A., 131, 132
Wehrli, E., 31
Weih, M., 264
Weiher, H., 231
Weiler, E. W., 238, 454, 469
Weinberger-Ohana, P., 265, 266
Weinhouse, C., 265
Weinhouse, H., 265, 266
Weir, B. S., 393
Weis, C., 52
Weis, E., 78, 79
Welinder, K. G., 206-9, 212, 214
Welle, B. J. H. ter, 188
Wells, B., 122, 124, 126, 133, 214
Welte, W., 30, 31
Werker, E., 188
Werr, W., 223, 224
Wessel, K., 226

West, C. A., 420-24, 441, 452-55
Westerhoff, H. V., 348
Westerhuis, W. H. J., 23
Westheimer, F. H., 352
Wetmore, R. H., 190, 193, 194, 196
Wettern, M., 14
Weyer, K. A., 27, 28, 354
Whitbread, V., 195
White, A. R., 264, 265
White, F. F., 469
Whitfeld, P. R., 392, 396, 399-405, 408-10
Whitmarsh, C. J., 377, 378
Whitmarsh, J., 19, 363
Wiborg, O., 243
Wick, S. M., 120, 121, 126-28, 130, 132
Widder, E. A., 8
Widell, S., 52
Wiebe, H. H., 181
Wielgat, B., 76
Wiemken, A., 96
Wiemken, V., 27, 31
Wienecke, K., 427
Wiessner, W., 270
Wildhaber, I., 31
Wildman, S. G., 305
Wilkins, M. B., 84, 85, 318, 335
Wilkinson, R. E., 456
Willemsen, R., 212
Willey, D. L., 356, 357
Williams, A. A., 158
Williams, J. C., 27
Williams, M., 81, 82
Williams, R. C., 24
Williams, R. J. P., 348
Williams, W. P., 349
Williamson, R. E., 50, 96, 99, 103, 104, 108, 109, 111, 127
Williams-Smith, D. L., 16
Willis, C. L., 433, 442-45
Willison, J. H. M., 264, 268
Willmitzer, L., 75
Wilson, J. E., 191
Wilson, T. M. A., 294, 295, 298, 305, 308
Wimpee, C. F., 224
Winans, S. C., 223, 469
Winer, J. A., 240
Wingate, V. P. M., 243
Winget, G. D., 358, 359
Winter, J. A., 483
Witt, H. T., 13, 358
Witt, J., 30
Witte, O., 133
Witz, J., 295
Witztum, A., 127, 134

Wobus, U., 235
Wodzicki, A. B., 184
Wodzicki, T. J., 184, 197
Woese, C. R., 33
Wojcik, S. J., 80, 86
Wolf, A., 191, 192
Wolniak, S. M., 51, 131
Wong, S. L., 243
Wong, S. S., 279, 280
Wood, E. A., 214
Wood, K. R., 303, 304, 306
Woodbury, N. W. T., 28, 29
Woodward, C. K., 350
Woolhouse, H. W., 159
Woolley, K. J., 30, 32
Worland, A. J., 143, 144, 146
Wraight, C. A., 14, 28, 29
Wright, K., 194
Wright, M., 475
Wright, S. T. C., 159
Wrigley, C. W., 142, 148, 150
Wu, C.-I., 394, 411
Wullems, G. J., 475
Wurtele, E. S., 427
Wylegalla, C., 77
Wyn Jones, R. G., 164, 166
Wysman, H. J. W., 206
Wyss, F., 30

X

Xinyling, Z., 188

Y

Yadav, N. S., 468, 477
Yaguzhinsky, L. E., 378
Yamagishi, A., 14, 24
Yamaguchi, I., 420, 428, 437, 438, 446, 451-53
Yamamoto, F., 192
Yamamoto, K. R., 231, 238
Yamamoto, K. T., 224
Yamamoto, R., 276
Yamanaka, G., 24
Yamane, H., 420, 428, 438
Yamashita, H., 426
Yamashita, T., 158
Yamaya, T., 53
Yang, S.-F., 155, 157-59, 163, 164, 169, 192
Yang, T., 324
Yano, M., 110, 134
Yanofsky, M. F., 223, 245, 468, 472
Yanshing, W., 86
Yaroshius, A. V., 319-21, 326-31, 333, 334
Yeo, A. R., 96
Yi, B.-Y., 246

Yi, W. J., 170
Yocum, R. R., 127
Yonetani, T., 207, 208
Yoshihisa, T., 275
Young, C., 468
Young, D. H., 55, 57, 63
Young, K., 166
Young, N. D., 298
Young, R. E., 158, 159, 164, 165, 170
Younis, L. Y., 48
Youvan, D. C., 27, 30, 32
Yuefen, D., 188

Z

Zaar, K., 264
Zabel, P., 298, 307
ZAITLIN, M., 291-315; 291-95, 297, 298, 300-5, 308
Zajaczkowski, S., 184, 197
Zakrzewski, J., 184
Zambryski, P., 223, 245, 468, 469, 470, 472, 474, 477
Zamski, E., 183, 184
Zander, M., 430
Zaug, A. J., 407
Zeeh, B., 452-54
Zeevaart, J. A. D., 426, 428, 430, 432, 436, 446, 447, 452
Zeiger, E., 233
Zeigler, R. S., 437
Zelechowska, M., 242
Zemel, E., 234
Zeroni, M., 155
Zeronian, S. H., 260
Zhang, W., 159
Zherelova, O. M., 100, 102
Zhvalikovskaya, V. P., 319, 320, 322, 326, 327, 331, 332, 334, 335
Zick, Y., 86
Ziervogel, U., 240
Zimanyi, L., 363
Zimmermann, J. I., 14
Zimmermann, M. H., 181, 182, 186, 187, 191, 195, 196
Zimmermann, U., 63, 96, 99
Zimmern, D., 300, 301
Zinth, W., 29
Zocchi, G., 80, 84, 106, 111
Zokolica, M., 224
Zuber, H., 27, 29-31, 33, 35
Zucchelli, G., 20
Zuidema, D., 300, 301
Zundel, G., 380
ZURAWSKI, G., 391-418; 396, 400-5, 407-10, 414
Zurfluh, L. L., 239

SUBJECT INDEX

A

Abscisic acid, 159, 237-38
 biosynthesis inhibition, 454
Acetic anhydride
 localized proton domains,
  360-61, 363-64
 labeling sites, 356
 probe for sequestered domains, 351-52
Acetobacter xylinum, 262
 cellulose biosynthesis, 260-
  61, 264-67
 degree of polymerization,
  262
 movement of cells, 271
Acetohydroxy acid synthase,
 473
Actomyosin, 134
Agrobacterium
 plasmid genes, 245-56
 rhizogenes, 243
 tumefaciens
  gene transfer intermediate,
   223
Agrobacterium-mediated plant
 transformation, 467-86
Agrobacterium biology,467-70
 beyond gene transfer, 476
 T-DNA use as mutagen,
  476-77
 gene rescue by shotgun
  cloning, 479-80
 mutagenesis with promoter-
  less markers, 477-79
 phytohormones, 480-82
 vectoring plant viruses, 480
 conclusions, 482-83
 hosts for transformation
  disarmed strains, 473-74
  host range, 474-75
  nondisarmed strains, 474
 plant transformation
  cocultivation, 475
  leaf disc transformation,
   475
 selectable markers, 471-73
 transgenic plants list, 472
 vectors, 470
  binary vectors, 471
  cointegrating vectors, 470-
   71
Alfalfa mosaic virus
 movement protein, 300
 replication phases, 299

Allophycocyanin, 19
1-Aminocyclopropane-1-
 carboxylic acid, 158
1-Aminocyclopropane-1-
 carboxylic acid synthase,
 158
Aminoethylisothiourea, 323
Amino levulinic dehydratase,
 210
Aminovinylglycine, 158
Amiprophos-methyl, 124,
 127
Amylase, 213, 238
Amyloplasts
 gravitational effects, 329-30,
  338
Ancymidol, 452, 455
Arabidopsis thaliana
 growth in space, 328-29
Ascorbate, 438, 440
ATP
 see Chloroplast bioenergetics,
  membrane-proton in-
  teractions
ATP synthetase, 348
Auxin
 Agrobacterium-transformed
  plants, 481, 483
 biosynthesis pathway, 468
 tracheid diameter, 187-88
 transport, 183-84
 vascular differentiation, 190-
  91, 192-94
  fiber differentiation, 194-96

B

B 1015 antenna complex, 29-31
 amino acid sequence of
  polypeptide components,
  30
Bacteriochlorophylls, 25-36
Bean golden mosaic virus, 301
Beta conglycinin, 236-37
Blinks, L., 4
Bouck, B., 7
Bread-making quality
 see Wheat, genetic basis for
  variation in bread-making
  quality
Brome mosaic virus, 298
 replication phases, 299
 translation strategies, 295-96
Brown, F., 5

C

Calcium, 111, 233, 318
 amylase, 213
 callose synthase activation,
  277-78, 283
 cellulose synthesis, 275-76,
  283
 cytoplasmic streaming control,
  108-10
 fruit ripening
  induction inhibition, 160
  softening, 170-71
 gravitropic perception-
  transduction-response
  mechanism, 333-34
 membrane excitation, 101-4
 mitosis control, 130
 weightlessness effect, 337-
  38
 osmoregulation
  osmotic pressure, 98
  turgor, 99-100
 peroxidase release, 212
Calcium-dependent regulation in
 metabolism
 active transport, 51-53
  biochemical characteristics,
   53
  mechanisms, 51-52
 callose deposition induction,
  56
 callose synthesis
  open questions and im-
   plications, 62-64
 callose synthesis and 1,3-β-
  glucan synthase
  callose formation by enforced calcium influx,
   55-58
  dependence of plasma
   membrane-located 1,3-
   β-glucan synthase, 58-
   62
 cell volume regulation, 68-69
  indirect vs direct action of
   calcium, 48-50
 introduction, 47-48
 maintenance and function of
  calcium gradients
  cytoplasmic calcium concentration, 50-51
 regulative proteinase generation, 64-65
  in cell homogenates, 65-68

507

uptake in chloroplasts, 54-55
uptake in mitochondria, 53-54
Callose
synthesis, 261-62
   open questions and im-
      plications, 62-64
   regulation model, 282-84
   relation to cellulose synthe-
      sis, 277-79
   synthesis and 1,3-β-glucan
      synthase
   calcium dependence of 1,3-
      β-glucan synthase, 58-
      62
   callose formation by en-
      forced calcium influx,
      55-58
Callose synthase, 276-79, 283
   activation, 277-78
Calmidazolium, 60, 67
Calmodulin, 48, 51, 59-60, 67,
   77, 80, 110-11, 126, 130-
   31
Cambium, 183-84, 190
Carnation etched ring virus
   movement protein, 300
Cauliflower mosaic virus
   genes, 244-45
   movement protein, 300-1
Cell walls
   gravitational effects, 333
Cellulase, 160, 164
   ripening fruit, 170
   synthesis, 168
      control, 161
Cellulose, 59
   deposition, 123-24
   synthesizing complexes,
      124-25
Cellulose biosynthesis, 259-90
   gravitational effect, 333
   in algae and higher plants
      biochemistry of polymeriza-
         tion process, 273-77
      globules, 268-71
      instability of enzyme com-
         plex, 276
      linear complexes, 267-71
      microtubules role in micro-
         fibril orientation, 271-
         73
      rosette complexes, 268-71
      terminal complexes, 267-71
   in bacteria and fungi, 264-67
      Acetobacter, 264-67
   introduction, 259-60
   lactose synthesis regulations
      in mammals, 279-80
   microfibril structure and
      biosynthesis mechanism,
      262
   crystallite size variation,
      262-63

degree of crystallinity, 264
degree of polymerization,
   262
parallel vs antiparallel
   chains, 263-64
overview, 260-62
receptor for 2,6-
   dichlorobenzonitrile, 280-
   81
regulation model, 282-85
relation to callose synthesis,
   277-79
Cellulose synthase, 265-66
Agrobacterium, 267
Chalcone, 243
Characeae
   see Membrane control in
      Characeae
Chitin, 124
Chitinase, 243
Chitosan, 55, 57, 63-64
Chloride
   chloroplast electron transport
      functions
   Cl⁻ release from oxygen-
      evolving complex,
      363-64
   osmotic pressure regulation,
      97-98
   plasmalemma excitation, 101-
      2
   tonoplast action potential,
      103
   turgor regulation, 98-101
Chlorobium limicola, 32
   photosynthetic apparatus
      organization, 32-34
   properties, 35
   reaction centers, 34
2-Chloroethylphosphonic acid,
   192
Chloroflexus aurantiacus, 33
   photosynthetic apparatus, 35
   properties, 35
Chloroisopropylphenylcarbamate,
   129
Chlorophyll a
   "core" antenna complex
      photosystem I, 18-18
      photosystem II, 14
Chlorophyll a/b
   light-harvesting complex, 78-
      79
   light-harvesting complex I, 18
   light-harvesting complex II,
      14-16
   gene families, 16
Chloroplasts
   calcium uptake, 54-55
   gravitational effects, 330-31
   interactions
      plasmalemma and
         mitochondria, 104-8

movement requirement, 134
photochemical apparatus
   organization, 20-21
photosystem stoichiometry,
   20-21
plants grown in space, 326,
   328
virus effects, 304-5
Chloroplast bioenergetics, membrane-proton interactions,
   347-89
   conclusions and speculations,
      379-81
   physiological role for local-
      ized and delocalized
      proton gradients, 381-
      83
   electron transpost functions,
      363
   active site of oxygen-
      evolving complex,
      364-65
   proton deposition into the
      lumen, 365-67
   release of Cl⁻ from oxygen-
      evolving complex,
      363-64
introduction, 348-49
localized domains, membrane
   energization for ATP
   formation, 367-68
bulk phase kinetics and
   localized proton gra-
   dients, 377-78
coupling mode and sample
   preparation, 368
dual proton gradient cou-
   pling, 368-75
hydroxyethane morpholine
   effect on phosphoryla-
   tion rates, 373
localized coupling via
   membrane domains,
   378-79
protons utilization to
   energize ATP forma-
   tion, 375-77
pyridine effect on single-
   turnover flash-initiated
   phosphorylation, 370
thylakoid membrane protein
   arrangement model,
   374-75, 380-81
localized domains, thylakoid
   H⁺ pumps, and site
   specificity
early experiments suggesting site specificity,
   358-60
experiments with acetic anhydride, 360-61
other photosystem-specific
   effects, 361-63

localized proton interaction
domain, 349-50
thylakoid membrane meta-
stable proton-buffering
domains, 350
acetic anhydride as probe
for sequestered do-
mains, 351-52
sequestering vs lumen-
exposure, 352-54
general properties, 350-51
organization of membrane
components, 355
proteins associated with
buried domains, 354-
57
thylakoid proteins contributing acetic anhydride
labeling sites, 356
Chloroplast DNA-encoded genes
evolution, 391-418
conclusions, 414-16
evolutionary filtering procedure, 392, 399-409
functionally important sequence identification,
399
atpB translation initiation
region, 406
atpBE promoter region,
403
comparative sequences of
rbcL promoter and
leader regions, 401
intervening sequences, 407-
9
leaders, 402
promoters, 399-402
protein-coding regions,
404-5, 407
ribosome-binding sites,
402-4
rbcL transcription termination region, 410
rpl2-rps19 intergenic region, 405
sequences proximal to the
rpl2 coding region,
408
transcription terminators,
409-11
introduction, 392-93
nature of nucleotide sub-
stitutions, 393-94
comparative structural
organization, 395-96
estimation of numbers of
substitutions for *rbcL*,
atpBE, and *rpl2*, 394-
99
per-nucleotide number of
substitution events,
398

systematic relationships
among plant genera,
397
phylogenetic topologies estimation, 411-12
divergence times and sub-
stitution estimates, 412
noncoding regions, 413-14
protein-coding regions,
412-1
Chlorotetracycline, 131
Chromosomes
gravitational effects, 322-26
Chrysolaminarin, 65
Colchicine, 321
Compactin, 455
Concanavalin A, 211
Coumarin, 274
Cowpea chlorotic mottle virus,
298
Cowpea mosaic virus, 297
replication phases, 299
symptom amelioration by satellites, 309
translation strategies, 295-96
Cremart, 134
Cyanobacteria
excitation energy distribution
in thylakoid membrane,
19-20
overview, 24-25
Cyclic diguanylic acid, 265-66
Cysteine, 323
Cytochalasin, 134-35
Cytochrome *f*, 357
Cytochrome *c* peroxidase, 207
Cytokinin, 449
Agrobacterium-transformed
plants, 481-82
biosynthesis pathway, 468
ent-kaurene oxidation inhibition, 455
vascular differentiation, 191
fiber differentiation, 196-
97
Cytokinin isopentenyl adenosine,
469
Cytoplasmic streaming control,
108-10
Cytoskeleton, fluorescence microscopy impact, 119-39
interphase microtubule array
cortical array establishment,
125-26
helical cortical arrays behavior, 122
microtubule nucleation,
120-21
organization, 121-26
proteins co-localizing with
the cortical array, 126-
28
introduction, 119-20

mitotic spindle, 130
proteins co-localizing with
the spindle, 130-31
organelle movement, 134-35
phragmoplast, 131-32
co-localizing proteins, 132
relationship with pre-
prophase band zone,
132-33
preprophase band, 128-29
co-localizing proteins, 130
prophase bands, 129
relationship with phragmoplast, 132-33
prospects, 135-36
tip-growing cells, 133-34

D

2,4-D, 76, 86-87, 125
DCMU, 366-67
1-*n*-Decylimidazole, 453-54
2-Deoxyglucose, 274-75
2,6-Dichlorobenzonitrile
receptor in plants, 280-81
2,6-Dichlorophenylazide, 280
Dictyosomes
gravitational effects, 332-33,
338
Dicyclohexylcarbodiimide, 360-
61
Digitonin, 57-58, 62-63, 265,
278
Dihydrofolate reductase, 473
Dithioerythritol, 440
Dithiothreitol, 440
T-DNA, 467, 469-71, 473-80
genes
auxin and cytokinin
biosynthesis, 468, 480-
81
tumor formation induction,
468
use as mutagen, 476-77
DNA-encoded genes
see Chloroplast DNA-encoded
genes evolution
Dwarf plants, 441

E

Echinocandin B, 60
EGTA, 49-50, 57, 61, 66, 68
Embryonic organs
gravitational effects, 327
Emerson, L., 4-5
Endoplasmic reticulum
active calcium transport, 51-
53
fruit ripening, 164
gibberellin biosynthesis, 427
gravitational effects, 331-32
Endopolygalacturonase, 170

Epinasty
  gravitational effects, 337
Ethylene, 164, 243
  effect
    microtubules, 122-23
    fruit ripening, 157, 163, 166-
      69, 171
    control, 157-59
    vascular differentiation, 192
Evolution
  see Chloroplast DNA-encoded
    genes evolution

F

F-actin, 127, 133-35
Fibers, 182, 185
  differentiation
    role of leaves, auxin and
      gibberellin, 194-96
    role of roots and cytokinin,
      196-97
Fluence, 226, 233
Fluorescence microscopy
  see Cytoskeleton, fluorescence
    microscopy impact
Fruit ripening, 155-78
  concluding remarks, 171-72
  developmental regulation
    control by growth regula-
      tors, 157-60
    gene expression, 160-63
    ripening as senescence,
      156-57
    ripening mutants, 163-64
  introduction, 155-56
  membranes and organizational
    resistance, 164-66
  oxidative reactions
    climacteric rise and protein
      synthesis, 166-68
    other oxidative reactions,
      169
    respiratory climacteric, 166
  softening process, 169
    calcium role, 170-71
    chemistry and enzymology,
      169-70
  regulator of ripening, 171
Fungicides
  relationships to plant growth
    retardants, 454-56

G

Galactose, 276
Gene expression regulation,
  221-57
  concluding remarks, 246-47
  constitutive genes
    agrobacterial genes, 245-46
    cellular genes, 244
    rRNA genes, 244

viral genes, 244-45
  gene transfer systems, 222-24
    Agrobacterium, 223
    direct transfer, 223
    GT sequence motif com-
      parison, 232
    heat shock genes, 239-41
    introduction, 221-22
    plant-microbe interactions
      defense genes, 243
      nodulin genes, 242-43
    regulation by light, 224-25
      cis-acting elements for
        light-regulated
        transcription, 227-230
      fluence, 226
      nature of light response,
        225-27
      signal transduction, 233-34
      trans-acting elements for
        light-regulated
        transcription, 230-33
    tissue-specific expression
      auxin-regulated genes, 238-
        39
      hormonal regulation of seed
        protein expression,
        237-38
      light-regulated genes, 234-
        35
      seed storage protein genes,
        235-37
Geranylgeranyl pyrophosphate,
  420-22
Gibberella fujikuroi
  see Gibberellin biosynthesis
    and control
Gibberellic acid, 76, 187, 195-
  96, 237-38
  effect on microtubules, 125
  phloem differentiation, 194
Gibberellin
  fiber differentiation, 194-96
Gibberellin biosynthesis and
  control, 419-65
  concluding remarks, 456-57
  from ent-kaurene to gib-
    berellin A$_{12}$-aldehyde
    other cell-free systems,
      427-28
    pathway and enzymes, 424-
      27
  introduction, 419-20
  ent-kaurene biosynthesis
    new cell-free systems and
      other aspects, 422-24
    pathway and enzymes, 420-
      22
  pathways after gibberellin
    A$_{12}$-aldehyde
    enzymes, 438-41
    new features of general
      pathway, 430-31

overview of general path-
  way, 428-30
pathway in other species,
  436-37
  Phaseolus spp. pathway,
    434-36
  Pisum sativum pathway,
    431-34
  summary, 438
physiology of control
  conjugation, 449-51
  genetic control of internode
    elongation, 441-46
  internode elongation in
    Pisum sativum, 442-45
  internode elongation in
    other species, 445-46
  internode elongation in Zea
    mays, 445
  photoperiodic control, 446-
    47
  plant growth retardants,
    452-56
  retardants and fungitoxicity,
    454-56
  retardants effect on
    biosynthesis, 452-53
  seed development and
    germination, 447-49
  sterol biosynthesis inhibi-
    tion, 453-54
  structure-activity rela-
    tionships, 451-52
  tissue cultures, 451
Gibberellin A$_{12}$-aldehyde, 428-
  41
  biosynthesis
    other cell-free systems,
      427-28
    pathway and enzymes, 424-
      27
Gliadin, 142-43
  genes, 148-49
  LMW subunit genes, 146-
    48
Gloeobacter violaceus, 36
1,4-β-Glucan, 275-76
1,3-β-Glucan synthase
  activity
    stimulation by polyethylene
      glycol, 265
  callose synthesis
    calcium dependence, 58-62
    formation by enforced cal-
      cium influx, 55-58
1,4-β-Glucan synthase, 275
1,3-β-Glucanase, 239
1:4-β-Glucanhydrolase, 160,
  273
Glucose, 194
β-Glucoside, 278
Glutamic acid, 210
Glutamine synthase, 242

Glutenin, 142-43
  HMW subunit genes, 143-
    46
Glycosamine, 211
Glyceollin, 55, 64
Glycine, 210
Golgi apparatus, 214
  peroxidase, 211-12
Gramicidin, 366-67
Guanyl cylase, 265

H

H⁺-ATPase
  membrane control, 104-6
Hastings, W., 5-6, 8
Haxo, F., 4, 6, 8
HCO₃⁻
  transport, 106-7
Heat shock genes, 239-41
Heliobacterium chlorum
  reaction centers, 36
Heliothrix oregonensis, 36
Heme synthesis, 209-10
Hexylamine, 374
Histone H1, 75
Hydrogen peroxide, 165
Hydroxyethane morpholine, 373
C-20 Hydroxylase, 440-41
2β-Hydroxylase, 439, 441
3β-Hydroxylase, 440-41
Hygromycin, 473
Hypersensitive reaction, 305-6
Hypocotyls
  graviperception thresholds,
    336
  microgravity effect, 319
HZE particles, 324

I

Indole-3-acetic acid, 159, 169,
  190, 195-96, 238-39, 469
Indoleacetamide hydrolase, 468-
  69
Intracytoplasmic membranes, 25
Invertase, 160
Iron, 438-40
Isofluroside, 65
Isofluroside phosphate-synthase,
  65-68
Isopentenyl transferase, 468-69

K

Karakashian, M., 6
ent-Kaurene
  genetic regulation of synthe-
    sis, 445
  new cell-free systems and
    other aspects, 422-24
  oxidation inhibition, 453, 455
  pathway and enzymes, 420-22

to gibberellin A₁₂-aldehyde
  other cell-free systems,
    427-28
  pathway and enzymes, 424-
    27
ent-Kaurene synthetase, 421-23,
    452
Kinetin, 191, 197

L

Lactose
  synthesis regulation in mam-
    mals, 179-80
Lanosterol, 455
Legumins, 235
Light
  gene expression regulation,
    224-25
  defense genes, 243
  fluence, 226
  nature of light response,
    225-27
  tissue-specific expression,
    234-35
  protein phosphorylation, 85-
    86
Light-harvesting complex I
  see also Chlorophyll *a/b*
Light-harvesting complex II
  phosphorylation
    effect on PSI, 22-24
    see also Chlorophyll *a/b*
Light-induced hyperpolarization
  of plasmalemma, 107-8
Lignin,
  peroxidase, 213-14
  synthesis
    gravitational effect, 333
    peroxidase, 208
Liposomes, 165
Lipoxygenase, 169
Luciferin-luciferase technique,
  369, 371
Lycopene, 163
Lysine, 141, 352

M

Magnesium, 333, 364-65
  callose synthase activation,
    278-79
Maltose, 193
Manganese, 364, 438
Mannose, 211
Membrane control in Characeae,
  95-117
  cell motility control, 108-10
  concluding remarks, 110-11
  introduction, 95-96
  membrane excitation, 101
  plasmalemma, 101-3
  tonoplast, 103-4

osmoregulation, 96
  osmotic pressure, 96-
    98
  turgor, 98-101
plasmalemma, chloroplasts,
  and mitochondria in-
    teractions, 104
  electrogenic H⁺-ATPase,
    104-6
  HCO₃⁻ and OH⁻ transport,
    106-7
  light-induced hyperpolariza-
    tion, 107-8
  K⁺-channel activation, 107-
    8
Methionine, 192
5-Methoxytryptamine, 323
Mevalonate, 420-21, 423-24,
  427, 455-56
Mevinolin, 455
Mg-ATP, 103-4
Microgravity
  see Plants in space
Microsomes
  gibberellin biosynthesis, 425,
    427
  membranes
    protein phosphorylation,
      79-81
Microtubules
  role in microfibril orientation,
    271-73
  see also Cytoskeleton,
    fluorescence microscopy
    impact
Mitochondria
  calcium uptake, 53-54
  fruit ripening, 164
  gravitational effects, 332
  heme synthesis, 210
  interactions
    plasmalemma and chloro-
      plasts, 104-8
  protein phosphorylation, 77-
    79
Mitosis
  gravitational effect, 320-22
Mitotic spindle
  gravitational effect, 321
  immunofluorescence studies,
    130
  proteins co-localizing with
    the spindle, 130-31
Monensin, 212
Monogalactosyldiacylglycerol,
  362
Mudrick, M., 7

N

Neomycin phosphotransferase,
  type II, 472-73, 478

Neutral red
  proton
    deposition into the lumen
      studies, 365-67
    release in internal aqueous
      phase, 368
Nicolas, M.-T., 8
Nifedipine, 58, 63
Nodulin genes expression, 242-
  43
Nopaline synthase, 223, 245
Nuclei
  gravitational effects, 331
  phosphorylation of proteins,
    77
Nutation, 336

O

OH⁻
  transport, 106-7
ORF1 protein, 300
Osmoregulation
  calcium function, 64-65
  osmotic pressure, 96-98
  turgor, 98-101
2-Oxoglutarate, 439-40

P

P515 electrochromic absorption
  change, 362-63, 367
P680, 13
P700, 16-18, 23
Paclobutrazol, 448, 453-55
Pectins, 170
PEP carboxylase
  modifications of regulatory
    properties, 94-85
Peroxidase, 169, 205-19
  apoprotein moiety
    in vitro synthesis, 209
  biosynthesis and enzyme
    activity, 207-8
  extraction, purification and
    antibody elicitation, 208-
    9
  function, 212-14
  glycosidic prosthetic groups
    and transport, 211-12
  heme synthesis, 209
  immunological studies, 214
  introduction, 205-6
  intra- and extracellular pro-
    teins and isozymes, 212-
    13
  isozymes and their implica-
    tion, 206
  fractions, 211-12
  molecular structure, 207
  summary, 214-15

Phalloidin, 130-31
Phallacidin, 127, 131
Phaseolin, 235
Phaseolus spp.
  gibberellin metabolism
    after gibberellin A₁₂-
      aldehyde, 434-36
Phloem
  see Vascular differentiation
Phosphoserine, 88
Phosphothreonine, 88
Phosphotyrosine, 88
Phosphorylation of proteins
  cascade systems as regulatory
    devices, 74-75
  enzymes, 81
    pyruvate dehydrogenase
      activity regulation, 81-
      82
    quinate:NAD⁺ oxidoreduc-
      tase activity regulation,
      82-83
    PEP carboxylase regulatory
      properties modifica-
      tion, 84-85
    proton-pumping system
      regulation, 84
    pyruvate, Pi dikinase activ-
      ity regulation, 83-84
  general conclusions, 89
  introduction, 73-74
  plant organelles
    microsomal membranes,
      79-81
    mitochondria, 77-79
    nuclei, 75-77
    plastids, 77-79
  reflections and prospects,
    87
  receptor phosphorylation,
    87-88
  transmembrane signalling,
    88
  stimulus/response coupling,
    85-87
Photochemical reaction centers,
  11-45
  cyanobacteria
    overview, 24-25
    enzyme transport regulation,
      22
    LHC II phosphorylation
      effect on PSI, 22-24
    physiological insignificance
      of LHC II
        phosphorylation, 24
    excitation energy distribution
      in thylakoid membrane,
      19
    cyanobacteria, 19-20
    mutations and LHC II, 21-
      22

  wild-type higher plant
    chloroplasts, 20-21
  green bacteria
    antenna pigment organiza-
      tion, 33-34
    overview, 32-33
    reaction centers of Chlor-
      obium, 34
    reaction centers of Chloro-
      flexus, 35
    reaction centers of
      Heliobacterium, 36
  introduction, 12-13
  photosystem I
    chlorophyll a "core" an-
      tenna complex, 17-18
    chlorophyll a/b light-
      harvesting complex I,
      18
    photochemical reaction cen-
      ter complex, 16-17
  photosystem II
    chlorophyll a "core" an-
      tenna complex, 14
    chlorophyll a/b light-
      harvesting complex II,
      14-16
    photochemical reaction cen-
      ter complex, 13-14
  photosystem stoichiometry,
    18-19
  photochemical apparatus
    organization, 18
  purple bacteria
    overview, 25-26
    reaction centers general
      properties, 27-28
    Rhodopseudomonas viridis
      B1015 antenna com-
      plex, 29-31
    Rhodopseudomonas viridis
      photosynthetic mem-
      brane organization, 31-
      32
    Rhodopseudomonas viridis
      reaction center com-
      plex, 28-29
Photosynthesis
  virus effect, 304-5
Photosystem I, 78-79
  chlorophyll a "core"antenna
    complex, 17-18
  chlorophyll a/b light-
    harvesting complex I,
    18
  LHC II phosphorylation
    effect, 22-24
  photochemical reaction center
    complex, 16-17
Photosystem II, 78-79
  chlorophyll a "core" antenna
    complex, 14

chlorophyll *a/b* light-
   harvesting complex II,
   14-16
photochemical reaction center
   complex, 13-14
Phragmoplast
   fluorescence microscopy, 131-
   32
   proteins co-localizing with
      the phragmoplast, 132
   relationship with prepro-
      phase band zone, 132-
      33
Phragmosome, 132-33
Phycobilisome, 19, 24-25
Phycocyanin, 19
Phytochrome, 86
   gene expression regulation,
      224-26
Pilson, M., 6
Pisum sativum
   gibberellin
      internode elongation con-
         trol, 442-45
      metabolism after gibberellin
         A_{12}-aldehyde, 431-34
Plant growth retardants, 452
   effect
      fungitoxicity, 454-56
      gibberellin biosynthesis,
         452-53
      sterol biosynthesis inhibi-
         tion, 453-54
Plants in space, 317-45
   cellular and subcellular
      changes, 329
   cell walls, 333
   dictyosomes, 332-33
   endoplasmic reticulum and
      ribosomes, 331-32
   essential elements, 333-
      34
   mitochondria, 332
   nuclei, 331
   plastids, 329-31
   conclusions, 337-39
   development, growth and
      reproduction
      aging, 327-28
      cell division, 320-22
      cell size and shape, 326
      chromosomal effects, 322-
         26
      differentiation, 326-28
      flowering, 320
      germination and growth,
         319-20
      radiation effects, 323-24
      reproduction, 328-29
   introduction, 318-19
   plant orientation and move-
      ment, 334

epinasty, 337
graviperception thresholds,
   335-36
nutation, 336
orientation, 334-35
tropisms, 335
Plasma membrane
   active calcium transport, 51-
      53
   1,3-β-glucan synthase
      calcium dependence, 58-
         62
Plasmalemma, 333-34
   fruit ripening, 164-65
   see also Membrane control in
      Characeae
Plasmids
   pMON557
      insertional mutagenesis,
         478-79
   Ri, 467, 473
      T-DNA, 474
   Ti, 467
      cointegrating vectors, 470-
         71
      T-DNA, 467-69
      hormone genes removal,
         473
      vir region, 469-70
Plasmodesmata
   virus passage, 299
Plastids
   gravitational effects, 329-31,
      338
   protein phosphorylation, 77-
      79
Plastocyanin, 301
Polyethylene glycol, 265
Polygalacturonase, 160
   ripening fruit, 162-64, 170
Polymyxin B, 60
Potassium, 333
   channel activation, 107-8
   osmotic pressure regulation,
      97-98
   turgor regulation, 98-101
Poterioochromonas malhamensis
   calcium and metabolic regula-
      tion, 64-65
   cell volume regulation, 68-
      69
   proteinase generation in cell
      homogenates, 65-68
Procambium, 183
Prolamins
   see Wheat, genetic basis for
      variation in bread-making
      quality
Prosthecochloris aestuarii, 32
Proteins
   see Phosphorylation of pro-
      teins

Protein kinases, 75, 77-78
   characterization procedure,
      87-88
   mitochondria, 77
Protein kinase C, 50
   transmembrane signalling, 88
Proteinase
   calcium dependent generation,
      64-65
   cell homogenates, 65-68
Protoheme, 210
Protons
   see Chloroplast bioenergetics,
      membrane-proton in-
      teractions
Proton-pumping system regulation, 84-85
Purple bacteria
   overview, 25-26
      bacteriochlorophylls, 26
   reaction centers
      general properties, 27-28
Pyridine, 369-71, 377-78
Pyruvate, Pi dikinase
   activity regulation, 83-84
Pyruvate dehydrogenase
   activity regulation, 81-82

Q

Q_A, 13-14
Q_B, 14
Quinate:NAD^+ reductase
   activity regulation, 82-83

R

Rhizobium, 242-43
Rhodobacter sphaeroides, 26
   reaction center complex, 27,
      32
Rhodospirillum
   rubrum, 27, 30-32
   tenue, 36
Rhodopseudomonas viridis
   B1015 antenna complex, 29-
      31
   photosynthetic membrane
      organization, 31-32
   reaction center complex, 27-
      29
      chromophore arrangement,
         28
Ribosomes
   gravitational effects, 332-32
RNA
   see Virus-host interactions
   rRNA genes, 244
   RNA-dependent RNA
      polymerases, 306-8

Roots
  differentiation in space,
    327
  graviperception thresholds,
    336
  orientation in space, 334-
    35
Root cap, 327
  microgravity effect, 319,
    335

S

Samuelsson, G., 8-9
Sargent, M., 5
Scholander, P., 6
Seed development and germina-
  tion
  gibberellin, 447-49
Silver, 159-60, 192
Skoog, F., 3
Snodgrass, J., 5
Spermidine, 61
Spermine, 61, 87
Squash mosaic virus, 305
Starch grains, 330
Sterol
  biosynthesis inhibition, 453-
    54
Sucrose, 193-94
Sweeney, Beatrice M., 1-9
  early life, 1-2
  formal studies, 2-3
  new knowledge, 3-4
  science, 2
  problems of woman scientist,
    8-9
  Scripps Institution of
    Oceanography, 4
  Acetabularia, 6-7
  algal photosynthesis,
    4-5
  dinoflagellates, 5-6
  University of California at
    Santa Barbara, 7-8

T

Taxol, 123, 134
Tetcyclacis, 452-54
Thaumatin, 306
Thimann, K., 3
Thylakoid membrane
  excitation energy distribution,
    19
  cyanobacteria, 19-20
  mutations and LHC II, 21-
    22
  wild-type higher plant
    chloroplasts, 20-21

organization
  Rhodopseudomonas viridis,
    31-32
  see also :Chloroplast
    bioenergetics, membrane-
    proton interactions
Ti plasmid genes, 245-46
Tobacco etch virus, 297-98
Tobacco mosaic virus, 297-98
  association with chloroplasts,
    304-5
  movement protein, 299-302
  replication phases, 299
  transgenic plants and resis-
    tance, 308
  translation strategies, 295-
    96
  uncoating, 294
Tobacco rattle virus
  movement protein, 300
Tobacco streak virus
  movement protein, 300
Tobacco vein mottle virus, 297
  translation strategies, 295-96
Tonoplast
  active calcium transport, 51-
    53
  protein phosphorylation, 80
  see also Membrane control in
    Characeae
Tracheids
  see Xylem differentiation
Trehalose, 193
Trifluoperazine, 60
Tiphenyltin chloride, 360
Troponin T, 130-31
Tryptophan monooxygenase,
    468-69
Tseng, C. K., 4
Tunicamycin, 212, 274
Turnip yellow mosaic virus,
    298
  association with chloroplasts,
    304
  RNA release, 295
  translation strategies, 295-96

U

Uricase, 242

V

Vascular differentiation, 179-
    204
  cell division role, 184-86
    tracheary elements, 185
  concluding remarks, 197-98
  conduit size and density con-
    trol

auxin, 187-88
  general patterns of vascular
    elements, 186-87
  problem of size control,
    187
  effects of pressure and
    ethylene, 192
  fiber differentiation control
    role of leaves, auxin, and
    gibberellin, 194-96
    role of roots and cytokinin,
    196-97
  introduction, 179-80
  relation between phloem and
    xylem differentiation,
    192-94
  role of roots and cytokinin,
    191
  tissues structure and function,
    180
    compartmentalization, 181-
    82
    fibers, 182
    phloem cell types, 181
    xylem, 181-82
  vascular adaptation control,
    188-90
    hypothesis, 188-89
  vascular meristems, 182-84
    cambium, 183
    procambium, 183
  xylem and phloem differentia-
    tion control
  vascular tissues induction,
    190-91
Vicilins, 235
Vir genes, 469, 475-76
Virus-host interactions, 291-
    315
  conclusions, 309-10
  early events
    gene expression and
    replication, 295-99
    uncoating, 294-95
  effects on host, 303
    host RNA-dependent RNA
    polymerases, 306-8
    hypersensitive reaction,
    305-6
    photosynthesis and the
    chloroplast, 304-5
  host range, 292-93
  introduction, 291-92
    virus nucleic acid types,
    292
  plant-to-plant spread, 293-94
  spread within the plant, 299-
    303
    long distance spread, 302-
    3
    movement protein, 299-302
    subliminal infections, 302

transgenic plants and virus resistance, 308-9
coat protein protection
mode, 308
negative strand capture,
308-9
satellites effect, 309
Vlasov, V., 6

W

Wheat, genetic basis for variation in bread-making quality, 141-53
allelic variation, 150
exploitation of variation in
plant breeding, 150-51

genes for α- and β-gliadins,
148-49
evolution, 148-49
homologies, 148
gliadin and glutenin, 142-43
HMW subunit genes, 143-
46
chromosomal position,
144
evolution, 146
genes per locus, 145
x and y types, 145
introduction, 141-42
LMW subunits genes, 146-48
chromosomal position, 144,
146-47
evolution, 147
gene number, 147

recombination, 146
minor gene loci, 149

X

Xanthine dehydrogenase, 242
Xylanase, 170
Xylem
see Vascular differentiation

Z

Zea mays
gibberellin biosynthesis,
445
Zeatin, 191

CUMULATIVE INDEXES

CONTRIBUTING AUTHORS, VOLUMES 30–38

A

Abeles, F. B., 37:49–72
Akazawa, T., 36:441–72
Albersheim, P., 35:243–75
Aloni, R., 38:179–204
Amasino, R. M., 35:387–413
Anderson, J. M., 37:93–136
Appleby, C. A., 35:443–78

B

Badger, M. R., 36:27–53
Bandurski, R. S., 33:403–30
Barber, J., 33:261–95
Bauer, W. D., 32:407–49
Beale, S. I., 34:241–78
Beard, W. A., 38:347–89
Bedbrook, J. R., 30:593–620
Beevers, H., 30:159–93
Bell, A. A., 32:21–81
Benson, D. R., 37:209–32
Benveniste, P., 37:275–308
Berry, J., 31:491–543
Bewley, J. D., 30:195–238
Bickel-Sandkötter, S., 35:97–120
Björkman, O., 31:491–543
Boller, T., 37:137–64
Bottomley, W., 34:279–310
Boudet, A. M., 38:73–93
Boyer, J. S., 36:473–516
Brady, C. J., 38:155–78
Brenner, M. L., 32:511–38
Buchanan, B. B., 31:341–74
Burnell, J. N., 36:255–86

C

Castelfranco, P. A., 34:241–78
Chapman, D. J., 31:639–78
Chua, N., 38:221–57
Clarke, A. E., 34:47–70
Clarkson, D. T., 31:239–98; 36:77–115
Clegg, M. T., 38:391–418
Cogdell, R. J., 34:21–45
Cohen, J. D., 33:403–30
Conn, E. E., 31:433–51

Cosgrove, D., 37:377–405
Craigie, J. S., 30:41–53
Cronshaw, J., 32:465–84
Cullis, C. A., 36:367–96

D

Darvill, A. G., 35:243–75
Davies, D. D., 30:131–58
Delmer, D. P., 38:259–90
Dennis, D. T., 33:27–50
Diener, T. O., 32:313–25
Digby, J., 31:131–48
Dilley, R. A., 38:347–89
Dutcher, F. R., 38:317–45

E

Edwards, G. E., 36:255–86
Eisbrenner, G., 34:105–36
Eisinger, W., 34:225–40
Elbein, A. D., 30:239–72
Ellis, R. J., 32:111–37
Elstner, E. F., 33:73–96
Etzler, M. E., 36:209–34
Evans, H. J., 34:105–36
Evans, L. T., 32:485–509
Evenari, M., 36:1–25

F

Farquhar, G. D., 33:317–45
Feldman, J. F., 33:583–608
Feldman, L. J., 35:223–42
Fincher, G. B., 34:47–70
Firn, R. D., 31:131–48
Flavell, R., 31:569–96
Fork, D. C., 37:335–61
Freeling, M., 35:277–98
French, C. S., 30:1–26
Fry, S. C., 37:165–86
Furuya, M., 35:349–73

G

Galston, A. W., 32:83–110
Galun, E., 32:237–66
Gantt, E., 32:327–47
Giaquinta, R. T., 34:347–87

Gifford, E. M. Jr., 34:419–40
Gifford, R. M., 32:485–509
Glass, A. D. M., 34:311–26
Glazer, A. N., 38:11–45
Good, N. E., 37:1–22
Goodwin, T. W., 30:369–404
Gordon, M. P., 35:387–413
Graebe, J. E., 38:419–65
Graham, D., 33:347–72
Gray, M. W., 33:373–402
Green, P. B., 31:51–82
Green, P. J., 38:221–57
Greenway, H., 31:149–90
Grisebach, H., 30:105–30
Guerrero, M. G., 32:169–204
Gunning, B. E. S., 33:651–98

H

Haehnel, W., 35:659–93
Hahlbrock, K., 30:105–30
Halstead, T. W., 38:317–45
Hanson, A. D., 33:163–203
Hanson, J. B., 31:239–98
Hara-Nishimura, I., 36:441–72
Hardham, A. R., 33:651–98
Harding, R. W., 31:217–38
Harris, N., 37:73–92
Hatch, M. D., 36:255–86
Haupt, W., 33:205–33
Heath, R. L., 31:395–431
Heber, U., 32:139–68
Heidecker, G., 37:439–66
Heldt, H. W., 32:139–68
Hepler, P. K., 36:397–439
Higgins, T. J. V., 35:191–221
Hirel, B., 36:345–65
Hitz, W. D., 33:163–203
Ho, T.-H. D., 37:363–76
Hoffman, N. E., 35:55–89
Horsch, R., 38:467–86
Howell, S. H., 33:609–50
Huber, S. C., 37:233–46
Hull, R., 38:291–315

J

Jackson, M. B., 36:145–74

K

Kamiya, N., 32:205–36
Kaplan, A., 35:45–83
Kauss, H., 38:47–72
King, R. W., 36:517–68
Klee, H., 38:467–86
Kolattukudy, P. E., 32:539–67
Kolodner, R., 30:593–620
Kowallik, W., 33:51–72
Kuhlemeier, C., 38:221–57

L

Labavitch, J. M., 32:385–406
Lang, A., 31:1–28
Laties, G. G., 33:519–55
Leaver, C. J., 33:373–402
Leong, S. A., 37:187–208
Letham, D. S., 34:163–97
Lieberman, M., 30:533–91
Lin, W., 37:309–34
Lloyd, C. W., 38:119–39
Loewus, F. A., 34:137–61
Loewus, M. W., 34:137–61
Loomis, R. S., 30:339–67
Lorimer, G. H., 32:349–83
Losada, M., 32:169–204
Lucas, W. J., 34:71–104

M

Maliga, P., 35:519–42
Malkin, R., 33:455–79
Marrè, E., 30:273–88
Marx, G. A., 34:389–417
McCandless, E. L., 30:41–53
McCarty, R. E., 30:79–104
Meins, F. Jr., 34:327–46
Melis, A., 38:11–45
Messing, J., 37:439–66
Miernyk, J. A., 33:27–50
Mimura, T., 38:95–117
Minchin, P. E. H., 31:191–215
Møller, I. M., 37:309–34
Moore, T. S. Jr., 33:235–59
Moreland, D. E., 31:597–638
Morgan, J. M., 35:299–319
Morris, R. O., 37:509–38
Munns, R., 31:149–90

N

Nakamoto, H., 36:255–86
Neilands, J. B., 37:187–208
Nester, E. W., 35:387–413
Ng, E., 30:33–67

O

Oaks, A., 36:345–65
Ogren, W. L., 35:415–42

O'Leary, M. H., 33:297–315
Outlaw, W. H. Jr., 31:299–311

P

Padan, E., 30:27–40
Palni, L. M. S., 34:163–97
Pate, J. S., 31:313–40
Patterson, B. D., 33:347–72
Payne, P. I., 38:141–53
Pharis, R. P., 36:517–68
Phillips, D. A., 31:29–49
Pickard, B. G., 36:55–75
Possingham, J. V., 31:113–29
Powles, S. B., 35:15–44
Pradet, A., 34:199–224
Pratt, L. H., 33:557–82
Preiss, J., 33:431–54
Preston, R. D., 30:55–78

Q

Quail, P. H., 30:425–84

R

Rabbinge, R., 30:339–67
Ragan, M. A., 31:639–78
Ranjeva, R., 38:73–93
Raven, J. A., 30:289–311
Raymond, P., 34:199–224
Reinhold, L., 35:45–83
Rennenberg, H., 35:121–53
Roberts, J. A., 33:133–62
Roberts, J. K. M., 35:375–86
Rogers, S., 38:467–86
Roughan, P. G., 33:97–132
Rubery, P. H., 32:569–96

S

Sachs, M. M., 37:363–76
Satoh, K., 37:335–61
Satter, R. L., 32:83–110
Schnepf, E., 37:23–47
Schubert, K. R., 37:539–74
Schulze, E.-D., 37:247–74
Schwintzer, C. R., 37:209–32
Sexton, R., 33:133–62
Sharkey, T. D., 33:317–45
Shepherd, R. J., 30:405–23
Shimmen, T., 38:95–117
Shininger, T. L., 30:313–37
Shropshire, W. Jr., 31:217–38
Silk, W. K., 35:479–518
Silverthorne, J., 36:569–93
Slack, C. R., 33:97–132
Smith, F. A., 30:289–311

Smith, H., 33:481–518
Smith, T. A., 36:117–43
Snell, W. J., 36:287–315
Somerville, C. R., 37:467–507
Spanswick, R. M., 32:267–89
Spiker, S., 36:235–53
Steponkus, P. L., 35:543–84
Stocking, C. R., 35:1–14
Stoddart, J. L., 31:83–111
Stone, B. A., 34:47–70
Strotmann, H., 35:97–120
Sweeney, B. M., 38:1–9
Sze, H., 36:175–208

T

Taiz, L., 35:585–657
Tang, P.-S., 34:1–19
Tazawa, M., 38:95–117
Theg, S. M., 38:347–89
Theologis, A., 37:407–38
Thomas, H., 31:83–111
Thomson, W. W., 31:375–94
Thorne, J. H., 36:317–43
Ting, I. P., 36:595–622
Tjepkema, J. D., 37:209–32
Tobin, E. M., 36:569–93
Tran Thanh Van, K. M., 32:291–311
Trelease, R. N., 35:321–47
Troughton, J. H., 31:191–215

V

van Huystee, R. B., 38:205–19
Vega, J. M., 32:169–204
Velthuys, B. R., 31:545–67
Vennesland, B., 32:1–20
Virgin, H. I., 32:451–63

W

Walbot, V., 36:367–96
Walton, D. C., 31:453–89
Wareing, P. F., 33:1–26
Wayne, R. O., 36:397–439
Weiler, E. W., 35:85–95
Whatley, J. M., 31:375–94
Whitfeld, P. R., 34:279–310
Wiemken, A., 37:137–64

Y

Yang, S. F., 35:155–89
Yanofsky, M. F., 35:387–413

Z

Zaitlin, M., 38:291–315
Zeiger, E., 34:441–75
Zurawski, G., 38:391–418

CHAPTER TITLES, VOLUMES 30–38

PREFATORY CHAPTERS

Fifty Years of Photosynthesis	C. S. French	30:1–26
Some Recollections and Reflections	A. Lang	31:1–28
Recollections and Small Confessions	B. Vennesland	32:1–20
A Plant Physiological Odyssey	P. F. Wareing	33:1–26
Aspirations, Reality, and Circumstances: The Devious Trail of a Roaming Plant Physiologist	P.-S. Tang	34:1–19
Reminiscences and Reflections	C. R. Stocking	35:1–14
A Cat Has Nine Lives	M. Evenari	36:1–25
Confessions of a Habitual Skeptic	N. E. Good	37:1–22
Living in the Golden Age of Biology	B. M. Sweeney	38:1–9

MOLECULES AND METABOLISM

Bioenergetics

Facultative Anoxygenic Photosynthesis in Cyanobacteria	E. Padan	30:27–40
Roles of a Coupling Factor for Photophosphorylation in Chloroplasts	R. E. McCarty	30:79–104
The Central Role of Phosphoenolpyruvate in Plant Metabolism	D. D. Davies	30:131–58
Efficiency of Symbiotic Nitrogen Fixation in Legumes	D. A. Phillips	31:29–49
Role of Light in the Regulation of Chloroplast Enzymes	B. B. Buchanan	31:341–74
Mechanisms of Electron Flow in Photosystem II and Toward Photosystem I	B. R. Velthuys	31:545–67
The Carboxylation and Oxygenation of Ribulose 1,5-Bisphosphate: The Primary Events in Photosynthesis and Photorespiration	G. H. Lorimer	32:349–83
The Physical State of Protochlorophyll(ide) in Plants	H. I. Virgin	32:451–63
Blue Light Effects on Respiration	W. Kowallik	33:51–72
Oxygen Activation and Oxygen Toxicity	E. F. Elstner	33:73–96
Photosystem I	R. Malkin	33:455–79
The Cyanide-Resistant, Alternative Path in Higher Plant Respiration	G. G. Laties	33:519–55
Photosynthetic Reaction Centers	R. J. Cogdell	34:21–45
Adenine Nucleotide Ratios and Adenylate Energy Charge in Energy Metabolism	A. Pradet, P. Raymond	34:199–224
Structure, Function, and Regulation of Chloroplast ATPase	H. Strotmann, S. Bickel-Sandkötter	35:97–120
Photorespiration: Pathways, Regulation, and Modification	W.L. Ogren	35:415–42
Photosynthetic Electron Transport in Higher Plants	W. Haehnel	35:659–93
Photosynthetic Oxygen Exchange	M. R. Badger	36:27–53
Pyruvate,P$_i$ Dikinase and NADP-Malate Dehydrogenase in C$_4$ Photosynthesis: Properties and Mechanism of Light/Dark Regulation	G. E. Edwards, H. Nakamoto, J. N. Burnell, M. D. Slack	36:255–86

Crassulacean Acid Metabolism | I. P. Ting | 36:595–622
Photoregulation of the Composition, Function,
and Structure of Thylakoid Membranes | J. M. Anderson | 37:93–136
Physiology of Actinorhizal Nodules | J. D. Tjepkema, C. R. Schwintzer,
| D. R. Benson | 37:209–32
Membrane-Bound NAD(P)H Dehydrogenases
in Higher Plant Cells | I. M. Møller, W. Lin | 37:309–34
The Control by State Transitions of the
Distribution of Excitation Energy in
Photosynthesis | D. C. Fork, K. Satoh | 37:335–61
Products of Biological Nitrogen Fixation in
Higher Plants: Synthesis, Transport, and
Metabolism | K. R. Schubert | 37:539–74
Membrane-Proton Interactions in Chloroplast
Bioenergetics: Localized Proton Domains | R. A. Dilley, S. M. Theg,
| W. A. Beard | 38:347–89
Photochemical Reaction Centers: Structure,
Organization, and Function | A. N. Glazer, A. Melis | 38:11–45

Small Molecules
Enzymic Controls in the Biosynthesis of
Lignin and Flavonoids | K. Hahlbrock, H. Grisebach | 30:105–30
The Role of Lipid-Linked Saccharides in the
Biosynthesis of Complex Carbohydrates | A. D. Elbein | 30:239–72
Biosynthesis of Terpenoids | T. W. Goodwin | 30:369–404
Biosynthesis and Action of Ethylene | M. Lieberman | 30:533–91
Photocontrol of Carotenoid Biosynthesis | R. W. Harding, W. Shropshire, Jr. | 31:217–38
A Descriptive Evaluation of Quantitative
Histochemical Methods Based on Pyridine
Nucleotides | W. H. Outlaw, Jr. | 31:299–311
Cyanogenic Compounds | E. E. Conn | 31:433–51
Biochemistry and Physiology of Abscisic
Acid | D. C. Walton | 31:453–89
Mechanisms of Action of Herbicides | D. E. Moreland | 31:597–638
Modern Methods for Plant Growth Substance
Analysis | M. L. Brenner | 32:511–38
Structure, Biosynthesis, and Biodegradation of
Cutin and Suberin | P. E. Kolattukudy | 32:539–67
Compartmentation of Nonphotosynthetic
Carbohydrate Metabolism | D. T. Dennis, J. A. Miernyk | 33:27–50
Cellular Organization of Glycerolipid
Metabolism | P. G. Roughan, C. R. Slack | 33:97–132
Phospholipid Biosynthesis | T. S. Moore, Jr. | 33:235–59
Chemistry and Physiology of the Bound
Auxins | J. D. Cohen, R. S. Bandurski | 33:403–30
myo-Inositol: Its Biosynthesis and Metabolism | F. A. Loewus, M. W. Loewus | 34:137–61
The Biosynthesis and Metabolism of
Cytokinins | D. S. Letham, L. M. S. Palni | 34:163–97
Chlorophyll Biosynthesis: Recent Advances
and Areas of Current Interest | P. A. Castelfranco, S. I. Beale | 34:241–78
Immunoassay of Plant Growth Regulators | E. W. Weiler | 35:85–95
The Fate of Excess Sulfur in Higher Plants | H. Rennenberg | 35:121–53
Ethylene Biosynthesis and its Regulation in
Higher Plants | S. F. Yang, N. E. Hoffman | 35:155–89
Polyamines | T. A. Smith | 36:117–43
Nitrogen Metabolism in Roots | A. Oaks, B. Hirel | 36:345–65
Siderophores in Relation to Plant Growth and
Disease | J. B. Neilands, S. A. Leong | 37:187–208
Sterol Biosynthesis | P. Benveniste | 37:275–308
Gibberellin Biosynthesis and Control | J. E. Graebe | 38:419–65

Macromolecules
Sulfated Polysaccharides in Red and Brown
Algae | E. L. McCandless, J. S. Craigie | 30:41–53

The Structure of Chloroplast DNA   J. R. Bedbrook, R. Kolodner   30:593–620
The Molecular Characterization and Organization of Plant Chromosomal DNA Sequences   R. Flavell   31:569–96
Chloroplast Proteins: Synthesis, Transport, and Assembly   R. J. Ellis   32:111–37
The Assimilatory Nitrate-Reducing System and Its Regulation   M. G. Guerrero, J. M. Vega, M. Losada   32:169–204
Cell Wall Turnover in Plant Development   J. M. Labavitch   32:385–406
Phosphoenolpyruvate Carboxylase: An Enzymologist's View   M. H. O'Leary   33:297–315
Regulation of the Biosynthesis and Degradation of Starch   J. Preiss   33:431–54
Phytochrome: The Protein Moiety   L. H. Pratt   33:557–82
Arabinogalactan-Proteins: Structure, Biosynthesis, and Function   G. B. Fincher, B. A. Stone, A. E. Clarke   34:47–70
Synthesis and Regulation of Major Proteins in Seeds   T. J. V Higgins   35:191–221
Leghemoglobin and *Rhizobium* Respiration   C. A. Appleby   35:443–78
Plant Lectins: Molecular and Biological Aspects   M. E. Etzler   36:209–34
Plant Chromatin Structure   S. Spiker   36:235–53
Rapid Genomic Change in Higher Plants   V. Walbot, C. A. Cullis   36:367–96
Topographic Aspects of Biosynthesis, Extracellular Secretion, and Intracellular Storage of Proteins in Plant Cells   T. Akazawa, I. Hara-Nishimura   36:441–72
Alteration of Gene Expression During Environmental Stress in Plants   M. M. Sachs, T.-H. D. Ho   37:363–76
Cellulose Biosynthesis   D. P. Delmer   38:259–90
Some Aspects of Calcium-Dependent Regulation in Plant Metabolism   H. Kauss   38:47–72
Some Molecular Aspects of Plant Peroxidase Biosynthetic Studies   R. B. van Huystee   38:205–19

ORGANELLES AND CELLS

Function
Intracellular pH and its Regulation   F. A. Smith, J. A. Raven   30:289–311
DNA Plant Viruses   R. J. Shepherd   30:405–23
The Cell Biology of Plant-Animal Symbiosis   R. K. Trench   30:485–53
The Chloroplast Envelope: Structure, Function, and Role in Leaf Metabolism   U. Heber, H. W. Heldt   32:139–68
Physical and Chemical Basis of Cytoplasmic Streaming   N. Kamiya   32:205–36
Plant Protoplasts as Physiological Tools   E. Galun   32:237–66
Electrogenic Ion Pumps   R. M. Spanswick   32:267–89
Viroids: Abnormal Products of Plant Metabolism   T. O. Diener   32:313–25
Light-Mediated Movement of Chloroplasts   W. Haupt   33:205–33
Influence of Surface Charges on Thylakoid Structure and Function   J. Barber   33:261–95
Mitochondrial Genome Organization and Expression in Higher Plants   C. J. Leaver, M. W. Gray   33:373–402
Plant Molecular Vehicles: Potential Vectors for Introducing Foreign DNA into Plants   S. H. Howell   33:609–50
Photosynthetic Assimilation of Exogenous HCO_3^- by Aquatic Plants   W. J. Lucas   34:71–104
Aspects of Hydrogen Metabolism in Nitrogen-Fixing Legumes and Other Plant-Microbe Associations   G. Eisbrenner, H. J. Evans   34:105–36
Organization and Structure of Chloroplast Genes   P. W. Whitfeld, W. Bottomley   34:279–310

The Biology of Stomatal Guard Cells	E. Zeiger	34:441–75
Membrane Transport of Sugars and Amino Acids	L. Reinhold, A. Kaplan	35:45–83
Plant Transposable Elements and Insertion Sequences	M. Freeling	35:277–98
Study of Plant Metabolism in vivo Using NMR Spectroscopy	J. K. M. Roberts	35:375–86
H+-Translocating ATPases: Advances Using Membrane Vesicles	H. Sze	36:175–208
Light Regulation of Gene Expression in Higher Plants	E. M. Tobin, J. Silverthorne	36:569–93
Dynamics of Vacuolar Compartmentation	T. Boller, A. Wiemken	37:137–64
Fructose 2,6-Bisphosphate as a Regulatory Metabolite in Plants	S. C. Huber	37:233–46
Structural Analysis of Plant Genes	G. Heidecker, J. Messing	37:439–66
Phosphorylation of Proteins in Plants: Regulatory Effects and Potential Involvement in Stimulus Response Coupling	R. Ranjeva, A. M. Boudet	38:73–93
Membrane Control in the Characeae	M. Tazawa, T. Shimmen, T. Mimura	38:95–117
Agrobacterium-Mediated Plant Transformation and Its Further Applications to Plant Biology	H. Klee, R. Horsch, S. Rogers	38:467–86
Regulation of Gene Expression in Higher Plants	C. Kuhlemeier, P. J. Green, N. Chua	38:221–57

Organization

Polysaccharide Conformation and Cell Wall Function	R. D. Preston	30:55–78
Microbodies in Higher Plants	H. Beevers	30:159–93
Plant Cell Fractionation	P. H. Quail	30:425–84
Phycobilisomes	E. Gantt	32:327–47
Microtubules	B. E. S. Gunning, A. R. Hardham	33:651–98
Biogenesis of Glyoxysomes	R. N. Trelease	35:321–47
Plant Cell Expansion: Regulation of Cell Wall Mechanical Properties	L. Taiz	35:585–657
Organization of the Endomembrane System	N. Harris	37:73–92
Cross-Linking of Matrix Polymers in the Growing Cell Walls of Angiosperms	S. C. Fry	37:165–86
The Plant Cytoskeleton: The Impact of Fluorescence Microscopy	C. W. Lloyd	38:119–39

Development

Plastid Replication and Development in the Life Cycle of Higher Plants	J. V. Possingham	31:113–29
Development of Nongreen Plastids	W. W. Thomson, J. M. Whatley	31:375–94
Auxin Receptors	P. H. Rubery	32:569–96
Heritable Variation in Plant Cell Culture	F. Meins, Jr.	34:327–46
Calcium and Plant Development	P. K. Hepler, R. O. Wayne	36:397–439
Cellular Polarity	E. Schnepf	37:23–47
Biophysical Control of Plant Cell Growth	D. Cosgrove	37:377–405
Genes Specifying Auxin and Cytokinin Biosynthesis in Phytopathogens	R. O. Morris	37:509–38

TISSUES, ORGANS, AND WHOLE PLANTS

Function

Fusicoccin: A Tool in Plant Physiology	E. Marrè	30:273–88
Quantitative Interpretation of Phloem Translocation Data	P. E. H. Minchin, J. H. Troughton	31:191–215
The Mineral Nutrition of Higher Plants	D. T. Clarkson, J. B. Hanson	31:239–98
Transport and Partitioning of Nitrogenous Solutes	J. S. Pate	31:313–40

Infection of Legumes by *Rhizobia*   W. D. Bauer   32:407–49
Phloem Structure and Function   J. Cronshaw   32:465–84
Photosynthesis, Carbon Partitioning, and
   Yield   R. M. Gifford, L. T. Evans   32:485–509
Regulation of Pea Internode Expansion by
   Ethylene   W. Eisinger   34:225–40
Regulation of Ion Transport   A. D. M. Glass   34:311–26
Phloem Loading of Sucrose   R. T. Giaquinta   34:347–87
Phytoalexins and Their Elicitors: A Defense
   Against Microbial Infection in Plants   A. G. Darvill, P. Albersheim   35:243–75
Factors Affecting Mineral Nutrient
   Acquisition by Plants   D. T. Clarkson   36:77–115
Ethylene and Responses of Plants to Soil
   Waterlogging and Submergence   M. B. Jackson   36:145–74
Cell-Cell Interactions in *Chlamydomonas*   W. J. Snell   36:287–315
Phloem Unloading of C and N Assimilates in
   Developing Seeds   J. H. Thorne   36:317–43
Water Transport   J. S. Boyer   36:473–516
Plant Chemiluminescence
F. B. Abeles   37:49–72
Fruit Ripening   C. J. Brady   38:155–78

Development
The Control of Vascular Development   T. L. Shininger   30:313–37
Biosynthesis and Action of Ethylene   M. Lieberman   30:533–91
Organogenesis—A Biophysical View   P. B. Green   31:51–82
Leaf Senescence   H. Thomas, J. L. Stoddart   31:83–111
The Establishment of Tropic Curvatures in
   Plants   R. D. Firn, J. Digby   31:131–48
Mechanisms of Control of Leaf Movements   R. L. Satter, A. W. Galston   32:83–110
Control of Morphogenesis in In Vitro
   Cultures   K. M. Tran Thanh Van   32:291–311
Cell Biology of Abscission   R. Sexton, J. A. Roberts   33:133–62
Genetic Approaches to Circadian Clocks   J. F. Feldman   33:583–608
Developmental Mutants in Some Annual Seed
   Plants   G. A. Marx   34:389–417
Concept of Apical Cells in Bryophytes and
   Pteridophytes   E. M. Gifford, Jr.   34:419–40
Regulation of Root Development   L. J. Feldman   35:223–42
Osmoregulation and Water Stress in Higher
   Plants   J. M. Morgan   35:299–319
Cell Division Patterns in Multicellular Plants   M.Furuya   35:349–73
Quantitative Descriptions of Development   W. K. Silk   35:479–518
Early Events in Geotropism of Seedling
   Shoots   B. G. Pickard   36:55–75
Gibberellins and Reproductive Development
   in Seed Plants   R. P. Pharis, R. W. King   36:517–68
Rapid Gene Regulation by Auxin   A. Theologis   37:407–38
Plants in Space   T. W. Halstead, F. R. Dutcher   38:317–45
Differentiation of Vascular Tissues   R. Aloni   38:179–204

POPULATION AND ENVIRONMENT

Physiological Ecology
Physiological Aspects of Desiccation
   Tolerance   J. D. Bewley   30:195–238
Explanatory Models in Crop Physiology   R. S. Loomis, R. Rabbinge, E. Ng   30:339–67
Mechanisms of Salt Tolerance in
   Nonhalophytes   H. Greenway, R. Munns   31:149–90
Photosynthetic Response and Adaptation to
   Temperature in Higher Plants   J. Berry, O. Björkman   31:491–543
Metabolic Responses of Mesophytes to Plant
   Water Deficits   A. D. Hanson, W. D. Hitz   33:163–203

Stomatal Conductance and Photosynthesis   G. D. Farquhar, T. D. Sharkey   33:317–45
Responses of Plants to Low, Nonfreezing
  Temperatures: Proteins, Metabolism, and
  Acclimation   D. Graham, B. D. Patterson   33:347–72
Light Quality, Photoperception, and Plant
  Strategy   H. Smith   33:481–518
Photoinhibition of Photosynthesis Induced by
  Visible Light   S. B. Powles   35:15–44
Carbon Dioxide and Water Vapor Exchange
  in Response to Drought in the Atmosphere
  and in the Soil   E.-D. Schulze   37:247–74

Genetics and Plant Breeding
Isolation and Characterization of Mutants in
  Plant Cell Culture   P. Maliga   35:519–42
Analysis of Photosynthesis with Mutants of
  Higher Plants and Algae   C. R. Somerville   37:467–507
Genetics of Wheat Storage Proteins and the
  Effect of Allelic Variation on
  Bread-Making Quality   P. I. Payne   38:141–53

Pathology and Injury
Initial Events in Injury to Plants by Air
  Pollutants   R. L. Heath   31:395–431
Biochemical Mechanisms of Disease
  Resistance   A. A. Bell   32:21–81
Crown Gall: A Molecular and Physiological
  Analysis   E.W. Nester, M. P. Gordon, R. M.
                                                            Amasino, M. F. Yanofsky   35:387–413
Role of the Plasma Membrane in Freezing
  Injury and Cold Acclimation   P. L. Steponkus   35:543–84
Plant Virus-Host Interactions   M. Zaitlin, R. Hull   38:291–315

Evolution
Evolution of Biochemical Pathways: Evidence
  from Comparative Biochemistry   D. J. Chapman, M. A. Ragan   31:639–78
Evolution of Higher Plant Chloroplast
  DNA-Encoded Genes: Implications for
  Structure-Function and Phylogenetic Studies   G. Zurawski, M. T. Clegg   38:391–418

Annual Reviews Inc.

NONPROFIT SCIENTIFIC PUBLISHER

4139 El Camino Way
P.O. Box 10139
Palo Alto, CA 94303-0897 • USA

ORDER FORM

Now you can order
TOLL FREE
1-800-523-8635
(except California)

Annual Reviews Inc. publications may be ordered directly from our office by mail or use our Toll Free telephone line (for orders paid by credit card or purchase order, and customer service calls only); through booksellers and subscription agents, worldwide; and through participating professional societies. Prices subject to change without notice. ARI Federal I.D. #94-1156476

Individuals: Prepayment required on new accounts by check or money order (in U.S. dollars, check drawn on U.S. bank) or charge to credit card — American Express, VISA, MasterCard.

Institutional buyers: Please include purchase order number.

Students: $10.00 discount from retail price, per volume. Prepayment required. Proof of student status must be provided (photocopy of student I.D. or signature of department secretary is acceptable). Students must send orders direct to Annual Reviews. Orders received through bookstores and institutions requesting student rates will be returned.

Professional Society Members: Members of professional societies that have a contractual arrangement with Annual Reviews may order books through their society at a reduced rate. Check with your society for information.

Toll Free Telephone orders: Call 1-800-523-8635 (except from California) for orders paid by credit card or purchase order and customer service calls only. California customers and all other business calls use 415-493-4400 (not toll free). Hours: 8:00 AM to 4:00 PM, Monday-Friday, Pacific Time.

Regular orders: Please list the volumes you wish to order by volume number.

Standing orders: New volume in the series will be sent to you automatically each year upon publication. Cancellation may be made at any time. Please indicate volume number to begin standing order.

Prepublication orders: Volumes not yet published will be shipped in month and year indicated.

California orders: Add applicable sales tax.

Postage paid (4th class bookrate/surface mail) by **Annual Reviews Inc.** Airmail postage or UPS, extra.

ANNUAL REVIEWS SERIES		Prices Postpaid per volume USA/elsewhere	Regular Order Please send:	Standing Order Begin with:
			Vol. number	Vol. number
Annual Review of ANTHROPOLOGY				
Vols. 1-14	(1972-1985)	$27.00/$30.00		
Vol. 15	(1986)	$31.00/$34.00		
Vol. 16	(avail. Oct. 1987)	$31.00/$34.00	Vol(s). _____	Vol. _____
Annual Review of ASTRONOMY AND ASTROPHYSICS				
Vols. 1-2, 4-20	(1963-1964; 1966-1982)	$27.00/$30.00		
Vols. 21-24	(1983-1986)	$44.00/$47.00		
Vol. 25	(avail. Sept. 1987)	$44.00/$47.00	Vol(s). _____	Vol. _____
Annual Review of BIOCHEMISTRY				
Vols. 30-34, 36-54	(1961-1965; 1967-1985)	$29.00/$32.00		
Vol. 55	(1986)	$33.00/$36.00		
Vol. 56	(avail. July 1987)	$33.00/$36.00	Vol(s). _____	Vol. _____
Annual Review of BIOPHYSICS AND BIOPHYSICAL CHEMISTRY				
Vols. 1-11	(1972-1982)	$27.00/$30.00		
Vols. 12-15	(1983-1986)	$47.00/$50.00		
Vol. 16	(avail. June 1987)	$47.00/$50.00	Vol(s). _____	Vol. _____
Annual Review of CELL BIOLOGY				
Vol. 1	(1985)	$27.00/$30.00		
Vol. 2	(1986)	$31.00/$34.00		
Vol. 3	(avail. Nov. 1987)	$31.00/$34.00	Vol(s). _____	Vol. _____

ANNUAL REVIEWS SERIES		Prices Postpaid per volume USA/elsewhere	Regular Order Please send:	Standing Order Begin with:
			Vol. number	Vol. number

Annual Review of COMPUTER SCIENCE
Vol. 1	(1986)	**$39.00/$42.00**		
Vol. 2	(avail. Nov. 1987)	**$39.00/$42.00**	Vol(s). _____	Vol. _____

Annual Review of EARTH AND PLANETARY SCIENCES
Vols. 1-10	(1973-1982)	**$27.00/$30.00**		
Vols. 11-14	(1983-1986)	**$44.00/$47.00**		
Vol. 15	(avail. May 1987)	**$44.00/$47.00**	Vol(s). _____	Vol. _____

Annual Review of ECOLOGY AND SYSTEMATICS
Vols. 1-16	(1970-1985)	**$27.00/$30.00**		
Vol. 17	(1986)	**$31.00/$34.00**		
Vol. 18	(avail. Nov. 1987)	**$31.00/$34.00**	Vol(s). _____	Vol. _____

Annual Review of ENERGY
Vols. 1-7	(1976-1982)	**$27.00/$30.00**		
Vols. 8-11	(1983-1986)	**$56.00/$59.00**		
Vol. 12	(avail. Oct. 1987)	**$56.00/$59.00**	Vol(s). _____	Vol. _____

Annual Review of ENTOMOLOGY
Vols. 10-16, 18-30	(1965-1971, 1973-1985)	**$27.00/$30.00**		
Vol. 31	(1986)	**$31.00/$34.00**		
Vol. 32	(avail. Jan. 1987)	**$31.00/$34.00**	Vol(s). _____	Vol. _____

Annual Review of FLUID MECHANICS
Vols. 1-4, 7-17	(1969-1972, 1975-1985)	**$28.00/$31.00**		
Vol. 18	(1986)	**$32.00/$35.00**		
Vol. 19	(avail. Jan. 1987)	**$32.00/$35.00**	Vol(s). _____	Vol. _____

Annual Review of GENETICS
Vols. 1-19	(1967-1985)	**$27.00/$30.00**		
Vol. 20	(1986)	**$31.00/$34.00**		
Vol. 21	(avail. Dec. 1987)	**$31.00/$34.00**	Vol(s). _____	Vol. _____

Annual Review of IMMUNOLOGY
Vols. 1-3	(1983-1985)	**$27.00/$30.00**		
Vol. 4	(1986)	**$31.00/$34.00**		
Vol. 5	(avail. April 1987)	**$31.00/$34.00**	Vol(s). _____	Vol. _____

Annual Review of MATERIALS SCIENCE
Vols. 1, 3-12	(1971, 1973-1982)	**$27.00/$30.00**		
Vols. 13-16	(1983-1986)	**$64.00/$67.00**		
Vol. 17	(avail. August 1987)	**$64.00/$67.00**	Vol(s). _____	Vol. _____

Annual Review of MEDICINE
Vols. 1-3, 6, 8-9	(1950-1952, 1955, 1957-1958)			
11-15, 17-36	(1960-1964, 1966-1985)	**$27.00/$30.00**		
Vol. 37	(1986)	**$31.00/$34.00**		
Vol. 38	(avail. April 1987)	**$31.00/$34.00**	Vol(s). _____	Vol. _____

Annual Review of MICROBIOLOGY
Vols. 18-39	(1964-1985)	**$27.00/$30.00**		
Vol. 40	(1986)	**$31.00/$34.00**		
Vol. 41	(avail. Oct. 1987)	**$31.00/$34.00**	Vol(s). _____	Vol. _____